Dynamics of Structures

"Non multa, sed multum"

Pliny, Epistolae

Texts on Computational Mechanics

Volume V

Series Editor:
JOHN ARGYRIS, F.R.S.

Presently planned Volumes:

- I Foundations of Simulation
- II Linear Systems
- III Non-Linear Systems, Part I
- IV Non-Linear Systems, Part II
- **V Dynamics of Structures**
- VI A Survey of Aerolasticity
- VII The Dynamics of Chaos
- VIII The Philosophy of Coupling and Decoupling

Dynamics of Structures

JOHN ARGYRIS

HANS-PETER MLEJNEK

N·H

1991

NORTH-HOLLAND

AMSTERDAM • NEW YORK • OXFORD • TOKYO

ELSEVIER SCIENCE PUBLISHERS B.V.
Sara Burgerhartstraat 25
P.O. Box 211,
1000 AE Amsterdam,
The Netherlands

Distributors for the USA and Canada:

ELSEVIER SCIENCE PUBLISHING COMPANY, INC.
655 Avenue of the Americas
New York, NY 10010, USA

Library of Congress Cataloging-in-Publication Data

Argyris, J. H. (John H.)
Dynamics of structures / John Argyris, Hans-Peter Mlejnek.
p. cm. -- (Texts on computational mechanics ; 5)
Includes bibliographical references and index.
ISBN 0-444-89045-9 (hardbound). -- ISBN 0-444-89112-9 (paperback)
1. Structural dynamics. I. Mlejnek, Hans-Peter. II. Title.
III. Series.
TA654.A74 1991
624. 1'7--dc20 91-12355
CIP

ISBN 0 444 89045 9 (Hardbound)
ISBN 0 444 89112 9 (Paperback)

CAT - Computer Aided Typography
KETEX
Institute for Computer Applications
University of Stuttgart
Germany

This volume is printed on acid-free paper
Transferred to digital printing 2005

"Neither seeking nor avoiding mathematical exercitations
we enter into problems solely with a view
to possible usefullness for physical science"

Lord Kelvin and Peter Cuthrie Tait
Treatise on Natural Philosophy, Part II

To posses	**g**enerosity	in	**g**eneralship
To offer	**e**quity	in	**e**xcellence
To seek	**r**edemption	in	**r**egeneration
To overcome	**d**espondency	in	**d**edication

Preface

Τῆς δ'ἀρετῆς ἱδρῶτα θεοὶ προπάροιθεν ἔθηκαν
Ἀθάνατοι, μακρὸς δὲ καὶ ὄρθιος οἶμος ἐς αὐτὴν.

But before virtue the immortal gods set sweat;
it is a long and steep path to her.

Hesiod, *Ἔργα καὶ ἡμέραι*

Prologue

This series of self-contained textbooks is designed to present a modern, but still elementary account of computational mechanics as first conceived in the Forties and early Fifties in the Department of Aeronautics at Imperial College, London, and developed further at the Institute of Statics and Dynamics of Aerospace Structures (ISD) and subsequently at the Institute for Computer Applications (ICA) of the University of Stuttgart. In order to give the authors some freedom in exploring the form and contents of such a major undertaking, it was decided to issue first a restricted and in many ways provisional edition in German extending over 2300 pages. This work indeed appeared in three volumes in 1986-88 and proved a considerable success. Its compilation within three years starting from a rough draught necessitated an intensive intellectual and organisational effort. The first two volumes comprise the linear and non-linear (stationary) theory of structures whilst the last volume presents an *aperçu* of computational dynamics but also contains substantial accounts on modern aeroelasticity and the theory of chaos. In this volume, only brief references to instationary aerodynamics could be incorporated as also on planetary motion. The reader will observe that these rapidly conceived volumes were not intended to accomodate a large number of modern topics like metal forming processes, aerodynamic and hydrodynamic problems as well as many other non-structural topics extensively researched in the aforementioned three institutes. This was immediately recognised as a serious omission which had to be remedied in the final English edition. We decided to incorporate in this authoritative work an overview over nearly all relevant activities of the Department of Aeronautics at Imperial College and the ISD and its successor institute, ICA, in Stuttgart. The organisation founded for the initial German edition had to be reconstituted and strengthened on the scientific and editorial side. We shall refer to the novel editorial techniques evolved at the ICA in what follows. Here, we shall concentrate on the scientific side. But before accounting for such aspects, we should mention that we decided at an early stage of this project to initiate the process of publication with Volume III of the original publication and to issue it in three companion Volumes V, VI and VII. In particular, the present Volume V offers an overview of linear and non-linear dynamics of structural systems. The companion Volume VI, whose authors are John Argyris, Kari Appa and Joachim Bühlmeier, is to give a modern account of aeroelasticity and expands greatly – especially in the aerodynamic context – on the account given in Chapter 26 of the original Volume III. Finally, the second companion Volume VII on chaotic manifestations in dynamic systems is composed by John Argyris, Gunter Faust and Maria Haase and extends massively in scope and contents the presentation given in Chapter 25 of Volume III of the German edition.

In eleven chapters, the present Volume V covers the computational dynamics for linear and non-linear engineering systems and conservative as well as non-conservative loads. The treatise has not been conceived as a translation of the original German edition but as an adaption to the higher exigencies of the English-speaking world. It includes a great number of significant clarifications, corrections, modifications and extensions. In addition to new items, we took the opportunity to improve greatly on the presentation, especially in connection with previously woolly arguments, infelicities of style, but also many downright errors. The team for this difficult but so sensitive work included John Argyris, Kari Appa, Prudence Lawday-Gienger and Karl Straub. Volume V as also Volume VI on aeroelasticity cover jointly the field of computer dynamics of structural systems as taught for over forty years in the Department of Aeronautics at Imperial College to under- and postgraduate students, with specific emphasis on the demands arising in the astronautical domain. These high standards could not be realised completely within the period between 1959 and 1984 in the original ISD in Stuttgart. The instruction was nevertheless greatly appreciated by industry since it provided a more or less unique curriculum on the Continent of Europe. The author hopes that these textbooks will not only communicate an interesting but complex subject to eager students of engineering sciences and applied physics, but also help the endeavours of practitioners in industry. The undersigned attempted to present the fundamental theory in a physically and mathematically attractive form, hopefully without forgetting the underlying engineering requirements.

Before detailing the contents of the present Volume V, we feel tempted to communicate to the reader our philosophy of work. As we already remarked on other occasions, many historical and scientific factors originally made it unavoidable for us to portray this theory from the point of view of aeronautics. It is aeronautics and astronautics which made – and still make – the highest intellectual demands on structural mechanics and aerodynamics. Economic necessities impose the highest pay load and the minimum fuel demands for an aircraft. These considerations and the stringent requirements on factors of safety demand extremely refined computational techniques for the aerodynamic and structural design of aeronautical vehicles. The creative efforts required in this field exceed by far the intellectual exertions in other branches of engineering design. The aeronautical background of the team of authors explains the rest. It is a fact that for many years, the aeronautical and astronautical specialist was faced with a greater intellectual challenge than the engineer working in other areas of structural design. This meant that the aviation engineer had to possess a specific philosophical attitude and an aptitude for generalisation which facilitated greatly his understanding of complex problems and their solution arising in other fields of engineering science. As the great scientist von Kàrmàn repeatedly observed, the reverse does not, in general, apply. Of course, this does not mean that it is a pre-requisite to study aeronautics and astronautics in order to shine in this field. The study of technical physics or fundamental engineering sciences as taught in Anglo-Saxon countries offers an alternative background; an expertise in civil or mechanical engineering is, in contrast, hardly helpful. What is important is a keen analytical, creative and markedly flexible mind such as most of us possess in our younger years. This drive for universality which is characteristic of aeronautics and astronautics within the domain of structural and aerodynamic engineering has left its stamp on this work. On the whole, only an aeronautical or astronautical engineer who has a mastery of engineering physics can be a "universalist". It is, however, necessary to be such a universalist in order to lend a comprehensible framework to the flood of problems which are and which can be solved by computer nowadays. This again renders it possible for us to undertake further excursions into unexplored areas. With these thoughts in mind, it will not surprise the intelligent reader that the Finite Element Method could only be conceived in the domain of aeronautics. In fact, the germ of the idea emerged in World War II as a *deus ex machina* through the intrinsic difficulties we aeronautical engineers had to face in the design and calculation of the swept-back wings of the first high-speed jet fighters.

For the benefit of our readers, we also have to refer briefly to our perception of teaching and research in an engineering faculty. This concerns the mathematical formalism of a lecture or paper. It goes without saying that we support clear and exact thinking (as far as it can be conceived) in our writings. At the same time, however, we insist with vehemence that a theory must be presented, even to our most gifted students, in a physically attractive form and not in an abstract mathematical frame. Mathematics are imperative to this purpose and an inventive mathematical skill is essential. But the emphasis in our argument must rest on the physical background and must be devoid of unnecessary abstractions (or rather obstructions) which are unattractive to able engineering students. Do not let us forget what a great mathematician like Constantine Carathéodory proclaimed: excessive abstraction stifles inventiveness. In this context, we should remember that practically all progress in computational mechanics has been achieved through theoretically gifted engineers and physicists and not, in our opinion, through applied mathematical exertions. At the same time, we physicists and engineers should admit that we continuously have to develop and perfect our mathematical expertise and let our inventive genes develop new mathematical techniques even if we are not always in a position to supply proofs of existence. Let us recall what a genius like Heaviside announced: shall I refuse my dinner because I do not understand the process of digestion?

With such an engineering philosophy as our capital, we aim in this first companion volume to give a systematic description of the analysis of the dynamic processes in complex structures on the basis of our computer-orientated research and our experience in the solution of engineering problems. We shall not, however, restrict ourselves to structures, but shall also occasionally refer to the motion of stars and satellites when depicting the use of direct numerical integration according to the method of time discretisation. As we already remarked in Volume I (G)*, this confirms the fact that the triumphant progress of the computer, parallel to the success of computer-orientated numerics, has forced us not only to have a profounder fundamental knowledge, but also to recast the theoretical principles in a form ideally suited to the computer, thus making new applications possible. This way of thinking has also effected dynamics and conditions contents and style of the present work. Our wish is to demonstrate to the gifted student as well as the experienced practician the theoretical and algorithmic tools in a form suitable

* The letter G refers to the original German edition

for immediate application. In this way, we hope to help prevent a catastrophe like Chernobyl in the future and, as far as is humanly possible, guarantee the safety of atomic power stations even in dangerous circumstances (like burst pipes, leakage of cooling fluid, human error, aircraft crashes and earthquakes) so that a failure does not lead to a serious endangerment of the environment. This and similar questions pose an enormous intellectual challenge for the creative engineer when assessing the safety of bridges, high-rise buildings, vehicles as well as aircraft and spacecraft.

As a result of these challenges, the mathematical and physical development of computer-orientated dynamics has made striking progress. This particular field of mechanics, especially in the important non-linear domain, has been recast and universalised decisively by the most crucial technical development of this century, the computer. The fascinating speed with which the first Cray 2 to be installed in Europe – soon to be superseded by a more powerful successor machine – masters highly complex computational processes initiated a new era for the solution of difficult non-linear problems with multi-degrees of freedom which no one would have dared consider feasible only a few years ago. However, in order to reassure the younger creative minds, let us make the following observation. There is one thing the computer is not (yet) capable of: it can neither develop new theories and algorithms, nor can it critically assess the meaning and absolute correctness of the results. Nevertheless, it has supported and inspired our intellect successfully for more than forty years and is still on the advance.

However, the computer still requires suitable instructions or programmes in order to function. The necessity of creating workable and at the same time general computer programmes forces us for economic and intellectual reasons to redirect our thoughts to basic theoretical and physical principles. The systematic development of the matrix calculus has long established itself in dynamics, and especially in the field of aeronautics, as a particularly convenient mathematical concept for the computer, but also for an elegant mind. Referring to our preceding statement on the mathematical inventiveness of gifted engineers, we observe that this group of researchers developed the matrix calculus in many new directions well beyond the limits set by past mathematicians. At the same time, however, an analysis of complex structures and other systems in a matrix formulation is unthinkable without the intellectual revolution caused by the finite element method. Although this method, in itself the brainchild of the invention of the computer, takes up only a modest amount of space in this volume, it should be noted that the extensive development of algorithms in matrix form dealt with here would lose its *raison d'être* to a great extent without this basis.

The field of dynamics is one of the most intriguing for a creative engineer, the applied mathematician and the technical physicist. The techniques developed therein secure the practical applicability of modern theories of continua. However, we also have to draw conceptual consequences from the changes which have taken place due to this development; they involve, for example, a drastic departure from the policy of increasing specialisation traditional at most technical universities. As we stated before, a return to basic principles which are attaining a growing significance because of the development of new computer models is of primary importance. The deepest possible understanding of the fundamentals of aeronautics as well as of technical physics and of their reformulation in a manner suitable for the computer is an indispensible pre-requisite for the solution of a wide range of complex problems. It is our intention to facilitate the work of the student and the practician – if not to render it possible in the first place – and radically condense the increasing number of single and apparently disconnected entries in a never-ending catalogue of uninspiring special cases. In the interest of technological progress, the multitude of specialised subjects – which often lose their significance overnight – must be replaced and taught to the enthusiastic student in a comprehensive form emphasising the general overview of a subject.

We hope that this companion volume fulfils these expectations. It is directed to open-minded students and practicians who are interested in a broadly based field of knowledge. The only requirement is a solid grasp of mechanics and mathematics as has been or should have been required for profitable study at English and German universities these many years. In this respect, the book offers the successful, more mature student the opportunity of familiarising himself with more recent developments in computer mechanics.

No textbook can offer a complete picture of a branch of knowledge; this one is no exception. Due to pressure of time and space, some modern aspects of dynamics could not be included. We can only hope that we shall be given the opportunity to deal with them in a future edition. It is a matter of particular regret that the so-called modal synthesis technique which has achieved great significance could not be incorporated in this treatise. This method, which, to express it roughly, represents the dynamic analogon to the static sub-structure technique, is based on the use of the eigenoscillation characteristics of sub-structures in conjunction with suitable interface conditions in order to establish the dynamic response of a total structure. This mainly applies of course to linear systems. Here, concepts such as effective mass and stiffness are used; they offer a criterion as to whether sufficient eigenvectors are at our disposal in order to establish economically a response pattern of the total structure. Selective model adjustment with structural characteristics established in the course of experiments is a further important step in the practical application of theoretical structural analyses (parameter or system identification, sensitivity analysis) to secure the necessary extensive information and establish in this way the dynamic response of systems. The wide field concerning the interaction of fluid or gaseous media and structures, for example in the case of rocket propulsion, is also of great significance. Aeroelasticity is one aspect of this extensive range of phenomena and is discussed in detail from the point of view of structural dynamics and aerodynamics in the companion Volume VI. In this connection, the coupling of acoustic influences and the dynamical response of vehicles or the cell of an aircraft is also of interest. A further, highly important procedure is the so-called spectral response method, used especially in earthquake analysis. Unfortunately, all these aspects had to be neglected in our limited textbook; we can only apologise to the reader for these *lacunae*. Unfortunately, we must admit to a further serious omission which has occurred for organisational reasons. In the present work, we have not dealt with dynamic systems incorporating periodic coefficient matrices. As a matter of fact, this was included in Vol. III (G). The development of this theory and the inclusion of the associated stability investigation following the method of Floquet will, however, be found in Vol. VI.

In the following 11 chapters, a multitude of modern theories and calculation processes is described which will hopefully satisfy the expectations of universities and the practical needs of industry. The following review of the contents offers a resumée of part of the teaching and research carried out in the course of more than 35 years at Imperial College and at the former ISD at the University of Stuttgart. The high standards in computer mechanics set in 1949 to 1978 in the Department of Aeronautics of Imperial College have happily been maintained at least in this institution. Regrettably, the same cannot necessarily be said of other institutions.

In the last three decades or so, aeronautical engineers and scientists were increasingly confronted by unfamiliar and complex problems arising in non-linear dynamical systems. This manifests itself in outbursts of chaotic manifestations as has been observed, for example, in hypersonic panel flutter. Such questions are touched upon in Chapter 11 of the present treatise, applying an elementary computational argument. The specific aeronautical aspects will be considered in Volume VI of this series, but the theory of chaos itself will be discussed – as stated before – at some length in Volume VII. The researchers realised, of course, that the novel phenomena in non-linear dynamical systems also demanded a very sophisticated experimental enquiry and verification. The author attempted in the late Fifties and early Sixties to establish this philosophy of work in the Department of Aeronautics of Imperial College. Although only limited financial funds could be secured, it nevertheless proved possible – in particular thanks to the highly gifted Sydney Kelsey and the enthusiasm of other staff – to achieve significant experimental successes. At the same time, the author hoped to initiate in the late Sixties a parallel effort at the original ISD. In fact, a researcher, Hans Flüh, whilst working in this Institute for the "Sonderforschungsbereich weitgespannte Flächentragwerke" of the "Deutsche Forschungsgemeinschaft", investigated experimentally in the period from 1977 to 1982 the non-linear oscillations of flat and slightly curved pre-stressed networks with great skill and imagination. Interestingly enough, the results confirmed the author's theoretical predictions, see also references [11.56],[11.58] in Chapter 11. Had this promising line of intelligent experiments been pursued vigorously, we could have realised also at the former ISD first significant experiments on chaotic motion and complex non-linear aeroelastic phenomena as early as 1980. Unfortunately, Hans Flüh left the Institute for industry and the hope of the author that his work might have been continued by new members of staff could regrettably not be fulfilled, although large financial funds were secured for the ISD in order to develop the sophisticated technique for non-linear oscillations and chaos, and also for aeroelastic problems and high temperature response of materials.

What caused such lack of response? The reasons are complex, but we may not be wrong in assuming that such an attitude might be due to the intellectual barriers some well-trained scientists and engineers experience when asked to face new and complex tasks. It does not suffice to be proficient in one's daily work but to refuse to enquire into the physics of unexplored phenomena. For the latter occupation, we require a certain philosophical attitude and curiosity as well as a desire to expand our horizons. These elements are unfortunately sometimes missing. The undersigned offers this confession as a reminder that there is always another side to each medal of success.

Exposition

We begin in Chapter 1 with an elementary introduction to elastodynamics restricted to linear oscillators with a single degree of freedom, intended particularly for less advanced students, but also as preparatory work for Chapters 5 and 6 of this volume. Two significant insertions have been added to the original trial edition. These refer to the concept of fictitious negative damping which facilitates the understanding of the outburst of an aeroelastic phenomenon and the exemplary demonstration of the usefulness of Dirac impulses through the analysis of a periodic application of such impulses. Although we have deliberately restricted ourselves in this chapter to examples with a single degree of freedom, all the formulations are set up in such a way that the reader can move on to multi-degree-of-freedom systems with relative ease.

In view of the new dimension "time" arising in dynamics, we considered it essential to re-examine in Chapter 2 some important basic equations of mechanics and the principle of virtual work. In this brief chapter, we select the dynamics of the mass point as our initial concept and remind the reader of the lucid philosophy and logic of Newton's mechanics. We extend the principle of virtual displacements (PVD) to dynamics and, on this basis, discuss the discretisation of continuous systems. At the same time, by restricting ourselves to the PVD, we reflect our decision to build up computational dynamics on the basis of the displacement method alone. There are, of course, other methodologies: the force method, mixed methods and the boundary element method. At the moment, however, we consider the value of a general formulation of these methods in dynamics to be limited and have therefore dispensed with a discussion of them.

In Chapter 3, we discuss linear dynamics in more detail using the finite element theory and develop as examples mass matrices for some element types. With the element stiffnesses already set up in statics (see e.g. Volume I (G)), we are thus in a position to offer the practician the basic tools for the numerical interpretation of dynamics in undamped linear systems.

We turn to this field in Chapter 4 which focuses on the calculation of the eigenvalues and eigenvectors of symmetric matrix pairs (like stiffness and mass), including special cases such as a singular stiffness of a free-free system (aircraft, rocket, satellite) as well as a singular mass matrix. A series of spectral analysis methods for different applications is discussed, all of which have been and still are being used at ISD/ICA. The classic vector iteration, the simultaneous vector iteration, the Jacobi method, the methods of Givens and Householder and the Eberlein method – not discussed until Chapter 9 – are of importance here. At the same time, the Lanczos method is continually gaining in significance in modern spectral analysis. This is dealt with in sub-section 4.3.5 in a concise contribution by Karl Straub. The economic feasilibility of computational processes is an essential aspect of Chapter 4. Due to the new dimension time, dynamic calculations involve a great deal more effort than static ones. As a result, every opportunity should be taken to reduce the problem size.

Eigenvalues and eigenvectors form the basis of many calculations of the time response of linear damped systems. In Chapter 5, we discuss the motion of freely oscillating systems which are not subject to outside impositions of forces and displacements. We also explore a current method, the so-called dynamic or modal reduction, of achieving a more economical computational approach.

From a practical point of view, systems subject to external excitations, whether due to forces or prescribed displacements, are of more interest. These are considered in Chapter 6. Using the spectral information of a freely oscillating system, we may express the excited motion as an assembly of single-degree-of-freedom responses. This methodology represents an elementary special case of the so-called modal reduction methods.

No process of motion in nature takes place, however, without mechanical energy being dissipated into thermal energy which is thus lost to the activation of the motion. This process is due to the damping forces which are discussed in Chapter 7. Modal and non-modal viscous and hysteretic damping are particularly important, whereas it is extremely difficult to incorporate the effect of dry friction in large-scale systems. On the other hand, it proves particularly simple to adopt the widespread modal damping model; this makes a decoupling to single-degree-of-freedom systems possible, like in the undamped case.

All the external influences discussed in the preceding chapters are deterministic, i.e. functionally uniquely defined in space and time. However, if we consider a flight from London to Washington, for example, it would be both illusory and pointless to establish the actual dynamic load on the aircraft cell as generated by airflow and turbulence at every instant of the flight and input it in the form of an explicit excitation function into a dynamic calculation. The effort would be enormous, the practical use questionable. Instead, we have to take recourse to statistical and probability methods which we discuss in Chapter 8. As a result of this philosophy, we cannot guarantee that the aircraft will not fail under the prescribed load: however, we can assert, for example, that with a probability of 99.99%, a critical stress will not be exceeded. The remaining 0.01%, unpleasant as it is, is unfortunately one of the dangers life involves.

The analysis of structures with arbitrary viscous damping is one of the more complex tasks of linear dynamics and is discussed in Chapter 9 (see Chapter 7 as a starting point). The reasoning requires more effort, approximately one order of magnitude in the degrees of freedom higher than for modal damping: in fact, the characteristics of the eigenvalue problem change completely as compared with the undamped or modally damped case. To elucidate these interesting aspects, Chapter 9 includes a discussion of pertinent eigenvalue procedures for non-modal viscous damping. However, these methods have not yet been applied to large problems.

As the reader will realise, the mathematical formalism behind the methods of dynamic analysis and behind the concept of eigenvalue analysis is by no means inconsiderable. In fact, dynamic problems can be solved much more easily by direct integration. A detailed account of such techniques as applying to linear problems is developed in Chapter 10. The reader who has studied Volume I (G) thoroughly will already know that modern structural analysis is based on a discretisation in space. In this chapter, discretisation is now extended to time in order to account for instationary cases. Without such a technique, the successful application of large-scale dynamic analysis to problems with arbitrary viscous damping – not to mention non-linear structures – would be completely unthinkable. We describe a series of algorithms ranging from the well-known Newmark variants to the Hermitian algorithms of the third or higher order. On the one hand, these can be conditionally stable, and on the other hand unconditionally stable (at least in the case of linear systems); they were developed and tested successfully more than 15 years ago at the ISD/ICA institutions.

In the final Chapter 11 of the present volume, we attempt to familiarise the reader with the complex field of non-linear dynamics. A somewhat extensive introduction to semi-analytical procedures applied to the solution of Duffing's differential equation (i.e. of an oscillator with a single degree of freedom and a non-linear spring including a cubic term) illustrates the strange, unwonted phenomena which occur in this field and which can even lead to chaos. We demonstrate a number of chaotic manifestations of the Duffing oscillator but do not discuss the background theory. We refer, however, to Volume VII of this series which is designed to serve as a textbook for physicists and engineers enquiring into the intriguing theory of non-linear chaotic systems.

Chapter 11 also presents a detailed account of a wide range of time integration methods for the analysis of non-linear oscillations and wave propagation in solid bodies and structures. The development of time algorithms in Chapter 10 serves as an initiation to these techniques. Their stability and convergence in the non-linear domain is also discussed extensively. We demonstrate the efficiency of the proposed methods on a large number of problems, some of them complex. In this chapter, we also consider displacement-dependent and velocity-dependent loads which can be non-conservative; in this way, we form a link to the theory of aeroelasticity in Volume VI. A practically inclined reader will no doubt find the last two sub-sections 11.8 and 11.9 of this volume on the computational analysis of crash and explosive loading applications particularly interesting. Indeed, sub-section 11.8 considers in detail the crash process of the front part or motor case of a car, using for the sheet material the TRUMP elements. The impact causes very severe elasto-plastic deformations. The strain-hardening of the material due to increasing strain as well as the dependence of the material response on the strain rate must be taken into account. The model adopted here is based on a rigid visco-plastic idealisation of the material. This leads to transient equations in the velocities in place of the usual expressions in the displacements. A remarkable correspondence between the results of the calculation and those of the complex experimental test was obtained. It must be admitted, however, that the computer effort required for this idealisation is high. Consequently, the authors enquired in sub-section 11.9 into a more economical solution of such problems. This was indeed achieved by a novel idealisation of the bending characteristics of beams and sheet material which ignores rotational modes and considers only displacements at the nodes. This involves a finite difference approximation in a natural formulation in place of the familiar FEM idealisation for the bending response of triangular sheet elements. Thanks to this idealisation, the computational effort for the temporal integration is reduced by a factor of 1/20. A most satisfying agreement with experimental results on the severe impact loading of a cylindrical shell subject to an explosive sheet charge is demonstrated. This completes the subjects considered in the present volume.

Epilogue

As mentioned in the prologue, the intensive research and development which has been carried out at the Department of Aeronautics at Imperial College, at the original ISD and at the newly created ICA at the University of Stuttgart was dictated by the high demands of aeronautics and astronautics and by the fundamental philosophical attitude of the senior author. With this attitude as a motivating force, the ICA, a central and independent institute, is involved in a much wider field of activity than the original ISD. Computer-orientated mechanics with its many manifestations of solid, liquid and gaseous materials is only one of the pillars on which the new institute stands. Further examples are radiation and heat transmission problems as well as the theory of electromagnetic fields. The computer simulation of the behaviour of ceramic material is another unusual research activity at the ICA. Technical physics, the basic principles of engineering sciences as well as software developments in connection with a further expansion of the finite element method and the availability of vector and parallel computers lead to the conception of new numerical procedures for the solution of complex physical and technical problems. Possibly the most exciting research in the ICA is concerned with the theory and application of chaos, discussed in some detail in the aforementioned Volume VII of this series. This provides the engineer with formidable intellectual impulses and forces him to depart from a deterministic attitude and study on a novel computer-orientated base unsolved physical and mathematical problems like turbulence. Strange as these subjects may initially sound to many readers, in a few years, they will form part of a normal curriculum for engineers. These developments demonstrate in what a staggering way the original ISD has moved beyond the field of aeronautics and astronautics into new areas of research.

It would be impossible for one single person, or even two, to carry out the work of writing an extensive and complex book such as this from the moment of its conception through the numerous draughts up to the point where it reaches the printer. Only scientists who have themselves experienced this long creative process can grasp how many talented people, working individually and yet as a team, are necessary to give the author or authors the enthusiastic, but critical support needed to complete a final text for the publishers. One small aspect may serve as an example. One often comes up against the mistaken idea that a manuscript with which the author himself is familiar can be handed to the printer who then produces the completed work without the slightest difficulty. Far from it! It is an illusion to believe that a complex mathematical text which is both scientifically correct and aesthetically satisfying can be created in this manner. Each formula must be considered carefully beforehand in order to optimise overall display, compactness, spacing and comprehensibility. The senior author, who can look back on fifty years of writing experience, was privileged to have the excellent support of an outstanding team for the realisation of the trial three-volume work from January 1985 onwards. Karl Straub and Gerhard Frik, in particular, assisted the undersigned most generously with endless patience and conscientiousness, with critical expertise and rare linguistic ability on the laborious path from the inception of Volume I to the completion of Volume III. Without their help, the three manuscripts of over 2300 pages with more than 1100 figures could never have been finished in the short time available.

The present volume of this English series of treatises is based on a yet different technical expertise. As stressed before, this work is not a mere translation of the

German original, but an independent composition in English. The great effort required in this creative undertaking could not have been realised without the linguistic mastery of our resident writer Prudence Lawday-Gienger who supported the undersigned superbly and independently. As from its initial conception, it was planned to compose the complete text on the computer within the Institute. To this purpose, we decided to expand greatly on the type-setting software system TeX and to build up a programming tool for a professional reproduction of a mathematical text and the computer generated figures. In this way, we hope to reach the very high standards offered before World War II by some European publishing houses. At the same time, we expect to accelerate the process of publication greatly by dispensing with proof reading and extensive correspondence with the publishing house. Most important of all: we hope in this way to achieve a significant reduction of the price of the paperback edition intended for our students.

To achieve these high aims, a special team was constituted under Steffen Kernstock. This includes an enthusiastic array of unusual talents like Jörg Albrecht, Stefan-Alexander Haake, Paul Sylvester Kernstock, Prudence Lawday-Gienger, Christiane Reisert, Lutz Hagen Schäfer, Gerald Schneider and Christiane Walter who worked with dedication towards the realisation of the initial plan ably guided by the meticulous sense of symmetry and order of Karl Straub. In this connection, it should be stressed that the most intriguing ideas were provided by the indefatigable Paul Sylvester Kernstock whose unflagging motivation overcame all initial difficulties. At the same time, it must be acknowledged that Steffen Kernstock's organisational skill combined with an engineer's understanding of the complex problems we face in astronautics proved a significant factor in securing the success of the venture. We hope that the present product will satisfy the most critical reader.

The figures are an important pre-requisite for the legibility and comprehensibility of the work. The demands of this volume were particularly exacting. We had the privilege of being able to use a computer generated production process together with a plotter or with direct photographic reproduction from the computer screen. This was also handled by the aforementioned team and in particular by Jörg Albrecht, Thilo Bieser, Stefan-Alexander Haake, Rolf Krentz, Patrick Liebich, Christiane Reisert, Lutz Hagen Schäfer, Gerald Schneider and Christiane Walter whose enthusiasm and artistic sensitivity ensured the high standards of figures and tables.

Christa Hübler, the best librarian the senior author has ever met in his long career, takes her place in this long list of invaluable colleagues. She secured all the scientific literature necessary for this work in the shortest possible time, thus forestalling the author's legendary impatience.

Yet however inspired the support of all these people was, one observation should be made: the ultimate responsibility for the contents of this volume rests with the undersigned.

A final remark: this work could never have been completed had the ICA not been established. The ICA owes its foundation to the then progressive political resolve of the Ministries responsible and the University of Stuttgart, represented by the "Rektorat", Chancellor and the Senate. We remember with gratitude the energetic Rector Hartmut Zwicker who unfortunately passed away so early. His place was ably taken by Franz Effenberger who guided the University with skill and intuition.

Our grateful appreciation also goes out to Jürgen Blum, the former Chancellor of the University of Stuttgart, who regrettably, but understandably left us for higher positions. We should also like to express our gratitude to the colleagues Gerhard Kohn and Helmut Sorg who supported us steadily and imaginatively through many turbulent periods, believing unfalteringly in the revolutionary importance of our work. Our thanks are offered to these and other friends who still encourage academic originality and inventiveness in spite of the levelling influences in universities and the distracting incursions of our times.

Before closing this overview of the work, the undersigned would like to offer his thanks and sincere appreciation to the Publishing House North-Holland and in particular to drs. Arie Jongejan and Ineke Kolen for their inspired support and organisational skill in the realisation of our common goal. Having worked with North-Holland for over 20 years, one may be forgiven if one felt inclined to take the smooth perfection of their clockwork perfomance for granted.

December 24th, 1990 *John Argyris*

Bibliography

[1] *John Argyris, Hans-Peter Mlejnek*; Die Methode der finiten Elemente in der elementaren Strukturmechanik, Band I, Verschiebungsmethode in der Statik, Friedr. Vieweg & Sohn, Wiesbaden, 1986.

[2] *John Argyris, Hans-Peter Mlejnek*; Die Methode der finiten Elemente in der elementaren Strukturmechanik, Band II, Kraft- und gemischte Methoden, Nichtlinearitäten, Friedr. Vieweg & Sohn, Wiesbaden, 1987.

[3] *John Argyris, Hans-Peter Mlejnek*; Die Methode der finiten Elemente in der elementaren Strukturmechanik, Band III, Einführung in die Dynamik, Friedr. Vieweg & Sohn, Wiesbaden, 1988.

Table of Contents

1 Introduction to elasto-dynamics: linear oscillators

"One picture is worth a thousand words."
Emperor Sung

This first chapter is intended both as a rehearsal of elementary dynamics and as an introduction to the subsequent extensive discussion of computational techniques in the dynamics of deformable solid bodies. We shall deal with simple, linear oscillators with a single degree of freedom. The exact solutions which are possible for such simple cases will serve as guidelines to the approximate methods of complex systems to be discussed later.

1.1 The concept of kinematic degree of freedom and some simple examples

Previously, in the static analysis (see Vols. I and II of this series), we investigated extensively the behaviour of structures under slowly imposed, externally applied loads or deformations and strains. In this case, the degrees of freedom are identical to the prescribed deformation modes of the structure. In principle, every structure has an infinite number of deformation modes as may be observed from $\boldsymbol{u} = \{u \;\; v \;\; w\}$ which defines the deformation vector appertaining to each point whose coordinates are given by $\boldsymbol{x} = \{x \;\; y \;\; z\}$. If we now formulate the partial differential equations for arbitrary structures in accordance with the classical theory of elasticity, we immediately observe how difficult they are to solve, even in the case of static problems.

In dynamics, we have to consider the additional difficulty that the response of the structure is clearly time-dependent. Such a time-dependent effect is experienced, for example, when an aircraft flies through a turbulent region. In this situation, it is evident that we have to introduce a new variable, namely time.

The possible deformation of a structure originally at rest now has to be described in terms of $\boldsymbol{u}(\boldsymbol{x}, t)$ where $\boldsymbol{x}$ is the spatial vector of the point in question and t then defines the time. The classical (analytical) approach to the theoretical understanding of the dynamics of deformable solid bodies is based on the formation of partial differential equations in space and time for which the number of deformation modes or degrees of freedom becomes strictly infinite.

When applied first to the limiting case of a body at rest, this technique must, in principle, be applicable to all problems of equilibrium for deformable bodies. However, unless the model is greatly simplified, only very few highly idealised technical problems can be solved in practice in a purely analytical manner. The most fruitful idea of a physicist or engineer would be to reduce the infinite modes of deformation of a structure to an acceptable, finite number in such a way that the main characteristics of the response of the system can still be reproduced with sufficient accuracy. Thus, by reducing these modes of deformation (or degrees of freedom) to a finite number, the partial differential equations of static problems will be reduced to a system of linear, algebraic equations; these can then be solved numerically by the slave of modern science, the computer.

In the more complex domain of dynamics, however, the procedure is not quite so straightforward; this is because in this case, the relevant partial differential equations in space and time are reduced to a system of ordinary time-dependent differential equations, even for a restricted finite number of degrees of freedom in space. The dynamic problem is thus not only much more extensive than the corresponding static one, but also more time-consuming on the computer.

	statics	dynamics
problem definition	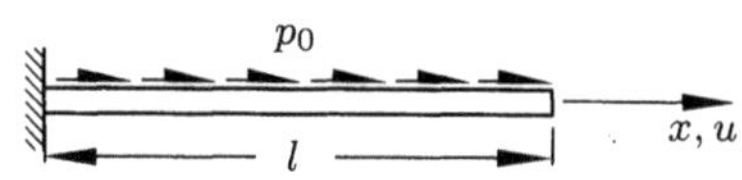	density $\varrho = \text{const.}$ Young's modulus $E = \text{const.}$ cross-section $A = \text{const.}$
	line load	
	$p(x) = p_0 = \text{const.}$	$p(x,t) = p_0(t) = \text{const. in } x$
	boundary conditions	
	$u\,(x=0) = 0$ $N(x=l\,) = 0$ (normal force)	$u(x=0,t) = 0$ $N(x=\,l,t) = 0$ (normal force)
	initial conditions	
		$u(x,t=0) = 0$ $\dot{u}(x,t=0) = 0$
formulation	$N = EA\dfrac{\partial u}{\partial x}$	
	equilibrium of forces 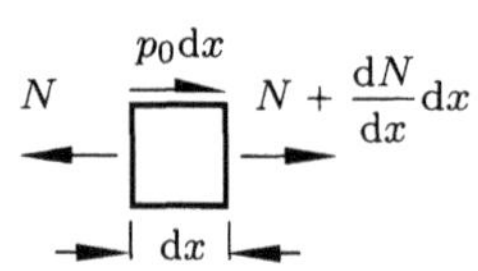 $\dfrac{\mathrm{d}N}{\mathrm{d}x} = -p_0$ ordinary differential equation (1D) $EA\dfrac{\mathrm{d}^2u}{\mathrm{d}x^2} = -p_0$	equilibrium of forces (d'Alembert) 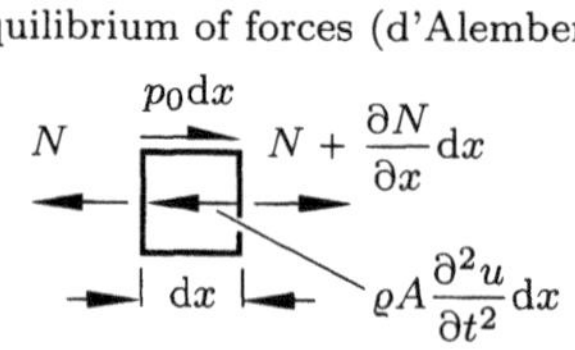 $\dfrac{\partial N}{\partial x} - \varrho A\dfrac{\partial^2 u}{\partial t^2} = -p_0$ partial differential equation $EA\dfrac{\partial^2 u}{\partial x^2} - \varrho A\dfrac{\partial^2 u}{\partial t^2} = -p_0$

Fig. 1.1.1 A simple static and dynamic problem, *Part 1*: classical formulation

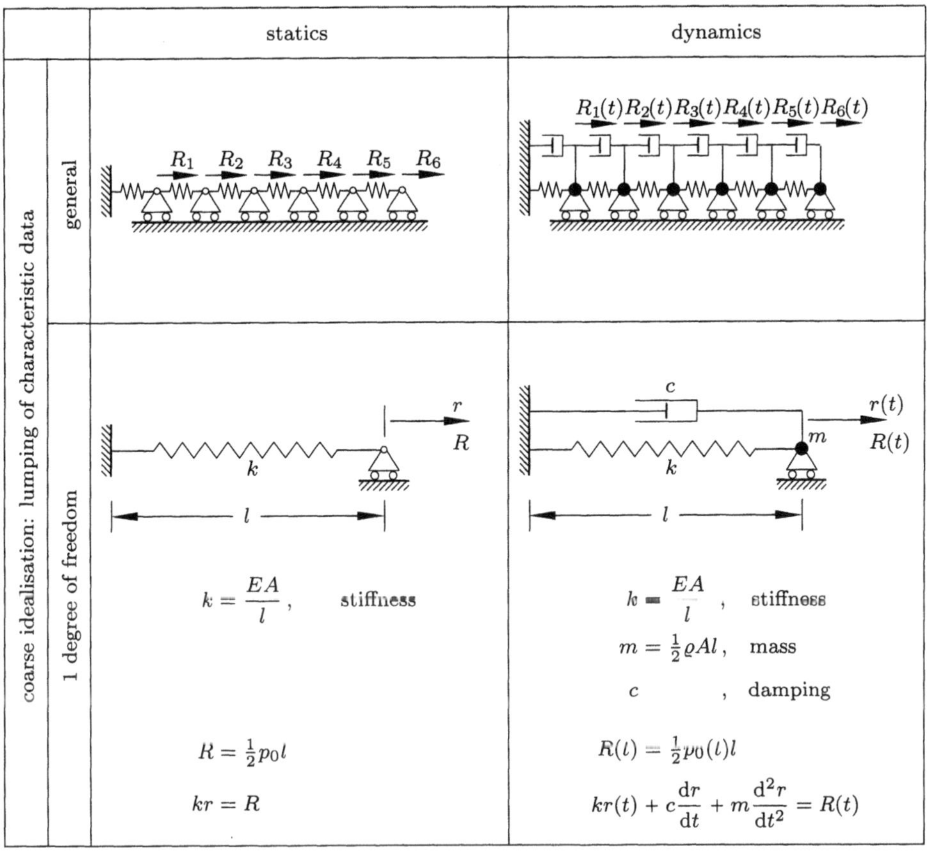

Fig. 1.1.1 A simple static and dynamic problem, *Part 2*: reduction to a single degree of freedom

In this introductory chapter, we propose the greatest simplification possible, i.e. the reduction of the infinite number of modes of a system to a single degree of freedom. Whilst on the one hand, systems with a single degree of freedom can obviously only represent a rough approximation of the real situation, on the other hand, this simple model and the study of its dynamics suggest a generalisation leading to an extension of this technique to multiple degrees of freedom. This will be discussed in subsequent chapters.

We shall now illustrate the above procedure with a simple example. Figure 1.1.1, *Part 1* shows a bar under a uniformly distributed, longitudinal load. In the formulation of the problem, the time-dependent loading and the initial conditions,neither of which appear in statics, are characteristic of a dynamic situation. In addition, a new concept, which will be described in more detail in section 1.2, has now to be introduced into the solution of a dynamic problem. This refers to the so-called inertial forces which have to be assigned to the masses of the moving system. These forces are deduced from Newton's Second Law of motion:

Force = mass × acceleration

By virtue of this relatively simple concept and d'Alembert's principle, the differential equations for static problems can easily be extended to dynamic situations. In our first analysis, we ignore damping. Since we have chosen a one-dimensional problem here, we obtain, in the static case, an ordinary differential equation in x (in general, partial in x, y, z). In dynamics, these equations become partial differential equations in x (in general x, y, z) and t. The number of degrees of freedom in our example is strictly speaking infinite. If we reduce these to a single degree of freedom, the choice of the approximate model may be in doubt. We could, for example, assume a linear variation of the longitudinal deformation, u, and select $u(x = l, t)$ as a free parameter. This, however, represents a generalisation of the concept of degrees of freedom and will be discussed later in section 1.7.

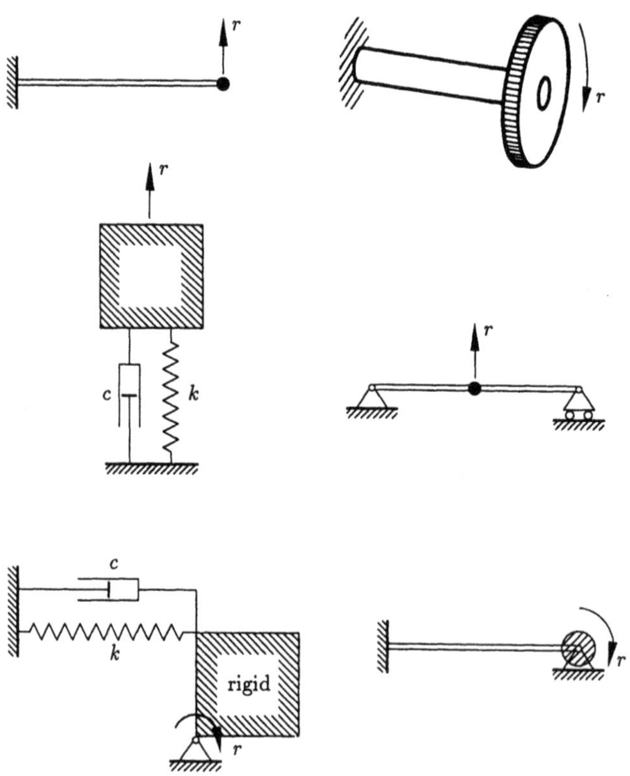

Fig. 1.1.2
Some simple models with a single degree of freedom and 'lumped' characteristics

Another simple way of obtaining a reduced number of degrees of freedom is to lump together the distributed characteristics of the structure such as elasticity, inertia and damping and thus to model the system with concentrated properties assigned intuitively to selected points. Historically, the simplest vibration models arose in this manner. In such a case, we may consider the system to be an assembly of point-masses joined by the elastic and damping ties as illustrated in fig. 1.1.1, *Part 2*. In the most extreme idealisation, we reduce the system to a single degree of freedom, the point-mass being assumed to move in one direction. For such a model, we obtain an algebraic equation in the static case and an ordinary time-dependent differential equation in the dynamic case. This is, however, a drastic simplification and the price we have to pay for it is by no means small. In fact, such a model cannot reproduce the dynamics of a continuum subject to a complex kinematic motion and thus will fail to account for the short wave phenomenon, for example.

In many technical problems, it is possible to lump together the characteristics, especially the mass and stiffness, in a meaningful way, so that the resulting model or idealisation is not as radical as shown in fig. 1.1.1. Thus, the wing of an aeroplane with large tip-tanks fits into the concept of a system with concentrated masses. Similarly, fly-wheel systems, car springs and machines on elastic foundations can also be idealised in this way, see fig. 1.1.2. In these simple problems, r unambiguously fixes the kinematics of the system and its inertial contribution. Our argument applies equally to r being a translation or rotation.

1.2 Forces in the dynamic equilibrium (stiffness, mass, damping)

The close relevance of dynamics to the study of the motion of structures immediàtely raises the question as to which reference system should be used to describe this motion. An analysis of the spring-mass system reveals how its response depends on the choice of the reference system. For example, an observer travelling on a lorry moving rectilinearly with a constant velocity notices that, in the absence of external forces, a spring-mass system does not deform in any direction, see fig. 1.2.1(a). However, if the vehicle accelerates, the spring stretches in the vehicle's direction of motion, see fig. 1.2.1(b), and if the vehicle travels at constant speed, but along a curved path, the spring-mass extends laterally, see fig. 1.2.1(c). If the acceleration case in (b) or the radius of the curve in case (c) vary, then the spring-mass system will start oscillating, even when it is initially at rest and the observer does not apply any forces. It is only in the case of (a) that there is no movement of the spring-mass system in the absence of external forces. Any forces are then referred to a special reference system known as the inertial reference or the Galilean system, after its original discoverer. In such a system, the acceleration and angular velocities possess absolute physical magnitudes and an object will only deviate from a uniform motion in the presence of external forces. It is also clear that the inertial property of a reference system can only be considered an approximation. Thus, in case (a), a lorry which is either standing still or moving along the Earth's surface is also subject to the Earth's rotation and experiences the known effects of a swinging pendulum (a reminder of physics lessons at school!).

In this introduction, we almost exclusively apply an inertial system. However, it is sometimes advantageous to introduce a different kind of reference system as is advisable, for example, in the case of the vibration of rotating turbine blades. In this case, special forces such as centrifugal and Coriolis' forces have to be considered and their effects can be clearly identified by an observer in a rotating non-Galilean system.

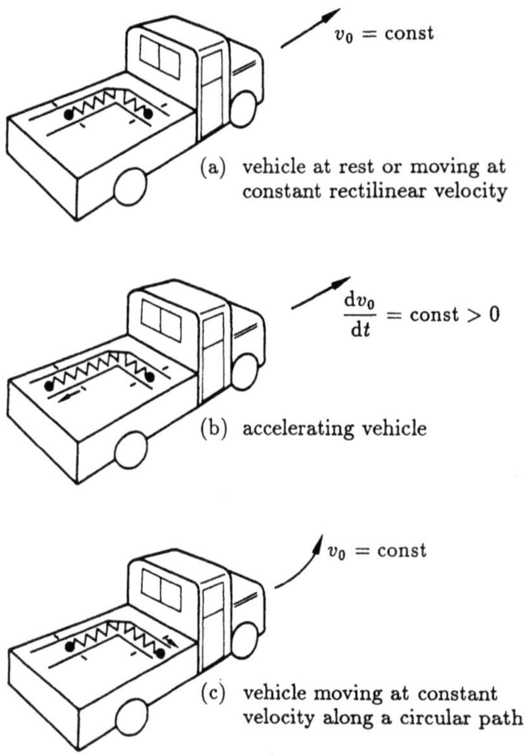

Fig. 1.2.1
Cargo platform of a truck as a reference system for the response of a spring-mass model

In a static analysis, a state of equilibrium is established between an external force, R, and a corresponding elastic restoring force, R_e (fig. 1.2.2, left). Similarly, in a dynamic analysis, an equilibrium state at any time, t, may be established using d'Alembert's principle. In realising this equilibrium, additional forces such as the inertial forces $R_I(t)$, have to be taken into account; all car drivers know that acceleration requires power. Also, in general, supplementary frictional forces $R_D(t)$ must be overcome. In statics, where any motion is by definition very slow, there are no mass forces because there is no acceleration and the frictional forces (at least in the case of fluid damping) are non-existent due to this absence of a relative velocity. In an inertial system with a single-degree-of-freedom model, the force equilibrium relations are illustrated in fig. 1.2.2

$$R_e(t) + R_I(t) + R_D(t) = R(t) \tag{1.2.1}$$

This is the fundamental equation of dynamics and its individual terms are described in more detail in the following paragraphs. In formulating the equations of motion for a non-inertial system, additional forces such as supplementary inertial forces, centrifugal and Coriolis' forces must also be taken into account.

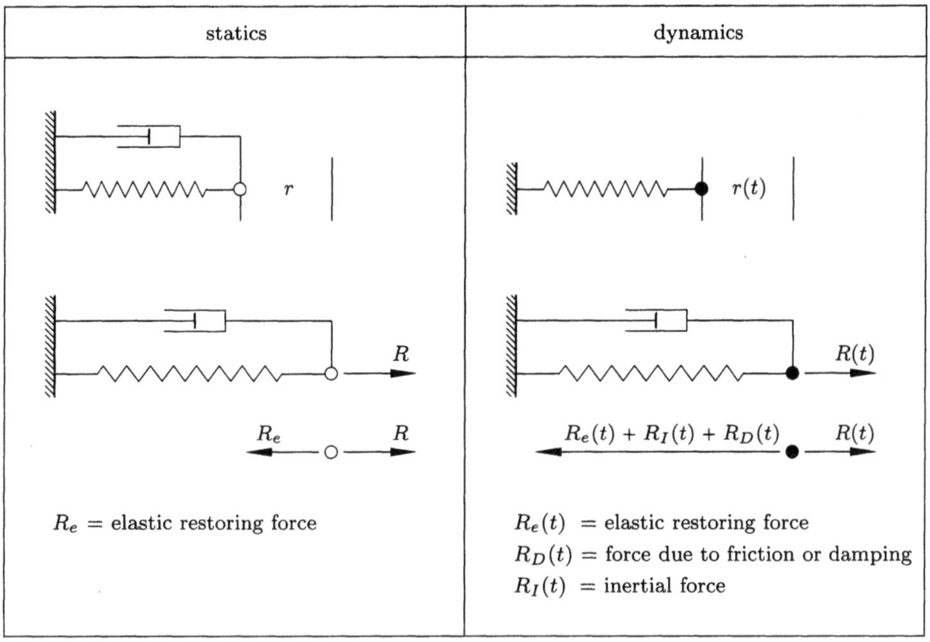

Fig. 1.2.2 Equilibrium in statics and dynamics

If a system is elastically deformed, elastic restoring forces arise which try to reduce the imposed deformation — the reader is already familiar with this behaviour in the static regime. For linear-elastic materials, the restoring forces are directly proportional to the deformation and may be determined solely by static considerations. Thus, the linear-elastic behaviour of the structure can be described by

$$R_e = kr \tag{1.2.2}$$

where k is both the linear elastic stiffness and the force required to stretch the structure by a unit length $r = 1$ (unit displacement method). Alternatively, we can write the above relation as

$$r = fR_e \tag{1.2.3}$$

where f is the flexibility of the structure and represents the displacement produced by a unit force $R_e = 1$ (unit load method).

From equs. (1.2.2) and (1.2.3), we deduce

$$\begin{aligned} k &= f^{-1} \\ f &= k^{-1} \end{aligned} \tag{1.2.4}$$

How k or f may be established has been discussed in detail in a number of chapters of Volumes I and II. Figure 1.2.3 shows several elementary examples.

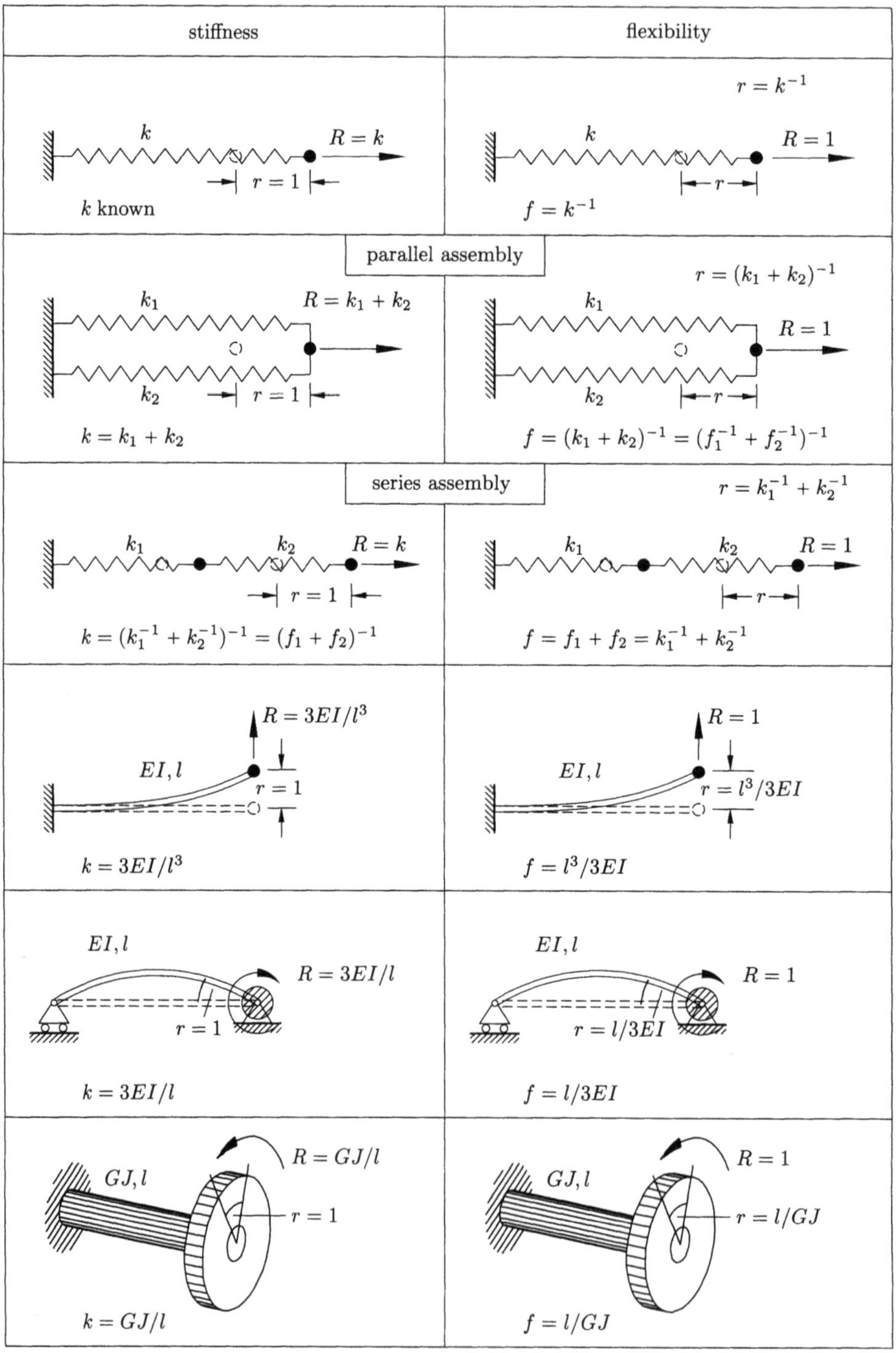

Fig. 1.2.3 A few simple examples for stiffness and flexibility

Having discussed the elastic forces, we next elucidate the inertia forces. Let us consider a spring-mass system which is set in motion under the influence of an external force $R(t)$. In this case, the external force must overcome not only the elastic but also the inertial force, fig. 1.2.4. Even if there are no elastic forces present, the acceleration of the mass requires a force. Conversely, it can be deduced that if $R(t)$ has to accelerate the mass, then $R_e < R(t)$ must hold in the initial phase of motion. Thus, the inertia forces initially reduce the effective load acting on the spring although this is not necessarily true in the later phases of the motion. Inertial forces can also magnify the stresses and strains in a system as opposed to a static loading.

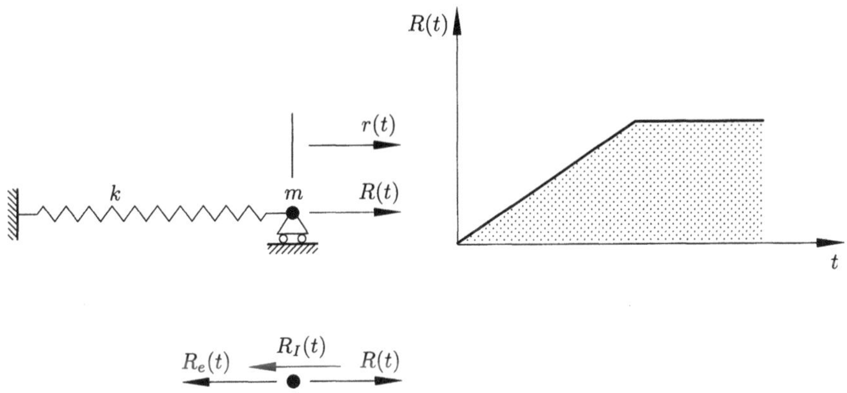

Fig. 1.2.4 When setting a mass in motion by an excitation force $R(t)$, both spring and inertia forces have to be overcome

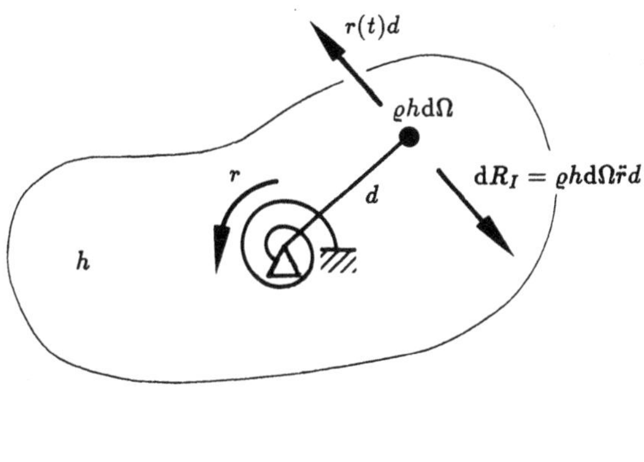

Fig. 1.2.5
Generation of rotational inertia in a disk of thickness h

From Newton's law, the inertia force $R_I(t)$ is:

$$R_I(t) = m\ddot{r}(t) \tag{1.2.5}$$

If r denotes a translational motion, then the unit of mass m is defined in [kg = $\mathrm{Ns^2/m}$]. If, on the other hand, r represents a rotation, R_I represents an angular inertial moment and m becomes the polar moment of mass inertia.

In order to calculate the polar moment of inertia of an arbitrary rotating disk of thickness h, we consider a local mass $\rho h \mathrm{d}\Omega$ [kg] placed at a distance d from the centre of rotation (fig. 1.2.5) where ρ is the density and $\mathrm{d}\Omega$ is an infinitesimal area of the disk. For an angular acceleration $\ddot{r}(t)$, the mass experiences an acceleration and hence an inertia force $\rho h \mathrm{d}\Omega d\ddot{r}(t)$. The mass polar moment of inertia is then easily obtained by integration

$$m = \rho h \int_\Omega d^2 \mathrm{d}\Omega = \rho h I_p \tag{1.2.6}$$

and has the dimension [$\mathrm{kg\,m^2 = Ns^2m}$]. Here, I_p denotes the standard or static polar moment of inertia about the axis of rotation. For example, for a solid circular disk of radius a, I_p is given by

$$I_p = \int_0^{2\pi}\int_0^{a} \bar{r}^3 \mathrm{d}\bar{r}\mathrm{d}\varphi = \tfrac{1}{2}\pi a^4$$

and for a solid square with an edge length a

$$I_p = \int_{-a/2}^{a/2}\int_{-a/2}^{a/2} (x^2 + y^2)\mathrm{d}x\mathrm{d}y = I_x + I_y = 2I_x$$

$$= \tfrac{1}{12}a^4$$

The third type of force listed in equ. (1.2.1), namely the damping force, will be discussed in more detail in Chapter 7. Here, we consider only briefly some elementary characteristics of damping models as applying to a simple oscillator.

In nature, we observe that the amplitude of any oscillating system, when set in motion, decays with time. Consequently, in addition to the interchange of elastic and inertia forces, globally represented by strain and kinetic energies respectively, there must be a mechanism at work dissipating mechanical energy. If, for example, we were to measure the temperature of the spring during the oscillation of a spring-mass system, a temperature rise would be detected, indicating that mechanical energy is converted into heat, resulting in a decrease of the kinetic energy. The cause of this dissipation needs to be clearly understood, although the usual damping models proposed so far only approximate the process with a reasonable degree of accuracy. This is due, *inter alia*, to the various sources of dissipation which act simultaneously upon the structure. Generally, the damping forces are classified according to the following groups:

— Viscous damping.

  This model is the most widely used. It is found to be a good approximation for the friction effect on an oscillator in oil or air.

— Structural damping.

  This is the result of internal material damping and the friction of the joints. The latter source is very difficult to specify as this friction depends, amongst other things, upon the bearing clearance and upon lubrication.

— Dry friction or Coulomb damping.

  Although this kind of damping describes the motion of a body on a dry surface quite well, it has not been generally applied up to the present time.

The friction force, $R_D(t)$, resulting from the viscous model, is taken to be proportional to the velocity, $\dot{r}(t)$, and a damping constant, c, is introduced as a constant of proportionality. This is dependent upon the system, but may also depend on the temperature. For a single one-dimensional oscillator, this mechanism may be represented by a dash-pot (as in a car suspension, see fig. 1.2.6).

$$R_D(t) = c\dot{r}(t) \tag{1.2.7}$$

Here, c cannot simply be calculated in terms of internal or external properties such as the mass and stiffness, but must generally be either measured experimentally or estimated.

The second cause of frictional, structural damping is also known as hysteretic damping (see fig. 1.4.6). A discussion of this kind of damping will be given in section 1.4.1 in conjunction with the steady-state oscillation. For a single-degree-of-freedom system, the process may formally be expressed as in equ. (1.2.7).

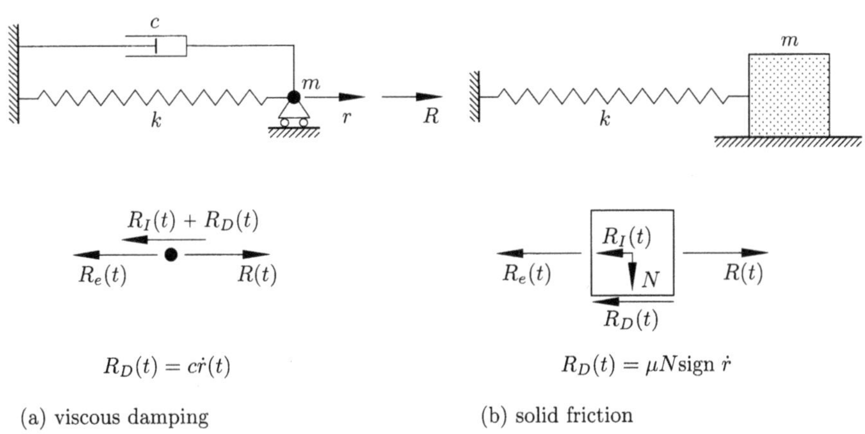

Fig. 1.2.6 Two models of friction in a moving system

For dry friction, the frictional force is proportional to the contact pressure N. In the case of our simple model in fig. 1.2.6(b), this is simply the weight of the body. Introducing the coefficient of friction μ, we have,

$$R_D(t) = \mu N \,\mathrm{sign}\, \dot{r} \tag{1.2.8}$$

Values of μ can be found for various combinations of materials in standard handbooks. It is different for sliding and adherent contact. For example, for the combination of steel on steel, the coefficient of static friction, μ_0, has a value ranging from 0.1 to 0.15 (oiled or dry surfaces) but for sliding friction, μ is reduced to a value between 0.01 and 0.15 (oiled or dry surfaces).

1.3 Free oscillations

Using the fundamental expression for the equilibrium of forces, equ. (1.2.1), and the definitions of the forces given in section 1.2, we are now in a position to establish the equation of motion of an oscillator with viscous damping

$$m\ddot{r} + c\dot{r} + kr = R(t) \tag{1.3.1}$$

Equation (1.3.1) is a linear, inhomogeneous, time-dependent differential equation whose solution, $r(t)$, depends upon the external force, $R(t)$. In contrast to the static analysis where, in the absence of an external force, there is only the trivial solution $r = 0$, in the dynamic case a non-trivial solution exists for the homogeneous differential equation ($R(t) = 0$). If the external force $R(t) = 0$, then equ. (1.3.2) reduces to the case of a free oscillation and the corresponding linear equation of motion for viscous damping becomes

$$m\ddot{r} + c\dot{r} + kr = 0 \tag{1.3.2}$$

Dividing by m and introducing the abbreviations ω and ζ, the significance of which will become clear later on,

$$\omega^2 = \frac{k}{m} = \frac{1}{fm} \tag{1.3.3}$$

$$2\zeta\omega = \frac{c}{m} = \frac{c}{k}\omega^2 \tag{1.3.4}$$

equ. (1.3.2) reduces to

$$\ddot{r} + 2\zeta\omega\dot{r} + \omega^2 r = 0 \tag{1.3.5}$$

A solution of equ. (1.3.5) is then given by

$$r = \mathrm{e}^{pt} \tag{1.3.6}$$

By substituting equ. (1.3.6) into equ. (1.3.5), the second-order differential equation is transformed into the quadratic algebraic equation

$$p^2 + 2\zeta\omega p + \omega^2 = 0 \tag{1.3.7}$$

The two solutions of this equation are

$$p_{1,2} = \omega\left(-\zeta \pm \sqrt{\zeta^2 - 1}\right) \tag{1.3.8}$$

where ω has a positive value

$$\omega = {}_{+}\sqrt{\omega^2} = {}_{+}\sqrt{k/m} \tag{1.3.9}$$

Thus, the general solution of the homogeneous differential equation (1.3.5) may be written in the form

$$r = A_1 \mathrm{e}^{p_1 t} + A_2 \mathrm{e}^{p_2 t} \tag{1.3.10}$$

where the constants of integration, A_1 and A_2, have to be determined from the initial values of $r(t_0)$ and $\dot{r}(t_0)$. In the case of equal roots, $p_1 = p_2$, the general solution reduces to the well-known expression

$$r = (A_1 + A_2 t)\, \mathrm{e}^{p_1 t} \tag{1.3.11}$$

Clearly, the magnitude of the dimensionless damping coefficient ζ plays a decisive role in determining the nature of oscillation. For $\zeta = 1$, the motion shows a critical limit. We distinguish between the following cases:

- $\zeta < 1$ sub-critical or light damping, (as a rule, metallic structures have values in the range of ζ=0.02 to 0.05).
- $\zeta = 1$ critical damping.
- $\zeta > 1$ over-critical or heavy damping (this is usually applied for measuring instruments).

As a matter of fact, systems exist for which ζ has a negative value indicating a continuous excitation or magnification of the oscillation. This case may arise, for example, in aeroelasticity and will be discussed in Vol. VI.

Sub-critical damping $\zeta < 1$

For $\zeta < 1$, the roots become complex and it is helpful to express them in the form

$$p_{1,2} = \omega\left(-\zeta \pm i\sqrt{1 - \zeta^2}\right)$$

The corresponding displacement, r, is then given by

$$r = \mathrm{e}^{-\zeta\omega t}\left(A_1 \mathrm{e}^{i\sqrt{1-\zeta^2}\omega t} + A_2 \mathrm{e}^{-i\sqrt{1-\zeta^2}\omega t}\right) \tag{1.3.12}$$

Using Euler's formula

$$\mathrm{e}^{\pm i\alpha} = \cos\alpha \pm i \sin\alpha \tag{1.3.13}$$

equ. (1.3.12) may now be written as

$$r(t) = e^{-\zeta\omega t}\left[(A_1 + A_2)\cos\sqrt{1-\zeta^2}\omega t + i(A_1 - A_2)\sin\sqrt{1-\zeta^2}\omega t\right] \quad (1.3.14)$$

It should be noted that in the general solution (1.3.10), the constants of integration, A_1 and A_2, are usually complex. They are furthermore complex conjugate since $r(t)$ must be real. Hence, the general solution can be rewritten as

$$r(t) = e^{-\zeta\omega t}\left(a\cos\sqrt{1-\zeta^2}\omega t + b\sin\sqrt{1-\zeta^2}\omega t\right) \quad (1.3.15)$$

where a and b are now real constants. This general solution describes for $\zeta > 0$ a decaying oscillatory motion. The frequency of oscillation for an undamped system with $\zeta = 0$ is ω, whereas for a damped system with $\zeta \neq 0$, the frequency is given by

$$\omega_D = \sqrt{1-\zeta^2}\omega \quad (1.3.16)$$

Here, ω_D denotes the damped angular eigenfrequency. For lightly damped systems, for example $\zeta = 0.05$, there is little difference between the two values ($\omega_D = 0.9987\omega$). The angular frequency ω, or ω_D, defines the number of equidirectional zero-crossings in 2π seconds. On the other hand, the engineering eigenfrequencies ν resp. ν_D denote the number of uni-directional zero-crossings in 1 sec

$$\nu = \frac{\omega}{2\pi}, \qquad \nu_D = \frac{\omega_D}{2\pi} \quad (1.3.17)$$

The reciprocals of ν and ν_D define the periods of the corresponding oscillations,i.e.

$$T = \frac{1}{\nu} = \frac{2\pi}{\omega}, \qquad T_D = \frac{1}{\nu_D} = \frac{2\pi}{\omega_D} \quad (1.3.18)$$

These expressions measure the time between two consecutive, equidirectional zero-crossings. With these definitions, we deduce from equ. (1.3.15) the alternative expressions

$$\begin{aligned} r(t) &= e^{-\zeta\omega t}\left(a\cos\omega_D t + b\sin\omega_D t\right) \\ &= e^{-2\pi\zeta t/T}\left(a\cos(2\pi t/T_D) + b\sin(2\pi t/T_D)\right) \\ &= \bar{a}e^{-2\pi\zeta t/T}\cos\left(2\pi t/T_D - \varphi\right) \end{aligned} \quad (1.3.19)$$

Following the addition theorem for trigonometric functions, we may use a single shifted sine or cosine function in place of a combination of cosine and sine functions.

In this case, we introduce as alternative constants the amplitude $\bar{a}$ and the phase shift φ. The new constants are related to the original ones via

$$\begin{aligned} a &= \bar{a}\cos\varphi \\ b &= \bar{a}\sin\varphi \\ \bar{a} &= \sqrt{a^2 + b^2} \\ \varphi &= \tan^{-1}(b/a) \end{aligned} \tag{1.3.20}$$

The determination of the unknown constants a and b (or $\bar{a}$ and φ) follows, as already mentioned, from the initial conditions. Our solution does not, of course, imply that an oscillatory state has occurred without any prior excitation, but simply that the motion defined by equ. (1.3.19) can take place when no external force, $R(t)$, is exerted. In addition, there is always the trivial solution $r(t) = 0$ (state of rest). The condition of excitation of a free oscillation is here realised by the imposition of initial conditions. For example, $r(0)$ and $\dot{r}(0)$ may be given. Then we have

$$\begin{aligned} a &= r(0) \\ b &= \frac{\dot{r}(0)}{\omega_D} + \frac{\omega}{\omega_D}\zeta r(0) \\ \bar{a} &= \left[r(0)^2 + \left(\frac{\dot{r}(0) + \zeta\omega r(0)}{\omega_D} \right)^2 \right]^{1/2} \\ \varphi &= \tan^{-1}\left(\frac{\omega}{\omega_D}\zeta + \frac{\dot{r}(0)}{\omega_D r(0)} \right) \end{aligned} \tag{1.3.21}$$

The kinematic interpretation of the constants $\bar{a}$ and φ is demonstrated in fig. 1.3.1. The amplitude of the motion is illustrated in a polar diagram on the left of the figure, in which the projection onto the vertical axis of the vector-pointer represents $r(t)$. Due to damping, the amplitude and the length of the vector decrease with time. Consequently, the tip of the travelling arrow describes a spiral.

Finally, let us discuss the decreasing amplitude of the free motion. To this end, we consider two displacements separated by a time interval $\mathrm{n}T_D$, where T_D is the period of the damped oscillator

$$\begin{aligned} r(t) &= \bar{a}\mathrm{e}^{-\zeta\omega t}\cos\left(\omega_D t - \varphi\right) \\ r(t + \mathrm{n}T_D) &= \bar{a}\mathrm{e}^{-\zeta\omega(t+\mathrm{n}T_D)}\cos\left(\omega_D t - \varphi + 2\mathrm{n}\pi\right) \end{aligned} \tag{1.3.22}$$

The cosine term has a period of 2π. Hence, dividing the first expression by the second, we get (assuming that $\cos \neq 0$)

$$\frac{r(t)}{r(t + \mathrm{n}T_D)} = \mathrm{e}^{\zeta\omega\mathrm{n}T_D} \tag{1.3.23}$$

or

$$\ln\frac{r(t)}{r(t + \mathrm{n}T_D)} = \zeta\omega\mathrm{n}T_D = \mathrm{n}2\pi\zeta(1 - \zeta^2)^{-1/2} \tag{1.3.24}$$

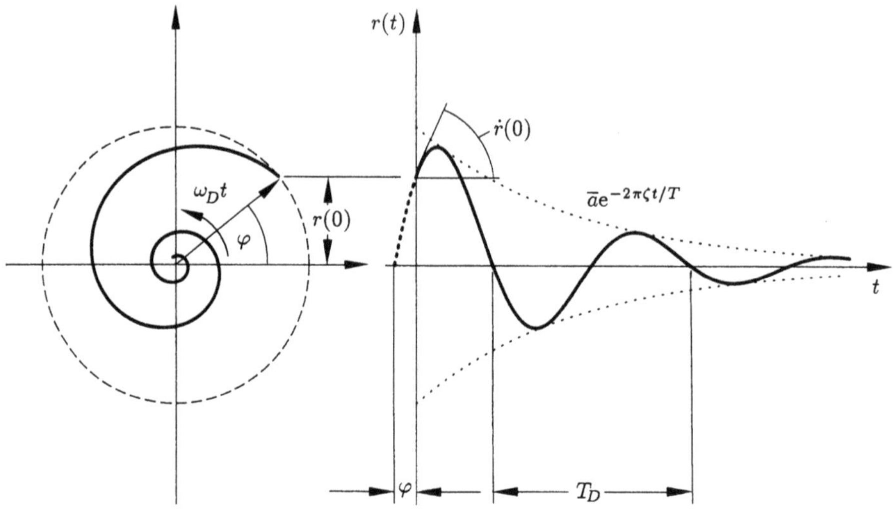

Fig. 1.3.1 Free response of an oscillator with viscous damping; left: vector diagram, right: r/t-diagram

For n=1, the equ. (1.3.24) represents the so-called logarithmic decrement

$$\delta = \ln \frac{r(t)}{r(t+T_D)} = 2\pi\zeta(1-\zeta^2)^{-1/2} \tag{1.3.25}$$

For lightly-damped systems

$$\delta \approx 2\pi\zeta, \qquad \text{because } (1-\zeta^2)^{-1/2} \approx 1 \tag{1.3.26}$$

This logarithmic decrement, and the corresponding damping coefficient, can be measured experimentally by observing the decay of the amplitude over one or more periods

$$\delta = \frac{1}{\mathrm{n}} \ln \frac{r(t)}{r(t+\mathrm{n}T_D)} \tag{1.3.27}$$

Thus, if the amplitude were to be reduced to half of its value during one period, then $\zeta \approx 0.1$, whereas if the same reduction occurred within two periods, $\zeta \approx 0.05$.

Critical damping $(\zeta = 1)$

For $\zeta = 1$, equ. (1.3.7) yields the real double root

$$p_1 = p_2 = -\zeta\omega = -\omega \tag{1.3.28}$$

The corresponding general solution is then (cf. also equ. (1.3.11))

$$r(t) = (a + bt)\,\mathrm{e}^{-\omega t} \tag{1.3.29}$$

This solution does not, of course, describe an oscillatory motion in the original sense of the word but rather a slowly decaying motion (fig. 1.3.2). From the initial conditions, such as $r_0 = r(0)$ and $v_0 = \dot{r}(0)$, the constants of integration, a and b, can be determined; the response of the system may then be expressed as

$$r(t) = \big(r_0 + (v_0 + \omega r_0)t\big)\mathrm{e}^{-\omega t} \tag{1.3.30}$$

Over-critical damping ($\zeta > 1$)

For $\zeta > 1$, both roots of equ. (1.3.8) are real and since

$$\sqrt{\zeta^2 - 1} < \zeta$$

both are negative

$$p_{1,2} = -\omega\left(\zeta \pm \sqrt{\zeta^2 - 1}\right) < 0$$

The response in the present case is simply

$$r(t) = a\,\mathrm{e}^{p_1 t} + b\,\mathrm{e}^{p_2 t} \tag{1.3.31}$$

Once again, this solution does not describe an oscillatory motion but rather a decaying motion consisting of two superimposed exponential functions both with negative exponents.

Figure 1.3.2 demonstrates the influence of the viscous damping on the motion over a wide range of undamped to heavily over-damped systems. The transition from a sub-critical to critical damping and from over-critical to critical damping can clearly be discerned and may also be derived analytically by a limiting process $\zeta \to 1$ from equs. (1.3.19), (1.3.21) and equ. (1.3.31). The fastest reduction to an equilibrium position (state of rest) can be achieved through critical damping, further increase of damping merely prolongs the decaying process of the amplitude.

Negative damping ($\zeta < 0$)

As has already been mentioned in the introduction, for computational and physical convenience, the engineer conceives models of complex processes, in which energy is not only dissipated through standard damping but also pumped into the system through action of the surrounding medium. This is, for example, the case in aeroelastic problems, in which the surrounding air may inject energy into a flying vehicle inducing an incited oscillation instead of a damped motion; see Vol. VI. If the inciting forces are taken to be proportional to the velocity, such oscillation processes can still be ascribed to the familiar viscous damping model as long as

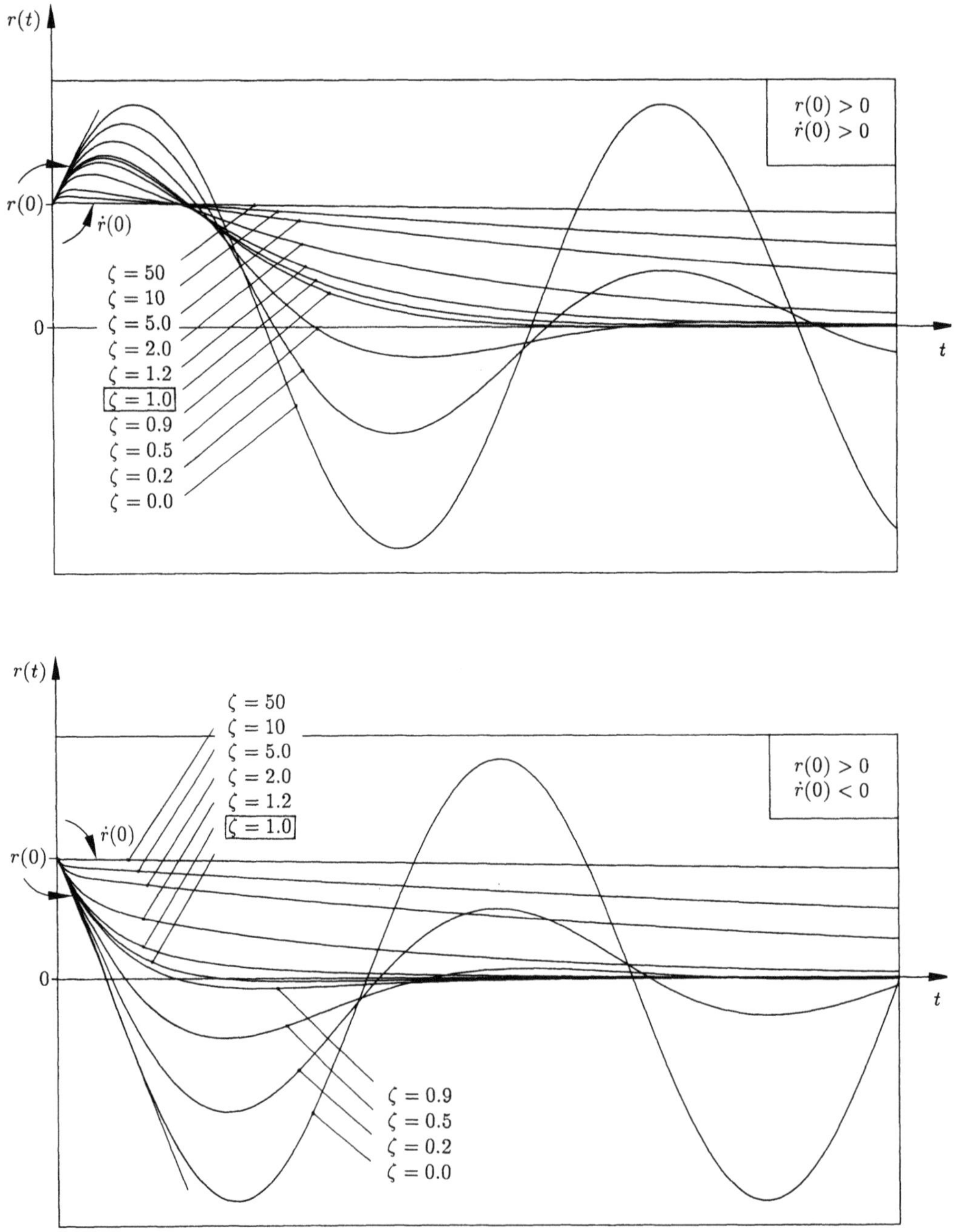

Fig. 1.3.2 Characteristic responses of an oscillator with viscous damping; *Part 1*: sub-critical, critical, over-critical damping

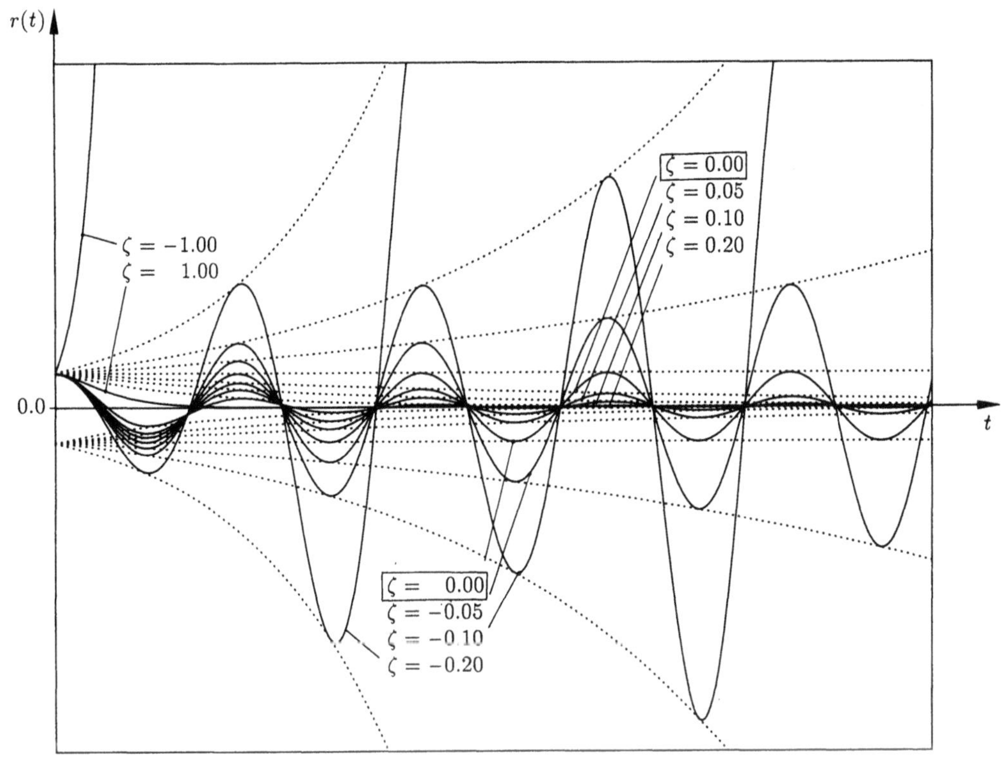

Fig. 1.3.2 Characteristic responses of an oscillator with viscous damping; *Part 2*: positive and negative viscous damping

negative values are assigned to the damping parameter ζ (hence the expression "negative damping").

The mathematical deduction developed for positive ζ-values need not be modified, nor do the distinctions made above, indeed the three ranges for ζ become now $-1 < \zeta < 0$, $\zeta = -1$, $\zeta < -1$). Since $p_{1,2}$ of equ. (1.3.8) is now always positive, the general response according to equs. (1.3.10) and (1.3.11) always contains terms $\exp(pt)$ with positive p, i.e. with increasing t, the response (theoretically) exceeds all limits.

Some time-history responses for not-too-large damping or incitement ratio ζ are shown in fig. 1.3.2, *Part 2*. The envelopes in the sub-critical cases ($|\zeta| < 1$) are also indicated; their mathematical expression for both positive and negative ζ-values is taken from equ. (1.3.19)

$$r_{\text{env}}(t) = \bar{a}\exp(-\zeta\omega t)$$

Coulomb damping or dry friction

In this case, the resisting force due to damping, apart from its sign, is independent of the velocity (see equ. (1.2.8)). Hence, the equation of motion is best set up separately over two ranges of validity

$$
\begin{aligned}
m\ddot{r} + kr &= -\mu mg, \qquad \text{for} \quad \dot{r} > 0 \\
m\ddot{r} + kr &= +\mu mg, \qquad \text{for} \quad \dot{r} < 0
\end{aligned} \tag{1.3.32}
$$

where μ is the coefficient of friction. It is assumed that the frictional force results solely from the weight of the mass, mg (see fig. 1.2.6, right). If the static friction force, $\mu_0 mg$, is not overcome at the dead-centre of motion, the movement will cease (μ_0, coefficient for static friction). The homogeneous solutions of both differential equations (1.3.32) are identical. In fact, transforming equ. (1.3.32) slightly, we have

$$\ddot{r} + \omega^2 r = 0 \tag{1.3.33}$$

Its standard solution is

$$r_h = a\cos\omega t + b\sin\omega t \tag{1.3.34}$$

(see also equ. (1.3.15) for $\zeta = 0$).

Furthermore, we may immediately deduce from

$$\ddot{r} + \omega^2 r = \mp\mu g \tag{1.3.35}$$

the particular solutions

$$r_p = \mp\mu\frac{g}{\omega^2} \tag{1.3.36}$$

Next, we must specify the initial conditions in order to determine the constants a and b. To this effect, we assume that the mass m is displaced statically by r_0 and then set free. The initial conditions are then

$$
\begin{aligned}
r(0) &= r_0 \\
\dot{r}(0) &= 0
\end{aligned} \tag{1.3.37}
$$

In order to initiate the motion of the mass, the static displacement should be sufficiently large so that the elastic restoring force can overcome the frictional force

$$kr_0 > \mu_0 mg$$

or

$$r_0 > \mu_0\frac{g}{\omega^2} \tag{1.3.38}$$

Assuming that this condition holds, the mass moves to the left ($\dot{r} < 0$) and hence, in equ. (1.3.36), the positive sign is valid. The complete solution is then given by

$$r(t) = a\cos\omega t + b\sin\omega t + \mu\frac{g}{\omega^2} \tag{1.3.39}$$

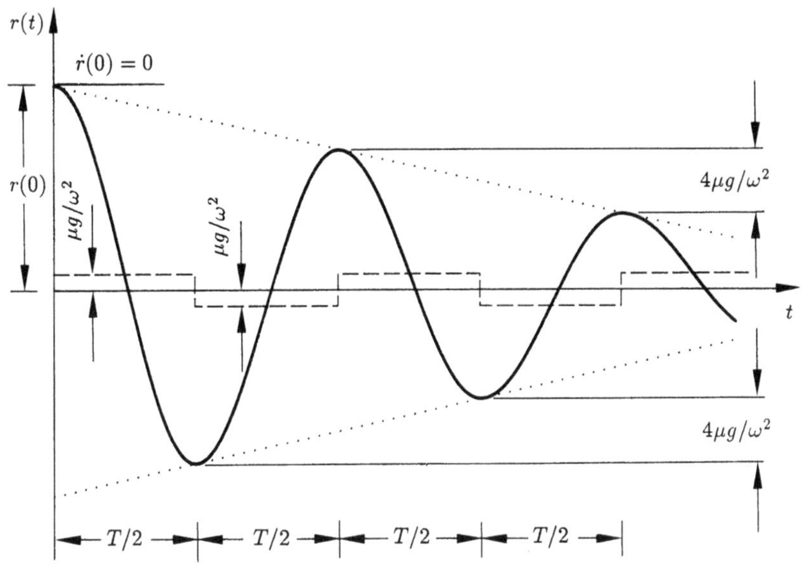

Fig. 1.3.3 Response of an oscillator with Coulomb damping or dry friction

Applying the initial conditions of equ. (1.3.37), we obtain the specific solution

$$r(t) = \left(r_0 - \mu\frac{g}{\omega^2}\right)\cos\omega t + \mu\frac{g}{\omega^2} \tag{1.3.40}$$

This represents a cosine oscillation shifted by $\mu g/\omega^2$, as shown in fig. 1.3.3. We derive the velocity from equ. (1.3.40) as

$$\dot{r}(t) = -\omega\left(r_0 - \mu\frac{g}{\omega^2}\right)\sin\omega t \tag{1.3.41}$$

The reversal point at $\dot{r}(t_1) = 0$ occurs for

$$t_1 = T/2 \tag{1.3.42}$$

The corresponding amplitude

$$|r(t_1)| = \left|-r_0 + 2\mu\frac{g}{\omega^2}\right| > \mu_0\frac{g}{\omega^2} \tag{1.3.43}$$

satisfies the condition given by equ. (1.3.38) and overcomes the static friction at the dead centre. The upper sign in equ. (1.3.35) is now applicable and the initial conditions at t_1 yield the solution

$$r(t) = \left(r_0 - 3\mu\frac{g}{\omega^2}\right)\cos\omega t - \mu\frac{g}{\omega^2} \tag{1.3.44}$$

The amplitude of the cosine term has decreased by $2\mu g/\omega^2$ as compared to the preceding solution of equ. (1.3.40). The shift is now $\mu g/\omega^2$ downwards, as shown in fig. 1.3.3. The next reversal-point occurs at

$$t_2 = T \tag{1.3.45}$$

and its displacement is given by

$$r(t_2) = r_0 - 4\mu \frac{g}{\omega^2} \tag{1.3.46}$$

This procedure can be continued until the system comes to rest. Further calculations can be dispensed with since the solution always consists of shifted cosine oscillations with the same frequency as that of the undamped system. The total amplitude falls off linearly (in the viscous case exponentially) with a rate of decrease of $4\mu g/\omega^2$ per period T.

1.4 Forced oscillations

Having discussed the dynamic motion in the absence of an external force, we now assume the presence of such a force $R(t)$. In place of equ. (1.3.5), the equation of motion for viscous damping now becomes

$$\ddot{r} + 2\zeta\omega\dot{r} + \omega^2 r = \omega^2 \frac{R(t)}{k} \tag{1.4.1}$$

If we are to limit ourselves to light-damping ($\zeta < 1$), which occurs most frequently in engineering, the homogeneous solution of the system is readily obtained from equ. (1.3.19) in the form

$$r_h(t) = \mathrm{e}^{-\zeta\omega t}\left(a\cos\omega_D t + b\sin\omega_D t\right) \tag{1.4.2}$$

The particular solution of equ. (1.4.1) depends upon the nature of the force, $R(t)$. In the following sections, we discuss the most important cases.

1.4.1 Harmonic excitation

Let

$$R(t) = R_0 \sin\omega_E t \tag{1.4.3}$$

where ω_E is the excitation frequency of the external force, fig. 1.4.1. Clearly, the static response r_0, caused by the amplitude R_0 of the load, is given by

$$r_0 = \frac{R_0}{k} \tag{1.4.4}$$

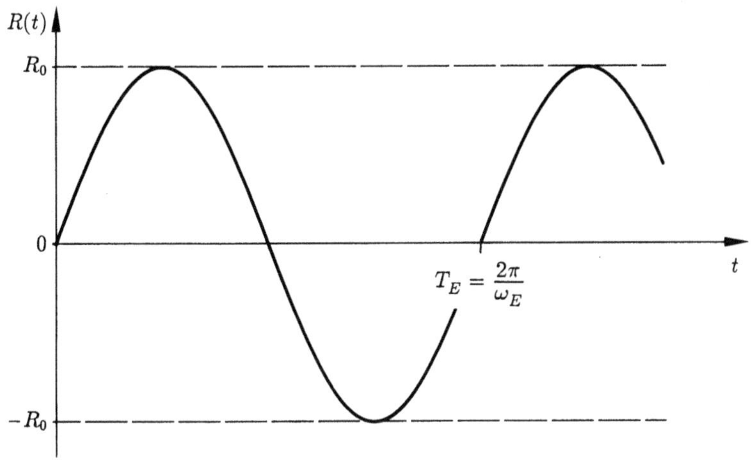

Fig. 1.4.1 Harmonic force excitation

Thus, for extremely slow oscillations, we would obtain a maximum deflection of $\pm r_0$. We now seek a particular solution r_p for

$$\ddot{r} + 2\zeta\omega\dot{r} + \omega^2 r = \omega^2 r_0 \sin \omega_E t \tag{1.4.5}$$

and assume that

$$r_p(t) = a_p \sin \omega_E t + b_p \cos \omega_E t \tag{1.4.6}$$

Substituting this expression into equ. (1.4.5) and comparing coefficients, we obtain upon introduction of the frequency ratio

$$\beta = \frac{\omega_E}{\omega} \tag{1.4.7}$$

two linear simultaneous equations for the constants a_p and b_p

$$\begin{bmatrix} 1-\beta^2 & -2\zeta\beta \\ 2\zeta\beta & 1-\beta^2 \end{bmatrix} \begin{bmatrix} a_p \\ b_p \end{bmatrix} = \begin{bmatrix} r_0 \\ 0 \end{bmatrix} \tag{1.4.8}$$

Their solution yields

$$a_p = \frac{1-\beta^2}{(2\zeta\beta)^2 + (1-\beta^2)^2} r_0$$

$$b_p = \frac{-2\zeta\beta}{(2\zeta\beta)^2 + (1-\beta^2)^2} r_0 \tag{1.4.9}$$

Thus, the general solution of equ. (1.4.5) becomes

$$
\begin{aligned}
r(t) &= r_h(t) + r_p(t) \\
&= \mathrm{e}^{-\zeta\omega t}\left(a\cos\omega_D t + b\sin\omega_D t\right) \\
&\quad + \frac{r_0}{(2\zeta\beta)^2 + (1-\beta^2)^2}\left[(1-\beta^2)\sin\omega_E t - 2\zeta\beta\cos\omega_E t\right]
\end{aligned}
\tag{1.4.10}
$$

The two constants, a and b, can be determined from the initial conditions. The solution, equ. (1.4.10), consists of two parts: the first part represents a damped oscillation with the damped eigenfrequency ω_D, and the second an undamped oscillation with the excitation frequency of ω_E. If the initial transient motion is required, then both terms of the solution are significant. However, if one is interested only in the ultimate steady-state response, as is frequently the case, it is clear that only the second term, i.e. the particular solution, is of importance since the first term decays with time. A more concise expression for r_p is

$$
r_p(t) = r_0 V(\beta,\zeta)\sin(\omega_E t - \varphi) \tag{1.4.11}
$$

Using simple trigonometry and comparing coefficients with equ. (1.4.10), we obtain

$$
\begin{aligned}
r_0 V\cos\varphi &= \frac{1-\beta^2}{(2\zeta\beta)^2 + (1-\beta^2)^2} r_0 \\
r_0 V\sin\varphi &= \frac{2\zeta\beta}{(2\zeta\beta)^2 + (1-\beta^2)^2} r_0
\end{aligned}
\tag{1.4.12}
$$

Here, V is the so-called amplification factor and is given by

$$
V = \left[(2\zeta\beta)^2 + (1-\beta^2)^2\right]^{-1/2} \tag{1.4.13}
$$

and the corresponding phase angle is

$$
\varphi = \tan^{-1}\frac{2\zeta\beta}{1-\beta^2} \tag{1.4.14}
$$

The steady-state response due to a sinusoidal excitation is thus sinusoidal with the excitation frequency ω_E and a phase shift of φ. The amplitude of this motion is magnified by the amplification factor V (sometimes also termed 'dynamic magnification factor') as against the static response r_0. Important limiting cases are:

(a) $\beta \to 0$, which gives

$$
V \to 1
$$
$$
\varphi \to 0
$$

Thus, if the excitation frequency, ω_E, is very low, the displacement follows the excitation without any phase shift or amplification and the damping plays no role.

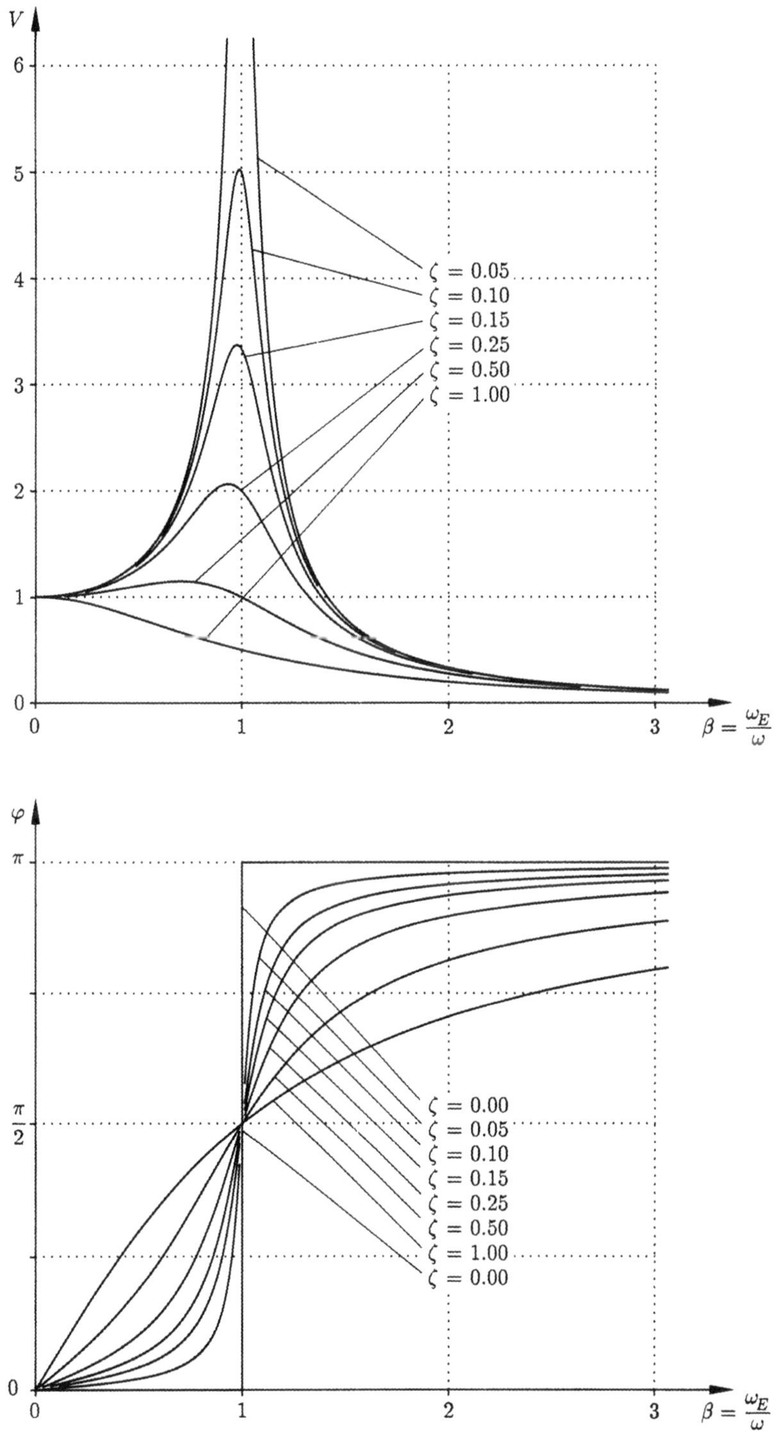

Fig. 1.4.2 Amplification factor and phase angle for the steady-state response of a harmonically excited system with viscous damping

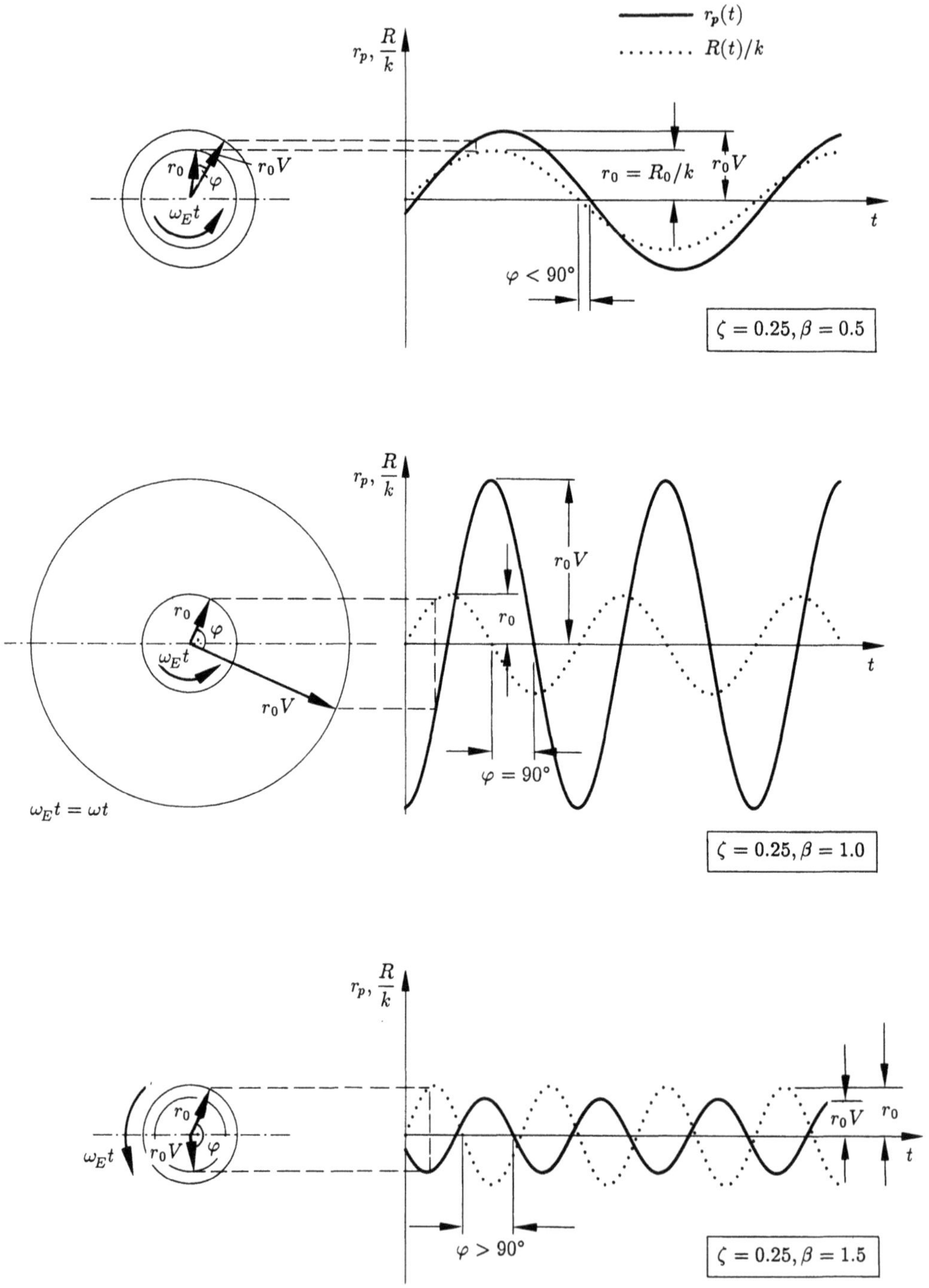

Fig. 1.4.3 Steady-state response for harmonic excitations; sub-critical, critical, over-critical

(b) $\beta = 1$ (resonance),

$V = (2\zeta)^{-1}$

$\varphi = \frac{\pi}{2} = 90°$

Here, the excitation frequency is equal to the natural frequency of the system and the response shows a phase shift of 90°, independent of the damping. However, the amplification factor, V, depends strongly on the damping. In particular, we obtain large values of amplification V for a lightly-damped system ($\zeta = 0.05 \Rightarrow V = 10, \zeta = 0 \Rightarrow V \to \infty$).

(c) $\beta \to \infty$, thus

$V \to 0$

$\varphi \to \pi = 180° \qquad (\cos\varphi = -1,\ \sin\varphi = 0)$

In this case, the excitation frequency is so high that the system cannot respond dynamically: hence, it remains at rest. Any fortuitous weak response of the system occurs out of phase with the excitation. Once again, the damping has no effect.

Figure 1.4.2 shows the variation of the functions V and φ for different frequency ratios and damping conditions. In general, we observe that the damping ratio, ζ, has a decisive influence on the amplification factor V in the neighbourhood of the resonance frequency, i.e. for $\beta = 1$. Detailed information on the straining of the structure thus demands more exact data on damping. Furthermore, we observe that the maximum of the amplification factor lies to the left of the resonance point, i.e. at $\beta < 1$, the shift becoming larger, the larger the value of ζ. The influence, however, is negligible for light damping. The damping ratio, ζ, has a striking influence on the phase angle, φ, and here the resonance range is especially sensitive.

Since ζ possesses only a limited influence on the steady-state response outside the range of resonance, approximations are frequently adopted by ignoring the effect of damping. In this case, we have

$$r_p(t) = \frac{r_0}{1-\beta^2}\sin\omega_E t \qquad (\zeta = 0) \tag{1.4.15}$$

This yields a conservative estimate of the amplitude. Typical oscillations for excitations at the critical point, and below and above the resonance point are shown in fig. 1.4.3. These diagrams clarify the significance of the physical data discussed so far.

In order to determine the resonance point in an experiment, the maximum amplitude is only of limited value since it depends upon the damping ratio and occurs for $\beta < 1$. The only striking measure is the phase angle, $\varphi = 90°$. If we were to display the relationship on an oscilloscope and assign the excitation to the horizontal axis (x) and the response to the vertical axis (y), we would obtain

$$x = R_0 \sin\omega_E t$$

$$y = \frac{R_0}{k} V \sin(\omega_E t - \varphi) \tag{1.4.16}$$

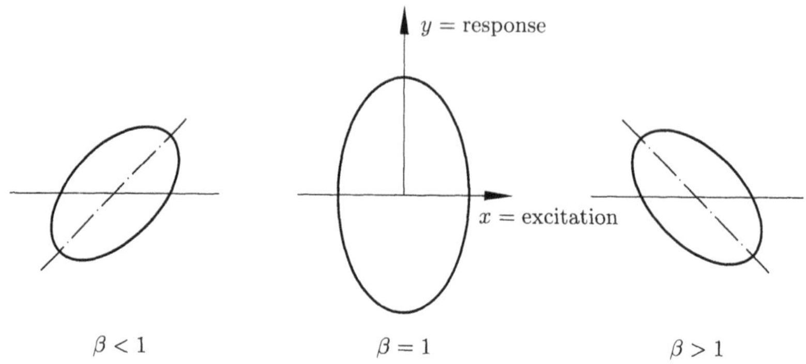

Fig. 1.4.4 Lissajous' figures for the determination of the point of resonance

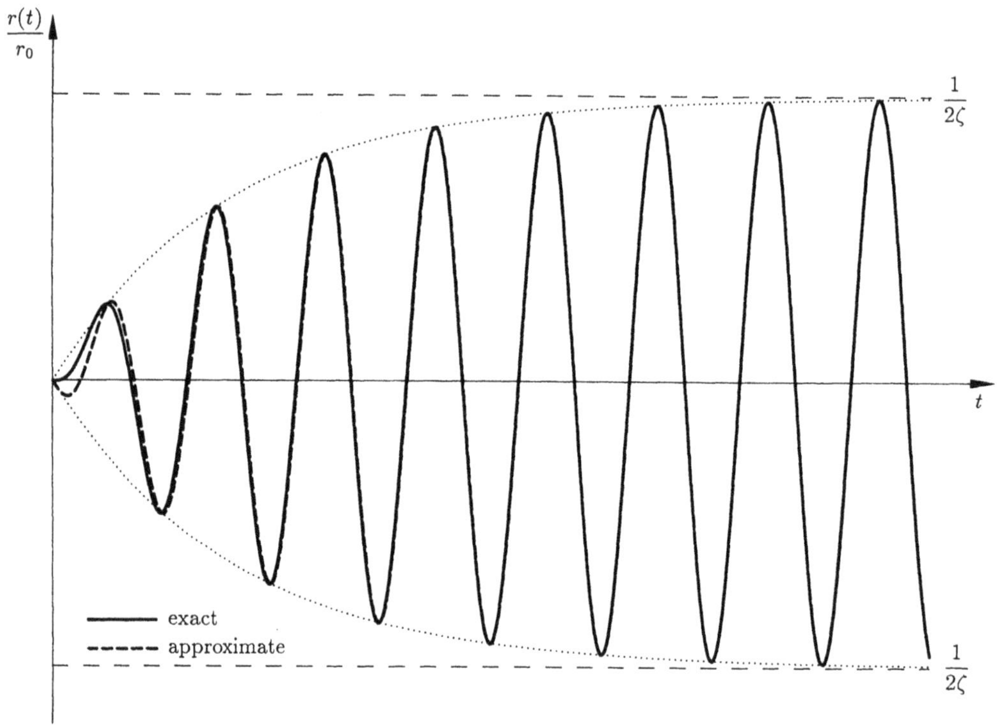

Fig. 1.4.5 Evolution of steady-state amplitude at resonance for weak damping

In this way, we observe the so-called Lissajous' figures on the oscilloscope. A vertical ellipse demonstrates the condition of resonance (fig. 1.4.4).

To conclude the discussion on harmonic excitations, we once again consider transient behaviour with the initial conditions $r(0) = 0$ and $\dot{r}(0) = 0$ and restrict ourselves to the resonance case ($\beta = 1$). From equ. (1.4.10), we deduce the transient response as

$$r(t) = \frac{r_0}{2\zeta}\left[\mathrm{e}^{-\zeta\omega t}\left(\frac{\zeta}{\sqrt{1-\zeta^2}}\sin\omega_D t + \cos\omega_D t\right) - \cos\omega t\right] \tag{1.4.17}$$

For lightly damped systems, the sinusoidal term contributes only insignificantly to the response. Omitting this term and setting $\omega_D \approx \omega$, we derive the approximate expression

$$\frac{r(t)}{r_0} = \frac{1}{2\zeta}(\mathrm{e}^{-\zeta\omega t} - 1)\cos\omega t \tag{1.4.18}$$

This solution indicates that the smaller the value of ζ, the longer it takes the amplitude to reach the final steady-state value. Figure 1.4.5 compares the exact solution to equ. (1.4.17) with the approximate one of equ. (1.4.18).

Hysteretic or Structural Damping

Up to now, we have ignored this particular type of damping which is closely connected with the steady-state response of real structures only when subject to harmonic excitation. Having now built up the necessary background, we turn our attention to structural damping.

In a harmonic excitation of an oscillator, mechanical energy is continuously supplied to the system, whereas damping converts part of the mechanical energy into heat. If a steady-state response is achieved, then there will be a balance between the energy input and the energy dissipated. Thus, the energy lost by damping in one cycle can easily be established by calculating the work done by an excitation force, $R(t)$, over the corresponding displacement, $r(t)$. For example, consider the infinitesimal contribution to this work

$$\mathrm{d}A_D = R(t)\mathrm{d}r = R(t)\dot{r}\mathrm{d}t \tag{1.4.19}$$

Integrating over a cycle with a period T_E, we obtain

$$\Delta A_D = \int_0^{T_E} R(t)\dot{r}\mathrm{d}t \tag{1.4.20}$$

The dissipated work over a cycle is represented by the area of the hysteresis loop in the R-r diagram of fig. 1.4.6. For viscous damping and

$$\begin{aligned} R(t) &= R_0 \sin\omega_E t \\ r(t) &= \frac{R_0}{k}V\sin(\omega_E t - \varphi) \end{aligned} \tag{1.4.21}$$

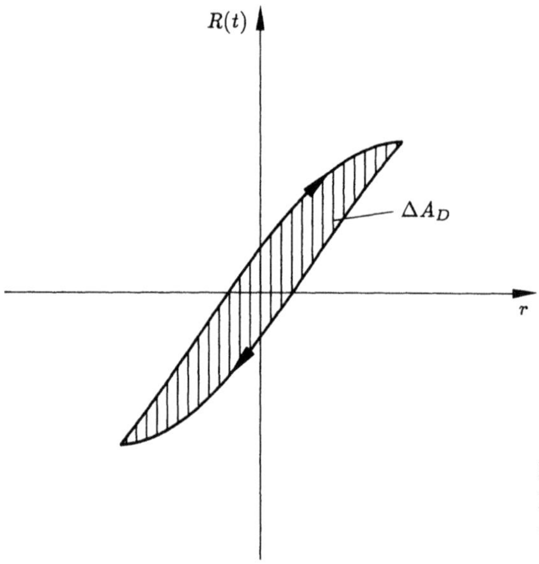

Fig. 1.4.6
Loss of energy due to damping; R/r-diagram

we deduce immediately from equ. (1.4.20)

$$\Delta A_D = \frac{R_0^2}{k} V \omega_E \int_0^{T_E} \sin(\omega_E t) \cos(\omega_E t - \varphi) \mathrm{d}t = \frac{R_0^2}{k} V \pi \sin \varphi \tag{1.4.22}$$

Using equs. (1.4.12), (1.4.13) and equ. (1.3.4), we obtain

$$\Delta A_D = \frac{R_0^2}{k} V \pi 2\zeta \left(\frac{\omega_E}{\omega} \right) V = c \pi \omega_E r_{\max}^2 \tag{1.4.22a}$$

where the amplitude of the steady-state response, $r_{\max}$, is given by

$$r_{\max} = r_0 V = \frac{R_0}{k} V \tag{1.4.23}$$

Consequently, the energy dissipated per cycle ought to be dependent on the frequency ω_E of the excitation. However, experiments conducted on a whole series of materials show that this is not the case when internal friction or structural damping is predominant. The observed experimental relationship can be expressed as

$$\Delta A_D = c_H r_{\max}^2 \tag{1.4.24}$$

where c_H is a constant and the energy dissipated is proportional to the square of the amplitude, as before. Comparing equs. (1.4.22a) and (1.4.24), we can see that structural damping can be converted into an equivalent viscous damping

$$c_{eq} = \frac{c_H}{\pi \omega_E} \tag{1.4.25}$$

This, however, presupposes that the excitation is harmonic. Applying equ. (1.4.25), the equation of motion of a system with hysteretic damping can be written as

$$\ddot{r} + \gamma \frac{\omega}{\omega_E} \omega \dot{r} + \omega^2 r = \omega^2 r_0 \sin \omega_E t \tag{1.4.26}$$

where

$$\gamma = \frac{c_H}{\pi k} \tag{1.4.27}$$

is the so-called structural damping coefficient (as an indication, $\gamma \approx 0.05$ for metal structures). We should emphasise once more that expression (1.4.26) applies strictly to a harmonic excitation only. This fact may easily be overlooked when using the complex notation which is frequently adopted in dynamics. This will be illustrated below by an example. We first observe that a harmonic excitation may be expressed in the form

$$\begin{aligned} R(t) &= R_0 \mathrm{e}^{i\omega_E t} \\ &= R_0 (\cos \omega_E t + i \sin \omega_E t) \end{aligned} \tag{1.4.28}$$

The real part of the complex excitation thus expresses the cosine excitation and the imaginary part the sine excitation. For hysteretic damping, the corresponding differential equation becomes

$$\ddot{r} + \gamma \frac{\omega}{\omega_E} \omega \dot{r} + \omega^2 r = \omega^2 r_0 \mathrm{e}^{i\omega_E t} \tag{1.4.29}$$

and has the following particular (i.e. steady-state) solution

$$r_p(t) = a_p \mathrm{e}^{i\omega_E t} \tag{1.4.30}$$

Substituting this expression into equ. (1.4.29), we find

$$a_p = \frac{r_0}{1 + i\gamma - (\omega_E/\omega)^2} \tag{1.4.31}$$

where the real part in equ. (1.4.30) is the solution for the cosine excitation and the imaginary part is the solution for the sine excitation. From equ. (1.4.30), we also deduce

$$\begin{aligned} \dot{r}_p(t) &= i\omega_E a_p \mathrm{e}^{i\omega_E t} \\ &= i\omega_E r_p(t) \end{aligned} \tag{1.4.32}$$

In place of equ. (1.4.29), we can then write

$$\ddot{r}_p + \omega^2 (1 + i\gamma) r_p = \omega^2 r_0 \mathrm{e}^{i\omega_E t} \tag{1.4.33}$$

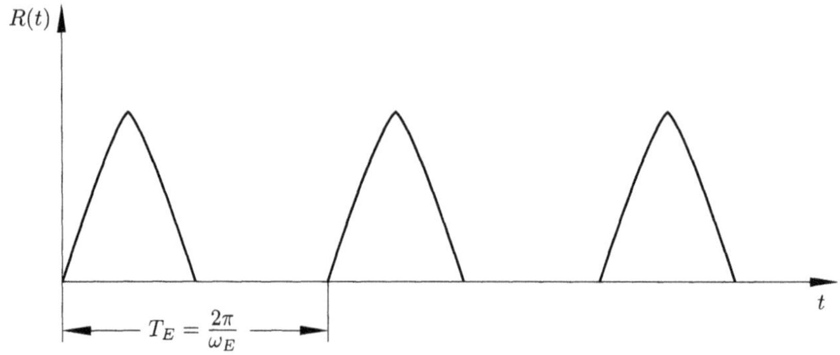

Fig. 1.4.7 Example of a periodic excitation

Here, any reference to the presupposed harmonic excitation no longer appears explicitly on the left-hand side and we must not be misled into believing that the following equation holds in general when structural damping is present

$$\ddot{r} + \omega^2(1 + i\gamma)r = R(t)$$

1.4.2 Periodic excitation

We now investigate more general histograms of forces and consider, in particular, periodic excitations. Such a histogram repeats itself at regular intervals of time with an excitation period, T_E, but, apart from that, the time-dependence of the force may be quite arbitrary. A typical example is shown in fig. 1.4.7. The most important step in the analysis of such problems, in which the loading has a period of $T_E = 2\pi/\omega_E$, is the representation of the load, $R_F(t)$, by a Fourier series . The relevant technique can be found in any mathematical handbook for engineers and yields

$$R_F(t) = \tfrac{1}{2}A_0 + \sum_{\mathrm{n}=1}^{\infty} \left(A_\mathrm{n} \cos(\mathrm{n}\omega_E t) + B_\mathrm{n} \sin(\mathrm{n}\omega_E t)\right) \tag{1.4.34}$$

where

$$A_\mathrm{n} = \frac{2}{T_E} \int_0^{T_E} R(t) \cos(\mathrm{n}\omega_E t)\mathrm{d}t \ ; \qquad \mathrm{n} = 0,\, 1,\, 2,\, \ldots$$

$$B_\mathrm{n} = \frac{2}{T_E} \int_0^{T_E} R(t) \sin(\mathrm{n}\omega_E t)\mathrm{d}t \ ; \qquad \mathrm{n} = 1,\, 2,\, \ldots \tag{1.4.35}$$

We must now clarify one important characteristic of our relevant differential equation. Let $r_1(t)$ be a solution for $R_1(t)$, and $r_2(t)$ be a solution for $R_2(t)$, i.e.

$$\ddot{r}_1 + 2\zeta\omega\dot{r}_1 + \omega^2 r_1 = R_1(t)$$

$$\ddot{r}_2 + 2\zeta\omega\dot{r}_2 + \omega^2 r_2 = R_2(t) \tag{1.4.36}$$

If c_1 and c_2 are arbitary constants, then $c_1 r_1 + c_2 r_2$ is also a solution of the differential equation for the loading $c_1 R_1(t) + c_2 R_2(t)$. This may easily be confirmed by substitution

$$\begin{aligned} &c_1\ddot{r}_1 + c_2\ddot{r}_2 + 2\zeta\omega(c_1\dot{r}_1 + c_2\dot{r}_2) + \omega^2(c_1 r_1 + c_2 r_2) \\ &\quad = c_1(\ddot{r}_1 + 2\zeta\omega\dot{r}_1 + \omega^2 r_1) + c_2(\ddot{r}_2 + 2\zeta\omega\dot{r}_2 + \omega^2 r_2) \\ &\quad = c_1 R_1(t) + c_2 R_2(t) \end{aligned} \tag{1.4.37}$$

Since the differential equation is linear, the superposition principle is seen to be applicable. Thus, for general periodic excitations as given by equ. (1.4.34), it is sufficient to know what the response will be for a single sine or cosine excitation, i.e. for a purely harmonic excitation. The sinusoidal excitation case has already been discussed (see section 1.4.1). The cosine excitation need not be considered separately because it can be deduced from a sinusoidal variation simply by a shift in the origin. Observing equs. (1.4.11), (1.4.13) and (1.4.14), we notice that the steady-state response to the excitation of equ. (1.4.34) can be written as

$$r_p(t) = \frac{1}{k}\left\{\tfrac{1}{2}A_0 + \sum_{n=1}^{\infty}\left[A_n V_n \cos(n\omega_E t - \varphi_n) + B_n V_n \sin(n\omega_E t - \varphi_n)\right]\right\} \tag{1.4.38}$$

where

$$V_n = \left[(2n\zeta\beta)^2 + (1 - n^2\beta^2)^2\right]^{-1/2} \tag{1.4.39}$$

is the amplification factor and

$$\varphi_n = \tan^{-1}\frac{2n\zeta\beta}{1 - n^2\beta^2} \tag{1.4.40}$$

is the phase shift of the n-th mode. Since $V_n \to 0$ as $n \to \infty$, the amplification factor for the response decreases with increasing frequency. Even if a particular Fourier expression for $R(t)$ does not possess an optimal convergence, the response can still converge rapidly. This proves valuable in practical applications and ensures satisfactory approximations when working with a limited number of terms in the series.

A simple example will clarify this point. In fig. 1.4.8a, we show a block-shaped excitation function with a period interval $0 \leqslant t \leqslant T_E$. Its mathematical definition is

$$\begin{aligned} R(t) &= R_0 \qquad &\text{for} \quad 0 \leqslant t \leqslant T_E/2 \\ R(t) &= -R_0 \qquad &\text{for} \quad T_E/2 < t \leqslant T_E \end{aligned}$$

Since $R(t)$ is antisymmetric in the interval of integration $(0, T_E)$, it follows that the coefficients of the cosine terms, A_n, in the Fourier series vanish

$$A_n = 0 \qquad n = 0,\ 1,\ 2,\ \ldots$$

From equ. (1.4.35), the sine terms are derived as

$$B_n = \frac{2R_0}{n\pi}(1 - \cos n\pi)$$

i.e.

$$B_n = 0 \qquad \text{for} \quad n = 2,\ 4,\ 6,\ \ldots$$

$$B_n = \frac{4R_0}{n\pi} \qquad \text{for} \quad n = 1,\ 3,\ 5,\ \ldots$$

The Fourier expansion for the loading function, $R(t)$, up to order n_{max} is then given by the following expression

$$R_F(t) = \frac{4R_0}{\pi} \sum_{n=1,3,5}^{n_{max}} \frac{1}{n} \sin(n\omega_E t)$$

This function does not possess a satisfactory convergence as is shown in fig. 1.4.8b. The deviation of the Fourier expression from the rectangular form using 10 terms ($n_{max} = 19$) is significant and even with 50 terms in the series ($n_{max} = 99$), it is still noticeable. The approximation to the steady-state response according to equ. (1.4.38) is

$$r_p(t) = \frac{4r_0}{\pi} \sum_{n=1,3,5}^{n_{max}} \frac{V_n}{n} \sin(n\omega_E t - \varphi_n)$$

where V_n and φ_n are given by equs. (1.4.39), (1.4.40). For physical clarity, we have introduced the 'static' deflection, r_0,

$$r_0 = \frac{R_0}{k}$$

For the case when $\beta = \omega_E/\omega = 1$ and $\zeta = 0.05$, the approximations to the response corresponding to the same number of terms used for the excitation are shown in fig. 1.4.8c. We immediately recognise that the first term in the series of the response yields practically the exact solution. Figure 1.4.8d further illustrates the order of magnitude of the respective Fourier coefficients. The former are proportional to $(1/n)$ whereas the latter are proportional to (V_n/n).

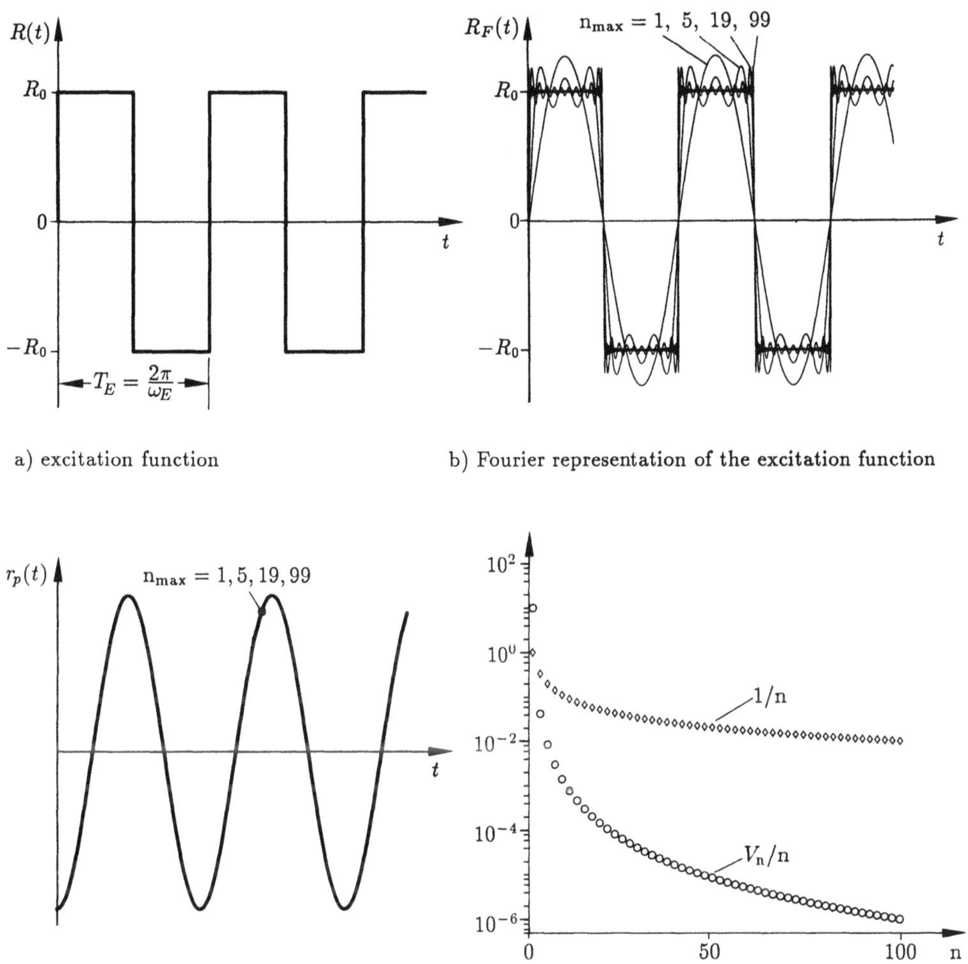

a) excitation function

b) Fourier representation of the excitation function

c) steady-state response

d) Fourier coefficients of excitation (◇) and response (○)

Fig. 1.4.8 Periodic rectangular excitation and steady-state response ($\beta = 1$, $\zeta = 0.05$)

1.4.3 General excitation and Fourier integral

We have emphasised that the time-dependence of the load within a period T_E can be arbitrary. Now we assume that the time interval, T_E, is continuously extended and simultaneously $\omega_E = 2\pi/T_E$ is correspondingly reduced. At the same time, the discrete values, $n\omega_E$, get progressively closer and the summation in equ. (1.4.34) converges to an integral. In this way, we simulate a histogram for an arbitrary excitation (see fig. 1.4.9). We have

$$\begin{aligned} T_E &\to \infty \\ \omega_E &\to d\Omega \\ n\omega_E &\to \Omega \end{aligned} \tag{1.4.41}$$

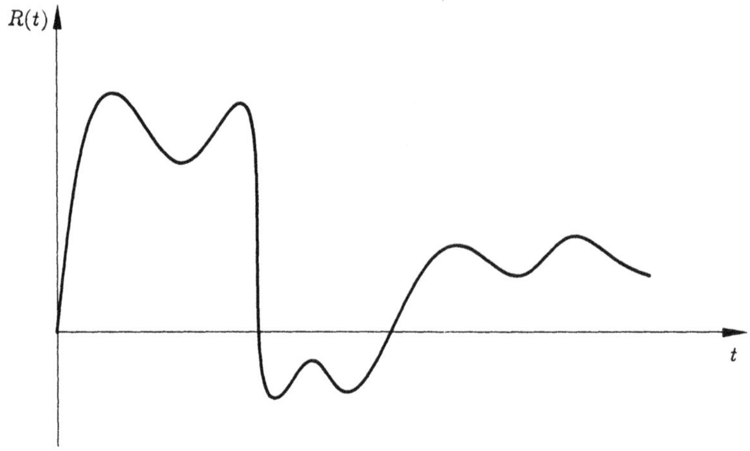

Fig. 1.4.9 Arbitrary force histogram

In order to formulate the limit, the factor $2/T_E = \omega_E/\pi$, as given in equ. (1.4.35), has to be entered into equ. (1.4.34). We then obtain

$$R_{FI} = \int_0^\infty \Big(A(\Omega)\cos\Omega t + B(\Omega)\sin\Omega t\Big)\mathrm{d}\Omega \tag{1.4.42}$$

with

$$A(\Omega) = \frac{1}{\pi}\int_0^\infty R(t)\cos\Omega t\mathrm{d}t$$

$$B(\Omega) = \frac{1}{\pi}\int_0^\infty R(t)\sin\Omega t\mathrm{d}t \tag{1.4.43}$$

For the validity of the above procedure, we have to presume that $R(t)$ is piecewise continuous within a given range and that the integral

$$\int_0^\infty |R(t)|\mathrm{d}t$$

is finite. The excitation defined by equ. (1.4.42) can also be thought of as a summation of the basic harmonic contributions in Ω, i.e. $A(\Omega)\cos\Omega t$ and $B(\Omega)\sin\Omega t$. Consequently, the corresponding response may also be written in the form (cf. equ. (1.4.38))

$$r_p(t) = \frac{1}{k}\int_0^{\infty} V(\Omega)\Big[A(\Omega)\cos\{\Omega t - \varphi(\Omega)\} + B(\Omega)\sin\{\Omega t - \varphi(\Omega)\}\Big]\mathrm{d}\Omega \tag{1.4.44}$$

where

$$V(\Omega) = \left[\left\{2\zeta\,(\Omega/\omega)\right\}^2 + \left\{1 - (\Omega/\omega)^2\right\}^2\right]^{-1/2} \tag{1.4.45}$$

and

$$\varphi(\Omega) = \tan^{-1}\frac{2\zeta\Omega/\omega}{1-(\Omega/\omega)^2} \tag{1.4.46}$$

The analytical evaluation of equs. (1.4.42) to (1.4.46) generally proves to be tedious and depends upon the existence of the integrals. In practice, these integrals are best computed numerically, either by discrete Fourier transformation (DFT) or fast Fourier transformation (FFT). Further discussion of these methods, however, is beyond the scope of this introductory textbook.

1.4.4 Pulse-type excitation (Dirac's impulse function)

By pulse-type loading, we mean an application of an external load to a structure over a very short interval of time. A general excitation is reproduced in fig. 1.4.10. The time over which the load is applied is called the impulse duration, t_1, and the area under the load-time curve is denoted as the strength of the impulse, I

$$I = \int_0^{t_I} R(t)\mathrm{d}t \tag{1.4.47}$$

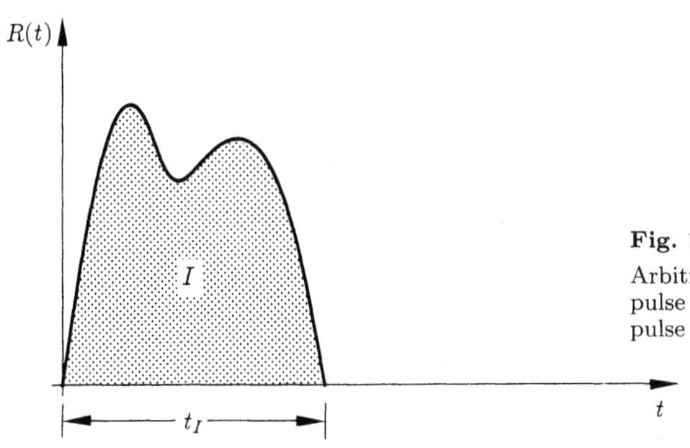

Fig. 1.4.10

Arbitrary pulse-type excitation; impulse period t_I, and strength of impulse I

Considering once more the general differential equation for a viscously damped oscillator subject to a force excitation (1.4.1), we obtain the following expression for the velocity at the end of the impulse period with initial conditions $r(0) = 0$ and $\dot{r}(0) = 0$

$$\begin{aligned}\dot{r}(t_I) &= \frac{\omega^2}{k}\int_0^{t_I} R(t)\,\mathrm{d}t - 2\zeta\omega\int_0^{t_I}\dot{r}(t)\,\mathrm{d}t - \omega^2\int_0^{t_I} r(t)\,\mathrm{d}t \\ &= \frac{\omega^2 I}{k} - 2\zeta\omega\int_0^{t_I}\dot{r}(t)\,\mathrm{d}t - \omega^2\int_0^{t_I} r(t)\,\mathrm{d}t \end{aligned} \tag{1.4.48}$$

Assuming the duration of the impulse, t_I, to be very short, no appreciable displacement can have developed before the end of the impulse. Hence, the integrals in equ. (1.4.48) can be neglected and we obtain

$$\begin{aligned} r(t_I) &\approx 0 \\ \dot{r}(t_I) &\approx \frac{\omega^2 I}{k} = \frac{I}{m} \end{aligned} \tag{1.4.49}$$

These values then serve as the initial conditions for the ensuing free oscillation.

These approximations for an impulse with an arbitrary time-dependence of the force and a finite duration are in the limit exact for a Dirac impulse. By Dirac impulse, we mean an impulse of infinitely short duration but of finite strength

$$I = \lim_{t_I \to 0}\left(\int_0^{t_I} R(t)\,\mathrm{d}t\right) \tag{1.4.50}$$

The response to a Dirac impulse applied at $t = 0$ is a damped, free oscillation which satisfies exactly the initial conditions of equ. (1.4.49) at $t = 0$. Noting equ. (1.3.19), we immediately obtain

$$r(t) = \frac{I}{m\omega_D}\mathrm{e}^{-\zeta\omega t}\sin\omega_D t \tag{1.4.51}$$

Thus, the approximate maximum amplitude for a not too heavily damped system is

$$r_{\max} \approx \frac{I}{m\omega_D} \tag{1.4.52}$$

An impulse of finite duration can be replaced in practice by a Dirac impulse acting at the centroid t_S of the force-time histogram. The resulting approximate response is

$$r(t) \approx \frac{I}{m\omega_D}\mathrm{e}^{-\zeta\omega(t-t_S)}\sin\omega_D(t - t_S) \tag{1.4.53}$$

Equations (1.4.52) and (1.4.53) should, however, only be applied for

$$t_I/T < 1/4 \qquad (T = \text{period of free oscillation})$$

since, for impulses of longer duration, the specific form of the load distribution develops a significant influence. In addition, the maximum deflection need not take place in the free oscillatory phase but can occur at an earlier instant. Our approximation does not provide any information about this range, however.

An example: periodic application of a Dirac impulse

Authors: John Argyris, Karl Straub

Let us conclude this short introduction to the Dirac impulse with a simple but instructive example of industrial relevance and familiarise the reader in this way with the at first sight rather mathematical concept of the Dirac function. We continue our discussion in section 1.4.2 on "periodic excitations" and consider a single-mass oscillator (mass m, period of oscillation T, undamped circular eigenfrequency ω, sub-critical damping ratio ζ) which is subjected to a regular succession (taken as infinite) of impulses of very short duration. In this case, "regular" means that each impulse acts with the same intensity I and that the intervals between the impulses are of equal duration T_E, i.e. the excitation is periodic; by denoting the impulses as "short", we are justified in treating them as Dirac impulses, equ. (1.4.50). Thus, the excitation function (see fig. 1.4.11) can be formally expressed in the form

$$R(t) = I \sum_{\mathrm{n}} \delta(t - \mathrm{n}T_E) \tag{1}$$

if, in analogy to equ. (1.4.50), we define a standard impulse $\delta(t - t_p)$ at the instant t_p by means of the following condition

$$\int_{-\infty}^{+\infty} \delta(t - t_p)\, dt = 1 \qquad \text{and} \qquad \delta(t - t_p) = 0 \qquad \text{for } t \neq t_p \tag{2}$$

We explore now the possibility of a steady-state response with the period T_E which should set in after a transient phase.

In order to establish the solution, we make use of elementary observations already familiar to us:

— when observing any single time interval of the duration T_E, the postulated periodicity implies that the states of motion (displacement, velocity) must be the same at the beginning and the end of the interval;

— each impulse induces a velocity jump $\Delta\dot{r} = I/m$ in the response but the displacement does not change at this instant, see equ. (1.4.49);

— between two consecutive impulses, the response is described by a free damped oscillation already familiar to us from section 1.3.

Without limiting the general validity of the argument, we can shift the time axis so that the two impulses act at $t = 0$ and $t = T_E$ and observe the motion in this interval. For clarity's sake, we define that at

$$t = 0 - \epsilon = 0_-$$

the impulse has not been applied, but has been at

$$t = 0 + \epsilon = 0_+$$

Analogously, we interpret the terms T_{E-} and T_{E+}. At the beginning of the interval under observation, we have the initial conditions

$$\begin{aligned} &\text{for} \quad t = 0 - \epsilon: \quad && r(0_-), && \dot{r}(0_-) \\ &\text{for} \quad t = 0 + \epsilon: \quad && r(0_+) = r(0_-) \equiv r_0, && \dot{r}(0_+) = \dot{r}(0_-) + I/m \equiv \dot{r}_0 \end{aligned} \tag{3}$$

The periodicity condition requires for the instant $t = T_{E_-}$

$$\begin{aligned} r(T_{E_-}) &= r(0_-) = r_0 \\ \dot{r}(T_{E_-}) &= \dot{r}(0_-) = \dot{r}_0 - I/m \end{aligned} \tag{3'}$$

At the instant $t = 0_+$, a free, sub-critically damped oscillation with the (yet unknown) initial conditions r_0 and $\dot{r}_0$ sets in. We can deduce its course directly from equ. (1.3.19)

$$r(t) = \mathrm{e}^{-\zeta\omega t}\big(a\cos(\omega_D t) + b\sin(\omega_D t)\big) \tag{4}$$

where

$$\omega_D = \omega\sqrt{1-\zeta^2} \qquad \text{, see equ. (1.3.16)}$$

and

$$\begin{aligned} a &= r_0 \\ b &= (\dot{r}_0 + \zeta\omega r_0)/\omega_D \qquad \text{, see equ. (1.3.21)} \end{aligned} \tag{4'}$$

Simple differentiation yields the velocity

$$\dot{r}(t) = \zeta\omega r(t) + \omega_D \mathrm{e}^{-\zeta\omega t}(-a\sin(\omega_D t) + b\cos(\omega_D t))$$

If we now use this information to determine displacement and velocity at the instant $t = T_{E-}$ and apply the condition of equ. (3′), we obtain two linear equations for the unknown interval initial conditions r_0 and $\dot{r}_0$

$$\begin{aligned} (\omega_D - \omega_D ec - \zeta\omega es)r_0 - \quad (es)\dot{r}_0 &= 0 \\ (\zeta\omega + \omega_D es - \zeta\omega ec)r_0 + (1 - ec)\dot{r}_0 &= I/m \end{aligned}$$

In these expressions, we apply for the sake of typographical brevity

$$
\begin{aligned}
e &= \mathrm{e}^{-\zeta\omega T_E} \\
c &= \cos(\omega_D T_E) \\
s &= \sin(\omega_D T_E)
\end{aligned} \tag{5}
$$

We thus obtain for the initial conditions of the free motion between two consecutive impulses

$$
\begin{aligned}
r_0 &= \frac{I}{m\omega_D}\frac{es}{1-2ec+e^2} \\
\dot{r}_0 &= \frac{I}{m\omega_D}\frac{\omega_D(1-ec)-\zeta\omega es}{1-2ec+e^2}
\end{aligned} \tag{6}
$$

Substituting these expressions into equ. (4), we establish a piecewise analytical formulation of the steady-state solution, this being a function of the time t and dependent on the system parameters m, ω, ζ and the excitation parameters I and T_E. The terms e, c and s incorporate the period ratio T_E/T for which we now introduce the notation α (this is the inverse value of the frequency ratio β defined in equ. (1.4.7))

$$
\alpha = T_E/T = \omega/\omega_E = \beta^{-1}
$$

The steady-state response $r(t)$ for various α-values and a 5% critical damping ratio ($\zeta = 0.05$) is shown in fig. 1.4.11. It should be noted that each α-value possesses in this diagram its own time-scale since T_E is dependent on α: the abscissa scale should in fact be t/T_E. For this reason, the constant velocity jump $\Delta\dot{r} = I/m$ is represented in different scales, depending on the α-value of the individual response.

An alternative expression to equ. (4) which has already been given in equ. (1.3.19) is

$$
r(t) = \bar{a}\mathrm{e}^{-\zeta\omega t}\cos(\omega_D t - \varphi)
$$

where

$$
\bar{a} = (a^2 + b^2)^{1/2}
$$

This leads to a clear formulation of the strength of the response by an amplification factor. With equs. (4′) and (6), we obtain

$$
\bar{a} = \frac{I}{m\omega_D}\frac{1}{(1-2ec+e^2)^{1/2}}
$$

for the amplitude. The second denominator is only dependent on α and ζ. Introducing now the amplification factor

$$
V(\alpha, \zeta) = (1 - 2ec + e^2)^{-1/2} \tag{7}
$$

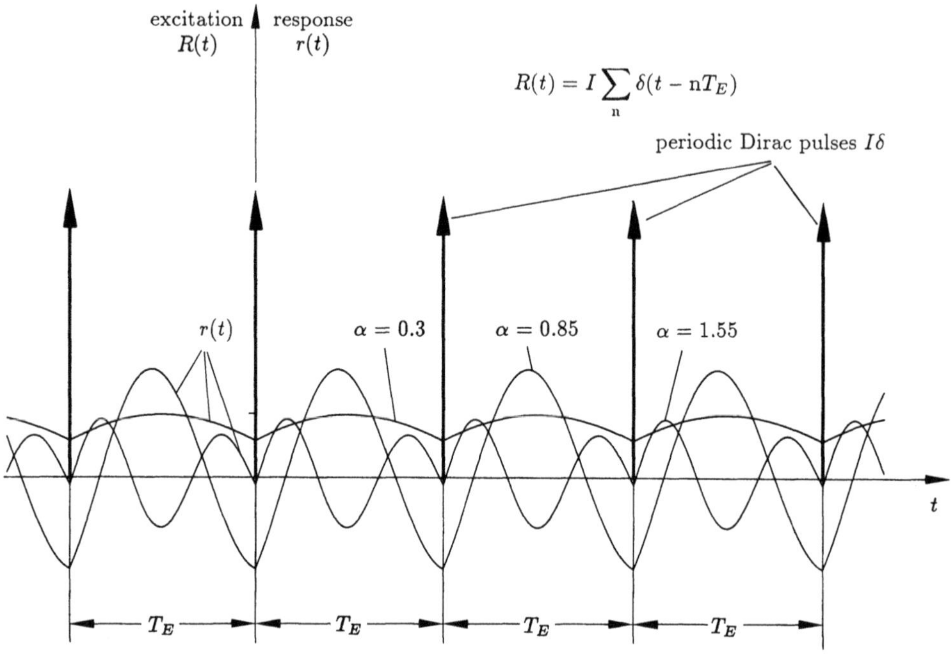

Fig. 1.4.11 Excitation by a periodic succession of Dirac impulses and steady-state response for various α-values; $\zeta = 0.05$

we find the alternative expression for the steady-state response

$$r(t) = \frac{I}{m\omega_D} V(\alpha, \zeta)\, e^{-\zeta\omega t} \cos(\omega_D t - \varphi) \tag{8}$$

By elementary transformations, we can express the dependence on α and ζ in the elegant form

$$V(\alpha, \zeta) = \frac{1}{\sqrt{2}} \frac{e^{\pi\alpha\zeta}}{[\cosh(2\pi\alpha\zeta) - \cos(2\pi\alpha\sqrt{1-\zeta^2})]^{1/2}} \tag{9}$$

In the case of near integral α-values, the course of this family of curves (see fig. 1.4.12) displays peaks which become points of discontinuity in an undamped system

$$V(\alpha, \zeta = 0) = \frac{1}{\sqrt{2}} \frac{1}{(1 - \cos(2\pi\alpha))^{1/2}}$$

Once again by means of elementary transformations, we deduce from equ. (7)

$$V(\alpha, \zeta) = \frac{(1-e)^{-1}}{\left(1 + \frac{2e(1-c)}{(1-e)^2}\right)^{1/2}}$$

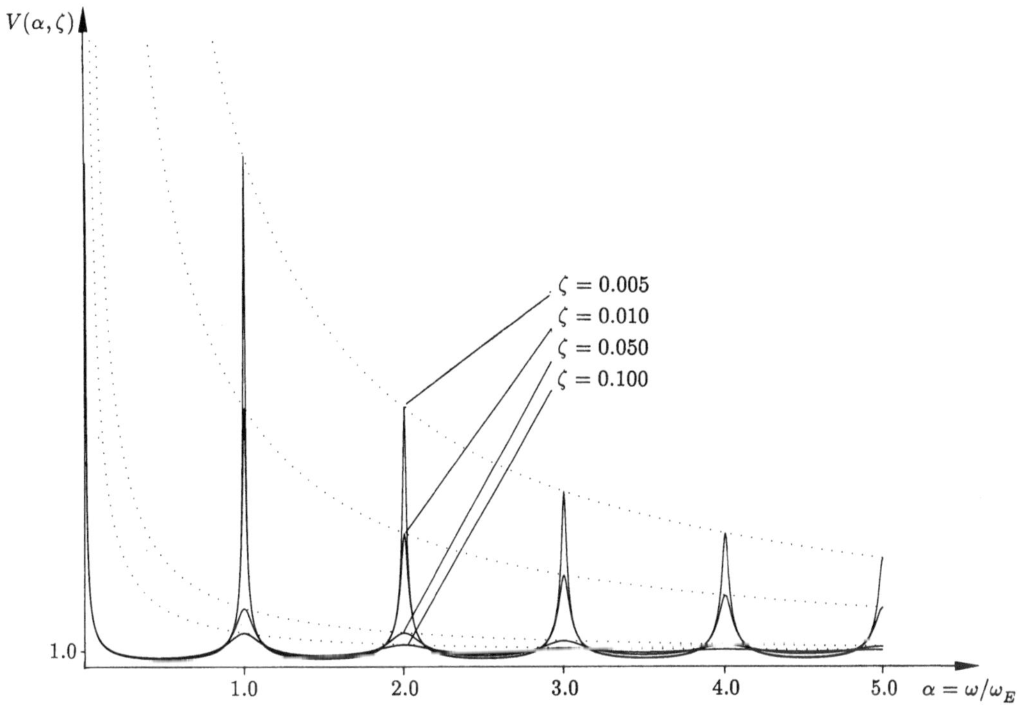

Fig. 1.4.12 Amplification factor of the steady-state response of a single-mass oscillator excited by a periodic sequence of Dirac impulses (see fig. 1.4.11)

As $|c| \leqslant 1$ applies, we note for the second term in the denominator

$$2e(1-c)/(1-e)^2 \geqslant 0$$

i.e. the denominator itself is always $\geqslant 1$. The lower limit of 1 is obtained for $c = +1$. The family of curves

$$\begin{aligned} H(\alpha,\zeta) &= (1-e)^{-1} \\ &= (1-\mathrm{e}^{-2\pi\alpha\zeta})^{-1} \end{aligned} \tag{10}$$

thus represents the envelopes in the V-α-diagram: this is demonstrated by the dotted lines in fig. 1.4.12. As mentioned above, the amplification factor V touches the envelope when $c = 1$, i.e. when $\omega_D T_E = 2\pi\alpha\sqrt{1-\zeta^2}$ is a multiple of 2π or

$$\alpha\sqrt{1-\zeta^2} = \text{integral}$$

This applies for $\zeta \ll 1$ with a good approximation if α itself has an integral value, i.e. $\alpha =$n.

In the introduction to this example, we already referred to section 1.4.2 on periodic excitations; it is only natural to apply the general results obtained there to the present example. Thus, we attempt to expand the Dirac excitation function $R(t)$, equ. (1), into a Fourier series and to deduce hence the steady-state response.

According to equs. (1.4.34) and (1.4.35), the Fourier series of the excitation function is

$$R_F(t) = \frac{1}{2}A_0 + \sum_{\mathrm{n}=1}^{\infty}(A_{\mathrm{n}}\cos(\mathrm{n}\omega_E t) + B_{\mathrm{n}}\sin(\mathrm{n}\omega_E t))$$

with

$$A_{\mathrm{n}} = \frac{2}{T_E}\int_0^{T_E} R(t)\cos(\mathrm{n}\omega_E t)\,\mathrm{d}t \qquad \mathrm{n} = 0, 1, 2, ...$$

$$B_{\mathrm{n}} = \frac{2}{T_E}\int_0^{T_E} R(t)\sin(\mathrm{n}\omega_E t)\,\mathrm{d}t \qquad \mathrm{n} = 1, 2, ...$$

Let the impulse be applied at the left boundary of the interval under observation; with the expressions introduced above, the integration interval is thus $(0_-, T_{E_-})$. In the range $0_+ \leqslant t \leqslant T_{E_-}$, $R(t) = 0$ applies; for the integration, the interval is thus reduced to $0 - \epsilon \leqslant t \leqslant 0 + \epsilon$ with an arbitrarily small ϵ

$$A_{\mathrm{n}} = \frac{2}{T_E}\lim_{\epsilon\to 0}\int_{0-\epsilon}^{0+\epsilon} R(t)\cos(\mathrm{n}\omega_E t)\,\mathrm{d}t$$

In this ϵ-proximity to the t-origin, the continuous cosine function can be assumed to be constant, i.e. we have

$$\begin{aligned} A_{\mathrm{n}} &= \frac{2}{T_E}\cos(\mathrm{n}\omega_E t)\bigg|_{t=0}\lim_{\epsilon\to 0}\int_{0-\epsilon}^{0+\epsilon} R(t)\,\mathrm{d}t \\ &= \frac{2}{T_E}\cdot 1\cdot I \end{aligned}$$

With the same argumentation, all B_{n}-constants are seen to vanish

$$B_{\mathrm{n}} = 0$$

The result of our rather heuristic reflections also follows directly from an alternative definition of the Dirac function which is equivalent to the earlier expression, equ. (1). To this purpose, we define the δ-function directly by the condition

$$\int_{-\infty}^{+\infty} \delta(t - t_p) f(t)\,\mathrm{d}t = f(t - t_p) \tag{11}$$

where $f(t)$ is an arbitrary function continuous at the instant $t = t_p$.
The formal Fourier expansion of the excitation function is thus

$$R_F(t) = \frac{2I}{T_E}\left(\frac{1}{2} + \sum_{\mathrm{n}=1}^{\infty} \cos(\mathrm{n}\omega_E t)\right) \tag{12}$$

Since the terms of the series clearly do not tend to zero with increasing n, the series is not convergent. In fact, the sequence of functions $R_F^{(\mathrm{N})}(t)$

$$R_F^{(\mathrm{N})}(t) = \frac{2I}{T_E}\left(\frac{1}{2} + \sum_{\mathrm{n}=1}^{\mathrm{N}} \cos(\mathrm{n}\omega_E t)\right)$$

does not converge towards the original function $R(t)$. Figure 1.4.13 reproduces the periodic course of the curves for some approximations $R_F^{(\mathrm{N})}(t)(\mathrm{N} = \mathrm{n_{max}} = 1, 10, 100)$. We clearly observe a certain similarity with the periodic sequence of Dirac impulses $R(t)$. With increasing n, the boundary values $R_F(t = 0) = R_F(t = T_E)$ continue to rise (because $\cos(\mathrm{n} \cdot 0) = +1$); in the inner range of the interval, the approximations remain "small", but do not tend to zero (the cosine function possesses values between -1 and $+1$).

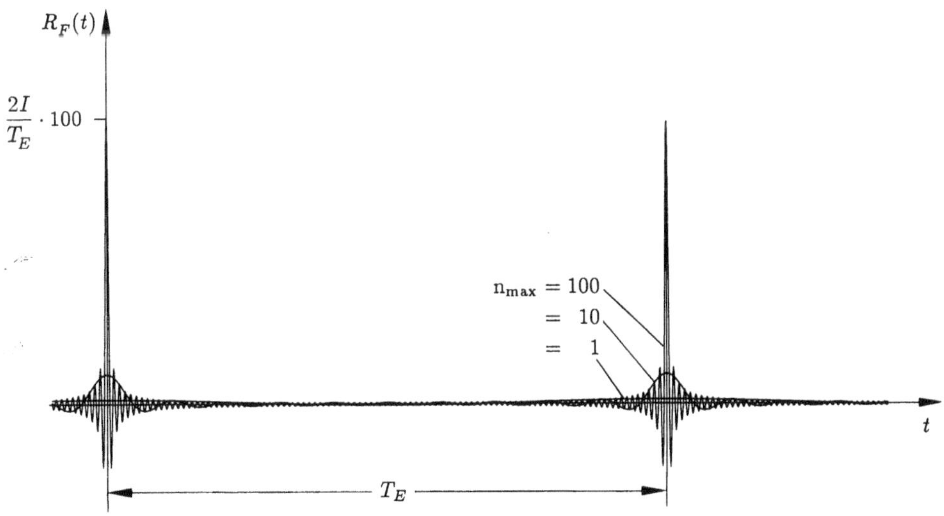

Fig. 1.4.13 Fourier approximation for the periodic pulse excitation

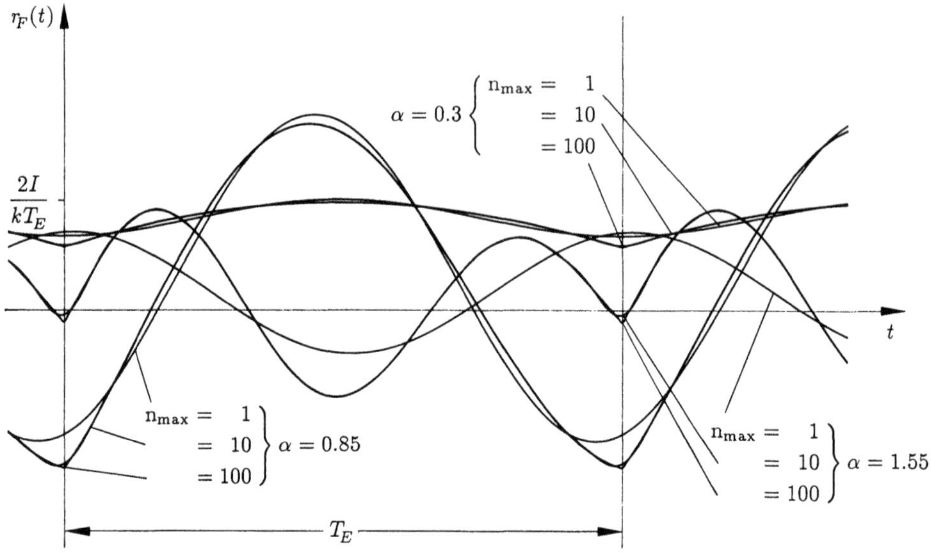

Fig. 1.4.14 Approximation for the steady-state response due to the periodic Dirac impulse excitation of fig. 1.4.12; damping ratio $\zeta = 0.05$

Without entering into the mathematical details of the handling of discontinuous functions and their non-converging series, we trust in our experience which instructs us that due to the inertia of the mass, the higher frequency components of the system response become increasingly weak. Consequently, we adopt formally the steady-state response from equ. (1.4.38) and apply the specific values for A_n and B_n determined above

$$r_F(t) = \frac{2I}{kT_E}\left(\frac{1}{2} + \sum_{n=1}^{\infty} V_n \cos(n\omega_E t - \varphi_n)\right) \tag{13}$$

where the amplification factor and the phase angle of the individual solution components are given by

$$\begin{aligned} V_n &= \left[(2n\zeta\beta)^2 + (1 - n^2\beta^2)^2\right]^{-1/2} \\ \varphi_n &= \tan^{-1}\frac{2n\zeta\beta}{1 - n^2\beta^2} \end{aligned} \tag{14}$$

For large values of n, V_n is approximately proportional to $1/n^2$, the series for the steady-state response is thus convergent – in fact, it converges very rapidly as can be seen in fig. 1.4.14. We achieve a satisfactory degree of accuracy for technical purposes with ten terms in the series expansion; further terms merely increase the accuracy with which the kink (tangential discontinuity) at the instant the impact acts (velocity jump) is described. The physical parameters of the examples in this figure were chosen so as to enable a direct comparison with fig. 1.4.11. It goes without saying that we may sum – with some effort – the series of equ. (13) and confirm in this way the formula for the amplification factor V of equ. (9).

One final remark: in the numerical analysis of equs. (13) and (14), we should not forget that equs. (14) are only consistent if both signs $\pm$ are taken into consideration in the root expression for V_{n} or if the inverse function $\tan^{-1}$ is not restricted to its standard principle value $-\pi/2 \leqslant \varphi_{\mathrm{n}} \leqslant +\pi/2$. The expressions defining V_{n} and φ_{n} are two equations which contain $\cos\varphi_{\mathrm{n}}$ and $\sin\varphi_{\mathrm{n}}$ (in analogy to equ. (1.4.12)); these must be satisfied either by the correct choice of the sign of V_{n} or by the correct φ_{n}-value: for example, $V_{\mathrm{n}} < 0$ applies if $1 - \mathrm{n}^2\beta^2 < 0$ holds.

1.4.5 General excitation and Duhamel integral

Clearly, an arbitrary load distribution can be represented by a series of successive impulses (fig. 1.4.15). Typically, such an impulse at time $t = \bar{t}$ has a strength of

$$\mathrm{d}I = R(\bar{t})\mathrm{d}\bar{t} \tag{1.4.54}$$

and a corresponding response for $t > \bar{t}$

$$\mathrm{d}r(t) = \frac{1}{m\omega_D} R(\bar{t})\, \mathrm{e}^{-\zeta\omega(t-\bar{t})} \sin\omega_D(t - \bar{t})\, \mathrm{d}\bar{t} \tag{1.4.55}$$

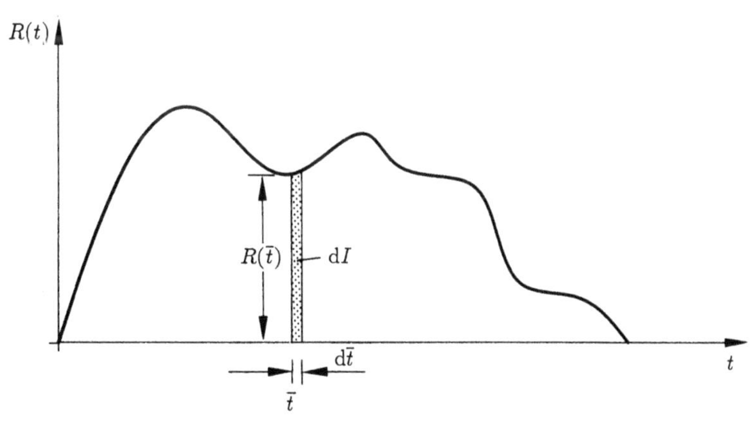

Fig. 1.4.15 Representation of an arbitrary force histogram by a sequence of infinitesimal impulses

We assume an initial state of rest and sum (integrate) the contributions of all impulses up to a time, t, to find

$$r(t) = \frac{1}{m\omega_D} \int_0^t R(\bar{t}\,)\,\mathrm{e}^{-\zeta\omega(t-\bar{t}\,)} \sin\omega_D(t-\bar{t}\,)\,\mathrm{d}\bar{t} \tag{1.4.56}$$

This is the well-known Duhamel integral. Next, we apply the trigonometric addition theorem and integrate equ. (1.4.56) to obtain

$$r(t) = A(t)\sin\omega_D t - B(t)\cos\omega_D t \tag{1.4.57}$$

where

$$\begin{aligned} A(t) &= \frac{1}{m\omega_D} \int_0^t R(\bar{t}\,)\,\mathrm{e}^{-\zeta\omega(t-\bar{t}\,)} \cos\omega_D(\bar{t}\,)\,\mathrm{d}\bar{t} \\ B(t) &= \frac{1}{m\omega_D} \int_0^t R(\bar{t}\,)\,\mathrm{e}^{-\zeta\omega(t-\bar{t}\,)} \sin\omega_D(\bar{t}\,)\,\mathrm{d}\bar{t} \end{aligned} \tag{1.4.58}$$

These integrals can easily be evaluated numerically. In order to do this, we divide the time axis, $\bar{t}$, into equal intervals and compute the integrands in equ. (1.4.58) at a sequence of times $\mathrm{n}\Delta\bar{t}$ ($\mathrm{n} = 0, 1, 2, \ldots$), i.e.

$$\begin{aligned} a_\mathrm{n} &= R(\mathrm{n}\Delta\bar{t}\,)\,\mathrm{e}^{-\zeta\omega(\mathrm{m}-\mathrm{n})\Delta\bar{t}} \cos(\omega_D \mathrm{n}\Delta\bar{t}\,) \\ b_\mathrm{n} &= R(\mathrm{n}\Delta\bar{t}\,)\,\mathrm{e}^{-\zeta\omega(\mathrm{m}-\mathrm{n})\Delta\bar{t}} \sin(\omega_D \mathrm{n}\Delta\bar{t}\,) \end{aligned} \tag{1.4.59}$$

Here, the terminal time of integration is fixed at $t = \mathrm{m}\Delta\bar{t}$ (see fig. 1.4.16). Using the trapezoidal rule, we find

$$\begin{aligned} A(t) &\approx \frac{\Delta\bar{t}}{2m\omega_D}(a_0 + 2a_1 + 2a_2 + \ldots + 2a_{\mathrm{m}-1} + a_\mathrm{m}) \\ B(t) &\approx \frac{\Delta\bar{t}}{2m\omega_D}(b_0 + 2b_1 + 2b_2 + \ldots + 2b_{\mathrm{m}-1} + b_\mathrm{m}) \end{aligned} \tag{1.4.60}$$

These formulae can easily be applied for an increasing time, t, so that the response of the system can be established without difficulty in conjunction with equ. (1.4.57) and in dependence on $\Delta\bar{t}$.

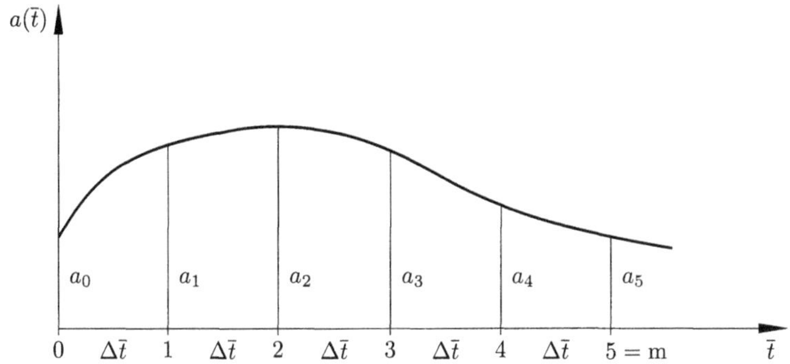

Fig. 1.4.16 On the numerical evaluation of the Duhamel integral

1.4.6 Oscillation induced by moving foundations

An important class of vibration problems is encountered in the study of recording apparatus and in motor vehicles. Here, the oscillator is represented by a simple spring-mass system mounted on a base (fig. 1.4.17) which is subjected to an excitation. Let $r_B(t)$ denote the excitation motion of the foundation and $r(t)$ the corresponding response motion of the mass, m, in an inertial system. Then, the elastic force is given by

$$R_e = k(r - r_B) \tag{1.4.61}$$

and the viscous damping force is

$$R_D = c(\dot{r} - \dot{r}_B) \tag{1.4.62}$$

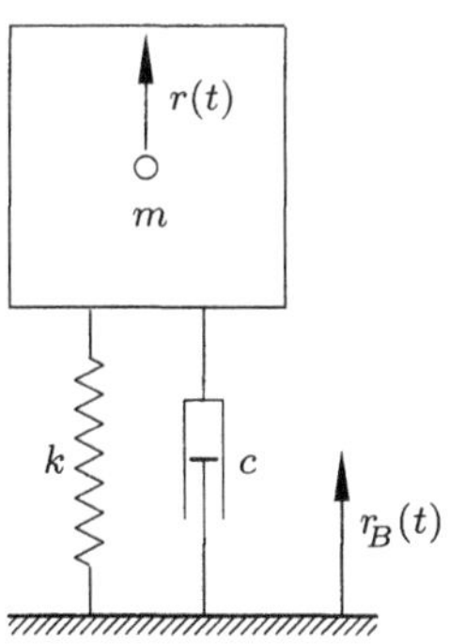

Fig. 1.4.17
Model of an oscillator excited by motion of a foundation

In these two equations, the relative motion between the base and the mass determines the restoring forces. On the other hand, the inertia force is determined by the absolute motion

$$R_I = m\ddot{r} \tag{1.4.63}$$

In the absence of an external force $R(t)$, the equation of motion is given by

$$m\ddot{r} + c\dot{r} + kr = kr_B + c\dot{r}_B \tag{1.4.64}$$

Dividing by m yields

$$\ddot{r} + 2\zeta\omega\dot{r} + \omega^2 r = \omega^2 r_B + 2\zeta\omega\dot{r}_B \tag{1.4.65}$$

Here, it is assumed that $r_B(t)$ is known. The solution of this equation proceeds by the usual technique employed for an external excitation force. A case of great practical importance arises when the base executes a harmonic motion

$$r_B(t) = r_0 \sin\omega_E t \tag{1.4.66}$$

for which the equivalent force is given by

$$\begin{aligned} R_B(t) &= k\left(r_B + \frac{2\zeta}{\omega}\dot{r}_B\right) \\ &= kr_0(\sin\omega_E t + 2\zeta\beta\cos\omega_E t) \end{aligned} \tag{1.4.67}$$

Hence, the principle of superposition yields the steady-state response

$$r_p(t) = r_0 V\left[\sin(\omega_E t - \varphi) + 2\zeta\beta\cos(\omega_E t - \varphi)\right] \tag{1.4.68}$$

where the amplification function, V, and phase shift, φ, are already known for a harmonic force excitation (see equs. (1.4.13), (1.4.14)). Using the trigonometric addition theorem, we can then easily deduce

$$r_p(t) = r_0 V_B \sin(\omega_E t - \varphi_B) \tag{1.4.69}$$

where

$$V_B = V\left[1 + (2\zeta\beta)^2\right]^{1/2} \tag{1.4.70}$$

Here, V_B is also denoted as the transfer function. The corresponding phase angle is

$$\varphi_B = \varphi + \tan^{-1}(-2\zeta\beta) \tag{1.4.71}$$

The functions V_B and φ_B are shown as in dependence on β in fig. 1.4.18. As expected, at resonance ($\beta = 1$), the amplitude becomes significantly large. On the other hand, for $\beta > \sqrt{2}$, the magnitude of the dynamic response is less than the static value.

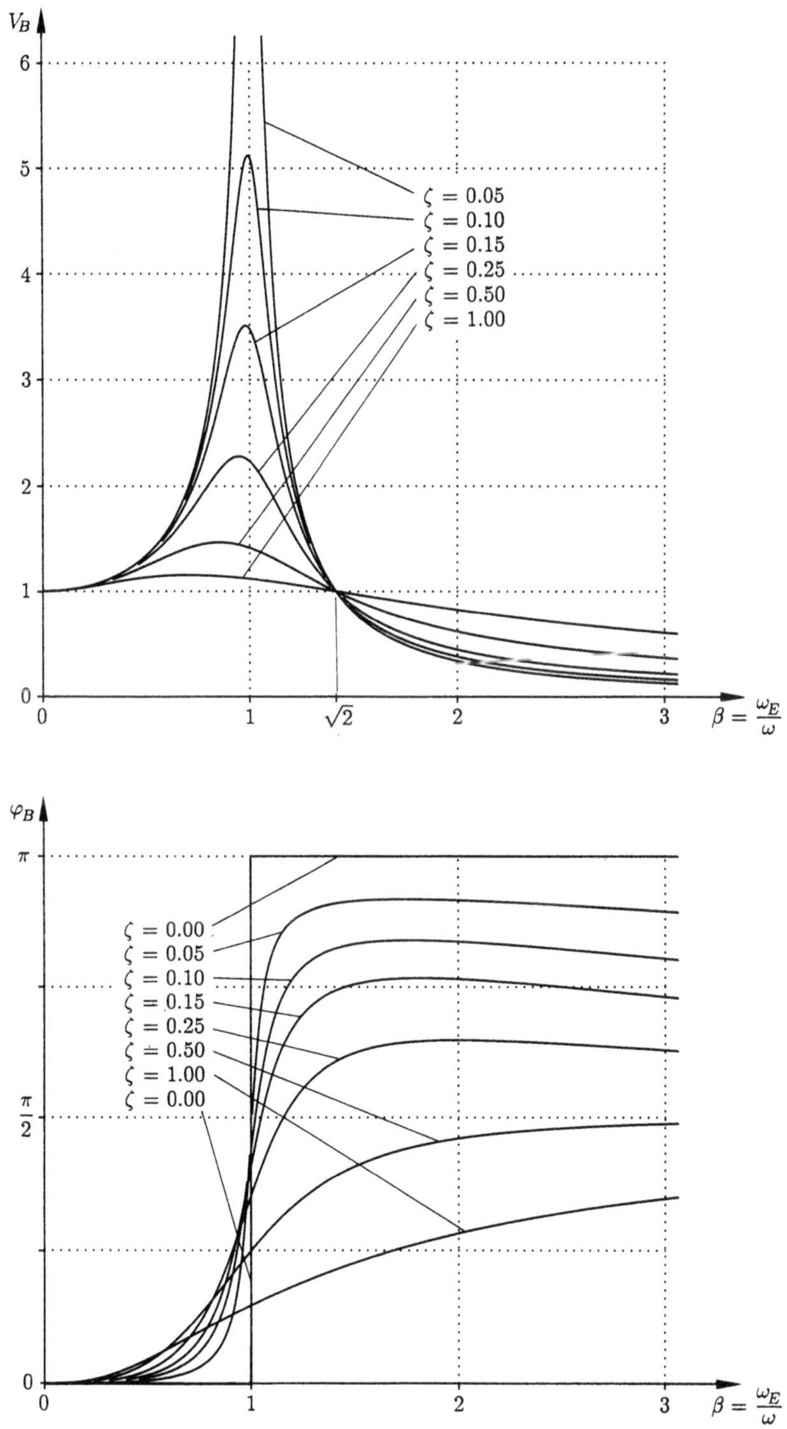

Fig. 1.4.18 Transfer function and phase variation due to an excitation by harmonic motion of a foundation

1.5 Oscillator with rigid body freedoms of motion

The system shown in fig. 1.5.1 has two masses and generally thus possesses two freedoms of motion. Since the translational movement of the model is free, the sum of the horizontal external forces in the absence of an excitation must be zero. This is certainly not the case if the movement of the two masses is fully arbitrary. Even though there are two degrees of freedom, the movement is constrained in such a way that only one independent degree of freedom remains. Thus, the system effectively possesses a single degree of freedom. The forces acting on the first mass are

$$\begin{aligned} R_{1I} &= m\ddot{r}_1 \\ R_{1D} &= c(\dot{r}_1 - \dot{r}_2) \\ R_{1e} &= k(r_1 - r_2) \end{aligned} \tag{1.5.1}$$

and on the second mass

$$\begin{aligned} R_{2I} &= m\ddot{r}_2 \\ R_{2D} &= c(\dot{r}_2 - \dot{r}_1) \\ R_{2e} &= k(r_2 - r_1) \end{aligned} \tag{1.5.2}$$

Each of these groups of forces must be in equilibrium

$$\begin{aligned} &m\ddot{r}_1 + c(\dot{r}_1 - \dot{r}_2) + k(r_1 - r_2) = 0 \\ &m\ddot{r}_2 + c(\dot{r}_2 - \dot{r}_1) + k(r_2 - r_1) = 0 \end{aligned} \tag{1.5.3}$$

If no external force acts upon the system as a whole, the sum of all the forces must also be zero. Consequently, the damping and elastic forces, considered as internal forces, are eliminated (as confirmed by summation of equ. (1.5.3)) and we obtain

$$m\ddot{r}_1 + m\ddot{r}_2 = 0 \tag{1.5.4}$$

This auxiliary condition effectively reduces the problem to a single degree of freedom.

Instead of the coupled equs. (1.5.3) and (1.5.4) , which are somewhat cumbersome to solve, we can apply an alternative description of the movement of the system and introduce distinct rigid and elastic components as shown in fig. (1.5.2). In this way, we encounter a first, very simple concept of generalised degrees of freedom.

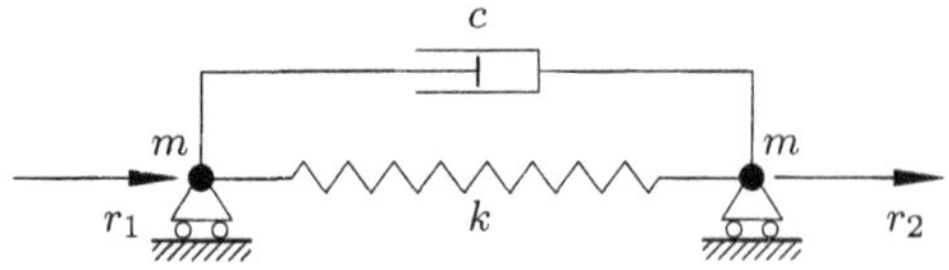

Fig. 1.5.1
Oscillator with rigid body freedom of motion (free-free oscillator)

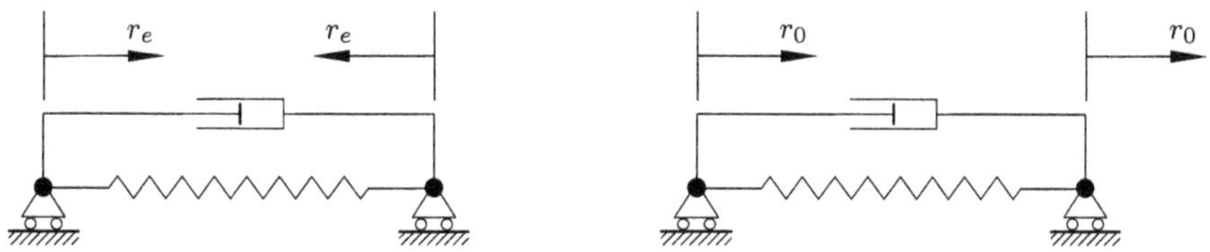

Fig. 1.5.2 Elastic and rigid body mode shape for the oscillator with rigid body freedom

We can express the relation between the original degrees of freedom, r_1 and r_2, and the new components, r_e and r_0, as

$$\begin{aligned} r_1 &= r_e + r_0 \\ r_2 &= -r_e + r_0 \end{aligned} \tag{1.5.5}$$

Substituting these into equs. (1.5.3) and (1.5.4), the subsidiary condition, equ. (1.5.4) reduces to

$$\ddot{r}_0 = 0 \tag{1.5.6}$$

Moreover, both equations (1.5.3) merge into the identical expression

$$m\ddot{r}_e + 2c\dot{r}_e + 2kr_e = 0 \tag{1.5.7}$$

We have thus de-coupled two components of motion and immediately note the solution of equ. (1.5.6)

$$r_0 = a_0 + b_0 t \tag{1.5.8}$$

This simply means that both masses may experience a uniform rigid body motion. Next, we superimpose the solution of equ. (1.5.7) which has already been established in equ. (1.3.19) and reads in the present case

$$r_e(t) = \mathrm{e}^{-2\zeta\omega t}\left(a_e \cos\omega_D t + b_e \sin\omega_D t\right) \tag{1.5.9}$$

with

$$\omega_D = \sqrt{1-(2\zeta)^2}\sqrt{\frac{2k}{m}} \tag{1.5.10}$$

In fact, the elastic component of the motion describes a damped oscillation. The four coefficients a_0, b_0, a_e and b_e can be determined from the four initial conditions $r_1(0)$, $\dot{r}_1(0)$, $r_2(0)$ and $\dot{r}_2(0)$. Figure 1.5.3 shows an undamped oscillation of the system subject only to internal elastic forces; as expected, the centre of gravity of the system does not move in the absence of any external forces.

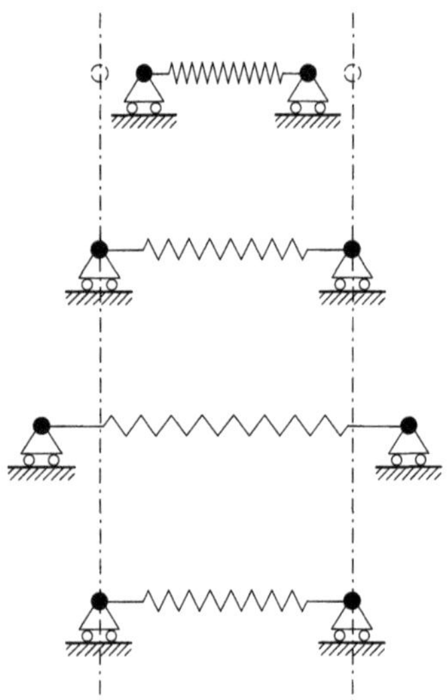

Fig. 1.5.3
Sequential motion of a free-free oscillator

1.6 Generalised degrees of freedom and corresponding generalised stiffnesses, masses and damping (Rayleigh-Ritz approach)

The simplest way of reducing the number of possible modes of oscillation of a technical system is to lump together typical structural properties such as stiffness, inertia and damping. Without doubt, the engineer's intuition plays an important role here.

An alternative and mathematically substantiated method of reducing the number of degrees of freedom was developed by Rayleigh and Ritz. According to this procedure, we postulate an assumed spatial distribution $\boldsymbol{\omega}(\boldsymbol{x})$ for the displacement function $\boldsymbol{u}$. The degree of freedom, r, appears then as a kind of time-dependent amplitude. Thus we write

$$\underset{(3\times1)}{\boldsymbol{u}(\boldsymbol{x},t)} = \begin{bmatrix} u(x,\, y,\, z;\, t) \\ v(x,\, y,\, z;\, t) \\ w(x,\, y,\, z;\, t) \end{bmatrix} = \underset{(3\times1)}{\boldsymbol{\omega}(\boldsymbol{x})} \cdot r(t) \tag{1.6.1}$$

In this way, the displacement function only has to satisfy so-called primary or essential boundary conditions and not necessarily secondary or natural boundary conditions. In the present case, this means that the displacement function must satisfy the kinematic boundary conditions, but not necessarily the static ones. This statement holds here since our dynamic problem is described in terms of displacements.

For the moment, we will consider the simple example shown in fig. 1.6.1 and seek an approximate estimate of the oscillatory characteristics of a cantilever beam. The kinematic boundary conditions are given by

$$\begin{aligned} v(0) &= 0 \\ v'(0) &= 0 \end{aligned} \tag{1.6.2}$$

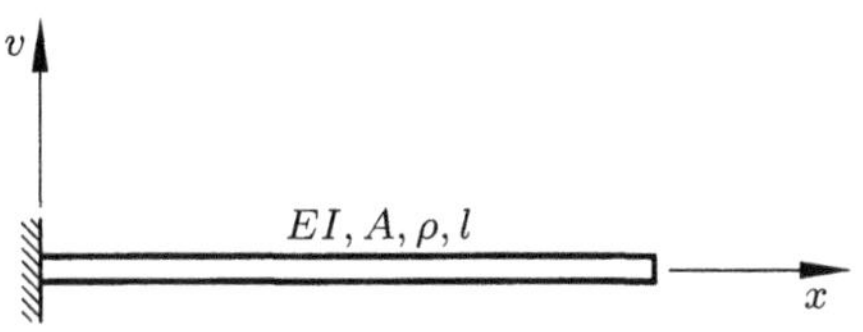

Fig. 1.6.1 Oscillatory model of a cantilever beam

The static boundary conditions state that both the moment ($\sim v''$) and shear force ($\sim v'''$) must be equal to zero at the free end of the beam. In accordance with the Rayleigh-Ritz procedure, we select the displacement function

$$v(x) = \left(\frac{x}{l}\right)^2 r \tag{1.6.3}$$

This is permissible, although the expression does not yield a zero moment at the free end. From equ. (1.6.1), we deduce the bending strain γ according to the engineers' theory of bending (see e.g. Vol. I(G) of this treatise)

$$\begin{aligned} \gamma &= \boldsymbol{D}^{\mathrm{t}}\boldsymbol{u} = \boldsymbol{D}^{\mathrm{t}}\boldsymbol{\omega} r \\ &= \boldsymbol{e}_1 r \end{aligned} \tag{1.6.4}$$

Using the constitutive equation defining the elastic properties of the material

$$\boldsymbol{\sigma} = \boldsymbol{\kappa\varepsilon} = \boldsymbol{\kappa}[\gamma - \boldsymbol{\eta}] \tag{1.6.5}$$

and substituting the strain from equ. (1.6.4), the stiffness matrix of the beam is given simply by

$$k = \int_V \boldsymbol{e}_1^{\mathrm{t}} \boldsymbol{\kappa} \boldsymbol{e}_1 \mathrm{d}V \tag{1.6.6}$$

Next, we calculate the inertial and damping forces. For an infinitesimal element of volume, $\mathrm{d}V$, of a continuous system, the mass is $\rho \mathrm{d}V$ (ρ = mass density). From Newton's law, the inertial force is then

$$\mathrm{d}\boldsymbol{R}_I = \rho \mathrm{d}V \ddot{\boldsymbol{u}} = \rho \boldsymbol{\omega}(\boldsymbol{x}) \mathrm{d}V \ddot{r} \tag{1.6.7}$$

If we assume that the energy dissipation takes place uniformly over the length of the beam, then we obtain for viscous damping

$$\mathrm{d}\boldsymbol{R}_D = \mu \mathrm{d}V \dot{\boldsymbol{u}} = \mu \boldsymbol{\omega}(\boldsymbol{x}) \mathrm{d}V \dot{r} \tag{1.6.8}$$

where $\mu(\boldsymbol{x})$ is the local damping coefficient. The inertia and damping forces are distributed over the entire structure. In order to replace them by equivalent discrete (i.e. concentrated or lumped) forces, R_I and R_D, we evaluate the work done by these forces in moving through the modal amplitude, r, and in order to do this, we apply the principle of virtual work. Thus, we require that R_I or R_D execute the same work when acting over the time-dependent amplitude r as $\boldsymbol{R}_I$ or $\boldsymbol{R}_D$ respectively over the local displacement vector $\boldsymbol{u}$. In fact,

$$rR_I = \int_V \boldsymbol{u}^{\mathrm{t}} \mathrm{d}\boldsymbol{R}_I = r \int_V \rho \boldsymbol{\omega}^{\mathrm{t}} \boldsymbol{\omega} \mathrm{d}V \ddot{r}$$

$$rR_D = \int_V \boldsymbol{u}^{\mathrm{t}} \mathrm{d}\boldsymbol{R}_D = r \int_V \mu \boldsymbol{\omega}^{\mathrm{t}} \boldsymbol{\omega} \mathrm{d}V \ddot{r} \tag{1.6.9}$$

Hence, the equivalent mass (in general a mass matrix) becomes

$$m = \int_V \rho \boldsymbol{\omega}^{\mathrm{t}} \boldsymbol{\omega} \mathrm{d}V \tag{1.6.10}$$

Also, the equivalent damping constant (more generally, a damping matrix) is given by

$$c = \int_V \mu \boldsymbol{\omega}^{\mathrm{t}} \boldsymbol{\omega} \mathrm{d}V \tag{1.6.11}$$

If our continuous oscillator is subject to distributed excitation forces such as body forces, $\boldsymbol{p}_V(\boldsymbol{x}, t)$, and/or surface tractions, $\boldsymbol{p}_S(\boldsymbol{x}, t)$, the work done by these forces is

$$rR(t) = \int_V \boldsymbol{u}^{\mathrm{t}} \boldsymbol{p}_V \mathrm{d}V + \int_S \boldsymbol{u}^{\mathrm{t}} \boldsymbol{p}_S \mathrm{d}S = r\Big(\int_V \boldsymbol{\omega}^{\mathrm{t}} \boldsymbol{p}_V \mathrm{d}V + \int_S \boldsymbol{\omega}^{\mathrm{t}} \boldsymbol{p}_S \mathrm{d}S\Big) \tag{1.6.12}$$

from which the equivalent external force is found to be

$$R(t) = \int_V \boldsymbol{\omega}^{\mathrm{t}} \boldsymbol{p}_V \mathrm{d}V + \int_S \boldsymbol{\omega}^{\mathrm{t}} \boldsymbol{p}_S \mathrm{d}S \tag{1.6.13}$$

If this technique is applied consistently, then the calculated fundamental characteristic frequency will always be higher than the true value. In fact, the Rayleigh-Ritz method yields an eigenfrequency which always possesses a lower bound defined by the 'exact' analytical value. We thus observe that we are now in a position to improve upon the methods developed in the preceding sections, this time using the Rayleigh-Ritz procedure in place of the rough 'lumping' technique.

Some further comments here should clarify the methodology applied to the example of fig. 1.6.1. The so-called exact result for the undamped first eigenfrequency can be obtained from any handbook of mechanics (rotational inertia and shear deformation being ignored here). Thus, we find

$$\omega_{analytic} = \frac{3.5160}{l^2}\sqrt{\frac{EI}{\rho A}}$$

Now we first try the simplest possible displacement function of equ. (1.6.3)

$$v = \left(\frac{x}{l}\right)^2 r$$

The elementary theory of bending yields the longitudinal strain

$$\epsilon_{xx} = -\frac{\mathrm{d}^2 v}{\mathrm{d}x^2} y = -\frac{2}{l^2} y r$$

As a result, the stiffness becomes

$$k = \int\limits_V \left(-\frac{2y}{l^2}\right) E \left(-\frac{2y}{l^2}\right) \mathrm{d}V = \frac{4E}{l^4} \int\limits_0^l \left(\int\limits_A y^2 \mathrm{d}A\right) \mathrm{d}x$$

$$= 4\frac{EI}{l^3}$$

Similarly, the equivalent mass becomes

$$m = \int\limits_V \rho \left(\frac{x}{l}\right)^4 \mathrm{d}V = \frac{\rho A}{l^4} \int\limits_0^l x^4 \mathrm{d}x$$

$$= \frac{1}{5}\rho A l$$

We deduce the approximate eigenfrequency

$$\omega = \sqrt{\frac{k}{m}} = \frac{4.472}{l^2}\sqrt{\frac{EI}{\rho A}} \qquad (+27.2\%)$$

This result is rather disappointing. However, we must remember that we have attempted to describe the behaviour of a system with an infinite number of degrees of freedom by a primitive system with a single degree of freedom.

Next, we apply a lumping procedure and partition the total mass into two components, one half each at the two ends of the beam; the two individual masses are then

$$m = \frac{1}{2}\rho A l$$

The stiffness for a discrete load at the free end of the cantilever beam is

$$k = 3\frac{EI}{l^3}$$

From this approximation, it follows that

$$\omega = \frac{\sqrt{6}}{l^2}\sqrt{\frac{EI}{\rho A}} = \frac{2.4495}{l^2}\sqrt{\frac{EI}{\rho A}} \qquad (-30.3\%)$$

This result is even worse than the previous one and the theory of the upper bound valid for the Rayleigh-Ritz procedure is obviously not applicable to the lumping technique.

We now conclude the discussion by selecting a more refined modal function for the Rayleigh-Ritz technique

$$v = \left[\left(\frac{x}{l}\right)^2 - \frac{1}{3}\left(\frac{x}{l}\right)^3\right] r$$

which, in addition to satisfying the kinematic boundary conditions, also fulfils the condition of zero moment at the free end of the beam ($M = 0$). The corresponding bending strain becomes

$$\epsilon_{xx} = \left[-\frac{2}{l^2} + \frac{2}{l^2}\left(\frac{x}{l}\right)\right] yr$$

We deduce the stiffness as

$$\begin{aligned} k &= \int_V \left(-\frac{2}{l^2} + \frac{2}{l^2}\frac{x}{l}\right) E \left(-\frac{2}{l^2} + \frac{2}{l^2}\frac{x}{l}\right) \mathrm{d}V \\ &= 4\frac{E}{l^4}\int_0^l \left(1 - \frac{x}{l}\right)^2 \left(\int_A y^2 \mathrm{d}A\right) \mathrm{d}x \\ &= \frac{4}{3}\frac{EI}{l^3} \end{aligned}$$

Further, the equivalent mass can be found to be

$$m = \int_V \rho \left[\left(\frac{x}{l}\right)^2 - \frac{1}{3}\left(\frac{x}{l}\right)^3\right]^2 \mathrm{d}V$$

$$= \rho A l \int_0^1 \left(\xi^2 - \tfrac{1}{3}\xi^3\right)^2 \mathrm{d}\xi$$

$$= \frac{11}{105}\rho A l$$

The last two values yield the approximate eigenfrequency

$$\omega = \sqrt{\frac{4}{3}\frac{105}{11}}\sqrt{\frac{EI}{\rho A l^4}} = \frac{3.5675}{l^2}\sqrt{\frac{EI}{\rho A}} \qquad (+1.5\%)$$

As engineers, we can now be satisfied with this result which also complies with the upper bound rule.

In this and the following sections, we have limited ourselves to positive values of damping ζ. However, an oscillatory system surrounded by a medium such as air can extract energy, not only to overcome internal and external damping, but also to excite critical and over-critical destructive vibrations. Such behaviour can be expressed by negative damping ratios and occurs for example in the form of an aeroelastic oscillation of a wing. A first introduction into this subject will be given in Volume VI to this treatise.

Bibliography for Chapter 1

Literature dealing with the subject matter of Chapter 1 is very extensive. More or less detailed discussions on the 1-mass-oscillator can be found in all textbooks on technical mechanics or physics. This modest selection of a few easily obtainable monographs is intended to stimulate the reader to supplementary studies.

[1.1] *K. Magnus*; Schwingungen (Teubner, Stuttgart, 3.Aufl. 1976).
[1.2] *K. Klotter*; Technische Schwingungslehre, Band 1 (Springer, Berlin, 1980).
[1.3] *P.C. Müller, W.O. Schiehlen*; Lineare Schwingungen (Akademische Verlagsgesellschaft, Wiesbaden, 1976).
[1.4] *S. Timoshenko*; Vibration problems in engineering (van Nostrand, Princeton, 1955).
[1.5] *J.P. den Hartog*; Mechanical vibrations (McGraw-Hill, New York, 1956).
[1.6] *R.W. Clough, J. Penzien*; Dynamics of structures (McGraw-Hill, New York, 1975).
[1.7] *W.C. Hurty, M.F. Rubinstein*; Dynamics of structures (Prentice-Hall, Englewood Cliffs., 1964).

2 The equations of motion and virtual work methods in dynamics

"They that loathe the Fountaines heade,
will neuer drinke of the lyttle Brookes."
*John Lyly, Euphues and this England:
To the Gentlemen Readers, p. 223 (1580)*

The main aim of this chapter is the formulation of the general equations of motion of a deformable body based on an arbitrary reference system. Starting from Newton's law, the principle of virtual displacement is extended to dynamics. By then applying a partial discretisation restricted to space, we obtain a system of ordinary differential equations which can then be integrated numerically or analytically to yield the motion of the structure in time and space. Assuming conservative loads and approximations to the damping characteristics, we obtain a number of simplifications which in their turn facilitate the solution of the equations of motion.

2.1 A short reminder of the dynamics of a point-mass

In our preceding account of statics of a structure (see Vol. I), we have already noted the position and displacement vectors $\boldsymbol{x}$ (equ. 2.2.1 of Vol. I, statics) and $\boldsymbol{u}(\boldsymbol{x})$ (equ. 2.2.2 of Vol. I, statics) which define the individual points of a structure and their displacements. In a static case, however, the deflection is assumed to take place at a very slow (strictly, infinitely slow) rate. In contrast, a dynamic loading implies a finite speed of deformation. Consequently, the deformation $\boldsymbol{u}(\boldsymbol{x}, t)$ is then not only a function of the spatial position $\boldsymbol{x}$ but also of time t. From our point of view, it proves more convenient to initiate the introduction of this new dimension by means of point-mass dynamics for, in the limit, the continuum can be represented as an accumulation of infinitesimal point-masses.

For the purpose of our treatise, we assume the existence of a basic system of coordinates which is used to measure movement. Independently of this, there should also be a means of measuring time. The Newtonian principles of independence of time and space and the presumed existence of an absolute time and a Euclidean space form a theoretical edifice based on many experimental earth-bound observations. At the beginning of the century, these had to be revised in order to explain, *inter alia*, some phenomena in electromagnetical theory. However, the Newtownian space-time framework will, in general, be adequate for our needs.

We now continue to follow the Newtonian edifice and, using its first law of motion, attempt to describe the natural motion of a point-mass. Amongst all possible types of reference systems, we seek a coordinate system for which a point-mass moving with uniform vectorial velocity retains this motion in the absence of external forces. Such a distinctive reference system is known as an inertial system and can be traced back to Galilei. It must also be mentioned that it is in principle impossible to specify a system which strictly fulfils these conditions. For most earth-bound

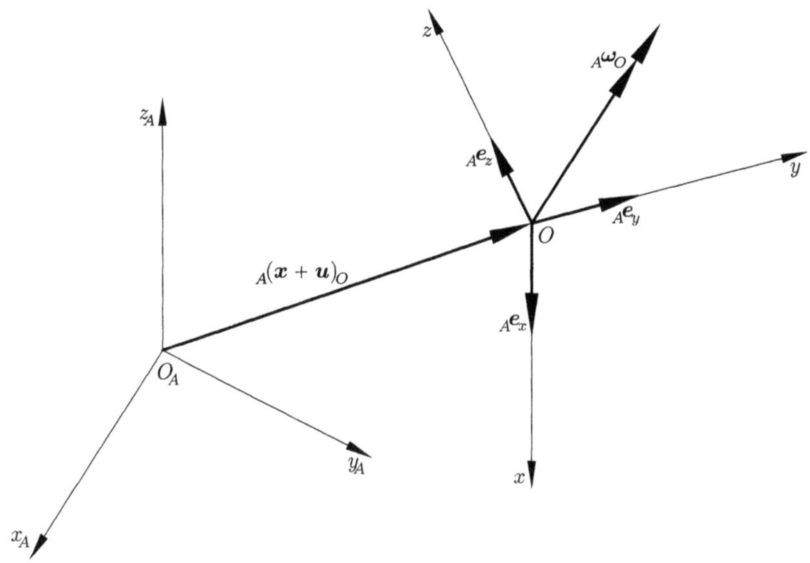

Fig. 2.1.1 Position of the moving system O with reference to an inertial system O_A

problems, however, we may accept the existence of such a system for engineering purposes.

It will not always be convenient to work with an inertial system. Thus, if we were to investigate the oscillatory response of a turbine blade, helicopter blade or a wind turbine, we would quickly discover that a co-rotational or convected frame of reference is preferable. The same applies to aeroplanes or rockets for which a moving system of coordinates will come in very useful.

In general, we need to distinguish between the Cartesian coordinate systems $O(x,\ y,\ z)$ and $O_A(x_A,\ y_A,\ z_A)$. The former defines the system selected for the current problem and may incidentally be a convected or moving one; the latter denotes the absolute or inertial one. If we work *a priori* in an inertial system, then O and O_A are coincident. A standard vector can be described by its components in the x, y, z directions or alternatively in the x_A, y_A, z_A directions. In order to be explicit, the first set of vectors may be denoted by $\boldsymbol{a}$ (with no subscripts for the coordinate system O) and the second set by $_A\boldsymbol{a}$ (pre-set subscript for the coordinate system O_A).

Starting with the inertial system O_A, we denote the position vector of a moving coordinate system O at time $t = 0$ by $_A\boldsymbol{x}_O$, and the same vector after a time interval t by $_A(\boldsymbol{x} + \boldsymbol{u}(t))_O$ (fig. 2.1.1). The direction cosines of the moving coordinate system O relative to that at O_A can be assembled in the matrix $_A[\boldsymbol{e}_x\ \boldsymbol{e}_y\ \boldsymbol{e}_z]$, where $\boldsymbol{e}_x$, $\boldsymbol{e}_y$, $\boldsymbol{e}_z$ are unit vectors along the O_x, O_y, O_z axes. With the help of these unit vectors, any arbitrary vector $\boldsymbol{a}$ (referred to O) can be transformed to a vector $_A\boldsymbol{a}$ (referred to O_A) and *vice versa.* Thus, we form the transformation matrix

$$\boldsymbol{T} = {}_A[\boldsymbol{e}_x \quad \boldsymbol{e}_y \quad \boldsymbol{e}_z] \tag{2.1.1}$$

and obtain the transformation rule

$$ {}_A\boldsymbol{a} = \boldsymbol{T}\boldsymbol{a} \tag{2.1.2}$$

The inverse transformation reads

$$ \boldsymbol{a} = \boldsymbol{T}^{\mathrm{t}}({}_A\boldsymbol{a}) \tag{2.1.3}$$

in which we take account of the orthogonality of the unitary transformation matrix $\boldsymbol{T}$

$$ \boldsymbol{T}^{\mathrm{t}}\boldsymbol{T} = \boldsymbol{I} \tag{2.1.4}$$

Next, we assume that the moving coordinate system O rotates with an angular velocity ω_O about a unit vector ${}_A\boldsymbol{e}_\omega$. Then

$$ {}_A\boldsymbol{\omega}_O = \omega_O({}_A\boldsymbol{e}_\omega) \tag{2.1.5}$$

(a right-handed screw is taken as positive). In the moving coordinate system (referred to O) a point, P, has the position

$$ \boldsymbol{x} + \boldsymbol{u} = \{(x+u) \quad (y+v) \quad (z+w)\} \tag{2.1.6}$$

An observer in O records the relative velocity of P as

$$ \frac{\partial \boldsymbol{u}}{\partial t} = \dot{\boldsymbol{u}} = \{\dot{u} \quad \dot{v} \quad \dot{w}\} \tag{2.1.7}$$

and the corresponding relative acceleration as

$$ \frac{\partial^2 \boldsymbol{u}}{\partial t^2} = \ddot{\boldsymbol{u}} = \{\ddot{u} \quad \ddot{v} \quad \ddot{w}\} \tag{2.1.8}$$

Simultaneously, an observer in an absolute or inertial coordinate system, ${}_AO$, monitors the position of the same point P (fig. 2.1.2) in the form

$$ {}_A[\boldsymbol{x}+\boldsymbol{u}]_P = {}_A[\boldsymbol{x}+\boldsymbol{u}]_O + {}_A[\boldsymbol{x}+\boldsymbol{u}] = {}_A[\boldsymbol{x}+\boldsymbol{u}]_O + \boldsymbol{T}[\boldsymbol{x}+\boldsymbol{u}] \tag{2.1.9}$$

The change of position with time is obtained by differentiating the positon vector with respect to t. Here, we must remember that the unit vectors $\boldsymbol{e}_x$, $\boldsymbol{e}_y$ and $\boldsymbol{e}_z$ in the matrix $\boldsymbol{T}$ are also functions of time

$$ {}_A\dot{\boldsymbol{u}}_P = {}_A\dot{\boldsymbol{u}}_O + \dot{u}\,{}_A\boldsymbol{e}_x + \dot{v}\,{}_A\boldsymbol{e}_y + \dot{w}\,{}_A\boldsymbol{e}_z + (x+u)\,{}_A\dot{\boldsymbol{e}}_x + (y+v)\,{}_A\dot{\boldsymbol{e}}_y + (z+w)\,{}_A\dot{\boldsymbol{e}}_z \tag{2.1.10}$$

The derivative of a unit vector, ${}_A\boldsymbol{e}_i$ ($i = x, y, z$), with respect to time can easily be determined (see fig. 2.1.3). We note that the vector $\dot{\boldsymbol{e}}_i$ stands normal to a plane containing the vectors ${}_A\boldsymbol{\omega}_O$ and ${}_A\boldsymbol{e}_i$, i.e. ${}_A\boldsymbol{e}_i$ must point in the direction of the

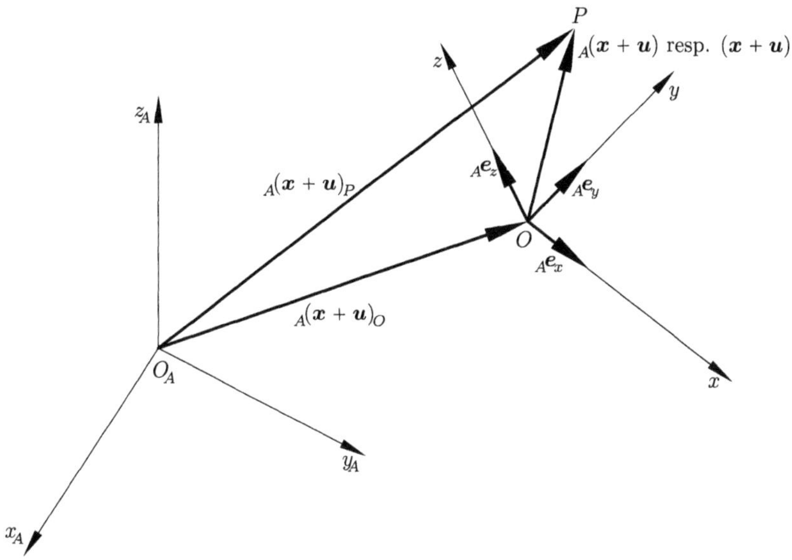

Fig. 2.1.2 Position of a point P with reference to an inertial system O_A

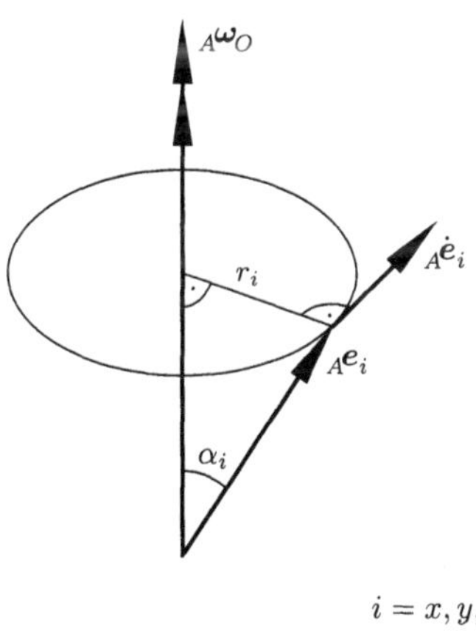

Fig. 2.1.3
Temporal variation of a unit vector $_A\boldsymbol{e}_i$ in a rotating system O

vector product $_A\boldsymbol{\omega}_O \times {_A\boldsymbol{e}_i}$. The magnitude of the vector $_A\dot{\boldsymbol{e}}_i$ can be deduced from the increment generated by the circular motion

$$|_A\dot{\boldsymbol{e}}_i| = \omega_O r_i \tag{2.1.11}$$

where the measure r_i is given by

$$r_i = |\boldsymbol{e}_i| \sin\alpha_i = \sin\alpha_i \tag{2.1.12}$$

Here, α_i is the angle between ${}_A\boldsymbol{e}_i$ and ${}_A\boldsymbol{\omega}_O$. From the definition of a vector product, it follows that

$$ {}_A\dot{\boldsymbol{e}}_i = {}_A\boldsymbol{\omega}_O \times {}_A\boldsymbol{e}_i, \qquad i = x, y, z \tag{2.1.13}$$

Hence, the time derivative of the matrix $\boldsymbol{T}$ (equ. (2.1.1)) can be written formally as

$$\dot{\boldsymbol{T}} = {}_A\boldsymbol{\omega}_O \times \boldsymbol{T} \tag{2.1.13a}$$

where the cross product (vector) × (matrix) is to be understood as the assembly of the cross products (vector) × (column vectors of the matrix) and is collected once more in a matrix form.

Using this definition, equ. (2.1.10) can be rewritten as

$$\begin{aligned} {}_A\dot{\boldsymbol{u}}_P &= {}_A\dot{\boldsymbol{u}}_O + \boldsymbol{T}\dot{\boldsymbol{u}} + \dot{\boldsymbol{T}}[\boldsymbol{x} + \boldsymbol{u}] \\ &= {}_A\dot{\boldsymbol{u}}_O + \boldsymbol{T}\dot{\boldsymbol{u}} + {}_A\boldsymbol{\omega}_O \times (\boldsymbol{T}[\boldsymbol{x} + \boldsymbol{u}]) \\ &= {}_A\dot{\boldsymbol{u}}_O + {}_A[\boldsymbol{e}_x \quad \boldsymbol{e}_y \quad \boldsymbol{e}_z]\dot{\boldsymbol{u}} + {}_A\boldsymbol{\omega}_O \times ({}_A[\boldsymbol{e}_x \quad \boldsymbol{e}_y \quad \boldsymbol{e}_z][\boldsymbol{x} + \boldsymbol{u}]) \end{aligned} \tag{2.1.14}$$

If we refer the vectors to the system O, i.e. if we describe the vectors by their components in the direction x, y, z, of the convected system, we obtain the simpler expression

$$\dot{\boldsymbol{u}}_P = \dot{\boldsymbol{u}} + \dot{\boldsymbol{u}}_O + \boldsymbol{\omega}_O \times [\boldsymbol{x} + \boldsymbol{u}] \tag{2.1.15}$$

where the vectors ${}_A\dot{\boldsymbol{u}}_O$ and ${}_A\boldsymbol{\omega}_O$ are expressed in components of the reference system O in accordance with equ. (2.1.3).

In equ. (2.1.15), $\dot{\boldsymbol{u}}$ defines the relative velocity to which we referred earlier in equ. (2.1.7). Furthermore,

$$\dot{\boldsymbol{u}}_O + \boldsymbol{\omega}_O \times [\boldsymbol{x} + \boldsymbol{u}]$$

is the so-called transport velocity of O with respect to ${}_AO$. The absolute velocity is thus the relative velocity plus the transport velocity.

To calculate the absolute acceleration, we differentiate equ. (2.1.14) once again with respect to time and obtain

$$\begin{aligned} {}_A\ddot{\boldsymbol{u}}_P =& {}_A\ddot{\boldsymbol{u}}_O + \dot{\boldsymbol{T}}\dot{\boldsymbol{u}} + \boldsymbol{T}\ddot{\boldsymbol{u}} + {}_A\dot{\boldsymbol{\omega}}_O \times (\boldsymbol{T}[\boldsymbol{x} + \boldsymbol{u}]) \\ &+ {}_A\boldsymbol{\omega}_O \times (\dot{\boldsymbol{T}}[\boldsymbol{x} + \boldsymbol{u}]) + {}_A\boldsymbol{\omega}_O \times (\boldsymbol{T}\dot{\boldsymbol{u}}) \end{aligned}$$

or, using equ. (2.1.13a),

$$\begin{aligned} {}_A\ddot{\boldsymbol{u}}_P =& \boldsymbol{T}\ddot{\boldsymbol{u}} + {}_A\ddot{\boldsymbol{u}}_O + {}_A\dot{\boldsymbol{\omega}}_O \times (\boldsymbol{T}[\boldsymbol{x} + \boldsymbol{u}]) \\ &+ {}_A\boldsymbol{\omega}_O \times \left[{}_A\boldsymbol{\omega}_O \times (\boldsymbol{T}[\boldsymbol{x} + \boldsymbol{u}])\right] + 2\,{}_A\boldsymbol{\omega}_O \times (\boldsymbol{T}\dot{\boldsymbol{u}}) \end{aligned} \tag{2.1.16}$$

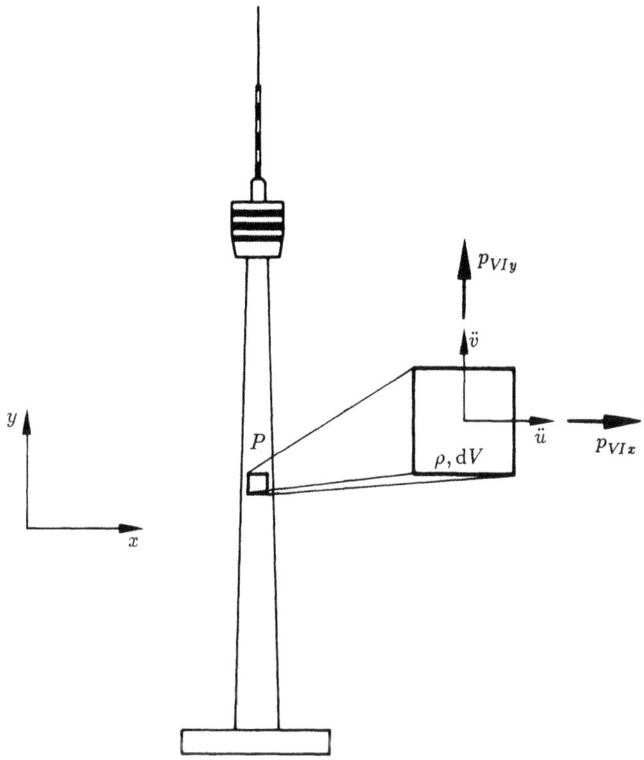

Fig. 2.1.4
Inertial forces at a point P of a structure

Multiplying the above expression by $\boldsymbol{T}^{\mathrm{t}}$, the acceleration in the moving coordinate system reduces to a simple expression

$$\ddot{\boldsymbol{u}}_P = \ddot{\boldsymbol{u}} + \ddot{\boldsymbol{u}}_O + \dot{\boldsymbol{\omega}}_O \times [\boldsymbol{x} + \boldsymbol{u}] + \boldsymbol{\omega}_O \times (\boldsymbol{\omega}_O \times [\boldsymbol{x} + \boldsymbol{u}]) + 2\boldsymbol{\omega}_O \times \dot{\boldsymbol{u}} \tag{2.1.17}$$

Here, $\ddot{\boldsymbol{u}}$ is the relative acceleration and

$$\ddot{\boldsymbol{u}}_O + \dot{\boldsymbol{\omega}}_O \times [\boldsymbol{x} + \boldsymbol{u}] + \boldsymbol{\omega}_O \times (\boldsymbol{\omega}_O \times [\boldsymbol{x} + \boldsymbol{u}])$$

represents the so-called transport acceleration whilst

$$2\boldsymbol{\omega}_O \times \dot{\boldsymbol{u}}$$

describes the so-called Coriolis acceleration. The absolute acceleration thus appears as the sum of the relative, transport and Coriolis accelerations.

Having discussed the kinematics of the motion, we next consider the dynamics of a system subject to a number of forces. As mentioned earlier, Newton's first law of motion states that a particle initially at rest or in uniform motion in an inertial system continues to remain on an undisturbed course unless otherwise acted upon by external forces. In a somewhat loose formulation, Newton's second law of motion states that the acceleration of a particle is directly proportional to the external force and inversely proportional to its mass. Let us assume that a point P in a structure is assigned an infinitesimally small mass (fig. 2.1.4)

$$\mathrm{d}m = \rho \mathrm{d}V \tag{2.1.18}$$

where ρ is the mass density and $\mathrm{d}V$ an infinitesimal volume element enclosing the point P. The force exerted on this mass is a volume force arising from the effects of inertia and is expressed here as $\boldsymbol{p}_{VI}\mathrm{d}V$ (I for inertia), and resolved into the components $p_{VIx}, p_{VIy}, p_{VIz}$ in the directions of the axes x, y, z

$$\boldsymbol{p}_{VI} = \{p_x \quad p_y \quad p_z\}_{VI} \tag{2.1.19}$$

According to Newton's second law and with reference to an inertial system, the force can be written as

$$\boldsymbol{p}_{VI}\mathrm{d}V = \rho \mathrm{d}V \ddot{\boldsymbol{u}}$$

or

$$\boldsymbol{p}_{VI} = \rho \ddot{\boldsymbol{u}} \tag{2.1.20}$$

In a convected coordinate system , the analogous expression is obtained by simply replacing $\ddot{\boldsymbol{u}}$ by the absolute acceleration $\ddot{\boldsymbol{u}}_P$ defined in equ. (2.1.17), i.e.

$$\boldsymbol{p}_{VI} = \rho \ddot{\boldsymbol{u}}_P \tag{2.1.20a}$$

2.2 The principle of virtual displacements (P.V.D.) in dynamics

We first refer to the principle of virtual displacements in statics and in particular to equ. (3.1.1) of Volume I of this series

$$\int_V \underline{\boldsymbol{\gamma}}^{\mathrm{t}} \boldsymbol{\sigma} \mathrm{d}V = \int_V \underline{\boldsymbol{u}}^{\mathrm{t}} \boldsymbol{p}_V \mathrm{d}V + \int_{S_\sigma} \underline{\boldsymbol{u}}^{\mathrm{t}} \boldsymbol{p}_S \mathrm{d}S$$

$$\underline{\boldsymbol{u}} = \boldsymbol{o} \quad \text{on} \quad S_u \tag{2.2.1}$$

This principle asserts that a system of applied body forces $\boldsymbol{p}_V$ and surface forces $\boldsymbol{p}_S$ (acting on the surface S_σ) must be in equilibrium (or static compatibility) with the corresponding internal stresses $\boldsymbol{\sigma}$. This state of equilibrium can now be linked with an arbitrary kinematically consistent virtual displacement field $\underline{\boldsymbol{u}}$. The associated virtual strain $\underline{\boldsymbol{\gamma}}$ has, of course, to be expressed in terms of the virtual displacement $\underline{\boldsymbol{u}}$. Thus

$$\underline{\boldsymbol{\gamma}} = \boldsymbol{D}^{\mathrm{t}} \underline{\boldsymbol{u}} \tag{2.2.2}$$

where $\boldsymbol{D}$ is a matrix cum differential operator and is detailed in equ. (2.2.23) of Volume I. The virtual field $\underline{\boldsymbol{u}}$ must not contain any gaps or overlaps and is necessarily zero on the part of the surface (S_u) over which the actual displacements $\boldsymbol{u}$ are prescribed

$$\underline{\boldsymbol{u}} = \boldsymbol{o}, \qquad \text{on} \quad S_u \tag{2.2.3}$$

Applying the Gaussian divergence integral formula to the left-hand side of equ. (2.2.1), the equilibrium state can be rewritten as

$$\int_V \underline{\boldsymbol{u}}^{\mathrm{t}}[\boldsymbol{D\sigma} + \boldsymbol{p}_V]\mathrm{d}V - \int_{S_\sigma} \underline{\boldsymbol{u}}^{\mathrm{t}}[\boldsymbol{D}_S\boldsymbol{\sigma} - \boldsymbol{p}_S]\mathrm{d}S = 0 \tag{2.2.4}$$

Since this equation must be satisfied for any arbitrary, virtual displacement $\underline{\boldsymbol{u}}$, it follows that

$$\boldsymbol{D\sigma} + \boldsymbol{p}_V = \boldsymbol{o} \tag{2.2.5}$$

$$\boldsymbol{D}_S\boldsymbol{\sigma} - \boldsymbol{p}_S = \boldsymbol{o}, \qquad \text{on} \quad S_\sigma \tag{2.2.6}$$

are valid. These are, however, the well-known differential equations expressing the equilibrium of internal stresses with the internal volume forces (see, e.g., equ. (2.3.9) of Vol. I) and the external surface forces (see equ. (2.3.16) of Vol. I). Consequently, the principle of virtual displacements provides a fully equivalent alternative formulation of the equilibrium conditions.

In literature, three methods are employed in the formulation of the equations of motion in dynamics, each of which has its own particular advantage depending upon the nature of the problem. These are:

— direct formulation of the state of equilibrium

— generalised principle of virtual displacement, P.V.D.

— application of the Hamiltonian principle

Basically, any problem in dynamics can be solved by means of any one of the above methods, although the selection of a particular approach is usually determined by individual preference. For the approximate computation of complex structural systems, however, the application of the virtual displacement method or the Hamiltonian principle possesses some advantages. In the virtual displacement method, the applied forces appear explicitly, while their influence in the Hamiltonian principle is contained in the expression for the kinetic and potential energies. However, both methods are fully equivalent and yield identical equations of motion in all cases. Therefore, there is no reason – at least in this introductory treatise – not to choose the P.V.D. method since this preference was also demonstrated in our initial comments. Indeed, the P.V.D. involves the simplest methodology for the mind of an engineer. Further, we may also argue that if one can derive the forces directly from the principle of virtual displacement, then why should one use a potential energy function whose derivatives with respect to the displacements yield the same forces? Consequently, we feel encouraged to employ the simplest approach, namely the principle of virtual displacement (P.V.D.). The reader is referred, however, to Vol. VII which contains an elementary discussion of the Hamiltonian principle.

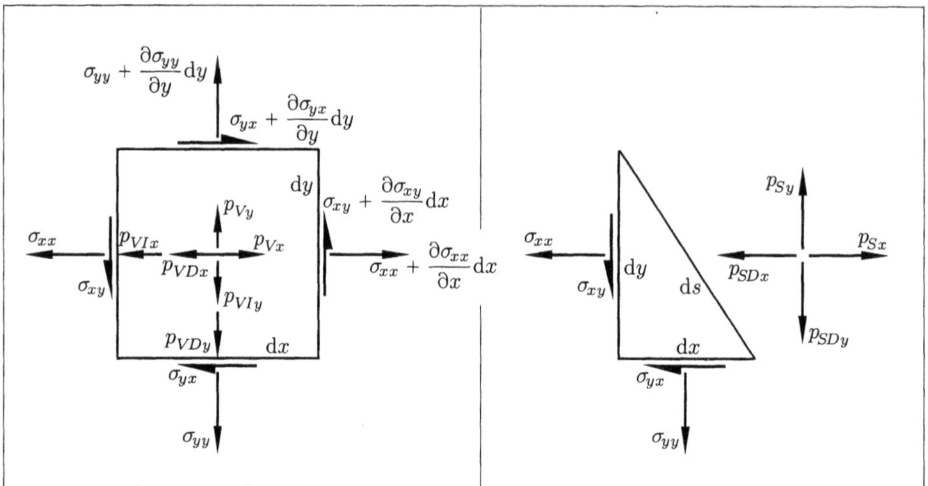

$$\begin{bmatrix} \frac{\partial}{\partial x} & 0 & \frac{1}{\sqrt{2}}\frac{\partial}{\partial x} \\ 0 & \frac{\partial}{\partial y} & \frac{1}{\sqrt{2}}\frac{\partial}{\partial y} \end{bmatrix} \begin{bmatrix} \sigma_{xx} \\ \sigma_{yy} \\ \sqrt{2}\sigma_{xy} \end{bmatrix} = \begin{bmatrix} p_{Vx} - p_{VIx} - p_{VDx} \\ p_{Vy} - p_{VIy} - p_{VDy} \end{bmatrix} \qquad \begin{bmatrix} l & 0 & \frac{1}{\sqrt{2}}m \\ 0 & m & \frac{1}{\sqrt{2}}l \end{bmatrix} \begin{bmatrix} \sigma_{xx} \\ \sigma_{yy} \\ \sqrt{2}\sigma_{xy} \end{bmatrix} = \begin{bmatrix} p_{Sx} - p_{SDx} \\ p_{Sy} - p_{SDy} \end{bmatrix}$$

in volume V	on the surface S

Fig. 2.2.1 Equilibrium in a dynamic situation; two-dimensional case

Following this preamble, we will now extend our horizon and proceed to the analysis of dynamic problems. In dynamic analysis, the system of forces on a structure must be in equilibrium at any particular instant, t, as long as all forces arising from the motion are taken into consideration. This conceptual approach is attributed to d'Alembert. Which forces do we need to consider in dynamics? Apart from the body forces $\boldsymbol{p}_V(t)$, we have to include the inertia forces $\boldsymbol{p}_{VI}$ already defined in equs. (2.1.19) and (2.1.20). Part of the external forces are required to overcome the inertial forces which are due to acceleration; another part of the external forces must overcome the volume or bulk damping forces, $\boldsymbol{p}_{VD}$ (D = damping). Thus, the effective volume forces become

$$\boldsymbol{p}_V \rightarrow \boldsymbol{p}_V(t) - \boldsymbol{p}_{VI}(t) - \boldsymbol{p}_{VD}(t) \tag{2.2.7}$$

In addition to the body forces, we also have the surface tractions. In dynamics, we assume that in addition to the applied surface tractions $\boldsymbol{p}_S(t)$, frictional forces also act on the surface $\boldsymbol{p}_{SD}(t)$. Depending on the sign of the overall force acting on a body, we obtain an acceleration or retardation of the system

$$\boldsymbol{p}_S \rightarrow \boldsymbol{p}_S(t) - \boldsymbol{p}_{SD}(t) \tag{2.2.8}$$

The system of forces which are in dynamic equilibrium in a two-dimensional case is shown in fig. 2.2.1. Using the expressions given by equs. (2.1.20), (2.2.7) and (2.2.8), the principle of virtual displacement in dynamics can be stated as

$$\int_V \rho \underline{\boldsymbol{u}}^{\mathrm{t}} \ddot{\boldsymbol{u}}_A(t)\mathrm{d}V + \int_V \underline{\boldsymbol{u}}^{\mathrm{t}} \boldsymbol{p}_{VD}(t)\mathrm{d}V + \int_S \underline{\boldsymbol{u}}^{\mathrm{t}} \boldsymbol{p}_{SD}(t)\mathrm{d}S + \int_V \underline{\boldsymbol{\gamma}}^{\mathrm{t}} \boldsymbol{\sigma}(t)\mathrm{d}V$$

$$= \int_V \underline{\boldsymbol{u}}^{\mathrm{t}} \boldsymbol{p}_V(t)\mathrm{d}V + \int_S \underline{\boldsymbol{u}}^{\mathrm{t}} \boldsymbol{p}_S(t)\mathrm{d}S \tag{2.2.9}$$

The principle is valid for any arbitrary reference system since we have used the absolute acceleration $\ddot{\boldsymbol{u}}_A(t)$. The virtual displacement, $\underline{\boldsymbol{u}}$ (strain $\underline{\boldsymbol{\gamma}}$), is independent of time. In fact, before applying $\underline{\boldsymbol{u}}$, we assume that the deformed state is frozen in an equilibrium state at the instant t. Thus, the procedure corresponds closely to that in statics. Equation (2.2.9) forms the basis for the derivation of the equations of motion in dynamics. It does not entail an approximation, but is exact in the analytical sense.

2.3 Discretisation of continua and arbitrary structures

As the reader could observe, the application of the principle of virtual displacements in dynamics is basically straightforward. As a matter of fact, it is relatively simple to establish the differential equations to describe the dynamic motion of a system. On the other hand, however, difficulties may arise when attempting to satisfy this principle for complex systems with arbitrary virtual displacement, $\underline{\boldsymbol{u}}$. Closed-form solutions for relatively simple geometries, stiffness and mass distributions can only be determined in a few cases and, naturally, are welcome because they supply a yardstick of excellence and provide a valuable insight into the dynamic behaviour of mechanical systems. It is, of course, more than unlikely that such an analytical solution can be obtained for complex systems like an aircraft or car.

When seeking a closed form solution for a problem in a two- or three-dimensional system, a major difficulty arises through the necessity of satisfying the boundary conditions of the partial differential equations. This state of affairs also occurs in static problems but in dynamics, the additional variable, time, has to be included. The magic formulation in static problems is achieved by means of discretisation in space and an associated concentration of the characteristic properties of the system at the nodes. The latter may be of the lumped kind or involve a consistent concentration through a selective number of approximate but kinematically consistent modal sets. In any case, these discretisations reduce the partial differential equations of the static problem to a set of linear algebraic equations which can easily be solved with the aid of modern computers.

The same procedure may be applied for the solution of a dynamic problem. Here, the dimension of time, t, enlarges that of space, $\boldsymbol{x}$, leading to the functional distribution $\boldsymbol{u}(\boldsymbol{x}, t)$ for the displacement $\boldsymbol{u}$ ([2.1], [2.2]). Recollecting what a formidable increase of computational effort arises in statics when switching from one- to two- or, even more so, to three-dimensional problems, we realise the magnitude of the problem when also incorporating the time domain. This forces us to look for simpler techniques. A simpler and commonly employed method of solving time-dependent problems involves a partial discretisation restricted to the spatial domain. Consequently, we are led to write the solution as a product of two functions, namely a space function, $\boldsymbol{\omega}_{\mathrm{i}}(\boldsymbol{x})$, and a temporal function, $r_{\mathrm{i}}(t)$,

$$\boldsymbol{u}(\boldsymbol{x},t)=\sum_{\mathrm{i}=1}^{\mathrm{n}}\boldsymbol{\omega}_{\mathrm{i}}(\boldsymbol{x})r_{\mathrm{i}}(t)=\underset{(3\times\mathrm{n})}{\boldsymbol{\omega}(\boldsymbol{x})}\,\underset{(\mathrm{n}\times 1)}{\boldsymbol{r}(t)} \tag{2.3.1}$$

where $\boldsymbol{r}(t)$ represents a column vector with the time-functions $r_{\mathrm{i}}(t)$ being entries and matrix $\boldsymbol{\omega}(\boldsymbol{x})$ contains the vectors $\boldsymbol{\omega}_{\mathrm{i}}(\boldsymbol{x})$ as columns

$$\boldsymbol{\omega}(\boldsymbol{x})=[\boldsymbol{\omega}_1(\boldsymbol{x})\ \ldots\ \boldsymbol{\omega}_{\mathrm{i}}(\boldsymbol{x})\ \ldots\ \boldsymbol{\omega}_{\mathrm{n}}(\boldsymbol{x})]$$

This technique can be traced back to Kantorovich's approach [2.3] and serves as the basis for most finite element solutions to time-dependent problems. This solution procedure possesses the great advantage of adopting and extending the analysis procedure employed in statics to dynamic problems. In the present context, we are therefore only concerned with those problems for which equ. (2.3.1) is applicable. The partial discretisation defined by equ. (2.3.1) reduces the partial differential equations to a system of ordinary differential equations. Moreover, if the domain of the structure is not time-dependent, fairly simple relations can be derived, the coefficients of the ordinary differential equations being independent of time. If the stiffness, mass and damping coefficients may also be assumed to be constant, the ordinary differential equations possess constant coefficients and can be solved by standard analytical methods. However, if this is not the case, then the solution may be accomplished numerically by time-marching methods.

Incidentally, the partial discretisation of a problem is not quite new to us. It was applied successfully in statics for rotationally symmetric structures subject to the influence of arbitrary or periodically distributed forces. There, the discretisation was restricted to the r,z-plane, the circumferential procedure being based on analytical techniques using Fourier series.

In equ. (2.3.1), the initial and boundary conditions of the problem are not satisfied *a priori*, but only after the formulation of the problem. We assume, however, that the given set of displacement functions complies with the kinematic boundary conditions through an appropriate selection of a number of functions $r_{\mathrm{i}}(t)$ in

$$\boldsymbol{u}(\boldsymbol{x},t)=\boldsymbol{\omega}(\boldsymbol{x})\boldsymbol{r}(t)=\boldsymbol{u}_p(\boldsymbol{x},t),\qquad \text{on}\quad S_u \tag{2.3.2}$$

Moreover the set of equ. (2.3.2) should satisfy the initial conditions, as for example

$$\boldsymbol{u}(\boldsymbol{x},0)=\boldsymbol{\omega}(\boldsymbol{x})\boldsymbol{r}(0)=\boldsymbol{u}_0(\boldsymbol{x}) \tag{2.3.3}$$

$$\dot{\boldsymbol{u}}(\boldsymbol{x},0)=\boldsymbol{\omega}(\boldsymbol{x})\dot{\boldsymbol{r}}(0)=\dot{\boldsymbol{u}}_0(\boldsymbol{x}) \tag{2.3.4}$$

Adopting this technique, we may for the time being ignore the not so easy problem of establishing kinematically consistent functional sets $\boldsymbol{\omega}_{\mathrm{i}}(\boldsymbol{x})$, as we were able to do in the static analogon. In general, the P.V.D. in dynamics, equ. (2.2.9), cannot be fulfilled exactly by the set of equ. (2.3.2); this could only be achieved by the true solution. Exact fulfilment does mean that the equilibrium state, defined by equ. (2.2.9), is satisfied for any virtual displacement $\underline{\boldsymbol{u}}$. Nevertheless, this does not imply that equ. (2.3.1) cannot be fulfilled at least for some virtual displacement sets. For this purpose, we may employ the set $\boldsymbol{\omega}_{\mathrm{i}}(\boldsymbol{x})$ already at hand. For the n unknown functions of time, $r_{\mathrm{i}}(t)$, in equ. (2.3.1), we then have n conditions. At the same time, we should recognise that the state of equilibrium is not satisfied exactly, i.e. at each point, but only for n specific virtual displacements. With this approach, we satisfy the condition of equilibrium in a global or overall sense. However, if the number of unknown time functions, n, is raised, the equilibrium state will be satisfied exactly in the limit.

By applying the P.V.D. of equ. (2.2.9) simultaneously n times for n virtual displacement functions, $\boldsymbol{\omega}_{\mathrm{i}}(\boldsymbol{x})$, we obtain

$$\int_V \rho \boldsymbol{\omega}^{\mathrm{t}} \ddot{\boldsymbol{u}}_A(t) \mathrm{d}V + \int_V \boldsymbol{\omega}^{\mathrm{t}} \boldsymbol{p}_{VD}(t) \mathrm{d}V + \int_S \boldsymbol{\omega}^{\mathrm{t}} \boldsymbol{p}_{SD}(t) \mathrm{d}S + \int_V \boldsymbol{\alpha}^{\mathrm{t}} \boldsymbol{\sigma}(t) \mathrm{d}V$$
$$= \int_V \boldsymbol{\omega}^{\mathrm{t}} \boldsymbol{p}_V(t) \mathrm{d}V + \int_S \boldsymbol{\omega}^{\mathrm{t}} \boldsymbol{p}_S(t) \mathrm{d}S \tag{2.3.5}$$

where (see also equ. (5.1.18), Vol. I)

$$\boldsymbol{\alpha} = \boldsymbol{D}^{\mathrm{t}} \boldsymbol{\omega} \tag{2.3.6}$$

is the strain function already introduced in statics. The system of equs. (2.3.5) still expresses a very general statement on the discrete equilibrium of forces in dynamics. This includes non-linearities of all kinds and allows a time-dependent boundary (in this case, the volume V is the domain of integration at the instant t). In equ. (2.3.5), the contribution

$$\boldsymbol{R}_I(t) = \int_V \rho \boldsymbol{\omega}^{\mathrm{t}} \ddot{\boldsymbol{u}}_A(t) \mathrm{d}V \tag{2.3.7}$$

denotes the inertial forces in an arbitrary reference system,

$$\boldsymbol{R}_D(t) = \int_V \boldsymbol{\omega}^{\mathrm{t}} \boldsymbol{p}_{VD}(t) \mathrm{d}V + \int_S \boldsymbol{\omega}^{\mathrm{t}} \boldsymbol{p}_{SD}(t) \mathrm{d}S \tag{2.3.8}$$

the damping forces (without any specification of the actual damping mode),

$$\boldsymbol{R}_e(t) = \int_V \boldsymbol{\alpha}^{\mathrm{t}} \boldsymbol{\sigma}(t) \mathrm{d}V \tag{2.3.9}$$

the elastic forces, and

$$\boldsymbol{R}(t) = \int_V \boldsymbol{\omega}^{\mathrm{t}} \boldsymbol{p}_V(t) \mathrm{d}V + \int_S \boldsymbol{\omega}^{\mathrm{t}} \boldsymbol{p}_S(t) \mathrm{d}S \tag{2.3.10}$$

the external forces , already familiar from the static analogon. Thus, the equilibrium of the discretised system of forces can simply be expressed as

$$\boldsymbol{R}_I(t) + \boldsymbol{R}_D(t) + \boldsymbol{R}_e(t) = \boldsymbol{R}(t) \tag{2.3.11}$$

This generalises the elementary condition of equilibrium in a system with a single degree of freedom as discussed in Chapter 1. Furthermore, equ. (2.3.11) in conjunction with equ. (2.3.9) demonstrates the formulation of the unit displacement method (UDM) in dynamics. As in statics, it is also valid for non-linear behaviour and serves as a starting-point for the derivation of the equations of motion.

We now examine in more detail the formation of the individual groups of forces. Adopting an inertial system, the acceleration of a point-mass at P is

$$\ddot{\boldsymbol{u}}_P = \ddot{\boldsymbol{u}} = \boldsymbol{\omega}\ddot{\boldsymbol{r}} \tag{2.3.12}$$

and the inertial force from equ. (2.3.7) becomes

$$\boldsymbol{R}_I(t) = \Big(\int_V \rho\boldsymbol{\omega}^{\mathrm{t}}\boldsymbol{\omega}\,\mathrm{d}V\Big)\ddot{\boldsymbol{r}} = \underset{(\mathrm{n}\times\mathrm{n})}{\boldsymbol{M}}\,\underset{(\mathrm{n}\times 1)}{\ddot{\boldsymbol{r}}} \tag{2.3.13}$$

Here, $\boldsymbol{M}$ is the symmetric mass matrix of the structure and will be discussed comprehensively later. For the moment, however, we turn our attention to a moving coordinate system and, in conjunction with equ. (2.3.1), rewrite the acceleration of equ. (2.1.17) in the form

$$\begin{aligned}\ddot{\boldsymbol{u}}_P = \ddot{\boldsymbol{u}}_O &+ \dot{\boldsymbol{\omega}}_O\times\boldsymbol{x} + \boldsymbol{\omega}_O\times(\boldsymbol{\omega}_O\times\boldsymbol{x}) + \big[(\dot{\boldsymbol{\omega}}_O\times\boldsymbol{\omega}) + \boldsymbol{\omega}_O\times(\boldsymbol{\omega}_O\times\boldsymbol{\omega})\big]\boldsymbol{r}\\ &+ [2\boldsymbol{\omega}_O\times\boldsymbol{\omega}]\dot{\boldsymbol{r}} + \boldsymbol{\omega}\ddot{\boldsymbol{r}}\end{aligned} \tag{2.3.14}$$

Here, $\boldsymbol{\omega}_O$ is the angular velocity vector with respect to a moving coordinate system and should not be confused with $\boldsymbol{\omega}(\boldsymbol{x})$ which is the set of modal functions of an element in a structure. We observe that the absolute acceleration contains derivatives of the amplitude vector $\boldsymbol{r}(t)$ up to the second order in time. In the present case, the corresponding inertial forces of equ. (2.3.7) become

$$\begin{aligned}\boldsymbol{R}_I(t) &= \int_V \rho\boldsymbol{\omega}^{\mathrm{t}}\big[\ddot{\boldsymbol{u}}_O + (\dot{\boldsymbol{\omega}}_O\times\boldsymbol{x}) + \boldsymbol{\omega}_O\times(\boldsymbol{\omega}\times\boldsymbol{x})\big]\mathrm{d}V\\ &\quad+ \int_V \rho\boldsymbol{\omega}^{\mathrm{t}}\big[(\dot{\boldsymbol{\omega}}_O\times\boldsymbol{\omega}) + \boldsymbol{\omega}_O\times(\boldsymbol{\omega}_O\times\boldsymbol{\omega})\big]\mathrm{d}V\boldsymbol{r}\\ &\quad+ \int_V 2\rho\boldsymbol{\omega}^{\mathrm{t}}(\boldsymbol{\omega}_O\times\boldsymbol{\omega})\mathrm{d}V\dot{\boldsymbol{r}} + \int_V \rho\boldsymbol{\omega}^{\mathrm{t}}\boldsymbol{\omega}\,\mathrm{d}V\ddot{\boldsymbol{r}}\\ &= \boldsymbol{R}_{IO}(t) + [\boldsymbol{H} + \boldsymbol{Z}]\boldsymbol{r} + \boldsymbol{G}\dot{\boldsymbol{r}} + \boldsymbol{M}\ddot{\boldsymbol{r}}\end{aligned} \tag{2.3.15}$$

$\boldsymbol{R}_{IO}(t)$ is a force vector which results from the motion of the convected system O and is given by

$$\boldsymbol{R}_{IO}(t) = \int_V \rho\boldsymbol{\omega}^{\mathrm{t}}\left[\ddot{\boldsymbol{u}}_O + (\dot{\boldsymbol{\omega}}_O \times \boldsymbol{x}) + \boldsymbol{\omega}_O \times (\boldsymbol{\omega} \times \boldsymbol{x})\right]\mathrm{d}V \tag{2.3.16}$$

Even in the absence of an explicit excitation force $\boldsymbol{R}(t)$ operating in the system O, a response of the body arises due to the motion of the system O.

The matrix $[\boldsymbol{H} + \boldsymbol{Z}]$ generates a force proportional to the displacement $\boldsymbol{r}$ and hence may be termed as a pseudo-stiffness matrix arising from the rotation of the system. Its first component is

$$\boldsymbol{H} = \int_V \rho\boldsymbol{\omega}^{\mathrm{t}}(\dot{\boldsymbol{\omega}}_O \times \boldsymbol{\omega})\,\mathrm{d}V \tag{2.3.17}$$

The element h_{ij} of the $\boldsymbol{H}$ matrix is

$$h_{\mathrm{ij}} = \int_V \rho\boldsymbol{\omega}_{\mathrm{i}}^{\mathrm{t}}(\dot{\boldsymbol{\omega}}_O \times \boldsymbol{\omega}_{\mathrm{j}})\,\mathrm{d}V = \int_V -\rho\boldsymbol{\omega}_{\mathrm{j}}^{\mathrm{t}}(\dot{\boldsymbol{\omega}}_O \times \boldsymbol{\omega}_{\mathrm{i}})\,\mathrm{d}V \tag{2.3.18}$$

where $\boldsymbol{\omega}_{\mathrm{i}}$ (resp. $\boldsymbol{\omega}_{\mathrm{j}}$) is the i-th (resp. j-th) column of the spatial modal matrix $\boldsymbol{\omega}$.

By interchanging the indices i and j, we observe that the absolute value of the element is retained but its sign reversed. Thus, the matrix $\boldsymbol{H}$ is skew-symmetric and can be computed very economically. It is only generated in the presence of an accelerated rotating system.

We next consider the derivation of the second pseudo-stiffness component matrix $\boldsymbol{Z}$

$$\boldsymbol{Z} = \int_V \rho\boldsymbol{\omega}^{\mathrm{t}}\left[\boldsymbol{\omega}_O \times (\boldsymbol{\omega}_O \times \boldsymbol{\omega})\right]\mathrm{d}V \tag{2.3.19}$$

This contribution exists even in the absence of an accelerated rotation and presupposes only a steady rotation of the system. The coefficient z_{ij} of this matrix is

$$\begin{aligned} z_{\mathrm{ij}} &= \int_V \rho\boldsymbol{\omega}_{\mathrm{i}}^{\mathrm{t}}\left[\boldsymbol{\omega}_O \times (\boldsymbol{\omega}_O \times \boldsymbol{\omega}_{\mathrm{j}})\right]\mathrm{d}V \\ &= \int_V \rho\left[(\boldsymbol{\omega}_O^{\mathrm{t}}\boldsymbol{\omega}_{\mathrm{j}})\boldsymbol{\omega}_{\mathrm{i}}^{\mathrm{t}}\boldsymbol{\omega}_O - (\boldsymbol{\omega}_O^{\mathrm{t}}\boldsymbol{\omega}_O)\boldsymbol{\omega}_{\mathrm{i}}^{\mathrm{t}}\boldsymbol{\omega}_{\mathrm{j}}\right]\mathrm{d}V \end{aligned} \tag{2.3.20}$$

By interchanging the indicies i and j, we confirm that the matrix $\boldsymbol{Z}$ is symmetric.

Next, we consider the matrix $\boldsymbol{G}$ which derives from the Coriolis forces

$$\boldsymbol{G} = \int_V 2\rho\boldsymbol{\omega}^{\mathrm{t}}(\boldsymbol{\omega}_O \times \boldsymbol{\omega})\mathrm{d}V \tag{2.3.21}$$

This matrix is proportional to the velocity and hence can be termed a pseudo-viscous damping matrix. However, it does not entail any loss of energy since a corresponding physical mechanism is missing. This is also apparent from an inspection of the elements g_{ij} of the matrix

$$\begin{aligned} g_{\mathrm{ij}} &= \int_V 2\rho\boldsymbol{\omega}_{\mathrm{i}}^{\mathrm{t}}(\boldsymbol{\omega}_O \times \boldsymbol{\omega}_{\mathrm{j}})\,\mathrm{d}V \\ &= \int_V -2\rho\boldsymbol{\omega}_{\mathrm{j}}^{\mathrm{t}}(\boldsymbol{\omega}_O \times \boldsymbol{\omega}_{\mathrm{i}})\,\mathrm{d}V \end{aligned} \tag{2.3.22}$$

By interchanging the indices, the sign is reversed and the matrix $\boldsymbol{G}$ is seen to be skew-symmetric.

The mass matrix $\boldsymbol{M}$ is, of course, well-known to the reader and has already been detailed in equ. (2.3.13).

Having discussed the extensive list of various inertial forces in a moving coordinate system, we now turn our attention to an elaboration of the damping forces (equ. (2.3.8)). At this point, it is appropriate to comment upon the nature of the damping forces $\boldsymbol{p}_{VD}$ and $\boldsymbol{p}_{SD}$. The origin of damping indeed creates great difficulties and is still the subject of intensive research. It is not only a result of friction inside the structure, i.e. in the material itself and in the joints of the structure, but also on the surface. We have already become familiar with a few damping models in the case of an oscillator with a single degree of freedom. The damping forces of a dynamically deforming system need not depend, however, solely on the velocity or the deformations of the body alone. It is very difficult to conceive the true analytical formulation of damping for most real physical systems. Moreover, the complexity of such expressions would magnify the difficulty of solving the equations of motion. Consequently, a viscous damping model appears generally preferable for establishing a simple solution. It is thus common practice and, from the practical view point, fully justifiable to approximate the real damping forces by a viscous model. We will proceed in this way now, although we will explore other possibilities later on.

We assume that a volume $\mathrm{d}V$ of the structure experiences a velocity, $\dot{\boldsymbol{u}}$, and is decelerated by the damping forces. Assuming the most elementary form of an isotropic damping, we write

$$\boldsymbol{p}_{VD} = \mu\dot{\boldsymbol{u}} \tag{2.3.23}$$

Hence, the damping force vector is taken to be solely dependent on and proportional to the local velocity vector. A similar simplifying assumption can be adopted for an infinitesimal element of the surface dS experiencing a velocity $\dot{\boldsymbol{u}}$

$$\boldsymbol{p}_{SD} = \mu_S\dot{\boldsymbol{u}} \tag{2.3.24}$$

Applying equs (2.3.8), (2.3.23) and (2.3.24) in conjunction with equ. (2.3.1), the damping force may be expressed as

$$\boldsymbol{R}_D = \Big(\int_V \mu\boldsymbol{\omega}^{\mathrm{t}}\boldsymbol{\omega}\mathrm{d}V + \int_S \mu_S\boldsymbol{\omega}^{\mathrm{t}}\boldsymbol{\omega}\mathrm{d}S\Big)\dot{\boldsymbol{r}} = \underset{(\mathrm{n}\times\mathrm{n})}{\boldsymbol{C}}\,\dot{\boldsymbol{r}} \tag{2.3.25}$$

Clearly, the damping matrix $\boldsymbol{C}$ is symmetric. From a practical point of view, we have to admit that the damping matrix can seldom be set up as demonstrated here. Usually, estimated damping characteristics are inserted directly into the calculation and hence, if necessary, the matrix $\boldsymbol{C}$ has to be construed. We will return to this point later in the discussion of a modal damping model. This does not affect equ. (2.3.25) or the symmetry of $\boldsymbol{C}$, however.

Following this exploratory talk, we leave the difficult question associated with the modelling of the damping and turn our attention to the discussion of the elastic forces, $\boldsymbol{R}_e(t)$, which we have already studied for the static case in Vols. I and II. To eliminate the stress vector in equ. (2.3.9), we require a constitutive law. For this purpose, we assume a linear elasticity in which the initial strains may generally be neglected in time-dependent problems

$$\boldsymbol{\sigma}(t) = \boldsymbol{\kappa}\boldsymbol{\varepsilon}(t) \tag{2.3.26}$$

$$\boldsymbol{\eta}(t) = \boldsymbol{o} \tag{2.3.27}$$

The elastic strains $\boldsymbol{\varepsilon}(t)$ are deduced from equs. (2.3.1) and (2.3.6) in the form

$$\begin{aligned}\boldsymbol{\varepsilon}(t) &= \boldsymbol{D}^{\mathrm{t}}\boldsymbol{u}(\boldsymbol{x},t) = \big[\boldsymbol{D}^{\mathrm{t}}\boldsymbol{\omega}(\boldsymbol{x})\big]\boldsymbol{r}(t)\\ &= \boldsymbol{\alpha}(\boldsymbol{x})\boldsymbol{r}(t)\end{aligned} \tag{2.3.28}$$

Thus, the elastic restoring force becomes

$$\begin{aligned}\boldsymbol{R}_e(t) &= \Big[\int_V \boldsymbol{\alpha}^{\mathrm{t}}\boldsymbol{\kappa}\boldsymbol{\alpha}\mathrm{d}V\Big]\boldsymbol{r}(t)\\ &= \boldsymbol{K}\boldsymbol{r}(t)\end{aligned} \tag{2.3.29}$$

as in statics. The stiffness matrix, $\boldsymbol{K}$, is naturally symmetric just as in the static case.

Finally, we come to the last group of forces, namely the kinematically equivalent force $\boldsymbol{R}(t)$, discussed in equ. (2.3.10). This vector arises from distributed body forces $\boldsymbol{p}_V(t)$ and surface forces $\boldsymbol{p}_S(t)$. In the simplest of cases, these forces are conservative in nature. Forces independent of the magnitude and direction of the displacements and their temporal derivatives have such a conservative character. We can, of course, also include different types of forces. For example, we may introduce a pressure distribution depending on time, $p(t)$, which is always acting normal to the deformed surface of a structure. The resulting load, $\boldsymbol{p}_S(t)$, depends upon the instantaneous position of the surface S, and consequently on the displacement $\boldsymbol{u}$. In the case of a downward motion, a lifting surface subject to an

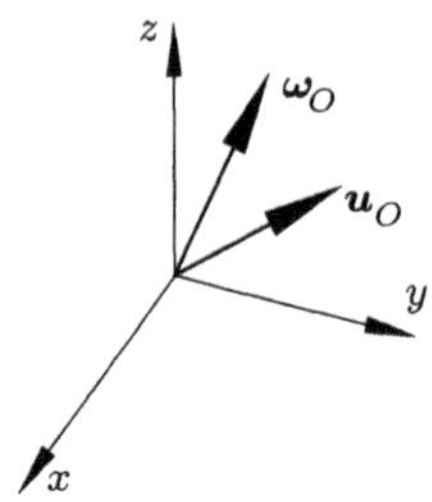

moving reference system
linear oscillator
viscous damping
conservative excitation

initial statement	$\boldsymbol{u}(\boldsymbol{x},t) = \boldsymbol{\omega}(\boldsymbol{x})\boldsymbol{r}(t)$ $\boldsymbol{\epsilon}(\boldsymbol{x},t) = \boldsymbol{\alpha}(\boldsymbol{x})\boldsymbol{r}(t)$
equation of motion	$\boldsymbol{M}\ddot{\boldsymbol{r}} + [\boldsymbol{C} + \boldsymbol{G}]\dot{\boldsymbol{r}} + [\boldsymbol{K} + \boldsymbol{Z} + \boldsymbol{H}]\boldsymbol{r} = \boldsymbol{R}(t) - \boldsymbol{R}_{IO}$
mass matrix	$\boldsymbol{M} = \int_V \rho\boldsymbol{\omega}^{\mathrm{t}}\boldsymbol{\omega}\mathrm{d}V$ (symmetric)
damping matrix	$\boldsymbol{C} = \int_V \mu\boldsymbol{\omega}^{\mathrm{t}}\boldsymbol{\omega}\mathrm{d}V + \int_S \mu_S\boldsymbol{\omega}^{\mathrm{t}}\boldsymbol{\omega}\mathrm{d}S$ (symmetric)
Coriolis matrix	$\boldsymbol{G} = \int_V 2\rho\boldsymbol{\omega}^{\mathrm{t}}(\boldsymbol{\omega}_O \times \boldsymbol{\omega})\mathrm{d}V$ (antisymmetric) (is non-existent in non-rotating systems)
stiffness matrix	$\boldsymbol{K} = \int_V \boldsymbol{\alpha}^{\mathrm{t}}\boldsymbol{\kappa}\boldsymbol{\alpha}\mathrm{d}V$ (symmetric)
centrifugal stiffness	$\boldsymbol{Z} = \int_V \rho\boldsymbol{\omega}^{\mathrm{t}}[\boldsymbol{\omega}_O \times (\boldsymbol{\omega}_O \times \boldsymbol{\omega})]\mathrm{d}V$ (symmetric) (is non-existent in non-rotating systems)
spin stiffness	$\boldsymbol{H} =$ (antisymmetric) (is non-existent in uniformly rotating systems)
direct force excitation	$\boldsymbol{R}(t) = \int_V \boldsymbol{\omega}^{\mathrm{t}}\boldsymbol{p}_V(t)\mathrm{d}V + \int_S \boldsymbol{\omega}^{\mathrm{t}}\boldsymbol{p}_S(t)\mathrm{d}S$ (vector)
indirect force excitation	$\boldsymbol{R}_{IO}(t) = \int_V \rho\boldsymbol{\omega}^{\mathrm{t}}[\ddot{\boldsymbol{u}}_O + (\dot{\boldsymbol{\omega}}_O \times \boldsymbol{x}) + \boldsymbol{\omega}_O \times (\boldsymbol{\omega}_O \times \boldsymbol{x})]\mathrm{d}V$ (vector) (is non-existent in non-rotating and non-accelerated systems)

Fig. 2.3.1 General equation of motion of a linear oscillator with viscous damping under conservative excitation

opposing horizontal flow experiences a relative upward velocity resulting in an increased angle of attack. Here, the forces again depend on the relative velocity of the lifting surface. In all these cases, correction force matrices arise in accordance with equ. (2.3.10); in the first case, this leads to a quasi-stiffness and in the second case to a quasi-damping matrix. Here, the reader's attention is drawn to Chapter 11 and to the companion Vol. VI which hopefully provide a better insight into these intriguing problems. We do not wish, however, to complicate matters here and restrict ourselves in the following to conservative loads. Equation (2.3.10) simply represents a known time-dependent vector.

We are now in a position to write the general equation of motion in an arbitrarily moving coordinate system

$$\boldsymbol{M}\ddot{\boldsymbol{r}} + [\boldsymbol{C} + \boldsymbol{G}]\dot{\boldsymbol{r}} + [\boldsymbol{K} + \boldsymbol{H} + \boldsymbol{Z}]\boldsymbol{r} = \boldsymbol{R}(t) - \boldsymbol{R}_{IO}(t) \tag{2.3.30}$$

This relation is discussed in more detail in fig. 2.3.1. The existence of non-symmetric coefficient matrices with respect to $\dot{\boldsymbol{r}}$ and $\boldsymbol{r}$ results in a costly solution procedure.

To simplify the problem, we assume a steady rotation of the system. Then, the matrix $\boldsymbol{H}$ is zero and we obtain the equations of motion for a damped, so-called gyroscopic system

$$\boldsymbol{M}\ddot{\boldsymbol{r}} + [\boldsymbol{C} + \boldsymbol{G}]\dot{\boldsymbol{r}} + [\boldsymbol{K} + \boldsymbol{Z}]\boldsymbol{r} = \boldsymbol{R}(t) - \boldsymbol{R}_{IO}(t) \tag{2.3.31}$$

Here, the quasi-stiffness matrix $[\boldsymbol{K} + \boldsymbol{Z}]$ is symmetric, but the quasi-damping matrix $[\boldsymbol{C} + \boldsymbol{G}]$ is still asymmetric.

A further simplification can be achieved by omitting either the damping matrix C or the rotation of the system. In the first case, we get an undamped gyroscopic system

$$\boldsymbol{M}\ddot{\boldsymbol{r}} + \boldsymbol{G}\dot{\boldsymbol{r}} + [\boldsymbol{K} + \boldsymbol{Z}]\boldsymbol{r} = \boldsymbol{R}(t) - \boldsymbol{R}_{IO}(t) \tag{2.3.31a}$$

All three matrices have special characteristics; matrices $\boldsymbol{M}$ and $[\boldsymbol{K} + \boldsymbol{Z}]$ are symmetric and $\boldsymbol{G}$ is skew-symmetric. In the case of a non-rotating system, the equations of motion are reduced to

$$\boldsymbol{M}\ddot{\boldsymbol{r}} + \boldsymbol{C}\dot{\boldsymbol{r}} + \boldsymbol{K}\boldsymbol{r} = \boldsymbol{R}(t) - \boldsymbol{R}_{IO}(t) \tag{2.3.32}$$

Here, the matrices $\boldsymbol{M}$, $\boldsymbol{C}$, and $\boldsymbol{K}$ are all symmetric and hence, the solution is greatly eased. For simplicity, one tends to prefer a study of the dynamic behaviour of an undamped system. By neglecting the acceleration of the reference system, one obtains the simplest possible equation of motion in dynamics

$$\boldsymbol{M}\ddot{\boldsymbol{r}} + \boldsymbol{K}\boldsymbol{r} = \boldsymbol{R}(t) \tag{2.3.33}$$

Throughout this treatise, we will generally be working with the type of problems defined by equs. (2.3.32) and (2.3.33).

Bibliography for Chapter 2

[2.1] *J.T. Oden*; A general theory of finite elements, Part 2, Applications, Int. J. Num. Meth. Eng. 1, no. 3 (1969) 254-264.

[2.2] *J. Argyris, D.W. Scharpf*; Finite elements in space and time, Aeron. J. Roy. Aeron. Soc., 73, no. 708 (1969) 1041-1044.

[2.3] *L.V. Kantorovich*; Bull. Accad. Sci. USSR, 5 (1933) 647.

[2.4] *J. Wittenburg*; Dynamics of systems of rigid bodies (Teubner, Stuttgart, 1977).

[2.5] *B.A. Finlayson*; The method of weighted residuals and variational principles (Academic Press, New York and London, 1972).

[2.6] *R. Wait, A.R. Mitchell*; Finite element analysis and applications (Wiley, New York, 1985).

[2.7] *T. Belytschko, T.J.R. Hughes* (Eds.); Computational methods for transient analysis (Volume 1 in: Mechanics and mathematical methods, North-Holland, Amsterdam, 1983).

3 The nature of the inertia forces and the mass matrix

"The love of indolence is universal"
S.T. Coleridge, Table-Talk, 8th June, 1833

In a dynamic analysis based on the finite element method, it is common practice to assume that the structural masses are concentrated at the structural nodes; this leads to the so-called lumped mass matrix. Such a concentration of the distributed mass results in the formation of point-masses which are assembled in diagonal elemental and global mass matrices and lead to inexpensive matrix operations. Unfortunately, however, the cross-coupling effects of the distributed mass are then ignored and entail a loss of the valuable property of an upper bound. This deficiency does not arise, however, when applying kinematically consistent mass matrices obtained by a systematic application of the principle of virtual displacements.

In the present chapter, we establish a general method for the computation of kinematically equivalent mass matrices and exemplify these techniques on a few elements of the simplex family. Based on this methodology, we may establish the consistent mass matrices of curved four-node membrane and hexahedron elements via our sub-element concept. At the same time, we discuss briefly the lumped point-mass concept and demonstrate it on an element.

3.1 General principles

In the present section, we examine an arbitrary structure (see fig. 3.1.1) and discretise it into a number of finite elements. Let us consider an infinitesimal cube of volume $\mathrm{d}V = \mathrm{d}x\,\mathrm{d}y\,\mathrm{d}z$, the mass density of which is ρ. If this cube is accelerated with $\ddot{u}$, $\ddot{v}$ and $\ddot{w}$ in the x, y and z directions respectively, then the inertia forces in the coordinate direction can be obtained from Newton's law.

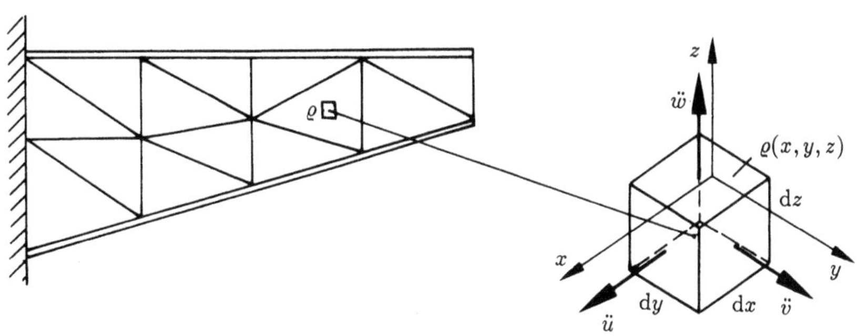

Fig. 3.1.1 Infinitesimal element in a discretised structure subject to dynamic excitation

These inertia forces are applied to the infinitesimal element in question by the surrounding structure. The structure itself experiences the distributed forces in a direction opposite to that of the local acceleration vector of the infinitesimal elements. The force per unit volume is then

$$\boldsymbol{p}_{VI} = -\rho(x, y, z)\begin{bmatrix} \ddot{u} \\ \ddot{v} \\ \ddot{w} \end{bmatrix} = -\rho\ddot{\boldsymbol{u}} \tag{3.1.1}$$

As in the case of statics, this distributed load is expressed globally in terms of kinematically equivalent nodal forces (see equ. (5.1.87), Vol. I, statics)

$$\boldsymbol{R}_I = \sum_{\mathrm{g}=1}^{\mathrm{m}} \mathbf{a}_{\mathrm{g}}^{\mathrm{t}} \int_{V_{\mathrm{g}}} \boldsymbol{\omega}^{\mathrm{t}} \boldsymbol{p}_{VI} \mathrm{d}V \tag{3.1.2}$$

Here, $\mathbf{a}_{\mathrm{g}}$ is the standard Boolean assembly matrix of the j-th element. By integration or summation over the elemental volumes V_{g}, we obtain the contribution from all elements, g=1 to m, and thus determine the total inertial load vector $\boldsymbol{R}_I$. The matrix $\boldsymbol{\omega}$ contains the assumed Cartesian modal functions of the element

$$\boldsymbol{u} = \boldsymbol{\omega}\boldsymbol{\rho}_{\mathrm{g}} = \boldsymbol{\omega}\mathbf{a}_{\mathrm{g}}\boldsymbol{r} \tag{3.1.3}$$

where $\boldsymbol{u}$ is the displacement vector at a point within the element and $\boldsymbol{\rho}$ is the nodal displacement vector. By differentiating equ. (3.1.3) twice with respect to time and assuming a time-independent interpolation function, we obtain

$$\ddot{\boldsymbol{u}} = \boldsymbol{\omega}\ddot{\boldsymbol{\rho}}_{\mathrm{g}} \tag{3.1.4}$$

and hence from equ. (3.1.2)

$$\boldsymbol{R}_I = -\sum_{\mathrm{g}=1}^{\mathrm{m}} \mathbf{a}_{\mathrm{g}}^{\mathrm{t}} \Big[\int_{V_{\mathrm{g}}} \rho \boldsymbol{\omega}^{\mathrm{t}} \boldsymbol{\omega} \mathrm{d}V \Big] \ddot{\boldsymbol{\rho}}_{\mathrm{g}} \tag{3.1.5}$$

Using the Boolean relation between the element displacement vector $\boldsymbol{\rho}$ and the structural displacement vector $\boldsymbol{r}$, equ. (3.1.3), the elemental acceleration vector is given by

$$\ddot{\boldsymbol{\rho}}_{\mathrm{g}} = \mathbf{a}_{\mathrm{g}}\ddot{\boldsymbol{r}} \tag{3.1.6}$$

The global inertial force vector is then

$$\boldsymbol{R}_I = -\left[\sum_{\mathrm{g}=1}^{\mathrm{m}} \mathbf{a}_{\mathrm{g}}^{\mathrm{t}} \Big[\int_{V_{\mathrm{g}}} \rho \boldsymbol{\omega}^{\mathrm{t}} \boldsymbol{\omega} \mathrm{d}V \Big] \mathbf{a}_{\mathrm{g}} \right] \ddot{\boldsymbol{r}} \tag{3.1.7}$$

where we denote the expression in the inner brackets as the kinematically equivalent element mass matrix of the element g

$$\boldsymbol{m}_{\mathrm{g}} = \int_{V_{\mathrm{g}}} \rho \boldsymbol{\omega}^{\mathrm{t}} \boldsymbol{\omega} \mathrm{d}V \tag{3.1.8}$$

The global mass matrix of the overall structure is obtained by a Boolean assembly of the elemental mass matrices in the standard form

$$\boldsymbol{M} = \sum_{\mathrm{g}=1}^{\mathrm{m}} \mathbf{a}_{\mathrm{g}}^{\mathrm{t}} \boldsymbol{m}_{\mathrm{g}} \mathbf{a}_{\mathrm{g}} \tag{3.1.9}$$

The assembly of the structural mass matrix thus follows the same pattern as that for the stiffness matrix of a structure. In fact, the derivation of the mass matrices requires hardly any additional effort since the same modal functions $\boldsymbol{\omega}$ have to be employed. We are now in a position to formulate the equations of motion of a dynamic system subject to negligible damping forces. This simplifying assumption cannot be maintained, however, near the resonance points. From equ. (2.3.11), we obtain

$$\boldsymbol{M}\ddot{\boldsymbol{r}} + \boldsymbol{K}\boldsymbol{r} = \boldsymbol{R}(t) \tag{3.1.10}$$

3.2 The mass matrix of a flange element (FLA2)

To illustrate the preceding theory, we consider a flange element with two nodes which experiences only axial forces (fig. 3.2.1). The modal function is linear and is given by

$$\underset{(3\times6)}{\boldsymbol{\omega}} = \tfrac{1}{2}\left[(1-\zeta)\boldsymbol{I}_3 \quad (1+\zeta)\boldsymbol{I}_3\right] \tag{3.2.1}$$

in a global coordinate system and refers to the element displacement vector $\boldsymbol{\rho}$ (see, e.g., equ. (5.3.11) of Vol. I, statics)

$$\boldsymbol{\rho} = \{u_1 \quad v_1 \quad w_1 \quad u_2 \quad v_2 \quad w_2\} \tag{3.2.2}$$

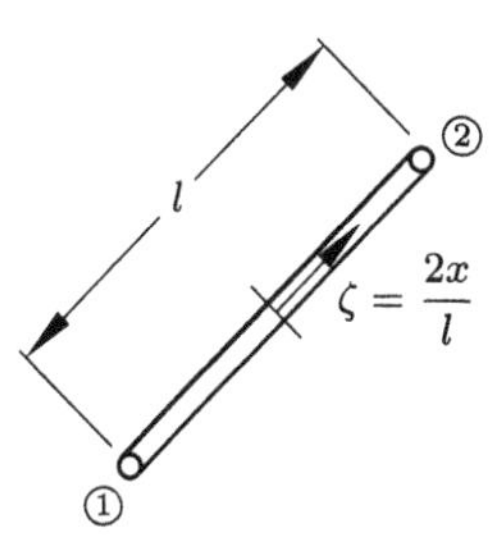

Fig. 3.2.1
Flange element FLA2 with linear displacement distribution

For a constant mass density ρ and a constant cross section A, the mass matrix of the bar element is found to be

$$\underset{(6\times 6)}{\boldsymbol{m}} = \tfrac{1}{8}\rho A l \int_{-1}^{+1} \begin{bmatrix} (1-\zeta)^2 \boldsymbol{I}_3 & (1-\zeta)(1+\zeta)\boldsymbol{I}_3 \\ (1-\zeta)(1+\zeta)\boldsymbol{I}_3 & (1+\zeta)^2 \boldsymbol{I}_3 \end{bmatrix} \mathrm{d}\zeta$$

$$= \tfrac{1}{6}\rho A l \begin{bmatrix} 2\boldsymbol{I}_3 & \boldsymbol{I}_3 \\ \boldsymbol{I}_3 & 2\boldsymbol{I}_3 \end{bmatrix} \qquad (3.2.3)$$

3.3 The mass matrix of a triangular membrane element (TRIM3) and a tetrahedron element (TET4)

The triangular element TRIM3 belongs to the same simplex family as FLA2. It is a membrane element with an assumed linear variation of the displacement. For the description of the displacement distribution, it is most convenient to apply the homogeneous triangular or natural coordinates ζ_i, (see fig. 3.3.1) which we repeatedly employed before (see Vol. I(G) of this treatise for the derivation of the stiffness matrix, see e.g. equ. (5.3.60)). Thus, the modal function in a global system becomes

$$\underset{(3\times 9)}{\boldsymbol{\omega}} = [\zeta_1 \boldsymbol{I}_3 \quad \zeta_2 \boldsymbol{I}_3 \quad \zeta_3 \boldsymbol{I}_3] \qquad (3.3.1)$$

and the corresponding displacement vector is

$$\boldsymbol{\rho} = \{u_1 \quad v_1 \quad w_1 \quad u_2 \quad v_2 \quad w_2 \quad u_3 \quad v_3 \quad w_3\} \qquad (3.3.2)$$

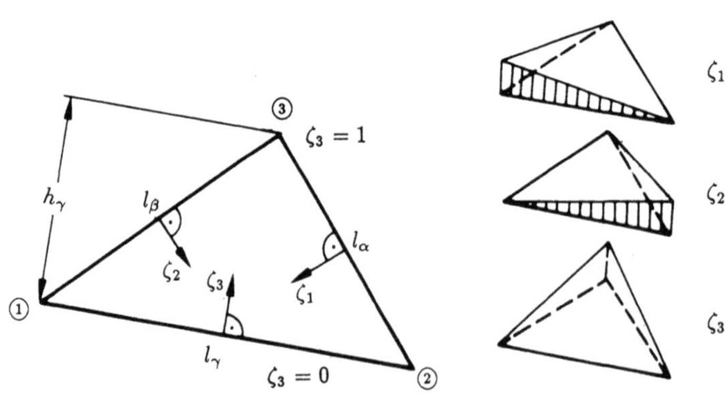

Fig. 3.3.1 Membrane element TRIM3 with linear displacement distribution; homogeneous triangular or natural coordinates

The general integration formula for the linear homogeneous coordinates over a triangle is (see equ. (5.3.62), Vol. I, statics)

$$\frac{1}{\Omega}\int \zeta_1^{\mathrm{p}}\zeta_2^{\mathrm{q}}\zeta_3^{\mathrm{r}}\mathrm{d}\Omega = \frac{2!\,\mathrm{p}!\,\mathrm{q}!\,\mathrm{r}!}{(2+\mathrm{p}+\mathrm{q}+\mathrm{r})!} \tag{3.3.3}$$

where Ω is the area of the triangle.

For constant mass density ρ and constant element thickness t, the element mass matrix of a triangular membrane element then becomes

$$\underset{(9\times9)}{\boldsymbol{m}} = \rho\Omega t\frac{1}{\Omega}\int_{\Omega}\begin{bmatrix} \zeta_1^2\boldsymbol{I}_3 & \zeta_1\zeta_2\boldsymbol{I}_3 & \zeta_1\zeta_3\boldsymbol{I}_3 \\ & \zeta_2^2\boldsymbol{I}_3 & \zeta_2\zeta_3\boldsymbol{I}_3 \\ \text{sym.} & & \zeta_3^2\boldsymbol{I}_3 \end{bmatrix}\mathrm{d}\Omega$$

$$= \tfrac{1}{12}\rho\Omega t\begin{bmatrix} 2\boldsymbol{I}_3 & \boldsymbol{I}_3 & \boldsymbol{I}_3 \\ & 2\boldsymbol{I}_3 & \boldsymbol{I}_3 \\ \text{sym.} & & 2\boldsymbol{I}_3 \end{bmatrix} \tag{3.3.4}$$

Even without a knowledge of the integral formula, equ. (3.3.3), the same result could have been derived by standard procedures, though admittedly less elegantly. Thus, we may employ the following linear relation between the redundant coordinates (see equ. (5.4.51) of Vol. I, statics)

$$\zeta_1 + \zeta_2 + \zeta_3 = 1 \tag{3.3.5}$$

in order to eliminate one of the three coordinates. In this way, we can integrate with respect to the two remaining coordinates, which are, however, skew to each other.

Analogous calculations may also be performed for the basic three-dimensional tetrahedron element TET 4 and lead to the element mass matrix

$$\underset{(12\times12)}{\boldsymbol{m}} = \frac{\rho V}{20}\begin{bmatrix} 2\boldsymbol{I}_3 & \boldsymbol{I}_3 & \boldsymbol{I}_3 & \boldsymbol{I}_3 \\ \boldsymbol{I}_3 & 2\boldsymbol{I}_3 & \boldsymbol{I}_3 & \boldsymbol{I}_3 \\ \boldsymbol{I}_3 & \boldsymbol{I}_3 & 2\boldsymbol{I}_3 & \boldsymbol{I}_3 \\ \boldsymbol{I}_3 & \boldsymbol{I}_3 & \boldsymbol{I}_3 & 2\boldsymbol{I}_3 \end{bmatrix} \tag{3.3.6}$$

With this analytical background and with the help of the sub-element principles (see Vol. I, statics), the reader can establish without difficulty the equivalent mass matrices for arbitrary two- or three-dimensional elements, such as curved quadrilateral membrane or hexahedron elements.

3.4 The mass matrix of a plane beam element

To simplify matters, we establish the mass matrix of a plane beam element with respect to a system of local coordinates, in which the x-axis is directed along the longitudinal axis of the element (fig. 3.4.1). Having obtained the associated mass matrix, we can transform it to a global coordinate system as we demonstrated in preceding volumes for the stiffness matrix. We start by assuming that the kinematics of the beam element are based on the elementary theory of bending (ETB) which ignores the shear deformations. The latter may have a significant contribution, but are better ignored in this introductory analysis.

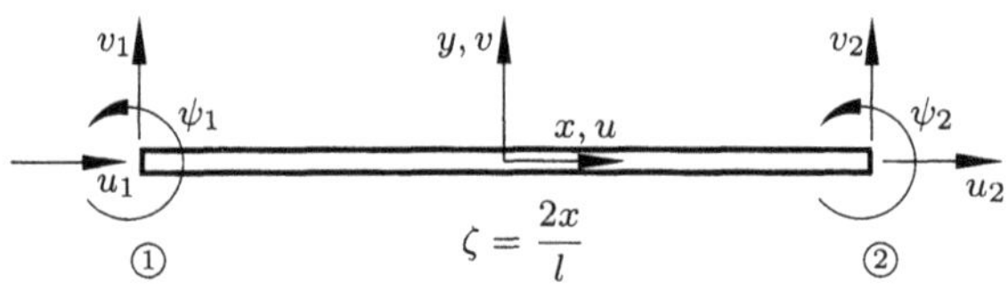

Fig. 3.4.1
The two-dimensional beam referred to a local system

There are three nodal degrees of freedom at each node, shown in fig. 3.4.1 for a two node element. At any point within an element, the longitudinal displacement is the sum of the extension of the middle axis and the straining of the cross-section due to bending

$$\begin{aligned} u &= u_m(x) - y\frac{\partial v(x)}{\partial x} \\ &= u_m(\zeta) - \frac{2y}{l}\frac{\partial v(\zeta)}{\partial \zeta} \end{aligned} \tag{3.4.1}$$

From this, we deduce the modal functional matrix for u and v in the form

$$\underset{(2\times 6)}{\boldsymbol{\omega}} = \tfrac{1}{8}\begin{bmatrix} 4(1-\zeta) & 12\tfrac{y}{l}(1-\zeta^2) & 2y(1+2\zeta-3\zeta^2) & 4(1+\zeta) & 12\tfrac{y}{l}(-1+\zeta^2) & 2y(1-2\zeta-3\zeta^2) \\ 0 & 2(2-3\zeta+\zeta^3) & l(1-\zeta-\zeta^2+\zeta^3) & 0 & 2(2+3\zeta-\zeta^3) & l(-1-\zeta+\zeta^2+\zeta^3) \end{bmatrix} \tag{3.4.2}$$

To simplify the integration, we split this matrix expression into primary and secondary parts. The secondary part is linear in y and determines the rotational inertia of the cross-section of the beam. Hence, we have

$$\begin{aligned} \boldsymbol{\omega} &= \tfrac{1}{8}\begin{bmatrix} 4(1-\zeta) & 0 & 0 & 4(1+\zeta) & 0 & 0 \\ 0 & 2(2-3\zeta+\zeta^3) & l(1-\zeta-\zeta^2+\zeta^3) & 0 & 2(2+3\zeta-\zeta^3) & l(-1-\zeta+\zeta^2+\zeta^3) \end{bmatrix} \\ &\quad + \tfrac{2}{8}y\begin{bmatrix} 0 & \tfrac{6}{l}(1-\zeta^2) & (1+2\zeta-3\zeta^2) & 0 & \tfrac{6}{l}(-1+\zeta^2) & (1-2\zeta-3\zeta^2) \\ 0 & 0 & 0 & 0 & 0 & 0 \end{bmatrix} \\ &= \tfrac{1}{8}\left[\boldsymbol{\omega}_t(\zeta) + 2y\boldsymbol{\omega}_r(t)\right] \end{aligned} \tag{3.4.3}$$

In this way, we obtain the mass matrix assuming here constant density

$$\boldsymbol{m} = \frac{\rho l}{128} \int_{-1}^{+1} \left[\boldsymbol{\omega}_t^{\mathrm{t}} \boldsymbol{\omega}_t A + 2(\boldsymbol{\omega}_t^{\mathrm{t}} \boldsymbol{\omega}_r + \boldsymbol{\omega}_r^{\mathrm{t}} \boldsymbol{\omega}_t) \int_A y \mathrm{d}A + 4 \boldsymbol{\omega}_r^{\mathrm{t}} \boldsymbol{\omega}_r \int_A y^2 \mathrm{d}A \right] \mathrm{d}\zeta \tag{3.4.4}$$

where A is the cross-section of the beam. If the coordinate system is located at the centre of gravity (C.G.) of the cross-section, then

$$\begin{aligned} \int_A y \mathrm{d}A = 0 &= \text{ first moment of area} \\ \int_A y^2 \mathrm{d}A = I &= \text{ moment of inertia} \end{aligned} \tag{3.4.5}$$

Thus, we find

$$\boldsymbol{m} = \frac{\rho l}{128} \int_{-1}^{+1} (A \boldsymbol{\omega}_t^{\mathrm{t}} \boldsymbol{\omega}_t + 4I \boldsymbol{\omega}_r^{\mathrm{t}} \boldsymbol{\omega}_r) \mathrm{d}\zeta \tag{3.4.6}$$

The remaining integrations are somewhat laborious but straightforward and can easily be confirmed by spot checks. For constant values of A and I, we obtain

$$\boldsymbol{m} = \frac{\rho A l}{420} \begin{bmatrix} 140 & 0 & 0 & 70 & 0 & 0 \\ & 156 & 22l & 0 & 54 & -13l \\ & & 4l^2 & 0 & 13l & -3l^2 \\ & & & 140 & 0 & 0 \\ & \text{sym.} & & & 156 & -22l \\ & & & & & 4l^2 \end{bmatrix} + \frac{\rho A l}{30} \left(\frac{i}{l} \right)^2 \begin{bmatrix} 0 & 0 & 0 & 0 & 0 & 0 \\ & 36 & 3l & 0 & -36 & 3l \\ & & 4l^2 & 0 & 3l & -1l^2 \\ & & & 0 & 0 & 0 \\ & \text{sym.} & & & 36 & -3l \\ & & & & & 4l^2 \end{bmatrix} \tag{3.4.7}$$

in which we introduce i, the radius of gyration of the cross-section of the beam defined by

$$i^2 = \frac{I}{A} \tag{3.4.8}$$

The second contributory matrix on the right-hand side of equ. (3.4.7) represents the effects of rotational inertia. We observe that its magnitude is primarily determined by the non-dimensional parameter l/i, i.e. the slenderness ratio of the beam.

For an I-profile I60.60.5, DIN 9712 (height 60mm, width 60mm, wall thickness 5mm), the radius of gyration i becomes 0.0245m. For slim beams, the second contribution in the mass matrix of equ. (3.4.7) hence becomes negligible. On the other hand, it can be important for short beams. In such cases, we should remember, however, that the simple beam theory is not applicable anyway unless we take shear deformations into consideration. In both this case and in a more precise analysis including higher-order effects in dynamics, the shear effect should be included since its influence becomes of the same order of magnitude as the rotational inertia of the mass matrix.

The problem of a uniform beam with shear deformation was discussed in Chapter 4 of Vol. I, statics. Here, we shall complement the discussion by demonstrating its effect on the formation of the mass matrix. The simplest technique of introducing the shear uses the force method and in particular its natural modes in place of the displacement method. The reader will recall that the shear force derives solely from the anti-symmetric moment distribution; hence, only the associated anti-symmetric kinematic mode exhibits shear deformations. For convenience, we use these natural modes to derive the mass matrix. In a plane, the beam element possesses six overall modes consisting of three rigid-body modes and three natural straining modes. The third straining mode defines the anti-symmetric contribution. To describe it, we introduce the shearing mode ρ_{N3}^S in addition to the pure bending mode ρ_{N3}^B (see fig. 3.4.2). The associated kinematic freedom is defined, however, solely by the sum of the two components

$$\rho_{N3} = \rho_{N3}^B + \rho_{N3}^S \tag{3.4.9}$$

From the elementary theory of bending (ETB), the pure bending contribution, ρ_{N3}^B, is given by

$$\rho_{N3}^B = \frac{l}{3EI}\mathrm{P}_{N3} \tag{3.4.10}$$

where P_{N3} is the anti-symmetric bending moment. The associated deformation due to shear, ρ_{N3}^S, is

$$\rho_{N3}^S = 2\frac{1}{GA_S}\frac{2\mathrm{P}_{N3}}{l} \tag{3.4.11}$$

Substituting equs. (3.4.10) and (3.4.11) into equ. (3.4.9), we obtain

$$\rho_{N3} = \frac{l}{3EI}(1+\alpha_S)\mathrm{P}_{N3} \tag{3.4.12}$$

(see also equ. (4.2.31) of Vol. I, statics). The non-dimensional parameter α_S is defined by

$$\alpha_S = \frac{12EI}{GA_S l^2} = 12\frac{EA}{GA_S}\left(\frac{i}{l}\right)^2 \tag{3.4.13}$$

The flexibility of the beam associated with the anti-symmetric mode is thus increased by an order of magnitude approximately equal to that of the rotational

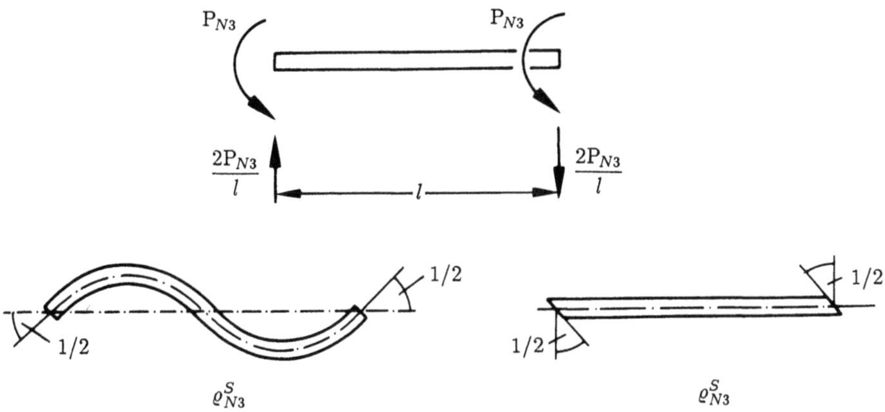

Fig. 3.4.2 Introduction of a shearing mode for an anti-symmetric loading

inertia of the cross-section. From equs. (3.4.12), (3.4.10) and (3.4.11), the breakdown of the total deformation ρ_{N3} in bending and shear components is given by

$$\rho_{N3}^{B} = \frac{1}{1+\alpha_S}\rho_{N3} \tag{3.4.14}$$

and

$$\rho_{N3}^{S} = \frac{\alpha_S}{1+\alpha_S}\rho_{N3} \tag{3.4.15}$$

The modal form $\overline{\boldsymbol{\omega}}$ corresponding to the generalised displacements $\boldsymbol{\rho}' = \{\boldsymbol{\rho}_0 \quad \boldsymbol{\rho}_N\}$ is simply

$$\begin{aligned}\overline{\boldsymbol{\omega}} &= [\boldsymbol{\omega}_0 \quad \boldsymbol{\omega}_N] \\ &= \left[\underset{(2\times3)}{\boldsymbol{\omega}_0} \quad \underset{(2\times1)}{\boldsymbol{\omega}_{N1}} \quad \underset{(2\times1)}{\boldsymbol{\omega}_{N2}} \quad \underset{(2\times1)}{\boldsymbol{\omega}_{N3}} + \frac{\alpha_S}{1+\alpha_S}\left[\begin{bmatrix}\omega_{N3}^{S} \\ 0\end{bmatrix} - \boldsymbol{\omega}_{N3}\right]\right]\end{aligned} \tag{3.4.16}$$

where the modal function due to the shear

$$\omega_{N3}^{S} = -\tfrac{1}{2}y \tag{3.4.17}$$

is the only new entry as against the beam element with zero shear deformations. All other interpolation functions derive from the classical beam theory without shear. According to equ. (5.2.13), Vol. I, statics, the Cartesian or local interpolation function $\boldsymbol{\omega}$ associated with the displacement vector $\boldsymbol{\rho}$ can be obtained from $\overline{\boldsymbol{\omega}}$ via the transformation

$$\begin{aligned}\boldsymbol{\omega} &= \overline{\boldsymbol{\omega}}\boldsymbol{a}_e \\ &= [\boldsymbol{\omega}_0 \quad \boldsymbol{\omega}_N]\{\boldsymbol{a}_0 \quad \boldsymbol{a}_N\}\end{aligned} \tag{3.4.18}$$

In equ. (3.4.16), only the term corresponding to the third natural mode $\boldsymbol{\omega}_{N3}$ has undergone a modification due to shear. Consequently, using the third row $\boldsymbol{a}_{N3}$, we immediately deduce

$$\boldsymbol{\omega}_{ETBS} = \boldsymbol{\omega}_{ETB} - \frac{\alpha_S}{1+\alpha_S}\left[\begin{bmatrix}\frac{1}{2}y \\ 0\end{bmatrix} + \boldsymbol{\omega}_{N3}\right]\boldsymbol{a}_{N3} \tag{3.4.19}$$

From this, the reader may easily obtain the additional terms in the mass matrix due to shear deformation.

3.5 The lumped mass matrix as a simplified model for the inertia forces

The mass matrices discussed so far can be denoted as kinematically consistent, since they are related to the same modal functions used to derive the stiffness matrix and fit consistently into the energy concept. Furthermore, these mass matrices may, in principle, also be derived from the Rayleigh-Ritz method, except that in the present case, the approximation extends over an element and not over the complete structure. Because the boundary and kinematic compatibility conditions are maintained in the Finite Element Method, the rule of bounds is also applicable. Thus, our approach always yields an over-estimation of the fundamental eigenfrequency. This is simply due to the fact that the limited number of system functions somewhat restricts the deformability of the structure. The success of linear dynamics was immediately achieved on its inception even though its historical development did not follow the consistent path presented here. Analogously to the static analysis, the Finite Element Method in the dynamic domain was based initially on lumped (e.g. inertia) forces acting at the nodal points of the structure. Distributed surface forces were simulated somewhat arbitrarily by a number of nodal forces.

Although from a theoretical point of view, the forces in nature are not concentrated but rather distributed, due to their inborn common sense, engineeers tend to successfully employ the resultant and not distributed forces. It was not until later years that the kinematically consistent or equivalent evaluation of nodal forces was conceived. This last concept is still being applied nowadays in modern FEM calculations.

As mentioned above, the use of lumped concentrated loads in statics is extended in dynamics to the inertia forces, which are lumped at the nodes. The inertia of the structure is thus represented by concentrated masses at the structural nodes. In this distribution, the total mass of each element has to be maintained. The lumped mass matrix of the element is clearly diagonal and for the FLA2 element we have

$$\boldsymbol{m}_l = \frac{\rho A l}{2} \boldsymbol{I}_6 \tag{3.5.1}$$

and for the membrane triangular element (TRIM3)

$$\boldsymbol{m}_l = \frac{\rho \Omega t}{3} \boldsymbol{I}_9 \tag{3.5.2}$$

A corresponding simple mass matrix for a beam, neglecting rotational inertia, is

$$\boldsymbol{m}_l = \frac{\rho A l}{2} \lceil 1 \quad 1 \quad 0 \quad 1 \quad 1 \quad 0 \rfloor \tag{3.5.3}$$

In the meantime, it has been proved theoretically that the results for a given number of degrees of freedom, using these lumped mass matrices, are in general less accurate. The concentration of inertial masses leads to a decrease in the calculated fundamental eigenfrequency and a loss of the rule of bounds. Any speculation that this lumped mass effect runs contrary to the stiffening effect in the discretisation and thus results in a smaller error is pointless as long as we know nothing of the magnitude of the errors involved.

Despite these misgivings, the lumped mass model can still be usefully applied in practice in certain cases. For example, it is often an economical necessity to employ the diagonal mass matrix concept in many non-linear algorithms.

Bibliography for Chapter 3

[3.1] *J. Argyris*; Some results on the free-free oscillations of aircraft type structures, Revue Française de Mécanique, 15 (1965).

[3.2] *J.S. Archer*; Consistent mass matrix for distributed systems, Proc. ASCE, 89 (1963).

[3.3] *F.A. Leckie, G.M. Lindberg*; The effect of lumped parameters on beam frequencies, Aero. Quart., 14 (1963).

[3.4] *R.J. Guyan*; Distributed mass matrix for plate elements in bending, AIAA Journal, 3 (1965).

[3.5] *P. Teng, T.H.H. Pian, L.L. Buciovelli*; Mode shapes and frequencies by the finite element method using consistent and lump matrices, J. Comp. Struct., 1 (1971).

[3.6] *I. Fried, D.S. Malkus*; Finite element mass matrix lumping by numerical integration with the convergence rate loss, Int. J. Solids Struct., 11 (1975).

[3.7] *E. Hinton, A. Rock, O.C. Zienkiewicz*; A note on mass lumping in related processes in the finite element method, Int. J. Earthq. Eng. Struct. Dyn., 4 (1976) 245–249.

4 Natural vibrations of undamped systems

"Step after step the ladder is ascended"
George Herbert, Jacula prudentum (1640)

It will not be immediately apparent to the reader why the authors have chosen to present such an extensive account of the eigenmodes or natural vibrations of undamped systems (we should refer to these more appropriately as the free oscillation modes of linear systems not subject to external forces or constraints) in this chapter. In practice, engineers seek to determine the dynamic response of structures under deterministic or stochastic external loads whilst natural or eigenmodes of oscillation are in fact characterised by the absence of such loads.

Natural oscillations are nevertheless of great importance to the experienced engineer. A careful determination of the natural oscillatory behaviour (eigenfrequencies and eigenmodes) will allow him a better insight into, and a clearer understanding of the dynamic response of a system under external excitation. In addition, eigenmodes are, in many cases, better suited as natural generalised coordinates for the description of the dynamic motion of a system than the original degrees of freedom ('nodal displacements'). This is especially true in the case of forced excitations. A knowledge of natural oscillations is a basic prerequisite for the application of the so-called modal methods employed in dynamics. These methods will be discussed in greater detail in subsequent chapters.

4.1 Eigenvalues and eigenvectors

We now proceed to a systematic account of the eigenvalues (natural frequencies) and eigenmodes of an elastic system. For compactness of terminology, we denote the process of the determination of eigenvalues and eigenmodes as spectral analysis.

Having already discussed the nature of inertial forces of a continuum in preceding chapters, we are now in a position to formulate the equations of motion of undamped structures. Omitting the external force vector in equ. (3.1.10), we have

$$\boldsymbol{M}\ddot{\boldsymbol{r}} + \boldsymbol{K}\boldsymbol{r} = \boldsymbol{o} \tag{4.1.1}$$

Applying once more the usual statement for the solution of linear homogeneous differential equations with constant coefficients, we write

$$\boldsymbol{r} = \boldsymbol{x}\,\mathrm{e}^{pt} \tag{4.1.2}$$

Substituting equ. (4.1.2) into equ. (4.1.1), we obtain the homogeneous algebraic system

$$[p^2\boldsymbol{M} + \boldsymbol{K}]\boldsymbol{x} = \boldsymbol{o} \tag{4.1.3}$$

The trivial solution is $\boldsymbol{x}=\boldsymbol{o}$ and represents the state of rest; it is of no interest to us here. Other non-trivial solutions arise only if columns (and consequently also rows) of the symmetric matrix $[p^2\boldsymbol{M} + \boldsymbol{K}]$ are linearly dependent. However, if linear dependence of the rows and columns of this matrix exists, its determinant has to be zero

$$\det[p^2\boldsymbol{M} + \boldsymbol{K}] = 0 \tag{4.1.4a}$$

or more explicitly

$$\begin{vmatrix} p^2m_{11}+k_{11} & p^2m_{12}+k_{12} & p^2m_{13}+k_{13} & \dots \\ & p^2m_{22}+k_{22} & p^2m_{23}+k_{23} & \dots \\ & & p^2m_{33}+k_{33} & \dots \\ \text{sym} & & & \ddots \end{vmatrix} = 0 \tag{4.1.4b}$$

The expansion of this determinant leads to an n-th order polynomial in p^2, which is described as the characteristic (or secular) equation of the system

$$b_n(p^2)^n + b_{n-1}(p^2)^{n-1} + \ldots + b_0 = 0 \tag{4.1.4c}$$

Anyone who is familiar with such determinants will at once confirm that the term with the highest exponent derives from the elements p^2m_{ij}. Thus, we have

$$b_n = \det[\boldsymbol{M}] \tag{4.1.5}$$

Furthermore, it can also be seen that the constant term in the characteristic equation is obtained from the stiffness coefficients k_{ij}. Hence, we find

$$b_0 = \det[\boldsymbol{K}] \tag{4.1.6}$$

We immediately infer that special conditions must prevail for a singular mass matrix and/or stiffness matrix. We will return to these special cases later on in the chapter. For the time being, however, let us assume that the mass matrix is positive definite and hence $\det[\boldsymbol{M}]$ is also positive. As we shall see, this assumption holds if consistent mass matrices are employed. Similarly, we presume that the stiffness matrix is also positive definite which entails that our structure is supported in an, at least, statically determinate manner and does not contain any kinematic mechanisms. Then, $\det[\boldsymbol{K}]$ is also positive. In this case, the characteristic equation possesses n non-zero solutions in p^2 (roots of the secular equation). For each solution of p^2, we obtain two roots: $\pm\sqrt{p^2}$; thus, there are 2n solutions in p.

Since the characteristic polynomial given in equ. (4.1.4c) is extremely complicated, especially for large values of n, it at first seems difficult to predict the characteristics of the solution p_i^2 (and hence also of $\boldsymbol{x}_i$).Some features of the solutions for simple oscillatory systems are shown in fig. 4.1.1. Both a single-mass oscillator and a system with two degrees of freedom yield negative real solutions for p_i^2 and hence purely imaginary values of p_i. The corresponding eigenvectors $\boldsymbol{x}_i$ are real and are listed in fig. 4.1.1, but for an arbitrary constant c_i. Our results are, of

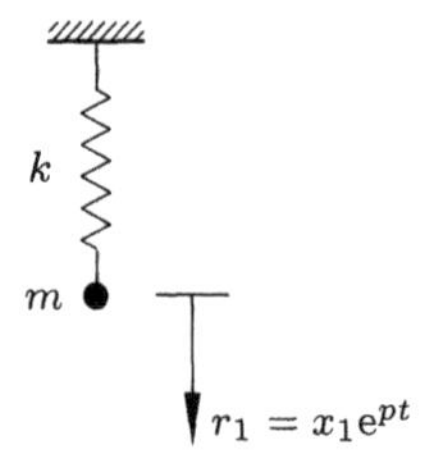

equation of motion:

$$\left[p^2\boldsymbol{M} + \boldsymbol{K}\right]\boldsymbol{x} = \boldsymbol{o} \qquad \boldsymbol{M} = m, \quad \boldsymbol{K} = k$$

$$\left(\frac{p^2 m}{k} + 1\right)x_1 = 0$$

$$\left(\frac{p^2 m}{k}\right)_{\mathrm{I}} = -1 \quad \Rightarrow \quad p_{\mathrm{I}-} = -i\sqrt{\frac{k}{m}}$$

$$p_{\mathrm{I}+} = +i\sqrt{\frac{k}{m}}$$

(a) oscillator with a single degree of freedom $\qquad x_1 = c_1 =$ arbitrary constant

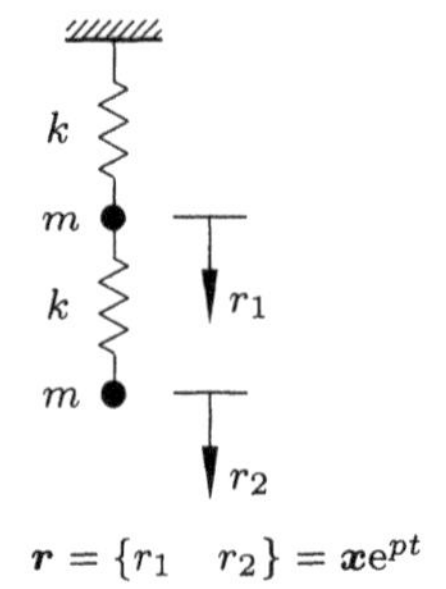

equation of motion:

$$\left[p^2\boldsymbol{M} + \boldsymbol{K}\right]\boldsymbol{x} = \boldsymbol{o} \qquad \boldsymbol{M} = m\lceil 1 \quad 1 \rfloor$$

$$\begin{bmatrix} \frac{p^2 m}{k} + 2 & -1 \\ -1 & \frac{p^2 m}{k} + 1 \end{bmatrix}\begin{bmatrix} x_1 \\ x_2 \end{bmatrix} = \begin{bmatrix} 0 \\ 0 \end{bmatrix} \qquad \boldsymbol{K} = k\begin{bmatrix} 2 & -1 \\ -1 & 1 \end{bmatrix}$$

$$\left(\frac{p^2 m}{k}\right)^2 + 3\left(\frac{p^2 m}{k}\right) + 1 = 0$$

$$\left(\frac{p^2 m}{k}\right)_{\mathrm{I}} = -\frac{1}{2}\left(3 - \sqrt{5}\right) = -0.38 \qquad \left(\frac{p^2 m}{k}\right)_{\mathrm{II}} = -\frac{1}{2}\left(3 + \sqrt{5}\right) = -2.62$$

$$p_{\mathrm{I}-} = -0.62i\sqrt{\frac{k}{m}} \qquad p_{\mathrm{II}-} = -1.62i\sqrt{\frac{k}{m}}$$

$$p_{\mathrm{I}+} = +0.62i\sqrt{\frac{k}{m}} \qquad p_{\mathrm{II}+} = +1.62i\sqrt{\frac{k}{m}}$$

$$\frac{1}{2}\begin{bmatrix} 1 + \sqrt{5} & -2 \\ -2 & -1 + \sqrt{5} \end{bmatrix}\begin{bmatrix} x_1 \\ x_2 \end{bmatrix}_{\mathrm{I}} = \begin{bmatrix} 0 \\ 0 \end{bmatrix} \qquad \frac{1}{2}\begin{bmatrix} 1 - \sqrt{5} & -2 \\ -2 & -1 - \sqrt{5} \end{bmatrix}\begin{bmatrix} x_1 \\ x_2 \end{bmatrix}_{\mathrm{II}} = \begin{bmatrix} 0 \\ 0 \end{bmatrix}$$

$$\boldsymbol{x}_{\mathrm{I}} = c_1\left\{1 \quad \tfrac{1}{2}\left(1 + \sqrt{5}\right)\right\} \qquad \boldsymbol{x}_{\mathrm{II}} = c_2\left\{1 \quad \tfrac{1}{2}\left(1 - \sqrt{5}\right)\right\}$$

(b) oscillator with two degrees of freedom $\qquad c_1, c_2 =$ arbitrary constant

Fig. 4.1.1 Spring-mass oscillator: the p_{i}^2 solutions of the characteristic equations are negative real, the pairs of p_{i} solutions are purely imaginary. The corresponding solution vectors $\boldsymbol{x}_{\mathrm{i}}$ are real and determinable but for an arbitrary (complex or real) constant c_{i}

course, not coincidental but apply also to systems with any number of degrees of freedom. To prove this in general, we substitute the complex vector

$$\boldsymbol{x}_{\mathrm{i}} = \boldsymbol{a}_{\mathrm{i}} + i\boldsymbol{b}_{\mathrm{i}} \tag{4.1.7}$$

into equ. (4.1.3) and pre-multiply the expression by the transpose of the complex conjugate vector

$$\boldsymbol{x}_{\mathrm{i}}^{*} = \boldsymbol{a}_{\mathrm{i}}^{\mathrm{t}} - i\boldsymbol{b}_{\mathrm{i}}^{\mathrm{t}} \tag{4.1.8}$$

to find

$$p_{\mathrm{i}}^{2}\boldsymbol{x}_{\mathrm{i}}^{*}\boldsymbol{M}\boldsymbol{x}_{\mathrm{i}} + \boldsymbol{x}_{\mathrm{i}}^{*}\boldsymbol{K}\boldsymbol{x}_{\mathrm{i}} = 0 \tag{4.1.9}$$

More explicitly for the term involving $\boldsymbol{K}$, we obtain

$$\boldsymbol{x}_{\mathrm{i}}^{*}\boldsymbol{K}\boldsymbol{x}_{\mathrm{i}} = \boldsymbol{a}_{\mathrm{i}}^{\mathrm{t}}\boldsymbol{K}\boldsymbol{a}_{\mathrm{i}} + \boldsymbol{b}_{\mathrm{i}}^{\mathrm{t}}\boldsymbol{K}\boldsymbol{b}_{\mathrm{i}} - i(\boldsymbol{b}_{\mathrm{i}}^{\mathrm{t}}\boldsymbol{K}\boldsymbol{a}_{\mathrm{i}} - \boldsymbol{a}_{\mathrm{i}}^{\mathrm{t}}\boldsymbol{K}\boldsymbol{b}_{\mathrm{i}}) \tag{4.1.10}$$

Since $\boldsymbol{K}\,(\boldsymbol{M})$ is symmetrical, the imaginary component is seen to vanish, even for complex values of $\boldsymbol{x}_{\mathrm{i}}$; consequently, the scalars $\boldsymbol{x}_{\mathrm{i}}^{*}\boldsymbol{K}\boldsymbol{x}_{\mathrm{i}}\,(\boldsymbol{x}_{\mathrm{i}}^{*}\boldsymbol{M}\boldsymbol{x}_{\mathrm{i}})$ are real. If we assume for the moment that both $\boldsymbol{K}$ and $\boldsymbol{M}$ are positive definite, then $\boldsymbol{a}_{\mathrm{i}}^{\mathrm{t}}\boldsymbol{K}\boldsymbol{a}_{\mathrm{i}}\,(\boldsymbol{a}_{\mathrm{i}}^{\mathrm{t}}\boldsymbol{M}\boldsymbol{a}_{\mathrm{i}})$ and $\boldsymbol{b}_{\mathrm{i}}^{\mathrm{t}}\boldsymbol{K}\boldsymbol{b}_{\mathrm{i}}\,(\boldsymbol{b}_{\mathrm{i}}^{\mathrm{t}}\boldsymbol{M}\boldsymbol{b}_{\mathrm{i}})$ – as well as $\boldsymbol{x}_{\mathrm{i}}^{*}\boldsymbol{K}\boldsymbol{x}_{\mathrm{i}}$ (or $\boldsymbol{x}_{\mathrm{i}}^{*}\boldsymbol{M}\boldsymbol{x}_{\mathrm{i}}$) – are in any case positive real numbers. It then follows from equ. (4.1.9)

$$p_{\mathrm{i}}^{2} = -\frac{\boldsymbol{x}_{\mathrm{i}}^{*}\boldsymbol{K}\boldsymbol{x}_{\mathrm{i}}}{\boldsymbol{x}_{\mathrm{i}}^{*}\boldsymbol{M}\boldsymbol{x}_{\mathrm{i}}} = -\omega_{\mathrm{i}}^{2} = \text{negative real} \tag{4.1.11}$$

or

$$\begin{aligned} p_{\mathrm{i}-} &= -i\omega_{\mathrm{i}} \\ p_{\mathrm{i}+} &= +i\omega_{\mathrm{i}} \\ \omega_{\mathrm{i}} &= +\sqrt{\frac{\boldsymbol{x}_{\mathrm{i}}^{*}\boldsymbol{K}\boldsymbol{x}_{\mathrm{i}}}{\boldsymbol{x}_{\mathrm{i}}^{*}\boldsymbol{M}\boldsymbol{x}_{\mathrm{i}}}} > 0 \end{aligned} \tag{4.1.12}$$

Having established the characteristics of the solution p_{i}^{2}, equ. (4.1.3) represents a system of equations with a real symmetric matrix of coefficients

$$[p_{\mathrm{i}}^{2}\boldsymbol{M} + \boldsymbol{K}]\boldsymbol{x}_{\mathrm{i}} = \boldsymbol{o} \tag{4.1.13}$$

However, since p_{i}^{2} satisfies the determinant condition of equ. (4.1.4a), the matrix $[p_{\mathrm{i}}^{2}\boldsymbol{M} + \boldsymbol{K}]$ must be singular. In order to avoid complexities, we assume at this point that p_{i}^{2} represents a simple root of the characteristic polynomial equ. (4.1.4c). Knowing the roots $p_{1}^{2}, p_{2}^{2}, \ldots p_{\mathrm{n}}^{2}$, we obtain an alternative expression of the characteristic equation (4.1.4c) in the form

$$(p^{2} - p_{\mathrm{n}}^{2})(p^{2} - p_{\mathrm{n}-1}^{2})\ldots(p^{2} - p_{2}^{2})(p^{2} - p_{1}^{2}) = 0 \tag{4.1.14}$$

If we restrict the argument to a simple eigenvalue p_i^2, the factor $(p^2 - p_i^2)$ can indeed only appear once. In this case, one equation in equ. (4.1.13) is linearly dependent and hence superfluous. This is also evident from the examples in fig. 4.1.1. Hence, a unique solution for $\boldsymbol{x}_i$ cannot be deduced from equ. (4.1.13). If, for example, $\boldsymbol{x}_i$ satisfies equ. (4.1.13), then it is evident that $c_i\boldsymbol{x}_i$ is also an admissible solution whereby c_i can be either real or complex. To obtain a unique solution for $\boldsymbol{x}_i$, we have to set up a subsidiary condition. A possible condition would be to assign a definite value to one of the elements, e.g. $x_{ij} = 1$. Then we may eliminate the j-th row of the system of equ. (4.1.13) since it is superfluous, and transfer the j-th column to the right-hand side. The resulting inhomogeneous real system is of the order (n−1) and has a solution because the remaining (n−1) equations are linearly independent if p_i^2 represents a simple root. The solution of the system is naturally real.

Let us now assume that we have n solutions p_i^2 or ω_i^2 and the corresponding n eigenvectors $\boldsymbol{x}_i$. The two arbitrary constants c_{i-} and c_{i+} associated with the eigenvector $\boldsymbol{x}_i$ appertain to the solution pair $p_{i-} = -i\omega$ and $p_{i+} = +i\omega$. For simple solutions ω_i^2, the 2n solutions determined from equ. (4.1.2) can now be superimposed. Thus, we may express the displacement as the linear combination

$$\boldsymbol{r} = \sum_{i=1}^{n} \boldsymbol{x}_i (c_{i+} e^{i\omega_i t} + c_{i-} e^{-i\omega_i t}) \tag{4.1.15}$$

This solution is certainly still complex and of little practical relevance. As a matter of fact, the displacement vector must be expressible in real terms. Rearranging the above expression in terms of real numbers, we obtain with the help of Euler's well-known formula

$$\begin{aligned} e^{i\psi} + e^{-i\psi} &= 2\cos\psi \\ -i(e^{i\psi} - e^{-i\psi}) &= 2\sin\psi \end{aligned} \tag{4.1.16}$$

the real expression

$$\begin{aligned} \boldsymbol{r} &= \sum_{i=1}^{n} \boldsymbol{x}_i \left[\frac{c_{i+} + c_{i-}}{2} (e^{i\omega_i t} + e^{-i\omega_i t}) + \frac{c_{i+} - c_{i-}}{2} (e^{i\omega_i t} - e^{-i\omega_i t}) \right] \\ &= \sum_{i=1}^{n} \boldsymbol{x}_i (a_i \sin\omega_i t + b_i \cos\omega_i t) \\ &= \sum_{i=1}^{n} \boldsymbol{x}_i c_i \sin(\omega_i t + \varphi_i) \end{aligned} \tag{4.1.17}$$

The general solution, $\boldsymbol{r}$, still contains 2n unknown coefficients, c_i and φ_i (or a_i and b_i), i=1,2,...n. The values of these constants can be determined from the initial conditions imposed on the system.

Equation (4.1.17) provides information on the physical meaning of the 2n characteristic values ω_i and $\boldsymbol{x}_i$. The general motion of a free system is, in fact, described by the superposition of several oscillations with angular frequencies ω_i. We denote these oscillations as natural or eigenoscillations , with the corresponding frequencies ω_i as the so-called natural angular frequencies or eigenvalues, and the vectors of the amplitudes, $\boldsymbol{x}_i$, as the natural modes or eigenvectors. If a free, undamped system oscillates in a definite natural mode, then all point-masses oscillate in phase. The modal shapes are determined by the eigenvector $\boldsymbol{x}_i$. The function $\sin(\omega_i t + \varphi_i)$ operates simply as a mapping scale. The resulting vibration nodes and anti-nodes of the oscillation are stationary and one thus refers to such vibrations as stationary oscillations. Such a natural vibration mode repeats itself with a period

$$T_i = \frac{2\pi}{\omega_i} \tag{4.1.18}$$

The reciprocal of this period is the frequency and gives the number of oscillations per second performed by an elastic system

$$\nu_i = \frac{1}{T_i} \tag{4.1.19}$$

In addition to the displacements defined by equ. (4.1.17), we are also often interested in the velocities and the accelerations

$$\begin{aligned}\dot{\boldsymbol{r}} &= \sum_{i=1}^{n} \boldsymbol{x}_i c_i \omega_i \cos(\omega_i t + \varphi_i) \\ &= \sum_{i=1}^{n} \boldsymbol{x}_i c_i \omega_i \sin(\omega_i t + \varphi_i - 90°)\end{aligned} \tag{4.1.20}$$

$$\begin{aligned}\ddot{\boldsymbol{r}} &= \sum_{i=1}^{n} -\boldsymbol{x}_i c_i \omega_i^2 \sin(\omega_i t + \varphi_i) \\ &= \sum_{i=1}^{n} \boldsymbol{x}_i c_i \omega_i^2 \sin(\omega_i t + \varphi_i - 180°)\end{aligned} \tag{4.1.21}$$

Apart from the factors ω_i and ω_i^2, velocities and accelerations of a natural vibration possess the same amplitude vectors as the displacements. Their phase angles, however, lag behind the displacement vector by 90^0 and 180^0 respectively. These relations are reproduced in fig. 4.1.2 in the form of a vector diagram.

Whether a single mode, i, exists independently (stationary oscillation) or whether a general motion takes place involving the superposition of a number of eigenmodes depends solely upon the initial conditions. These initial conditions may be defined by the displacements and the corresponding velocities at a definite instant t_0. We will discuss this in greater detail in Chapter 5. At this point, we would like to return to the determination of ω_i^2 and $\boldsymbol{x}_i$ and discuss some important characteristics which are particularly significant in the mathematical formulation of the problem.

Let $\boldsymbol{x}_i$ and $\boldsymbol{x}_j$ be two eigenvectors which satisfy equ. (4.1.13)

$$\begin{aligned}\omega_i^2 \boldsymbol{M}\boldsymbol{x}_i &= \boldsymbol{K}\boldsymbol{x}_i \\ \omega_j^2 \boldsymbol{M}\boldsymbol{x}_j &= \boldsymbol{K}\boldsymbol{x}_j\end{aligned} \tag{4.1.22}$$

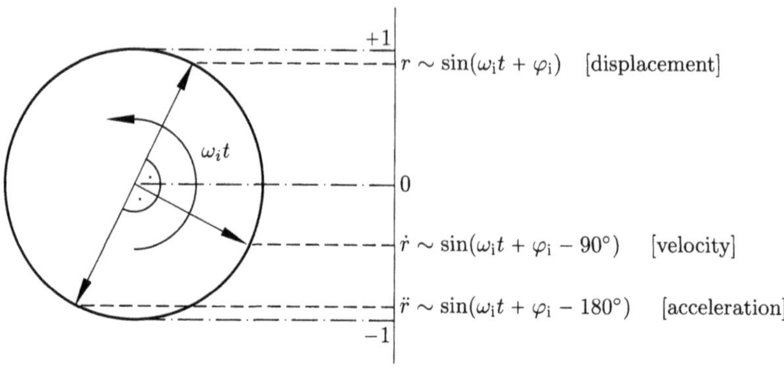

Fig. 4.1.2 Time-phases between displacement, velocity and acceleration of an undamped eigen-oscillation

We multiply the first equation by $\boldsymbol{x}_j^t$ and the second one by $\boldsymbol{x}_i^t$ and, because of the symmetry of the $\boldsymbol{M}$ and $\boldsymbol{K}$ matrices

$$\begin{aligned} \overline{m}_{ij} &= \boldsymbol{x}_i^t \boldsymbol{M} \boldsymbol{x}_j = \boldsymbol{x}_j^t \boldsymbol{M} \boldsymbol{x}_i \\ \overline{k}_{ij} &= \boldsymbol{x}_i^t \boldsymbol{K} \boldsymbol{x}_j = \boldsymbol{x}_j^t \boldsymbol{K} \boldsymbol{x}_i \end{aligned} \tag{4.1.23}$$

deduce the relations

$$\begin{aligned} \omega_i^2 \overline{m}_{ij} &= \overline{k}_{ij} \\ \omega_j^2 \overline{m}_{ij} &= \overline{k}_{ij} = \omega_i^2 \overline{m}_{ij} \end{aligned} \tag{4.1.24}$$

If $\omega_i^2 \neq \omega_j^2$, then the second line of equation (4.1.24) is satisfied only if

$$\overline{m}_{ij} = 0 \tag{4.1.25}$$

Consequently, it then follows from the first or second line that

$$\overline{k}_{ij} = 0 \tag{4.1.26}$$

Next, we collect all possible formulations of equ. (4.1.22) in a matrix form

$$\boldsymbol{M}[\boldsymbol{x}_1 \;\; \boldsymbol{x}_2 \;\; \dots \;\; \boldsymbol{x}_n] = \boldsymbol{K}[\boldsymbol{x}_1 \;\; \boldsymbol{x}_2 \;\; \dots \;\; \boldsymbol{x}_n] \begin{bmatrix} \frac{1}{\omega_1^2} & \frac{1}{\omega_2^2} & \cdots & \frac{1}{\omega_n^2} \end{bmatrix} \tag{4.1.27}$$

and introduce the following abbreviations

$$\underset{(n \times n)}{\boldsymbol{X}} = [\boldsymbol{x}_1 \;\; \boldsymbol{x}_2 \;\; \dots \;\; \boldsymbol{x}_n] \tag{4.1.28}$$

for the eigenvectors and the diagonal matrix

$$\boldsymbol{\Lambda} = \lceil \lambda_1 \ \lambda_2 \ \ldots \ \lambda_n \rfloor = \left\lceil \frac{1}{\omega_1^2} \ \frac{1}{\omega_2^2} \ \cdots \ \frac{1}{\omega_n^2} \right\rfloor \tag{4.1.29}$$

for the spectrum of the eigenvalues (= reciprocal of the square of the angular frequencies). With these notations, equ. (4.1.27) can be rewritten in the compact form

$$\boldsymbol{MX}=\boldsymbol{KX\Lambda} \tag{4.1.30}$$

If the above equation is pre-multiplied by $\boldsymbol{X}^t$, then both $\boldsymbol{X}^t\boldsymbol{MX}$ and $\boldsymbol{X}^t\boldsymbol{KX}$ are diagonalised as a consequence of equs. (4.1.25) and (4.1.26). If all eigenvalues are hence different, the eigenvectors diagonalise both the mass and the stiffness matrices.

We are, of course, by now aware that the eigenvectors may only be determined up to an arbitrary scalar factor. The possibility of prescribing the value of a suitable element, $x_{ij} = 1$, has already been mentioned but is only a variant of many possibilities. It proves much more useful to normalise the eigenvector $\boldsymbol{x}_i$ so that e.g.

$$\bar{k}_{ii} = \boldsymbol{x}_i^t \boldsymbol{K} \boldsymbol{x}_i = 1 \quad ; \qquad i = 1, 2, \ldots, n \tag{4.1.31a}$$

or

$$\boldsymbol{X}^t\boldsymbol{KX}=\boldsymbol{I} \tag{4.1.31b}$$

This process is known as the normalisation of the eigenvectors with respect to the stiffness matrix. Alternatively, it is possible to normalise the eigenvectors with respect to the mass matrix.

When the eigenvectors are normalised with respect to a stiffness matrix, we deduce from equ. (4.1.30)

$$\boldsymbol{X}^t\boldsymbol{MX} = \boldsymbol{\Lambda} \tag{4.1.32}$$

Thus, eigenvectors orthonormalised with respect to the stiffness matrix also reduce the mass matrix into the diagonal matrix of the eigenvalues. This, of course, also applies for the case of equal eigenvalues, provided the eigenvectors satisfy the normalisation condition of equ. (4.1.31b). We shall return to this case in connection with the special cases to be discussed in section 4.2.

At this stage, it may serve as light relief to consider fig. 4.1.1(b) once more. The following expression reproduces the eigenvectors that are initially not yet normalised with respect to $\boldsymbol{K}$

$$\boldsymbol{X} = \begin{bmatrix} 1 & 1 \\ \frac{1}{2}(1+\sqrt{5}) & \frac{1}{2}(1-\sqrt{5}) \end{bmatrix}$$

The mass matrix, $\boldsymbol{M}$, is already diagonal and remains so after the congruent transformation with the matrix $\boldsymbol{X}$. We obtain

$$\boldsymbol{X}^{\mathrm{t}}\boldsymbol{M}\boldsymbol{X} = \frac{m}{2}\left\lceil (5+\sqrt{5}) \quad (5-\sqrt{5}) \right\rfloor$$

The fully populated stiffness matrix will also be reduced to a diagonal form by the congruent transformation with the matrix $\boldsymbol{X}$

$$\boldsymbol{X}^{\mathrm{t}}\boldsymbol{K}\boldsymbol{X} = \frac{k}{2}\left\lceil (5-\sqrt{5}) \quad (5+\sqrt{5}) \right\rfloor$$

As $\boldsymbol{X}$ is not orthonormal with respect to $\boldsymbol{K}$, the squares of the angular frequencies are deduced from equ. (4.1.11). This expression is known as the Rayleigh quotient

$$\begin{aligned}
\boldsymbol{\Lambda}^{-1} &= \lceil \omega_1^2 \quad \omega_2^2 \rfloor \\
&= [\boldsymbol{X}^{\mathrm{t}}\boldsymbol{K}\boldsymbol{X}][\boldsymbol{X}^{\mathrm{t}}\boldsymbol{M}\boldsymbol{X}]^{-1} \\
&= \frac{k}{m}\left\lceil \frac{5-\sqrt{5}}{5+\sqrt{5}} \quad \frac{5+\sqrt{5}}{5-\sqrt{5}} \right\rfloor \\
&= \frac{k}{2m}\left\lceil (3-\sqrt{5}) \quad (3+\sqrt{5}) \right\rfloor
\end{aligned}$$

However, the eigenvector can also be orthonormalised easily with respect to $\boldsymbol{K}$ by dividing the elements of the eigenvector $\boldsymbol{x}_{\mathrm{i}}$ by the scalar $(\boldsymbol{x}_{\mathrm{i}}^{\mathrm{t}}\boldsymbol{K}\boldsymbol{x}_{\mathrm{i}})^{1/2}$. The result of this procedure is

$$\boldsymbol{X}_{Ko} = (2k)^{-1/2}\begin{bmatrix} 2(5-\sqrt{5})^{-1/2} & 2(5+\sqrt{5})^{-1/2} \\ (1+\sqrt{5})(5-\sqrt{5})^{-1/2} & (1-\sqrt{5})(5+\sqrt{5})^{-1/2} \end{bmatrix}$$

The fact that the eigenvectors $\boldsymbol{X}$ diagonalise the mass and stiffness matrices simultaneously is essential for the solution of the eigenvalue problem, i.e. the determination of $\boldsymbol{X}$ and $\boldsymbol{\Lambda}$, in equ. (4.1.30). The application of computers for the determination of the roots of the determinant of $[\boldsymbol{M} - \lambda\boldsymbol{K}]$ ($\widehat{=}$ $[p^2\boldsymbol{M} + \boldsymbol{K}]$ $\widehat{=}$ $[\omega^2\boldsymbol{M} - \boldsymbol{K}]$), for example by variation of λ and subsequent analysis of the function $\det[\boldsymbol{M} - \lambda K]$, is not recommended and leads to an inordinate computational expenditure.

Even more general forms exist in addition to the important orthogonalisation conditions of equs. (4.1.31b) and (4.1.32). Once again, we require that $\boldsymbol{X}$ is orthonormal with respect to $\boldsymbol{K}$, thus satisfying equ. (4.1.31b). Pre-multiplying equ. (4.1.30) by $\boldsymbol{X}^{\mathrm{t}}\boldsymbol{M}\boldsymbol{K}^{-1}$, we find

$$\begin{aligned}
\boldsymbol{X}^{\mathrm{t}}\boldsymbol{M}\boldsymbol{K}^{-1}\boldsymbol{M}\boldsymbol{X} &= \boldsymbol{X}^{\mathrm{t}}\boldsymbol{M}\boldsymbol{K}^{-1}\boldsymbol{K}\boldsymbol{X}\boldsymbol{\Lambda} = \boldsymbol{X}^{\mathrm{t}}\boldsymbol{M}\boldsymbol{X}\boldsymbol{\Lambda} \\
&= \boldsymbol{\Lambda}^2
\end{aligned} \tag{4.1.33}$$

where $\boldsymbol{K}$ naturally has to be positive definite.

Still more general orthogonalisation characteristics can be obtained by using multiple products of stiffness and mass. Assuming that the eigenvectors $\boldsymbol{X}$ are normalised as in equ. (4.1.31) or (4.1.32) results directly in

$$[\boldsymbol{K}^{-1}\boldsymbol{M}]^{\mathrm{p}}\boldsymbol{X}=\boldsymbol{X}\boldsymbol{\Lambda}^{\mathrm{p}} \quad \text{or} \quad [\boldsymbol{M}^{-1}\boldsymbol{K}]^{\mathrm{p}}\boldsymbol{X}=\boldsymbol{X}\boldsymbol{\Lambda}^{-\mathrm{p}} \tag{4.1.34}$$

(in which we assume that the corresponding inverses exist).

Hence, we may deduce further orthogonality relations, e.g.

$$\begin{aligned} \boldsymbol{X}^{\mathrm{t}}\boldsymbol{K}[\boldsymbol{K}^{-1}\boldsymbol{M}]^{\mathrm{p}}\boldsymbol{X} &= \boldsymbol{\Lambda}^{\mathrm{p}} \\ \boldsymbol{X}^{\mathrm{t}}\boldsymbol{M}[\boldsymbol{K}^{-1}\boldsymbol{M}]^{\mathrm{p}}\boldsymbol{X} &= \boldsymbol{\Lambda}^{\mathrm{p}+1} \\ \boldsymbol{X}^{\mathrm{t}}\boldsymbol{K}[\boldsymbol{M}^{-1}\boldsymbol{K}]^{\mathrm{p}}\boldsymbol{X} &= \boldsymbol{\Lambda}^{-\mathrm{p}} \end{aligned} \tag{4.1.35}$$

Important special expressions are obtained for p=0 and p=1.

The formulation in equ. (4.1.30) is known as the *general* eigenvalue problem and serves the purpose of finding the vectors $\boldsymbol{X}$ which can diagonalise the *two* matrices $\boldsymbol{K}$ and $\boldsymbol{M}$ simultaneously. In contrast, when only *one* matrix has to be diagonalised, one talks of a *special* eigenvalue problem. The conversion of equ. (4.1.30) into a 'special' eigenvalue problem, is obtained, for example, by a premultiplication with $\boldsymbol{M}^{-1}$ ($\boldsymbol{M}$ positive definite), or with $\boldsymbol{K}^{-1}$ ($\boldsymbol{K}$ positive definite). The resulting 'dynamic' matrix $\boldsymbol{M}^{-1}\boldsymbol{K}$ or $\boldsymbol{K}^{-1}\boldsymbol{M}$ is unfortunately unsymmetric. The 'special' eigenvalue problem is, however, easier to solve if the matrix to be diagonalised is symmetric. This can be achieved by means of a simple transformation, as follows.

Firstly, we assume that the stiffness matrix $\boldsymbol{K}$ is positive definite. Applying Cholesky's triangular decomposition, we obtain

$$\boldsymbol{K} \to \boldsymbol{U}^{\mathrm{t}}\boldsymbol{U} \tag{4.1.36}$$

Next, we define the transformed eigenvector

$$\overline{\boldsymbol{X}} = \boldsymbol{U}\boldsymbol{X} \tag{4.1.37}$$

and multiply equ. (4.1.30) by $[\boldsymbol{U}^{\mathrm{t}}]^{-1} = \boldsymbol{U}^{-\mathrm{t}}$. The resulting eigenvalue formulation is

$$\boldsymbol{A}\overline{\boldsymbol{X}} = \overline{\boldsymbol{X}}\boldsymbol{\Lambda} \tag{4.1.38}$$

in which $\boldsymbol{A}$ is known as the dynamic matrix and is given by

$$\boldsymbol{A} = [\boldsymbol{U}^{\mathrm{t}}]^{-1}\boldsymbol{M}\boldsymbol{U}^{-1} \tag{4.1.39}$$

All that now remains is to determine the eigenvectors $\overline{\boldsymbol{X}}$ which diagonalise the symmetric matrix $\boldsymbol{A}$. If one chooses to normalise the eigenvectors $\overline{\boldsymbol{X}}$ such that

$$\overline{\boldsymbol{X}}^{\mathrm{t}}\overline{\boldsymbol{X}} = \boldsymbol{I} \tag{4.1.40}$$

we have

$$\overline{\boldsymbol{X}}^{\mathrm{t}}\boldsymbol{A}\overline{\boldsymbol{X}} = \boldsymbol{\Lambda} \tag{4.1.41}$$

Once again, $\boldsymbol{\Lambda}$ contains the inverses of the squares of the angular frequencies. Having obtained the solution $\overline{\boldsymbol{X}}$ to equ. (4.1.38), the eigenvectors $\boldsymbol{X}$ of the original problem can be determined by the following transformation

$$\boldsymbol{X} = \boldsymbol{U}^{-1}\overline{\boldsymbol{X}} \tag{4.1.42}$$

An alternative method of generating a special eigenvalue problem proceeds by triangularisation of the mass matrix (if positive definite). This approach is feasible even in the presence of rigid-body freedoms of the system. However, the method is numerically sensitive when using kinematically equivalent mass matrices (see [4.1], p. 34).

4.2 Special cases in spectral analysis : Singular mass and stiffness matrices and multiple eigenvalues

To begin with, we concentrate upon a discussion of a singular stiffness matrix $\boldsymbol{K}$. Such a case occurs quite regularly in practice and is of special relevance to oscillations of free-free or semi-free structures (i.e. aeroplanes, rockets, satellites or hinged helicopter blades). Such systems possess rigid-body freedoms, defined by $\boldsymbol{x}_0$ (see fig. 4.2.1). A rigid-body motion cannot induce any elastic forces, i.e. we have

$$\boldsymbol{K}\boldsymbol{x}_0 = \boldsymbol{o} \qquad (\boldsymbol{x}_0 \neq \boldsymbol{o}) \tag{4.2.1}$$

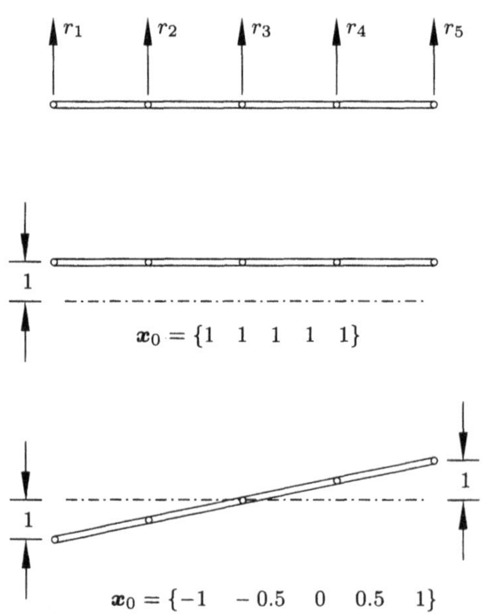

Fig. 4.2.1
Rigid-body modes $\boldsymbol{x}_0$ contained in the $\boldsymbol{r}$-system

It is immediately evident from the eigenvalue problem

$$\boldsymbol{K}\boldsymbol{x}_{\mathrm{i}} = \boldsymbol{M}\boldsymbol{x}_{\mathrm{i}}\lambda_{\mathrm{i}}^{-1} \tag{4.2.2}$$

that the vector $\boldsymbol{x}_0$ represents a non-trivial solution of the system, which appertains to

$$\lambda_0^{-1} = \omega_0^2 = 0 \tag{4.2.3}$$

provided $\boldsymbol{M}\boldsymbol{x}_0$ is a non-zero vector. In fact, it follows from $\boldsymbol{M}\boldsymbol{x}_0 = \boldsymbol{o}$ that $\boldsymbol{x}_0$ is not simply a rigid body vector but that it also describes an inertialess motion. This aspect will be discussed briefly at the end of the section.

The solution specified by equ. (4.2.3) can also be derived either from equ. (4.1.11), the Rayleigh quotient, or taken directly from the characteristic equation (4.1.4c) which we write here in the form

$$b_{\mathrm{n}}\left(\frac{1}{\lambda}\right)^{\mathrm{n}} + b_{\mathrm{n}-1}\left(\frac{1}{\lambda}\right)^{\mathrm{n}-1} + \ldots + b_0 = 0 \tag{4.2.4}$$

Since $\boldsymbol{K}$ is singular, $\det\boldsymbol{K}$ is zero and consequently, in accordance with equ. (4.1.6), it follows that b_0 is also zero. Thus, we have at least one solution of the form (4.2.3). On the other hand, however, if we base our solution procedure for the eigenvalue problem on the formulation of equs. (4.1.30) or (4.1.38), we are left with the unfortunate consequence that a few values of λ_{i} become infinitely large. What is more, the triangular decomposition of equ (4.1.36) cannot be realised. Thus, we are obliged to extract the rigid body modes from the eigenvalue calculation.

In order to resolve this problem, we can split the displacements, as was the case in the natural approach to the finite element method, into elastic and rigid-body motions. The most obvious procedure here would be to suppress the rigid-body modes by a constraint of the structure through statically determinate supports and subsequently to correct the error by the reintroduction of missing rigid-body modes ([4.2], [4.3], fig. 4.2.2). Thus, we obtain

$$\boldsymbol{r} = [\boldsymbol{T}_0 \quad \boldsymbol{T}_e]\{\boldsymbol{r}_0 \quad \boldsymbol{r}_e\} \tag{4.2.5}$$

If this procedure is carried out by a suppression of individual degrees of freedom, as is illustrated in fig. 4.2.2, $\boldsymbol{T}_e$ becomes a simple Boolean matrix and describes the elastic displacement of individual nodes when the selected supports are fixed. Each column of the $\boldsymbol{T}_e$ matrix contains a '1' element (the ij indices indicate that $r_{e\mathrm{j}}$ contributes to r_{i}). Moreover, every row in the $\boldsymbol{T}_e$ matrix contains either solely zero-values (r_{i} being suppressed by the supports) or a single entry of 1 (r_{i} existing in such a case, even in the presence of supports). A number of other mathematical approaches, which will be discussed later on, exist for the determination of the $\boldsymbol{T}_0$ and $\boldsymbol{T}_e$ matrices of general structures. As a rule, however, the simple pattern of the matrices is then lost. Although the matrix of the rigid-body vectors $\boldsymbol{T}_0$ is generally fully populated, it can easily be generated automatically by means of the nodal coordinates stored in the computer. Thus, in the case of the example in fig. 4.2.1, the first and second columns of the $\boldsymbol{T}_0$ matrix can be determined by means

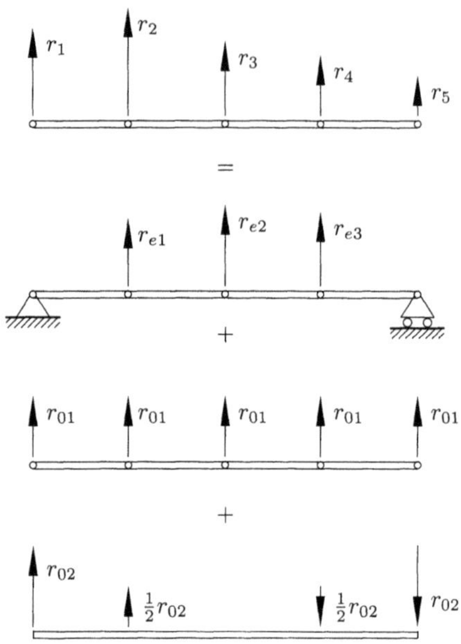

$$\begin{bmatrix} r_1 \\ r_2 \\ r_3 \\ r_4 \\ r_5 \end{bmatrix} = \left[\begin{array}{cc|ccc} 1 & 1 & 0 & 0 & 0 \\ 1 & 0.5 & 1 & 0 & 0 \\ 1 & 0 & 0 & 1 & 0 \\ 1 & -0.5 & 0 & 0 & 1 \\ 1 & -1 & 0 & 0 & 0 \end{array}\right] \begin{bmatrix} r_{01} \\ r_{02} \\ r_{e1} \\ r_{e2} \\ r_{e3} \end{bmatrix}$$

$$\boldsymbol{r} = [\boldsymbol{T}_0 \mid \boldsymbol{T}_e]\{\boldsymbol{r}_0 \mid \boldsymbol{r}_e\}$$

Fig. 4.2.2
Free and partly free oscillators: complete motion r is divided into rigid and elastic components, $(\boldsymbol{r}_0)$ and $(\boldsymbol{r}_e)$

of this technique – for example, the first column is generated by the instruction: "vertical translation" and the second by the instruction: "rotation about node 3 in a clockwise direction". As a general rule, the $\boldsymbol{T}_0$ matrix comprises only a small number of columns and this should not cause any storage problems in the computer. It is, of course, necessary that the columns in $\boldsymbol{T}_0$ and $\boldsymbol{T}_e$ be linearly independent, a requirement which can be fulfilled without difficulty.

We are now in a position to put the above process into practice and reformulate the dynamic problem of equ. (4.1.1) in terms of the two new degrees of freedom, $\boldsymbol{r}_0$ and $\boldsymbol{r}_e$. To this purpose, we substitute equ. (4.2.5) into equ. (4.1.1), pre-multiply by $\{\boldsymbol{T}_0^{\mathrm{t}} \quad \boldsymbol{T}_e^{\mathrm{t}}\}$ and note that columns of $\boldsymbol{T}_0$ represent rigid body modes

$$\boldsymbol{K}\boldsymbol{T}_0 = \boldsymbol{O} \tag{4.2.6}$$

We obtain the transformed system

$$\begin{bmatrix} \boldsymbol{M}_{00} & \boldsymbol{M}_{0e} \\ \boldsymbol{M}_{e0} & \boldsymbol{M}_{ee} \end{bmatrix} \begin{bmatrix} \ddot{\boldsymbol{r}}_0 \\ \ddot{\boldsymbol{r}}_e \end{bmatrix} + \begin{bmatrix} \boldsymbol{O} & \boldsymbol{O} \\ \boldsymbol{O}^{\mathrm{t}} & \boldsymbol{K}_{ee} \end{bmatrix} \begin{bmatrix} \boldsymbol{r}_0 \\ \boldsymbol{r}_e \end{bmatrix} = \begin{bmatrix} \boldsymbol{o} \\ \boldsymbol{o} \end{bmatrix} \tag{4.2.7}$$

in which the following compact expressions are introduced

$$\begin{aligned} \boldsymbol{M}_{00} &= \boldsymbol{T}_0^{\mathrm{t}} \boldsymbol{M} \boldsymbol{T}_0 \\ \boldsymbol{M}_{0e} &= \boldsymbol{T}_0^{\mathrm{t}} \boldsymbol{M} \boldsymbol{T}_e = \boldsymbol{M}_{e0}^{\mathrm{t}} \\ \boldsymbol{M}_{ee} &= \boldsymbol{T}_e^{\mathrm{t}} \boldsymbol{M} \boldsymbol{T}_e \\ \boldsymbol{K}_{ee} &= \boldsymbol{T}_e^{\mathrm{t}} \boldsymbol{K} \boldsymbol{T}_e \end{aligned} \tag{4.2.8}$$

When using simple support conditions in the generation of matrix $\boldsymbol{T}_e$, then $\boldsymbol{M}_{ee}$ ($\boldsymbol{K}_{ee}$) represents the mass (stiffness) matrix of the supported structure. Both matrices are sparsely populated and the $\boldsymbol{M}_{00}$ matrix, whilst generally being fully populated, is only of very small dimension. Moreover, if the rigid-body motions are referred to the structure's principal axes of inertia, $\boldsymbol{M}_{00}$ will be diagonal. The coupling matrix $\boldsymbol{M}_{eo}$ will, in general, be fully populated but this should not cause any problems, since it is a very narrow matrix. Of course, we may adopt a special matrix $\boldsymbol{T}_e$ for which $\boldsymbol{M}_{0e}$ is zero. As a result, we obtain a complete decoupling of the two equations in equ. (4.2.7), albeit at the expense of the sparsity of the matrices $\boldsymbol{M}_{ee}$ and $\boldsymbol{K}_{ee}$.

We are now in a position to use the first hyperequation in equ. (4.2.7) to eliminate the rigid-body motions. Since $\boldsymbol{M}_{00}$ is non-singular (this is also reflected in its diagonalisable form), the rigid-body acceleration is given by

$$\ddot{\boldsymbol{r}}_0 = -\boldsymbol{M}_{00}^{-1} \boldsymbol{M}_{0e} \ddot{\boldsymbol{r}}_e \tag{4.2.9}$$

From equ. (4.2.9) and the first set of equations in (4.2.7), it can be deduced that the total sum of the inertia forces in the direction of the rigid-body motions must be zero. Substituting now equ. (4.2.9) into the second set of equ. (4.2.7), we obtain a system of equations solely in terms of the elastic degrees of freedom

$$[\boldsymbol{M}_{ee} - \boldsymbol{M}_{e0} \boldsymbol{M}_{00}^{-1} \boldsymbol{M}_{0e}] \ddot{\boldsymbol{r}}_e + \boldsymbol{K}_{ee} \boldsymbol{r}_e = \boldsymbol{o} \tag{4.2.10}$$

The sole distinguishing factor between the present formulation and the standard oscillatory analysis of a supported structure is a modified mass matrix. The eigenvalue problem can be solved here as in section 4.1 since the stiffness matrix $\boldsymbol{K}_{ee}$ is positive definite and all eigenvalues are finite since the zero eigenvalues have been eliminated. Equation (4.2.10), however, only yields the elastic components of the eigenvectors $\boldsymbol{X}_e$ of the system, which nevertheless determine the stress-strain behaviour of the structure. By superposition of the non-degenerate solutions (see also equ. (4.1.17)), we immediately deduce the elastic components of the motion

$$\boldsymbol{r}_e = \sum_{\mathrm{i}=1}^{\mathrm{ne}} c_{\mathrm{i}} \boldsymbol{x}_{e\mathrm{i}} \sin(\omega_{\mathrm{i}} t + \varphi_{\mathrm{i}}) \tag{4.2.11}$$

in which ne is the number of elastic degrees of freedom.

Integrating equ. (4.2.9) twice with respect to time, we obtain

$$\boldsymbol{r}_0 = -\boldsymbol{M}_{00}^{-1}\boldsymbol{M}_{0e}\boldsymbol{r}_e + \boldsymbol{a}_0 + \boldsymbol{b}_0 t \tag{4.2.12}$$

The entries in the vectors $\boldsymbol{a}_0$ and $\boldsymbol{b}_0$ are still unknown. The total response of the system expressed in the original degrees of freedom, $\boldsymbol{r}$, becomes in conjunction with equ. (4.2.5) simply

$$\begin{aligned}\boldsymbol{r} &= [\boldsymbol{T}_e - \boldsymbol{T}_0\boldsymbol{M}_{00}^{-1}\boldsymbol{M}_{0e}]\boldsymbol{r}_e + \boldsymbol{T}_0[\boldsymbol{a}_0 + \boldsymbol{b}_0 t] \\ &= [\boldsymbol{T}_e - \boldsymbol{T}_0\boldsymbol{M}_{00}^{-1}\boldsymbol{M}_{0e}]\sum_{\mathrm{i}=1}^{\mathrm{ne}} c_\mathrm{i}\boldsymbol{x}_{e\mathrm{i}}\sin(\omega_\mathrm{i} t + \varphi_\mathrm{i}) + \boldsymbol{T}_0[\boldsymbol{a}_0 + \boldsymbol{b}_0 t]\end{aligned} \tag{4.2.13}$$

The first part represents the oscillatory motion, whereby the eigenvectors in terms of $\boldsymbol{r}$ are clearly given by

$$\underset{(\mathrm{n}\times\mathrm{ne})}{\boldsymbol{X}} = [\boldsymbol{T}_e - \boldsymbol{T}_0\boldsymbol{M}_{00}^{-1}\boldsymbol{M}_{0e}]\boldsymbol{X}_e \tag{4.2.14}$$

The $\boldsymbol{X}$ matrix belongs to the non-degenerate solutions λ; thus, it is an upright rectangular matrix. The full rank of the eigenvectors is established by superposition of the n_0 rigid-body modes. The second part of the solution derives from the conditions associated with partial support or lack of support of the structure. If such a structure now experiences a constant velocity, then this state of motion will remain unaffected since no external forces are acting upon the structure. Furthermore, because the displacements now vary linearly with respect to time, we obtain for the special initial conditions

$$\begin{aligned}\boldsymbol{r}_0(t=0) &= \boldsymbol{o} \\ \dot{\boldsymbol{r}}_0(t=0) &= \boldsymbol{o}\end{aligned} \tag{4.2.15}$$

simply

$$\boldsymbol{a}_0 = \boldsymbol{b}_0 = \boldsymbol{o} \tag{4.2.16}$$

i.e. terms linear with respect to time drop out in equ. (4.2.13).

The two examples illustrated in fig. 4.2.3 demonstrate the technique discussed above. On the left, we analyse a spring-mass oscillator whilst on the right hand side, we consider the oscillatory motion of a beam with two rigid-body modes. Since the rigid-body modes of the beam model refer to the principal axes of the mass of the beam, the rigid-body inertia matrix $\boldsymbol{M}_{00}$ is diagonal. There is, however, no decoupling between the rigid-body and elastic modes, since the coupling matrix $\boldsymbol{M}_{e0}$ is non-zero in both examples. Consequently, the rigid-body motion enters in the calculation of the total motion of the examples shown in fig. 4.2.3. The oscillatory modes expressed in the degrees of freedom $\boldsymbol{r}$ highlight the balance of forces in the course of the motion. In order to achieve an uncoupling of the inertia matrices between the rigid-body and the elastic modes and in order to achieve an

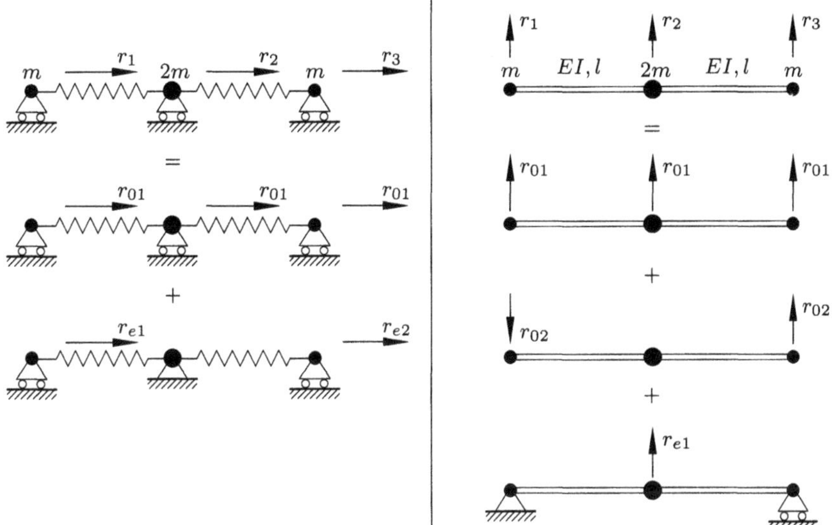

$$T_0 = \begin{bmatrix} 1 \\ 1 \\ 1 \end{bmatrix} \qquad T_e = \begin{bmatrix} 1 & 0 \\ 0 & 0 \\ 0 & 1 \end{bmatrix}$$

$$\boldsymbol{M} = m\lceil 1 \quad 2 \quad 1 \rfloor$$

$$\boldsymbol{M}_{00} = 4m$$

$$\boldsymbol{M}_{e0} = m\{1 \quad 1\}$$

$$\boldsymbol{M}_{ee} = m\lceil 1 \quad 1 \rfloor$$

$$\boldsymbol{M}_{ee} - \boldsymbol{M}_{e0}\boldsymbol{M}_{00}^{-1}\boldsymbol{M}_{0e} = \frac{m}{4}\begin{bmatrix} 3 & -1 \\ -1 & 3 \end{bmatrix}$$

$$\boldsymbol{K}_{ee} = k\lceil 1 \quad 1 \rfloor$$

$$\begin{bmatrix} \frac{3}{4}m - k\lambda & -\frac{1}{4}m \\ -\frac{1}{4}m & \frac{3}{4}m - k\lambda \end{bmatrix} \begin{bmatrix} x_{e1} \\ x_{e2} \end{bmatrix} = \boldsymbol{o}$$

$$\lambda_1 = \frac{m}{2k} \qquad \boldsymbol{x}_{e1} = \{1 \quad 1\}$$

$$\lambda_2 = \frac{m}{k} \qquad \boldsymbol{x}_{e2} = \{1 \quad -1\}$$

$$T_0\boldsymbol{M}_{00}^{-1}\boldsymbol{M}_{0e} = \frac{1}{4}\begin{bmatrix} 1 & 1 \\ 1 & 1 \\ 1 & 1 \end{bmatrix}$$

$$\boldsymbol{x}_1 = \{1 \quad -1 \quad 1\}$$

$$\boldsymbol{x}_2 = \{1 \quad 0 \quad -1\}$$

$$T_0 = \begin{bmatrix} 1 & -1 \\ 1 & 0 \\ 1 & 1 \end{bmatrix} \qquad T_e = \begin{bmatrix} 0 \\ 1 \\ 0 \end{bmatrix}$$

$$\boldsymbol{M} = m\lceil 1 \quad 2 \quad 1 \rfloor$$

$$\boldsymbol{M}_{00} = 2m\lceil 2 \quad 1 \rfloor$$

$$\boldsymbol{M}_{e0} = m[2 \quad 0]$$

$$\boldsymbol{M}_{ee} = 2m$$

$$\boldsymbol{M}_{ee} - \boldsymbol{M}_{e0}\boldsymbol{M}_{00}^{-1}\boldsymbol{M}_{0e} = m$$

$$\boldsymbol{K}_{ee} = \frac{6EI}{l^3}$$

$$\left(m - \frac{6EI}{l^3}\lambda\right)x_{e1} = 0$$

$$\lambda_1 = \frac{ml^3}{6EI} \qquad \boldsymbol{x}_{e1} = 1$$

$$T_0\boldsymbol{M}_{00}^{-1}\boldsymbol{M}_{0e} = \frac{1}{4}\begin{bmatrix} 1 \\ 1 \\ 1 \end{bmatrix}$$

$$\boldsymbol{x}_1 = \{-1 \quad 1 \quad -1\}$$

Fig. 4.2.3 Two simple examples demonstrating the analysis of oscillators with singular stiffness, *Part 1*

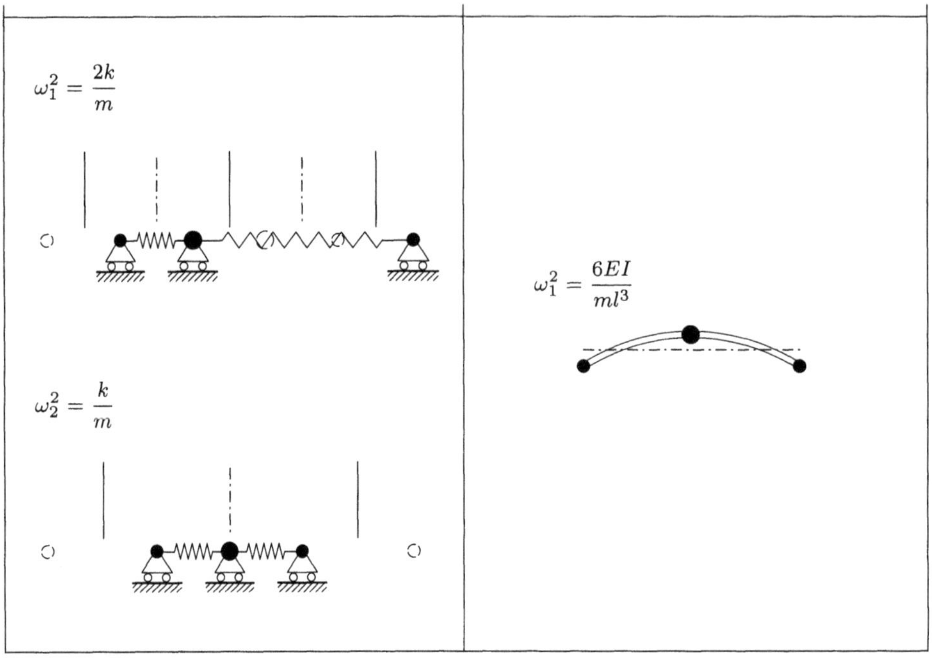

Fig. 4.2.3 Two simple examples demonstrating the analysis of oscillators with singular stiffness, *Part 2*

orthogonality between them, the $\boldsymbol{T}_e$ matrix cannot be set up by a specific system of supports but must instead be construed by the imposed condition

$$\boldsymbol{M}_{e0} = \boldsymbol{T}_e^{\mathrm{t}} \boldsymbol{M} \boldsymbol{T}_0 = \boldsymbol{O} \tag{4.2.17}$$

This means that for the spring-mass system, we seek two linearly independent vectors, orthogonal to $\boldsymbol{MT}_0$

$$\boldsymbol{MT}_0 = m\{1 \quad 2 \quad 1\}$$

These vectors must therefore satisfy

$$v_1 + 2v_2 + v_3 = 0$$

Obviously, this holds if we adopt the transformation matrix

$$\boldsymbol{T}_e = \begin{bmatrix} 1 & 1 \\ 0 & -1 \\ -1 & 1 \end{bmatrix}$$

The reader should now reanalyse the eigenvalue problem and also attempt to solve the beam problem in a similar manner.

Let us now focus our attention upon the second critical case, namely that of a singular mass matrix. In this case, inertialess motions $\boldsymbol{x}_\infty$ exist which satisfy

$$\boldsymbol{M}\boldsymbol{x}_\infty = \boldsymbol{o} \qquad (\boldsymbol{x}_\infty \neq \boldsymbol{o}) \tag{4.2.18a}$$

or also

$$\boldsymbol{x}_\infty^{\mathrm{t}}\boldsymbol{M}\boldsymbol{x}_\infty = 0 \tag{4.2.18b}$$

If we employ the kinematically consistent mass concept, then we obtain from equ. (4.2.18b)

$$\begin{aligned}\boldsymbol{x}_\infty^{\mathrm{t}}\boldsymbol{M}\boldsymbol{x}_\infty &= \boldsymbol{x}_\infty^{\mathrm{t}}\sum_{\mathrm{g}=1}^{\mathrm{m}}\mathbf{a}_{\mathrm{g}}^{\mathrm{t}}\Big[\int_{V_{\mathrm{g}}}\rho\boldsymbol{\omega}^{\mathrm{t}}\boldsymbol{\omega}\,\mathrm{d}V\Big]\mathbf{a}_{\mathrm{g}}\boldsymbol{x}_\infty\\ &= \sum_{\mathrm{g}=1}^{\mathrm{m}}\int_{V_{\mathrm{g}}}\rho\boldsymbol{u}_\infty^{\mathrm{t}}\boldsymbol{u}_\infty\,\mathrm{d}V\end{aligned} \tag{4.2.19}$$

where $\boldsymbol{u}_\infty$ denotes the nodal displacement vector corresponding to the vector $\boldsymbol{x}_\infty$. To satisfy the condition of equ. (4.2.18b), either the density $\rho = 0$ or the displacement vector $\boldsymbol{u}_\infty = \boldsymbol{o}$ must hold at all points. An inertialess motion $\boldsymbol{x}_\infty$ of the system is only possible if certain regions of the structure are mass-free, clearly implying an unrealistic idealisation.

A slightly different issue arises for simple point-mass matrices. As is shown in fig. 4.2.4, inertialess motions are possible for such point-mass models. The eigenvalue problem

$$\boldsymbol{M}\boldsymbol{x}_{\mathrm{i}} = \boldsymbol{K}\boldsymbol{x}_{\mathrm{i}}\lambda_{\mathrm{i}} \tag{4.2.20}$$

indicates immediately that $\boldsymbol{x}_\infty$ represents a non-trivial solution which appertains to the eigenvalue

$$\lambda_\infty = \frac{1}{\omega_\infty^2} = 0 \tag{4.2.21}$$

It is assumed here that $\boldsymbol{K}\boldsymbol{x}_\infty$ is not a null vector, i.e. that $\boldsymbol{x}_\infty$ does not stand for a rigid-body motion. The result of equ. (4.2.21) can also be deduced from the Rayleigh quotient (equ. (4.1.11)) or the characteristic equation (4.1.4c) if written in the form

$$b_{\mathrm{n}} + b_{\mathrm{n}-1}\lambda + b_{\mathrm{n}-2}\lambda^2 + \ldots + b_0\lambda^{\mathrm{n}} = 0 \tag{4.2.22}$$

It follows from equ. (4.1.5) that for a singular mass matrix at least $b_{\mathrm{n}} = 0$, and hence not less than one simple solution $\lambda = 0$ exists. The latter can be tolerated in our scheme of eigenvalue analysis. In analogy to the degenerate solutions of the rigid body modes, this degenerate solution can, however, be eliminated at all

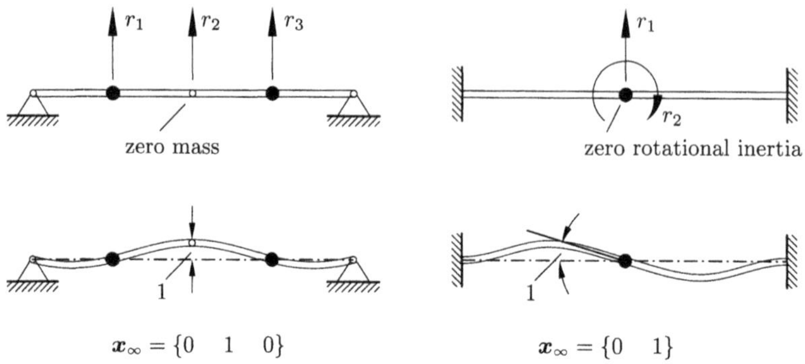

Fig. 4.2.4 Inertialess motions in point-mass models

times. For this purpose, we split the motion into inertia-dependent and inertialess modes, described by $\boldsymbol{T}_m$ and $\boldsymbol{T}_\infty$ respectively

$$\boldsymbol{r} = [\boldsymbol{T}_\infty \quad \boldsymbol{T}_m]\{\boldsymbol{r}_\infty \quad \boldsymbol{r}_m\} \tag{4.2.23}$$

If we consider simple massless degrees of freedom, as in the examples of fig. 4.2.4, then $\boldsymbol{T}_\infty$ and $\boldsymbol{T}_m$ are simply Boolean matrices. In this scheme, $\boldsymbol{r}_\infty$ represents degrees of freedom associated with zero mass and $\boldsymbol{r}_m$ those associated with a finite mass. Since

$$\boldsymbol{M}\boldsymbol{T}_\infty = \boldsymbol{O} \tag{4.2.24}$$

we obtain the transformed system in $\{\boldsymbol{r}_\infty \quad \boldsymbol{r}_m\}$ as

$$\begin{bmatrix} \boldsymbol{O} & \boldsymbol{O} \\ \boldsymbol{O}^{\mathrm{t}} & \boldsymbol{M}_{mm} \end{bmatrix} \begin{bmatrix} \ddot{\boldsymbol{r}}_\infty \\ \ddot{\boldsymbol{r}}_m \end{bmatrix} + \begin{bmatrix} \boldsymbol{K}_{\infty\infty} & \boldsymbol{K}_{\infty m} \\ \boldsymbol{K}_{m\infty} & \boldsymbol{K}_{mm} \end{bmatrix} \begin{bmatrix} \boldsymbol{r}_\infty \\ \boldsymbol{r}_m \end{bmatrix} = \begin{bmatrix} \boldsymbol{o} \\ \boldsymbol{o} \end{bmatrix} \tag{4.2.25}$$

By virtue of the first row, the inertialess motions can be eliminated

$$\boldsymbol{r}_\infty = -\boldsymbol{K}_{\infty\infty}^{-1}\boldsymbol{K}_{\infty m}\boldsymbol{r}_m \tag{4.2.26}$$

and we are left with a system of equations solely dependent on $\boldsymbol{r}_m$

$$\boldsymbol{M}_{mm}\ddot{\boldsymbol{r}}_m + [\boldsymbol{K}_{mm} - \boldsymbol{K}_{m\infty}\boldsymbol{K}_{\infty\infty}^{-1}\boldsymbol{K}_{\infty m}]\boldsymbol{r}_m = \boldsymbol{o} \tag{4.2.27}$$

where $\boldsymbol{M}_{mm}$ is now positive definite. This process of elimination is identical to static condensation; see equ. (3.2.33) of Vol. I(G).

This system offers nm=n$-$n$_\infty$ non-degenerate solutions $\lambda_{\mathrm{i}} = 1/\omega_{\mathrm{i}}^2$ and vectors $\boldsymbol{x}_{m\mathrm{i}}$ which once more may reproduce by superposition the total solution $\boldsymbol{r}_m$. Thus,

$$\boldsymbol{r}_m = \sum_{\mathrm{i}=1}^{\mathrm{nm}} c_{\mathrm{i}}\boldsymbol{x}_{m\mathrm{i}} \sin(\omega_{\mathrm{i}}t + \varphi_{\mathrm{i}}) \tag{4.2.28}$$

In terms of the original degrees of freedom, the response is then

$$\boldsymbol{r} = [\boldsymbol{T}_m - \boldsymbol{T}_\infty \boldsymbol{K}_{\infty\infty}^{-1} \boldsymbol{K}_{\infty m}] \sum_{\mathrm{i}=1}^{\mathrm{nm}} c_\mathrm{i} \boldsymbol{x}_{m\mathrm{i}} \sin(\omega_\mathrm{i} t + \varphi_\mathrm{i}) \tag{4.2.29}$$

Hence, the eigenvectors corresponding to the freedoms $\boldsymbol{r}$ are obtained by the reverse transformation

$$\underset{(\mathrm{n}\times\mathrm{nm})}{\boldsymbol{X}} = [\boldsymbol{T}_m - \boldsymbol{T}_\infty \boldsymbol{K}_{\infty\infty}^{-1} \boldsymbol{K}_{\infty m}] \underset{(\mathrm{nm}\times\mathrm{nm})}{\boldsymbol{X}_m} \tag{4.2.30}$$

Here, $\boldsymbol{X}$ is an upright rectangular matrix and contains the nm eigenvectors corresponding to the non-degenerate solutions $\lambda_\mathrm{i} \neq 0$. In place of equ. (4.2.29), the vector $\boldsymbol{r}$ is given by

$$\boldsymbol{r} = \sum_{\mathrm{i}=1}^{\mathrm{nm}} c_\mathrm{i} \boldsymbol{x}_\mathrm{i} \sin(\omega_\mathrm{i} t + \varphi_\mathrm{i}) \tag{4.2.31}$$

Thus, if we dispense with the reduction of the order of the system in the case of a singular mass matrix – which is possible by the technique suggested in the preceding section –, we observe that the eigenvalue problem

$$\boldsymbol{M}\boldsymbol{X} = \boldsymbol{K}\boldsymbol{X}\boldsymbol{\Lambda} \tag{4.2.32}$$

possesses n_∞ zero eigenvalues λ_i and corresponding vectors $\boldsymbol{x}_\mathrm{i}$; however, these can be ignored as being of no significance for the description of the free motion of a structure (equ. (4.2.31)). These solutions will, though, become relevant if external forces are applied to the system, an aspect to which we shall return later.

Up to now, we have treated the cases of a singular stiffness matrix and mass matrix separately. In practice, however, it is not unlikely that the two matrices may be both singular (e.g. free point-mass models; fig. 4.2.5). As a matter of fact, this should not cause any difficulties since the preceding two methods may be coupled. To this purpose, we split the motion into rigid-body ($\boldsymbol{T}_0$), inertialess ($\boldsymbol{T}_\infty$) and independent ($\boldsymbol{T}_i$) components. Thus,

$$\boldsymbol{r} = [\boldsymbol{T}_0 \quad \boldsymbol{T}_\infty \quad \boldsymbol{T}_i]\{\boldsymbol{r}_0 \quad \boldsymbol{r}_\infty \quad \boldsymbol{r}_i\} \tag{4.2.33}$$

The transformed equations of motion are then

$$\begin{bmatrix} \boldsymbol{M}_{00} & \boldsymbol{O} & \boldsymbol{M}_{0i} \\ \boldsymbol{O}^\mathrm{t} & \boldsymbol{O} & \boldsymbol{O} \\ \boldsymbol{M}_{i0} & \boldsymbol{O}^\mathrm{t} & \boldsymbol{M}_{ii} \end{bmatrix} \begin{bmatrix} \ddot{\boldsymbol{r}}_0 \\ \ddot{\boldsymbol{r}}_\infty \\ \ddot{\boldsymbol{r}}_i \end{bmatrix} + \begin{bmatrix} \boldsymbol{O} & \boldsymbol{O} & \boldsymbol{O} \\ \boldsymbol{O}^\mathrm{t} & \boldsymbol{K}_{\infty\infty} & \boldsymbol{K}_{\infty i} \\ \boldsymbol{O}^\mathrm{t} & \boldsymbol{K}_{i\infty} & \boldsymbol{K}_{ii} \end{bmatrix} \begin{bmatrix} \boldsymbol{r}_0 \\ \boldsymbol{r}_\infty \\ \boldsymbol{r}_i \end{bmatrix} = \begin{bmatrix} \boldsymbol{o} \\ \boldsymbol{o} \\ \boldsymbol{o} \end{bmatrix} \tag{4.2.34}$$

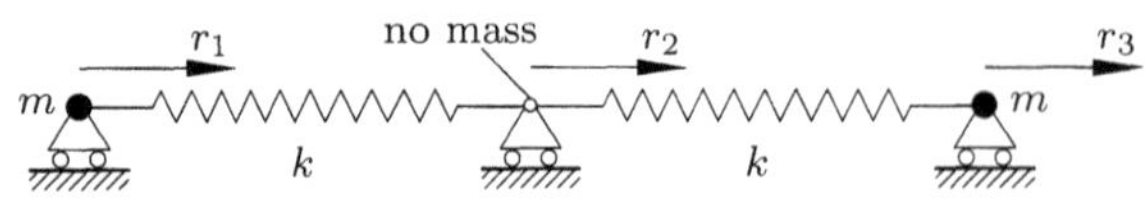

Fig. 4.2.5
Structure with singular mass and stiffness matrix arising in a meaningful dynamic problem

Provided that the rigid-body modes involve mass motions ($\boldsymbol{MT}_0 \neq \boldsymbol{O}, \boldsymbol{M}_{00}$ positive definite) and that the inertialess modes are not associated with rigid-body modes ($\boldsymbol{KT}_\infty \neq \boldsymbol{O}, \boldsymbol{K}_{\infty\infty}$ positive definite), we observe that $\boldsymbol{r}_0$ and $\boldsymbol{r}_\infty$ can be eliminated, reducing the system of equations of motion to a set of equations with positive definite matrices solely dependent on the independent degrees of freedom $\boldsymbol{r}_i$

$$[\boldsymbol{M}_{ii} - \boldsymbol{M}_{i0}\boldsymbol{M}_{00}^{-1}\boldsymbol{M}_{0i}]\ddot{\boldsymbol{r}}_i + [\boldsymbol{K}_{ii} - \boldsymbol{K}_{i\infty}\boldsymbol{K}_{\infty\infty}^{-1}\boldsymbol{K}_{\infty i}]\boldsymbol{r}_i = \boldsymbol{o} \tag{4.2.35}$$

In this manner, the degree of freedom r_1 in the example of fig. 4.2.5 could be suppressed in order to remove the rigid-body motion and r_2 could be omitted to eliminate the inertialess motion, leaving only a single independent degree of freedom r_{1i} in the direction of r_3. The corresponding transformation matrices are

$$\begin{aligned} \boldsymbol{T}_0 &= \{1 \quad 1 \quad 1\} \\ \boldsymbol{T}_\infty &= \{0 \quad 1 \quad 0\} \\ \boldsymbol{T}_i &= \{0 \quad 0 \quad 1\} \end{aligned}$$

The mass and stiffness matrices of the original system are

$$\boldsymbol{M} = m\lceil 1 \quad 0 \quad 1 \rfloor$$

and

$$\boldsymbol{K} = k\begin{bmatrix} 1 & -1 & 0 \\ & 2 & -1 \\ \text{sym} & & 1 \end{bmatrix}$$

Following the aforementioned transformation, these matrices become

$$\boldsymbol{T}^{\mathrm{t}}\boldsymbol{MT} = m\begin{bmatrix} 2 & 0 & 1 \\ & 0 & 0 \\ \text{sym} & & 1 \end{bmatrix}$$

and

$$\boldsymbol{T}^{\mathrm{t}}\boldsymbol{KT} = k\begin{bmatrix} 0 & 0 & 0 \\ & 2 & -1 \\ \text{sym} & & 1 \end{bmatrix}$$

Finally, the reduced vibration problem is

$$m\ddot{r}_{1i} + kr_{1i} = 0$$

with the non-degenerate solution

$$\lambda_1 = \frac{m}{k}$$
$$\boldsymbol{x}_{1i} = \{1\}$$
$$\boldsymbol{x}_1 = \{-1 \quad 0 \quad 1\}$$

Since this example contains one rigid body mode and one independent inertialess mode, the number of independent degrees of freedom can only be 1.

As a useful exercise for the reader, we propose that he may experiment with the suppression of the degree of freedom r_2, thus obtaining – seemingly – a standard system with two degrees of freedom. However, two possibilities of reduction are feasible in this case. One may either employ the transformation $\boldsymbol{T}_0$ (noting that the support is introduced to suppress a rigid-body motion) or the transformation $\boldsymbol{T}_\infty$ (here, the support is introduced on account of the inertialess motion). The reader will discover that in both cases, one of the reduced matrices is still singular and hence further reduction is possible.

Finally, we pose the question as to what is entailed if the system is qualified to execute all at once a combined rigid-body and an inertialess motion

$$\boldsymbol{M}\boldsymbol{x}_{0,\infty} = \boldsymbol{o} \qquad\qquad \boldsymbol{K}\boldsymbol{x}_{0,\infty} = \boldsymbol{o} \tag{4.2.30}$$

Then, naturally, the eigenvalue problem reads

$$\boldsymbol{M}\boldsymbol{x} = \boldsymbol{K}\boldsymbol{x}\lambda$$

and can be solved for any arbitrary real or complex root λ. Let us consider the simple example in fig. 4.2.6; the massless part of the structure can rotate freely as a rigid-body about the pin-joint and does not impose any elastic or inertial forces on the remaining part of the structure. Consequently, it can be removed. Such cases, which we classify as an inconsistent problem formulation, need not be discussed any further here.

Figure 4.2.7 summarises the degenerate solutions obtained so far.

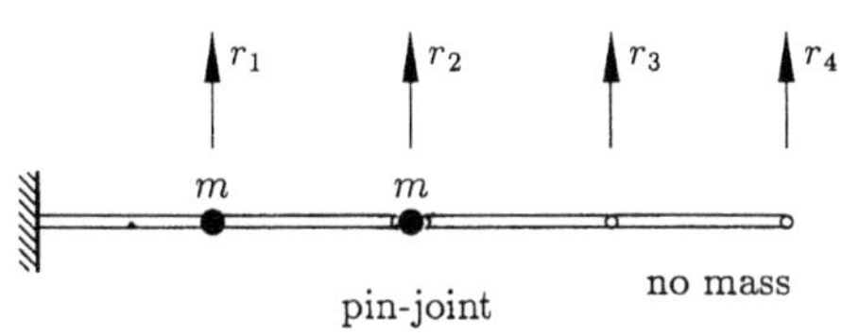

Fig. 4.2.6

Structure with a singular mass and stiffness matrix resulting from an unacceptable dynamic specification; the part linked to the joint can be removed as it is meaningless

<table>
<tr><th>singular stiffness matrix</th><th>singular mass matrix</th></tr>
<tr><td>structure has n_0 linearly independent rigid body modes</td><td>structure has n_∞ linearly independent inertialess modes</td></tr>
<tr><td>there exists an n_0-fold degenerate solution
$\lambda_i^{-1} = \omega_i^2 = 0$</td><td>there exists an n_∞-fold degenerate solution
$\lambda_i = 1/\omega_i^2 = 0$</td></tr>
<tr><td>explicit representation of rigid body modes
$\underset{(n\times n_0)}{\boldsymbol{T}_0}$ where $\boldsymbol{K}\boldsymbol{T}_0 = \boldsymbol{O},\ \boldsymbol{M}\boldsymbol{T}_0 \neq \boldsymbol{O}$</td><td>explicit representation of inertialess modes
$\underset{(n\times n_\infty)}{\boldsymbol{T}_\infty}$ where $\boldsymbol{M}\boldsymbol{T}_\infty = \boldsymbol{O},\ \boldsymbol{K}\boldsymbol{T}_\infty \neq \boldsymbol{O}$</td></tr>
<tr><td colspan="2">number of independent degrees of freedom
$n_i = n - n_0 - n_\infty$
$= \text{number of non-degenerate solutions}$</td></tr>
<tr><td colspan="2">transformation
$\underset{(n\times 1)}{\boldsymbol{r}} = [\underset{(n\times n_0)}{\boldsymbol{T}_0}\ \underset{(n\times n_\infty)}{\boldsymbol{T}_\infty}\ \underset{(n\times n_i)}{\boldsymbol{T}_i}]\{\underset{(n_0\times 1)}{\boldsymbol{r}_0}\ \underset{(n_\infty\times 1)}{\boldsymbol{r}_\infty}\ \underset{(n_i\times 1)}{\boldsymbol{r}_i}\}$</td></tr>
<tr><td colspan="2">reduced system
$[\boldsymbol{M}_{ii} - \boldsymbol{M}_{i0}\boldsymbol{M}_{00}^{-1}\boldsymbol{M}_{0i}]\ddot{\boldsymbol{r}}_i + [\boldsymbol{K}_{ii} - \boldsymbol{K}_{i\infty}\boldsymbol{K}_{\infty\infty}^{-1}\boldsymbol{K}_{\infty i}]\boldsymbol{r}_i = \boldsymbol{o}$</td></tr>
<tr><td colspan="2">backward transformation
$\boldsymbol{r} = [\boldsymbol{T}_0\ \ \boldsymbol{T}_\infty\ \ \boldsymbol{T}_i]\begin{bmatrix} -\boldsymbol{M}_{00}^{-1}\boldsymbol{M}_{0i} \\ -\boldsymbol{K}_{\infty\infty}^{-1}\boldsymbol{K}_{\infty i} \\ \boldsymbol{I} \end{bmatrix}\boldsymbol{r}_i + \begin{bmatrix} \boldsymbol{a}_0 + \boldsymbol{b}_0 t \\ \boldsymbol{o} \\ \boldsymbol{o} \end{bmatrix}$</td></tr>
</table>

Fig. 4.2.7 Degenerate solutions for eigenoscillations

Let us now turn to the final special case, namely that of equal (non-degenerate) eigenvalues. We have already discussed multiple eigenvalues of the degenerate type. If a characteristic polynomial possesses multiple solutions λ_i, then not one, but in general several equations of the system

$$[\boldsymbol{M} - \lambda_i \boldsymbol{K}]\boldsymbol{x}_i = \boldsymbol{o} \tag{4.2.37}$$

will become linearly dependent. If the eigenvalue λ_i has a multiplicity of p_i, then there are p_i superfluous linearly dependent equations. Mathematically, we say that the system $[\boldsymbol{M} - \lambda_i \boldsymbol{K}]$ has a rank deficiency of p_i. This characteristic attribute does not apply to arbitrary matrices, but holds for symmetric matrices $\boldsymbol{M}$ and $\boldsymbol{K}$, one of which at least must be positive definite and the other positive semi-definite. If we eliminate p_i over-determined equations in equ. (4.2.37), we obtain $(n - p_i)$ linearly independent equations. We thus obtain a flat, rectangular system

$$\underset{(n-p_i\times n)}{\boldsymbol{A}}\ \underset{(n\times 1)}{\boldsymbol{x}_i} = \underset{(n-p_i\times 1)}{\boldsymbol{o}} \tag{4.2.38}$$

In the case of a simple eigenvalue λ_i, we were free to prescribe one element in the vector $\boldsymbol{x}_i$. But here, in contrast, p_i elements in $\boldsymbol{x}_i$ can be chosen arbitrarily, while the remaining $(n - p_i)$ values can be determined through equ. (4.2.38). Instead of selecting p_i free variables in the vector $\boldsymbol{x}_i$ arbitrarily, we may alternatively select p_i linearly independent vectors and prescribe in each of them *one* entry x_{ij} and set the remaining free entries equal to zero, the remainder $(n - p_i)$ values being determined from equ. (4.2.38). Then, any linear combination of these vectors is a solution

$$\boldsymbol{x}_i = c_1 \boldsymbol{x}_i^{(1)} + c_2 \boldsymbol{x}_i^{(2)} + \ldots + c_{p_i} \boldsymbol{x}_i^{(p_i)} \tag{4.2.39}$$

We can thus say that the p_i-fold eigenvalue λ_i also possesses p_i linearly independent eigenvectors $\boldsymbol{x}_i^{(k)}$. As the reader will remember, the eigenvectors belonging to a single eigenvalue have one useful characteristic which is that they can simultaneously diagonalise both $\boldsymbol{M}$ and $\boldsymbol{K}$. The question is: does the same apply to multiple eigenvalues? To this purpose, let us consider the special eigenvalue problem

$$\boldsymbol{U}^{-t} \boldsymbol{M} \boldsymbol{U}^{-1} \overline{\boldsymbol{X}} = \overline{\boldsymbol{X}} \boldsymbol{\Lambda} \tag{4.2.40}$$

with

$$\overline{\boldsymbol{X}} = \boldsymbol{U} \boldsymbol{X} \tag{4.2.41}$$

The spectral analysis of equ. (4.2.40) yields a complete set of $(n \times n)$ eigenvectors $\overline{\boldsymbol{X}}$ and n eigenvalues λ_i, of which some may be equal (multiple). The eigenvectors $\overline{\boldsymbol{x}}_i$ and $\overline{\boldsymbol{x}}_j$, appertaining to different eigenvalues λ_i and λ_j, are orthogonal since

$$\overline{\boldsymbol{x}}_i^t \overline{\boldsymbol{x}}_j = \boldsymbol{x}_i^t \boldsymbol{U}^t \boldsymbol{U} \boldsymbol{x}_j = \boldsymbol{x}_i^t \boldsymbol{K} \boldsymbol{x}_j = 0 \tag{4.2.42}$$

which we have already proved in equ. (4.1.26). In the case of multiple eigenvalues λ_i, we are at liberty to recombine the linearly independent vectors $\overline{\boldsymbol{x}}_i^{(k)}$ in such a form that they are also orthogonal (for example using Schmidt's method). Hence, it is always possible to satisfy the condition

$$\overline{\boldsymbol{X}}^t \overline{\boldsymbol{X}} = \boldsymbol{I} = \boldsymbol{X}^t \boldsymbol{K} \boldsymbol{X} \tag{4.2.43}$$

from which we deduce

$$\overline{\boldsymbol{X}}^t \boldsymbol{U}^{-t} \boldsymbol{M} \boldsymbol{U}^{-1} \overline{\boldsymbol{X}} = \boldsymbol{\Lambda} = \boldsymbol{X}^t \boldsymbol{M} \boldsymbol{X} \tag{4.2.44}$$

This proves that the eigenvectors appertaining to multiple roots may be orthonormalised with respect to the stiffness matrix $\boldsymbol{K}$ and then possess the same characteristics as standard eigenvectors belonging to single eigenvalues. What, however, happens to the solution to the differential equation in $\boldsymbol{r}$? Indeed, we may confirm that the technique is the same as in the case of simple eigenvalues. We demonstrate this by introducing new variables

$$\underset{(n\times 1)}{\boldsymbol{r}(t)} = \underset{(n\times n)}{\boldsymbol{X}} \; \underset{(n\times 1)}{\boldsymbol{\eta}(t)} \tag{4.2.45}$$

In place of the standard equation

$$\boldsymbol{M}\ddot{\boldsymbol{r}} + \boldsymbol{K}\boldsymbol{r} = \boldsymbol{o} \tag{4.2.46a}$$

we obtain after pre-multiplication by $\boldsymbol{X}^{\mathrm{t}}$ the alternative system

$$\boldsymbol{\Lambda}\ddot{\boldsymbol{\eta}} + \boldsymbol{\eta} = \boldsymbol{o} \tag{4.2.46b}$$

assuming, of course, that $\boldsymbol{X}$ is orthonormal to $\boldsymbol{K}$. The above equation represents a system of n uncoupled differential equations

$$\lambda_{\mathrm{i}}\ddot{\eta}_{\mathrm{i}} + \eta_{\mathrm{i}} = 0, \qquad \mathrm{i} = 1 \dots \mathrm{n} \tag{4.2.46c}$$

For each of them, the solution can immediately be written down

$$\eta_{\mathrm{i}} = c_{\mathrm{i}} \sin(\omega_{\mathrm{i}} t + \varphi_{\mathrm{i}}) \tag{4.2.47}$$

It makes no difference whether $\lambda_{\mathrm{i}} = \omega_{\mathrm{i}}^{-2}$ is a multiple or a single eigenvalue. The overall solution of $\boldsymbol{r}$ becomes hence

$$\boldsymbol{r} = \sum_{\mathrm{i}=1}^{\mathrm{n}} c_{\mathrm{i}} \boldsymbol{x}_{\mathrm{i}} \sin(\omega_{\mathrm{i}} t + \varphi_{\mathrm{i}}) \tag{4.2.48}$$

As an instructive exercise, we consider the simple example shown in fig. 4.2.8. Here, the mass matrix is diagonal

$$\boldsymbol{M} = m\boldsymbol{I}_4$$

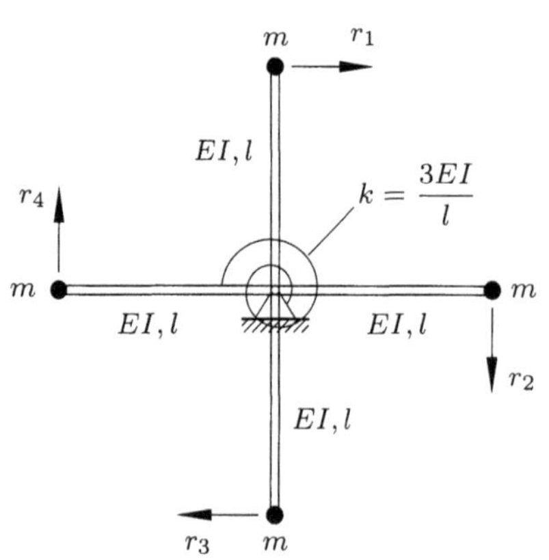

Fig. 4.2.8
Oscillator with multiple eigenfrequencies (eigenvalues)

but the stiffness matrix is fully populated

$$\boldsymbol{K} = \frac{k}{5l^2}\begin{bmatrix} 4 & -1 & -1 & -1 \\ -1 & 4 & -1 & -1 \\ -1 & -1 & 4 & -1 \\ -1 & -1 & -1 & 4 \end{bmatrix}$$

Introducing the abbreviation

$$\overline{\lambda} = \frac{k}{5ml^2\omega^2} = \frac{k\lambda}{5ml^2}$$

we obtain a homogeneous system of equations for the eigenvalues $\overline{\lambda}_i$ and eigenvectors $\boldsymbol{x}_i$

$$\begin{bmatrix} \overline{\lambda}^{-1} - 4 & 1 & 1 & 1 \\ 1 & \overline{\lambda}^{-1} - 4 & 1 & 1 \\ 1 & 1 & \overline{\lambda}^{-1} - 4 & 1 \\ 1 & 1 & 1 & \overline{\lambda}^{-1} - 4 \end{bmatrix}\begin{bmatrix} x_1 \\ x_2 \\ x_3 \\ x_4 \end{bmatrix} = \boldsymbol{o}$$

Setting the determinant of the matrix to zero, we deduce the eigenvalues

$$\overline{\lambda}_{1,2,3}^{-1} = 5 \quad \text{or} \quad \lambda_{1,2,3} = \frac{1}{\omega^2} = \frac{ml^2}{k} \qquad \text{(triple)}$$

$$\overline{\lambda}_4^{-1} = 1 \qquad \lambda_4 = \frac{1}{\omega^2} = \frac{5ml^2}{k} \qquad \text{(single)}$$

The validity of these solutions may immediately be proved by substitution of the roots into the homogeneous system. Indeed, we obtain linearly dependent equations. Thus, for $\overline{\lambda}_{1,2,3}$

$$\begin{bmatrix} 1 & 1 & 1 & 1 \\ 1 & 1 & 1 & 1 \\ 1 & 1 & 1 & 1 \\ 1 & 1 & 1 & 1 \end{bmatrix}\begin{bmatrix} x_1 \\ x_2 \\ x_3 \\ x_4 \end{bmatrix}_{1,2,3} = \boldsymbol{o}$$

We observe that as a result of its triple multiplicity, three of the equations are supernumerary. We recognise that the matrix is three-fold degenerate (loss of rank 3). Each eigenvector can be put in the form

$$\boldsymbol{x}_{1,2,3} = \{c_1 \quad c_2 \quad c_3 \quad -(c_1 + c_2 + c_3)\}$$

with arbitrary constants c_1, c_2, c_3. In fact, the following three linearly independent vectors may also be considered eigenvectors, as well as linear combinations thereof

$$\boldsymbol{x}_1 = \{1 \quad 0 \quad 0 \quad -1\}$$

$$\boldsymbol{x}_2 = \{0 \quad 1 \quad 0 \quad -1\}$$

$$\boldsymbol{x}_3 = \{0 \quad 0 \quad 1 \quad -1\}$$

On the other hand, the single eigenvalue $\lambda_4^{-1} = 1$ yields

$$\begin{bmatrix} -3 & 1 & 1 & 1 \\ 1 & -3 & 1 & 1 \\ 1 & 1 & -3 & 1 \\ 1 & 1 & 1 & -3 \end{bmatrix} \begin{bmatrix} x_1 \\ x_2 \\ x_3 \\ x_4 \end{bmatrix}_4 = \boldsymbol{o}$$

Here, only one equation is superfluous (the first line is, for example, the negative sum of the last three). If x_1 is considered a free variable (and we put $x_1 = 1$), we obtain

$$\boldsymbol{x}_4 = \{1 \quad 1 \quad 1 \quad 1\}$$

which represents a point symmetric oscillation. Since $\lambda_4 \neq \lambda_{1,2,3}$ we have

$$\overline{m}_{\mathrm{i}4} = \overline{m}_{4\mathrm{j}} = 0 \qquad \text{(see equ. (4.1.23))}$$

$$\overline{k}_{\mathrm{i}4} = \overline{k}_{4\mathrm{j}} = 0, \quad \mathrm{i,j} \neq 4$$

in the transformed matrices. The remaining elements of the mass matrix and stiffness matrix can only be diagonalised if $\boldsymbol{x}_{1,2,3}$ are orthogonal to $\boldsymbol{K}$. This is not the case here. In fact, the corresponding matrices are

$$\boldsymbol{X}^{\mathrm{t}}\boldsymbol{M}\boldsymbol{X} = m \begin{bmatrix} 2 & 1 & 1 & 0 \\ 1 & 2 & 1 & 0 \\ 1 & 1 & 2 & 0 \\ 0 & 0 & 0 & 4 \end{bmatrix}$$

and

$$\boldsymbol{X}^{\mathrm{t}}\boldsymbol{K}\boldsymbol{X} = \frac{k}{5l^2} \begin{bmatrix} 10 & 5 & 5 & 0 \\ 5 & 10 & 5 & 0 \\ 5 & 5 & 10 & 0 \\ 0 & 0 & 0 & 4 \end{bmatrix}$$

In contrast, if we choose as eigenvectors the set

$$\boldsymbol{X}_{Ko} = \sqrt{\frac{l^2}{12k}} \begin{bmatrix} \sqrt{6} & -\sqrt{2} & -1 & \sqrt{15} \\ 0 & 2\sqrt{2} & -1 & \sqrt{15} \\ 0 & 0 & 3 & \sqrt{15} \\ -\sqrt{6} & -\sqrt{2} & -1 & \sqrt{15} \end{bmatrix}$$

we confirm its orthonormalisation with respect to $\boldsymbol{K}$

$$\boldsymbol{X}^{\mathrm{t}}\boldsymbol{K}\boldsymbol{X} = \boldsymbol{I}_4$$

and deduce with

$$\boldsymbol{X}^{\mathrm{t}}\boldsymbol{M}\boldsymbol{X} = \frac{ml^2}{k}\lceil 1 \quad 1 \quad 1 \quad 5 \rfloor = \boldsymbol{\Lambda} = \boldsymbol{\Omega}^{-2}$$

a diagonal matrix for the eigenvalues or the inverses of the angular frequencies. In this elementary example, it would have been much simpler still to work with $\boldsymbol{M}$-orthonormal eigenvectors. The reader should ponder why this is so and then execute the corresponding calculations.

Supplement to 4.2

*4.2 Extraction of an orthogonal complement $\boldsymbol{T}_e$ out of $\boldsymbol{M}\boldsymbol{T}_0$ using Householder's method

We have already demonstrated that the system in equ. (4.2.7) may be completely decoupled by the assumed introduction of statically determinate supports into a system of rigid body motions and a system of elastic deformations. As a result, the mass matrix $\boldsymbol{M}_{ee}$ and stiffness matrix $\boldsymbol{K}_{ee}$ become fully populated, whilst they are sparsely populated for the case of given supports. The decoupling requires that

$$\boldsymbol{M}_{e0} = \boldsymbol{T}_e^{\mathrm{t}}\boldsymbol{M}\boldsymbol{T}_0 = \underset{(\mathrm{ne}\times\mathrm{n})}{\boldsymbol{T}_e^{\mathrm{t}}}\underset{(\mathrm{n}\times\mathrm{n0})}{\boldsymbol{Y}} = \boldsymbol{O} \tag{*4.2.1}$$

where

$$\underset{(\mathrm{n}\times\mathrm{n0})}{\boldsymbol{Y}} = \underset{(\mathrm{n}\times\mathrm{n})}{\boldsymbol{M}}\underset{(\mathrm{n}\times\mathrm{n0})}{\boldsymbol{T}_0} \tag{*4.2.2}$$

In what follows, we propose a method of setting up $\boldsymbol{T}_e$ for known vectors $\boldsymbol{Y}$, so that equ. (*4.2.1) is satisfied. To this purpose, we assume that it is possible to pre-multiply $\boldsymbol{Y}$ by a matrix $\boldsymbol{Q}$, such that

$$\underset{(\mathrm{n}\times\mathrm{n})}{\boldsymbol{Q}}\underset{(\mathrm{n}\times\mathrm{n0})}{\boldsymbol{Y}} = \begin{bmatrix} \underset{(\mathrm{n0}\times\mathrm{n0})}{\boldsymbol{U}} \\ \boldsymbol{O} \end{bmatrix} \tag{*4.2.3}$$

whereby $\boldsymbol{U}$ becomes an upper triangular matrix. Were this operation feasible and should $\boldsymbol{Q}$ have linearly independent rows, then the lower ne=n−n0 rows of $\boldsymbol{Q}$ indeed form the matrix $\boldsymbol{T}_e^{\mathrm{t}}$.

Adopting the Householder method, we now attempt to establish the desired form in a step-by-step procedure. Assuming that (k−1) columns have already been computed in (k−1) steps, the column k is to be set up in the desired form within the k-th step

$$\boldsymbol{Y}^{(\mathrm{k}-1)} = \begin{bmatrix} \times & \times & \times & \times & \times & \times \\ 0 & \times & \times & \times & \times & \times \\ 0 & 0 & \times & \times & \times & \times \\ 0 & 0 & 0 & \times & \times & \times \\ 0 & 0 & 0 & \times & \times & \times \\ 0 & 0 & 0 & \times & \times & \times \\ 0 & 0 & 0 & \times & \times & \times \\ 0 & 0 & 0 & \times & \times & \times \end{bmatrix} \Rightarrow \boldsymbol{Y}^{(\mathrm{k})} = \begin{bmatrix} \times & \times & \times & \times & \times & \times \\ 0 & \times & \times & \times & \times & \times \\ 0 & 0 & \times & \times & \times & \times \\ 0 & 0 & 0 & \times & \times & \times \\ 0 & 0 & 0 & \boxed{0} & \times & \times \\ 0 & 0 & 0 & 0 & \times & \times \\ 0 & 0 & 0 & 0 & \times & \times \\ 0 & 0 & 0 & 0 & \times & \times \end{bmatrix} = \boldsymbol{Q}^{(\mathrm{k})}\boldsymbol{Y}^{(\mathrm{k}-1)} \tag{*4.2.4}$$

(column k marked above the fourth column of each matrix)

Providing that this procedure is possible, we complete our task in n steps and obtain

$$\boldsymbol{Q} = \boldsymbol{Q}^{(\mathrm{n0})}\boldsymbol{Q}^{(\mathrm{n0}-1)}\ldots\boldsymbol{Q}^{(2)}\boldsymbol{Q}^{(1)} = \begin{bmatrix} \boldsymbol{Q}_0 \\ \boldsymbol{T}_e^{\mathrm{t}} \end{bmatrix} \tag{*4.2.5}$$

In order to realise the operation outlined in equ. (*4.2.4), we employ the following matrix

$$\underset{(\mathrm{n}\times\mathrm{n})}{\boldsymbol{Q}^{(\mathrm{k})}} = \boldsymbol{I}_{\mathrm{n}} - 2\underset{(\mathrm{n}\times 1)(1\times\mathrm{n})}{\boldsymbol{w}\;\boldsymbol{w}^{\mathrm{t}}} \tag{*4.2.6}$$

in which

$$\boldsymbol{w}^{\mathrm{t}}\boldsymbol{w} = 1 \tag{*4.2.7}$$

Clearly, the matrix $\boldsymbol{Q}^{(\mathrm{k})}$ is symmetric and orthonormal

$$[\boldsymbol{Q}^{(\mathrm{k})}]^{\mathrm{t}} = \boldsymbol{Q}^{(\mathrm{k})} \tag{*4.2.8}$$

$$[\boldsymbol{Q}^{(\mathrm{k})}]^{\mathrm{t}}\boldsymbol{Q}^{(\mathrm{k})} = \boldsymbol{Q}^{(\mathrm{k})}\boldsymbol{Q}^{(\mathrm{k})} = \boldsymbol{Q}^{(\mathrm{k})}[\boldsymbol{Q}^{(\mathrm{k})}]^{\mathrm{t}} = \boldsymbol{I}_{\mathrm{n}} \tag{*4.2.9}$$

Substituting the matrix of equ. (*4.2.6) in the procedure of equ. (*4.2.4), we find

$$\begin{aligned} \boldsymbol{Y}^{(\mathrm{k})} &= \boldsymbol{Y}^{(\mathrm{k}-1)} - 2\boldsymbol{w}\boldsymbol{w}^{\mathrm{t}}\boldsymbol{Y}^{(\mathrm{k}-1)} \\ &= \boldsymbol{Y}^{(\mathrm{k}-1)} - 2\boldsymbol{w}\boldsymbol{p}^{\mathrm{t}} \end{aligned} \tag{*4.2.10}$$

where

$$\underset{(\mathrm{n0}\times 1)}{\boldsymbol{p}} = \underset{(\mathrm{n0}\times\mathrm{n})}{[\boldsymbol{Y}^{(\mathrm{k}-1)}]^{\mathrm{t}}}\;\underset{(\mathrm{n}\times 1)}{\boldsymbol{w}} \tag{*4.2.11}$$

Thus, for any column vector j of the array $\boldsymbol{Y}^{(\mathrm{k})}$, we have

$$\boldsymbol{y}_{\mathrm{j}}^{(\mathrm{k})} = \boldsymbol{y}_{\mathrm{j}}^{(\mathrm{k}-1)} - 2p_{\mathrm{j}}\boldsymbol{w}\;; \qquad \mathrm{j} = 1\ldots\mathrm{n0} \tag{*4.2.12}$$

and

$$p_j = [\boldsymbol{y}_j^{(k-1)}]^t \boldsymbol{w} \qquad (*4.2.13)$$

Hence, in this operation, a certain multiple of the vector $\boldsymbol{w}$ will simply be subtracted from every column vector. At the same time, we desire to leave the first (k−1) columns, once completed, unaffected. This can clearly be achieved if p_j ($j = 1, k-1$) is zero. The vectors $\boldsymbol{y}_j^{(k-1)}$ ($j = 1, k-1$) possess, at most up to the element (k−1), non-zero elements. Consequently, equ. (*4.2.13) demands that the first (k−1) terms in $\boldsymbol{w}$ must be equal to zero in order to exclude coupling. We have now achieved our purpose, namely

$$w_i = 0 \qquad i = 1 \ldots k-1$$

and thence

$$\boldsymbol{w} = \{0 \quad 0 \quad \ldots \quad 0 \quad w_k \quad w_{k+1} \quad \ldots \quad w_n\} \qquad (*4.2.14)$$

The k-th column is now our working vector, to which equ. (*4.2.12) applies. Here, though, the aim is to nullify all $y_{ik}^{(k)}$ elements for $i = k+1$ to n. This condition immediately yields

$$w_i = \frac{y_{ik}^{(k-1)}}{2p_k}\,; \qquad i = k+1 \ldots n \qquad (*4.2.15)$$

This implies that, with the exception of w_k, all non-zero elements of the $\boldsymbol{w}$ vector are proportional to the corresponding elements in the column vector $\boldsymbol{y}_k^{(k-1)}$. Although the terms p_k and w_k are still unknown, we have two additional conditions to determine these terms. Thus, we deduce immediately from equs. (*4.2.7) and (*4.2.15) that

$$1 = w_k^2 + \left(\frac{1}{2p_k}\right)^2 \sum_{i=k+1}^{n} \left(y_{ik}^{(k-1)}\right)^2 \qquad (*4.2.16)$$

Also from equs. (*4.2.13) and (*4.2.15) we obtain

$$\begin{aligned} p_k &= \left[\boldsymbol{y}_k^{(k-1)}\right]^t \boldsymbol{w} \\ &= y_{kk}^{(k-1)} w_k + \frac{1}{2p_k} \sum_{i=k+1}^{n} \left(y_{ik}^{(k-1)}\right)^2 \end{aligned} \qquad (*4.2.17)$$

For the printer's convenience, we introduce the following abbreviation

$$s^2 = \sum_{i=k}^{n} \left(y_{ik}^{(k-1)}\right)^2 \qquad (*4.2.18)$$

In this way we find the two additional equations

$$4p_{\mathrm{k}}^2 = (2p_{\mathrm{k}}w_{\mathrm{k}})^2 - \left(y_{\mathrm{kk}}^{(\mathrm{k}-1)}\right)^2 + s^2 \tag{*4.2.19}$$

$$4p_{\mathrm{k}}^2 = 2y_{\mathrm{kk}}^{(\mathrm{k}-1)}(2p_{\mathrm{k}}w_{\mathrm{k}}) - 2\left(y_{\mathrm{kk}}^{(\mathrm{k}-1)}\right)^2 + 2s^2 \tag{*4.2.20}$$

Subtracting (*4.2.19) from (*4.2.20), we have

$$\left(2p_{\mathrm{k}}w_{\mathrm{k}} - y_{\mathrm{kk}}^{(\mathrm{k}-1)}\right)^2 = s^2 \tag{*4.2.21}$$

or

$$2p_{\mathrm{k}}w_{\mathrm{k}} = y_{\mathrm{kk}}^{(\mathrm{k}-1)} \pm s \tag{*4.2.22}$$

Substituting this expression into equ. (*4.2.20), we obtain the ultimate expression

$$p_{\mathrm{k}} = s\sqrt{\frac{1}{2}\left(1 \pm \frac{y_{\mathrm{kk}}^{(\mathrm{k}-1)}}{s}\right)} \tag{*4.2.23}$$

The sign has to be selected in such a way that no entries are nullified. Since the proportionality factor $2p_{\mathrm{k}}$ appears in every w_{i} element, it can be extracted and we write

$$\begin{aligned}\boldsymbol{w} &= (2p_{\mathrm{k}})^{-1}\boldsymbol{u} \\ &= (2p_{\mathrm{k}})^{-1}\{0 \quad 0 \quad \ldots \quad 0 \quad y_{\mathrm{kk}}^{(\mathrm{k}-1)} \pm s \quad y_{\mathrm{k+1,k}}^{(\mathrm{k}-1)} \quad \ldots \quad y_{\mathrm{nk}}^{(\mathrm{k}-1)}\}\end{aligned} \tag{*4.2.24}$$

Applying the vector $\boldsymbol{u}$ instead of $\boldsymbol{w}$ in equ. (4.2.10), we obtain

$$\boldsymbol{Y}^{(\mathrm{k})} = \boldsymbol{Y}^{(\mathrm{k}-1)} - \frac{1}{2p_{\mathrm{k}}^2}\boldsymbol{u}\boldsymbol{u}^{\mathrm{t}}\boldsymbol{Y}^{(\mathrm{k}-1)} \tag{*4.2.25}$$

In this way, we circumvent the square root operation in equ. (*4.2.23).

Finally, we demonstrate the Householder method on the simple example already shown in fig. 4.2.3 (left-hand side). We have already computed

$$\boldsymbol{Y}^{(1)} = \boldsymbol{M}\boldsymbol{T}_0 = m\{1 \quad 2 \quad 1\}$$

in which the factor m is insignificant and can be omitted here. Clearly, we have

$$s^2 = 6$$

and therefore

$$2p_1^2 = 6 + \sqrt{6}$$

Also,

$$\boldsymbol{u} = \{(1+\sqrt{6}) \quad 2 \quad 1\}$$

The matrix $\boldsymbol{Q}$ is hence simply

$$\boldsymbol{Q} = \boldsymbol{Q}^{(1)} = \boldsymbol{I} - \frac{1}{2p_1^2}\boldsymbol{u}\boldsymbol{u}^{\mathrm{t}}$$

$$= \frac{1}{6+\sqrt{6}}\begin{bmatrix} -(1+\sqrt{6}) & -2(1+\sqrt{6}) & -(1+\sqrt{6}) \\ -2(1+\sqrt{6}) & 2+\sqrt{6} & -2 \\ -(1+\sqrt{6}) & -2 & 5+\sqrt{6} \end{bmatrix}$$

Consequently, the desired transformation matrix $\boldsymbol{T}_e$ is given by

$$\boldsymbol{T}_e = \frac{1}{6+\sqrt{6}}\begin{bmatrix} -2(1+\sqrt{6}) & -(1+\sqrt{6}) \\ 2+\sqrt{6} & -2 \\ -2 & 5+\sqrt{6} \end{bmatrix}$$

The reader can easily confirm that this matrix is orthogonal to $\boldsymbol{Y}$. The described procedure can also be extended to hypermatrices; see [4.4]

4.3 Solution methods in spectral analysis

4.3.1 Determination of the largest eigenvalue (the smallest eigenfrequency) and the corresponding eigenvector: vector iteration

Knowledge of the fundamental frequency (or the first characteristic vibration mode) of a structure is often a necessary requirement when assessing its dynamic behaviour. This can be achieved by means of a very simple vector iteration technique. To this purpose, we consider the special eigenvalue problem

$$\boldsymbol{A}\overline{\boldsymbol{X}} = \overline{\boldsymbol{X}}\boldsymbol{\Lambda} \tag{4.3.1a}$$

$$\boldsymbol{A}\overline{\boldsymbol{x}}_1 = \overline{\boldsymbol{x}}_1\lambda_1 \tag{4.3.1b}$$

where $\overline{\boldsymbol{X}}$ is assumed to be orthonormal (i.e. $\boldsymbol{X}$ is orthonormal with respect to $\boldsymbol{K}$)

$$\overline{\boldsymbol{X}}^{\mathrm{t}}\overline{\boldsymbol{X}} = \boldsymbol{I} \tag{4.3.2}$$

Hence, using equ. (4.3.1a), we establish the following expression for the symmetric dynamic matrix $\boldsymbol{A}$

$$\boldsymbol{A} = \overline{\boldsymbol{X}}\boldsymbol{\Lambda}\overline{\boldsymbol{X}}^{\mathrm{t}} \tag{4.3.3}$$

To comprehend this method, it may suffice to mention that n eigenvectors $\overline{\boldsymbol{x}}_i$ are linearly independent as corroborated by equ. (4.3.2). Any arbitrary (n × 1)-vector $\overline{\boldsymbol{x}}_1^{(k)}$ can thus be expressed as a linear combination of these eigenvectors $\overline{\boldsymbol{x}}_i^{(k)}$. Thus, we have

$$\overline{\boldsymbol{x}}_1^{(k)} = \sum_{i=1}^{n} c_i \overline{\boldsymbol{x}}_i = \overline{\boldsymbol{X}}\boldsymbol{c} \tag{4.3.4}$$

where $\boldsymbol{c}$ represents a column vector

$$\boldsymbol{c} = \{c_1 \quad c_2 \quad \dots \quad c_n\} \tag{4.3.5}$$

The scalar factors c_i are, of course, initially unknown because $\overline{\boldsymbol{X}}$ is not yet available. However, if we post-multiply the dynamic matrix $\boldsymbol{A}$ by the vector $\overline{\boldsymbol{x}}_1^{(k)}$ and use the formulations in equs. (4.3.3) and (4.3.2), we get

$$\begin{aligned}\overline{\boldsymbol{x}}_1^{(k+1)} = \boldsymbol{A}\overline{\boldsymbol{x}}_1^{(k)} &= \overline{\boldsymbol{X}}\boldsymbol{\Lambda}\overline{\boldsymbol{X}}^t\overline{\boldsymbol{X}}\boldsymbol{c} = \overline{\boldsymbol{X}}\boldsymbol{\Lambda}\boldsymbol{c} \\ &= \sum_{i=1}^{n} c_i \lambda_i \overline{\boldsymbol{x}}_i\end{aligned} \tag{4.3.6}$$

The resulting vector $\overline{\boldsymbol{x}}_1^{(k+1)}$ contains as factors $(c_i\lambda_i)$, which are amplified as against the factors c_i of $\overline{\boldsymbol{x}}_1^{(k)}$ by λ_i. It is thus evident that the eigenvector appertaining to the largest eigenvalue (e.g. λ_1) is much more predominant in $\overline{\boldsymbol{x}}_1^{(k+1)}$ than in $\overline{\boldsymbol{x}}_1^{(k)}$. By repeat multiplication with the matrix $\boldsymbol{A}$, we can augment the participation of $\overline{\boldsymbol{x}}_1$ and ultimately achieve in this way a convergent procedure yielding in the limit the eigenvector $\overline{\boldsymbol{x}}_1$. However, this procedure will fail if the largest eigenvalue possesses a multiplicity p, in which case the associated p eigenvectors are equally magnified. The result is

$$\overline{\boldsymbol{x}}_1^{(k+1)} = \boldsymbol{A}\overline{\boldsymbol{x}}_1^{(k)} = \lambda_1 \sum_{i=1}^{p} c_i \overline{\boldsymbol{x}}_i + \sum_{i=p+1}^{n} \lambda_i c_i \overline{\boldsymbol{x}}_i \tag{4.3.7}$$

The iterated vector $\overline{\boldsymbol{x}}_1^{(k+1)}$ thus contains a magnified combination of p eigenvectors belonging to the multiple eigenvalue λ_1 but with the same participating factors as in $\overline{\boldsymbol{x}}_1^{(k)}$. By a single iteration, we obtain only one of the p linearly independent eigenvectors. To find all other eigenvectors appertaining to λ_1, we must re-run the iteration process with a new vector which is orthogonal to the preceding eigenvectors already found.

Although it is always possible to obtain a computationally meaningful eigenvector appertaining to exactly equal maximum eigenvalues, the situation is much more critical for almost equal (close) maximum eigenvalues. Vector iteration separates the corresponding eigenvectors only very slowly for they are amplified with a quasi-identical intensity. In such a situation, the extension of the simple vector iteration technique to simultaneous vector iteration is an absolute necessity; here, the reader should also consult section 4.3.2.

The significance of vector iteration (and of the simultaneous vector iteration to be discussed in what follows) arises through a computational scheme that does not demand the explicit formation of the dynamical matrix $\boldsymbol{A}$. Otherwise, we would be confronted by a fully populated matrix which, in the case of large order matrices, would require a prohibitive amount of storage space (say about 50×10^6 computer words for n=10 000). It is extremely expensive to compute the eigenvalues and eigenvectors of such matrices. Bearing these difficulties in mind, we refer to the product representation of equ. (4.1.39) and divide the iteration into the following steps:

$$\begin{aligned}
&(1)\ \text{Normalisation of the initial input vector } \overline{\boldsymbol{x}}_1^{(k)},\ \text{i.e. } [\overline{\boldsymbol{x}}_1^{(k)}]^{t}\overline{\boldsymbol{x}}_1^{(k)} = 1\\
&(2)\ \text{Solution of the triangular system } \boldsymbol{U}\boldsymbol{s} = \overline{\boldsymbol{x}}_1^{(k)}\\
&(3)\ \text{Multiplication } \boldsymbol{t} = \boldsymbol{M}\boldsymbol{s} \qquad\qquad (4.3.8)\\
&(4)\ \text{Solution of the triangular system } \boldsymbol{U}^{t}\overline{\boldsymbol{x}}_1^{(k+1)} = \boldsymbol{t}\\
&(5)\ \text{Determination of the Rayleigh-quotient } \lambda_1^{(k)} = [\overline{\boldsymbol{x}}_1^{(k+1)}]^{t}\overline{\boldsymbol{x}}_1^{(k)} = (1/\omega_1^{(k)})^2
\end{aligned}$$

The result of the triangular decomposition $\boldsymbol{U}$, of the stiffness matrix $\boldsymbol{K}$ is already familiar to the reader from our preceding discussion on statics. As is the case for $\boldsymbol{K}$, it is sparsely populated. Hence, it is clear that the iteration scheme operates here with sparsely populated matrices. As a result, we should be in a position to solve problems with the same order of magnitude as in statics.

As we already know from equs. (4.2.10) and (4.2.35), the above procedure can be applied to a free-free structural system without any difficulty, once we have established a positive definite stiffness matrix.

The selection of a suitable starting vector is a significant problem when applying the simple vector iteration method. If the starting vector $\overline{\boldsymbol{x}}_1$ does not contain the vector $\overline{\boldsymbol{x}}_1^{(k)}$ at all, i.e. $c_1 = 0$, then the iteration technique cannot yield the correct vector. In reality, the rounding errors introduce a germ of the vector $\overline{\boldsymbol{x}}_1$, which can become dominant. However, if the coefficient c_1 is small, the iteration method converges slowly and the procedure is not computationally economical. In any case, an experienced engineer should be able to predict, at least approximately, the fundamental vibration mode which can be employed as a starting vector. He will at least be able to name one degree of freedom r_i, which plays a dominant role in this oscillatory form. Alternatively, one can use the static deformation due to a load $R_i = 1$, located near the maximum amplitude of the fundamental mode, [4.37]. We then have

$$\boldsymbol{U}^{t}\boldsymbol{U}\boldsymbol{x}_1^{(0)} = \{0 \quad 0 \quad \dots \quad 0 \quad R_i = 1 \quad 0 \quad \dots \quad 0\} \tag{4.3.9a}$$

or since we require the transformed starting vector

$$\boldsymbol{U}^{t}\overline{\boldsymbol{x}}_1^{(0)} = \{0 \quad 0 \quad \dots \quad 0 \quad R_i = 1 \quad 0 \quad \dots \quad 0\} \tag{4.3.9b}$$

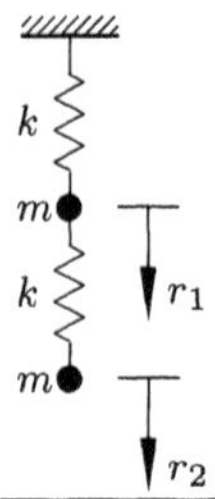

$$\boldsymbol{M} = m\boldsymbol{I}_2$$

$$\boldsymbol{K} = k\begin{bmatrix} 2 & -1 \\ -1 & 1 \end{bmatrix} \Rightarrow \boldsymbol{U} = \sqrt{\frac{k}{2}}\begin{bmatrix} 2 & -1 \\ 0 & 1 \end{bmatrix}$$

r_2 predominates in $\boldsymbol{x}_1$

k	0	1	2	...	exact
$\overline{\boldsymbol{x}}^{(k)}$ normalised	$\begin{bmatrix} 0 \\ 1 \end{bmatrix}$	$\begin{bmatrix} 0.20 \\ 0.98 \end{bmatrix}$	$\begin{bmatrix} 0.23 \\ 0.97 \end{bmatrix}$		
$\boldsymbol{s} = \boldsymbol{U}^{-1}\overline{\boldsymbol{x}}^{(k)} = \boldsymbol{x}_1^{(k)}$	$\sqrt{\frac{2}{k}}\begin{bmatrix} 0.5 \\ 1.0 \end{bmatrix}$	$\sqrt{\frac{2}{k}}\begin{bmatrix} 0.59 \\ 0.98 \end{bmatrix}$	$\sqrt{\frac{2}{k}}\begin{bmatrix} 0.60 \\ 0.97 \end{bmatrix}$		$\sqrt{\frac{2}{k}}\begin{bmatrix} 0.60 \\ 0.97 \end{bmatrix}$
$\boldsymbol{t} = \boldsymbol{M}\boldsymbol{s}$	$m\boldsymbol{s}$	$\leftarrow$	$\leftarrow$		
$\overline{\boldsymbol{x}}^{(k+1)} = \boldsymbol{U}^{-t}\boldsymbol{t}$	$\frac{2m}{k}\begin{bmatrix} 0.25 \\ 1.25 \end{bmatrix}$	$\frac{2m}{k}\begin{bmatrix} 0.30 \\ 1.28 \end{bmatrix}$	$\frac{2m}{k}\begin{bmatrix} 0.30 \\ 1.27 \end{bmatrix}$		
$\lambda_1^{(k)} = \left[\overline{\boldsymbol{x}}^{(k+1)}\right]^t \overline{\boldsymbol{x}}^{(k)}$	$2.50\frac{m}{k}$	$2.63\frac{m}{k}$	$2.60\frac{m}{k}$		
$\lvert\overline{\boldsymbol{x}}^{(k+1)}\rvert$	$1.63\frac{2m}{k}$	$1.31\frac{2m}{k}$	$1.30\frac{2m}{k}$		
$\left(\omega_1^{(k)}\right)^2 = \frac{1}{\lambda_1^{(k)}}$	$0.40\frac{k}{m}$	$0.38\frac{k}{m}$	$0.38\frac{k}{m}$		$0.38\frac{k}{m}$

Fig. 4.3.1 Simple vector iteration exemplified on the example of fig. 4.1.1

Consequently, the necessary starting vector can be obtained by an economical forward substitution method. All leading elements of $\overline{x}_{1j}^{(0)}$, for j < i are clearly zero.

After completion of the iteration, a reverse transformation of the eigenvector is called for. In this way, we obtain in conjunction with a normalised $\overline{\boldsymbol{x}}_1$ and in accordance with equ. (4.1.42) a $\boldsymbol{K}$-orthonormal eigenvector $\boldsymbol{x}_1$. This corresponds to the calculation of $\boldsymbol{s}$ in step 2 of equ. (4.3.8).

As an example, we first turn our attention to the two-mass oscillator described in fig. 4.1.1. The vector iteration procedure is detailed in fig. 4.3.1. Since the eigenvalues are clearly separated, the dominant eigenvalue is determined only after three iterations, to an accuracy of about two decimal digits. We also note that the convergence of the eigenvalue is achieved faster than that of the eigenvectors; this is not a coincidence.

Figure 4.2.3, left, shows a structure with a singular stiffness matrix, which is analysed again in fig. 4.3.2. Since we are working in the present case with purely

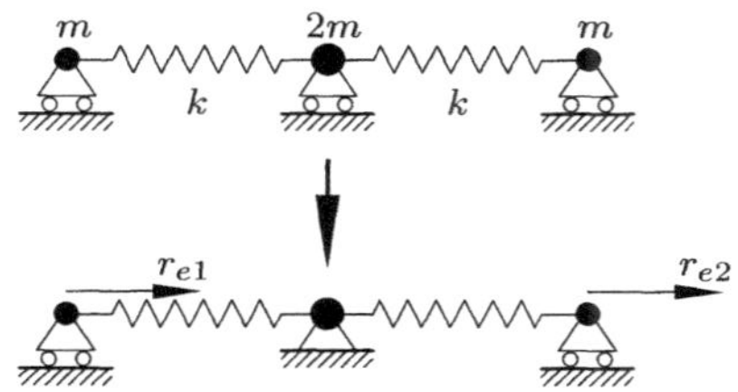

$$\boldsymbol{M}_{ee} - \boldsymbol{M}_{e0}\boldsymbol{M}_{00}^{-1}\boldsymbol{M}_{0e} = \frac{m}{4}\begin{bmatrix} 3 & -1 \\ -1 & 3 \end{bmatrix} = \boldsymbol{M}_e$$

$$\boldsymbol{K}_{ee} = k\boldsymbol{I}_2 \Rightarrow \boldsymbol{U}_e = \sqrt{k}\,\boldsymbol{I}_2$$

r_{e1} predominates in $\boldsymbol{x}_{e1}$

	0	1	2	3	4	5	exact
$\overline{\boldsymbol{x}}_e^{(k)}$ normalised	$\begin{bmatrix}1.0\\0\end{bmatrix}$	$\begin{bmatrix}0.95\\-0.32\end{bmatrix}$	$\begin{bmatrix}0.86\\-0.52\end{bmatrix}$	$\begin{bmatrix}0.79\\-0.62\end{bmatrix}$	$\begin{bmatrix}0.75\\-0.66\end{bmatrix}$	$\begin{bmatrix}0.73\\-0.68\end{bmatrix}$	
$\boldsymbol{s} = \boldsymbol{U}_e^{-1}\overline{\boldsymbol{x}}_e^{(k)} = \boldsymbol{x}_{e1}^{(k)}$	$\frac{1}{\sqrt{k}}\overline{\boldsymbol{x}}_e^{(k)}$	←	←	←	←	←	$\frac{1}{\sqrt{k}}\begin{bmatrix}0.71\\-0.71\end{bmatrix}$
$\boldsymbol{t} = \boldsymbol{M}_e\boldsymbol{s}$	$\frac{m}{4\sqrt{k}}\begin{bmatrix}3\\-1\end{bmatrix}$	$\frac{m}{4\sqrt{k}}\begin{bmatrix}3.17\\-1.91\end{bmatrix}$	$\frac{m}{4\sqrt{k}}\begin{bmatrix}3.10\\-2.42\end{bmatrix}$	$\frac{m}{4\sqrt{k}}\begin{bmatrix}2.99\\-2.65\end{bmatrix}$	$\frac{m}{4\sqrt{k}}\begin{bmatrix}2.91\\-2.73\end{bmatrix}$	$\frac{m}{4\sqrt{k}}\begin{bmatrix}2.87\\-2.77\end{bmatrix}$	
$\overline{\boldsymbol{x}}_e^{(k+1)} = \boldsymbol{U}_e^{-t}\boldsymbol{t}$	$\frac{1}{\sqrt{k}}\boldsymbol{t}$	←	←	←	←	←	
$\lambda_1^{(k)} = \left[\overline{\boldsymbol{x}}_e^{(k+1)}\right]^t\overline{\boldsymbol{x}}_e^{(k)}$	$0.75\frac{m}{k}$	$0.91\frac{m}{k}$	$0.98\frac{m}{k}$	$1.00\frac{m}{k}$	$1.00\frac{m}{k}$	$0.99\frac{m}{k}$	$\frac{m}{k}$
$\left\|\overline{\boldsymbol{x}}_e^{(k+1)}\right\|$	$3.16\frac{m}{4k}$	$3.70\frac{m}{4k}$	$3.93\frac{m}{4k}$	$4.00\frac{m}{4k}$	$3.99\frac{m}{4k}$	$3.99\frac{m}{4k}$	
$\left(\omega_1^{(k)}\right)^2 = \frac{1}{\lambda_1^{(k)}}$	$1.33\frac{k}{m}$	$1.10\frac{k}{m}$	$1.02\frac{k}{m}$	$1.00\frac{k}{m}$	$1.00\frac{k}{m}$	$1.01\frac{k}{m}$	$\frac{k}{m}$

Fig. 4.3.2 Simple vector iteration exemplified on the example of fig. 4.2.3, left (singular stiffness matrix)

elastic degrees of freedom, we obtain the largest non-degenerate eigenvalue without any difficulty. The example may thus be of little relevance. However, we should observe that the eigenvalues are not as well separated ($\lambda_1/\lambda_2 = 2$) as those in the example of fig. 4.3.1 ($\lambda_1/\lambda_2 \approx 7$). The convergence is consequently slower.

Next, we consider an oscillator (fig. 4.3.3) possessing through appropriate design two equal maximum eigenvalues. As shown in fig. 4.3.4, we employ simple vector iteration and obtain, because of well separated roots, ($\overline{\lambda}_{1,2}/\overline{\lambda}_3 = 76$), rapid convergence to the dominant eigenvalue. However, the iterated eigenvector necessarily takes the form of a linear combination of the two independent eigenvectors of fig. 4.3.3. In fact, in our case, we have

$$\boldsymbol{x}_1^{(2)} = \sqrt{\frac{l^3}{24EI}}\begin{bmatrix}4.5593\\1.7781\\-0.8529\end{bmatrix} = \sqrt{\frac{l^3}{24EI}}\left(1.8532\begin{bmatrix}1.\\0.9595\\1.\end{bmatrix} + 2.7061\begin{bmatrix}1.\\0.\\-1.\end{bmatrix}\right)$$

$$= c_1\boldsymbol{x}_1 + c_2\boldsymbol{x}_2$$

As an incidental exercise, the reader should prove that the equation

$$\boldsymbol{M}\boldsymbol{x}_\mathrm{i} = \lambda_\mathrm{i}\boldsymbol{K}\boldsymbol{x}_\mathrm{i}$$

is satisfied for $\lambda_1^{(2)}$ and $\boldsymbol{x}_1^{(2)}$. He should also perform anew the iterations using a starting vector, which is orthogonal to $\boldsymbol{x}_1^{(2)}$ or $\overline{\boldsymbol{x}}_1^{(2)}$. Orthogonalisation at every step of the iteration is thereby recommended. The reader may also consider how a repetition of the procedure may be made to yield the third eigenvalue and eigenvector.

In concluding this section, we summarise the pros and cons of the simple vector iteration method.

The advantages are clearly:

(1) The matrices $\boldsymbol{M}$ and $\boldsymbol{U}$ are sparse and hence permit the computation of very large problems.

(2) The computational effort of solving linear equations with triangular matrices (forward and backward substitutions) is relatively insignificant.

(3) The decomposition of the stiffness matrix may already be on hand through a preceding static analysis.

The disadvantages are as follows:

(1) In the case of multiple maximum eigenvalues λ_1, the procedure yields only *one* eigenvector

(2) In the case of poorly separated eigenvalues λ_1 (cluster of maximum eigenvalues) the convergence is very slow since it depends upon the ratio λ_1/λ_2.

4.3.2 Determination of the lower frequency spectrum and the corresponding eigenvectors: simultaneous vector iteration

There are, in principle, three possibilities of extending the simple vector iteration method to the computation of several natural frequencies in the lower domain.

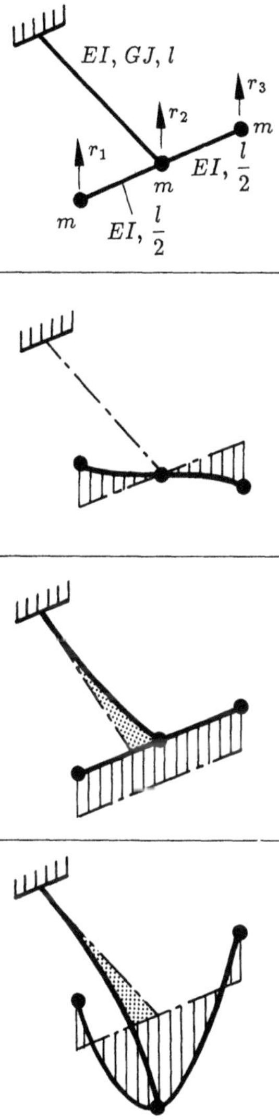

$$\alpha = \frac{EI}{GJ} = \frac{23 + \sqrt{593}}{24} = 1.9730$$

$$\bar{\lambda} = \frac{24EI}{ml^3}\lambda = \frac{24EI}{\omega^2 ml^3}$$

$$\bar{\lambda}_1 = 1 + 12\alpha = 1 + \frac{23 + \sqrt{593}}{2} = 24.6758$$

$$\boldsymbol{x}_1 = \{1 \quad 0 \quad -1\}$$

$$\bar{\lambda}_2 = \tfrac{1}{2}(25 + \sqrt{593}) = 24.6758$$

$$\boldsymbol{x}_2 = \left\{1 \quad \frac{32}{9 + \sqrt{593}} \quad 1\right\} = \{1 \quad 0.9595 \quad 1\}$$

$$\bar{\lambda}_3 = \tfrac{1}{2}(25 - \sqrt{593}) = 0.3242$$

$$\boldsymbol{x}_3 = \left\{1 \quad \frac{32}{9 - \sqrt{593}} \quad 1\right\} = \{1 \quad -2.0845 \quad 1\}$$

Fig. 4.3.3 Oscillator with equal maximum eigenvalues as a demonstration for the simple vector iteration (see fig. 4.3.4)

Firstly, the system's largest computed eigenvalue can be eliminated so that the second eigenvalue becomes the largest. This process, which is known as the deflation of the matrix, is, however, unsuitable for large, sparse matrices. The second approach has already been discussed in the previous section and involves the repetition of the iteration with orthogonalised starting vectors. The disadvantage of

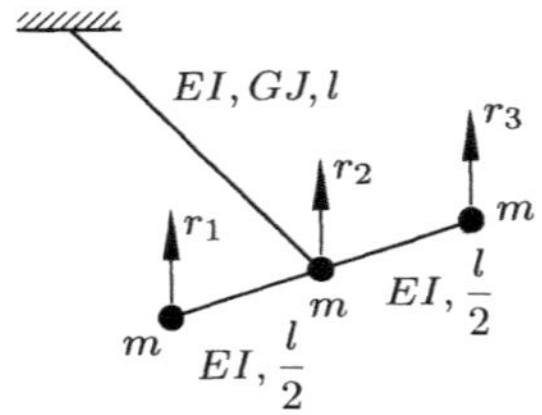

$$\alpha = \frac{EI}{GJ} = \frac{23+\sqrt{593}}{24}$$

r_2 predominates in initial vector

$$\boldsymbol{M} = m\boldsymbol{I}_3$$

$$\boldsymbol{K} = \frac{24\,EI}{l^3}\begin{bmatrix} \frac{1+6\alpha}{1+12\alpha} & -1 & \frac{6\alpha}{1+12\alpha} \\ -1 & \frac{17}{8} & -1 \\ \frac{6\alpha}{1+12\alpha} & -1 & \frac{1+6\alpha}{1+12\alpha} \end{bmatrix} = \frac{24\,EI}{l^3}\begin{bmatrix} 0.5203 & -1 & 0.4797 \\ -1 & 2.1250 & -1 \\ 0.4797 & -1 & 0.5203 \end{bmatrix}$$

$$\boldsymbol{U} = \sqrt{\frac{24\,EI}{l^3}}\begin{bmatrix} 0.7213 & -1.3864 & 0.6651 \\ 0 & 0.4504 & -0.1729 \\ 0 & 0 & 0.2191 \end{bmatrix}$$

k	0	1	2	3	exact
$\overline{\boldsymbol{x}}^{(k)}$ normalised	$\begin{bmatrix} 0.3037 \\ 0.9348 \\ -0.1842 \end{bmatrix}$	$\begin{bmatrix} 0.2568 \\ 0.9484 \\ -0.1868 \end{bmatrix}$	$\begin{bmatrix} 0.2562 \\ 0.9484 \\ -0.1868 \end{bmatrix}$	$\begin{bmatrix} 0.2562 \\ 0.9482 \\ -0.1868 \end{bmatrix}$	
$\boldsymbol{s} = \boldsymbol{U}^{-1}\overline{\boldsymbol{x}}^{(k)} = \boldsymbol{x}^{(k)}$	$\sqrt{\frac{l^3}{24EI}}\begin{bmatrix} 4.5649 \\ 1.7525 \\ -0.8407 \end{bmatrix}$	$\sqrt{\frac{l^3}{24EI}}\begin{bmatrix} 4.5594 \\ 1.7777 \\ -0.8528 \end{bmatrix}$	$\sqrt{\frac{l^3}{24EI}}\begin{bmatrix} 4.5593 \\ 1.7781 \\ -0.8529 \end{bmatrix}$		see fig. 4.3.3 and text
$\overline{\boldsymbol{x}}^{(k+1)} = \boldsymbol{U}^{-t}\boldsymbol{t} = m\boldsymbol{U}^{-t}\boldsymbol{s}$	$\frac{ml^3}{24EI}\begin{bmatrix} 6.3287 \\ 23.3698 \\ -4.6042 \end{bmatrix}$	$\frac{ml^3}{24EI}\begin{bmatrix} 6.3212 \\ 23.4026 \\ -4.6107 \end{bmatrix}$	$\frac{ml^3}{24EI}\begin{bmatrix} 6.3210 \\ 23.4026 \\ -4.6107 \end{bmatrix}$		
$\lambda_1^{(k)} = \left[\overline{\boldsymbol{x}}^{(k+1)}\right]^t \overline{\boldsymbol{x}}^{(k)}$	$\frac{ml^3}{24EI}$ 24.6162	$\frac{ml^3}{24EI}$ 24.6748	$\frac{ml^3}{24EI}$ 24.6757		$\frac{ml^3}{24EI}$ 24.6758
$\left\lvert\overline{\boldsymbol{x}}^{(k+1)}\right\rvert$	$\frac{ml^3}{24EI}$ 24.6455	$\frac{ml^3}{24EI}$ 24.6758	$\frac{ml^3}{24EI}$ 24.6758		

Fig. 4.3.4 Demonstration of simple vector iteration for structures with multiple maximum eigenvalues

this technique, though, is that in the case of not well-separated large eigenvalues, the cost of determining an eigenvector becomes relatively high. The third, most effective and economical method performs iterations using a set of several orthonormal vectors. In this case, orthonormalisation should be repeated at the end of each step otherwise the rounding errors cause re-emergence of the first eigenvector leading to a duplication as the ultimate result. Once this elementary aspect is realised, we are in a position to extend the simple vector iteration to an effective simultaneous iteration:

(1) Orthonormalisation of an approximation $\boldsymbol{V}$

$$\underset{(n\times ni)}{\boldsymbol{V}} \rightarrow \underset{(n\times ni)}{\overline{\boldsymbol{X}}^{(k)}} \boldsymbol{R}$$

where $[\overline{\boldsymbol{X}}^{(k)}]^t \overline{\boldsymbol{X}}^{(k)} = \boldsymbol{I}_{ni}$

(2) Solution $\boldsymbol{U}\boldsymbol{S} = \overline{\boldsymbol{X}}^{(k)}$

(3) Multiplication $\boldsymbol{T} = \boldsymbol{M}\boldsymbol{S}$

(4) Solution $\boldsymbol{U}^t\boldsymbol{V} = \boldsymbol{T}$

(5) Rayleigh quotients

$$\underset{(ni\times ni)}{\boldsymbol{\Lambda}} = \text{diag}\left[\boldsymbol{V}^t\overline{\boldsymbol{X}}^{(k)}\right] = \text{diag}\left[[\overline{\boldsymbol{X}}^{(k)}]^t \boldsymbol{A}\overline{\boldsymbol{X}}^{(k)}\right] \tag{4.3.10}$$

From the preceding discussion, it is evident that this convergence process also is dependent upon the ratio of adjoining eigenvalues. Reference [4.5] shows that after k iterations the norm of the deviation between the iterated vector $\overline{\boldsymbol{x}}_i^{(k)}$ (of $\overline{\boldsymbol{X}}^{(k)}$) and the actual true eigenvector $\overline{\boldsymbol{x}}_i$ is of the order of magnitude q^k. The expression for the norm is thereby defined by

$$\|\overline{\boldsymbol{x}}_i^{(k)} - \overline{\boldsymbol{x}}_i\| = \left[[\overline{\boldsymbol{x}}_i^{(k)} - \overline{\boldsymbol{x}}_i]^t[\overline{\boldsymbol{x}}_i^{(k)} - \overline{\boldsymbol{x}}_i]\right]^{1/2} \tag{4.3.11}$$

Furthermore, the ratio q is given by

$$q = \max\left(\frac{\lambda_{i+1}}{\lambda_i}, \frac{\lambda_i}{\lambda_{i-1}}\right) \tag{4.3.12}$$

The ratios of the current eigenvalue to the immediately neighbouring eigenvalues are the determining factor in the convergence of a vector. Thus, the procedure tabulated in equ. (4.3.10) does not present a perceptible improvement upon a repeated simple vector iteration method in conjunction with an orthonormalisation of the starting vector. It is hence necessary to seek methods of improving the rate of convergence.

For the computation of approximate eigenvalues, we established in step 5 the diagonal of the matrix operation

$$\underset{(ni\times ni)}{\boldsymbol{B}} = \boldsymbol{V}^t\overline{\boldsymbol{X}}^{(k)} = [\overline{\boldsymbol{X}}^{(k)}]^t \boldsymbol{A}\overline{\boldsymbol{X}}^{(k)} \tag{4.3.13}$$

in accordance with the basic procedure of equ. (4.3.10). In fact, we should not only be interested in the diagonal elements, but also in the off-diagonal elements of the symmetric matrix $\boldsymbol{B}$. Indeed, if $\overline{\boldsymbol{X}}^{(\mathrm{k})}$ were to contain the set of the true eigenvectors, then $\boldsymbol{B}$ would be reduced to a diagonal. Since $\overline{\boldsymbol{X}}^{(\mathrm{k})}$ is, however, only an approximation, the matrix $\boldsymbol{B}$ attains not more than a diagonally dominant form. In general, the number of iterated vectors ni is small compared to the total number of degrees of freedom of the system, n. Hence, the dimension of the matrix $\boldsymbol{B}$ is also restricted and $\boldsymbol{B}$ can be reduced to a diagonal form by inexpensive eigenvalue analysis methods

$$\boldsymbol{B} \rightarrow \underset{(\mathrm{ni}\times\mathrm{ni})}{\boldsymbol{X}_B} \underset{(\mathrm{ni}\times\mathrm{ni})}{\boldsymbol{\Lambda}_B} \underset{(\mathrm{ni}\times\mathrm{ni})}{\boldsymbol{X}_B^{\mathrm{t}}}$$

$$\text{where} \quad \boldsymbol{X}_B^{\mathrm{t}} \boldsymbol{X}_B = \boldsymbol{I}_{\mathrm{ni}} \tag{4.3.14}$$

Equation (4.3.14) yields all ni eigenvectors $\boldsymbol{X}_B$ of the matrix $\boldsymbol{B}$. A suitable technique for this purpose is given by the Jacobi algorithm described in section 4.3.3. If we are now to adopt a transformed vector instead of $\overline{\boldsymbol{X}}^{(\mathrm{k})}$

$$\underset{(\mathrm{n}\times\mathrm{ni})}{\overline{\boldsymbol{X}}_0^{(\mathrm{k})}} = \underset{(\mathrm{n}\times\mathrm{ni})}{\overline{\boldsymbol{X}}^{(\mathrm{k})}} \underset{(\mathrm{ni}\times\mathrm{ni})}{\boldsymbol{X}_B} \tag{4.3.15}$$

which is also orthogonal

$$\begin{aligned} [\overline{\boldsymbol{X}}_0^{(\mathrm{k})}]^{\mathrm{t}} \overline{\boldsymbol{X}}_0^{(\mathrm{k})} &= \boldsymbol{X}_B^{\mathrm{t}} [\overline{\boldsymbol{X}}^{(\mathrm{k})}]^{\mathrm{t}} \overline{\boldsymbol{X}}^{(\mathrm{k})} \boldsymbol{X}_B = \boldsymbol{X}_B^{\mathrm{t}} \boldsymbol{X}_B \\ &= \boldsymbol{I}_{\mathrm{ni}} \end{aligned} \tag{4.3.16}$$

then instead of $\boldsymbol{V}$, we obtain the iterated set of vectors

$$\boldsymbol{V}_0 = \boldsymbol{A} \overline{\boldsymbol{X}}_0^{(\mathrm{k})} = \boldsymbol{A} \overline{\boldsymbol{X}}^{(\mathrm{k})} \boldsymbol{X}_B = \boldsymbol{V} \boldsymbol{X}_B \tag{4.3.17}$$

with the characteristic property

$$\begin{aligned} \boldsymbol{V}_0^{\mathrm{t}} \overline{\boldsymbol{X}}_0^{(\mathrm{k})} &= [\overline{\boldsymbol{X}}_0^{(\mathrm{k})}]^{\mathrm{t}} \boldsymbol{A} \overline{\boldsymbol{X}}_0^{(\mathrm{k})} = \boldsymbol{X}_B^{\mathrm{t}} [\overline{\boldsymbol{X}}^{(\mathrm{k})}]^{\mathrm{t}} \boldsymbol{A} \overline{\boldsymbol{X}}^{(\mathrm{k})} \boldsymbol{X}_B \\ &= \boldsymbol{\Lambda}_B \end{aligned} \tag{4.3.18}$$

We employ $\boldsymbol{V}_0$ instead of $\boldsymbol{V}$ in the next iteration loop. Since $\overline{\boldsymbol{X}}_0^{(\mathrm{k})}$ is only a linear combination of vectors $\overline{\boldsymbol{X}}^{(\mathrm{k})}$, the step given in equ. (4.3.15) is not expected to bring about a sharper definition and separation of the ni iterated vectors from the remaining (n−ni) eigenvectors. However, if we focus our attention solely on ni vectors, the decoupling of equ. (4.3.15) will in any case improve the convergence. We can assert that a subset of nir first vectors (nir < ni) is always better than the preceding set. Accordingly, the rate of convergence for i ⩽ ni no longer depends on the eigenvalues in the immediate neighbourhood, but in place of equ. (4.3.12) on the ratio

$$q = \frac{\lambda_{\mathrm{ni}+1}}{\lambda_{\mathrm{i}}} \tag{4.3.19}$$

This means that the rate of convergence is conditioned by the first eigenvalue which appears outside the spectral range of iteration ([4.1], p.22). This at once leads us to introduce yet another improvement.

For example, consider that we wish to determine three eigenvectors and that the spectrum of the eigenvalues of the problem reads

$$\boldsymbol{\Lambda} = \lceil 100 \quad 100 \quad 99 \quad 98 \quad 10 \quad 9 \quad 3 \rfloor$$

According to the criteria of equ. (4.3.19), when iterating with three vectors, the frequency ratios (98/100, 98/100 and 98/99) are the determining factors and the rate of convergence appears unfavourable. On the other hand, a natural split of the spectrum lies between the 4th and the 5th eigenvalues. Consequently, if four iteration vectors were selected, the convergence rate of the first three vectors becomes very favourable with characteristic ratios 10/100, 10/100 10/99. The convergence rate may be increased still further if more vectors were to be considered in the iteration set. At the same time, we should realise that the procedure becomes more time-consuming when the vector set is increased. There is clearly an optimum choice in this procedure. Generally, in finite element models, the frequencies are distinctly separated – and thus determine a suitable split in the spectrum – only in their lower range. A simple technique of using the characteristic properties described above is as follows: we start the vector iteration with nis > ni eigenvectors (e.g. nis=2ni) and we seek a suitable natural split nie in the spectrum following the evaluation of the first relatively reliable eigenvalues. Subsequently, we continue the iteration with nie $\geqslant$ ni eigenvectors.

The simple expression in equ. (4.3.19) for the critical ratio of the eigenvalues suggests a further improvement [4.6]. If the iterated eigenvalues are well separated – for example, ni=3 and nie=4 and the eigenspectrum is given by

$$\boldsymbol{\Lambda} = [100 \quad 50 \quad 30 \underset{\text{ni}}{\uparrow} 15 \underset{\text{nie}}{\uparrow} 2 \quad 2 \quad 1.5]$$

– we find that the first eigenvector with the ratio 2/100 will converge more rapidly than the second and the third. Before the end of the iteration process designed for the three eigenvectors, we should have obtained the first eigenvector with sufficient accuracy and subsequently the second one and thus be in a position to extract these two vectors from the iteration procedure.

However, even with this technique, all possiblilites of improving the iteration process are not yet exhausted. Further speeding-up of the iteration is achieved by a purposive change in the eigenvalues of the dynamic matrix $\boldsymbol{A}$. A simple way of achieving this is by a shift of the spectrum of the eigenvalues. In particular, the matrix $[\boldsymbol{A} - s\boldsymbol{I}]$ possesses the eigenvalues

$$\boldsymbol{\Lambda}_s = \overline{\boldsymbol{X}}^{\,t}[\boldsymbol{A} - s\boldsymbol{I}]\overline{\boldsymbol{X}} = \boldsymbol{\Lambda} - s\boldsymbol{I} \tag{4.3.20}$$

This characterisitic can be utilised to accelerate the convergence. Thus, a spectral shift for the eigenvalues

$$\boldsymbol{\Lambda} = [100 \quad 50 \quad 30 \underset{\text{ni}}{\uparrow} 15 \underset{\text{nie}}{\uparrow} 2 \quad 2 \underset{\text{nis}}{\uparrow} 1.5]$$

with the shift

$$s = \frac{1}{2}(1.5 + 2) = 1.75$$

yields the new spectrum

$$\mathbf{\Lambda}_s = [98.25 \quad 48.25 \quad 28.25 \underset{\text{ni}}{\uparrow} \quad 13.25 \underset{\text{nie}}{\uparrow} \quad 0.25 \quad 0.25 \quad -0.25]$$

with a dramatically improved convergence. For the first eigenvector, the rate of convergence is increased nearly by a factor of eight. To achieve this improvement, we require, of course, the optimal spectral shift. This is in the present case

$$s = \frac{1}{2}(\lambda_{\text{nie}+1} + \lambda_{\text{n}}) \tag{4.3.21}$$

where neither λ_{n} nor the exact value of $\lambda_{\text{nie}+1}$ are known, however. All we know is that as a rule the value of λ_{n} is small and positive. Therefore, we approximate it to zero and obtain

$$s = \frac{1}{2}\lambda_{\text{nie}+1} \tag{4.3.22}$$

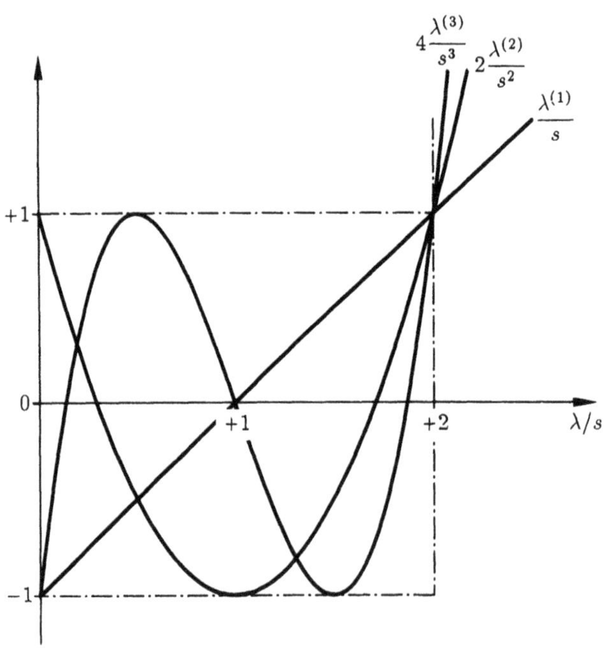

Fig. 4.3.5

Application of Chebychev polynomials to achieve a spreading of the eigenvalue spectra using the Chebychev iteration

Our small example thus yields a spectral shift of 1, which still results in a substantial improvement in convergence. Of course, we know at best an approximation to $\lambda_{\text{nie}+1}$ obtained at the start of the iteration for nis>nie. Continuous improvement can be achieved only for λ_{nie}. Unfortunately, the spectral shift

$$s = \tfrac{1}{2}\lambda_{\text{nie}} \tag{4.3.23}$$

which could be improved continuously does not guarantee any success. Indeed, our simple example demonstrates that $s = 7.5$ has unfavourable effects. Thus, it is not always easy to operate with a simple spectral shift.

However, these difficulties can be alleviated with a little thought. We have already seen that the matrix

$$\boldsymbol{A}^{(1)} = [\boldsymbol{A} - s\boldsymbol{I}\,] \tag{4.3.24}$$

possesses the modified eigenvalues $\lambda^{(1)}$

$$\frac{\lambda^{(1)}}{s} = \frac{\lambda}{s} - 1 \tag{4.3.25}$$

Application of fig. 4.3.5 yields a means of reevaluating the eigenvalues. We see from the graphical scheme that for $\lambda/s > 2$, the corresponding ratios of the modified eigenvalues exceed unity ($\lambda^{(1)}/s > 1$), while for $0 < \lambda/s < 2$ the ratio of the modified eigenvalues is $-1 < \lambda^{(1)}/s < 1$.

We now consider the somewhat more complicated matrix

$$\boldsymbol{A}^{(2)} = [\boldsymbol{A} - s\boldsymbol{I}\,]^2 - \tfrac{1}{2}s^2\boldsymbol{I} \tag{4.3.26}$$

which has the same eigenvectors as $\boldsymbol{A}$ and the eigenvalues $\lambda^{(2)}$

$$2\frac{\lambda^{(2)}}{s^2} = 2\left(\frac{\lambda}{s} - 1\right)^2 - 1 \tag{4.3.27}$$

The graphical interpretation of equ. (4.3.27) is also demonstrated in fig. 4.3.5. We see that all eigenvalues for $\lambda/s > 2$, compared to those for $0 < \lambda/s < 2$, are more strongly magnified than in the case of the matrix $\boldsymbol{A}^{(1)}$. The polynomials of the type given in equs. (4.3.25) and (4.3.27) are known as Chebychev polynomials. The series of such polynomials can be extended easily to arbitrarily high orders. We demonstrate this on the third order polynomial defined by the matrix

$$\boldsymbol{A}^{(3)} = [\boldsymbol{A} - s\boldsymbol{I}\,]^3 - \tfrac{3}{4}s^2[\boldsymbol{A} - s\boldsymbol{I}\,] \tag{4.3.28}$$

with the eigenvalue characteristic

$$4\frac{\lambda^{(3)}}{s^3} = 4\left(\frac{\lambda}{s} - 1\right)^3 - 3\left(\frac{\lambda}{s} - 1\right) \tag{4.3.29}$$

All eigenvalues for $\lambda/s > 2$ are seen to be amplified to a still higher order. Without much further elaboration, one can ascertain that with increasing order of the

Chebychev polynomial, the magnitude of the eigenvalues in the original range $0 < \lambda/s \leqslant 2$ recedes as against those in the range $\lambda/s > 2$. Consequently, for a suitable value of s, the following sequence

$$\begin{aligned}
\overline{\boldsymbol{X}}^{(1)} &= \boldsymbol{A}^{(1)}\overline{\boldsymbol{X}}^{(0)} \\
\overline{\boldsymbol{X}}^{(2)} &= \boldsymbol{A}^{(2)}\overline{\boldsymbol{X}}^{(0)} \\
\overline{\boldsymbol{X}}^{(3)} &= \boldsymbol{A}^{(3)}\overline{\boldsymbol{X}}^{(0)} \\
&\vdots
\end{aligned} \tag{4.3.30}$$

which *de facto* describes at any one step a vector iteration with a modified matrix will yield continuously improving results. The iteration given by equ. (4.3.26) is denoted the Chebychev iteration. Its particular advantage is that it can be expressed in a recursive form. Thus, we obtain from equ. (4.3.30) in conjunction with equs. (4.3.24), (4.3.26) and (4.3.28)

$$\begin{aligned}
\overline{\boldsymbol{X}}^{(1)} &= [\boldsymbol{A} - s\boldsymbol{I}]\overline{\boldsymbol{X}}^{(0)} \\
\overline{\boldsymbol{X}}^{(2)} &= [\boldsymbol{A} - s\boldsymbol{I}]\overline{\boldsymbol{X}}^{(1)} - \tfrac{1}{2}s^2\overline{\boldsymbol{X}}^{(0)} \\
\overline{\boldsymbol{X}}^{(3)} &= [\boldsymbol{A} - s\boldsymbol{I}]\overline{\boldsymbol{X}}^{(2)} - \tfrac{1}{4}s^2\overline{\boldsymbol{X}}^{(1)} \\
&\vdots \\
\overline{\boldsymbol{X}}^{(\mathrm{i})} &= [\boldsymbol{A} - s\boldsymbol{I}]\overline{\boldsymbol{X}}^{(\mathrm{i}-1)} - \tfrac{1}{4}s^2\overline{\boldsymbol{X}}^{(\mathrm{i}-2)}
\end{aligned} \tag{4.3.31}$$

All the eigenvalues to be amplified should lie as markedly as possible to the right of the limit $\lambda/s = 2$. In this group of large eigenvalues, λ_{ni} lies next to the critical limit. We postulate therefore

$$\lambda_{\mathrm{ni}}/s > 2 \qquad \text{or} \qquad s < \lambda_{\mathrm{ni}}/2 \tag{4.3.32}$$

On the other hand, we must prevent an amplification of the eigenvalues outside the range of interest. This is true for

$$\lambda_{\mathrm{nie}+1}/s \leqslant 2 \qquad \text{or} \qquad s \geqslant \lambda_{\mathrm{nie}+1}/2 \tag{4.3.33}$$

Thereby, the equals sign guarantees that the vector $\boldsymbol{x}_{\mathrm{nie}}$ falls just outside the magnification range. This means that for

$$s = \tfrac{1}{2}\lambda_{\mathrm{nie}+1} \tag{4.3.34}$$

the Chebychev iteration method is optimal. Since this information is generally unavailable in the vector iteration process, the next possible choice appears to be

$$s = \tfrac{1}{2}\lambda_{\mathrm{nie}}^{(\mathrm{k})} \leqslant \tfrac{1}{2}\lambda_{\mathrm{nie}} \tag{4.3.35}$$

where $\lambda_{\mathrm{nie}}^{(\mathrm{k})}$ is the current iterated approximation to the eigenvalue λ_{nie}. The necessity of a buffer zone between the significant eigenvalues and the delimitation of the iteration field – which proved so advantageous in the simple iteration technique – applies here too. To satisfy the requirement given by equ. (4.3.22), we should include at least one more additional vector $\overline{\boldsymbol{x}}_{\mathrm{ni}+1}$ corresponding to a smaller eigenvalue $\lambda_{\mathrm{ni}+1} < \lambda_{\mathrm{ni}}$. A prerequisite for the application of Chebychev's iteration is, naturally, an anterior standard iteration step followed by decoupling in order to determine the shift s. Further, we should take into consideration that the order of the Chebychev polynomial should not be too high because of the limitation imposed on the word length in the computer, which has a detrimental effect on the formation of powers of the matrix $\boldsymbol{A}$. Hence, it is recommended to scale the mass matrix $\boldsymbol{M}$ by a factor of the order of magnitude $(1/\lambda_1)$.

Two additional points, namely the selection of the initial vectors and the design of a suitable criterion for the iteration break-off, must be clarified to complete the discussion on simultaneous vector iteration. For the initial vectors, we may adopt a similar approach to that used in the case of the simple vector iteration. Assuming that we ultimately require ni<n eigenvectors, we initiate the procedure by prescribing nb degrees of freedom, ni⩽nb<n, which in our experience as engineers describe the dynamic behaviour of the system with sufficient accuracy. This permits the reduction of the problem size from n to nb by means of static condensation (see section 4.4). For this purpose, nb displacement vectors are computed due to a unit-load in the direction of the nb fundamental degrees of freedom. These serve as initial vectors. The load matrix is a Boolean matrix $\mathbf{a}_b(\mathrm{n}\times \mathrm{nb})$, which contains a single 1 in each column. The displacement vectors for the general eigenvalue problem are then given by $\boldsymbol{U}^{-1}\boldsymbol{U}^{-\mathrm{t}}\mathbf{a}_b$, and for the special restricted problem by

$$\boldsymbol{V} = \boldsymbol{U}^{-\mathrm{t}}\mathbf{a}_b \tag{4.3.36}$$

If the nb degrees of freedom indeed satisfactorily describe the required dynamic response, then $\boldsymbol{V}$ contains the major share of the desired eigenvectors. If this assertion still contains some uncertainties, then the vector set should be enlarged by additional linearly independent vectors which are orthogonal to $\boldsymbol{V}$ and are generated by a random process. A loop of the vector iteration with nb vectors is now performed, ultimately yielding the uncoupled matrix $\boldsymbol{B}(\mathrm{nb}\times\mathrm{nb})$. The result comprises an approximation to the first nb > ni eigenvalues. This vector set is then reduced to nie vectors such that ni < nie ⩽ nb, on which the simultaneous vector iteration is applied using the Chebychev algorithm.

We can now turn to a final consideration, namely the break-off criterion of the iteration procedure. Detailed information on this subject is beyond the scope of this book which is necessarily restricted to an outline of the method, but the reader is advised to consult the classical treatise of Wilkinson [4.7]. If we select a specific iterated eigenvalue $\lambda_{\mathrm{i}}^{(\mathrm{k})}$ and the corresponding iterated vector $\overline{\boldsymbol{x}}_{\mathrm{i}}^{(\mathrm{k})}$, we note that the residual vector

$$\boldsymbol{\eta}_{\mathrm{i}}^{(\mathrm{k})} = \boldsymbol{A}\overline{\boldsymbol{x}}_{\mathrm{i}}^{(\mathrm{k})} - \lambda_{\mathrm{i}}^{(\mathrm{k})}\overline{\boldsymbol{x}}_{\mathrm{i}}^{(\mathrm{k})} \tag{4.3.37}$$

provides some information on the accuracy of the result. If the iterated solution were exact, the vector $\boldsymbol{\eta}_{\mathrm{i}}^{(\mathrm{k})}$ would be zero. Wilkinson shows that the iterated

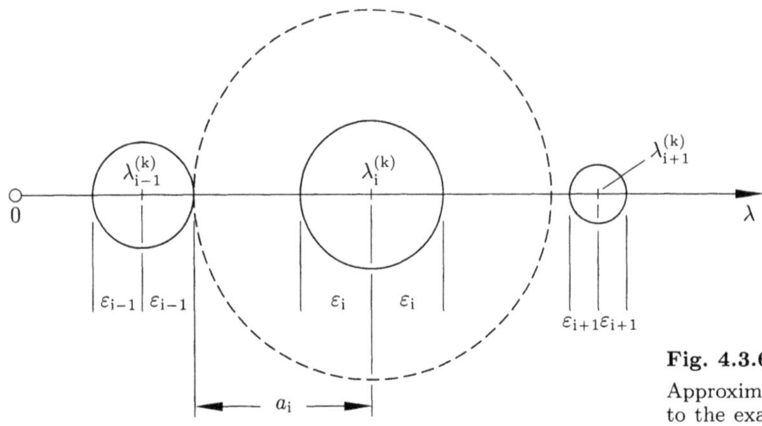

Fig. 4.3.6
Approximations and ranges to the exact eigensolutions

solution $\lambda_i^{(k)}$ can deviate at the most by ε_i from the true eigenvalue λ_i (naturally, λ_i can be a multiple root)

$$\min_i(|\lambda_i - \lambda_i^{(k)}|) \leqslant \varepsilon_i \tag{4.3.38}$$

where ε_i is simply

$$\varepsilon_i = \|\boldsymbol{\eta}_i^{(k)}\|_2 = (\boldsymbol{\eta}^{(k)^t}\boldsymbol{\eta}_i^{(k)})^{1/2} \tag{4.3.39}$$

Consequently, using the criterion of equ. (4.3.38), we can circumscribe about each approximation a circle, within which the true eigenvalue λ_i must be located (see fig. 4.3.6). In addition, to evaluate the error in $\overline{\boldsymbol{x}}_i^{(k)}$

$$\boldsymbol{e}_i = \overline{\boldsymbol{x}}_i^{(k)} - \overline{\boldsymbol{x}}_i \tag{4.3.40}$$

one must also consider the neighbouring eigenvalues. If there is only one eigenvalue in the range $\lambda_i^{(k)} - \varepsilon_i$ to $\lambda_i^{(k)} + \varepsilon_i$ and the remaining eigenvalues lie outside the interval $\lambda_i^{(k)} - a_i$ and $\lambda_i^{(k)} + a_i$ (see fig. 4.3.6), then the norm of the error vector is

$$\|\boldsymbol{e}_i\|_2 = (\boldsymbol{e}_i^t\boldsymbol{e}_i)^{1/2} \leqslant \frac{\varepsilon_i}{a_i} \tag{4.3.41}$$

The above expression is somewhat more complicated for multiple eigenvalues. Furthermore, it is evident that for close eigenvalues, ε_i must be very small to guarantee sufficient accuracy in $\overline{\boldsymbol{x}}_i^{(k)}$, since a_i is small in any case.

Finally, to conclude the theoretical discussion, we refer to fig. 4.3.7, which outlines the computational scheme discussed above. The simultaneous vector iteration scheme proves a very useful and practical tool if we wish to compute a restricted

extended simultaneous vector iteration

user selects nb basic degrees of freedom to describe the dynamic behaviour, with nb (> ni) = number of eigenvalues and -vectors

$$\underset{(\mathrm{n\times nb})}{\mathbf{a}_b} \quad \text{is generated}$$

initial vectorial set

$$\underset{(\mathrm{n\times nb})}{\boldsymbol{V}_0} = \underset{(\mathrm{n\times n})}{\boldsymbol{U}^{-\mathrm{t}}}\underset{(\mathrm{n\times nb})}{\mathbf{a}_b}$$

with $\boldsymbol{U}$ = triangular decomposition of the stiffness matrix $\boldsymbol{K}$

initialisation

reduced set of nie vectors (ni < nie ⩽ nb), where $\lambda_{\mathrm{nie}}/\lambda_{\mathrm{nie}+1}$ is as large as possible

break-off or convergence criterion

$$\left.\begin{aligned}\boldsymbol{\eta}_\mathrm{i} &= \boldsymbol{v}_{0\mathrm{i}} - \lambda_{B\mathrm{i}}\overline{\boldsymbol{x}}_{0\mathrm{i}} \\ \varepsilon_\mathrm{i} &= [\boldsymbol{\eta}_\mathrm{i}^\mathrm{t}\boldsymbol{\eta}_\mathrm{i}]^{1/2}\end{aligned}\right\} \mathrm{i}=1\ldots\mathrm{ni}$$

e.g. $\dfrac{\varepsilon_\mathrm{i}}{\lambda_\mathrm{i}} < \varepsilon$ (user), i=1...ni

no | yes

main loop

backward transformation

$$\underset{(\mathrm{n\times ni})}{\boldsymbol{X}} = \boldsymbol{U}^{-1}\underset{(\mathrm{n\times ni})}{\overline{\boldsymbol{X}}_0}$$

initialisation extended simultaneous vector iteration

orthonormalisation (Cholesky)

$$\boldsymbol{V}_0^\mathrm{t}\boldsymbol{V}_0 = \underset{(\mathrm{nb\times nb})}{\boldsymbol{G}} \Rightarrow \boldsymbol{U}_G^\mathrm{t}\boldsymbol{U}_G$$

$$\underset{(\mathrm{n\times nb})}{\overline{\boldsymbol{X}}} = \boldsymbol{V}_0\boldsymbol{U}_G^{-1}$$

power operation

$$\boldsymbol{S} = \boldsymbol{U}^{-1}\overline{\boldsymbol{X}}$$

$$\boldsymbol{T} = \boldsymbol{M}\boldsymbol{S}$$

$$\boldsymbol{V} = \boldsymbol{U}^{-\mathrm{t}}\boldsymbol{T}$$

Rayleigh-Ritz procedure

$$\boldsymbol{B} = \boldsymbol{V}^\mathrm{t}\overline{\boldsymbol{X}} = \overline{\boldsymbol{X}}^\mathrm{t}\boldsymbol{A}\overline{\boldsymbol{X}}$$

$$\boldsymbol{B} \Rightarrow \boldsymbol{X}_B\boldsymbol{\Lambda}_B\boldsymbol{X}_B^\mathrm{t} \quad \text{(Jacobi)}$$

$$\overline{\boldsymbol{X}}_0 = \overline{\boldsymbol{X}}\boldsymbol{X}_B$$

$$\boldsymbol{V}_0 = \boldsymbol{V}\boldsymbol{X}_B$$

Fig. 4.3.7 Survey of the process of simultaneous vector iteration (for the sake of clarity some details referred to in the text have been omitted), *Part 1*

main loop extended simultaneous vector iteration

loop

orthonormalisation

$$\underset{}{V_0^t}\ \underset{(n\times nie)}{V_0} = \underset{(nie\times nie)}{G} \Rightarrow U_G^t U_G$$

$$\underset{(n\times nie)}{\overline{X}} = V_0 U_G^{-1}$$

power operation

$$S = U^{-1}\overline{X}$$
$$T = MS$$
$$V = U^{-t}T$$

Rayleigh-Ritz procedure

$$B = V^t\overline{X} = \overline{X}^t A\overline{X}$$
$$B \Rightarrow X_B \Lambda_B X_B^t \quad \text{(Jacobi)}$$
$$\overline{X}_0 = \overline{X}X_B$$
$$V_0 = VX_B$$

break-off or convergence criterion

$$\left.\begin{aligned}\eta_i &= v_{0i} - \lambda_{Bi}\overline{x}_{0i}\\ \varepsilon_i &= \left[\eta_i^t\eta_i\right]^{1/2}\end{aligned}\right\}\ i=1...ni$$

e.g. $\dfrac{\varepsilon_i}{\lambda_i} < \varepsilon$ (user), i=1...ni

Chebychev iteration

$$s = \tfrac{1}{2}\lambda_{Bnie}$$

$$V_0^{(1)} = [U^{-t}MU^{-1} - sI]V_0$$
$$V_0^{(2)} = [U^{-t}MU^{-1} - sI]V_0^{(1)} - \tfrac{1}{2}s^2V_0$$

from j = 3 onwards:

$$V_0^{(j)} = [U^{-t}MU^{-1} - sI]V_0^{(j-1)} - \tfrac{1}{4}s^2V_0^{(j-2)}$$

up to $j = j_{max}$ according to user's instruction

Fig. 4.3.7 Survey of the process of simultaneous vector iteration (for the sake of clarity some details referred to in the text have been omitted), *Part 2*

number of eigenvalues, n_i, in the lower frequency spectrum (high eigenvalues) within a system of high order n.

Small examples are of course not suitable to demonstrate the merits of the simultaneous vector iteration scheme. They have, however, the advantage of clarity and transparency. To demonstrate the acceleration by the Chebychev scheme, we use the oscillatory problem of a chain of spring-masses shown in fig. 4.3.8 to compute the first three eigenvalues and eigenvectors. The slowest convergence can be expected for the third eigenvector, which is conditioned by the ratio of the eigenvalues $\lambda_3/\lambda_4 = 1.5$. We next consider the basic formulation of the simultaneous vector iteration involving orthogonalisation and power iteration steps (fig.

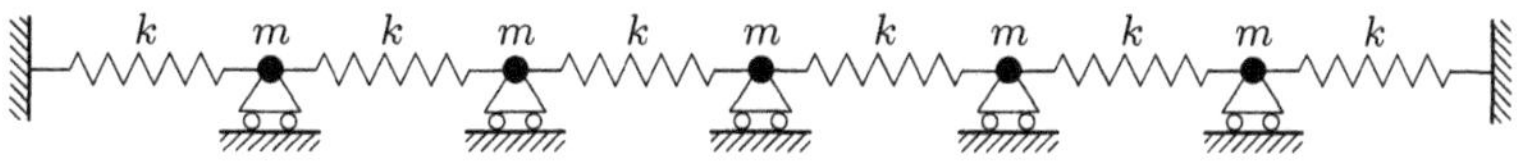

$$K = k\begin{bmatrix} 2 & -1 & 0 & 0 & 0 \\ & 2 & -1 & 0 & 0 \\ & & 2 & -1 & 0 \\ & \text{sym} & & 2 & -1 \\ & & & & 2 \end{bmatrix}, \qquad M = mI_5 \qquad \tilde{\lambda}_i = \lambda_i \frac{k}{m}$$

= eigenvalues of K, M-pair, factors k and m omitted

eigenvalues and eigenvectors:

i	1	2	3	4	5
$\tilde{\lambda}_i$	3.7321	1.0000	0.5000	0.3333	0.2679
x_i	0.5577 0.9659 1.1154 0.9659 0.5577	–0.5000 –0.5000 0.0000 0.5000 0.5000	0.4082 0.0000 –0.4082 0.0000 0.4082	–0.2887 0.2887 0.0000 –0.2887 0.2887	0.1494 –0.2588 0.2989 –0.2588 0.1494

Fig. 4.3.8 Example of an oscillatory chain for the demonstration of Chebychev's acceleration

4.3.9(a)) and notice that after eight iterations, the eigenvalue converges to a four decimal accuracy, while the third eigenvector has not as yet emerged with the same accuracy (this result is to be expected). The exact vector is incidentally

$$x_3 = \{0.4082 \quad 0. \quad -0.4082 \quad 0. \quad 0.4082\}$$

As the next alternative, we consider the Chebychev acceleration method using the third order polynomial. Since in our case the extension of the sub-space is not applicable (ni=nie), the criterion of equ. (4.3.34) for the spectral shift yields

$$s = \tfrac{1}{2}\tilde{\lambda}_{ni+1} = \tfrac{1}{2}\tilde{\lambda}_4 = \tfrac{1}{6} = 0.1667$$

We observe in fig. 4.3.9(b) that the Chebychev iteration step converges more rapidly than the power-iterative scheme. Often, in practice, one observes that only the first Chebychev iterations achieve a pronounced advantage. Therefore, we have limited the Chebychev method to the first power only (fig. 4.3.9(c)). In this case, we may prove without difficulty that a spectral shift of $s = 0.3$ and the modified spectrum

$$\tilde{\Lambda} = [3.43 \quad 0.70 \quad 0.20 \quad 0.03 \quad -0.03]$$

(a) standard simultaneous vector iteration

k	0	1	2	3	4	5	6	7	8	9
	0.0000	0.0587	0.2260	0.3199	0.3649	0.3866	0.3973	0.4026	0.4054	0.4068
	0.0000	−0.0293	0.0075	0.0119	0.0088	0.0056	0.0034	0.0020	0.0012	0.0007
$\boldsymbol{x}_3$	0.0000	−0.4107	−0.4762	−0.4611	−0.4406	−0.4266	−0.4183	−0.4137	−0.4112	−0.4098
	0.8165	0.4107	0.2020	0.1032	0.0534	0.0278	0.0145	0.0076	0.0040	0.0021
	0.4082	0.4107	0.4067	0.4090	0.4092	0.4087	0.4083	0.4082	0.4081	0.4081
$\tilde{\lambda}_3$	0.8333	0.5103	0.4841	0.4929	0.4977	0.4993	0.4998	0.4999	0.5000	0.5000

(b) three Chebychev iterations per loop ($s = 1/6 = 0.17$)

k	0	1				2			
		p.step	cheb 1	cheb 2	cheb 3	p.step	cheb 1	cheb 2	cheb 3
	0.0000	0.0587	0.2910	0.3870	0.4027	0.4057	0.4070	0.4075	0.4079
	0.0000	−0.0293	−0.0096	−0.0325	−0.0167	−0.0086	−0.0031	−0.0005	−0.0001
$\boldsymbol{x}_3$	0.0000	−0.4107	−0.4565	−0.3984	−0.3996	−0.4037	−0.4069	−0.4084	−0.4084
	0.8165	0.4107	0.1372	0.0181	0.0023	0.0009	0.0007	0.0008	0.0004
	0.4082	0.4107	0.4315	0.4373	0.4217	0.4152	0.4108	0.4088	0.4084
$\tilde{\lambda}_3$	0.8333	0.5103	0.4982	0.5011	0.5000	0.5000	0.5000	0.5000	0.5000

(c) one Chebychev iteration per loop ($s = 0.3$)

k	0	1		2		3		4	
		p.step	cheb 1	p.step	cheb 1	p.step	cheb 1	p.step	cheb 1
	0.0000	0.0587	0.3727	0.3918	0.3994	0.4038	0.4074	0.4078	0.4081
	0.0000	−0.0293	−0.0803	−0.0423	−0.0044	−0.0021	−0.0013	−0.0006	−0.0001
$\boldsymbol{x}_3$	0.0000	−0.4107	−0.3627	−0.3863	−0.4114	−0.4100	−0.4080	−0.4081	−0.4083
	0.8165	0.4107	0.0215	0.0083	0.0100	0.0051	0.0008	0.0004	0.0002
	0.4082	0.4107	0.4735	0.4423	0.4138	0.4110	0.4094	0.4088	0.4084
$\tilde{\lambda}_3$	0.8333	0.5103	0.5016	0.5002	0.5001	0.5000	0.5000	0.5000	0.5000

Fig. 4.3.9 Results of the Chebychev acceleration on the example in fig. 4.3.8., 3rd eigenvector (p.step = power step, cheb = Chebychev)

are conducive to a further acceleration of the iteration procedure. Indeed, the Chebychev step is now dependent on the ratio

$$\tilde{\lambda}_3/\tilde{\lambda}_4 = 0.2000/0.0333 = 6.61$$

which is more effective compared to the value of 2.00 (1.5) for a spectral shift of $s = 1/6$ ($s = 0$). This last computation provides only, of course, a demonstration artistry since in a real application, the eigenvalue spectrum is unknown and therefore cannot be exploited as in our demonstration.

Our next example is taken from [4.1], where the computations were performed on a VAX 785 computer using double precision accuracy. The numerical operations are based on the simultaneous vector iteration scheme implemented in ASKA, which is essentially the same as that discussed in the preceding section. Figure 4.3.10 shows the geometrical and material data of the structure as well as the complete spectrum of the eigenvalues. A consistent mass matrix was used in the analysis and

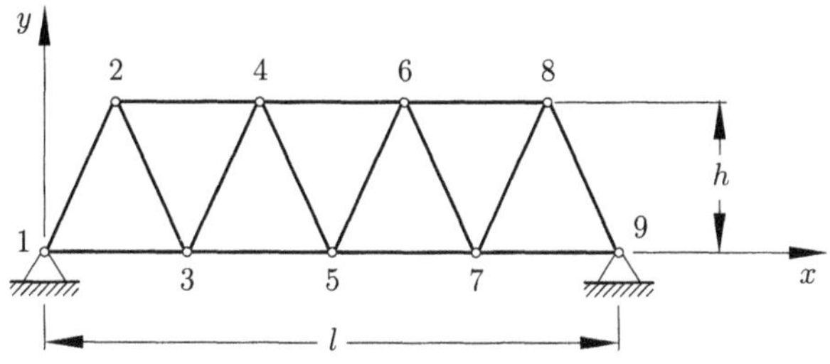

data:

span	$l = 8$ m
height	$h = 2$ m
cross-section area	$A = 0.001$ m^2
density	$\rho = 8000$ Ns^2/m^4
Young's modulus	$E = 2.1 \cdot 10^{11}$ N/m^2

i	ν_i [Hz]
1	73.0
2	135.7
3	187.5
4	309.0
5	338.6
6	361.0
7	455.0
8	523.4
9	562.0
10	598.3
11	618.3
12	640.0
13	772.5
14	785.0

Fig. 4.3.10 Structure idealised with FLA2 elements [4.1]

the eigenvalues were computed by the Jacobi method. Our aim was to determine the first three eigenvalues and eigenvectors by the method of simultaneous vector iteration. A perusal of the eigenvalue spectrum indicates that the choice of the desired eigenvalues is not unfavourable, because a marked discontinuity in the spectrum occurs between ν_3 and ν_4. The user is, however, generally unaware of such a situation and would therefore choose more vectors than required. For example, since ni=3 is the desired number of vectors, one would select nie=5 for the initial iteration vectors. With this, the rate of convergence always increases but, of course, so does the cost of computation. Next, we turn our attention to fig. 4.3.11, which displays the quality of the convergence, and the desired eigenvalues and eigenvectors. In the first iteration, the power iteration together with the Rayleigh-Ritz decoupling was employed. This step provides the first information on the eigenvalue spectrum, and hence permits the execution of a Chebychev acceleration step. As a consequence of the vectors carried along in our operations, only approximations for λ_1 to λ_5 are available. The general-purpose computer programme ASKA utilises $s = \lambda_5/2$ as the spectral shift. This guarantees the amplification of the roots λ_1 to λ_4 as compared to the eigenvalues λ_5 to λ_n. Since λ_5 is still inaccurate after the first iteration, it is not worth performing more than one Chebychev step. The accuracy is controlled by linear and quadratic bounds to the eigenvalues. In the main iteration step 3, nearly exact information on the eigenvalues is available, so that a two-step Chebychev iteration can be performed as in the standard technique. As we have said earlier, there is no need to adopt a power iteration scheme of order higher than two. However, the user has the option of exceeding this order of the power if required. After the main iteration loop 3, the first eigenvalue and eigenvectors have fully converged (as may be checked by the residues) and may be excluded from the subsequent iterations. The same procedure then applies to the 2nd eigenvalue and vector, and the scheme may be

k	1	2	3	4	5
	p.step	p.step + cheb 1	p.step + cheb 1+2	p.step + cheb 1+2	p.step + cheb 1+2
$10^6\lambda_i$	4.75233 1.37467 0.71410	4.75243 1.37486 0.72060	4.75243 1.37486 0.72072	* 1.37486 0.72072	* * 0.72072
$10^6 r_i$	0.02097 0.02875 0.16630	0.00040 0.00166 0.02177	0.00000 0.00005 0.00195	0.00000 0.00000 0.00001	0.00000 0.00000 0.00000

(a) convergence behaviour $\lambda_i = \dfrac{1}{\omega_i^2} = \dfrac{1}{(2\pi\nu_i)^2}$ r_i = vector residuals

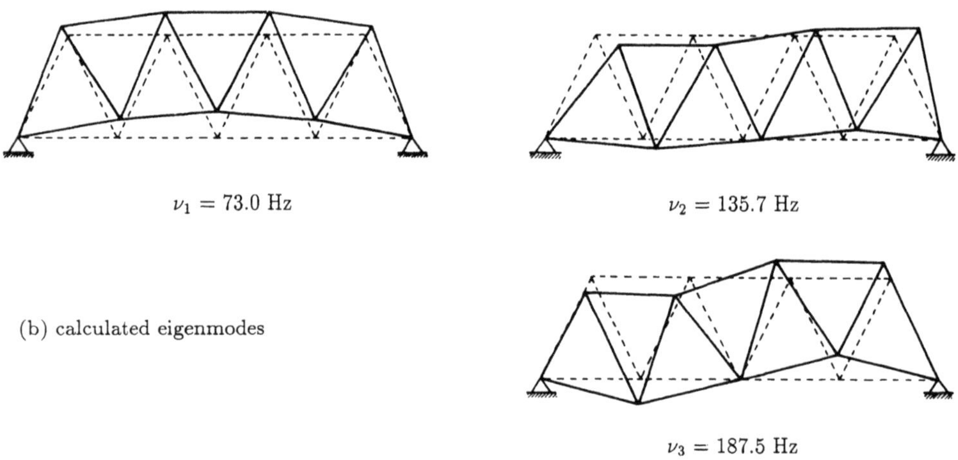

(b) calculated eigenmodes

Fig. 4.3.11 Calculation of the example in fig. 4.3.10 using simultaneous vector iteration and Chebychev's acceleration [4.1] (p. step = power step, cheb = Chebychev)

continued in this way until all five eigenvalues and vectors are established. For the norm of the vector residues we have

$$r_i = \left([\boldsymbol{A}\boldsymbol{x}_i - \lambda_i\boldsymbol{x}_i]^t[\boldsymbol{A}\boldsymbol{x}_i - \lambda_i\boldsymbol{x}_i]\right)^{1/2}$$

We restrict the presentation here to six decimal places in fixed format so as not to overload the figure, although the actual computations were carried out with floating-point arithmetics to fourteen decimal places.

4.3.3 Complete spectral analysis of a system: Jacobi algorithm

The iterative computation of eigenvalues using the Jacobi method also involves an indeterminate number of computational steps as was the case for the algorithms discussed in sections 4.3.1 and 4.3.2. These methods are hereafter classified as 'indirect methods'. The number of operations is in the order of n^3, where n is the order of the dynamical matrix.

The Jacobi method proceeds to the diagonalisation of the dynamical matrix $\boldsymbol{A}$ by a series of transformations. Each of these transformations is construed in such a manner that one element outside the diagonal becomes zero (this need not remain so in the course of subsequent transformations). The process is repeated until all off-diagonal elements are reduced to sufficiently small values.

Denoting the transformation matrix of a step by $\boldsymbol{Q}^{(k)}$, we obtain

$$\boldsymbol{A}^{(k+1)} = [\boldsymbol{Q}^{(k)}]^t \boldsymbol{A}^{(k)} \boldsymbol{Q}^{(k)} \tag{4.3.42}$$

which converges to

$$\boldsymbol{A}^{(k)} \to \boldsymbol{\Lambda} \tag{4.3.43}$$

Since the sequence of $\boldsymbol{Q}^{(k)}$ transformations diagonalises the original matrix $\boldsymbol{A}$, the resulting eigenvectors are given by

$$\overline{\boldsymbol{X}} = \boldsymbol{Q}^{(1)} \boldsymbol{Q}^{(2)} \boldsymbol{Q}^{(3)} \ldots \boldsymbol{Q}^{(k)} \tag{4.3.44}$$

The specific transformation matrix for the cancellation of the element a_{pq} takes the form

$$\boldsymbol{Q}^{(k)} = \begin{array}{c} \\ \left[\begin{array}{ccccccc} 1 & & & & & & \\ & 1 & & & & & \\ & & \boxed{c} & & & \boxed{s} & \\ & & & 1 & & & \\ & & & & 1 & & \\ & & & & & 1 & \\ & & \boxed{-s} & & & \boxed{c} & \\ & & & & & & 1 \end{array} \right] \end{array} \begin{array}{l} \\ \\ p \\ \\ \\ \\ q \\ \\ \end{array} \tag{4.3.45}$$

(column labels: p above the c/$-s$ column, q above the s/c column)

with

$$1 = c^2 + s^2 \tag{4.3.46}$$

As we confirm easily with the multiplication scheme of equ. (4.3.42), only rows and columns p and q in the $\boldsymbol{A}$ matrix are affected in the operation

$$\boldsymbol{A}^{(\mathrm{k+1})} = [\boldsymbol{Q}^{(\mathrm{k})}]^{\mathrm{t}} \boldsymbol{A}^{(\mathrm{k})} \boldsymbol{Q}^{(\mathrm{k})} = \tag{4.3.47}$$

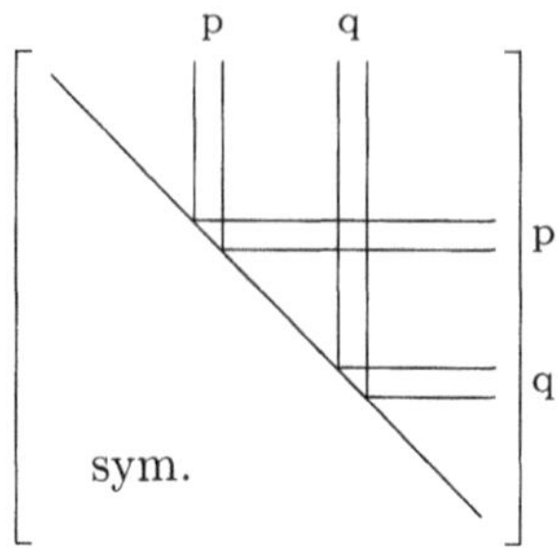

in which

$$a_{\mathrm{ip}}^{(\mathrm{k+1})} = a_{\mathrm{ip}}^{(\mathrm{k})} c - a_{\mathrm{iq}}^{(\mathrm{k})} s$$

$$a_{\mathrm{iq}}^{(\mathrm{k+1})} = a_{\mathrm{ip}}^{(\mathrm{k})} s + a_{\mathrm{iq}}^{(\mathrm{k})} c \; ; \qquad \mathrm{i} \neq \mathrm{p,q} \tag{4.3.48}$$

where the elements below the diagonal correspond to the rows on the right of the diagonal (symmetry). At the points of intersection, we have

$$a_{\mathrm{pp}}^{(\mathrm{k+1})} = a_{\mathrm{pp}}^{(\mathrm{k})} c^2 - 2a_{\mathrm{pq}}^{(\mathrm{k})} cs + a_{\mathrm{qq}}^{(\mathrm{k})} s^2$$

$$a_{\mathrm{qq}}^{(\mathrm{k+1})} = a_{\mathrm{pp}}^{(\mathrm{k})} s^2 + 2a_{\mathrm{pq}}^{(\mathrm{k})} cs + a_{\mathrm{qq}}^{(\mathrm{k})} c^2$$

$$a_{\mathrm{pq}}^{(\mathrm{k+1})} = \left(a_{\mathrm{pp}}^{(\mathrm{k})} - a_{\mathrm{qq}}^{(\mathrm{k})} \right) cs + a_{\mathrm{pq}}^{(\mathrm{k})} (c^2 - s^2) \stackrel{!}{=} 0 \tag{4.3.49}$$

The last equation in (4.3.49) determines c and s, in conjunction with equ. (4.3.46). We proceed thereby in the following steps

$$\begin{aligned} &(a) \qquad \vartheta = \frac{a_{\mathrm{qq}}^{(\mathrm{k})} - a_{\mathrm{pp}}^{(\mathrm{k})}}{2a_{\mathrm{pq}}^{(\mathrm{k})}} \\ &(b) \qquad t = \frac{1}{\vartheta \pm \sqrt{\vartheta^2 + 1}} \\ &(c) \qquad c = \frac{1}{\sqrt{1 + t^2}} \\ &(d) \qquad s = ct \end{aligned} \tag{4.3.50}$$

In step (b), the sign is chosen so that only a minimum loss of numerical accuracy is involved.

The fact that such a simple method as the Jacobi one can yield both eigenvalues and eigenvectors indeed appears so astonishing that the exploration of its background may be helpful to the reader. Further details on the Jacobi method may also be found in the excellent mathematical monograph by Schwarz/Rutishauser and Stiefel [4.8] or any other standard textbook on matrix computations.

Let us now look more closely at equ. (4.3.49). By summing all the diagonal elements of the matrix $\boldsymbol{A}^{(k+1)}$, we observe that only the pp and qq elements are modified. At the same time, equ. (4.3.49) shows that

$$a_{pp}^{(k+1)} + a_{qq}^{(k+1)} = a_{pp}^{(k)} + a_{qq}^{(k)} \tag{4.3.51}$$

From this, it follows immediately that the sum of all the diagonal elements is constant, i.e.

$$S_D^{(k+1)} = \sum_{i=1}^{n} a_{ii}^{(k+1)} = \sum_{i=1}^{n} a_{ii}^{(k)} = S_D^{(k)} \tag{4.3.52}$$

Having successfully diagonalised the dynamical matrix $\boldsymbol{A}$ by a series of Jacobi transformations, we deduce from equ. (4.3.52)

$$S_D = \sum_{i=1}^{n} a_{ii} = \sum_{i=1}^{n} \lambda_i \tag{4.3.53}$$

Consequently, the sum of the diagonal elements of $\boldsymbol{A}$ is equal to the sum of the eigenvalues.

Next, we consider the off-diagonal elements and form the sum of their squares. We notice that only the rows (columns) p and q are modified. By combining pairs of terms in equ. (4.3.48), we easily obtain

$$\left(a_{ip}^{(k+1)}\right)^2 + \left(a_{iq}^{(k+1)}\right)^2 = \left(a_{ip}^{(k)}\right)^2 + \left(a_{iq}^{(k)}\right)^2 \quad ; \qquad i \neq p, q \tag{4.3.54}$$

Thus, all these elements need not be considered when discussing the modification of the aforementioned sum. But the two elements $a_{pq}^{(k)}$ and $a_{qp}^{(k)}$, which do not appear in equ. (4.3.54), are reduced to zero. Therefore, we find

$$S_N^{(k+1)} = \sum_{i=1}^{n} \sum_{\substack{j=1 \\ j \neq i}}^{n} \left(a_{ij}^{(k+1)}\right)^2 = S_N^{(k)} - 2\left(a_{pq}^{(k)}\right)^2 \tag{4.3.55}$$

Obviously, a steady diagonalisation of $\boldsymbol{A}$ takes place because S_N decreases monotonically and S_D remains constant.

Although it would be theoretically advantageous to choose the largest possible a_{pq} element in order to reduce S_N rapidly, such a choice is by no means easy and therefore one generally prefers to process the matrix $\boldsymbol{A}$ with a systematic algorithm.

As a typical example, let us now consider the oscillator shown in fig. 4.3.3, which was analysed by the simple vector iteration scheme in fig. 4.3.4 and by the simultaneous vector iteration scheme in fig. 4.3.9. Figure 4.3.12 represents the dynamical matrix $\boldsymbol{A}$ of the problem which must now be formed explicitly. Since the matrix $\boldsymbol{A}$, unlike the $\boldsymbol{M}$ and $\boldsymbol{U}$ matrices, is fully populated, this requires a large memory space in the case of a high number of degrees of freedom. Consequently, the computation of large problems can no longer be carried out in the main memory of a computer, a problem to which we shall return later.

In our example, the dynamical matrix $\boldsymbol{A}$ possesses dominant diagonal elements, a feature which is most useful for the Jacobi method. Equation(4.3.53) can easily be verified with the help of the exact eigenvalues given in fig. 4.3.3. The extent to which the diagonal elements of this dynamical matrix approximate the true eigenvalues is of significant interest. Following [4.9], we have

$$\left|a_{\mathrm{ii}}^{(\mathrm{k})} - \lambda_{\mathrm{i}}\right| \leqslant \left(S_N^{(\mathrm{k})}\right)^{1/2} \quad ; \qquad \mathrm{i} = 1 \ldots \mathrm{n} \tag{4.3.56}$$

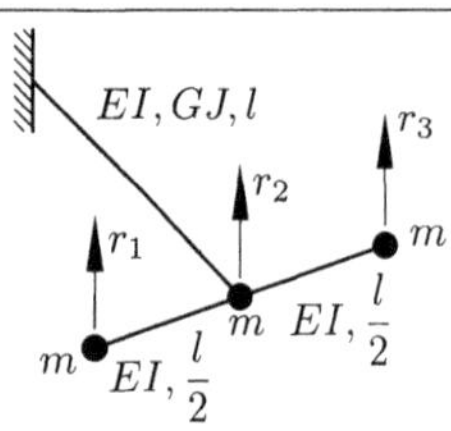

$$\frac{EI}{GJ} = \alpha = \frac{23 + \sqrt{593}}{24}$$

rotation	$\boldsymbol{A}^{(\mathrm{k})} \longrightarrow \boldsymbol{\Lambda}$	
	$\frac{ml^3}{24EI}\begin{bmatrix} 1.9221 & 5.9160 & -1.1656 \\ & 23.1376 & 0.3031 \\ \text{sym.} & & 24.6161 \end{bmatrix}$	$S_D = 49.6758$, $S_N = 72.8990$, $\sqrt{S_N} = 8.54$
p = 1 q = 2, $c = 0.9678$, $s = 0.2516$	$\frac{ml^3}{24EI}\begin{bmatrix} 0.3839 & 0 & -1.2044 \\ & 24.6758 & 0 \\ \text{sym.} & & 24.6161 \end{bmatrix}$	$S_N = 2.9010$, $\sqrt{S_N} = 1.70$
p = 1 q = 3, $c = 0.9988$, $s = -0.0495$	$\frac{ml^3}{24EI}\begin{bmatrix} 0.3244 & 0 & 0 \\ & 24.6758 & 0 \\ \text{sym.} & & 24.6758 \end{bmatrix}$	$S_N = 0$, $\sqrt{S_N} = 0$

$$\overline{\boldsymbol{X}} = \begin{bmatrix} 0.9678 & 0.2516 & 0 \\ -0.2516 & 0.9678 & 0 \\ 0 & 0 & 1 \end{bmatrix} \begin{bmatrix} 0.9988 & 0 & -0.0495 \\ 0 & 1 & 0 \\ 0.0495 & 0 & 0.9988 \end{bmatrix} = \begin{bmatrix} 0.9666 & 0.2516 & -0.0479 \\ -0.2513 & 0.9678 & 0.0125 \\ 0.0495 & 0 & 0.9988 \end{bmatrix} = \boldsymbol{UX}$$

$$\boldsymbol{X} = \boldsymbol{U}^{-1}\overline{\boldsymbol{X}} = \sqrt{\frac{l^3}{24EI}} \begin{bmatrix} 0.2263 & 4.4790 & -0.8529 \\ -0.4711 & 2.1488 & 1.7777 \\ 0.2259 & 0 & 4.5586 \end{bmatrix} \quad , \quad \boldsymbol{K}\text{-orthonormal}$$

Fig. 4.3.12 Application of the Jacobi method to the oscillator of fig. 4.3.3

which also defines a possible break-off criterion. If this is applied to our example, we observe that the diagonal elements of $\boldsymbol{A}$ differ from the eigenvalues by ≈ 8.5 at the most. Astonishingly enough, our example yields the exact solution in only two Jacobi iterations, a fact which is due to the special nature of the problem and does not generally apply. For example, in our present column-wise procedure, we accidentally pick the maximum off-diagonal element in each case. Usually, the off-diagonal elements of the $\boldsymbol{A}$ matrix must be worked through repeatedly, whereby (n-1)n/2 steps of computation are required per iteration. The modification of the p and q columns in accordance with equ. (4.3.48), for which 4n multiplications are required, is the most time-consuming process within each step. This illustrates that the Jacobi method is a process involving $\mathrm{n}_D 2\mathrm{n}^3$ multiplications in which n_D is the number of loops required and depends essentially upon the structure of the matrix $\boldsymbol{A}$ and the chosen break-off criterion. Let us now reconsider the eigenvectors in fig. 4.3.12. When comparing to the vectors in fig. 4.3.3, we should observe that in the present case, we obtain vectors orthonormal to the stiffness matrix $\boldsymbol{K}$. But for a scalar factor, the first vector agrees with $\boldsymbol{x}_3$. However, the second and third vectors given in fig. 4.3.12 are linear combinations of the vectors $\boldsymbol{x}_1$ and $\boldsymbol{x}_2$, as given in fig. 4.3.3. This is in fact not at all surprising for multiple eigenvalues.

Let us now turn finally to a brief discussion of the application of the Jacobi method to a large scale problem, in which the dynamic matrix $\boldsymbol{A}$ cannot be stored completely in the central memory of the computer [4.10]. Our aim here is not directed at merely eliminating an off-diagonal element of the matrix $\boldsymbol{A}$, but instead at cancelling an off-diagonal sub-matrix. Indeed, considering the hypermatrix $\boldsymbol{A}$ stored on the secondary memory as

$$\boldsymbol{A} = \begin{bmatrix} \boldsymbol{A}_{11} & \cdots & & \cdots & \boldsymbol{A}_{1\mathrm{N}} \\ & \ddots & & \vdots & \vdots \\ & & \boldsymbol{A}_{\mathrm{pp}} \cdots & \boldsymbol{A}_{\mathrm{pq}} \cdots & \vdots \\ & & \ddots & \vdots & \vdots \\ & & & \boldsymbol{A}_{\mathrm{qq}} & \vdots \\ & & & \ddots & \vdots \\ & & & & \boldsymbol{A}_{\mathrm{NN}} \end{bmatrix} \tag{4.3.57}$$

we now attempt to eliminate the sub-matrix $\boldsymbol{A}_{\mathrm{pq}}$. For this purpose, we form the small matrix

$$\begin{bmatrix} \boldsymbol{A}_{\mathrm{pp}} & \boldsymbol{A}_{\mathrm{pq}} \\ \mathrm{sym} & \boldsymbol{A}_{\mathrm{qq}} \end{bmatrix} \tag{4.3.58}$$

in the main memory and apply the Jacobi method, with which we are already familiar

$$\begin{bmatrix} \boldsymbol{A}_{\mathrm{pp}} & \boldsymbol{A}_{\mathrm{pq}} \\ \mathrm{sym} & \boldsymbol{A}_{\mathrm{qq}} \end{bmatrix} \Longrightarrow \begin{bmatrix} \boldsymbol{\Lambda}_{\mathrm{p}}^{(\mathrm{pq})} & \boldsymbol{O} \\ \boldsymbol{O}^{\mathrm{t}} & \boldsymbol{\Lambda}_{\mathrm{q}}^{(\mathrm{pq})} \end{bmatrix} , \begin{bmatrix} \overline{\boldsymbol{X}}_{\mathrm{pp}}^{(\mathrm{pq})} & \overline{\boldsymbol{X}}_{\mathrm{pq}}^{(\mathrm{pq})} \\ \overline{\boldsymbol{X}}_{\mathrm{qp}}^{(\mathrm{pq})} & \overline{\boldsymbol{X}}_{\mathrm{qq}}^{(\mathrm{pq})} \end{bmatrix} \tag{4.3.59}$$

This expression shows that a transformation of the global matrix in equ. (4.3.57) by

$$
\boldsymbol{Q}^{(\mathrm{pq})} = \begin{bmatrix}
\boldsymbol{I} & & & & & & & & \\
& \ddots & & & & & & & \\
& & \boldsymbol{I} & & & & & & \\
& & & \overline{\boldsymbol{X}}_{\mathrm{pp}}^{(\mathrm{pq})} & \cdots\cdots & & \overline{\boldsymbol{X}}_{\mathrm{pq}}^{(\mathrm{pq})} & & \\
& & & & \boldsymbol{I} & & & & \\
& & & & & \ddots & \vdots & & \\
& & & & & \boldsymbol{I} & & & \\
& & & & & & \overline{\boldsymbol{X}}_{\mathrm{qq}}^{(\mathrm{pq})} & & \\
& & & & & & & \boldsymbol{I} & \\
& & & & & & & & \ddots \\
& & & & & & & & & \boldsymbol{I}
\end{bmatrix} \tag{4.3.60}
$$

indeed leads to the cancellation of the sub-matrix $\boldsymbol{A}_{\mathrm{pq}}$. The hypercolumns and hyperrows p and q of the $\boldsymbol{A}$ matrix can be updated in the same way, and the corresponding eigenvectors can be obtained by a chain multiplication with the matrices $\boldsymbol{Q}^{(\mathrm{pq})}$. In analogy to the simple Jacobi alogrithm, the maximum 'element' or sub-matrix to be reduced to zero can be searched for. As a criterion for these sub-matrices, we use the Euclidean norm of the off-diagonal matrix in the form

$$
\underset{(\mathrm{np}\times\mathrm{nq})}{\| \boldsymbol{A}_{\mathrm{pq}} \|_E} = \left(\sum_{\mathrm{i}=1}^{\mathrm{np}} \sum_{\mathrm{j}=1}^{\mathrm{nq}} \left(a_{\mathrm{ij}}^{(\mathrm{pq})} \right)^2 \right)^{1/2} \tag{4.3.61}
$$

The total number of off-diagonal sub-matrices is generally not too large, so that the evaluation of the norms is not computationally expensive. Moreover, the norm of equ. (4.3.61) is required anyhow for the verification of the convergence (S_N). Consequently, there may be some advantages in applying the cyclic process.

A short summary of the main advantages and disadvantages of the Jacobi method completes our discussion.

The advantages of the method are:

(1) It is a simple algorithm

(2) A hypermatrix version also exists, thus enabling the solution of problems not fitting into the main memory of the computer.

(3) Multiple eigenvalues can be computed without any difficulty.

(4) Small eigenvalues (or higher natural frequencies) are obtained with a relatively high degree of accuracy.

(5) Problems with dominant diagonals, decoupled blocks or banded morphology converge rapidly.

The disadvantages of the Jacobi method are:

(1) The Jacobi method operates with the dynamical matrix $\boldsymbol{A}$, which is generally fully populated and thus requires a high storage capacity for large-scale problems.

(2) There is only the choice of computing all the eigenvectors or none at all. Although in theory it is possible to compute subsequently specific vectors, this is computationally expensive as the fully populated $\boldsymbol{A}$ matrix must be used.

4.3.4 Determination of a partial spectrum and the associated eigenvectors: The Givens/Householder method, combined with bisection and inverse iteration with shift (Wielandt)

Whilst also being founded on the special eigenvalue problem specified by the dynamical matrix $\boldsymbol{A}$, the Householder method does not attempt to diagonalise directly the $\boldsymbol{A}$ matrix by means of a series of transformations as the Jacobi method does, but instead, reduces a symmetric matrix $\boldsymbol{A}$ to a symmetric tridiagonal matrix matrix $\boldsymbol{T}$. The Householder method is also commonly known as a direct method since the process is completed in a finite number of steps. This is, however, not strictly correct as an undetermined number of steps is required to obtain the eigenvalue spectrum of the matrix $\boldsymbol{T}$. Alternative techniques of transforming an $\boldsymbol{A}$ matrix to a tridiagonal form are the Givens method and the Householder method. The two methods are essentially identical, although the Givens method tends to be simpler, whilst Householder's method is somewhat more economical. For the purpose of this discussion, the Givens method will be employed, but we also refer the reader to section *4.3.4. This method is very similar to that of Jacobi except that here, only the elements outside the tridiagonal band are reduced to zero and remain so in further transformations. The basic operation once again is

$$\boldsymbol{A}^{(\mathrm{k}+1)} = [\boldsymbol{Q}^{(\mathrm{k})}]^{\mathrm{t}} \boldsymbol{A}^{(\mathrm{k})} \boldsymbol{Q}^{(\mathrm{k})} \tag{4.3.62}$$

in which, proceeding column by column, the $a_{\mathrm{pq}}(\mathrm{q} \geqslant \mathrm{p} + 2)$ element is nullified with the transformation matrix

$$\boldsymbol{Q}^{(\mathrm{k})} = \begin{array}{c} \begin{matrix} & & \mathrm{p+1} & & & & \mathrm{q} & & \end{matrix} \\ \left[\begin{matrix} 1 & & & & & & & & \\ & 1 & & & & & & & \\ & & c & & & & s & & \\ & & & 1 & & & & & \\ & & & & \ddots & & & & \\ & & & & & 1 & & & \\ & & -s & & & & c & & \\ & & & & & & & 1 & \\ & & & & & & & & 1 \end{matrix}\right] \begin{matrix} \\ \\ \mathrm{p+1} \\ \\ \\ \\ \mathrm{q} \\ \\ \\ \end{matrix} \end{array} \tag{4.3.63}$$

in which

$$c = \frac{a_{p,p+1}}{\sqrt{a^2_{p,p+1} + a^2_{p,q}}}$$

$$s = \frac{a_{p,q}}{\sqrt{a^2_{p,p+1} + a^2_{p,q}}} \tag{4.3.64}$$

It is at once apparent that this single step is a great deal simpler than in the Jacobi method. It is helpful at this point to clarify the effects of the transformation. When the a_{pq} element is reduced to zero, all anteceding elements outside the tridiagonal band, operated upon in a column-wise sequence, have already been nullified. The transformation itself modifies only the rows (columns) p+1 and q of $\boldsymbol{A}^{(k)}$

$$\boldsymbol{A}^{(k)} = \begin{array}{c} \\ \left[\begin{array}{ccccccccc} \times & \times & 0 & 0 & 0 & 0 & 0 & \times & \times \\ & \times & \times & 0 & 0 & 0 & 0 & \times & \times \\ & & \times & \times & 0 & 0 & \otimes & \times & \times \\ & & & \times & \times & 0 & \times & \times & \times \\ & & & & \times & \times & \times & \times & \times \\ & & & & & \times & \times & \times & \times \\ & & & & & & \times & \times & \times \\ & & & & & & & \times & \times \\ & \text{sym.} & & & & & & & \times \end{array}\right] \end{array} \begin{array}{l} \\ \\ \text{p} \\ \text{p+1} \\ \\ \\ \text{q} \\ \\ \\ \end{array}$$

(columns p, p+1 and q are the 3rd, 4th and 7th columns)

At the points of intersection, which now contain the points (p+1, p+1), (p+1,q) and (q,q), the scheme in equ. (4.3.49) is, in principle, still applicable. In the present case, though, p is replaced by p + 1 and $a^{(k+1)}_{p+1,q}$ is no longer required to be zero. Thus, we have

$$a^{(k+1)}_{p+1,p+1} = a^{(k)}_{p+1,p+1}c^2 - 2a^{(k)}_{p+1,q}cs + a^{(k)}_{qq}s^2$$

$$a^{(k+1)}_{q,q} = a^{(k)}_{p+1,p+1}s^2 + 2a^{(k)}_{p+1,q}cs + a^{(k)}_{q,q}c^2$$

$$a^{(k+1)}_{p+1,q} = \left(a^{(k)}_{p+1,p+1} - a^{(k)}_{q,q}\right) cs + a^{(k)}_{p+1,q}(c^2 - s^2) \tag{4.3.65}$$

Outside these points of intersection, relations analogous to equ. (4.3.48) apply

$$a^{(k+1)}_{i,p+1} = a^{(k)}_{i,p+1}c - a^{(k)}_{i,q}s$$

$$a^{(k+1)}_{i,q} = a^{(k)}_{i,p+1}s + a_{i,q}c \qquad i \neq p + 1, q \tag{4.3.66}$$

Equation (4.3.66) indicates that already existent zeros in the columns (rows) p+1 and q remain unaltered. Unlike the Jacobi rotation scheme, the field of zeros once created by the Givens method is subsequently not affected. Since in a column-wise process the elements for i < p are already zero, it suffices subsequently to modify the index i from p to n. Hence, after (n-1)(n-2)/2 steps, the $\boldsymbol{A}$ matrix is reduced to a tridiagonal form. The overall transformation matrix is then given by

$$\underset{(n\times n)}{\boldsymbol{Q}} = \boldsymbol{Q}^{(1)}\boldsymbol{Q}^{(2)}\boldsymbol{Q}^{(3)}\dots \tag{4.3.67}$$

Also, $\boldsymbol{Q}$ is orthonormal

$$\boldsymbol{Q}^{t}\boldsymbol{Q} = \boldsymbol{I}_{n} = \boldsymbol{Q}\boldsymbol{Q}^{t} \tag{4.3.68}$$

since orthogonality applies to each factor of equ. (4.3.67)

$$[\boldsymbol{Q}^{(k)}]^{t}\boldsymbol{Q}^{(k)} = \boldsymbol{I}_{n} = \boldsymbol{Q}^{(k)}[\boldsymbol{Q}^{(k)}]^{t} \tag{4.3.69}$$

With respect to the eigenvalues of the dynamical matrix $\boldsymbol{A}$, we have

$$\begin{aligned}\Lambda &= \overline{\boldsymbol{X}}^{t}\boldsymbol{A}\overline{\boldsymbol{X}}\\ &= \overline{\boldsymbol{X}}^{t}\boldsymbol{Q}\boldsymbol{Q}^{t}\boldsymbol{A}\boldsymbol{Q}\boldsymbol{Q}^{t}\overline{\boldsymbol{X}} = \overline{\boldsymbol{X}}^{t}\boldsymbol{Q}\boldsymbol{T}\boldsymbol{Q}^{t}\overline{\boldsymbol{X}}\\ &= \boldsymbol{X}_{T}^{t}\boldsymbol{T}\boldsymbol{X}_{T}\end{aligned} \tag{4.3.70}$$

The vector set $\overline{\boldsymbol{X}}$ diagonalises the dynamic matrix $\boldsymbol{A}$, whilst the vector set $\boldsymbol{X}_T$ diagonalises the tridiagonal matrix $\boldsymbol{T}$, which possesses the same eigenvalues. Once the latter vector set has been found, it must naturally be transformed back to $\overline{\boldsymbol{X}}$ (which in turn will be transformed to $\boldsymbol{X}$)

$$\overline{\boldsymbol{X}} = \boldsymbol{Q}\boldsymbol{X}_{T} \tag{4.3.71}$$

Thus, if eigenvectors are also required, the $\boldsymbol{Q}$ matrix must be stored. Alternatively, we may secure the (n−1)(n−2)/2 pairs of c and s according to equ. (4.3.64).

At this point, we reconsider the example shown in fig. 4.2.8. The method of reducing the dynamic matrix $\boldsymbol{A}$ to the tridiagonal form $\boldsymbol{T}$ is demonstrated in fig. 4.3.13. Owing to the prevailing favourable factors, the task is performed in two steps only. As can be observed, the tridiagonal matrix is already decoupled and two eigenvalues appear directly on the diagonal. Consequently, the corresponding eigenvectors $\overline{\boldsymbol{x}}_3$ and $\overline{\boldsymbol{x}}_4$ are given by the third and fourth columns of the matrix $\boldsymbol{Q}$. Such a favourable circumstance for the solution is, of course, a rare exception.

Our next exercise consists of finding the eigenvalues of the tridiagonal matrix $\boldsymbol{T}$ (the elements of the $\boldsymbol{T}$ matrix are denoted by t_{ij}). For this purpose, an interval in which all eigenvalues have to be located, has to be selected. According to the Gershgorin theorem (see e.g. [4.11]), all eigenvalues of a tridiagonal matrix lie between the bounds a_0 and b_0

$$a_0 \leqslant \lambda_i \leqslant b_0\ ; \qquad i = 1\dots n \tag{4.3.72}$$

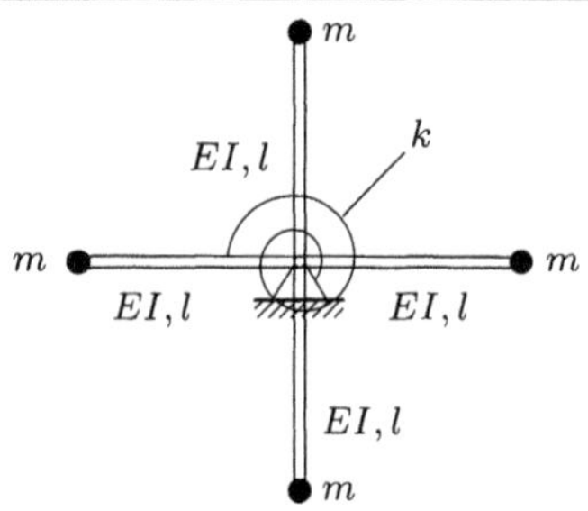

$$k = \frac{3EI}{l}$$

rotation	$A^{(k)} \longrightarrow T$
$S_D = 8 = \text{const}$	$\frac{ml^2}{k}\begin{bmatrix} \frac{5}{4} & \frac{1}{12}\sqrt{15} & \frac{1}{4}\sqrt{\frac{10}{3}} & \frac{1}{2}\sqrt{\frac{5}{2}} \\ & \frac{17}{12} & \frac{5}{12}\sqrt{2} & \frac{5}{6}\sqrt{\frac{3}{2}} \\ & & \frac{11}{6} & \frac{5}{6}\sqrt{3} \\ \text{sym.} & & & \frac{7}{2} \end{bmatrix}$
p = 1 , q = 3 p + 1 = 2 $c = \sqrt{1/3}$ $s = -\sqrt{2/3}$	$\frac{ml^2}{k}\begin{bmatrix} \frac{5}{4} & \frac{1}{4}\sqrt{5} & 0 & \frac{1}{2}\sqrt{\frac{5}{2}} \\ & \frac{9}{4} & 0 & \frac{5}{4}\sqrt{2} \\ & & 1 & 0 \\ \text{sym.} & & & \frac{7}{2} \end{bmatrix}$
p = 1 , q = 4 p + 1 = 2 $c = \sqrt{1/3}$ $s = -\sqrt{2/3}$	$\frac{ml^2}{k}\begin{bmatrix} \frac{5}{4} & \frac{3}{4}\sqrt{\frac{5}{3}} & 0 & 0 \\ & \frac{19}{4} & 0 & 0 \\ & & 1 & 0 \\ \text{sym.} & & & 1 \end{bmatrix}$

$$Q = Q_1 Q_2 = \frac{1}{3}\begin{bmatrix} 3 & 0 & 0 & 0 \\ 0 & 1 & -\sqrt{6} & -\sqrt{2} \\ 0 & \sqrt{2} & \sqrt{3} & -2 \\ 0 & \sqrt{6} & 0 & \sqrt{3} \end{bmatrix}$$

Fig. 4.3.13 Transformation of the dynamic matrix $\boldsymbol{A}$ of the problem in fig. 4.2.8 into the tridiagonal matrix $\boldsymbol{T}$

where

$$a_0 = \min_{\rm i}(t_{\rm ii} - r_{\rm i})$$
$$b_0 = \max_{\rm i}(t_{\rm ii} + r_{\rm i}) \tag{4.3.73}$$

in which

$$r_1 = |t_{21}|$$
$$r_{\rm i} = |t_{\rm i-1,i}| + |t_{\rm i+1,i}|$$
$$r_{\rm n} = |t_{\rm n-1,n}| \tag{4.3.74}$$

We thus sum up the moduli of the elements outside the diagonal within each column and form the difference to and the sum with the respective diagonal element. The minimum and maximum of these n numbers reproduce the lower and upper bounds a_0 and b_0 respectively. This information alone is very pertinent since in many cases, it suffices to know that the natural frequencies lie outside a certain frequency range.

Applying this scheme to the determination of the bounds in the example of fig. 4.3.13, we obtain the following enclosures for both the complete matrix and for the separable (2×2) sub-matrix

$$a_0 = \frac{1}{4}(5 - \sqrt{15})\frac{ml^2}{k} = 0.28\frac{ml^2}{k}$$

$$b_0 = (19 + \sqrt{15})\frac{ml^2}{k} = 5.72\frac{ml^2}{k}$$

which provides useful information considering our ignorance of the true eigenvalues $\lambda = 1ml^2/k$ and $\lambda = 5ml^2/k$.

Therefore, the zeros of the characteristic polynomial

$$P(\lambda) = \det \begin{bmatrix} \lambda - t_{11} & -t_{12} & 0 & 0 & \cdots \\ -t_{12} & \lambda - t_{22} & -t_{23} & 0 & \\ 0 & -t_{23} & \lambda - t_{33} & -t_{34} & \\ 0 & 0 & -t_{34} & \lambda - t_{44} & \\ \vdots & & & & \ddots \end{bmatrix} \tag{4.3.75}$$

must lie within the interval a_0, b_0. We now assume that the eigenvalues are arranged in descending order, i.e. the natural frequencies in ascending order, and our next step is to compute the g-th eigenvalue. For this purpose, we divide the interval $a_{\rm k}, b_{\rm k}$ by the mean value

$$c_{\rm k} = \tfrac{1}{2}(a_{\rm k} + b_{\rm k}); \qquad {\rm k} = 0, 1, \ldots \tag{4.3.76}$$

We begin with k=0 and inquire into the number of zeros of the characteristic polynomial $P(\lambda)$ for $\lambda > c_{\rm k}$. Let m be this number of zeros, and if m$\geqslant$g, then we know that the desired eigenvalue λ_{g} must lie in the right subinterval. Hence, the new interval is

$$\begin{aligned} a_{\mathrm{k}+1} &= c_{\mathrm{k}} \\ b_{\mathrm{k}+1} &= b_{\mathrm{k}} \qquad \text{if } \mathrm{m} \geqslant \mathrm{g} \end{aligned} \tag{4.3.77}$$

On the other hand, if $\mathrm{m} < \mathrm{g}$, then the desired eigenvalue must lie in the left subinterval

$$\begin{aligned} a_{\mathrm{k}+1} &= a_{\mathrm{k}} \\ b_{\mathrm{k}+1} &= c_{\mathrm{k}} \qquad \text{if } \mathrm{m} < \mathrm{g} \end{aligned} \tag{4.3.78}$$

The only question that now remains to be answered is what the magnitude of m is, i.e. the number of changes in sign of the polynomial $P(\lambda)$ in the range $\lambda > c_{\mathrm{k}}$. For this, we calculate the sequence of values of the polynomial

$$\begin{aligned} P_0(c_{\mathrm{k}}) &= 1 \\ P_1(c_{\mathrm{k}}) &= c_{\mathrm{k}} - t_{11} \\ P_2(c_{\mathrm{k}}) &= (c_{\mathrm{k}} - t_{11})(c_{\mathrm{k}} - t_{22}) - t_{12}^2 \\ &= (c_{\mathrm{k}} - t_{22})P_1(c_{\mathrm{k}}) - t_{12}^2 P_0(c_{\mathrm{k}}) \\ &\vdots \\ P_{\mathrm{i}}(c_{\mathrm{k}}) &= (c_{\mathrm{k}} - t_{\mathrm{ii}})P_{\mathrm{i}-1}(c_{\mathrm{k}}) - t_{\mathrm{i}-1,\mathrm{i}}^2 P_{\mathrm{i}-2}(c_{\mathrm{k}}) \end{aligned} \tag{4.3.79}$$

This series terminates with $P_{\mathrm{n}}(c_{\mathrm{k}})$, which reproduces the value of the characteristic polynomial for $\lambda = c_{\mathrm{k}}$. The other polynomials P_{i} represent the value of the determinant of a matrix formed by the first i rows and columns. Now, the required number is given by the number of changes in sign of the values of the polynomial series (4.3.79); for proof see [4.8]. In the trivial case where n=1, the assertion can easily be verified. Then the zero occurs simply at $\lambda_1 = t_{11}$ and if one were to choose $c_{\mathrm{k}} < \lambda_1$, then m=1 and one sign change should take place. This is indeed the case since $P_0 > 0$ and $P_1 = c_{\mathrm{k}} - t_{11} < 0$. Besides, in the general reasoning, it is assumed that the off-diagonal elements of the $\boldsymbol{T}$ matrix are different from zero. If this is not the case (as in our example, fig. 4.3.13), then the $\boldsymbol{T}$ matrix is decomposed into two or more smaller tridiagonal matrices, for which the above statement applies once more.

Figure 4.3.14 outlines the principle of the interval bisection method. To be more specific, we consider once more the problem of fig. 4.3.13, but limit ourselves to the decoupled submatrix

$$\boldsymbol{T}_{red} = \frac{ml^2}{k}\begin{bmatrix} \frac{5}{4} & \frac{3}{4}\sqrt{\frac{5}{3}} \\ \text{sym} & \frac{19}{4} \end{bmatrix}$$

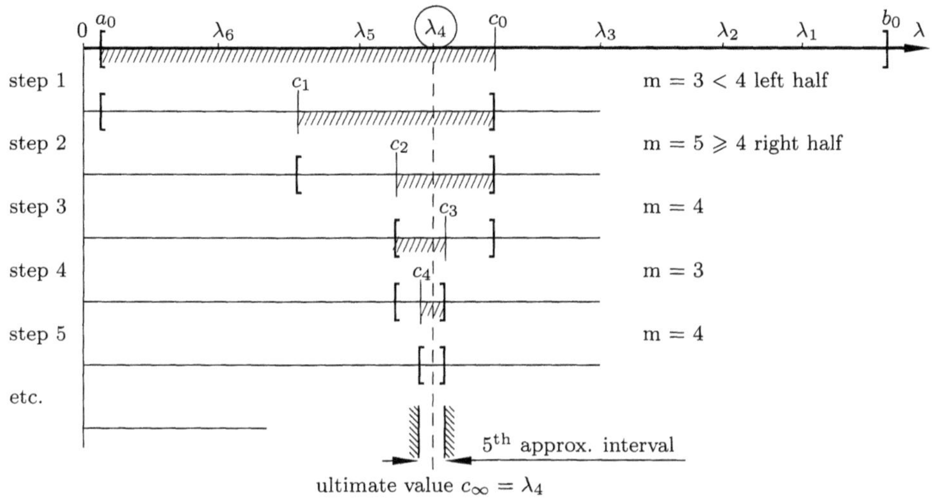

Fig. 4.3.14 Calculation of the eigenvalue λ_4 by the bisection method

Clearly, it is straightforward to calculate the eigenvalues directly from the quadratic secular equation. However, here we apply the bisection method using the following sequence of polynomials

$$P_0 = 1$$
$$P_1 = c_k - \tfrac{5}{4}$$
$$P_2 = (c_k - \tfrac{19}{4})P_1 - \tfrac{15}{16}$$

Let us attempt to find the largest eigenvalue λ_2 of the sub-system. The adopted procedure is reproduced in the following table (for simplicity, we have suppressed the factor ml^2/k):

k	lower limit a_k	centre c_k	upper limit b_k	P_0	P_1	P_2	m
0	0.28	3.00	5.72	+	+	−	1
1	3.00	4.36	5.72	+	+	−	1
2	4.36	5.04	5.72	+	+	+	0
3	4.36	4.70	5.04	+	+	−	1
4	4.70	4.87	5.04	+	+	−	1
5	4.87	4.96	5.04	+	+	−	1
6	4.96	5.00 solution	5.04				

The interval method also operates in the case of equal eigenvalues without difficulty. At all times, one obtains the correct magnitude and multiplicity of such eigenvalues.

Having determined the desired eigenvalues, we turn our attention to the eigenvectors which require some additional computations. For this purpose, we employ the inverse vector iteration with eigenvalue shift (Wielandt). We are already aware that vector iteration yields the eigenvector corresponding to the eigenvalue with the largest modulus, and the better the separation of the eigenvalues in question from all the others, the faster the convergence. It now remains to find the eigenvector for an arbitrary eigenvalue λ_k. The spectral shift with λ_k means that there is an eigenvalue shift from $\boldsymbol{\Lambda}$ to $[\boldsymbol{\Lambda} - \lambda_k \boldsymbol{I}]$. The shifted $\overline{\lambda}_k$ is thus zero and consequently the eigenvalue with the lowest modulus. We are already aware from our discussion on the simultaneous vector iteration that this procedure does not affect the eigenvectors

$$\boldsymbol{X}_T^t \boldsymbol{T} \boldsymbol{X}_T = \boldsymbol{\Lambda}$$

$$\boldsymbol{X}_T^t [\boldsymbol{T} - \lambda_k \boldsymbol{I}] \boldsymbol{X}_T = \boldsymbol{\Lambda} - \lambda_k \boldsymbol{I} = \overline{\boldsymbol{\Lambda}} \tag{4.3.80}$$

Our aim, however, is to make λ_k the eigenvalue with the largest modulus. To achieve this, we consider the inverse of the shifted matrix

$$[\boldsymbol{T} - \lambda_k \boldsymbol{I}]^{-1} = [\boldsymbol{X}_T \overline{\boldsymbol{\Lambda}} \boldsymbol{X}_T^t]^{-1} = \boldsymbol{X}_T^{-t} \overline{\boldsymbol{\Lambda}}^{-1} \boldsymbol{X}_T^{-1}$$

$$= \boldsymbol{X}_T \overline{\boldsymbol{\Lambda}}^{-1} \boldsymbol{X}_T^t \tag{4.3.81}$$

The inverse matrix is seen to possess the same eigenvectors as the original $\boldsymbol{T}$ matrix, but its eigenvalues are the inverse of the shifted spectrum. We observe that the eigenvalue corresponding to the original λ_k now has the largest modulus. Theoretically, it should be infinite. In fact, rounding-off errors inevitably produce a small number in place of a zero; as a result, $\overline{\lambda}_K^{-1}$ is merely very large. The following computational steps thus have to be taken:

(1) Shifting of the eigenvalues

$$\boldsymbol{T} \rightarrow \boldsymbol{T} - \lambda_k \boldsymbol{I} \tag{4.3.82}$$

(2) Operation with the inverse of $[\boldsymbol{T} - \lambda_k \boldsymbol{I}]$, in order to obtain by iteration the smallest eigenvalue of $[\boldsymbol{T} - \lambda_k \boldsymbol{I}]$

$$\boldsymbol{x}_T^{(\nu+1)} = [\boldsymbol{T} - \lambda_k \boldsymbol{I}]^{-1} \boldsymbol{x}_T^{(\nu)} \tag{4.3.83}$$

Since the convergence of the method is dependent upon the ratio between the second smallest and the smallest eigenvalue of $[\boldsymbol{T} - \lambda_k \boldsymbol{I}]$, the method converges within just a few iterations. A critical issue is the solution algorithm of equ. (4.3.83), since $[\boldsymbol{T} - \lambda_k \boldsymbol{I}]$ is nearly singular (for the exact value of λ_k, the determinant would be zero). In the following solution steps, this has to be taken into account:

(1) Triangularisation

$$[\boldsymbol{T} - \lambda_k \boldsymbol{I}] \rightarrow \boldsymbol{U}_T^t \boldsymbol{D}_T \boldsymbol{U}_T \tag{4.3.84}$$

(2) Forward substitution

$$\boldsymbol{U}_T^{\mathrm{t}}\boldsymbol{y} = \boldsymbol{x}_{T\mathrm{k}}^{(\nu)} \tag{4.3.85}$$

(3) Diagonal scaling

$$\boldsymbol{z} = \boldsymbol{D}_T^{-1}\boldsymbol{y} \tag{4.3.86}$$

(4) Back-substitution

$$\boldsymbol{U}_T\boldsymbol{x}_{T\mathrm{k}}^{(\nu+1)} = \boldsymbol{z} \tag{4.3.87}$$

Special measures, which cannot be detailed here, are necessary in the case of multiple eigenvalues. It still remains to transform the eigenvector deduced from equ. (4.3.87) into the form appertaining to the special eigenvalue problem

$$\overline{\boldsymbol{x}}_{\mathrm{k}} = \boldsymbol{Q}\boldsymbol{x}_{T\mathrm{k}} \tag{4.3.88}$$

and then further into the original general eigenvalue problem (see equ. (4.1.42)).

We now test our procedure on the example shown in fig. 4.3.13 and determine the eigenvector which corresponds to the eigenvalue $\lambda_{\mathrm{k}} = (5.00 + \varepsilon)ml^2/k$. Here, ε represents a kind of rounding error or the inaccuracy of the solution since we know the exact solution by chance. For the purpose of the iteration, we now consider once more the subsystem matrix of the order of 2×2 (fig. 4.3.15). The matrix $[\boldsymbol{T} - \lambda_{\mathrm{k}}\boldsymbol{I}\,]$ is tridiagonal and thus $\boldsymbol{U}_T$ contains only one secondary diagonal, which is not immediately apparent in our small example. It is necessary to extract the diagonal matrix $\boldsymbol{D}_T$, because $[\boldsymbol{T} - \lambda_{\mathrm{k}}\boldsymbol{I}\,]$ is symmetric, but not necessarily positive definite. Possibly, the diagonal matrix $\boldsymbol{D}_T$ will contain negative elements as is the case in the present example. Although $\boldsymbol{y}$ remains finite for small values of ε, the solutions $\boldsymbol{z}$ and the unnormalised vector $\boldsymbol{x}_T^{(1)}$ contain very large entries for small values of ε. However, by means of normalisation, finite entries, even for $\varepsilon = 0$, are obtained. For the purpose of convergence, it would be advantageous to select a very small value of ε. In the limit $\varepsilon = 0$, the iteration will evidently converge in one loop. In such a case, however, the computer imposes a limit since it is unable to represent arbitrarily large (or arbitrarily small) numbers. Thus, special attention must be paid to the practical implementation of the scheme.

To conclude, let us consider how to extend the method to the calculation of eigenvalues of large matrices which are stored in a hypermatrix scheme. Proceeding from the representation of the dynamic matrix $\boldsymbol{A}$ in equ. (4.3.57), all sub-matrices $\boldsymbol{A}_{\mathrm{pq}}$ outside the hyper-tridiagonal band can be eliminated, analogous to the simple Givens procedure. Finally, the hyper-tridiagonal matrix must be reduced to a tridiagonal matrix, an operation the Givens method is not suited to, since it destroys the established band structure. Nevertheless, we can call on another method, the Rutishauser process ([4.12], p.567) which executes the task in conjunction with the banded shape of the matrix. Alternatively, the Givens method can also be used to produce a tridiagonal matrix directly [4.4]. To this purpose, the submatrix $\boldsymbol{A}_{11}$ is tridiagonalised (fig. 4.3.16), this being a purely in-core process. The c and s terms

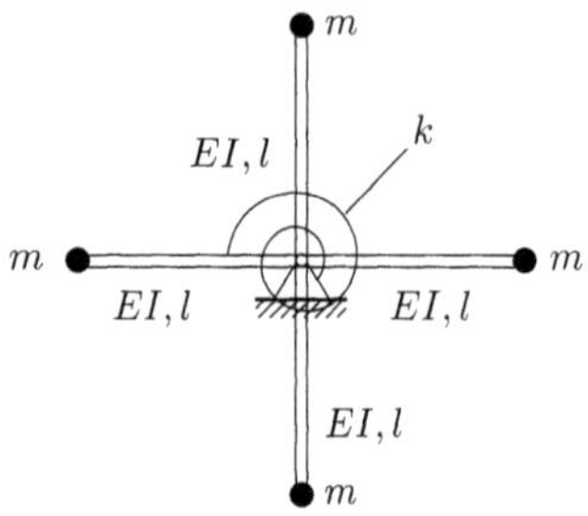

$$k = \frac{3EI}{l}$$

$$\lambda_k = (5.00 + \varepsilon)\frac{ml^2}{k} \qquad (\varepsilon = \text{small deviation})$$

see also figs. 4.3.13 and 4.2.8

$$\boldsymbol{T}_{\text{red}} - (\lambda_k + \varepsilon)\boldsymbol{I} = \frac{ml^2}{4k}\begin{bmatrix} -15 - 4\varepsilon & \sqrt{15} \\ \sqrt{15} & -1 - 4\varepsilon \end{bmatrix}$$ singular for $\varepsilon = 0$, tridiagonal

$$\boldsymbol{U}_T = \begin{bmatrix} 1 & -\sqrt{15}/(15 + 4\varepsilon) \\ 0 & 1 \end{bmatrix}$$ tridiagonal , non-singular

$$\boldsymbol{D}_T = \frac{ml^2}{4k}\left\lceil -(15 + 4\varepsilon) \quad - 16\varepsilon\frac{4 + \varepsilon}{15 + 4\varepsilon} \right\rfloor$$ diagonal , singular for $\varepsilon = 0$

$$\boldsymbol{x}_T^{(0)} = \{1 \quad 0\} \ , \qquad |\boldsymbol{x}_T^{(0)}| = 1$$

$$\boldsymbol{y} = \boldsymbol{U}_T^{-t}\boldsymbol{x}_T^{(0)} = \left\{1 \quad \frac{\sqrt{15}}{15 + 4\varepsilon}\right\}$$

$$\boldsymbol{z} = \boldsymbol{D}_T^{-1}\boldsymbol{y} = \frac{4k}{ml^2}\left\{-\frac{1}{15 + 4\varepsilon} \quad -\frac{\sqrt{15}}{16\varepsilon(4 + \varepsilon)}\right\}$$

$$\boldsymbol{x}_T^{(1)} = \boldsymbol{U}_T^{-1}\boldsymbol{z} = \frac{4k}{ml^2}\left\{-\frac{1 + 4\varepsilon}{16\varepsilon(4 + \varepsilon)} \quad -\frac{\sqrt{15}}{16\varepsilon(4 + \varepsilon)}\right\}$$ not normalised

$$\boldsymbol{x}_T^{(1)} = \frac{1}{4\sqrt{1 + 0.5\varepsilon + \varepsilon^2}}\{1 + 4\varepsilon \quad \sqrt{15}\}$$ normalised $|\boldsymbol{x}_T^{(1)}| = 1$

$$\lim_{\varepsilon \to 0} \boldsymbol{x}_T^{(1)} = \boldsymbol{x}_T(\lambda = 5) = \tfrac{1}{4}\{1 \quad \sqrt{15}\}$$

backward transformation:

1. $\overline{\boldsymbol{x}} = \boldsymbol{Q}\boldsymbol{x}_T = \frac{1}{4}\boldsymbol{Q}\left\{1 \quad \sqrt{15} \quad 0 \quad 0\right\} = \frac{1}{12}\left\{3 \quad \sqrt{15} \quad \sqrt{30} \quad \sqrt{90}\right\}$

2. $\boldsymbol{x} = \boldsymbol{U}^{-1}\overline{\boldsymbol{x}} = \sqrt{\frac{5l^2}{4k}}\{1 \quad 1 \quad 1 \quad 1\}$ $\quad$ $\boldsymbol{K}$-orthogonal

Fig. 4.3.15 Wielandt iteration for the calculation of the eigenvectors of a tridiagonal matrix defining the rotation must be stored because tridiagonalisation of the sub-matrix in question also affects other sub-matrices. This is taken into account in a so-called modification step. The tridiagonalisation of the sub-matrix $\boldsymbol{A}_{\mathrm{ii}}^{(\mathrm{k})}$ evidently only causes modifications to the i-th hyper-column and hyper-row. The process proceeds hence in hyper-columns. Two variants may be employed here: in variant 1 the matrix under consideration is located on the secondary hyper-diagonal. In this case, it will not be completely cancelled, but is reduced to an element of the tridiagonal matrix in the south-west corner. Variant 2, on the other hand, processes all the remaining sub-matrices which lie outside the hyper-tridiagonal.

Following the process outlined in fig. 4.3.16, we turn first to the sub-matrix $\boldsymbol{A}_{12}^{(1)}$ and thus to the variant 1. Again, we perform an in-core Givens process. This time, the field of activity comprises part of the $\boldsymbol{T}(\boldsymbol{T}_1^{(1)})$ matrix and the $\boldsymbol{A}_{12}^{(1)}$ matrix. The transformation procedure is already familiar to us. Once again, the c, s pairs must be stored for the computation of the subsequent modification step. Clearly, processing the secondary diagonal matrix $\boldsymbol{A}_{\mathrm{i,i+1}}^{(\mathrm{k})}$ affects hyper-rows and hyper-columns i and i + 1. We can miss out the next step in fig. 4.3.16 for it is a repetition of the first step. We next turn to variant 2 to cancel $\boldsymbol{A}_{13}^{(3)}$ completely. For this purpose, we form, including $\boldsymbol{T}_1^{(3)}$, a working area in the central memory and reactivate the procedure. Processing of a sub-matrix $\boldsymbol{A}_{\mathrm{ij}}^{(\mathrm{k})}$,(j>i+1) induces modifications in the hyper-columns and hyper-rows with indices i, i+1 and j.

This short insight into the hypermatrix technique concludes our discussion. The interested reader should consult the appendix which outlines the basic principles of the Householder method. Further information can be obtained from the special reference works, amongst others [4.8],[4.11],[4.12].

The principles of the methods described above may be applied with advantage when a considerable number of eigenvalues and eigenvectors is to be computed. In such a case, simultaneous vector iteration converges only slowly and in the Jacobi method, all eigenvectors must be included. Even when no eigenvectors are desired, this process is faster than the Jacobi method for a general dynamic matrix $\boldsymbol{A}$. In addition, it is possible to focus the operation on specific ranges of the eigenvalue spectrum.

The advantages of this procedure are as follows:

(1) It is a safe and economical method of determining the eigenvalues and – if desired – the eigenvectors; multiple eigenvalues do not cause any difficulty.

(2) The methodology has been extended to hypermatrices so that large problems can be solved.

(3) Specific operation on selected eigenvalue intervals is possible and some accuracy estimates can be made.

As a disadvantage, it is to be noted:

(1) The fully populated dynamic matrix must be formed and reduced to a tridiagonal form. For large problems, limitations of memory capacity may cause difficulties.

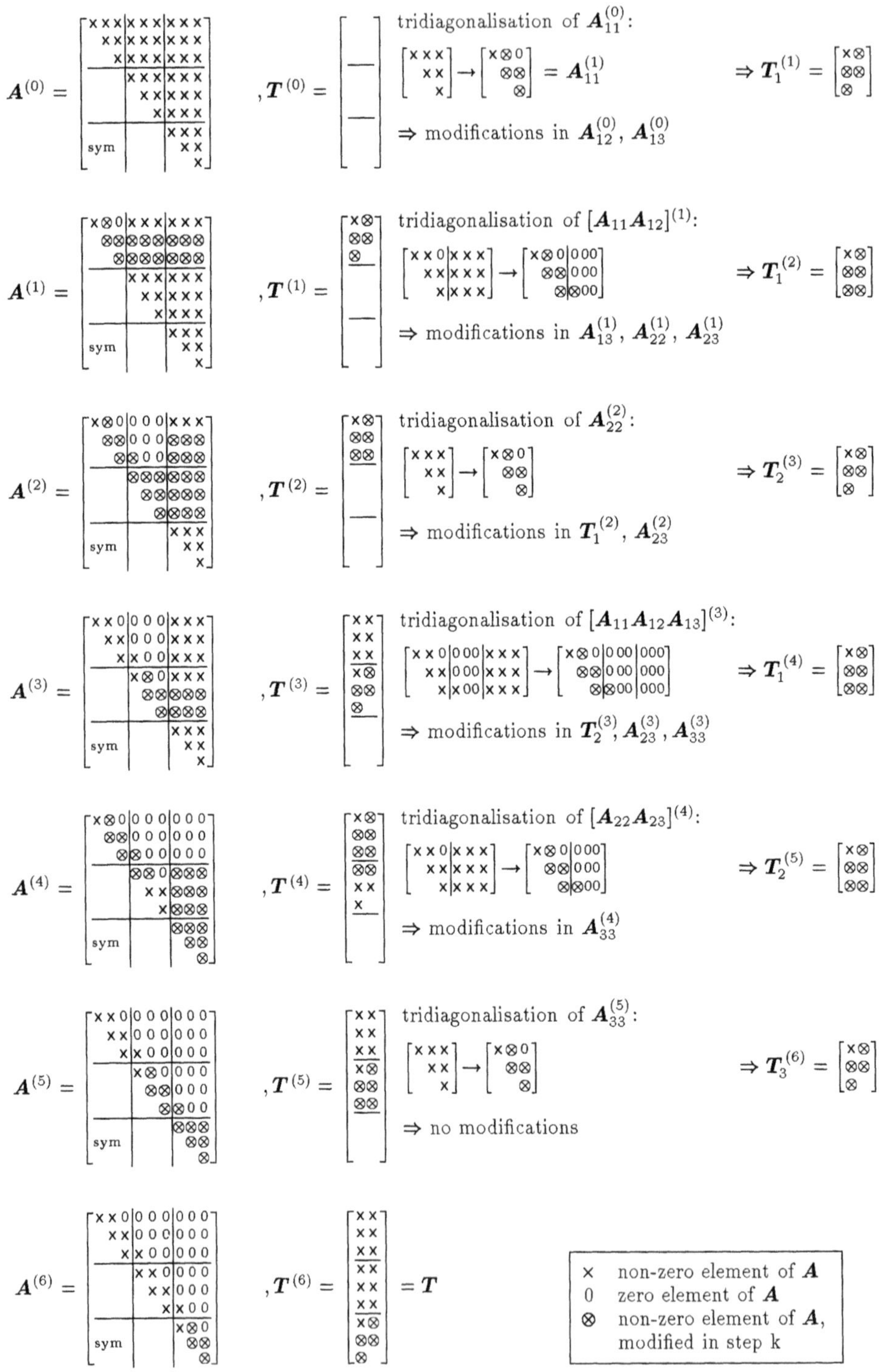

Fig. 4.3.16 A possible scheme for the transformation of a hypermatrix A into a tridiagonal matrix T according to the Givens method

Supplement to 4.3.4

*4.3.4 Transformation to tridiagonal form by the Householder method

It has already been mentioned that the Householder method is preferable to the Givens method, for both economical and numerical reasons. Thus, for the benefit of the especially interested reader, this method will be discussed briefly. More detailed discussions can be found in standard texts and in particular in [4.8] and [4.12]. The basic principles of the Givens and Householder methods are essentially the same. In the Givens method, the matrix $\boldsymbol{A}$ is reduced to a tridiagonal form by a series of orthogonal transformations $\boldsymbol{Q}^{(\mathrm{k})}$ whilst the Householder method reduces a complete column (and row) to the desired form in only one step. Accordingly, there are only (n – 2) steps in the Householder method, although they involve more effort. The Householder method (see [4.11]) proceeds in the k-th step to transform the k-th column resp. row into a tridiagonal form. All preceding columns and rows are, of course, already in tridiagonal form:

$$\boldsymbol{A}^{(\mathrm{k})} = \begin{bmatrix} \times & \times & 0 & 0 & 0 & 0 & 0 & 0 \\ \times & \times & \times & 0 & 0 & 0 & 0 & 0 \\ 0 & \times & \times & \times & \times & \times & \times & \times \\ 0 & 0 & \times & \times & \times & \times & \times & \times \\ 0 & 0 & \times & \times & \times & \times & \times & \times \\ 0 & 0 & \times & \times & \times & \times & \times & \times \\ 0 & 0 & \times & \times & \times & \times & \times & \times \\ 0 & 0 & \times & \times & \times & \times & \times & \times \end{bmatrix} \mathrm{k} \Longrightarrow \boldsymbol{A}^{(\mathrm{k}+1)} = \begin{bmatrix} \times & \times & 0 & 0 & 0 & 0 & 0 & 0 \\ \times & \times & \times & 0 & 0 & 0 & 0 & 0 \\ 0 & \times & \times & \times & 0 & 0 & 0 & 0 \\ 0 & 0 & \times & \times & \times & \times & \times & \times \\ 0 & 0 & 0 & \times & \times & \times & \times & \times \\ 0 & 0 & 0 & \times & \times & \times & \times & \times \\ 0 & 0 & 0 & \times & \times & \times & \times & \times \\ 0 & 0 & 0 & \times & \times & \times & \times & \times \end{bmatrix} \mathrm{k}$$

$$= [\boldsymbol{Q}^{(\mathrm{k})}]^{\mathrm{t}} \boldsymbol{A}^{(\mathrm{k})} \boldsymbol{Q}^{(\mathrm{k})} \qquad (^*4.3.1)$$

This can be achieved with

$$\underset{(\mathrm{n}\times\mathrm{n})}{\boldsymbol{Q}^{(\mathrm{k})}} = \boldsymbol{I}_{\mathrm{n}} - 2\,\underset{(\mathrm{n}\times 1)}{\boldsymbol{w}}\,\boldsymbol{w}^{\mathrm{t}} \qquad (^*4.3.2)$$

where $\boldsymbol{w}$ is a column vector with length 1

$$\boldsymbol{w}^{\mathrm{t}}\boldsymbol{w} = 1 \qquad (^*4.3.3)$$

We ascertain that $\boldsymbol{Q}^{(\mathrm{k})}$ is symmetric

$$[\boldsymbol{Q}^{(\mathrm{k})}]^{\mathrm{t}} = \boldsymbol{I}_{\mathrm{n}} - 2\boldsymbol{w}\boldsymbol{w}^{\mathrm{t}} = \boldsymbol{Q}^{(\mathrm{k})} \qquad (^*4.3.4)$$

and that it is an orthogonal matrix

$$\begin{aligned} [\boldsymbol{Q}^{(\mathrm{k})}]^{\mathrm{t}}\boldsymbol{Q}^{(\mathrm{k})} &= \boldsymbol{Q}^{(\mathrm{k})}\boldsymbol{Q}^{(\mathrm{k})} = \boldsymbol{Q}^{(\mathrm{k})}[\boldsymbol{Q}^{(\mathrm{k})}]^{\mathrm{t}} \\ &= \boldsymbol{I}_{\mathrm{n}} - 4\boldsymbol{w}\boldsymbol{w}^{\mathrm{t}} + 4\boldsymbol{w}\boldsymbol{w}^{\mathrm{t}}\boldsymbol{w}\boldsymbol{w}^{\mathrm{t}} \\ &= \boldsymbol{I}_{\mathrm{n}} \end{aligned} \qquad (^*4.3.5)$$

Hence, equ. (4.3.70) is still applicable and the eigenvalues are not modified by the orthogonal transformation. Naturally, we do not wish to reduce only the k-th column (row) to the desired form in the k-th step, but aim at retaining the first (k−1) columns (rows), already tridiagonalised, intact. Therefore, we examine the effects of the transformation (*4.3.1)

$$\begin{aligned} \boldsymbol{A}^{(k+1)} &= [\boldsymbol{Q}^{(k)}]^t \boldsymbol{A}^{(k)} \boldsymbol{Q}^{(k)} \\ &= [\boldsymbol{I} - 2\boldsymbol{w}\boldsymbol{w}^t]\boldsymbol{A}^{(k)}[\boldsymbol{I} - 2\boldsymbol{w}\boldsymbol{w}^t] \\ &= \boldsymbol{A}^{(k)} - 2\boldsymbol{w}\boldsymbol{w}^t\boldsymbol{A}^{(k)} - 2\boldsymbol{A}^{(k)}\boldsymbol{w}\boldsymbol{w}^t + 4\boldsymbol{w}\boldsymbol{w}^t\boldsymbol{A}^{(k)}\boldsymbol{w}\boldsymbol{w}^t \end{aligned} \tag{*4.3.6}$$

For typographical simplicity, the following abbreviations are introduced

$$\boldsymbol{p} = \boldsymbol{A}^{(k)}\boldsymbol{w} \tag{*4.3.7}$$

$$z = \boldsymbol{p}^t\boldsymbol{w} = \boldsymbol{w}^t\boldsymbol{A}^{(k)}\boldsymbol{w} \tag{*4.3.8}$$

$$\boldsymbol{q} = 2[\boldsymbol{p} - z\boldsymbol{w}] \tag{*4.3.9}$$

Using these symbols, equ. (*4.3.6) becomes

$$\boldsymbol{A}^{(k+1)} = \boldsymbol{A}^{(k)} - \boldsymbol{w}\boldsymbol{q}^t - \boldsymbol{q}\boldsymbol{w}^t \tag{*4.3.10}$$

If now we set in the k-th step the first k elements of the vector $\boldsymbol{w}$ to zero, then, as a result of the zero pattern in the $\boldsymbol{A}^{(k)}$ matrix, at least the first k−1 elements of $\boldsymbol{p}$ will also be zero (see equ. (*4.3.7)). The same applies to the vector $\boldsymbol{q}$ in equ. (*4.3.10). Thus, both $\boldsymbol{w}$ and $\boldsymbol{q}$ are vectors containing (k−1) leading zero elements, and as a result of equ. (*4.3.6), no change occurs in the first (k−1) rows and columns of the matrix $\boldsymbol{A}^{(k)}$. By the nullification of the first k elements in $\boldsymbol{w}$, the already established zero pattern remains unaffected. However, we wish to cancel n−(k+1) elements in the k-th step too. For this purpose, we require an equal number of freely disposable $\boldsymbol{w}$ elements. In view of equ. (*4.3.3), we can prescribe one non-zero element and in this way ensure for k zero-elements the availability of the correct number of freely disposable entries in $\boldsymbol{w}$. With

$$\boldsymbol{w} = \{0 \quad 0 \quad \dots \quad 0 \quad 0 \quad w_{k+1} \quad w_{k+2} \quad \dots \quad w_n\} \tag{*4.3.11}$$

$$\boldsymbol{q} = \{0 \quad 0 \quad \dots \quad 0 \quad q_k \quad q_{k+1} \quad q_{k+2} \quad \dots \quad q_n\} \tag{*4.3.12}$$

we deduce from equ. (*4.3.10) for the elements to be nullified in the k-th row (column)

$$a_{kj}^{(k+1)} = a_{kj}^{(k)} - q_k w_j = a_{kj}^{(k)} - 2p_k w_j \overset{!}{=} 0; \qquad j = k+2 \dots n \tag{*4.3.13}$$

from which follows

$$w_j = \frac{a_{kj}^{(k)}}{2p_k}; \qquad j = k+2 \dots n \tag{*4.3.14}$$

All non-zero elements of the $\boldsymbol{w}^{(\mathrm{k})}$ vector – with the exception of the first – are thus proportional to the corresponding elements of the row (column) k of the matrix $\boldsymbol{A}^{(\mathrm{k})}$. The proportionality factor is $1/(2p_{\mathrm{k}})$, which can be determined by application of the unit length property of $\boldsymbol{w}$ in equ. (*4.3.3). Thus,

$$\boldsymbol{w}^{\mathrm{t}}\boldsymbol{w} = 1 = w_{\mathrm{k+1}}^2 + \left(\frac{1}{2p_{\mathrm{k}}}\right)^2 \sum_{\mathrm{j=k+2}}^{\mathrm{n}} a_{\mathrm{kj}}^2 \tag{*4.3.15a}$$

or

$$4p_{\mathrm{k}}^2 = (2p_{\mathrm{k}}w_{\mathrm{k+1}})^2 + s^2 - a_{\mathrm{k,k+1}}^2 \tag{*4.3.15b}$$

where s^2 stands for

$$s^2 = \sum_{\mathrm{j=k+1}}^{\mathrm{n}} a_{\mathrm{kj}}^2 \tag{*4.3.16}$$

The formal calculation of p_{k} from equ. (*4.3.7) contributes another equation

$$\begin{aligned} p_{\mathrm{k}} &= \sum_{\mathrm{j=k+1}}^{\mathrm{n}} a_{\mathrm{kj}}w_{\mathrm{j}} \\ &= a_{\mathrm{k,k+1}}w_{\mathrm{k+1}} + \frac{1}{2p_{\mathrm{k}}} \sum_{\mathrm{j=k+2}}^{\mathrm{n}} a_{\mathrm{kj}}^2 \end{aligned} \tag{*4.3.17a}$$

or

$$2p_{\mathrm{k}}^2 = a_{\mathrm{k,k+1}}(2p_{\mathrm{k}}w_{\mathrm{k+1}}) - a_{\mathrm{k,k+1}}^2 + s^2 \tag{*4.3.17b}$$

Next, we eliminate p_{k}^2 by taking the difference between equs. (*4.3.15b) and (*4.3.17b) to obtain the following relation

$$(a_{\mathrm{k,k+1}} - 2p_{\mathrm{k}}w_{\mathrm{k+1}})^2 = s^2 \tag{*4.3.18a}$$

or

$$2p_{\mathrm{k}}w_{\mathrm{k+1}} = a_{\mathrm{k,k+1}} \pm s \tag{*4.3.18b}$$

Next, from equ. (*4.3.17b), we deduce

$$2p_{\mathrm{k}}^2 = s^2 \pm a_{\mathrm{k,k+1}}s \tag{*4.3.19a}$$

or

$$p_{\mathrm{k}} = s\sqrt{\frac{1}{2}\left(1 \pm \frac{a_{\mathrm{k,k+1}}}{s}\right)} \tag{*4.3.19b}$$

The final computation of w_{k+1} follows from equ. (*4.3.18b), once p_k is known. All other elements of $\boldsymbol{w}$ are obtained from equ. (*4.3.14). The Householder method involves less expenditure than the Givens method (the number of operations is halved). In addition, the root must be taken only once per columnal step, whilst this is necessary for every element in the Givens method. The theoretical background of the Householder method, however, is without doubt much more demanding. The effort of comprehending the technique does not, however, affect the implementation of the programme code.

For the Householder method, there are also hypermatrix variants that lean heavily on the in-core algorithm just outlined. Lack of space precludes the discussion of this matter, but the reader is referred to [4.4] and [4.13].

4.3.5 The determination of eigenvalues and eigenvectors by the Lanczos method

Author: Karl Straub

In addition to the simultaneous vector iteration with its distinct variants described in section 4.3.2, other iterative methods of spectral analysis for large sparse matrices have been widely developed in recent years. The Lanczos method is one example of this. It would be more correct here to refer to a family of algorithms, the central concept of which was proposed by C. Lanczos in 1950 for the approximate calculation of eigenvalues [4.23]. At first, this method was misunderstood to be nothing but a procedure of tridiagonalising a matrix, but in this task it proved inferior to the more direct methods of Householder and Givens discussed in section 4.3.4. In the course of the last few years since the publication of [4.24], however, there has been a significant change in the interpretation of this method, and it has in many respects been further developed in order to provide effective algorithms for the determination of eigenvalues and eigenvectors of large matrices, in particular in a specified frequency range (mostly at the upper or lower end of the frequency spectrum).

Although this method is routinely employed as a 'black box' in many general purpose finite element codes, it should be of interest to an inquiring mind to understand the fundamentals of the Lanczos method, forgoing a strict mathematical exposition (such as the accuracy and the rate of convergence). The relationship of the Lanczos method to other methods, such as the gradient one, cannot be discussed here, however, but the interested reader may consult [4.12], [4.28], [4.29] and [4.30] for additional information.

For the purpose of simplicity, we consider a symmetric matrix $\boldsymbol{A}$ and attempt to solve the special eigenvalue problem

$$\boldsymbol{A}\boldsymbol{X} = \boldsymbol{X}\boldsymbol{\Lambda} \tag{4.3.89}$$

The matrix $\boldsymbol{A}$ (n $\times$ n) represents the dynamic matrix formed by the stiffness matrix $\boldsymbol{K}$ and the mass matrix $\boldsymbol{M}$ of a system (see also equ. (4.1.39)). As we will see in the following, $\boldsymbol{A}$ need not necessarily be construed explicitly, only the operation '$\boldsymbol{A}$ multiplied by a vector' must be executable.

The iteration procedure begins with a starting vector $\boldsymbol{v}$ and, as is the case for vector iteration (section 4.3.1), additional vectors are established by successive multiplication with $\boldsymbol{A}$. Unlike in the vector iteration, the vector $\boldsymbol{A}^{\mathrm{i}-1}\boldsymbol{v}$ will not always be over-written by the new vector $\boldsymbol{A}^{\mathrm{i}}\boldsymbol{v}$, but will instead be inserted as a new column in a matrix $\boldsymbol{V}^{(\mathrm{i}+1)}$. In this way, we obtain in the m-th step

$$\boldsymbol{V}^{(\mathrm{m})} = [\boldsymbol{v} \quad \boldsymbol{A}\boldsymbol{v} \quad \boldsymbol{A}^2\boldsymbol{v} \quad \dots \quad \boldsymbol{A}^{(\mathrm{m}-1)}\boldsymbol{v}] \tag{4.3.90}$$

Matrices of this type are known as Krylov matrices, and the vector spaces formed by the columns are termed Krylov subspaces. A subspace basis $\boldsymbol{Q}$ can be built successively, by means of Gram-Schmidt's orthonormalisation method

$$\boldsymbol{Q}^{(\mathrm{m})} = [\boldsymbol{q}_1 \quad \boldsymbol{q}_2 \quad \boldsymbol{q}_2 \quad \dots \quad \boldsymbol{q}_\mathrm{m}] \tag{4.3.91}$$

such that

$$\boldsymbol{Q}^{(\mathrm{m})\mathrm{t}}\boldsymbol{Q}^{(\mathrm{m})} = \boldsymbol{I}_\mathrm{m} \tag{4.3.92}$$

(In the following $\boldsymbol{Q}$ may also be understood as the subspace itself formed by the basis vectors $\boldsymbol{q}_\mathrm{i}$. From the context, it should be discernable in each case whether the basis vectors or the subspace is being referred to.)

It is presumed here that the columns in equ. (4.3.90) are linearly independent. Because of the special form of the Krylov matrices (equ. (4.3.90)), the basis vectors $\boldsymbol{q}_\mathrm{i}$ (Lanczos vectors) have a distinctive characteristic in that they can reduce the $\boldsymbol{A}$ matrix to a tridiagonal form $\boldsymbol{T}(\mathrm{m} \times \mathrm{m})$

$$\boldsymbol{Q}^{(\mathrm{m})\mathrm{t}}\boldsymbol{A}\boldsymbol{Q}^{(\mathrm{m})} = \boldsymbol{T}^{(\mathrm{m})} \tag{4.3.93}$$

in which

$$\boldsymbol{T}^{(\mathrm{m})} = \begin{bmatrix} \alpha_1 & \beta_1 & & & \\ \beta_1 & \alpha_2 & \beta_2 & & \boldsymbol{O} \\ & \beta_2 & \ddots & \ddots & \\ & & \ddots & \ddots & \beta_{\mathrm{m}-1} \\ \boldsymbol{O} & & & \beta_{\mathrm{m}-1} & \alpha_\mathrm{m} \end{bmatrix} \tag{4.3.94}$$

The extension of the Krylov subspace by a new vector and the orthogonalisation are performed in one step, which means that instead of forming the complete Krylov matrix $\boldsymbol{V}^{(\mathrm{m})}$ and subsequently the $\boldsymbol{Q}^{(\mathrm{m})}$ matrix, the new $\boldsymbol{A}^{\mathrm{m}-1}\boldsymbol{v}$ (in symbolic form) will only be required when establishing $\boldsymbol{q}_\mathrm{m}$. The vector $\boldsymbol{A}^{\mathrm{m}-2}\boldsymbol{v}$ is contained in the subspace, $\boldsymbol{Q}^{(\mathrm{m}-1)}$

$$\boldsymbol{A}^{\mathrm{m}-2}\boldsymbol{v} = \sum_{\mathrm{k}=1}^{\mathrm{m}-1} c_\mathrm{k}\boldsymbol{q}_\mathrm{k} \tag{4.3.95}$$

and this $\boldsymbol{Q}^{(m-1)}$ can be extended to the new subspace $\boldsymbol{Q}^{(m)}$ only by the vector $\boldsymbol{A}\boldsymbol{q}_{m-1}$, obtained from the product $\boldsymbol{A}(\boldsymbol{A}^{m-2}\boldsymbol{v})$, because of

$$\begin{aligned} \boldsymbol{A}\boldsymbol{q}_1 &\subset \boldsymbol{Q}^{(2)} \\ \boldsymbol{A}\boldsymbol{q}_2 &\subset \boldsymbol{Q}^{(3)} \\ &\vdots \\ \boldsymbol{A}\boldsymbol{q}_{m-2} &\subset \boldsymbol{Q}^{(m-1)} \end{aligned}$$

When construing the new basis vector $\boldsymbol{q}_m$, we can thus use $\boldsymbol{A}\boldsymbol{q}_{m-1}$ instead of the vector $\boldsymbol{A}^{m-1}\boldsymbol{v}$.

Due to the orthogonalisation condition, we have

$$\boldsymbol{q}_m \perp \boldsymbol{Q}^{(m-1)} \supset \boldsymbol{Q}^{(m-2)} \supset \ldots \supset \boldsymbol{Q}^{(1)} \tag{4.3.96}$$

and

$$\boldsymbol{A}\boldsymbol{q}_i \subset \boldsymbol{Q}^{(i+1)} \tag{4.3.97}$$

whence immediately follows

$$\boldsymbol{q}_m^t \boldsymbol{A}\boldsymbol{q}_i = 0 \qquad \text{if } i+1 \leqslant m-1, \qquad \text{i.e.} \quad i < m-1 \tag{4.3.98}$$

Since $\boldsymbol{A}$ is symmetric, the transformed matrix $\boldsymbol{A}$ acquires the tridiagonal form given by equ. (4.3.94). Furthermore, we deduce from equs. (4.3.92) and (4.3.93) the following expression

$$\boldsymbol{A}\boldsymbol{Q}^{(m)} = \boldsymbol{Q}^{(m)}\boldsymbol{T}^{(m)} \tag{4.3.99}$$

Equating the j-th columns on the left and right of this matrix equation, we obtain

$$\boldsymbol{A}\boldsymbol{q}_j = \beta_{j-1}\boldsymbol{q}_{j-1} + \alpha_j\boldsymbol{q}_j + \beta_j\boldsymbol{q}_{j+1} \tag{4.3.100}$$

and from the orthogonality of the $\boldsymbol{q}$-vectors, we find

$$\alpha_j = \boldsymbol{q}_j^t \boldsymbol{A}\boldsymbol{q}_j \tag{4.3.101}$$

The elementary Lanczos iteration of the matrix $\boldsymbol{A}$ with the starting vector $\boldsymbol{v}$ may thus be represented by the following simple recursive algorithm:

$$\begin{aligned} &\text{set} \quad j = 0, \quad \beta_0 = 1, \quad \boldsymbol{q}_0 = \boldsymbol{o}, \qquad \boldsymbol{q}_1 = \boldsymbol{v}/|\boldsymbol{v}| \\ &j = j+1: \quad (\beta_j \neq 0 \text{ assumed}) \\ &\qquad \alpha_j = \boldsymbol{q}_j^t(\boldsymbol{A}\boldsymbol{q}_j) \\ &\qquad \overline{\boldsymbol{q}}_{j+1} = (\boldsymbol{A}\boldsymbol{q}_j) - \alpha_j\boldsymbol{q}_j - \beta_{j-1}\boldsymbol{q}_{j-1} \\ &\qquad \beta_j = |\overline{\boldsymbol{q}}_{j+1}| \\ &\qquad \boldsymbol{q}_{j+1} = \overline{\boldsymbol{q}}_{j+1}/\beta_j \end{aligned} \tag{4.3.102}$$

Beginning with a normalised starting vector $\boldsymbol{q}_1$, the elements of the $\boldsymbol{T}$-matrix and the Lanczos vectors $\boldsymbol{q}$ are computed successively:

$$\boldsymbol{q}_1,\ \alpha_1,\ \beta_1,\ \boldsymbol{q}_2,\ \alpha_2,\ \beta_2,\ \ldots$$

The only operation involving the matrix $\boldsymbol{A}$ is a multiplication by $\boldsymbol{q}_j$ within each iteration. Only three vectors of order n are required at each iteration step (n=order of the matrix $\boldsymbol{A}$). However, a careful memory managing strategy reduces this requirement to two vectors only (see [4.30]). Lanczos iteration is distinguished from simultaneous vector iteration (section 4.3.2) by its extremely economical central memory requirement and its relatively restricted number of computations.

The iteration step shown in equ. (4.3.102) also yields a theoretical indication of when the iteration scheme has been completed: in fact, when the off-diagonal element β_j is zero, no other orthogonal basis vector $\boldsymbol{q}_{j+1}$ can be generated. This is certainly true when j is equal to n since in an n-dimensional vector space, no more than n independent vectors $\boldsymbol{q}_i$ can be established. This natural termination of the iteration can occur even earlier, ($j < n$), namely if the starting vector $\boldsymbol{v}$ (or $\boldsymbol{q}_1$) lies completely within an eigenvector subspace of the matrix $\boldsymbol{A}$. For example, if we choose $\boldsymbol{v}$ as one of the eigenvectors $\boldsymbol{X}_k$, the resulting multiplications with $\boldsymbol{A}$ yield only a projection of $\boldsymbol{X}_k$ onto itself and an extension of the basis matrix $\boldsymbol{Q}$ cannot take place.

The determination of the tridiagonal matrix $\boldsymbol{T}^{(m)}$, in general with $m \ll n$, does not postulate that the eigenvalue problem of equ. (4.3.89) has been solved. From our experience of the vector iteration and the simultaneous vector iteration schemes, we know that the repeated multiplication with $\boldsymbol{A}$ continuously amplifies the convergence of the eigenvectors which appertain to the largest eigenvalues. From this, one can show that by increasing the dimension m of the matrix $\boldsymbol{T}^{(m)}$, better approximations to the eigensolutions of $\boldsymbol{A}$ can be achieved. The mathematical treatment of the problems of convergence and accuracy is beyond the scope of our introductory study. The reader should consult the references (Kaniel-Paige-Saad theorem, [4.24], [4.26], [4.28], [4.30], [4.31]). Since the dimension of the matrix $\boldsymbol{T}$ is generally small, analogous to the handling of the matrix $\boldsymbol{B}$ in equ. (4.3.13) in the simultaneous vector iteration scheme, its eigenvalues can be determined (partially or completely) by known methods such as the QR-alogrithm or the Jacobi alogrithm. The transformation of the eigenvectors from the subspace of the $\boldsymbol{T}$ matrix back to the n-dimensional space of the $\boldsymbol{A}$ matrix by means of $\boldsymbol{Q}$ as transformation matrix is known to us by now, see equs. (4.3.91), (4.3.71).

The Lanczos algorithm, in its simplest form for the determination of the p higher eigenvalues and eigenvectors of an n dimensional symmetric matrix $\boldsymbol{A}$, consists of the following steps:

— perform Lanczos iteration as detailed in equ. (4.3.102) to obtain tridiagonal matrices of dimension m ($\ll$n, $m \geqslant p$), in which m is possibly not specified *a priori*, but determined during the iteration process based on the progress of the convergence;

— perform the eigenvalue analysis of the matrix $\boldsymbol{T}^{(m)}$ and check the accuracy;

— transform all m or only the desired p eigenvectors from the subspace of $\boldsymbol{T}$ to the original $\boldsymbol{A}$-space, and, if necessary, also compute residua to check the accuracy.

Simple as the Lanczos method may appear, in the operational scheme discussed above, we should be aware that our presentation did not refer to the difficulties encountered in its application to practical problems of structural mechanics. Our introductory account is mainly concerned with the theoretical description of the Lanczos method. Also, our preceding discussion presumed exact arithmetical operations and did not consider round-off errors inevitably associated with the finite bit approximate specification of numbers in the computer. A few important concepts are discussed in what follows, but our presentation is by no means complete.

A first critical point already attained after relatively few iteration loops (4.3.102) concerns a deficiency in the orthogonality of the Lanczos vectors. This significantly affects the correspondence between the eigenvalues of the $\boldsymbol{T}$-matrix and those of the original matrix $\boldsymbol{A}$. The current iteration step ensures explicitly that the last vector $\boldsymbol{q}$ be orthogonal to the immediate two preceding vectors, while the orthogonality requirement with respect to other antecedent vectors is only contained implicitly (theoretically) in the algorithm. However, in a supplementary step, we may enforce on the new vector $\overline{\boldsymbol{q}}_{j+1}$ an orthogonality to all preceding vectors ('complete reorthogonalisation') but this is a very expensive process, since in realistic applications the number of Lanczos vectors may reach a few hundred. In addition, we require that all antecedent vectors are available in the central memory.

Another approach for safer orthogonalisation is more cost-effective but requires a subtle strategic concept of the whole iteration procedure. At the onset of a loss of orthogonality, reorthogonalisation is applied selectively with respect only to those vectors that represent sufficiently good approximations to the eigenvectors ('selective orthogonalisation'); see [4.27], [4.28], [4.30].

Particular attention must also be paid to multiple eigenvalues. One can show that the Lanczos algorithm with reorthogonalisation performed in an exact arithmetic operation cannot detect the multiplicity of roots. However, following a complete Lanczos iteration with a starting vector $\boldsymbol{v}_1$, another iteration cycle can be started with a starting vector $\boldsymbol{v}_2$ which is orthogonal to all the available eigenvectors of the first iteration, taken to be sufficiently accurate. If an eigenvector appertaining to a multiple eigenvalue is determined in the first iteration cycle, the second iteration will yield a second eigenvector orthogonal to the first one and corresponding to the same multiple eigenvalue. This process can be repeated until no new eigenvectors belonging to the multiple root can be found. Analogous to the extension of the simple vector iteration to the simultaneous vector iteration, one can arrange for the Lanczos iteration to start simultaneously with several starting vectors, and obtain in this way a block-tridiagonal matrix in place of a standard tridiagonal matrix. This is called a block or band Lanczos algorithm (see [4.28], [4.30], [4.31]). Here, the dimension of the blocks is uncertain since it depends upon the multiplicity (in general unknown) of the eigenvalues which can be detected in the selected eigenvalue interval. Apart from a 'trial' strategy, which is time consuming, the shift strategy described in the following can be applied.

To this purpose, we consider the determination of all eigenvalues in an interval (a, b) and their eigenvectors. This can be achieved by a modified version of the Lanczos algorithm. For this, we rewrite the original eigenvalue problem of equ. (4.3.89) in the following form involving a spectral shift

$$[\boldsymbol{A} - \mu\boldsymbol{I}]^{-1}\boldsymbol{X}^{(s)} = \boldsymbol{X}^{(s)}\boldsymbol{\Lambda}^{(s)} \tag{4.3.103}$$

whereby the eigenvalues λ of the original matrix $\boldsymbol{A}$ and the new eigenvalues $\lambda^{(s)}$ are simply related by

$$\lambda^{(s)} = 1/(\lambda - \mu) \tag{4.3.104}$$

By this 'spectral transformation' ([4.33]; see also equ. (4.3.20)), those eigenvalues λ closest to the shift parameter μ are mapped on the $\lambda^{(s)}$ values with the largest modulus. Consequently, application of the Lanczos method to $[\boldsymbol{A} - \mu\boldsymbol{I}]^{-1}$ ensures first convergence towards these extreme eigenvalues $\lambda^{(s)}$. The solution process requires the factorisation

$$\boldsymbol{A} - \mu\boldsymbol{I} = \boldsymbol{L}\boldsymbol{D}\boldsymbol{L}^{\mathrm{t}} \tag{4.3.105}$$

where $\boldsymbol{L}$ is a lower triangular matrix and $\boldsymbol{D}$ is a diagonal matrix. The number of negative elements in the diagonal matrix $\boldsymbol{D}$ indicates the number of eigenvalues λ which are smaller than the spectral shift μ. Since the Lanczos iteration scheme converges most rapidly towards the largest eigenvalues, all eigenvalues in the range a through to b can be extracted efficiently using a sequence of suitable shifts

$$\mu_1 = a, \mu_2, \mu_3, \ldots, \mu_{\mathrm{k}} = b \tag{4.3.106}$$

For any shift μ_{i}, the number of eigenvalues to the left of μ_{i} can be determined easily, and by comparison with already available eigenvalues it can be established whether sets of multiple eigenvalues have been determined completely. Any missing eigenvectors can also be computed, if required, in a corrector-iteration step (using new starting vectors). The interplay of spectral shifts μ_{i} and Lanczos iteration loops requires a subtle strategy. For additional information, the reader is advised to consult [4.33].

So far, we have considered the special eigenvalue problem of equ. (4.3.89). The handling of the general eigenvalue problem in structural mechanics in terms of the stiffness ($\boldsymbol{K}$) and mass ($\boldsymbol{M}$) matrices will be outlined next. If we use the spectral shift, the inverse of the dynamic matrix can be expressed as

$$[\boldsymbol{A} - \mu\boldsymbol{I}]^{-1} = \boldsymbol{M}^{1/2}[\boldsymbol{K} - \mu\boldsymbol{M}]^{-1}\boldsymbol{M}^{1/2} \tag{4.3.107}$$

in which the following decomposition

$$\boldsymbol{K} - \mu\boldsymbol{M} = \boldsymbol{L}\boldsymbol{D}\boldsymbol{L}^{\mathrm{t}} \tag{4.3.108}$$

may be considered. The factorisation in equ. (4.3.108) must be understood as symbolic since in large, sparsely populated matrices we can operate directly with the operand matrices in the triangularised and diagonalised form.

In conclusion, we would like to experiment on the Lanczos algorithm using some practical problems and compare the results with the simultaneous vector iteration method. The restricted number of computational operations required in the Lanczos method as against the simultaneous vector iteration method demonstrates its efficiency (an improvement in the order of 10 to 30 may be expected, see [4.34]). However, the required special strategies and the implementation of numerical control to ensure the reliability of the results reduces this efficiency bonus to three to five ([4.33], [4.34], [4.35]). Our first initial experiences with the Lanczos method when implemented in the general purpose finite element computer program, ASKA, did not, however, demonstrate such dramatic improvements. We assume this to be due to the fact that the Lanczos algorithm was not applied consistently with all its advantages or to a lack of optimal integration into the ASKA system. We do not believe that our first experiments indicate a deflation of the merits of the Lanczos method. We will return to this aspect after we have examined some examples.

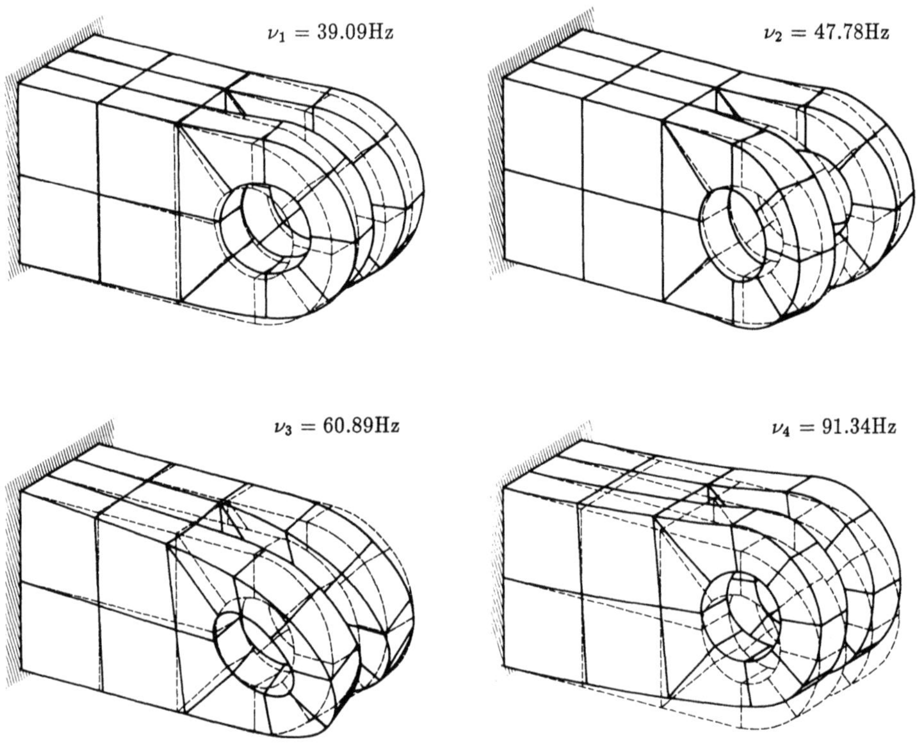

Fig. 4.3.17 Lowest eigenoscillations of an FE-idealised load bearing lug (n=744)

In the first example, we determine the lower eigenfrequency spectrum of a load bearing lug shown in fig. 4.3.17. The system in the adopted idealisation displays n=744 degrees of freedom and a half band width of b=744 (i.e. a full matrix). Another optimised idealisation yields b=315. The eigenfrequency spectrum ranges from 40 Hz to 800 Hz and contains 60 eigenfrequencies. The frequencies are more or less equally distributed, some, however, are very close to each other (only 0.5% apart). As compared with the simultaneous vector iteration method, the Lanczos method showed an improvement of 15% to 30% in the CPU for 60 eigenfrequencies and eigenvectors. In both the methods, approximately the same order of accuracy was achieved.

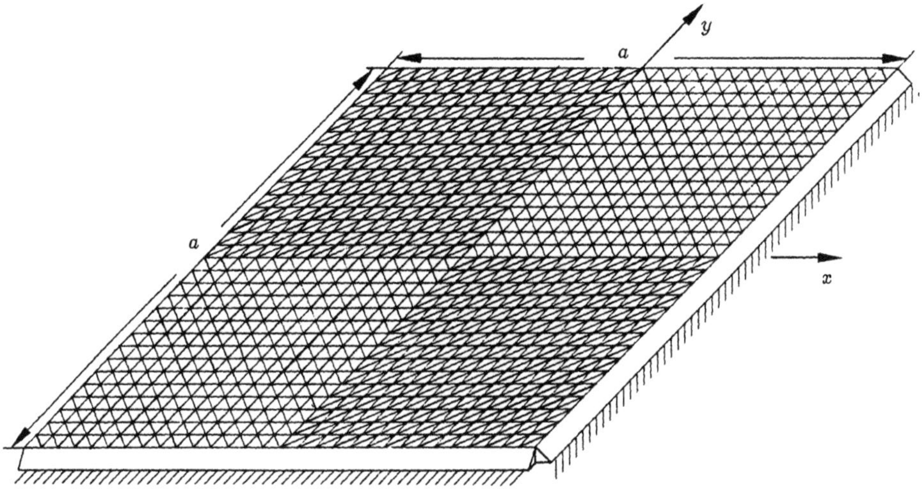

geometrical data:		material data:	
length	$a = 1$ m	Young's modulus	$E = 7 \cdot 10^{10}$ N/m^2
thickness	$t = 0.01$ m	Poisson's ratio	$\nu = 0.3$
		density	$\rho = 2700$ Ns^2/m^4

Fig. 4.3.18 FE-idealisation of a simply supported square plate ($n = 5406$)

The second example concerned a quadrilateral plate simply supported at all edges (fig. 4.3.18). Because of the double symmetry of the structure, there are many pairs of exactly equal eigenvalues. The chosen fine modelling of the structure results in n=5406 degrees of freedom and a band width of b $\approx$ 130. The thirty lowest eigenfrequencies were determined. The existence of equal eigenvalues permits simple control of the accuracy of the method and its ability to predict multiple roots of an eigenvalue problem. In the present realisation of the Lanczos method with an internal termination criterion, an accuracy of 2 to 3 digits was obtained in the eigenvalues. Equal precision for the simultaneous vector iteration demands a slightly increased (10%) computational effort. In our opinion, the possibility of combining the Lanczos and simultaneous vector iteration methods appears attractive. Thus, one may start in a subspace with Lanczos eigenvectors and subsequently attach a simultaneous vector iteration (see [4.36], FE-programme system ADINA).

In this summarised appraisal of their experience, the authors draw the reader's attention to the fact that their first experiments involved a difficult process of incorporating a subtle technique like that of Lanczos into a general programme system. This may have caused adaptation and implementation difficulties. Moreover, current developments and improvements on the Lanczos procedure are by no means complete, as 'The First World Congress on Computational Mechanics' (Austin/Texas, September 1986) made apparent: in the sessions on eigenvalue problems, most contributions were based upon 'Lanczos methods'. Finally, with reference to [4.4], we wish to state that the latest versions of Lanczos algorithms 'are too good to be ignored'.

4.4 Reducing the problem size (Guyan's reduction, condensation)

For a reliable analysis of large static problems, it is often necessary to discretise a given structure using several thousand degrees of freedom. For example, early in 1969, the F 104G aircraft had to be represented by 8000 degrees of freedom in order to obtain a satisfactory stiffness matrix. It would, however, be somewhat unrealistic to consider the analysis of an eigenvalue problem of this magnitude if we desire to establish more than a few eigenfrequencies and eigenvectors.

Generally, one is aware of the nature of the vibration modes and knows that in the lower frequency range, they may be described by simple shapes not involving thousands of degrees of freedom. Sometimes, it is only nessesary to determine the vibration characteristics of a component of structure, for example, the stiffened side panel of an automobile; the complete structure need only be considered in order to establish the elastic boundary conditions. Here, it suffices to idealise only the substructure in question (i.e. the stiffened side panel) with sufficient degrees of freedom. The specification of the relevant problem also plays a significant role in the selection of the starting vectors in the simultaneous vector iteration method. In the course of our subsequent analysis, we shall observe that the previously outlined technique in the simultaneous vector iteration is, indeed, identical to the procedure commonly known as the condensation or reduction method. This is described in an *addendum* to this section.

We demonstrate first the concept of condensation on an example of somewhat academic nature (fig. 4.4.1) and then expound the theory and demonstrate it on some practically significant examples. To apply this technique successfully, we require some practical experience in dynamics. The selection of significant degrees of freedom to describe the desired vibration modes is designated in fig. 4.4.1 by the subscript c (for coupling). If we are interested in the longitudinal oscillation of model (a), we have to select 4 longitudinal coupling or primary degrees of freedom, as shown in (b), in order to obtain all four eigenforms of oscillation of model (a). For an approximation to the first two longitudinal modes, however, it suffices to restrict ourselves to two degrees of freedom as shown in model (c). If we wish to establish a wider spectrum of eigenvalues, we have to progressively refine the idealisation of the system. On the other hand, if we concentrate more on the transverse vibration characteristics and desire to establish all modes of the idealised system in (a), the necessary degrees of freedom are shown in model (d). If we are to restrict our interest to an approximate computation of four or two vibration modes only, we should select models (e) or (f) respectively. Selection of the so-called primary degrees of freedom as in models (d), (e) and (f) causes all the characteristics of the longitudinal vibration to be lost.

In order to reduce the total number of degrees of freedom to a primary set, we require a functional relation between the two groups of degrees of freedom. Guyan [4.14] achieves this by assuming the absence of forces on the degrees of freedom which are to be eliminated. In fact, this method interpolates the deformation by

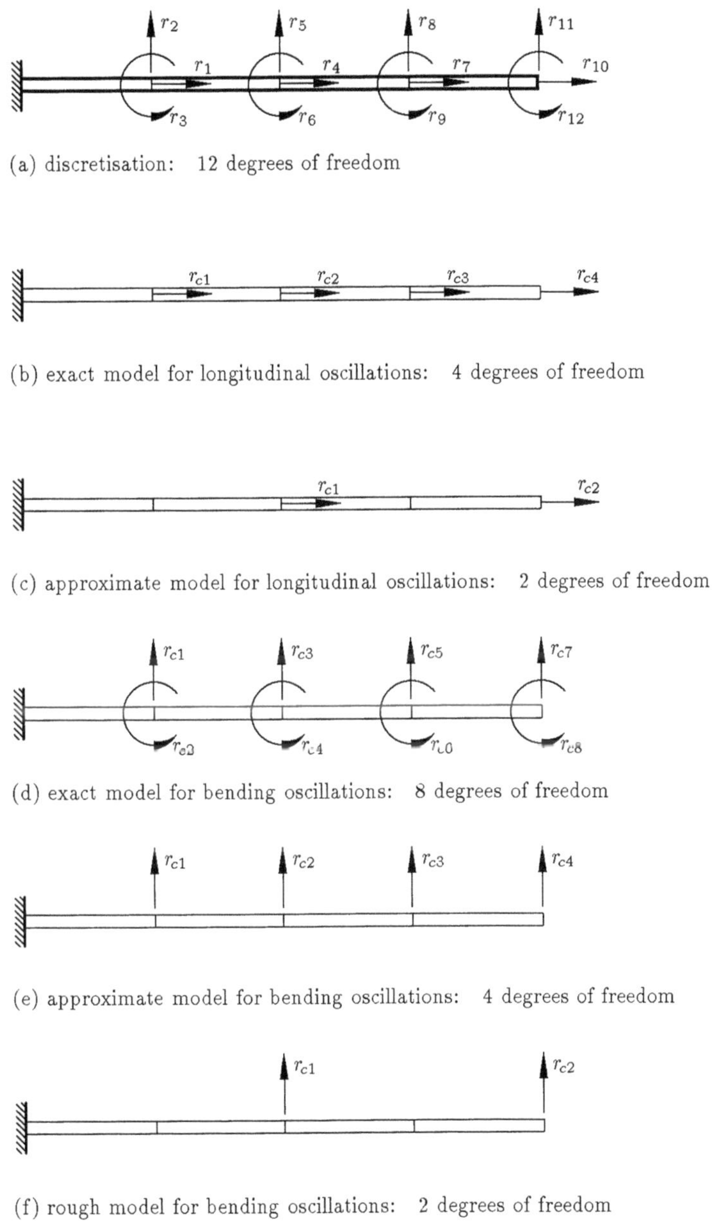

Fig. 4.4.1 Selection of the dynamically significant degrees of freedom, illustrated on a simple example

means of static deflection curves. Initiating our discussion with equ. (3.1.10), we divide the degrees of freedom into the two groups

$\boldsymbol{r}_\sigma$ = secondary degrees of freedom (to be eliminated)

and

$\boldsymbol{r}_c$ = primary degrees of freedom (to be maintained)

The particular indices are chosen such that there is a clear analogy to the substructure technique (see also Vol. I of statics, section 6.4). Accordingly, the following equilibrium relation applies

$$\begin{bmatrix} \boldsymbol{K}_{cc} & \boldsymbol{K}_{c\sigma} \\ \boldsymbol{K}_{\sigma c} & \boldsymbol{K}_{\sigma\sigma} \end{bmatrix} \begin{bmatrix} \boldsymbol{r}_c(t) \\ \boldsymbol{r}_\sigma(t) \end{bmatrix} = \begin{bmatrix} \boldsymbol{R}_c(t) \\ \boldsymbol{R}_\sigma(t) = \boldsymbol{o} \end{bmatrix} \tag{4.4.1}$$

The right-hand side of the above equation contains inertial, damping and excitation forces. It would thus be erroneous to incorporate a degree of freedom in the $\boldsymbol{r}_\sigma$ group if it were to contain a large mass or an excitation force. The second subsystem of equ. (4.4.1) can now be used for the elimination of the secondary degrees $\boldsymbol{r}_\sigma$. In fact, we obtain

$$\boldsymbol{r}_\sigma(t) = -\boldsymbol{K}_{\sigma\sigma}^{-1}\boldsymbol{K}_{\sigma c}\boldsymbol{r}_c(t) \tag{4.4.2}$$

and thus

$$\boldsymbol{r}(t) = \begin{bmatrix} \boldsymbol{r}_c(t) \\ \boldsymbol{r}_\sigma(t) \end{bmatrix} = \begin{bmatrix} \boldsymbol{I} \\ -\boldsymbol{K}_{\sigma\sigma}^{-1}\boldsymbol{K}_{\sigma c} \end{bmatrix} \boldsymbol{r}_c(t) = \boldsymbol{T}_c \boldsymbol{r}_c(t) \tag{4.4.3}$$

By substituting this expression, for example, into the equation for a free, undamped oscillation, we find

$$[\omega^2 \boldsymbol{M}\boldsymbol{T}_c - \boldsymbol{K}\boldsymbol{T}_c]\boldsymbol{x}_c = \boldsymbol{o} \tag{4.4.4}$$

Pre-multiplying this equation by $\boldsymbol{T}_c^{\mathrm{t}}$, we deduce a reduced system for the eigenvalue problem

$$[\omega^2 \overline{\boldsymbol{M}}_{cc} - \overline{\boldsymbol{K}}_{cc}]\boldsymbol{x}_c = \boldsymbol{o} \tag{4.4.5}$$

in which the following notations are used

$$\overline{\boldsymbol{M}}_{cc} = \boldsymbol{T}_c^{\mathrm{t}} \boldsymbol{M} \boldsymbol{T}_c \tag{4.4.6}$$

$$\overline{\boldsymbol{K}}_{cc} = \boldsymbol{T}_c^{\mathrm{t}} \boldsymbol{K} \boldsymbol{T}_c \tag{4.4.7}$$

This method is also applicable if the structure contains rigid body modes, i.e. if $\boldsymbol{K}$ is singular. We have only to ensure that the structure is at least supported in a statically determinate manner (in which case the degrees of freedom $\boldsymbol{r}_c$ are

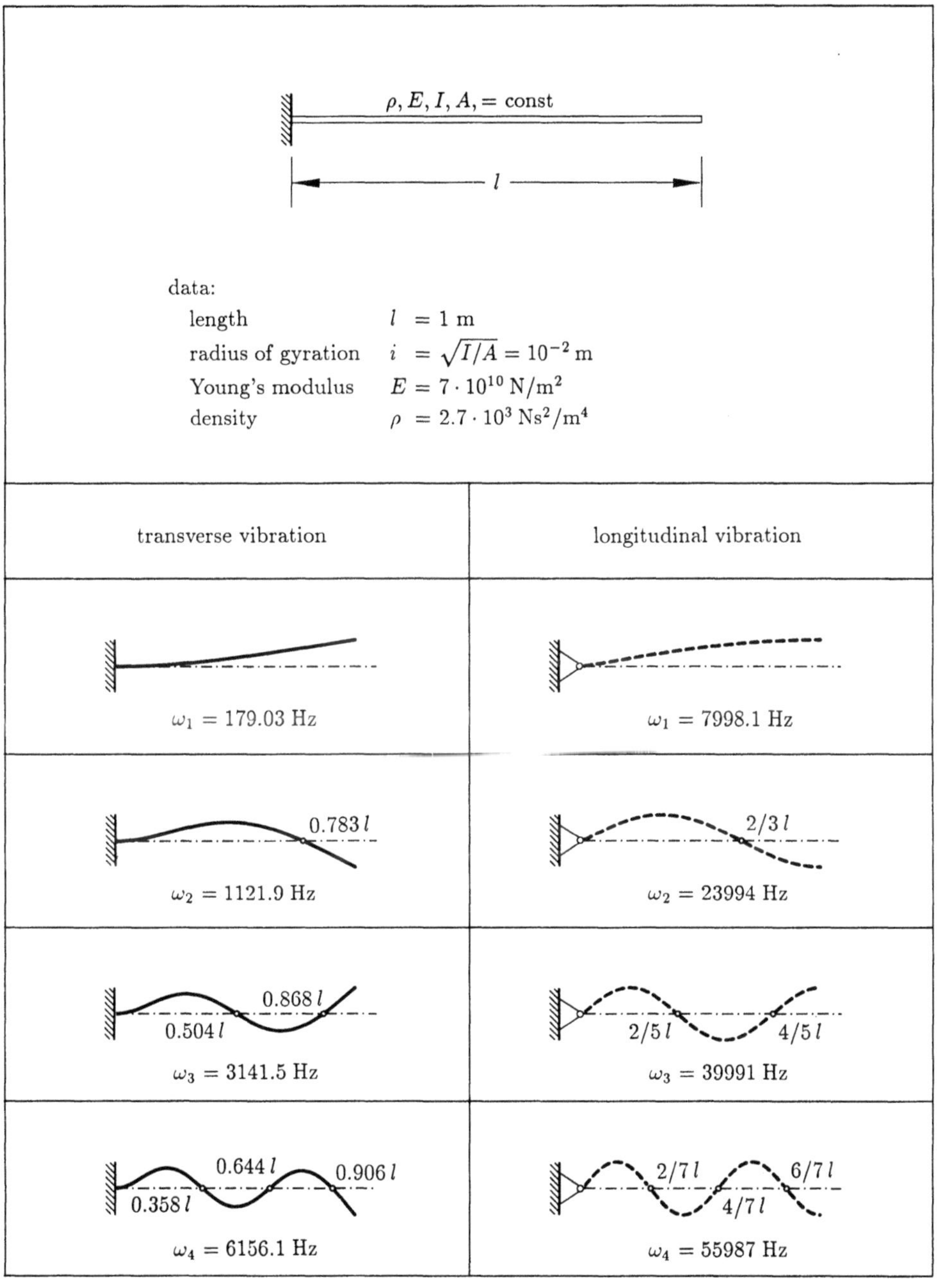

Fig. 4.4.2 Analytical solutions for the eigenoscillations of a cantilever beam (see e.g. [4.15], [4.16], [4.17])

model of fig. 4.4.1	transverse oscillations			longitudinal oscillations		
	ω_i[Hz]	e_D[%]		ω_i[Hz]	e_D[%]	
(a) 12 degrees of freedom	179.03 1123.20 3165.80 6245.40	0.00 0.12 0.77 1.45		8049.60 25393.60 46128.20 66705.40	0.64 5.83 15.35 19.14	
		e_c[%]	e_t[%]		e_c[%]	e_t[%]
(c) 2 degrees of freedom				8204.90 28663.00	1.93 12.87	2.59 19.46
(e) 4 degrees of freedom	179.04 1125.00 3206.30 6290.60	0.01 0.16 1.28 0.72	0.01 0.28 2.06 2.18			
(f) 2 degrees of freedom	179.33 1134.40	0.17 1.00	0.17 1.11			

discretisation error:

$$e_D = \frac{\text{uncondensed FEM solution} - \text{analytical solution}}{\text{analytical solution}} \cdot 100$$

condensation error:

$$e_c = \frac{\text{condensed FEM solution} - \text{uncondensed FEM solution}}{\text{uncondensed FEM solution}} \cdot 100$$

total error (condensed – analytical):

$$e_t = e_D + e_c + \frac{e_D e_c}{100}$$

Fig. 4.4.3 Finite element calculation of the beam model of fig. 4.4.1 with and without condensation

thought of as being suppressed)and thus possesses a non-singular $\boldsymbol{K}_{\sigma\sigma}$. Before the development of the simultaneous vector iteration, Guyan's reduction represented the only simple possibility of analysing large complex structures. When applying this method, we should nevertheless bear in mind that it is only an approximation. In many cases, excellent results are obtained and the errors conditioned by the reduction process are of the order of 1% or less for the lower eigenforms. When insufficiently experienced, however, one should not be surprised to meet some disillusionments. We now illustrate the technique by means of a few examples.

Once again, we turn to the example shown in fig. 4.4.1 and apply Guyan's reduction to it. The data used in this example and the classical analytical solutions are given in fig. 4.4.2. The corresponding results for a finite element model are reproduced in fig. 4.4.3. The transverse motion of the beam is represented by eight degrees of freedom with cubic interpolation, whilst the longitudinal motion is idealised by four degrees of freedom and linear interpolation. As to be expected, the accuracy of the computed frequencies in bending is higher than that for the longitudinal modes. Since the transverse and longitudinal degrees of freedom are uncoupled here, there is no need to consider the models (b) and (d) shown in fig. 4.4.1. The condensation technique is applied to models (c), (e) and (f), and the results are given in fig. 4.4.3. We clearly observe that the condensation technique has hardly any effect upon the bending frequencies. Even for the rough approximation with geometry and idealisation (simply supported quadratic plate)

5 degrees of freedom at a node: u , v , w , φ_x , φ_y

TRIB3 plate elements

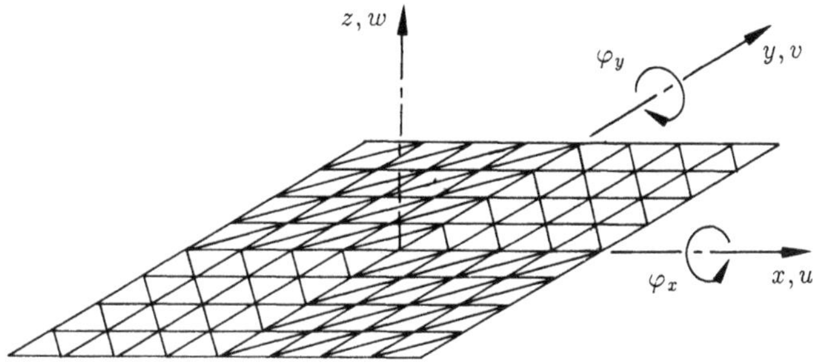

data:

length of side	$l = 1$ m
thickness	$t = 0.01$ m
Young's modulus	$E = 7 \cdot 10^{10}\text{N/m}^2$
Poisson's ratio	$\nu = 0.3$
density	$\rho = 2.7 \cdot 10^3$ Ns/m^4

calculation A: non condensed
81 x 5 = 405 degrees of freedom
6 rigid body modes

calculation B: condensed to 81 w degrees of freedom
+ u, v in center
+ v or u at a corner
= 84 primary degrees of freedom
6 rigid body modes of condensed system

Fig. 4.4.4 Calculation of a free-free square plate according to [4.1]

two degrees of freedom, the error hardly exceeds 1%. On the other hand, we observe for the model on longitudinal oscillations – that is deficient in its accuracy anyhow – a pronounced negative influence of the process of condensation. For two degrees of freedom, the error reaches nearly 20%. This is not surprising since the condensation imposes a linear variation of the displacement within two elements only. Hence, the results effectively correspond to an idealisation with two FLA2 elements which indeed represents a very rough approximation. All results indicate higher frequencies for the FEM models than for the exact analytical model. The condensation imposes a stiffening over the whole frequency spectrum. This is in accordance with the Rayleigh-Ritz method. The reader should notice that kinematic compatibility is maintained by the condensation procedure.

The idealisation of our beam model was relatively rough. Our next example is a somewhat more realistic example of a free-free square plate (fig. 4.4.4) and has already been discussed in [4.1]. A computation based on all prescribed degrees

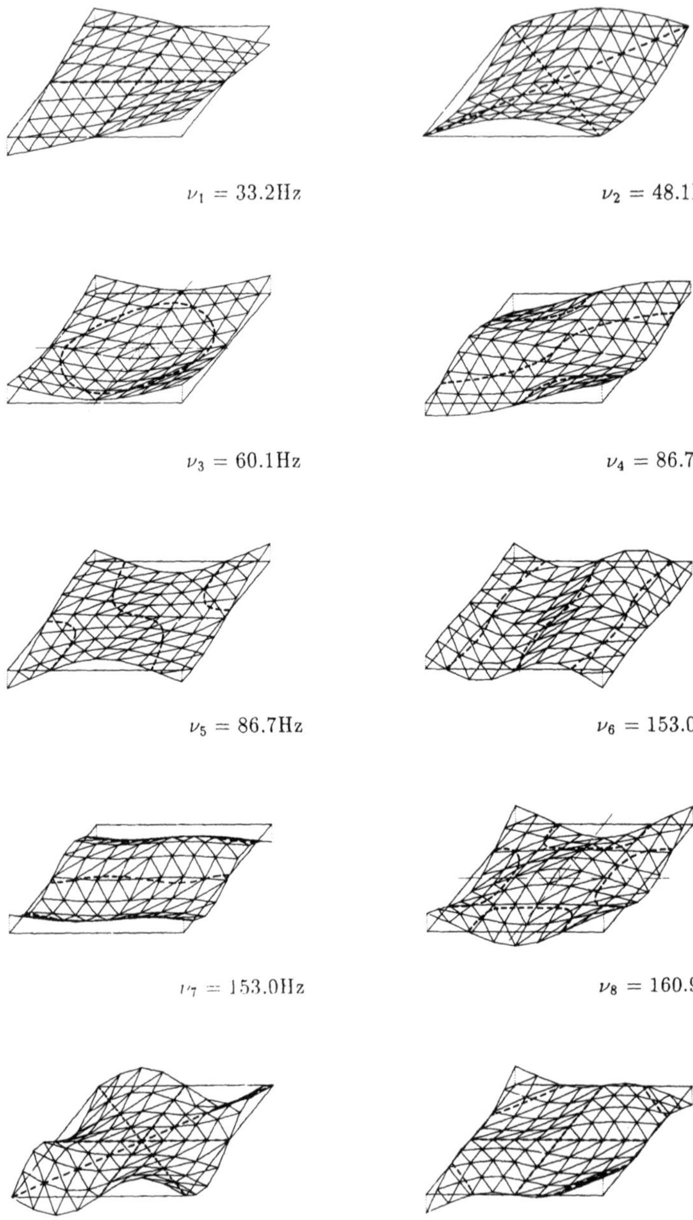

$\nu_1 = 33.2\text{Hz}$ $\nu_2 = 48.1\text{Hz}$

$\nu_3 = 60.1\text{Hz}$ $\nu_4 = 86.7\text{Hz}$

$\nu_5 = 86.7\text{Hz}$ $\nu_6 = 153.0\text{Hz}$

$\nu_7 = 153.0\text{Hz}$ $\nu_8 = 160.9\text{Hz}$

$\nu_9 = 175.1\text{Hz}$ $\nu_{10} = 194.2\text{Hz}$

Fig. 4.4.5 Eigenmode shapes of the free-free square plate according to [4.1]. The calculation was carried out with 84 primary degrees of freedom. The tracer programme used a linear interpolation within the elements

order of bending modes	analytical solution [17.18]	calculation A* 405 degrees of freedom [17.1]		calculation B* 84 primary degrees of freedom [17.1]		
i	ν_i[Hz]	ν_i[Hz]	$e_D = e_t$[%]	ν_i[Hz]	e_c[%]	e_t[%]
1	33.0395	33.1918	0.46	33.1933	0.0045	0.47
2	48.0557	48.1084	0.11	48.1121	0.0077	0.12
3	59.5182	60.1039	0.98	60.1111	0.0120	1.00
4	85.3432	86.7211	1.61	86.7477	0.0307	1.65
5	85.3432	86.7211	1.61	86.7477	0.0307	1.65
6	149.2800	152.9380	2.08	135.0560	0.0772	2.16
7	149.2800	152.9380	2.08	135.0560	0.0772	2.16
8	156.1800	160.7583	2.93	160.9150	0.0975	3.03
9	170.4410	174.8724	2.60	175.1220	0.1427	2.75
10	190.2740	193.9359	1.92	194.1900	0.1310	2.06

$\nu = \frac{1}{T} = \frac{\omega}{2\pi}$ (frequency)

e_D, e_c, e_t are defined in fig. 4.4.3

* identical results are obtained if the u, v degrees of freedom are suppressed, yielding 243 or 81 degrees of freedom

Fig. 4.4.6 Results of the dynamic analysis of the free-free square plate in fig. 4.4.4

of freedom yields the highly satisfactory results shown in figs. 4.4.5 and 4.4.6 for the first ten modes. In fig. 4.4.6, the results of the model based alternatively on a complete set of degrees of freedom and one with a condensed number are compared with the analytical results of ref. [4.18]. The excellent agreement is not effected if the 405 (243) degrees of freedom are condensed to 84(81); see fig. 4.4.6.

As a relevant practical example, we next consider the FS 28 power glider (fig. 4.4.7), designed and built by the Academic Flying Association of Stuttgart. The technical details of this analysis are reported in [4.19]. Since the aircraft is symmetric, only half of the structure need be idealised, a point to which we shall return later in the following chapter. The discretisation of the complete structure is represented by shell, membrane, beam and flange elements and encompasses 5515 degrees of freedom. A direct analysis of the eigenvalue problem of this magnitude is uneconomical unless the method of simultaneous vector iteration or Lanczos' method is employed. The engineer's intuition plays an important role in the selection of the primary degrees of freedom. For the short compact fuselage, one would expect a motion similar to a rigid body one, which can be approximated by four points. On the other hand, a reasonable number of primary degrees of freedom must be adopted to adequately represent the main oscillatory characteristics of the lifting surfaces (wing and vertical fin) involving torsion and bending modes. The boom which carries the vertical fin is so slender that the torsional modes are of secondary importance. However, the model represents satisfactorily the expected beam-like motion of the boom. The validity of our calculations is demonstrated by comparison with experimental results.

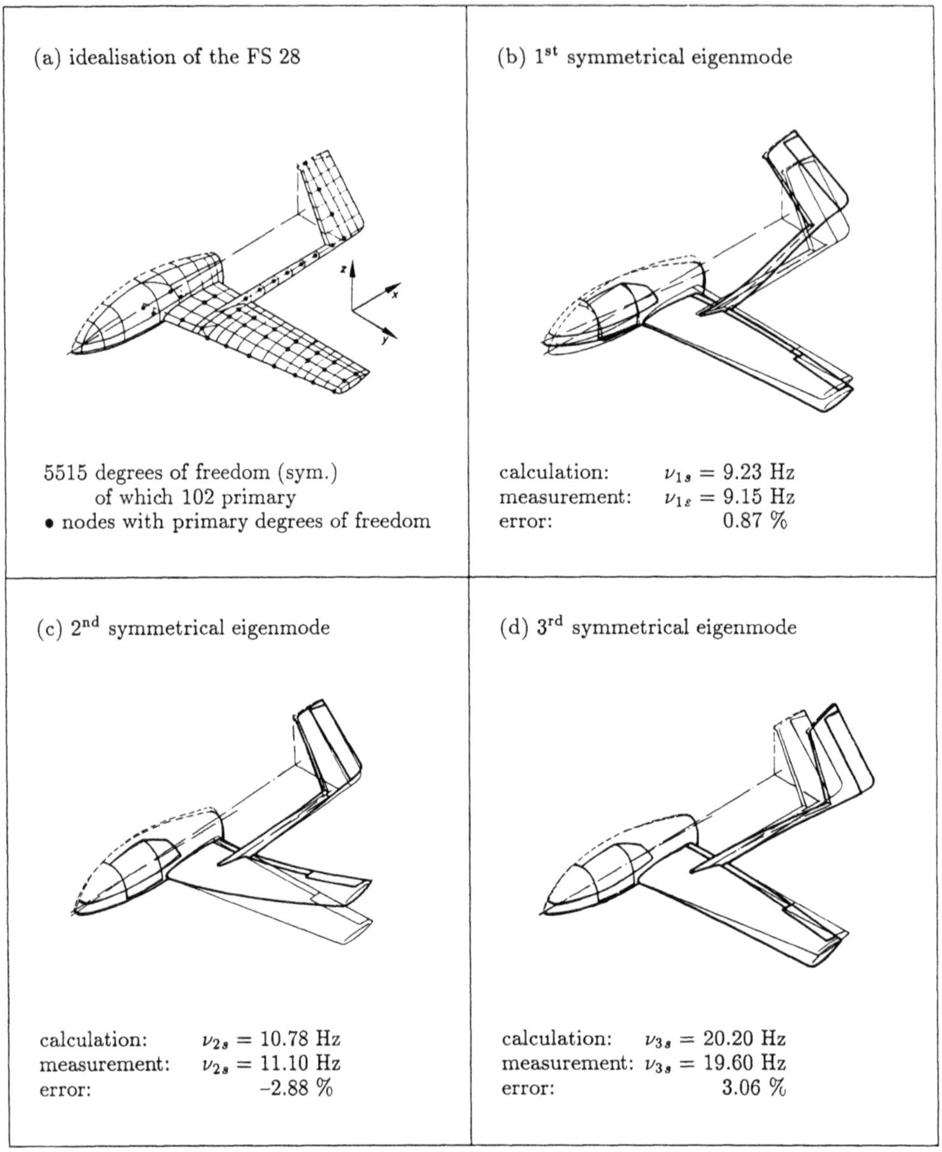

Fig. 4.4.7 Dynamic analysis of the motor glider FS 28, see [4.19]

As a final example, we consider the dynamic analysis of a pre-stressed concrete reactor pressure vessel, quoted in [4.20]. The same vessel was analysed statically in [4.21] and [4.22] where the data are taken from. Condensation of thick-walled containers and massive structures is somewhat problematic. If we observe the first five natural modes of the structure (fig. 4.4.9), it may appear that the response of the container involves mainly bending deformations which can be described satisfactorily by primary points on the surfaces of the container.However, since we desire to determine the first 20 frequencies and modes, this proves a deceptive

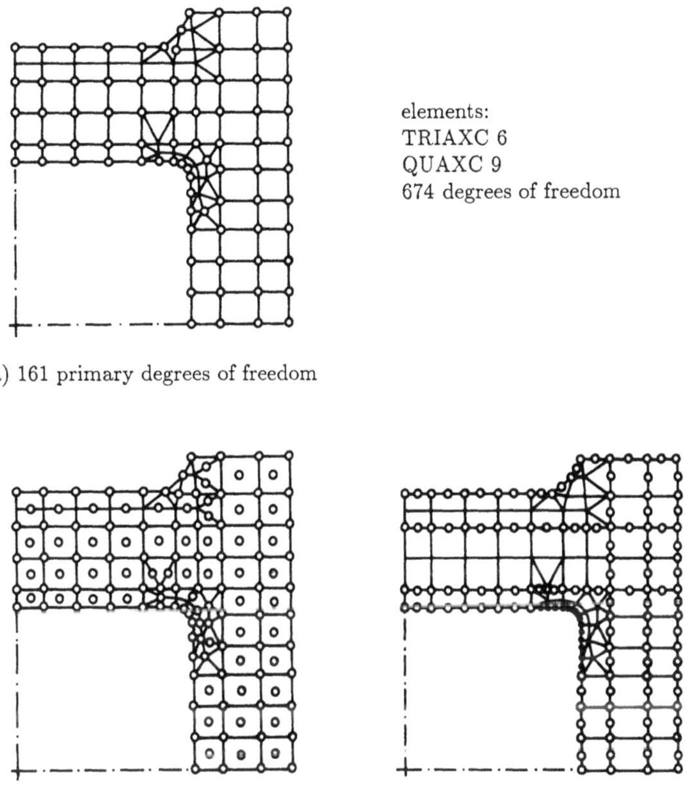

(a) 161 primary degrees of freedom

(b) 277 primary degrees of freedom

(c) 284 primary degrees of freedom

Fig. 4.4.8 Systems of primary degrees of freedom for dynamic analysis of a prestressed concrete reactor pressure vessel, see [4.20]

conclusion. Figure 4.4.8 reproduces three alternative idealisations of the structure with axial symmetry and indicates the nodal points with primary degrees of freedom. One would generally assume that model (a) possesses sufficient primary degrees of freedom as assigned to all corner nodes of the elements. However, the results for the frequencies given in fig. 4.4.10 indicate that the fifth eigenfrequency and higher ones are inaccurate. An increase in the primary degrees of freedom will, in principle, always bring an improvement. The degree of improvement, however, can vary considerably from idealisation to idealisation. The calculations for the variants (b) and (c) substantiate this statement. Variant (c), which contains the highest number of primary degrees of freedom, yields unsatisfactory results, at least from the tenth eigenfrequency onwards. This takes place although we have provided for four dense arrays of primary nodes in the calculation. In contrast, variant (b) proves optimal. Here, in comparison to (a), we have only included an additional primary nodal point in the centre of each element. If the reader surveys his knowledge of dynamics, he will recollect that in the calculation of kinematically

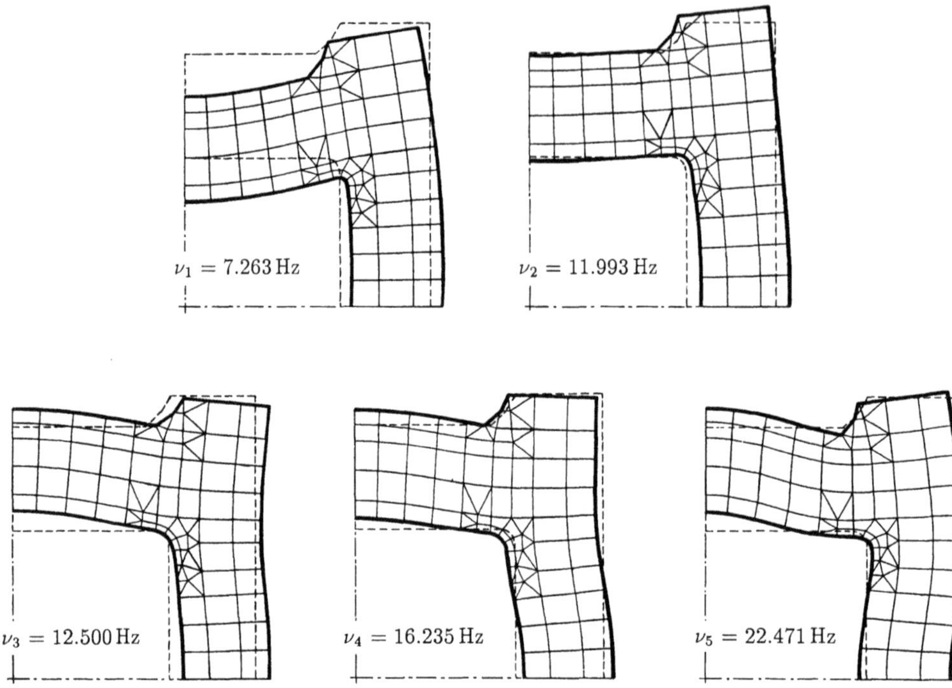

Fig. 4.4.9 The first 5 mode shapes of a pre-stressed concrete reactor pressure vessel, see [4.20] (axial symmetry)

equivalent volume forces, the central nodes and thereafter the midside nodes are dominantly loaded. It is thus advisable to assign primary degrees of freedom to these points.

Those readers interested in deepening their knowledge on primary degrees of freedom are referred to the supplement to this chapter.

Supplement to 4.4

*4.4 Guyan's reduction as a transformed dynamics problem

In Guyan's condensation technique of section 4.4, the degrees of freedom $\boldsymbol{r}_\sigma$ are assigned a statical dependence on the primary degrees of freedom $\boldsymbol{r}_c$. However, in order to improve the accuracy, we split the displacements into a basic group, identical to that of Guyan, and an additional group, which serves as a correction to our approximation. The corrected equilibrium hence becomes

$$\begin{bmatrix} \boldsymbol{r}_c \\ \boldsymbol{r}_\sigma \end{bmatrix} = \begin{bmatrix} \boldsymbol{I} & \boldsymbol{O} \\ -\boldsymbol{K}_{\sigma\sigma}^{-1}\boldsymbol{K}_{\sigma c} & \boldsymbol{I} \end{bmatrix} \begin{bmatrix} \boldsymbol{r}_c \\ \Delta\boldsymbol{r}_\sigma \end{bmatrix} = \begin{bmatrix} \boldsymbol{T}_c & \begin{matrix} \boldsymbol{O} \\ \boldsymbol{I} \end{matrix} \end{bmatrix} \{\boldsymbol{r}_c \quad \Delta\boldsymbol{r}_\sigma\}$$

$$= \boldsymbol{T}\{\boldsymbol{r}_c \quad \Delta\boldsymbol{r}_\sigma\} \tag{*4.4.1}$$

	basic model 674 DOFs	(a) 161 primary DOFs		(b) 277 primary DOFs		(c) 284 primary DOFs	
i	eigenfrequencies ν_i[Hz]	ν_{ic}[Hz]	error e_c [%]	ν_{ic} [Hz]	error e_c [%]	ν_{ic} [Hz]	error e_c [%]
1	7.263	7.281	0.252	7.266	0.037	7.270	0.102
2	11.993	12.084	0.757	12.016	0.189	12.017	0.199
3	12.500	12.601	0.809	12.522	0.180	12.525	0.200
4	16.235	16.434	1.223	16.278	0.262	16.300	0.403
5	22.471	22.982	2.273	22.583	0.497	22.618	0.654
6	31.480	32.838	4.315	31.849	1.171	31.865	1.223
7	34.140	36.007	5.468	34.487	1.015	34.917	2.275
8	37.580	39.902	6.180	38.079	1.327	38.276	1.851
9	43.111	46.114	6.965	44.161	2.436	44.161	2.436
10	44.701	48.890	9.370	45.481	1.745	47.304	5.822
15	60.097	68.621	14.184	62.118	3.363	62.758	4.428
20	71.195	82.987	16.564	73.786	3.640	74.633	4.829

$$e_c = \frac{\nu_{ic} - \nu_i}{\nu_i} \cdot 100$$

Fig. 4.4.10 Eigenfrequencies and errors for the alternative models of fig. 4.4.8, see [4.20] (axial symmetry)

Our transformed representation of the unknown displacements is thus given by $\boldsymbol{r}_c$ and $\Delta\boldsymbol{r}_\sigma$. Providing that we initially made an effective choice of primary degrees of freedom, the correction $\Delta\boldsymbol{r}_\sigma$ should prove small. In the original Guyan method, this correction was set to zero. These new unknowns are now introduced into the dynamic equation and we observe that this operation leads to a static decoupling (in fact, a stiffness decoupling) of the problem

$$\boldsymbol{T}^{\mathrm{t}}\boldsymbol{K}\boldsymbol{T}\begin{bmatrix} \boldsymbol{r}_c \\ \Delta\boldsymbol{r}_\sigma \end{bmatrix} + \boldsymbol{T}^{\mathrm{t}}\boldsymbol{M}\boldsymbol{T}\begin{bmatrix} \ddot{\boldsymbol{r}}_c \\ \Delta\ddot{\boldsymbol{r}}_\sigma \end{bmatrix} = \begin{bmatrix} \boldsymbol{o} \\ \boldsymbol{o} \end{bmatrix} \tag{*4.4.2}$$

Applying the usual exponential solution statement, we find

$$\begin{bmatrix} \overline{\boldsymbol{M}}_{cc} & \boldsymbol{M}_{c\sigma} - \boldsymbol{K}_{c\sigma}\boldsymbol{K}_{\sigma\sigma}^{-1}\boldsymbol{M}_{\sigma\sigma} \\ \boldsymbol{M}_{\sigma c} - \boldsymbol{M}_{\sigma\sigma}\boldsymbol{K}_{\sigma\sigma}^{-1}\boldsymbol{K}_{\sigma c} & \boldsymbol{M}_{\sigma\sigma} \end{bmatrix}\begin{bmatrix} \boldsymbol{X}_{cc} & \boldsymbol{X}_{c\sigma} \\ \Delta\boldsymbol{X}_{\sigma c} & \Delta\boldsymbol{X}_{\sigma\sigma} \end{bmatrix}$$

$$= \begin{bmatrix} \overline{\boldsymbol{K}}_{cc} & \boldsymbol{O} \\ \boldsymbol{O}^{\mathrm{t}} & \boldsymbol{K}_{\sigma\sigma} \end{bmatrix}\begin{bmatrix} \boldsymbol{X}_{cc} & \boldsymbol{X}_{c\sigma} \\ \Delta\boldsymbol{X}_{\sigma c} & \Delta\boldsymbol{X}_{\sigma\sigma} \end{bmatrix}\begin{bmatrix} \boldsymbol{\Lambda}_c & \boldsymbol{O} \\ \boldsymbol{O}^{\mathrm{t}} & \boldsymbol{\Lambda}_\sigma \end{bmatrix} \tag{*4.4.3}$$

The matrices $\overline{\boldsymbol{M}}_{cc}$ and $\overline{\boldsymbol{K}}_{cc}$ represent the reduced mass and stiffness matrices according to the Guyan method and are given in equs. (4.4.6) and (4.4.7). Since the condensation procedure involves only primary degrees of freedom, our calculations take place in the subspace of the $\boldsymbol{\Lambda}_c$ eigenvalues. A decoupled consideration of the stiffness matrix is possible; the mass matrix remains coupled, however. Indeed, we observe that Guyan employs as an approximation to the real set of eigenvectors

$$\begin{bmatrix} \boldsymbol{X}_{cc} \\ \Delta\boldsymbol{X}_{\sigma c} \end{bmatrix} \Longrightarrow \begin{bmatrix} \widetilde{\boldsymbol{X}}_{cc} \\ \boldsymbol{O} \end{bmatrix} \tag{*4.4.4}$$

Although the solution to the condensed problem

$$\overline{\boldsymbol{M}}_{cc}\widetilde{\boldsymbol{X}}_{cc} = \overline{\boldsymbol{K}}_{cc}\widetilde{\boldsymbol{X}}_{cc}\widetilde{\boldsymbol{\Lambda}}_{c} \tag{*4.4.5}$$

fulfils the first subset of equ. (*4.4.3), we are left in the second subset with the residue

$$\boldsymbol{H}_{\sigma c} = [\boldsymbol{M}_{\sigma c} - \boldsymbol{M}_{\sigma\sigma}\boldsymbol{K}_{\sigma\sigma}^{-1}\boldsymbol{K}_{\sigma c}]\widetilde{\boldsymbol{X}}_{cc} \tag{*4.4.6}$$

From the point of view of the residuum, mass concentrations should, if possible, be assigned primary degrees of freedom in order to avoid large entries in $\boldsymbol{M}_{\sigma c}$ and $\boldsymbol{M}_{\sigma\sigma}$. At this point, we remind the reader of the error bounds in equs. (4.3.37) and (4.3.48). A transcription to the present formulation using equ. (*4.4.6) leads to an estimate of the condensation error. For this purpose, we first transform the general eigenvalue problem in equ. (*4.4.3) into an equivalent special one and obtain with the triangularisation

$$\begin{bmatrix} \overline{\boldsymbol{K}}_{cc} & \boldsymbol{K}_{\sigma\sigma} \end{bmatrix} = \begin{bmatrix} \overline{\boldsymbol{U}}_{cc}^{\mathrm{t}} & \boldsymbol{U}_{\sigma\sigma}^{\mathrm{t}} \end{bmatrix}\begin{bmatrix} \overline{\boldsymbol{U}}_{cc} & \boldsymbol{U}_{\sigma\sigma} \end{bmatrix} \tag{*4.4.7}$$

at once

$$\begin{bmatrix} \overline{\boldsymbol{U}}_{cc}^{-\mathrm{t}}\overline{\boldsymbol{M}}_{cc}\overline{\boldsymbol{U}}_{cc}^{-1} & \overline{\boldsymbol{U}}_{cc}^{-\mathrm{t}}[\boldsymbol{M}_{c\sigma} - \boldsymbol{K}_{c\sigma}\boldsymbol{K}_{\sigma\sigma}^{-1}\boldsymbol{M}_{\sigma\sigma}]\boldsymbol{U}_{\sigma\sigma}^{-1} \\ \boldsymbol{U}_{\sigma\sigma}^{-\mathrm{t}}[\boldsymbol{M}_{\sigma c} - \boldsymbol{M}_{\sigma\sigma}\boldsymbol{K}_{\sigma\sigma}^{-1}\boldsymbol{K}_{\sigma c}]\overline{\boldsymbol{U}}_{cc}^{-1} & \boldsymbol{U}_{\sigma\sigma}^{-1}\boldsymbol{M}_{\sigma\sigma}\boldsymbol{U}_{\sigma\sigma}^{-1} \end{bmatrix}\begin{bmatrix} \overline{\boldsymbol{X}}_{cc} & \overline{\boldsymbol{X}}_{c\sigma} \\ \Delta\overline{\boldsymbol{X}}_{\sigma c} & \Delta\overline{\boldsymbol{X}}_{\sigma\sigma} \end{bmatrix}$$

$$= \begin{bmatrix} \overline{\boldsymbol{X}}_{cc} & \overline{\boldsymbol{X}}_{c\sigma} \\ \Delta\overline{\boldsymbol{X}}_{\sigma c} & \Delta\overline{\boldsymbol{X}}_{\sigma\sigma} \end{bmatrix}\begin{bmatrix} \boldsymbol{\Lambda}_{c} & \boldsymbol{O} \\ \boldsymbol{O}^{\mathrm{t}} & \boldsymbol{\Lambda}_{\sigma} \end{bmatrix} \tag{*4.4.8}$$

where

$$\begin{bmatrix} \overline{\boldsymbol{X}}_{cc} & \overline{\boldsymbol{X}}_{c\sigma} \\ \Delta\overline{\boldsymbol{X}}_{\sigma c} & \Delta\overline{\boldsymbol{X}}_{\sigma\sigma} \end{bmatrix} = \begin{bmatrix} \overline{\boldsymbol{U}}_{cc} & \boldsymbol{U}_{\sigma\sigma} \end{bmatrix}\begin{bmatrix} \boldsymbol{X}_{cc} & \boldsymbol{X}_{c\sigma} \\ \Delta\boldsymbol{X}_{\sigma c} & \Delta\boldsymbol{X}_{\sigma\sigma} \end{bmatrix} \tag{*4.4.9}$$

are the eigenvectors of the special eigenvalue problem. The matrix

$$\overline{\boldsymbol{A}}_{cc} = \overline{\boldsymbol{U}}_{cc}^{-\mathrm{t}}\overline{\boldsymbol{M}}_{cc}\overline{\boldsymbol{U}}_{cc}^{-\mathrm{t}} \tag{*4.4.10}$$

is, as a matter of fact, the dynamic matrix of the condensed problem

$$\overline{\boldsymbol{A}}_{cc}\widetilde{\overline{\boldsymbol{X}}}_{cc} = \widetilde{\overline{\boldsymbol{X}}}_{cc}\widetilde{\boldsymbol{\Lambda}}_{c} \tag{*4.4.11}$$

This eigenvalue equation yields, as usual, an estimate of the approximations $\widetilde{\boldsymbol{\Lambda}}_c$ and $\{\widetilde{\overline{\boldsymbol{X}}}_{cc} \quad \boldsymbol{O}\}$ for the higher eigenvalues and eigenvectors appertaining to the primary subspace. The first subset of the exact formulation is obviously fulfilled, but the second remains unsatisfied. We thus obtain

$$\overline{\boldsymbol{H}} = \begin{bmatrix} \boldsymbol{O} \\ \overline{\boldsymbol{H}}_{\sigma c} \end{bmatrix} = \begin{bmatrix} \boldsymbol{O} \\ \boldsymbol{U}_{\sigma\sigma}^{-\mathrm{t}}[\boldsymbol{M}_{\sigma c} - \boldsymbol{M}_{\sigma\sigma}\boldsymbol{K}_{\sigma\sigma}^{-1}\boldsymbol{K}_{\sigma c}]\overline{\boldsymbol{U}}_{cc}^{-1}\widetilde{\overline{\boldsymbol{X}}}_{cc} \end{bmatrix} \tag{*4.4.12}$$

Let us now consider the residue $\overline{\boldsymbol{\eta}}_{\mathrm{i}}^{(\sigma c)}$ corresponding to an eigenvalue $\widetilde{\lambda}_{\mathrm{i}}^{(c)}$

$$\overline{\boldsymbol{\eta}}_{\mathrm{i}}^{(\sigma c)} = \boldsymbol{U}_{\sigma\sigma}^{-\mathrm{t}}[\boldsymbol{M}_{\sigma c} - \boldsymbol{M}_{\sigma\sigma}\boldsymbol{K}_{\sigma\sigma}^{-1}\boldsymbol{K}_{\sigma c}]\overline{\boldsymbol{U}}_{cc}^{-1}\widetilde{\overline{\boldsymbol{x}}}_{\mathrm{i}}^{(c)} \tag{*4.4.13}$$

where the eigenvector $\widetilde{\overline{\boldsymbol{x}}}_{\mathrm{i}}^{(c)}$ is normalised in accordance with

$$\left[\widetilde{\boldsymbol{x}}_{\mathrm{i}}^{(c)}\right]^{\mathrm{t}}\widetilde{\boldsymbol{x}}_{\mathrm{i}}^{(c)} = 1 \tag{*4.4.14}$$

From

$$\overline{\boldsymbol{\eta}}_{\mathrm{i}} = \boldsymbol{A}\begin{bmatrix} \widetilde{\overline{\boldsymbol{x}}}_{\mathrm{i}}^{(c)} \\ \boldsymbol{o} \end{bmatrix} - \begin{bmatrix} \widetilde{\overline{\boldsymbol{x}}}_{\mathrm{i}}^{(c)} \\ \boldsymbol{o} \end{bmatrix}\widetilde{\lambda}_{\mathrm{i}}^{(c)} = \begin{bmatrix} \boldsymbol{o} \\ \overline{\boldsymbol{\eta}}_{\mathrm{i}}^{(\sigma c)} \end{bmatrix} \tag{*4.4.15}$$

we deduce the following statement in association with an exact complete set $\overline{\boldsymbol{X}}$

$$\overline{\boldsymbol{X}}^{\mathrm{t}}\overline{\boldsymbol{\eta}}_{\mathrm{i}} = \overline{\boldsymbol{X}}^{\mathrm{t}}\boldsymbol{A}\overline{\boldsymbol{X}}\,\overline{\boldsymbol{X}}^{\mathrm{t}}\begin{bmatrix} \widetilde{\overline{\boldsymbol{x}}}_{\mathrm{i}}^{(c)} \\ \boldsymbol{o} \end{bmatrix} - \overline{\boldsymbol{X}}^{\mathrm{t}}\begin{bmatrix} \widetilde{\overline{\boldsymbol{x}}}_{\mathrm{i}}^{(c)} \\ \boldsymbol{o} \end{bmatrix}\widetilde{\lambda}_{\mathrm{i}}^{(c)} = [\boldsymbol{\Lambda} - \widetilde{\lambda}_{\mathrm{i}}^{(c)}\boldsymbol{I}]\overline{\boldsymbol{X}}^{\mathrm{t}}\begin{bmatrix} \widetilde{\overline{\boldsymbol{x}}}_{\mathrm{i}}^{(c)} \\ \boldsymbol{o} \end{bmatrix} \tag{*4.4.16}$$

where

$$\overline{\boldsymbol{X}}^{\mathrm{t}}\overline{\boldsymbol{X}} = \overline{\boldsymbol{X}}\,\overline{\boldsymbol{X}}^{\mathrm{t}} = \boldsymbol{I} \tag{*4.4.17}$$

applies. Either one of the exact eigenvalues of $\boldsymbol{\Lambda}$ is identical to the calculated approximation $\overline{\lambda}_{\mathrm{i}}^{(c)}$ of the condensed system or

$$\overline{\boldsymbol{X}}^{\mathrm{t}}\begin{bmatrix} \widetilde{\overline{\boldsymbol{x}}}_{\mathrm{i}}^{(c)} \\ \boldsymbol{o} \end{bmatrix} = \left[\boldsymbol{\Lambda} - \widetilde{\lambda}_{\mathrm{i}}^{(c)}\boldsymbol{I}\right]^{-1}\overline{\boldsymbol{X}}^{\mathrm{t}}\begin{bmatrix} \boldsymbol{o} \\ \overline{\boldsymbol{\eta}}_{\mathrm{i}}^{(\sigma c)} \end{bmatrix} \tag{*4.4.18}$$

holds. According to the rules of the matrix norms, though, we have

$$\|\boldsymbol{Q}\boldsymbol{y}\| = \|\boldsymbol{y}\| \tag{*4.4.19}$$

if $\boldsymbol{Q}$ is an orthogonal matrix as, for example, $\overline{\boldsymbol{X}}$. Furthermore,

$$\|\boldsymbol{B}\boldsymbol{C}\| \leqslant \|\boldsymbol{B}\|\,\|\boldsymbol{C}\| \tag{*4.4.20}$$

applies which is true for any matrix norm. We also have

$$\left\| \begin{bmatrix} \widetilde{\overline{\boldsymbol{x}}}_{\rm i}^{(c)} \\ \boldsymbol{o} \end{bmatrix} \right\| = \|\widetilde{\overline{\boldsymbol{x}}}_{\rm i}^{(c)}\| = 1 \tag{*4.4.21}$$

since equ. (*4.4.14) applies. We next introduce the following abbreviation

$$\varepsilon_{\rm i} = \left\| \begin{bmatrix} \boldsymbol{o} \\ \overline{\boldsymbol{\eta}}_{\rm i}^{(\sigma c)} \end{bmatrix} \right\| = \|\overline{\boldsymbol{\eta}}_{\rm i}^{(\sigma c)}\| = \left(\overline{\boldsymbol{\eta}}_{\rm i}^{(\sigma c)\rm t}\overline{\boldsymbol{\eta}}_{\rm i}^{(\sigma c)}\right)^{1/2} \tag{*4.4.22}$$

Consequently, we observe immediately from equ. (*4.4.18) the result

$$1 \leqslant \max_{\rm k=1,n} \left(|\lambda_{\rm k} - \widetilde{\lambda}_{\rm i}^{(c)}|^{-1}\right)\varepsilon_{\rm i} \tag{*4.4.23}$$

or

$$\min_{\rm k=1,n} |\lambda_{\rm k} - \widetilde{\lambda}_{\rm i}^{(c)}| \leqslant \varepsilon_{\rm i} \tag{*4.4.24}$$

We are thus informed that a true eigenvalue $\lambda_{\rm k}$ can differ from the approximation $\lambda_{\rm i}^{(c)}$ by $\pm\varepsilon_{\rm i}$ at the most. In general, the lower limit is relevant since the condensation produces a stiffening of the system with a consequent increase of the eigenfrequencies respectively a decrease of the eigenvalues. From equ. (*4.4.24), we obtain the following estimate of the condensation error in [%]

$$e_c \leqslant \frac{\varepsilon_{\rm i}}{\widetilde{\lambda}_{\rm i}^{(c)} - \varepsilon_{\rm i}} 100$$

This error estimation can be useful in practice if we possess only a limited information on the natural modes of interest. In addition to the deviation in the eigenvalues, the deviations of eigenvectors are also of interest to us. In fact, these are of vital importance, especially for the calculation of the dynamic stresses since the associated differentiation process increases the error. Thus, if we are interested in precise stresses, it is necessary to turn to simultaneous vector iteration.

4.5 Exploitation of structural symmetry in eigenvalue analysis

Whilst Guyan's reduction only offers an approximation by a reduction of the problem size, the use of symmetry in dynamics provides an exact reduction procedure. In comparison to linear statics, some modifications must, however, be considered. In what follows we expect, as in linear statics, the structure to possess a symmetric geometry and distribution of material data (symmetric stiffness characteristics). In addition, we have to assume in dynamics a symmetric mass distribution (symmetric mass matrix).

We now consider the simple oscillator with symmetric properties of fig. 4.5.1 and introduce the new grouping of the degrees in $\boldsymbol{r}_s$ (pure symmetrical) and $\boldsymbol{r}_a$ (pure

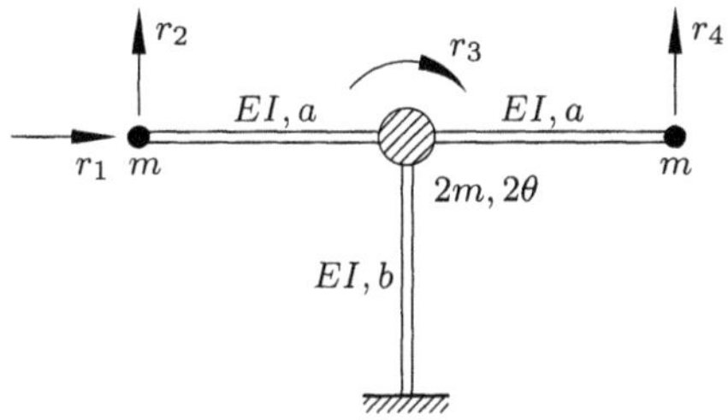

(a) symmetrical oscillator with 4 degrees of freedom (restricted to pure bending deformations)

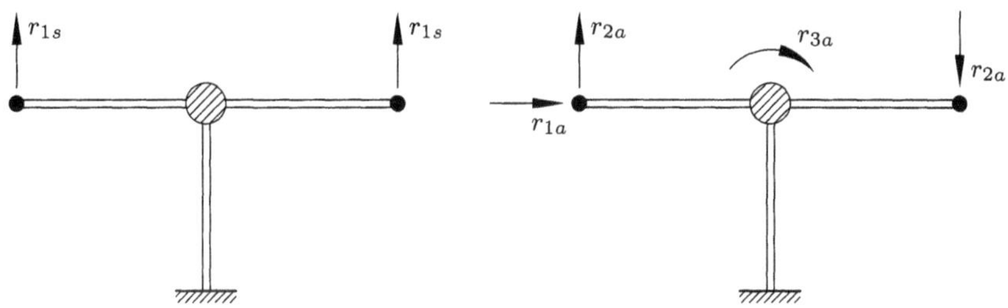

● masses having only translatory inertia

masses having also rotational inertia

$$\boldsymbol{r} = \boldsymbol{T}_s \boldsymbol{r}_s + \boldsymbol{T}_a \boldsymbol{r}_a$$

$$\begin{bmatrix} r_1 \\ r_2 \\ r_3 \\ r_4 \end{bmatrix} = \begin{bmatrix} 0 \\ 1 \\ 0 \\ 1 \end{bmatrix} \begin{bmatrix} r_{1s} \end{bmatrix} + \begin{bmatrix} 1 & 0 & 0 \\ 0 & 1 & 0 \\ 0 & 0 & 1 \\ 0 & -1 & 0 \end{bmatrix} \begin{bmatrix} r_{1a} \\ r_{2a} \\ r_{3a} \end{bmatrix}$$

(b) decomposition of the motion in symmetrical and antisymmetrical components

Fig. 4.5.1 Symmetrical oscillator

antisymmetrical) into the eigenvalue problem in place of the standard degrees of freedom $\boldsymbol{r}$. Thus,

$$\boldsymbol{r} = \boldsymbol{T}_s \boldsymbol{r}_s + \boldsymbol{T}_a \boldsymbol{r}_a \tag{4.5.1}$$

A grouping of the degrees of freedom into symmetric and antisymmetric components is possible for any symmetric structure. Substituting the expression (4.5.1) in equ. (4.1.1), we obtain the new equations of motion

$$\begin{bmatrix} \boldsymbol{T}_s^{\mathrm{t}}\boldsymbol{M}\boldsymbol{T}_s & \boldsymbol{T}_s^{\mathrm{t}}\boldsymbol{M}\boldsymbol{T}_a \\ \boldsymbol{T}_a^{\mathrm{t}}\boldsymbol{M}\boldsymbol{T}_s & \boldsymbol{T}_a^{\mathrm{t}}\boldsymbol{M}\boldsymbol{T}_a \end{bmatrix} \begin{bmatrix} \ddot{\boldsymbol{r}}_s \\ \ddot{\boldsymbol{r}}_a \end{bmatrix} + \begin{bmatrix} \boldsymbol{T}_s^{\mathrm{t}}\boldsymbol{K}\boldsymbol{T}_s & \boldsymbol{T}_s^{\mathrm{t}}\boldsymbol{K}\boldsymbol{T}_a \\ \boldsymbol{T}_a^{\mathrm{t}}\boldsymbol{K}\boldsymbol{T}_s & \boldsymbol{T}_a^{\mathrm{t}}\boldsymbol{K}\boldsymbol{T}_a \end{bmatrix} \begin{bmatrix} \boldsymbol{r}_s \\ \boldsymbol{r}_a \end{bmatrix} = \boldsymbol{o} \tag{4.5.2}$$

Considering the matrices $\boldsymbol{T}_s$ and $\boldsymbol{T}_a$ in fig. 4.5.1, we observe that each column of $\boldsymbol{T}_s$ describes a pure symmetrical motion and each column of $\boldsymbol{T}_a$ a pure antisymmetrical one. Consequently, the matrix product $\boldsymbol{M}\boldsymbol{T}_a$ can be interpreted as defining inertia forces due to a unit acceleration in the direction of an antisymmetric mode shape. The term $\boldsymbol{T}_s^{\mathrm{t}}\boldsymbol{M}\boldsymbol{T}_a$ then denotes the work of the antisymmetric inertia forces performed over the symmetric modes of motion. Since the contributions to the work arising left and right from the plane of symmetry are equal but of opposite sign, we find

$$\begin{aligned} \boldsymbol{T}_s^{\mathrm{t}}\boldsymbol{M}\boldsymbol{T}_a &= \boldsymbol{O} \\ \boldsymbol{T}_a^{\mathrm{t}}\boldsymbol{M}\boldsymbol{T}_s &= \boldsymbol{O}^{\mathrm{t}} \end{aligned} \tag{4.5.3}$$

Correspondingly, $\boldsymbol{K}\boldsymbol{T}_a$ may be considered elastic forces arising from the antisymmetric deformations $\boldsymbol{T}_a$. In a symmetric distribution of the stiffness characteristics, these elastic forces are also antisymmetric. They do not perform any work when acting upon the symmetric displacements $\boldsymbol{T}_s$

$$\begin{aligned} \boldsymbol{T}_s^{\mathrm{t}}\boldsymbol{K}\boldsymbol{T}_a &= \boldsymbol{O} \\ \boldsymbol{T}_a^{\mathrm{t}}\boldsymbol{K}\boldsymbol{T}_s &= \boldsymbol{O}^{\mathrm{t}} \end{aligned} \tag{4.5.4}$$

Thus, the dynamic problem of a symmetric structure (equ. (4.5.2)) and consequently also the associated eigenvalue problem can be uncoupled and written as

$$\begin{bmatrix} \boldsymbol{T}_s^{\mathrm{t}}\boldsymbol{M}\boldsymbol{T}_s & \boldsymbol{O} \\ \boldsymbol{O}^{\mathrm{t}} & \boldsymbol{T}_a^{\mathrm{t}}\boldsymbol{M}\boldsymbol{T}_a \end{bmatrix} \begin{bmatrix} \ddot{\boldsymbol{r}}_s \\ \ddot{\boldsymbol{r}}_a \end{bmatrix} + \begin{bmatrix} \boldsymbol{T}_s^{\mathrm{t}}\boldsymbol{K}\boldsymbol{T}_s & \boldsymbol{O} \\ \boldsymbol{O}^{\mathrm{t}} & \boldsymbol{T}_a^{\mathrm{t}}\boldsymbol{K}\boldsymbol{T}_a \end{bmatrix} \begin{bmatrix} \boldsymbol{r}_s \\ \boldsymbol{r}_a \end{bmatrix} = \begin{bmatrix} \boldsymbol{o} \\ \boldsymbol{o} \end{bmatrix} \tag{4.5.5}$$

or

$$\begin{aligned} \left[\boldsymbol{T}_s^{\mathrm{t}}\boldsymbol{M}\boldsymbol{T}_s\right]\boldsymbol{X}_s &= \left[\boldsymbol{T}_s^{\mathrm{t}}\boldsymbol{K}\boldsymbol{T}_s\right]\boldsymbol{X}_s\boldsymbol{\Lambda}_s \\ \left[\boldsymbol{T}_a^{\mathrm{t}}\boldsymbol{M}\boldsymbol{T}_a\right]\boldsymbol{X}_a &= \left[\boldsymbol{T}_a^{\mathrm{t}}\boldsymbol{K}\boldsymbol{T}_a\right]\boldsymbol{X}_a\boldsymbol{\Lambda}_a \end{aligned} \tag{4.5.6}$$

From the last equation, we conclude that a symmetric oscillator can only possess pure symmetric and pure antisymmetric eigenmodes. This knowledge allows us

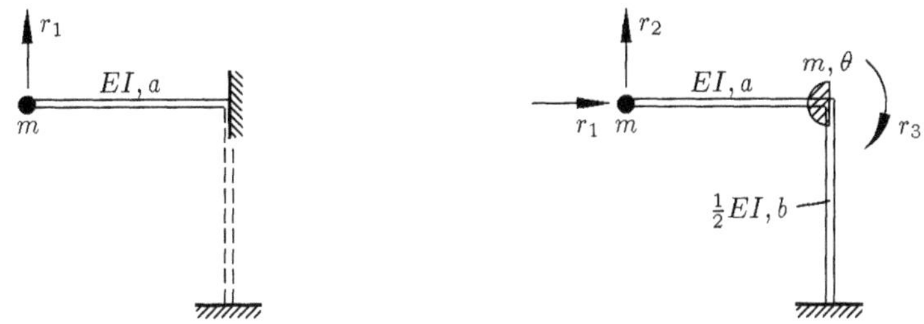

(a) half-model for symmetrical eigenoscillations

(b) half-model for antisymmetrical eigenoscillations

Fig. 4.5.2 Half-model of the oscillator in fig. 4.5.1 (only bending deformations admitted)

<table>
<tr><td colspan="3">three-fold symmetry
one eigth model</td></tr>
<tr><td colspan="2">two-fold symmetry
quarter model</td><td></td></tr>
<tr><td>simple symmetry
half model</td><td colspan="2"></td></tr>
<tr><td>plane of symmetry 1</td><td>plane of symmetry 2</td><td>plane of symmetry 3</td></tr>
<tr><td>s</td><td>s</td><td>s</td></tr>
<tr><td>s</td><td>s</td><td>a</td></tr>
<tr><td>s</td><td>a</td><td>s</td></tr>
<tr><td>s</td><td>a</td><td>a</td></tr>
<tr><td>a</td><td>s</td><td>s</td></tr>
<tr><td>a</td><td>s</td><td>a</td></tr>
<tr><td>a</td><td>a</td><td>s</td></tr>
<tr><td>a</td><td>a</td><td>a</td></tr>
</table>

s = symmetrical motion only in the plane of symmetry

a = antisymmetrical motion only in the plane of symmetry

Fig. 4.5.3 Computations and boundary conditions when taking advantage of symmetry

to circumvent the introduction of the generalised displacements $\boldsymbol{r}_s$ and $\boldsymbol{r}_a$, and to operate directly, as is seen from fig. 4.5.2, with a half-model and the appropriate boundary conditions. In this manner, the problem is reduced to approximately half-size (when multiple symmetries, 1/4 or 1/8) and can be solved for various boundary conditions. Since an eigenvalue calculation is a process involving n^3 operations, the effort is reduced to $2 * \mathrm{n}^3/8$. For multiple symmetry, the scheme illustrated in fig. 4.5.3 proves useful. For triple symmetry, only an eighth of the structure has to be analysed but for this, we require 8 runs of calculations with different boundary conditions. The calculation of a rectangular plate by means of the double symmetry is illustrated in fig. 4.5.4.

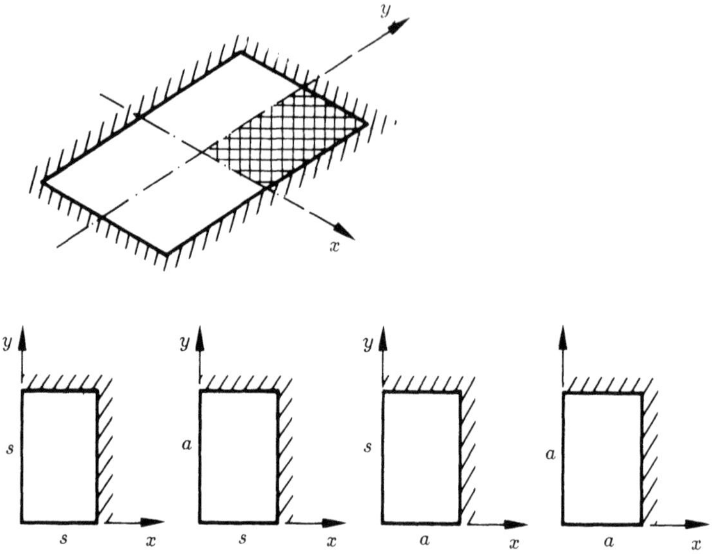

Fig. 4.5.4 Application of symmetry considerations in the spectral analysis of a uniform doubly-symmetrical rectangular plate

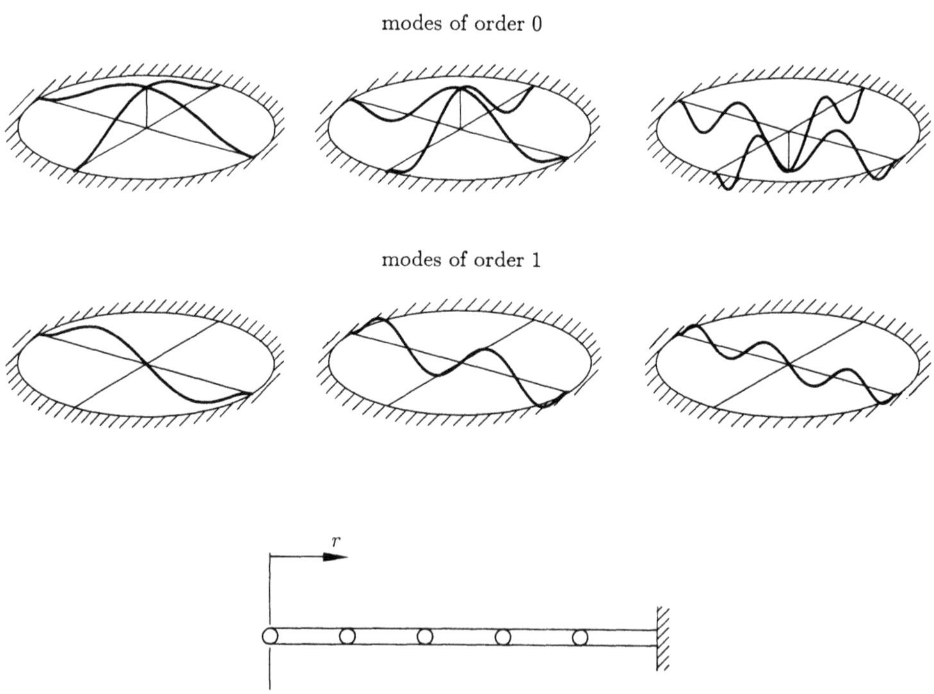

Fig. 4.5.5 Oscillation modes of a body with rotational symmetry (clamped circular plate). Each order demands its own calculation based on the idealisation shown below

Structures with point symmetry can also be analysed most economically using an analogous technique. However, we have to remember that a point symmetrical body possesses eigenmodes with an arbitrary number of waves in the circumferential direction (fig. 4.5.5). If, as is usual in statics, one idealises only a cross-section of the rotationally symmetric structure, then the determination of the eigenforms up to the order k demands k+1 separate eigenvalue analyses, but these computations are most economical in contrast to a complete three-dimensional calculation.

Bibliography for Chapter 4

[4.1] *K. A. Braun*,et al.; Linear dynamic analysis – Lecture notes and example problems, ASKA UM 212, Stuttgart (1982); and: ISD-Bericht Nr. 155, Stuttgart (1974).

[4.2] *J. Argyris, S.Kelsey*; Modern fuselage analysis and the elastic aircraft, Butterworths, London (1963) 93-100.

[4.3] *J. Argyris*; Some results on the free-free oscilations of aircraft type structures, Revue Française de Mécanique, 15 (1965); 3e trimèstre 59-73; and: ISD-Bericht Nr. 1, Stuttgart (1964).

[4.4] *O.E. Brønlund, Th.L. Johnsen*; QR-factorisation of partitioned matrices, Comp. Meths. Appl. Mech. Eng.,3 (1974) 153-172.

[4.5] *H. Rutishauser*; Computational aspects of F. L. Bauer's simultaneous iteration method, Numerische Mathematik, 13 (1969) 4-13.

[4.6] *G. Dietrich*; Improved simultaneous vector iteration, in: ASKA extensions '78, ISD-Bericht Nr. 239, Stuttgart (1978).

[4.7] *J.H. Wilkinson*; Rounding errors in algebraic processes, Notes on Applied Sciences No. 32, Her Majesty's Stationery Office, London; Prentice-Hall, New Jersey (1963).

[4.8] *H.R. Schwarz, H. Rutishauser, E.Stiefel*; Numerik symmetrischer Matrizen (Teubner, Stuttgart, 1968).

[4.9] *P. Henrici*; On the speed of convergenceof cyclic and quasicyclic Jacobi methods for computing eigenvalues of Hermitian matrices, J. Soc. Industr. Appl. Math., 6 (1958) 144-162.

[4.10] *K.A. Braun, et al.*; Some hypermatrix algorithms in linear algebra, in: Lecture notes in economics and mathematical systems, Vol. 134, Springer, Berlin-Heidelberg-New York (1976).

[4.11] *A.S. Householder*; Principles of numerical analysis (McGraw-Hill, New York, 1953).

[4.12] *J.H. Wilkinson*; The algebraic eigenvalue problem (Clarendon Press, Oxford, 1965).

[4.13] *G. Dietrich*; A new formulation of the hypermatrix Housholder - QR decomposition, Comp. Meths. Appl. Eng.,9 (1976) 273-280.

[4.14] *R.J. Guyan*; Reduction of stiffness and mass, AIAA Journal, Vol. 3 (1965).

[4.15] *C.M. Harris, Ch.E. Crede*; Shock and vibration handbook, Vol. 1 (basic theory and measurements) (McGraw-Hill, New York, 1961).

[4.16] *K. Klotter*; Technische Schwingungslehre, Band 2 (Schwinger von mehreren Freiheitsgraden) (Springer, Berlin,1960).

[4.17] *J.P. den Hartog*; Mechanical vibrations (McGraw-Hill, New York, 1956).

[4.18] *S. Iguchi*; Die Eigenschwingungen und Klangfiguren der vierseitig freien rechteckigen Platte, Ing. Archiv, 21 (1953) 303-322.

[4.19] *J. Argyris, et al.*; Aspects of the finite element method as applied to aero-space structures, XXIIIrd Intern. Astronautical Congress, Wien (1972), ISD-Bericht Nr. 138, Stuttgart (1973).

[4.20] *K.A. Braun, TH.L. Johnsen*; Eigencomputation of symmetric hypermatrices using a generalisation of the Householder method, 2nd Int. Conf. SMIRT, Berlin (1973).

[4.21] Forschungsgruppe Reaktordruckbehälter des ISD; Rotationssymmetrische Berechnung des 1:5 THTR Spannbetonbehälter - Modells, ISD-Bericht Nr. 102,Stuttgart (1971).

[4.22] *J. Argyris, et al.*; Finite Elemente zur Berechnung von Spannbeton-Reaktordruckbehälter n, ISD-Bericht Nr. 137, Stuttgart (1973).

[4.23] *C. Lanczos*; An iteration method for the solution of the eigenvalue problem of linear differential and integral operators, J. Res. Nat. Bur. Standards, 45 (1950) 255-282.

[4.24] *C.C. Paige*; The computation of eigenvalues and eigenvectors of very large sparse matrices, Ph. D. Thesis, University of London (1971).

[4.25] *C.C. Paige*; Computational variants of the Lanczos method for the eigenproblem, J. Inst. Math. Appl., 10 (1972) 373-381.

[4.26] *C.C. Paige*; Error analysis of the Lanczos algorithm for tridiagonalising a symmetric matrix, J. Inst. Math. Appl., 18 (1976) 341-349.

[4.27] *B.N. Parlett, D.S. Scott*; The Lanczos algorithm with selective orthogonalisation, Math. Comp., 33 (1979) 217-238.

[4.28] *B.N. Parlett*; The symmetric eigenvalue problem (Prentice-Hall, Englewood Cliffs, N.J., 1980).

[4.29] *D.K. Faddeev, W.N. Faddeeva*; Computational methods of linear algebra, W.H. Freeman & Co., San Francisco (1963).

[4.30] *G.H. Gloub, C.F. van Loan*; Matrix computations (North Oxford Academic, Oxford, 1983).

[4.31] *Y. Saad*; On the rate of convergence of the Lanczos and the block-Lanczos methods, SIAM J. Num. Anal., 17 (1980) 687-706.

[4.32] *A. Ruhe*; Computation of eigenvalues and eigenvectors, in: Sparse Matrix Techniques (ed. V.A. Baker), Lecture Notes in Mathematics 572, Springer-Verlag, Berlin (1977) 130-184.

[4.33] *T. Ericsson, A. Ruhe*; The spectral transformation Lanczos method for the numerical solution of large sparse generalised symmetric eigenvalue problems, Math. Comp., 35 (1980) 1251-1268.

[4.34] *B. Nour-Omid, B.N. Parlett, R.L.Taylor*; Lanczos versus subspace iteration for the solution of eigenvalue problems, Int. J. Num. Meth. Eng., 19 (1983) 859-871.

[4.35] *H.G. Matthies*; A subspace Lanczos method for the generalised symmetric eigenproblem, Comp. & Struct., 21 (1985) 319-325.

[4.36] *K.-J. Bathe, S. Ramaswamy*; An accelerated subspace iteration method, Comp. Meth. Appl. Mech. Eng., 23 (1980) 313-331.

[4.37] *Kari Appa*; Theoretische Bestimmung der Frei-Frei-Schwingungen eines Delta-Flügels, Dr.Ing. doctoral thesis, University of Stuttgart (1966).

5 Free vibrations of undamped systems

"You must strike in measure, when there are many to strike on one anvil"
George Herbert, Jacula prudentum (1640)

In Chapter 4, we inquired into the fundamental problem of what kind of oscillatory motion an elastic body can execute when no loads or specific initial conditions are imposed. This quest led us to the concept of eigenoscillations. In the present chapter, we extend our understanding of oscillatory motion by considering the influence of initial conditions. In particular, we examine the nature of oscillations of a system assuming that they are initiated not by forces but by prescribed initial conditions (displacement, velocity at the onset of motion).

5.1 Diagonalisation and integration of the system; dynamic reduction

The motion of the system is described as before by the set of differential equations given in equ. (4.1.1). For computational convenience, however, we introduce a classification of the degrees of freedom in the following three groups:

(a) $\boldsymbol{r}_c(t)$ primary degrees of freedom, which are required for the description of the dynamic response of the system;

(b) $\boldsymbol{r}_\sigma(t)$ secondary degrees of freedom, which arise from static considerations. These can, however, be eliminated by an interpolation procedure (Guyan reduction);

(c) $\boldsymbol{r}_s(t) = \boldsymbol{o}$ boundary displacements, which represent the suppressed degrees of freedom at the supports.

With this classification, the equation of motion of the system is expressed as

$$\begin{bmatrix} \boldsymbol{M}_{cc} & \boldsymbol{M}_{c\sigma} & \boldsymbol{M}_{cs} \\ \boldsymbol{M}_{\sigma c} & \boldsymbol{M}_{\sigma\sigma} & \boldsymbol{M}_{\sigma s} \\ \boldsymbol{M}_{sc} & \boldsymbol{M}_{s\sigma} & \boldsymbol{M}_{ss} \end{bmatrix} \begin{bmatrix} \ddot{\boldsymbol{r}}_c \\ \ddot{\boldsymbol{r}}_\sigma \\ \ddot{\boldsymbol{r}}_s = \boldsymbol{o} \end{bmatrix} + \begin{bmatrix} \boldsymbol{K}_{cc} & \boldsymbol{K}_{c\sigma} & \boldsymbol{K}_{cs} \\ \boldsymbol{K}_{\sigma c} & \boldsymbol{K}_{\sigma\sigma} & \boldsymbol{K}_{\sigma s} \\ \boldsymbol{K}_{sc} & \boldsymbol{K}_{s\sigma} & \boldsymbol{K}_{ss} \end{bmatrix} \begin{bmatrix} \boldsymbol{r}_c \\ \boldsymbol{r}_\sigma \\ \boldsymbol{r}_s = \boldsymbol{o} \end{bmatrix} = \begin{bmatrix} \boldsymbol{o} \\ \boldsymbol{o} \\ \boldsymbol{R}_s \end{bmatrix} \tag{5.1.1}$$

We first eliminate the degrees of freedom at the supports using the third row in equ. (5.1.1) and in this way, deduce the supporting forces or reactions assuming for the moment that $\boldsymbol{r}_c(t)$ and $\boldsymbol{r}_\sigma(t)$ are known

$$\boldsymbol{R}_s = \boldsymbol{M}_{sc}\ddot{\boldsymbol{r}}_c + \boldsymbol{M}_{s\sigma}\ddot{\boldsymbol{r}}_\sigma + \boldsymbol{K}_{sc}\boldsymbol{r}_c + \boldsymbol{K}_{s\sigma}\boldsymbol{r}_\sigma \tag{5.1.2}$$

Hereafter, the system is condensed (reduced) to the primary degrees of freedom by means of Guyan's interpolation technique. In this way, we find

$$\overline{\boldsymbol{M}}_{cc}\ddot{\boldsymbol{r}}_c + \overline{\boldsymbol{K}}_{cc}\boldsymbol{r}_c = \boldsymbol{o} \tag{5.1.3}$$

Here, the reduced mass matrix $\overline{\boldsymbol{M}}_{cc}$ and the associated stiffness matrix $\overline{\boldsymbol{K}}_{cc}$ have already been defined in equs. (4.4.6) and (4.4.7). If condensation is not applied according to the Guyan procedure, then these matrices are of course identical to $\boldsymbol{M}_{cc}$ and $\boldsymbol{K}_{cc}$ ($\boldsymbol{r}_\sigma$ being non-existent in this case). Once $\boldsymbol{r}_{\rm c}(t)$ has been calculated, $\boldsymbol{r}_\sigma(t)$ can be obtained from equ. (4.4.2).

We may now assume that the eigenanalysis or spectral analysis of the system (5.1.3) has been established by the methods expounded in Chapter 4. The response of the system can then be described by a linear combination of the eigenvectors (this process is known as modal superposition; $\eta_{\rm i}$ are the modal coordinates)

$$\boldsymbol{r}_{\rm c}(t) = \boldsymbol{x}_1\eta_1(t) + \boldsymbol{x}_2\eta_2(t) + \ldots + \boldsymbol{x}_{\rm nx}\eta_{\rm nx}(t) = \boldsymbol{X}\boldsymbol{\eta}(t) \tag{5.1.4}$$

If the nx vectors include the total number of eigenvectors of the system defined by equ. (5.1.3), the representation of equ. (5.1.4) is exact. But if we assume, as is often the case in practice, that the dynamic response can be reproduced with sufficient engineering accuracy by a restricted set of vectors, this being known as the dynamic reduction, then equ. (5.1.4) only offers an approximation.

Substituting next equ. (5.1.4) into equ. (5.1.3) and pre-multiplying by $\boldsymbol{X}^{\rm t}$, we obtain a decoupled system with nx independent, second order linear differential equations, since $\overline{\boldsymbol{M}}_{cc}$ as well as $\overline{\boldsymbol{K}}_{cc}$ are diagonalised by the congruent transformation with the set $\boldsymbol{X}$

$$\boldsymbol{\Lambda}\ddot{\boldsymbol{\eta}} + \boldsymbol{\eta} = \boldsymbol{o} \tag{5.1.5}$$

Here, $\boldsymbol{\Lambda}$ is defined in equ. (4.1.29).

A typical constituting equation has the form

$$\ddot{\eta}_{\rm i} + \omega_{\rm i}^2\eta_{\rm i} = 0, \qquad {\rm i} = 1\ldots{\rm nx} \tag{5.1.6}$$

the general solution of which is already known (see also equs. (1.3.5) and (1.3.15), for $\zeta = 0$)

$$\eta_{\rm i}(t) = a_{\rm i}\sin\omega_{\rm i}t + b_{\rm i}\cos\omega_{\rm i}t, \qquad {\rm i} = 1\ldots{\rm nx} \tag{5.1.7}$$

Adapting this expression to the imposed initial conditions at the instant $t = 0$

$$\begin{aligned} \eta_{\rm i}(0) &= \eta_{0{\rm i}} = b_{\rm i} \\ \dot{\eta}_{\rm i}(0) &= \dot{\eta}_{0{\rm i}} = a_{\rm i}\omega_{\rm i} \end{aligned} \tag{5.1.8}$$

the explicit solution becomes

$$\eta_{\rm i}(t) = \frac{\dot{\eta}_{0{\rm i}}}{\omega_{\rm i}}\sin\omega_{\rm i}t + \eta_{0{\rm i}}\cos\omega_{\rm i}t \tag{5.1.9}$$

However, one problem still remains to be solved. Since it is not generally common practice to seek the solution for initial values $\eta_{\rm i}(0)$, resp. $\dot{\eta}_{\rm i}(0)$, but rather for $\boldsymbol{r}_{\rm c}(0)$, resp. $\dot{\boldsymbol{r}}_{\rm c}(0)$, we have to deduce the corresponding initial values for $\boldsymbol{\eta}$ from the latter values. This is straightforward when working with a complete set of eigenvectors, since these are linearly independent ([5.2]). Consequently, we obtain

$$
\begin{aligned}
\boldsymbol{\eta}(0) &= \underset{(\mathrm{n\times n})}{\boldsymbol{X}^{-1}}\boldsymbol{r}_c(0) = \boldsymbol{X}^{\mathrm{t}}\overline{\boldsymbol{K}}_{cc}\boldsymbol{r}_c(0) \\
\dot{\boldsymbol{\eta}}(0) &= \underset{(\mathrm{n\times n})}{\boldsymbol{X}^{-1}}\dot{\boldsymbol{r}}_c(0) = \boldsymbol{X}^{\mathrm{t}}\overline{\boldsymbol{K}}_{cc}\dot{\boldsymbol{r}}_c(0) \\
&= \boldsymbol{\Lambda}^{-1}\boldsymbol{X}^{\mathrm{t}}\overline{\boldsymbol{M}}_{cc}\dot{\boldsymbol{r}}_c(0)
\end{aligned}
\tag{5.1.10}
$$

A difficulty arises, however, when applying a dynamic reduction. In this case, $\boldsymbol{X}$ becomes a rectangular matrix of the order n×nx, (nx<n), and is hence not uniquely invertible. The associated system of equations

$$
\underset{(\mathrm{n\times nx})}{\boldsymbol{X}}\;\underset{(\mathrm{nx\times 1})}{\boldsymbol{\eta}(0)} = \underset{(\mathrm{n\times 1})}{\boldsymbol{r}_c(0)}
\tag{5.1.11}
$$

does not possess a unique or single-valued solution, since any selected set of nx equations determines a specific $\boldsymbol{\eta}(0)$ but does not satisfy the remaining equations (the problem is analogous to that arising in the calculus of observations). In fact, since it proves impossible to satisfy the complete set of equations, we demand instead that the adopted displacement transformation should minimise the deviation of the potential energy and analogously that the corresponding velocity transformation should minimise the deviation of the kinetic energy. In this way, we obtain the two expressions

$$
\begin{aligned}
&[\boldsymbol{r}_c(0) - \boldsymbol{X}\boldsymbol{\eta}(0)]^{\mathrm{t}}\overline{\boldsymbol{K}}_{cc}[\boldsymbol{r}_c(0) - \boldsymbol{X}\boldsymbol{\eta}(0)] \stackrel{!}{=} \min \\
&[\dot{\boldsymbol{r}}_c(0) - \boldsymbol{X}\dot{\boldsymbol{\eta}}(0)]^{\mathrm{t}}\overline{\boldsymbol{K}}_{cc}[\dot{\boldsymbol{r}}_c(0) - \boldsymbol{X}\dot{\boldsymbol{\eta}}(0)] \stackrel{!}{=} \min
\end{aligned}
\tag{5.1.12}
$$

In equs. (5.1.12), $\boldsymbol{\eta}(0)$ resp. $\dot{\boldsymbol{\eta}}(0)$ represent the variables that have to fulfil the two conditions. We obtain the two vectors by differentiation of the two expressions in equ. (5.1.12) in turn by $\boldsymbol{\eta}(0)$ resp. $\dot{\boldsymbol{\eta}}(0)$ and subsequent nullification of the corresponding results. In this way, we deduce the simple statements

$$
\begin{aligned}
\boldsymbol{\eta}(0) &= \boldsymbol{X}^{\mathrm{t}}\overline{\boldsymbol{K}}_{cc}\boldsymbol{r}_c(0) \\
\dot{\boldsymbol{\eta}}(0) &= \boldsymbol{\Lambda}^{-1}\boldsymbol{X}^{\mathrm{t}}\overline{\boldsymbol{M}}_{cc}\dot{\boldsymbol{r}}_c(0)
\end{aligned}
\tag{5.1.13}
$$

which are formally identical to the expressions in equ. (5.1.10).

Having thus completed our discussion on the oscillatory motion of a deformable body subject to specific initial conditions, we note that the essential information for the description of the oscillation derives from the eigenoscillations as specified in equ. (5.1.9). Both amplitude and phase shift are determined by the initial conditions. We realise this more clearly if we apply the alternative formulation of equ. (4.1.17)

$$
\eta_{\mathrm{i}}(t) = c_{\mathrm{i}}\sin(\omega_{\mathrm{i}}t + \varphi_{\mathrm{i}})
\tag{5.1.14}
$$

Using the trigonometrical addition theorems, we easily obtain for the eigenoscillation i the relations

$$
\begin{aligned}
a_i &= c_i \cos\varphi_i \\
b_i &= c_i \sin\varphi_i
\end{aligned}
\tag{5.1.15}
$$

Consequently, the amplitude of the eigenoscillation i becomes

$$
c_i = \sqrt{a_i^2 + b_i^2} = \sqrt{\eta_{0i}^2 + (\dot{\eta}_{0i}/\omega_i)^2}
\tag{5.1.16}
$$

and the corresponding phase shift

$$
\varphi_i = \tan^{-1}\frac{b_i}{a_i} = \tan^{-1}\frac{\omega_i \eta_{0i}}{\dot{\eta}_{0i}}
\tag{5.1.17}
$$

As a result, the ensuing phase angle vanishes when the deflection at time $t = 0$ is $\eta_{0i} = 0$. At this instant, we run through a so-called zero passage with a velocity $\dot{\eta}_{0i}$. If, on the other hand, we prescribe the velocity $\dot{\eta}_{0i} = 0$ at time $t = 0$ and a deflection η_{0i}, we run through a so-called reversal point. The phase angle then becomes $\pm 90^0$, and we obtain a pure cosine oscillation (see fig. 5.1.1). In particular the pointers in the vector diagrams illustrate the various phase conditions.

At this stage of the discussion, we are concerned not only with the functions $\eta_i(t)$ but also with the specification of the actual motion $\boldsymbol{r}_c(t)$ of the system (see equ. (5.1.4)). We observe first that the description of an arbitrary free motion of the system involves all eigenoscillations. At the same time, we note that for certain specific initial conditions the motion may still be expressed by a single eigenoscillation. For example, in order to obtain a pure eigenoscillation with the eigenfrequency ω_j we have to prescribe

$$
\begin{aligned}
\eta_{0i} &= 0 \\
\dot{\eta}_{0i} &= 0 \qquad i = 1 \dots nx; \quad i \neq j
\end{aligned}
\tag{5.1.18}
$$

This can also be expressed in terms of $\boldsymbol{r}_c$ and $\dot{\boldsymbol{r}}_c$ in the form

$$
\begin{aligned}
\boldsymbol{r}_c(0) &= \alpha \boldsymbol{x}_j \\
\dot{\boldsymbol{r}}_c(0) &= \beta \boldsymbol{x}_j
\end{aligned}
\tag{5.1.19}
$$

But for the arbitrary constants α and β, the initial deflection and velocity vectors are hence proportional to the eigenvector $\boldsymbol{x}_j$. In the case of the full set of eigenvectors, application of equ. (5.1.1) then yields with equ. (5.1.10) the initial vectors

$$
\begin{aligned}
\boldsymbol{\eta}(0) &= \{0 \quad 0 \quad \dots \quad 0 \quad \overset{j}{\alpha} \quad 0 \quad \dots \quad \overset{nx}{0}\} \\
\dot{\boldsymbol{\eta}}(0) &= \{0 \quad 0 \quad \dots \quad 0 \quad \beta \quad 0 \quad \dots \quad 0\}
\end{aligned}
\tag{5.1.20}
$$

If a dynamic reduction is applied as in equ. (5.1.3), the same result is obtained.

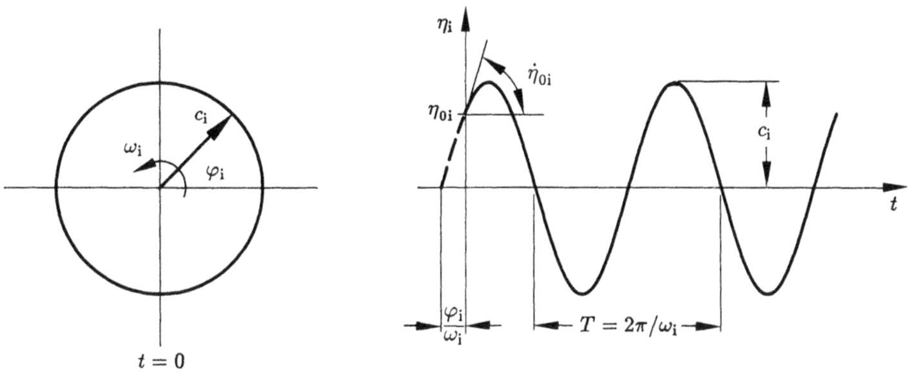

a) arbitrary initial conditions: η_{0i} and $\dot{\eta}_{0i}$

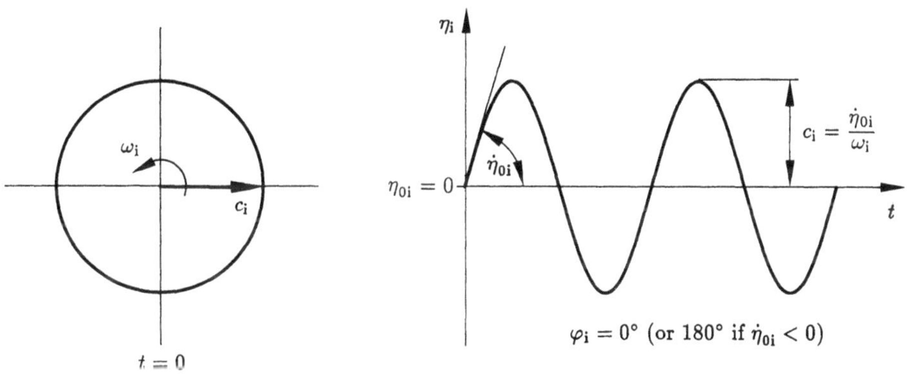

b) initial condition: $\eta_{0i} = 0$ (sine response)

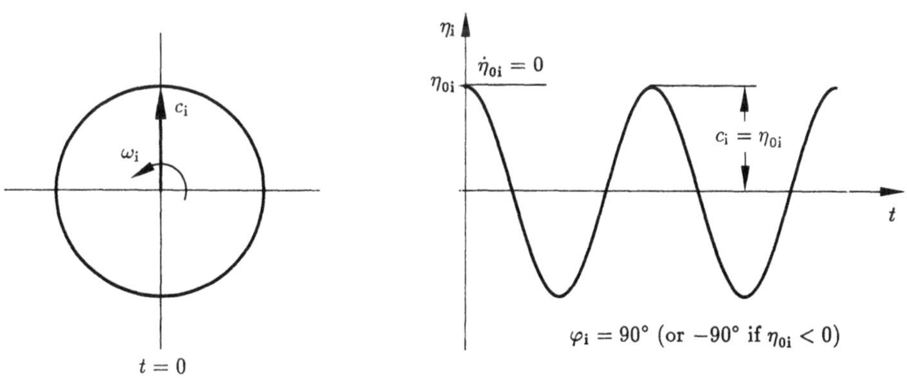

c) initial condition: $\dot{\eta}_{0i} = 0$ (cosine response)

Fig. 5.1.1 Oscillatory behaviour of the η_i-system for various initial conditions

model

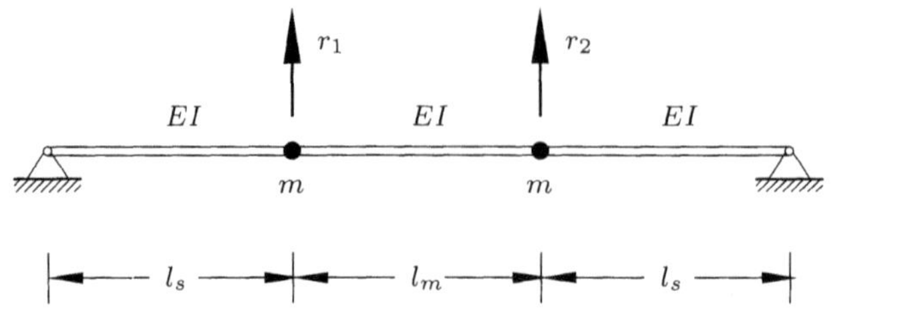

$l_m = l_s = l$

mass matrix	$\boldsymbol{M} = m\boldsymbol{I}_2$
stiffness matrix	$\boldsymbol{K} = \frac{6EI}{5l^3}\begin{bmatrix} 8 & -7 \\ -7 & 8 \end{bmatrix}$
eigenvectors (orthonormal to $\boldsymbol{K}$)	$\boldsymbol{X} = \frac{l}{6}\sqrt{\frac{l}{EI}}\begin{bmatrix} \sqrt{15} & 1 \\ \sqrt{15} & -1 \end{bmatrix}$
eigenvalues	$\boldsymbol{\Lambda} = \boldsymbol{X}^{\mathrm{t}}\boldsymbol{M}\boldsymbol{X} = \frac{ml^3}{18EI}\lceil 15 \quad 1 \rfloor = \left\lceil \frac{1}{\omega_1^2} \quad \frac{1}{\omega_2^2} \right\rfloor$

Fig. 5.1.2 Vibrational model and results of the eigenvalue analysis

Such a special excitation hence yields a stationary or steady-state oscillation. The motion of the system repeats itself with a period T_{j}. The shape of the response is defined by the eigenvector $\boldsymbol{x}_{\mathrm{j}}$ and a time-dependent global factor of proportionality. Thus, the points of maximum deflection (inter- or antinodes) and the zero-passage (nodal) points remain stationary.

In contrast, an arbitrary response of a free body generally involves all eigenmodes and progresses in an instationary manner. No time interval or period can be determined for which the motion repeats itself unless the eigenfrequencies are integral multiples of a fundamental frequency. Hence, in general, there are no stationary nodes and antinodes.

The discussion is best illustrated by means of a simple example. Figure 5.1.2 shows the model of a massless beam with two degrees of freedom and associated concentrated masses whose eigenvectors can be established simply by considerations of symmetry. In order to apply the procedure established in this chapter, we must first orthonormalise the eigenvectors with respect to $\boldsymbol{K}$. Once all the necessary preparatory steps have been completed, we proceed to the selection of the initial conditions. First, we assume a deflection proportional to the 2nd eigenvector (fig. 5.1.3). Allowing the structure to oscillate freely, we obtain a pure eigenoscillation with a frequency ω_2. Following equ. (5.1.10), we deduce

$$\boldsymbol{\eta}(0) = \boldsymbol{X}^{\mathrm{t}}\boldsymbol{K}_{cc}\boldsymbol{r}_c(0) = \sqrt{\frac{EI}{l^3}}\{0 \quad 6\}$$

$$\dot{\boldsymbol{\eta}}(0) = \boldsymbol{o}_2$$

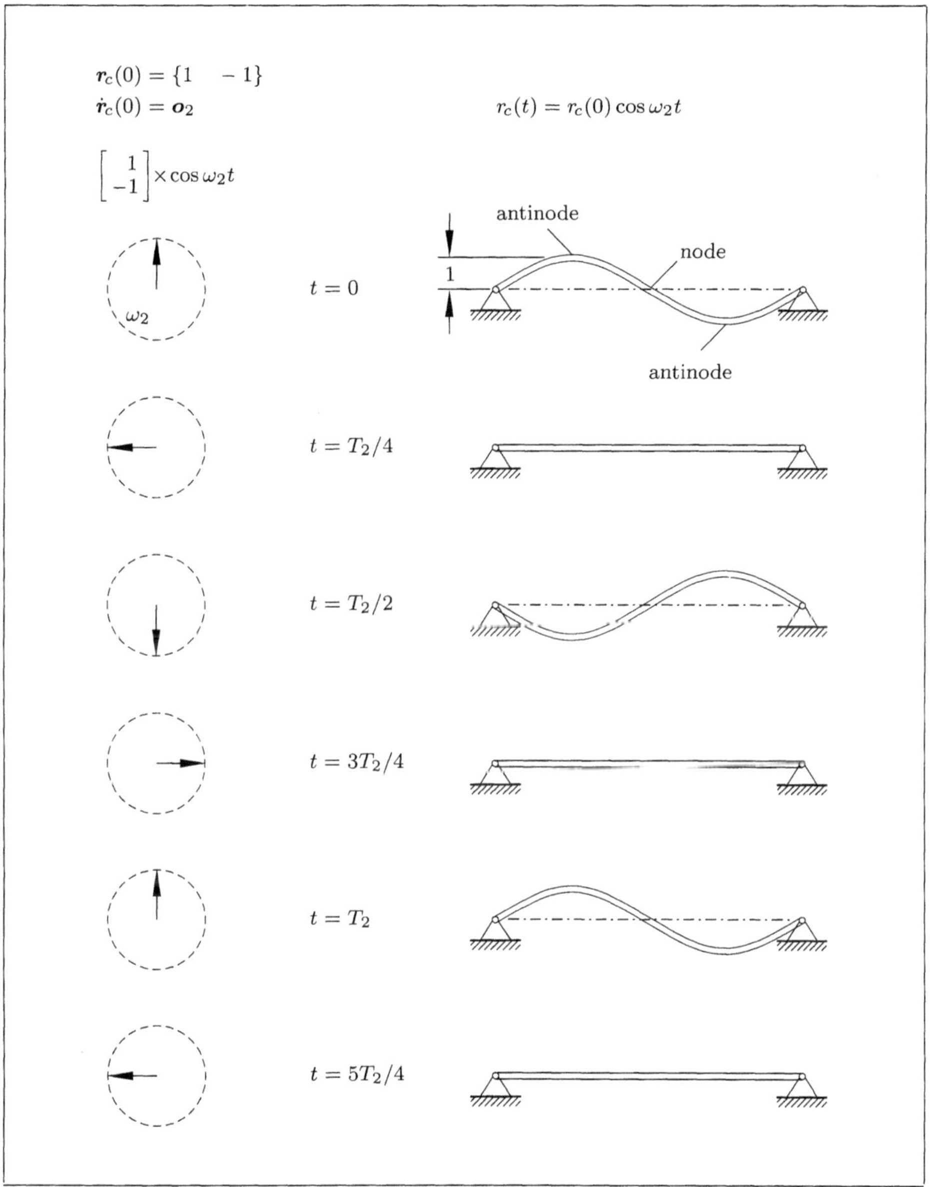

Fig. 5.1.3 Initial conditions, entailing a pure steady-state eigenoscillation

Applying equ. (5.1.9), we obtain at an arbitrary instant t

$$\boldsymbol{\eta}(t) = \begin{bmatrix} 0 \\ 6\sqrt{\frac{EI}{l^3}}\cos\omega_2 t \end{bmatrix}$$

Introducing next the reverse transformation of equ. (5.1.4), we obtain

$$\boldsymbol{r}_c(t) = \begin{bmatrix} 1 \\ -1 \end{bmatrix} \cos\omega_2 t = \begin{bmatrix} 1 \\ -1 \end{bmatrix} \cos\sqrt{\frac{18EI}{ml^3}}t$$

The initial conditions are now modified and, at time $t = 0$, solely the degree of freedom r_1 is deflected as shown in fig. 5.1.4. From equ. (5.1.10), we deduce

$$\boldsymbol{\eta}(0) = \frac{1}{5}\sqrt{\frac{EI}{l^3}}\{\sqrt{15} \quad 15\}$$

$$\dot{\boldsymbol{\eta}}(0) = \boldsymbol{o}_2$$

and immediately observe that in the present case, all the system's eigenoscillations participate in the motion. From equ. (5.1.9), we obtain at a time t

$$\boldsymbol{\eta} = \frac{1}{5}\sqrt{\frac{EI}{l^3}}\begin{bmatrix} \sqrt{15}\cos\omega_1 t \\ 15\cos\omega_2 t \end{bmatrix}$$

The reverse transformation of equ. (5.1.4) yields

$$\boldsymbol{r}_c = \frac{1}{2}\begin{bmatrix} \cos\omega_1 t & +\cos\omega_2 t \\ \cos\omega_1 t & -\cos\omega_2 t \end{bmatrix} = \frac{1}{2}\begin{bmatrix} \cos 2\pi\frac{t}{\sqrt{15}T_2} & +\cos 2\pi\frac{t}{T_2} \\ \cos 2\pi\frac{t}{\sqrt{15}T_2} & -\cos 2\pi\frac{t}{T_2} \end{bmatrix}$$

In the vector diagram, ω_2 is represented by a rapidly rotating pointer, whilst ω_1 is determined by a more slowly rotating pointer. The vertical components of these motions must then be superposed. Clearly, it is impossible to establish a specific time interval for a periodic repetition of the motion. The differing angular rotations of the pointers do not in principle exclude the possibility of achieving the same pattern of the motion, but this occurs at irregular intervals. The points of zero passage and the points of maximum deflection (antinodes) shift continuously.

An interesting variant of the present problem is obtained for the example of fig. 5.1.2 if we introduce the condition

$$l_m = 1.72076\, l_s$$

In this case, the eigenfrequency ratio is reduced to the integer

$$\frac{\omega_2}{\omega_1} = 3$$

This ensures after each time interval $3T_2$ a re-emergence of the initial configuration imposed at the instant $t = 0$. However, the oscillation is instationary, since the zero passage points shift in between these time intervals.

Special conditions prevail when the two eigenfrequencies coincide and are excited simultaneously. Particularly when phase coincidence holds for the two components, no phase shift occurs in the total motion. The interested reader may find

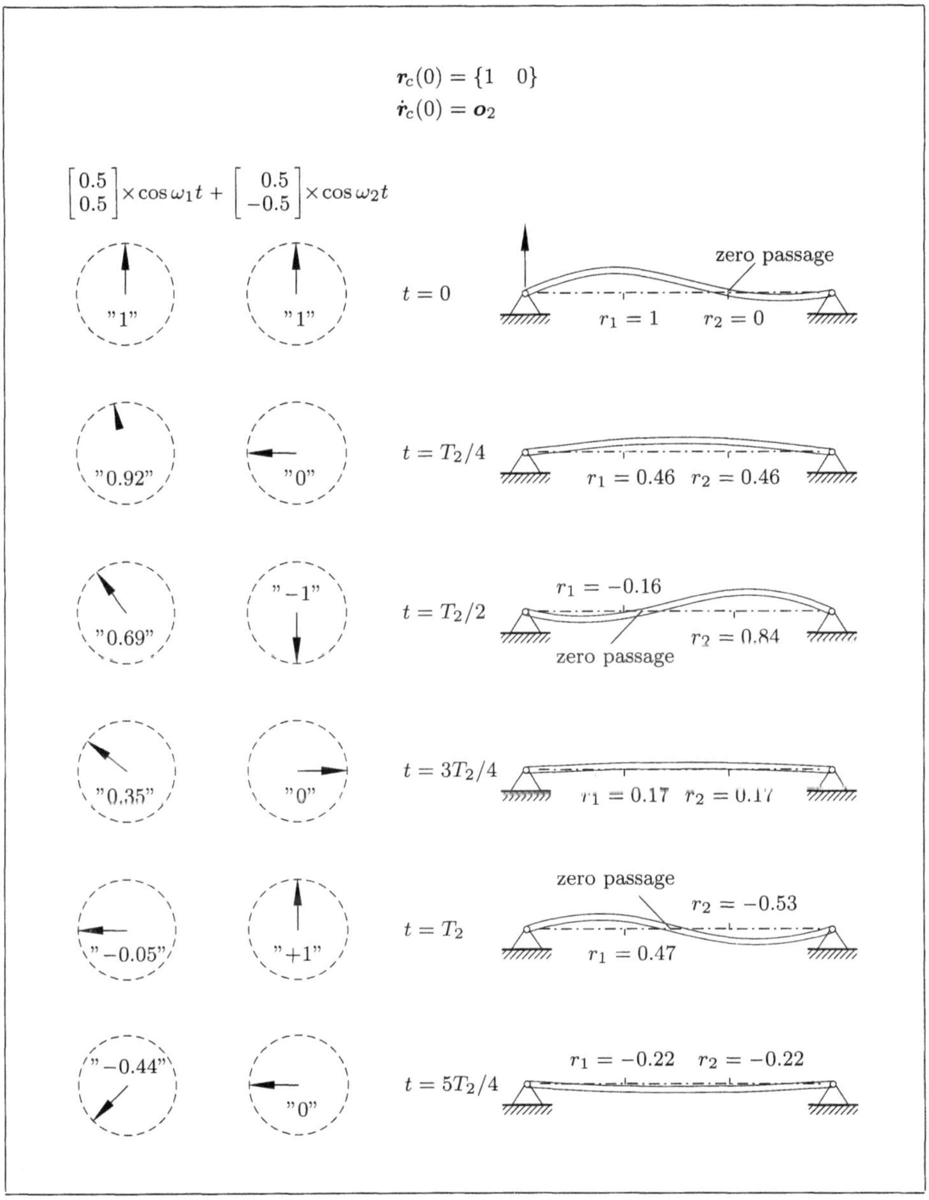

Fig. 5.1.4 Initial conditions, entailing an instationary vibration

it instructive to look once more at the example of fig. 4.3.3 and consider the two alternative sets of initial conditions

$$\boldsymbol{r}\,(t=0) = \left\{1 \quad \frac{16}{9+\sqrt{553}} \quad 0\right\}$$

$$\dot{\boldsymbol{r}}\,(t=0) = \boldsymbol{o}_3$$

or

$$\boldsymbol{r}\,(t=0) = \{0.5 \quad 0 \quad -0.5\}$$

$$\dot{\boldsymbol{r}}\,(t=0) = \frac{1}{\omega_{1,2}}\left\{0.5 \quad \frac{16}{9+\sqrt{553}} \quad 0.5\right\}$$

We recommend plotting and studying the oscillatory motion, in particular within the first period.

5.2 Analysis of systems with rigid body modes (free-free oscillations)

In this section, particular attention is paid to systems which possess rigid body freedoms. We should first note that all rigid body freedoms of motion of a system must be included in the group of the $\boldsymbol{r}_c$ degrees of freedom. Indeed, should a rigid body freedom be expressible in terms of $\boldsymbol{r}_\sigma$ freedoms alone, the $\boldsymbol{K}_{\sigma\sigma}$ would be singular and a Guyan reduction impossible. If, on the other hand, $\boldsymbol{r}_c$ includes all the rigid body modes, the system defined by equ. (5.1.3) can be established without difficulty, the matrix $\overline{\boldsymbol{K}}_{cc}$ now being singular, however. The reader is reminded of the representation of the degrees of freedom given in equ. (4.2.5) which we adopt here

$$\boldsymbol{r}_c = \boldsymbol{T}\{\boldsymbol{r}_{c0} \quad \boldsymbol{r}_{ce}\} = [\boldsymbol{T}_0 \quad \boldsymbol{T}_e]\{\boldsymbol{r}_{c0} \quad \boldsymbol{r}_{ce}\} \tag{5.2.1}$$

Using the catalogue of conversions

$$\begin{aligned}
\boldsymbol{r}_c &\rightarrow \boldsymbol{r}\\
\boldsymbol{r}_{c0} &\rightarrow \boldsymbol{r}_0\\
\boldsymbol{r}_{ce} &\rightarrow \boldsymbol{r}_e\\
\boldsymbol{K} &\rightarrow \overline{\boldsymbol{K}}_{cc}\\
\boldsymbol{M} &\rightarrow \overline{\boldsymbol{M}}_{cc}
\end{aligned} \tag{5.2.2}$$

all the results obtained in section 4.2, in particular in equs. (4.2.6) to (4.2.16), can be applied. In this case, the initial conditions can be prescribed only in the form of the primary degrees of freedom, $\boldsymbol{r}_c$, i.e. by $\boldsymbol{r}_c(0)$ and $\dot{\boldsymbol{r}}_c(0)$. It should be noted that all initial conditions of $\boldsymbol{r}_\sigma$ are then automatically fixed through equ. (4.4.2). Following these comments, we refer once more to the conversion catalogue of equ. (5.2.2), this time omitting the 'c' index for the sake of clarity.

Our first step is to re-interpret the initial conditions $\boldsymbol{r}\,(\boldsymbol{r}_c)$ in terms of $\boldsymbol{r}_0(\boldsymbol{r}_{c0})$ and $\boldsymbol{r}_e(\boldsymbol{r}_{ce})$. There is a simple rule for this: since the matrix $\boldsymbol{T}$ in equ. (5.2.1) resp. equ. (4.2.5) is quadratic and by definition contains linearly independent columns, it can hence be inverted. Thus,

$$\begin{bmatrix}\boldsymbol{r}_0\\ \boldsymbol{r}_e\end{bmatrix} = \underset{(\mathrm{n}\times\mathrm{n})}{\boldsymbol{T}^{-1}}\boldsymbol{r} \tag{5.2.3}$$

This procedure involving $\boldsymbol{T}$ of order (n×n) is, of course, very inconvenient and should be simplified. To this purpose, we draw the reader's attention to the important orthonormality property

$$\boldsymbol{T}_e^{\mathrm{t}}\boldsymbol{T}_e = \boldsymbol{I}_e \tag{5.2.4}$$

The simplest conception of the elastic sub-system is achieved by means of a system of supports. In this case, $\boldsymbol{T}_e$ is a Boolean matrix, containing a 'one' in each column but at most a 'one' or only zero-elements in each row. The consequence of this is that the matrix, $\boldsymbol{T}_e$, automatically fulfils the orthogonality condition of equ. (5.2.4). An alternative assembly of the $\boldsymbol{T}_e$ matrix is obtained by the device of making its field an orthogonal complement to $\boldsymbol{M}\boldsymbol{T}_0$, thereby completely decoupling the elastic and rigid parts of the system (see equ. (*4.2.1)). In accordance with equ. (*4.2.3), $\boldsymbol{T}_e$ is then construed as part of an orthogonal matrix $\boldsymbol{Q}$ (equ. (*4.2.5)) in the form

$$\boldsymbol{Q}\boldsymbol{Q}^{\mathrm{t}} = \boldsymbol{I} = \begin{bmatrix} \boldsymbol{Q}_0 \\ \boldsymbol{T}_e^{\mathrm{t}} \end{bmatrix} [\boldsymbol{Q}_0^{\mathrm{t}} \quad \boldsymbol{T}_e] \tag{5.2.5}$$

We observe that equ. (5.2.4) is once more automatically fulfilled. Pre-multiplication of equ. (4.2.5) resp. equ. (5.2.1) by $\boldsymbol{T}^{\mathrm{t}}$ yields

$$\begin{aligned} \boldsymbol{T}_0^{\mathrm{t}}\boldsymbol{r} &= \boldsymbol{T}_0^{\mathrm{t}}\boldsymbol{T}_0\boldsymbol{r}_0 + \boldsymbol{T}_0^{\mathrm{t}}\boldsymbol{T}_e\boldsymbol{r}_e \\ \boldsymbol{T}_e^{\mathrm{t}}\boldsymbol{r} &= \boldsymbol{T}_e^{\mathrm{t}}\boldsymbol{T}_0\boldsymbol{r}_0 + \boldsymbol{r}_e \end{aligned} \tag{5.2.6}$$

Using the second set of this system, we can eliminate the large number of elastic degrees of freedom

$$\boldsymbol{r}_e = \boldsymbol{T}_e^{\mathrm{t}}[\boldsymbol{r} - \boldsymbol{T}_0\boldsymbol{r}_0] \tag{5.2.7}$$

Substituting this result into the first set, we obtain a very small system of equations in the unknown $\boldsymbol{r}_0$

$$\boldsymbol{r}_0 = \left[\boldsymbol{T}_0^{\mathrm{t}}\boldsymbol{T}_0 - \boldsymbol{T}_0^{\mathrm{t}}\boldsymbol{T}_e\boldsymbol{T}_e^{\mathrm{t}}\boldsymbol{T}_0\right]^{-1}\boldsymbol{T}_0^{\mathrm{t}}\left[\boldsymbol{I} - \boldsymbol{T}_e\boldsymbol{T}_e^{\mathrm{t}}\right]\boldsymbol{r} \tag{5.2.8}$$

Apart from the orthonormality of $\boldsymbol{T}_e$, no special conditions need be fulfilled by $\boldsymbol{T}_0$. Orthogonality of $\boldsymbol{T}_0$ itself would indeed be completely insignificant for the solution process. In the event of $\boldsymbol{T}_e$ being generated as an orthogonal complement to $\boldsymbol{M}\boldsymbol{T}_0$, we can employ the matrix product representation of $\boldsymbol{Q}$, which also contains $\boldsymbol{T}_e^{\mathrm{t}}$.

Now using equs. (5.2.8) and (5.2.7), we can deduce the initial conditions $\boldsymbol{r}_0(0)$, $\dot{\boldsymbol{r}}_0(0)$ and $\boldsymbol{r}_e(0)$, $\dot{\boldsymbol{r}}_e(0)$ from $\boldsymbol{r}(0)$ and $\dot{\boldsymbol{r}}(0)$ and are thus in a position to analyse the elastic system of equ. (4.2.10). Applying in this equation the transformation

$$\boldsymbol{r}_e = \underset{(\mathrm{n}\times\mathrm{nx})}{\boldsymbol{X}_e}\boldsymbol{\eta}_e \tag{5.2.9}$$

where $\boldsymbol{X}_e$ is orthonormal with respect to $\boldsymbol{K}_{ee}$, i.e.

$$\boldsymbol{X}_e^{\mathrm{t}}\boldsymbol{K}_{ee}\boldsymbol{X}_e = \boldsymbol{I}_{\mathrm{nx}} \tag{5.2.10}$$

Guyan's reduction	full system
$\underset{(n\times 1)}{\boldsymbol{r}} = \underset{(n\times nc)}{\boldsymbol{T}_c}\underset{(nc\times 1)}{\boldsymbol{r}_c}$ $\overline{\boldsymbol{M}}_{cc}\ddot{\boldsymbol{r}}_c + \overline{\boldsymbol{K}}_{cc}\boldsymbol{r}_c = \boldsymbol{o}$	$\boldsymbol{M}\ddot{\boldsymbol{r}} + \boldsymbol{K}\boldsymbol{r} = \boldsymbol{o}$

decomposition of deformation into rigid and elastic components

$$\boldsymbol{r} = [\boldsymbol{T}_0 \quad \boldsymbol{T}_e]\{\boldsymbol{r}_0 \quad \boldsymbol{r}_e\}$$

transformed matrices

$$\boldsymbol{M}_{e0} = \boldsymbol{T}_e^t\boldsymbol{M}\boldsymbol{T}_0 = \boldsymbol{M}_{0e}^t$$
$$\boldsymbol{M}_{00} = \boldsymbol{T}_0^t\boldsymbol{M}\boldsymbol{T}_0$$
$$\boldsymbol{K}_{ee} = \boldsymbol{T}_e^t\boldsymbol{K}\boldsymbol{T}_e$$

new dynamic system

$$\ddot{\boldsymbol{r}}_0 = -\boldsymbol{M}_{00}^{-1}\boldsymbol{M}_{0e}\ddot{\boldsymbol{r}}_e$$
$$\left[\boldsymbol{M}_{ee} - \boldsymbol{M}_{e0}\boldsymbol{M}_{00}^{-1}\boldsymbol{M}_{0e}\right]\ddot{\boldsymbol{r}}_e + \boldsymbol{K}_{ee}\boldsymbol{r}_e = \boldsymbol{o}$$

Note: $\boldsymbol{T}_e$ can be generated in such a way that $\boldsymbol{M}_{e0} = \boldsymbol{O}$

eigenvalue analysis and uncoupling of the elastic response of the system

eigenvalue analysis $\left[\boldsymbol{M}_{ee} - \boldsymbol{M}_{e0}\boldsymbol{M}_{00}^{-1}\boldsymbol{M}_{0e}\right]$ and $\boldsymbol{K}_{ee} \quad \rightarrow \boldsymbol{\Lambda},\ \boldsymbol{X}_e$

decoupling transformation $\boldsymbol{r}_e = \underset{(ne\times nx)}{\boldsymbol{X}_e}\,\boldsymbol{\eta}_e$

yields independent equations $\boldsymbol{\Lambda}\ddot{\boldsymbol{\eta}}_e + \boldsymbol{\eta}_e = \boldsymbol{o}$

initial conditions for decomposed rigid-elastic deformations

given $\boldsymbol{r}(0), \dot{\boldsymbol{r}}(0)$ and $\boldsymbol{T}_e^t\boldsymbol{T}_e = \boldsymbol{I}$, then

$$\boldsymbol{r}_0(0) = \left[\boldsymbol{T}_0^t\boldsymbol{T}_0 - \boldsymbol{T}_0^t\boldsymbol{T}_e\boldsymbol{T}_e^t\boldsymbol{T}_0\right]^{-1}\boldsymbol{T}_0^t\left[\boldsymbol{I} - \boldsymbol{T}_e\boldsymbol{T}_e^t\right]\boldsymbol{r}(0)$$
$$\dot{\boldsymbol{r}}_0(0) = \left[\boldsymbol{T}_0^t\boldsymbol{T}_0 - \boldsymbol{T}_0^t\boldsymbol{T}_e\boldsymbol{T}_e^t\boldsymbol{T}_0\right]^{-1}\boldsymbol{T}_0^t\left[\boldsymbol{I} - \boldsymbol{T}_e\boldsymbol{T}_e^t\right]\dot{\boldsymbol{r}}(0)$$

also
$$\boldsymbol{r}_e(0) = \boldsymbol{T}_e^t\left[\boldsymbol{r}(0) - \boldsymbol{T}_0\boldsymbol{r}_0(0)\right]$$
$$\dot{\boldsymbol{r}}_e(0) = \boldsymbol{T}_e^t\left[\dot{\boldsymbol{r}}(0) - \boldsymbol{T}_0\dot{\boldsymbol{r}}_0(0)\right]$$

Fig. 5.2.1 Free vibrations of undamped dynamic systems with rigid-body modes, *Part 1*

we obtain the decoupled system

$$\boldsymbol{\Lambda}\ddot{\boldsymbol{\eta}}_e + \boldsymbol{\eta}_e = \boldsymbol{o} \tag{5.2.11}$$

A typical component of this system is simply

$$\ddot{\eta}_{ei} + \omega_i^2\eta_{ei} = 0, \qquad i = 1\ldots nx \tag{5.2.12}$$

<table>
<tr><td colspan="2">integration of elastic contributions to response
initial conditions $\boldsymbol{\eta}_e(0) = \boldsymbol{X}_e^t \boldsymbol{K}_{ee} \boldsymbol{r}_e(0)$
$\dot{\boldsymbol{\eta}}_e(0) = \boldsymbol{\Lambda}^{-1}\boldsymbol{X}_e^t \left[\boldsymbol{M}_{ee} - \boldsymbol{M}_{e0}\boldsymbol{M}_{00}^{-1}\boldsymbol{M}_{0e}\right]\dot{\boldsymbol{r}}_e(0)$
then $\eta_{ei}(t) = \frac{\dot{\eta}_{ei}(0)}{\omega_i} \sin\omega_i t + \eta_{ei}(0)\cos\omega_i t \qquad i = 1 \ldots nx$
$\boldsymbol{r}_e(t) = \boldsymbol{X}_e \boldsymbol{\eta}_e(t)$</td></tr>
<tr><td colspan="2">integration of rigid body contributions to response
$\boldsymbol{r}_0(t) = \boldsymbol{r}_0(0) + \boldsymbol{M}_{00}^{-1}\boldsymbol{M}_{0e}\boldsymbol{r}_e(0) + \left[\dot{\boldsymbol{r}}_0(0) + \boldsymbol{M}_{00}^{-1}\boldsymbol{M}_{0e}\dot{\boldsymbol{r}}_e(0)\right]t - \boldsymbol{M}_{00}^{-1}\boldsymbol{M}_{0e}\boldsymbol{r}_e$
Note: $\boldsymbol{M}_{0e} = \boldsymbol{O}$ leads to considerable simplifications</td></tr>
<tr><td colspan="2">superposition of rigid body and elastic contributions
$\boldsymbol{r}(t) = \begin{bmatrix}\boldsymbol{T}_0 & \boldsymbol{T}_e\end{bmatrix}\begin{Bmatrix}\boldsymbol{r}_0(t) & \boldsymbol{r}_e(t)\end{Bmatrix}$</td></tr>
<tr><td>in the case of Guyan's reduction,
$\boldsymbol{r}(t)$ is to be replaced by $\boldsymbol{r}_c(t)$

reversed transformation
$\boldsymbol{r}(t) = \boldsymbol{T}_c \boldsymbol{r}_c(t)$</td><td>otherwise no additional operation necessary</td></tr>
</table>

Fig. 5.2.1 Free vibrations of undamped dynamic systems with rigid-body modes, *Part 2*

for which the general solution is

$$\begin{aligned}\eta_{ei}(t) &= a_i \sin\omega_i t + b_i \cos\omega_i t \\ &= c_i \sin(\omega_i t + \varphi_i)\end{aligned} \tag{5.2.13}$$

The coefficients, a_i and b_i, resp. c_i and φ_i, can now be deduced from equs. (5.1.8) and (5.1.9). We find

$$\eta_{ei}(t) = \frac{\dot{\eta}_{ei}(0)}{\omega_i} \sin\omega_i t + \eta_{ei}(0)\cos\omega_i t \tag{5.2.14}$$

respectively

$$\begin{aligned}c_i &= \sqrt{\left(\eta_{ei}(0)\right)^2 + \left(\dot{\eta}_{ei}(0)/\omega_i\right)^2} \\ \varphi_i &= \tan^{-1}\frac{\eta_{ei}(0)\omega_i}{\dot{\eta}_{ei}(0)}\end{aligned} \tag{5.2.15}$$

The above scheme presumes a knowledge of the initial conditions, $\boldsymbol{\eta}_e(0)$, resp. $\dot{\boldsymbol{\eta}}_e(0)$. Since the analysis of the elastic component is in principle identical – but for the index e – to the overall analysis proposed in section 5.1, we are in a position to deduce from equs. (5.1.10) and (5.1.30) the modified result

$$\boldsymbol{\eta}_e(0) = \boldsymbol{X}_e^{\mathrm{t}}\boldsymbol{K}_{ee}\boldsymbol{r}_e(0)$$
$$\dot{\boldsymbol{\eta}}_e(0) = \boldsymbol{\Lambda}^{-1}\boldsymbol{X}_e^{\mathrm{t}}[\boldsymbol{M}_{ee} - \boldsymbol{M}_{e0}\boldsymbol{M}_{00}^{-1}\boldsymbol{M}_{0e}]\dot{\boldsymbol{r}}_e(0) \tag{5.2.16}$$

With this, our discussion has turned a complete circle and the specific, purely elastic motion has been established. Let us now turn to the rigid-body motion, using equ. (4.2.12) to determine the unknown vector columns, $\boldsymbol{a}_0$ and $\boldsymbol{b}_0$. In fact, we find

$$\boldsymbol{a}_0 = \boldsymbol{r}_0(0) + \boldsymbol{M}_{00}^{-1}\boldsymbol{M}_{0e}\boldsymbol{r}_e(0)$$
$$\boldsymbol{b}_0 = \dot{\boldsymbol{r}}_0(0) + \boldsymbol{M}_{00}^{-1}\boldsymbol{M}_{0e}\dot{\boldsymbol{r}}_e(0) \tag{5.2.17}$$

This completes the evaluation of the rigid-body motion. Particularly simple conditions are obtained when $\boldsymbol{M}_{0e} = \boldsymbol{O}$.

Finally, using equ. (5.2.1) and reversing the transformation, we obtain $\boldsymbol{r}\,(\boldsymbol{r}_c)$ from $\boldsymbol{r}_0(\boldsymbol{r}_{c0})$ and $\boldsymbol{r}_e(\boldsymbol{r}_{ce})$. In the case of the Guyan reduction, we transform the system $\boldsymbol{r}_c$ into $\boldsymbol{r}$ with the help of $\boldsymbol{T}_c$ and in this way deduce the response of the system in terms of the original degrees of freedom. The complete procedure is summarised in fig. 5.2.1.

To conclude this chapter, we consider a small example and reconsider the beam analysed on the right half of fig. 4.2.3. We assume at time $t = 0$ that the beam is bent into an arch and subsequently released as shown in fig. 5.2.2.

For the calculations, we employ the definitions, $\boldsymbol{T}_0$ and $\boldsymbol{T}_e$, given in fig. 4.2.3, and note that here, $\boldsymbol{M}_{e0}$ is a non-zero matrix. The procedure outlined in fig. 5.2.1 yields an eigenoscillation about a displaced equilibrium position. If $\dot{\boldsymbol{r}}\,(0) \neq 0$ the system may move with a linear dependence in t under certain conditions.

The advantages of making $\boldsymbol{T}_e$ orthogonal to $\boldsymbol{M}\boldsymbol{T}_0$ have already been mentioned. In such a situation, $\boldsymbol{M}_{e0} = \boldsymbol{O} = \boldsymbol{M}_{0e}^{\mathrm{t}}$ and integration of the rigid-body motion of the system yields the simple result

$$\boldsymbol{r}_0(t) = \boldsymbol{r}_0(0) + t\dot{\boldsymbol{r}}_0(0), \qquad \text{when } \boldsymbol{M}_{e0} = \boldsymbol{O} \tag{5.2.18}$$

Through transformation of the initial conditions $\boldsymbol{r}\,(0)$ and $\dot{\boldsymbol{r}}\,(0)$ into $\boldsymbol{r}_0(0)$ and $\dot{\boldsymbol{r}}_0(0)$ we are hence in a position to verify directly whether the entire structure moves as a rigid body or not. For the purpose of our simple example in fig. 5.2.2, the construction of such an alternative $\boldsymbol{T}$ leads to the transformation

$$\boldsymbol{r} = \frac{1}{\sqrt{6}}\left[\begin{array}{cc|c} \sqrt{2} & -\sqrt{3} & \sqrt{2} \\ \sqrt{2} & 0 & -\sqrt{2} \\ \sqrt{2} & \sqrt{3} & \sqrt{2} \end{array}\right]\begin{bmatrix} \boldsymbol{r}_0 \\ \boldsymbol{r}_e \end{bmatrix}$$

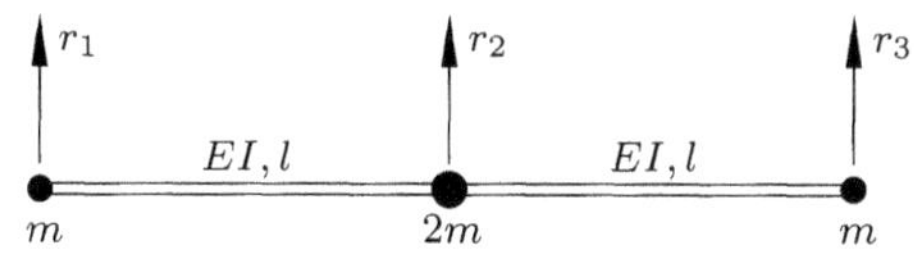

initial conditions	$\boldsymbol{r}(0) = \begin{Bmatrix} 0 & 1 & 0 \end{Bmatrix}$ $\dot{\boldsymbol{r}}(0) = \begin{Bmatrix} 0 & 1 & 0 \end{Bmatrix}$
kinematics	$\boldsymbol{r} = \left[\begin{array}{cc\|c} 1 & -1 & 0 \\ 1 & 0 & 1 \\ 1 & 1 & 0 \end{array}\right] \begin{bmatrix} \boldsymbol{r}_0 \\ \boldsymbol{r}_e \end{bmatrix}$
or inverted relation	$\begin{bmatrix} \boldsymbol{r}_0 \\ \boldsymbol{r}_e \end{bmatrix} = \frac{1}{2} \begin{bmatrix} 1 & 0 & 1 \\ -1 & 0 & 1 \\ -1 & 2 & -1 \end{bmatrix} \boldsymbol{r}$
eigenvalue analysis (see fig.4.2.3)	$\omega_1^2 = \frac{6EI}{ml^3} \qquad \boldsymbol{x}_{e1} = \begin{Bmatrix} 1 \end{Bmatrix}$
elastic displacements	$\boldsymbol{r}_e(t) = \cos \omega_1 t$
rigid body displacements	$\boldsymbol{r}_0(t) = \frac{1}{2} \begin{Bmatrix} 1 & 0 \end{Bmatrix} \left(1 - \cos \omega_1 t\right)$
total motion	$\boldsymbol{r}(t) = \begin{bmatrix} 0.5 \\ 0.5 \\ 0.5 \end{bmatrix} + \begin{bmatrix} -0.5 \\ 0.5 \\ -0.5 \end{bmatrix} \cos \omega_1 t$

Fig. 5.2.2 Free motion of a free-free structure: the beam oscillates around a displaced equilibrium position which contains the centre of gravity at the instant $t = 0$, *Part 1*

This indeed fulfils the conditions

$$
\begin{aligned}
\boldsymbol{T}_0^{\mathrm{t}} \boldsymbol{T}_0 &= \boldsymbol{I}_0 \\
\boldsymbol{T}_e^{\mathrm{t}} \boldsymbol{T}_e &= \boldsymbol{I}_e \\
\boldsymbol{T}_e^{\mathrm{t}} \boldsymbol{M} \boldsymbol{T}_0 &= \boldsymbol{O}
\end{aligned}
$$

$$r(t) = \begin{bmatrix} 0.5 \\ 0.5 \\ 0.5 \end{bmatrix} + \begin{bmatrix} -0.5 \\ 0.5 \\ -0.5 \end{bmatrix} \times \cos \omega_1 t$$

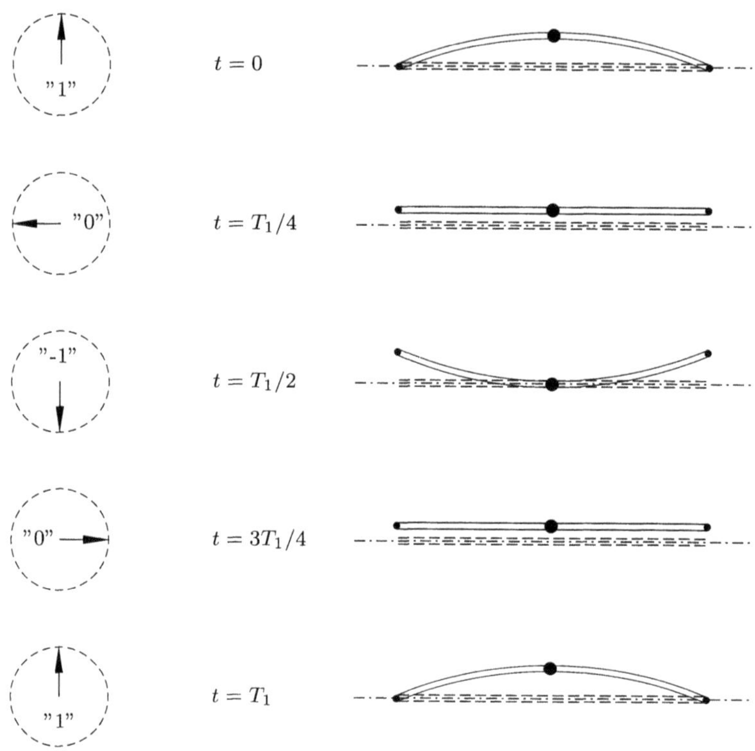

Fig. 5.2.2 Free motion of a free-free structure: the beam oscillates around a displaced equilibrium position which contains the centre of gravity at the instant $t = 0$, *Part 2*

An inversion of the kinematic transformation yields

$$\begin{bmatrix} \boldsymbol{r}_0 \\ \boldsymbol{r}_e \end{bmatrix} = \frac{1}{\sqrt{6}} \begin{bmatrix} \frac{3}{4}\sqrt{2} & \frac{6}{4}\sqrt{2} & \frac{3}{4}\sqrt{2} \\ -\sqrt{3} & 0 & \sqrt{3} \\ \frac{3}{4}\sqrt{2} & -\frac{6}{4}\sqrt{2} & \frac{3}{4}\sqrt{2} \end{bmatrix} \boldsymbol{r}$$

and this determines the corresponding initial conditions

$$\begin{bmatrix} \boldsymbol{r}_0(0) \\ \boldsymbol{r}_e(0) \end{bmatrix} = \frac{\sqrt{3}}{2} \begin{bmatrix} 1 \\ 0 \\ -1 \end{bmatrix}$$

and

$$\begin{bmatrix} \dot{\boldsymbol{r}}_0(0) \\ \dot{\boldsymbol{r}}_e(0) \end{bmatrix} = \boldsymbol{o}_3$$

Hence, we may immediately deduce the rigid body component of the displaced system as

$$\boldsymbol{r}_0(t) = \boldsymbol{r}_0(0) = \left\{ \tfrac{1}{2}\sqrt{3} \quad 0 \right\} = \text{const}$$

Its contribution to the overall response is thus

$$\boldsymbol{T}_0 \boldsymbol{r}_0(t) = \left\{ \tfrac{1}{2} \quad \tfrac{1}{2} \quad \tfrac{1}{2} \right\}$$

and is obtained without any prior knowledge of the elastic response. A detailed analysis of the problem may serve as a useful exercise for the reader.

Bibliography for Chapter 5

[5.1] *J. Argyris*; Some results on the free-free oscillations of aircraft type structures, Revue Française de Mécanique, 15 (1965), 3e trimestre, 59–73

[5.2] *G.A. Malejannakis*; Anwendung der Matrizenverschiebungsmethode auf erzwungene Schwingungen proportional gedämpfter elastischer Systeme, ISD-Bericht Nr. 99, Stuttgart (1971).

[5.3] *J. Argyris, D.W. Scharpf*; Finite elements in space and time, Aeron. J. Roy. Aeron. Soc., 73, (1969) 1041–1044.

[5.4] *R.W. Clough, J. Penzien*; Dynamics of structures (McGraw-Hill, New York, 1975).

[5.5] *W.C. Hurty, M.F. Rubinstein*; Dynamics of structures (Prentice-Hall, Englewood Cliffs., 1964).

[5.6] *J. Wittenburg*; Dynamics of systems of rigid bodies (Teubner, Stuttgart, 1977).

[5.7] *S. Timoshenko*; Vibration problems in engineering (van Nostrand, Princeton, 1955).

[5.8] *T. Belytschko, T.J.R. Hughes* (Eds.); Computational methods for transient analysis (Volume 1 in: Mechanics and mathematical methods, North-Holland, Amsterdam, 1983).

6 Forced vibrations of undamped systems

"Force without forecast is of little avail"
Thomas Fuller, Gnomologia No. 1589 (1732)

In the previous chapter, we discussed the eigenoscillations of an unconstrained elastic body as induced by specific initial conditions. In this case, no prescribed external forces were presumed to act upon the body, its temporal deformation being determined soleley by the assumed initial conditions but by no other external cause.

No such restrictions apply in this present chapter where external forces are assumed to be imposed on the oscillatory system during a set observation period.

The principles outlined earlier in section 1.4 of Chapter 1 for the calculation of a one-mass oscillator are once again applied, but this time to the more general case of a multi-mass oscillator subject to periodic, non-periodic and arbitrary load applications.

6.1 Diagonalisation and integration of the system

Bearing in mind that a forced oscillation can generally be generated by means of imposed forces and/or deflections (fig. 6.1.1), we now extend the family of the degrees of freedom as first discussed in section 5.1 by also incorporating time-dependent, prescribed degrees of freedom, $\boldsymbol{r}_u(t)$, the corresponding forces $\boldsymbol{R}_u(t)$ still being unknown, however.

The equation of motion of the dynamic system can now be expressed as

$$\begin{bmatrix} \boldsymbol{M}_{cc} & \boldsymbol{M}_{c\sigma} & \boldsymbol{M}_{cu} & \boldsymbol{M}_{cs} \\ \boldsymbol{M}_{\sigma c} & \boldsymbol{M}_{\sigma\sigma} & \boldsymbol{M}_{\sigma u} & \boldsymbol{M}_{\sigma s} \\ \boldsymbol{M}_{uc} & \boldsymbol{M}_{u\sigma} & \boldsymbol{M}_{uu} & \boldsymbol{M}_{us} \\ \boldsymbol{M}_{sc} & \boldsymbol{M}_{s\sigma} & \boldsymbol{M}_{su} & \boldsymbol{M}_{ss} \end{bmatrix} \begin{bmatrix} \ddot{\boldsymbol{r}}_c \\ \ddot{\boldsymbol{r}}_\sigma \\ \ddot{\boldsymbol{r}}_u \\ \ddot{\boldsymbol{r}}_s = \boldsymbol{o} \end{bmatrix} + \begin{bmatrix} \boldsymbol{K}_{cc} & \boldsymbol{K}_{c\sigma} & \boldsymbol{K}_{cu} & \boldsymbol{K}_{cs} \\ \boldsymbol{K}_{\sigma c} & \boldsymbol{K}_{\sigma\sigma} & \boldsymbol{K}_{\sigma u} & \boldsymbol{K}_{\sigma s} \\ \boldsymbol{K}_{uc} & \boldsymbol{K}_{u\sigma} & \boldsymbol{K}_{uu} & \boldsymbol{K}_{us} \\ \boldsymbol{K}_{sc} & \boldsymbol{K}_{s\sigma} & \boldsymbol{K}_{su} & \boldsymbol{K}_{ss} \end{bmatrix} \begin{bmatrix} \boldsymbol{r}_c \\ \boldsymbol{r}_\sigma \\ \boldsymbol{r}_u \\ \boldsymbol{r}_s = \boldsymbol{o} \end{bmatrix} = \begin{bmatrix} \boldsymbol{R}_c \\ \boldsymbol{R}_\sigma \\ \boldsymbol{R}_u \\ \boldsymbol{R}_s \end{bmatrix} \tag{6.1.1}$$

As in section 5.1, we first eliminate the supporting degrees of freedom, $\boldsymbol{r}_s$, the corresponding reacting forces being calculated from

$$\boldsymbol{R}_s = \boldsymbol{M}_{sc}\ddot{\boldsymbol{r}}_c + \boldsymbol{M}_{s\sigma}\ddot{\boldsymbol{r}}_\sigma + \boldsymbol{M}_{su}\ddot{\boldsymbol{r}}_u + \boldsymbol{K}_{sc}\boldsymbol{r}_c + \boldsymbol{K}_{s\sigma}\boldsymbol{r}_\sigma + \boldsymbol{K}_{su}\boldsymbol{r}_u \tag{6.1.2}$$

The same technique is then used for the evaluation of the group of forces $\boldsymbol{R}_u$ presuming that $\boldsymbol{r}_c(t)$ and $\boldsymbol{r}_\sigma(t)$ are known

$$\boldsymbol{R}_u = \boldsymbol{M}_{uc}\ddot{\boldsymbol{r}}_c + \boldsymbol{M}_{u\sigma}\ddot{\boldsymbol{r}}_\sigma + \boldsymbol{M}_{uu}\ddot{\boldsymbol{r}}_u + \boldsymbol{K}_{uc}\boldsymbol{r}_c + \boldsymbol{K}_{u\sigma}\boldsymbol{r}_\sigma + \boldsymbol{K}_{uu}\boldsymbol{r}_u \tag{6.1.3}$$

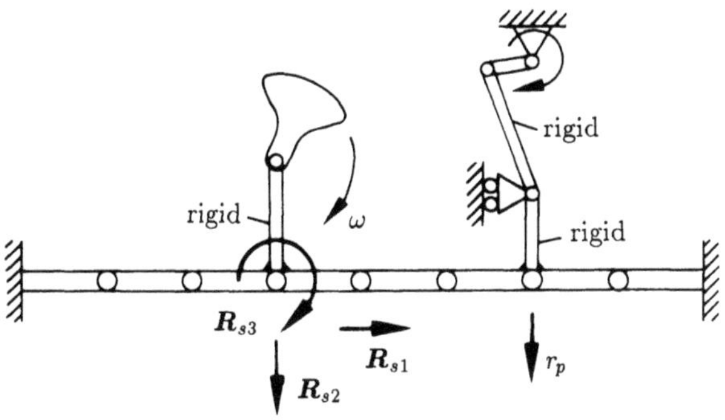

Fig. 6.1.1
Example of a dynamic excitation due to imposed forces and/or prescribed displacements

Applying next the method already set down in equ. (4.4.2), the residual system defined by $\boldsymbol{r}_c$ and $\boldsymbol{r}_\sigma$ can be further reduced to $\boldsymbol{r}_c$ alone. The final result reads

$$\overline{\boldsymbol{M}}_{cc}\ddot{\boldsymbol{r}}_c + \overline{\boldsymbol{K}}_{cc}\boldsymbol{r}_c = \overline{\boldsymbol{R}}_c(t) \tag{6.1.4}$$

where $\overline{\boldsymbol{M}}_{cc}$ and $\overline{\boldsymbol{K}}_{cc}$ have already been given in equs. (4.4.6) and (4.4.7) and

$$\overline{\boldsymbol{R}}_c(t) = \boldsymbol{R}_c - [\boldsymbol{M}_{cu}\ddot{\boldsymbol{r}}_u + \boldsymbol{K}_{cu}\boldsymbol{r}_u] - \boldsymbol{K}_{c\sigma}\boldsymbol{K}_{\sigma\sigma}^{-1}[\boldsymbol{R}_\sigma - \boldsymbol{M}_{\sigma u}\ddot{\boldsymbol{r}}_u - \boldsymbol{K}_{\sigma u}\boldsymbol{r}_u] \tag{6.1.5}$$

We now assume once more that equ. (6.1.4) has already undergone a spectral analysis and we express the corresponding solution $\boldsymbol{r}_c(t)$ in terms of the eigenvectors as in equ. (5.1.4). The procedure described in section 5.1 then yields a diagonal system of order nx

$$\boldsymbol{\Lambda}\ddot{\boldsymbol{\eta}} + \boldsymbol{\eta} = \boldsymbol{X}^{\mathrm{t}}\overline{\boldsymbol{R}}_c(t) \tag{6.1.6}$$

A typical individual equation has the form

$$\ddot{\eta}_{\mathrm{i}} + \omega_{\mathrm{i}}^2\eta_{\mathrm{i}} = \omega_{\mathrm{i}}^2 R_{\eta\mathrm{i}}(t), \qquad \mathrm{i} = 1\ldots\mathrm{nx} \tag{6.1.7}$$

and is formally identical with the expression for a single-mass oscillator of equ. (1.4.1). The homogeneous solution of this equation coincides with that of equ. (5.1.7). Whether a particular closed-form solution can be established depends on the nature of the excitation, $R_{\eta\mathrm{i}}(t)$. This can easily be realised for excitations described by circular functions and polynomials but not, in general, for other functional dependences of technical interest. These must then be solved by means of a numerical integration.

6.2 Periodic excitation functions

Periodic excitation functions and their simplest variants, harmonic functions, play a significant role in vibration analysis. We next consider a sinusoidal unit excitation

$$R_{\eta\mathrm{i}}(t) = 1 \cdot \sin \Omega_\mathrm{i} t \tag{6.2.1}$$

and establish the system's response, $\eta_\mathrm{i}(t)$, for the initial conditions

$$\begin{aligned} \eta_\mathrm{i}(0) &= 0 \\ \dot{\eta}_\mathrm{i}(0) &= 0 \end{aligned} \tag{6.2.2}$$

The particular solution to the differential equation (6.1.7) is also sinusoidal in the present case. Setting

$$\eta_{p\mathrm{i}} = \alpha_\mathrm{i} \sin \Omega_\mathrm{i} t \tag{6.2.3}$$

we immediately obtain

$$\alpha_\mathrm{i} = \frac{\omega_\mathrm{i}^2}{\omega_\mathrm{i}^2 - \Omega_\mathrm{i}^2} = \frac{1}{1 - (\Omega_\mathrm{i}/\omega_\mathrm{i})^2} \tag{6.2.4}$$

In conjunction with equ. (5.1.7), the complete general solution is then

$$\eta_\mathrm{i} = a_\mathrm{i} \sin \omega_\mathrm{i} t + b_\mathrm{i} \cos \omega_\mathrm{i} t + \frac{1}{1 - (\Omega_\mathrm{i}/\omega_\mathrm{i})^2} \sin \Omega_\mathrm{i} t \tag{6.2.5}$$

The reader will notice the link to the more general solution of equ. (1.4.10) for a simple oscillator.

The initial conditions of equ. (6.2.2) result in the specific solution

$$\eta_\mathrm{i} = \frac{1}{1 - (\Omega_\mathrm{i}/\omega_\mathrm{i})^2} \left(\sin \Omega_\mathrm{i} t - \frac{\Omega_\mathrm{i}}{\omega_\mathrm{i}} \sin \omega_\mathrm{i} t \right) \tag{6.2.6}$$

The expression comprises an oscillatory term with the frequency of excitation, Ω_i, and a term with the eigenfrequency, ω_i. As the reader yet remains to be informed, the first term represents the steady-state response whilst the second is relevant for the transient phase alone and vanishes rapidly as a result of the damping effects inevitably present in technical systems. For an undamped body, the lasting component of the response is in phase with the excitation, but this is no longer true when damping is acting. The amplitude of equ. (6.2.6) is greatly dependent upon the ratio $\Omega_\mathrm{i}/\omega_\mathrm{i}$ of excitation frequency to eigenfrequency. Equation (6.2.6) does not hold when $\Omega_\mathrm{i} = \omega_\mathrm{i}$, i.e. for an excitation in tune with the eigenfrequency.

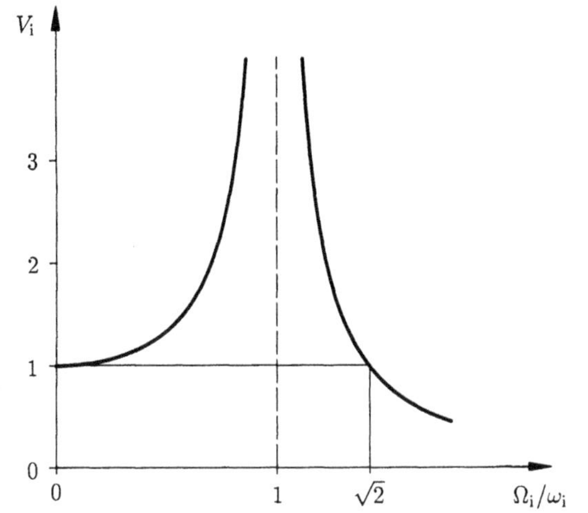

Fig. 6.2.1
Amplification factor V_i for undamped structures

Here, we have to apply a limiting process according to the Bernoulli-l'Hospital rule. Using Ω_i as a variable, we have

$$\lim_{\Omega_i \to \omega_i} \eta_i(t) = \lim_{\Omega_i \to \omega_i} \frac{t\cos\Omega_i t - \omega_i^{-1}\sin\omega_i t}{-2\Omega_i/\omega_i^2}$$

$$= \tfrac{1}{2}(\sin\Omega_i t - \Omega_i t\cos\Omega_i t) \tag{6.2.7}$$

This result suggests that the amplitude $\eta_i(t)$ increases steadily although in reality, damping will prevent an unrestricted growth; however, large amplitudes may destroy the system in certain cases.

It is advantageous to relate the amplitude of the term oscillating in tune with the excitation to the static deflection due to a unit load. In fact, from the differential equ. (6.1.7), the static response is immediately found to be $\eta_i = 1$. The ratio V_i of the dynamic amplitude to the static amplitude is known as the amplification factor (the reader should also consult section 1.4.1, in particular equ. (1.4.13)). Thus,

$$V_i(\Omega_i) = \left|\frac{1}{1 - (\Omega_i/\omega_i)^2}\right| \tag{6.2.8}$$

Consequently, in equ. (6.2.6), the steady-state amplitude is V_i and the 'transient' one $(\Omega_i/\omega_i)V_i$. The amplification factor, V_i, is plotted in fig. 6.2.1. It is characteristic of the sub-critical regime ($\Omega_i/\omega_i < 1$) that an augmentation of the amplitude always takes place. In contrast, the magnification in the over-critical regime tends to zero. In fact, in this regime, the inertia conditions the body to follow the rapidly oscillating unit load increasingly less pronouncedly. At a frequency ratio of $\sqrt{2}$, the amplification factor becomes smaller than 1 since not only the magnification factor itself, but also $(\Omega_i/\omega_i)V_i$ approaches zero in the over-critical regime. Thus, the overall response decreases with increasing frequency of excitation. On the other hand, the total response of the system approaches the static solution, $\eta_i = 1$, as the frequency of excitation decreases . This is because $(\Omega_i/\omega_i)V_i$ becomes significantly smaller than V_i. Figure 6.2.2 depicts three typical cases of sub-critical, critical and over-critical excitation for an oscillator with zero damping. We observe that in a sub-critical regime, the steady-state response takes place in phase with the excitation. In contrast, an over-critical excitation causes a phase shift of 180°. This is demonstrated in more detail in fig. 6.2.3.

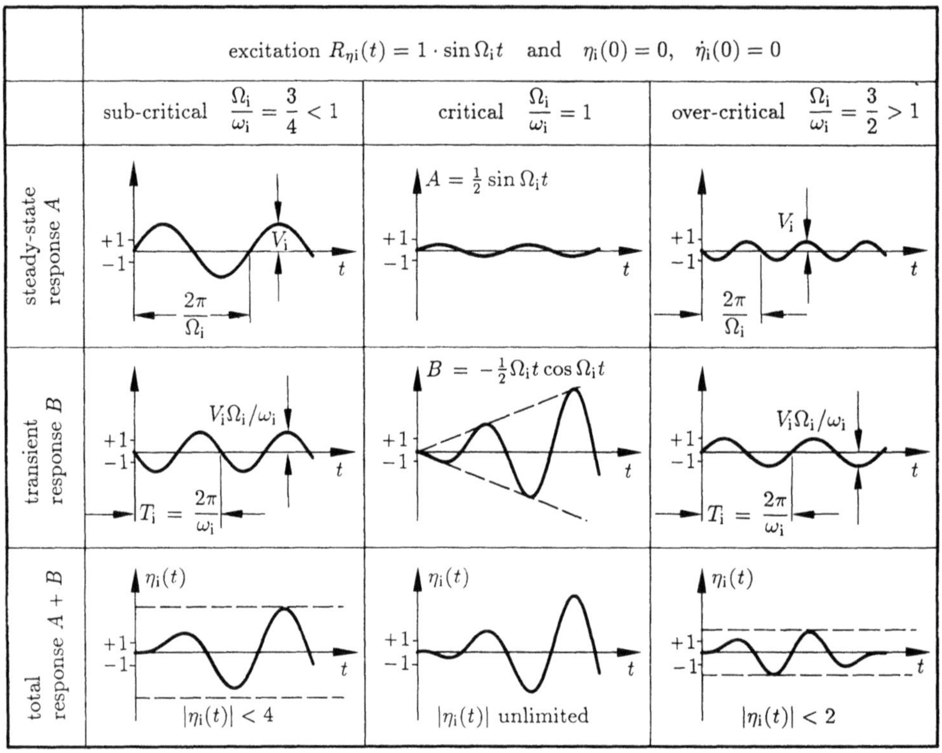

Fig. 6.2.2 Harmonic excitation and response (for zero damping)

The specific response due to a sinusoidal excitation with a unit amplitude can easily be generalised as the differential equation (6.1.7) is linear. Thus, the excitation

$$R_{\eta i}(t) = C_i \sin(\Omega_i t + \psi_i) \tag{6.2.9}$$

represents a sinusoidal mode with an amplitude of C_i instead of 1 and a time shift of ψ_i/Ω_i. The steady-state response possesses the same amplification factor, V_i, and the same phase shift, φ_i (0° or 180°), as before whilst the transient response will be modified depending, of course, upon the initial conditions.

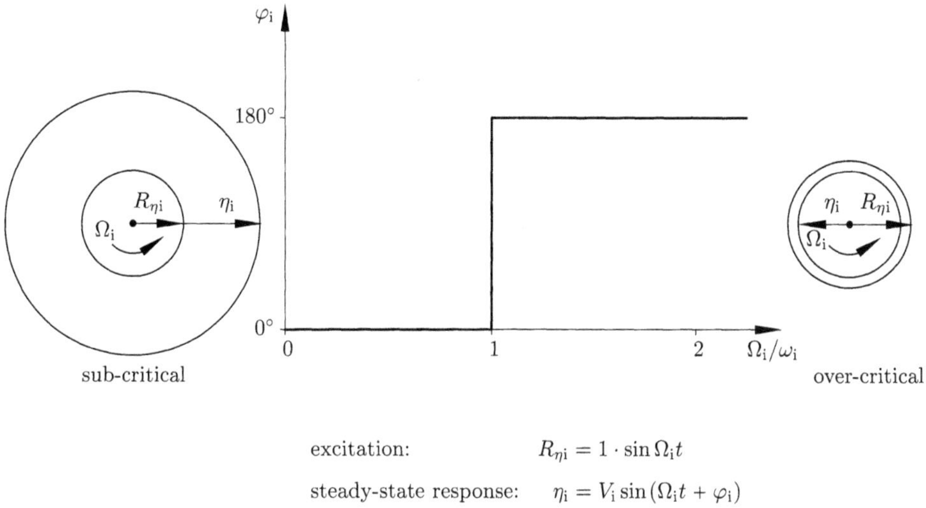

Fig. 6.2.3 Phase shift between excitation and steady-state response of an undamped structure

Equation (6.2.5) is now replaced by the new expression

$$\eta_i = a_i \sin \omega_i t + b_i \cos \omega_i t + \frac{C_i}{1 - (\Omega_i/\omega_i)^2} \sin(\Omega_i t + \psi_i) \tag{6.2.10}$$

For the initial conditions of equ. (6.2.2), the total response becomes

$$\eta_i = \frac{C_i}{1 - (\Omega_i/\omega_i)^2} \left[\sin(\Omega_i t + \psi_i) - \frac{\Omega_i}{\omega_i} \cos \psi_i \sin \omega_i t - \sin \psi_i \cos \omega_i t \right] \tag{6.2.11}$$

For $\psi_i = 0$, the solution reduces clearly to that of equ. (6.2.6). Another special case arises for a pure cosine excitation which corresponds to $\psi_i = 90°$. Here we have

$$\eta_i = \frac{C_i}{1 - (\Omega_i/\omega_i)^2} (\cos \Omega_i t - \cos \omega_i t) \tag{6.2.12}$$

An equivalent excitation to that defined by equ. (6.2.9) may be expressed as

$$R_{\eta i}(t) = A_i \sin \Omega_i t + B_i \cos \Omega_i t \tag{6.2.13}$$

where

$$C_i = \sqrt{A_i^2 + B_i^2}$$
$$\psi_i = \tan^{-1} \frac{B_i}{A_i} \tag{6.2.14}$$

The response to the excitation of equ. (6.2.13) derives in conjunction with equ. (6.2.14) either from equ. (6.2.11) or from a superposition of equs. (6.2.6) and (6.2.12). We find

$$\eta_{\mathrm{i}} = \frac{A_{\mathrm{i}}}{1-(\Omega_{\mathrm{i}}/\omega_{\mathrm{i}})^2}\left(\sin\Omega_{\mathrm{i}}t - \frac{\Omega_{\mathrm{i}}}{\omega_{\mathrm{i}}}\sin\omega_{\mathrm{i}}t\right) + \frac{B_{\mathrm{i}}}{1-(\Omega_{\mathrm{i}}/\omega_{\mathrm{i}})^2}(\cos\Omega_{\mathrm{i}}t - \cos\omega_{\mathrm{i}}t) \tag{6.2.15}$$

Euler's formula yields an interesting variant to the harmonic excitation functions

$$\mathrm{e}^{i\Omega t} = \cos\Omega t + i\sin\Omega t \tag{6.2.16}$$

Hence, the complex excitation

$$R_{\eta\mathrm{i}} = \mathrm{e}^{i\Omega_{\mathrm{i}}t} \tag{6.2.17}$$

comprises the real part $1\cdot\cos\Omega_{\mathrm{i}}t$ and the imaginary part $1\cdot\sin\Omega_{\mathrm{i}}t$. Because the differential equation is linear, the real part of η_{i} corresponds to the excitation $\cos\Omega_{\mathrm{i}}t$ and the imaginary part corresponds to $\sin\Omega_{\mathrm{i}}t$. We are thus in a position to analyse both types of harmonic excitation simultaneously.

A particular solution for the excitation of equ. (6.2.17), which reproduces the steady-state condition, is given by

$$\eta_{p\mathrm{i}} = F_{\mathrm{i}}\mathrm{e}^{i\Omega_{\mathrm{i}}t} \tag{6.2.18}$$

where F_{i} is known as the frequency response characteristic. Substitution into the differential equation (6.1.7) yields the expression

$$F_{\mathrm{i}} = \frac{1}{1-(\Omega_{\mathrm{i}}/\omega_{\mathrm{i}})^2} \tag{6.2.19}$$

This result appears also in equs. (6.2.4) and (6.2.12) or (6.2.15). The absolute value of the frequency response concurs with the amplification factor V_{i} already defined

$$V_{\mathrm{i}} = |F_{\mathrm{i}}| \tag{6.2.20}$$

Hence we have

$$\begin{aligned} F_{\mathrm{i}} &= V_{\mathrm{i}} \qquad &&\text{for } \Omega_{\mathrm{i}} < \omega_{\mathrm{i}} \quad \text{(sub-crtitical)} \\ F_{\mathrm{i}} &= -V_{\mathrm{i}} \qquad &&\text{for } \Omega_{\mathrm{i}} > \omega_{\mathrm{i}} \quad \text{(over-critical)} \end{aligned} \tag{6.2.21}$$

If F_{i} is defined as a complex number (as is indeed the case in damped structures), then

$$F_{\mathrm{i}} = |F_{\mathrm{i}}|(\cos\psi_{\mathrm{i}} + i\sin\psi_{\mathrm{i}}) = V_{\mathrm{i}}\mathrm{e}^{i\psi_{\mathrm{i}}} \tag{6.2.22}$$

where ψ_{i} is the associated phase angle.

In the present case, F_i is, of course, purely real, i.e.

$$\sin\psi_\mathrm{i} = 0 \tag{6.2.23}$$

or

$$\psi_\mathrm{i} = 0^\circ \quad \text{or} \quad 180^\circ \tag{6.2.24}$$

Noting equ. (6.2.21), we obtain hence

$$\begin{aligned} \psi_\mathrm{i} &= 0^\circ && \text{if} \quad \Omega_\mathrm{i} < \omega_\mathrm{i} \\ \psi_\mathrm{i} &= 180^\circ && \text{if} \quad \Omega_\mathrm{i} > \omega_\mathrm{i} \end{aligned} \tag{6.2.25}$$

where ψ_i indicates the phase shift for the steady-state response. Indeed, following equ. (6.2.18), we confirm the expression

$$\begin{aligned} \eta_{p\mathrm{i}} &= F_\mathrm{i} \mathrm{e}^{i\Omega_\mathrm{i} t} = V_\mathrm{i} \mathrm{e}^{i\psi_\mathrm{i}} \mathrm{e}^{i\Omega_\mathrm{i} t} = V_\mathrm{i} \mathrm{e}^{i(\Omega_\mathrm{i} t + \psi_\mathrm{i})} \\ &= V_\mathrm{i} \big[\cos(\Omega_\mathrm{i} t + \psi_\mathrm{i}) + i \sin(\Omega_\mathrm{i} t + \psi_\mathrm{i}) \big] \end{aligned} \tag{6.2.26}$$

The analysis of harmonic excitations is hence greatly facilitated by a complex formulation.

We conclude this section with the modest example of fig. 6.2.4. Let a sinusoidally varying force (e.g. due to an unbalanced rotating mass) be applied on a point of the beam. In the figure, we observe deflections which are far in excess of those achieved under static conditions for the maximal force. What is more, we note that the second eigenoscillation exercises hardly any influence on the response, although the system is unsymmetrically excited. A dynamic reduction to the first eigenvector alone hence appears permissible. With respect to the ever-present damping, this particular solution represents a valid approximation to the steady-state condition.

An arbitrary periodic excitation function, $R_{\eta\mathrm{i}}$, is generally understood to represent a force-time relation, which, although arbitrary, repeats itself with a constant time interval of $T_{E\mathrm{i}}$. Such a force function is shown in fig. 6.2.5 and may be developed into a Fourier series (see also 1.4.2). Introducing the forcing frequency

$$\Omega_0 = \frac{2\pi}{T_{E\mathrm{i}}} \tag{6.2.27}$$

we obtain the expansion

$$R_{\eta\mathrm{i}}^{(F)} = \tfrac{1}{2} a_0 + \sum_{\mathrm{k}=1}^{\infty} (a_\mathrm{k} \cos \mathrm{k}\Omega_0 t + b_\mathrm{k} \sin \mathrm{k}\Omega_0 t) \tag{6.2.28}$$

where

$$a_\mathrm{k} = \int_{-1}^{+1} R_{\eta\mathrm{i}}(\tau) \cos(\mathrm{k}\pi\tau)\, \mathrm{d}\tau \tag{6.2.29}$$

model

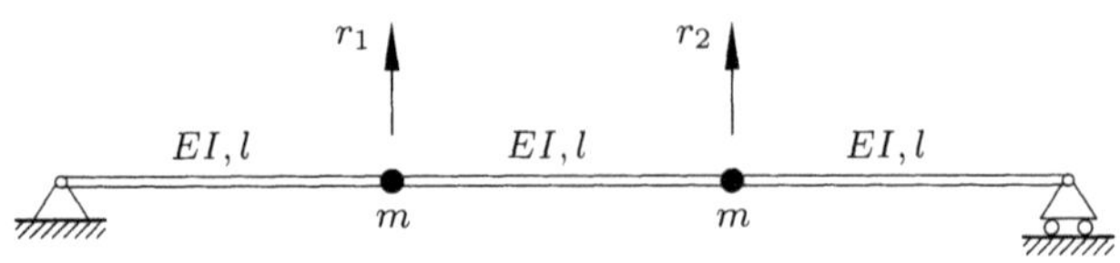

eigenvalue analysis: see fig. 5.1.2

excitation $\boldsymbol{R} = \begin{bmatrix} R_1 \\ R_2 \end{bmatrix} = \begin{bmatrix} \frac{1}{10}\frac{EI}{l^2}\sin\Omega t \\ 0 \end{bmatrix}$ for $t \geqslant 0$ and $\Omega = \left(\dfrac{EI}{ml^3}\right)^{\frac{1}{2}}$

initial conditions $\left.\begin{array}{l} \boldsymbol{r}\,(t=0) = \boldsymbol{o} \\ \dot{\boldsymbol{r}}\,(t=0) = \boldsymbol{o} \end{array}\right\}$ state of rest

excitation of decoupled system $\boldsymbol{R}_\eta = \boldsymbol{X}^{\mathrm{t}}\boldsymbol{R} = \dfrac{1}{60}\left(\dfrac{EI}{l}\right)^{\frac{1}{2}}\left\{\sqrt{15} \quad 1\right\}\sin\Omega t$

response of decoupled system $\boldsymbol{\eta} = \begin{bmatrix} (\frac{3}{2})^{\frac{1}{2}}\left(\frac{EI}{l}\right)^{\frac{1}{2}}\left(\sin\Omega t - (\frac{5}{6})^{\frac{1}{2}}\sin(\frac{6}{5})^{\frac{1}{2}}\Omega t\right) \\ \frac{3}{170}\left(\frac{EI}{l}\right)^{\frac{1}{2}}\left(\sin\Omega t - (\frac{1}{18})^{\frac{1}{2}}\sin(18)^{\frac{1}{2}}\Omega t\right) \end{bmatrix}$

response of original system

$$\boldsymbol{r} = \boldsymbol{X}\boldsymbol{\eta} = l\begin{bmatrix} 0.253\sin\Omega t - 0.228\sin 1.095\Omega t - 0.003\sin 4.243\Omega t \\ 0.247\sin\Omega t - 0.228\sin 1.095\Omega t + 0.003\sin 4.243\Omega t \end{bmatrix}$$

Fig. 6.2.4 Example of a harmonic excitation of an undamped oscillator (sub-critical), *Part 1*

$$b_{\mathrm{k}} = \int_{-1}^{+1} R_{\eta\mathrm{i}}(\tau)\sin(\mathrm{k}\pi\tau)\,\mathrm{d}\tau \tag{6.2.30}$$

The dimensionless time τ is here defined as

$$\tau = \frac{2t}{T_{\mathrm{Ei}}} = \Omega_0\frac{t}{\pi} \tag{6.2.31}$$

The usefulness of a complex representation has already been demonstrated in section 1.4 when considering a single-mass oscillator and in the first part of the present section for a pure harmonic excitation. If we adopt it here and modify static response

$$\boldsymbol{R} = \left\{ \frac{1}{10}\frac{EI}{l^2} \quad 0 \right\}$$

$$\boldsymbol{r} = l\{0.044 \quad 0.039\}$$

initial transient response for $\quad \Omega = \left(\dfrac{EI}{ml^3}\right)^{\frac{1}{2}}$

$\Omega t = 0 \qquad \boldsymbol{r} = l\begin{bmatrix} 0 \\ 0 \end{bmatrix}$

$\Omega t = \dfrac{\pi}{2} \qquad \boldsymbol{r} = l\begin{bmatrix} +0.025 \\ +0.025 \end{bmatrix}$

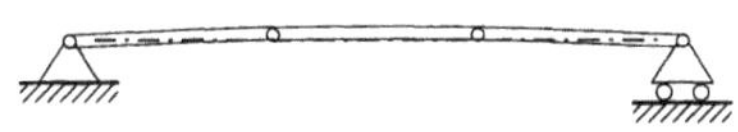

$\Omega t = \pi \qquad \boldsymbol{r} = l\begin{bmatrix} +0.067 \\ +0.067 \end{bmatrix}$

$\Omega t = \dfrac{3\pi}{2} \qquad \boldsymbol{r} = l\begin{bmatrix} -0.044 \\ -0.044 \end{bmatrix}$

$\Omega t = 2\pi \qquad \boldsymbol{r} = l\begin{bmatrix} -0.128 \\ -0.128 \end{bmatrix}$

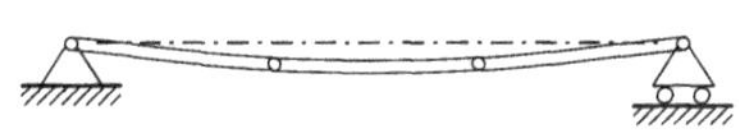

steady state response:
maximum deflection

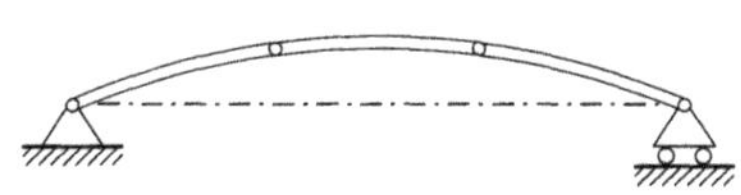

Fig. 6.2.4 Example of a harmonic excitation of an undamped oscillator (sub-critical), *Part 2*

equ. (6.2.28) accordingly, we obtain

$$\begin{aligned} R_{\eta \mathrm{i}}^{(F)} &= \tfrac{1}{2}a_0 + \sum_{\mathrm{k}=1}^{\infty} \big[\tfrac{1}{2}(a_\mathrm{k} - ib_\mathrm{k})(\cos \mathrm{k}\Omega_0 t + i \sin \mathrm{k}\Omega_0 t) \\ &\qquad + \tfrac{1}{2}(a_\mathrm{k} + ib_\mathrm{k})(\cos \mathrm{k}\Omega_0 t - i \sin \mathrm{k}\Omega_0 t)\big] \\ &= \tfrac{1}{2}a_0 + \sum_{\mathrm{k}=1}^{\infty} \big[\tfrac{1}{2}(a_\mathrm{k} - ib_\mathrm{k})\mathrm{e}^{i\mathrm{k}\Omega_0 t} + \tfrac{1}{2}(a_\mathrm{k} + ib_\mathrm{k})\mathrm{e}^{-i\mathrm{k}\Omega_0 t}\big] \end{aligned} \tag{6.2.32}$$

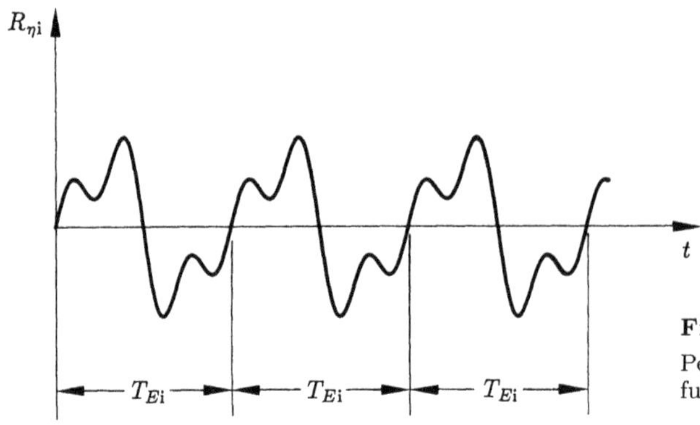

Fig. 6.2.5
Periodic variation of an excitation function, $R_{\eta i}$

The individual terms can be simplified if we allow for negative k-values. Introducing the definitions

$$c_k = \tfrac{1}{2}\left[a_k - \mathrm{sign}(k) i b_k\right] , \quad k = -\infty \cdots +\infty$$
$$c_k = \tfrac{1}{2} a_0 , \qquad k = 0 \tag{6.2.33}$$

we find the concise result

$$R_{\eta i}^{(F)} = \sum_{k=-\infty}^{+\infty} c_k e^{ik\Omega_0 t} \tag{6.2.34}$$

This is the most compact possible Fourier representation of an excitation. Naturally, it is also feasible to evaluate c_k directly instead of using equs. (6.2.29) and (6.2.30). Reapplication of Euler's formula yields

$$c_k = \tfrac{1}{2} \int_{-1}^{+1} R_{\eta i}(\tau) e^{-ik\pi\tau} d\tau \tag{6.2.35}$$

At this point, the reader should be reminded of the frequency response characteristic which determines the response for a harmonic excitation in a very simple manner, see equ. (6.2.26). In fact, in this way, we obtain the particular solution or steady-state part of the response as

$$\eta_{pi} = \sum_{k=-\infty}^{+\infty} F_{ik} c_k e^{ik\Omega_0 t}$$
$$= \sum_{k=-\infty}^{+\infty} V_{ik} c_k e^{i(k\Omega_0 t + \psi_i)} \tag{6.2.36}$$

where

$$V_{\text{ik}} = \left| \frac{1}{1 - (\text{k}\Omega_0/\omega_\text{i})^2} \right| \tag{6.2.37}$$

and

$$\psi_{\text{ik}} = 0 \; , \qquad \text{if } |\text{k}\Omega_0| < \omega_\text{i}$$

$$\psi_{\text{ik}} = 180^\circ = \pi \; , \qquad \text{if } |\text{k}\Omega_0| > \omega_\text{i} \tag{6.2.38}$$

are to be noted. Equations (6.2.34) and (6.2.36) describe the excitation and the response of the system strictly for an infinite Fourier series. In practice, the series has to be terminated after, say, nf number of terms. The decisive issue here is the convergence characteristics not of the excitation but of the system's response, which is in general better (see the example in fig. 1.4.8).

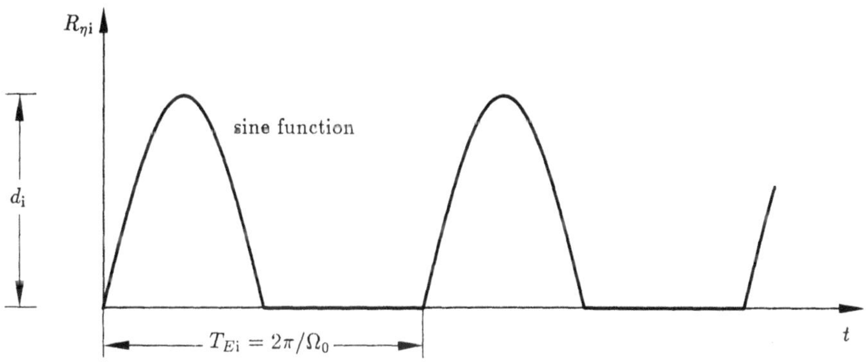

Fig. 6.2.6 Expansion of an excitation function into a Fourier series

Let us now conclude this section by applying the aforementioned technique to a simple example. Figure 6.2.6 illustrates a periodic excitation. In order to familiarise the reader with complex operations, we apply this technique in what follows. Consequently, we write the excitation function as

$$R_{\eta\text{i}}(\tau) = 0 \qquad -1 \leqslant \tau \leqslant 0$$

$$R_{\eta\text{i}}(\tau) = d_\text{i} \sin \pi\tau = \frac{d_\text{i}}{2i}(\text{e}^{i\pi\tau} - \text{e}^{-i\pi\tau}) \qquad 0 \leqslant \tau \leqslant 1$$

The coefficients c_k of the complex Fourier series derive from equ. (6.2.35)

$$c_\text{k} = \frac{d_\text{i}}{4i} \int_0^1 (\text{e}^{i\pi\tau(1-\text{k})} - \text{e}^{-i\pi\tau(1+\text{k})}) \text{d}\tau$$

When integrating, the special conditions for $k = \pm 1$ have to be taken into account. We thus focus our attention first upon this special case and obtain

$$c_{-1} = \frac{d_i}{4i}\int_0^1 (e^{i2\pi\tau} - 1)d\tau = -\frac{d_i}{4i}$$

Analogously, we have

$$c_{+1} = +\frac{d_i}{4i}$$

All other coefficients can be obtained by means of direct integration

$$c_k = \frac{d_i}{4i}\left[\frac{1}{i\pi(1-k)}(e^{i\pi(1-k)} - 1) + \frac{1}{i\pi(1+k)}(e^{-i\pi(1+k)} - 1)\right]$$

We also note

$$\begin{aligned} e^{i\pi(1-k)} &= e^{i\pi}e^{-i\pi k} = -e^{-i\pi k} \\ &= -\cos\pi k \end{aligned}$$

and correspondingly

$$e^{-i\pi(1+k)} = -\cos\pi k$$

For all odd-numbered k-values, $\cos\pi k$ is equal to -1 and for all even-numbered values it is equal to $+1$. This yields

$$c_k = 0, \qquad \text{if } |k| \text{ odd and } |k| \neq 1$$

$$c_k = \frac{d_i}{2\pi}\left(\frac{1}{1-k} + \frac{1}{1+k}\right) = \frac{d_i}{\pi}\frac{1}{1-k^2}, \qquad \text{if } |k| \text{ even}$$

Thus, the Fourier series of equ. (6.2.34) becomes

$$\begin{aligned} R_{\eta i}^{(F)} &= \sum_{k=-\infty}^{+\infty} c_k e^{ik\Omega_0 t} \\ &= \frac{d_i}{\pi} + \frac{d_i}{4i}\left(e^{i\Omega_0 t} - e^{-i\Omega_0 t}\right) + \frac{d_i}{\pi}\sum_{k=2,4,}^{\text{even}} \frac{1}{1-k^2}\left(e^{ik\Omega_0 t} + e^{-ik\Omega_0 t}\right) \\ &= \frac{d_i}{\pi} + \frac{d_i}{2}\sin\Omega_0 t + 2\frac{d_i}{\pi}\sum_{k=2,4,}^{\text{even}} \frac{\cos k\Omega_0 t}{1-k^2} \\ &= \frac{d_i}{\pi}\left(1 + \frac{\pi}{2}\sin\Omega_0 t - \frac{2}{3}\cos 2\Omega_0 t - \frac{2}{15}\cos 4\Omega_0 t \ldots\right) \end{aligned}$$

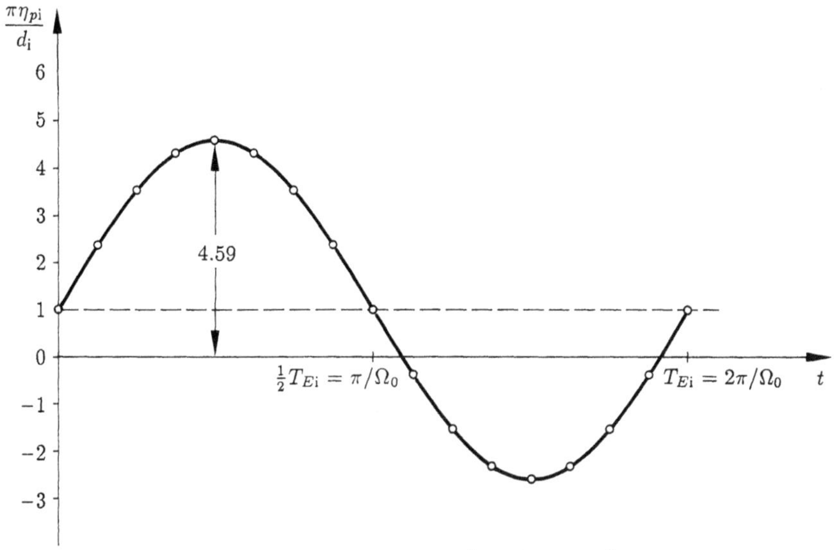

(a) response for 2 series terms $\frac{\pi\eta_{pi}}{d_i} = \left(1 + \frac{8\pi}{7}\sin\Omega_0 t\right)$

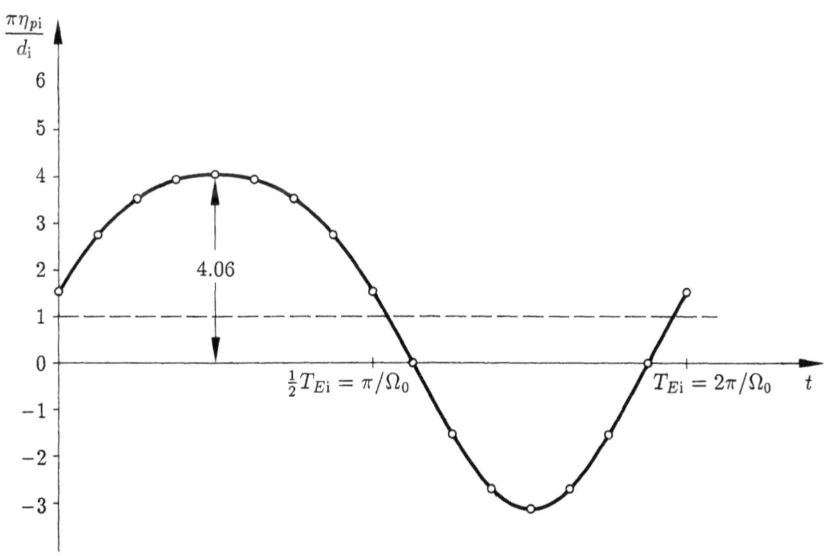

(b) response for 3 series terms $\frac{\pi\eta_{pi}}{d_i} = \left(1 + \frac{8\pi}{7}\sin\Omega_0 t + \frac{8}{15}\cos 2\Omega_0 t\right)$

Fig. 6.2.7 Influence of the number of expansion terms on the response for the excitation shown in fig. 6.2.6

The particular response of the system derives from equ. (6.2.36) for a given ratio of excitation frequency Ω_0 to eigenfrequency ω_i. For undamped structures, $k\Omega_0$ should obviously not coincide with the eigenfrequency ω_i, otherwise resonance would occur. Adopting the ratio

$$\frac{\Omega_0}{\omega_i} = \frac{3}{4}$$

we find

$$\begin{array}{lll} F_{i0} = 1 \;, & V_{i0} = 1 \;, & \psi_{i0} = 0 \\ F_{i1} = \frac{16}{7} \;, & V_{i1} = \frac{16}{7} \;, & \psi_{i1} = 0 \\ F_{i2} = -\frac{4}{5} \;, & V_{i2} = \frac{4}{5} \;, & \psi_{i2} = \pi \\ F_{i4} = -\frac{1}{8} \;, & V_{i4} = \frac{1}{8} \;, & \psi_{i4} = \pi \qquad \text{etc.} \end{array}$$

This indicates that the excitation is in part sub-critical and in part over-critical. Thus, in the present case, the steady-state response (particular solution) becomes

$$\eta_{pi}(t) = \frac{d_i}{\pi}\left(1 + \frac{8\pi}{7}\sin\Omega_0 t + \frac{8}{15}\cos 2\Omega_0 t + \frac{1}{60}\cos 4\Omega_0 t \ldots\right)$$

The first term reproduces the static response due to the applied excitation. The second term is clearly too significant to be ignored. Even if the series is truncated at this point, the maximum error in the deflection is still limited to approximately 10%. The contribution of the third term is still noticeable. The influence of the fourth term is, however, no longer visually perceptible in the graphic representation of fig. 6.2.7.

6.3 Non-periodic excitation functions (Fourier integral)

The transition from arbitrary periodic excitations to more general (non-periodic) excitations, discussed briefly in section 1.4.3, is demonstrated here with a complex formulation. In fig. 6.3.1 (a), we reproduce such a general non-periodic excitation function $R_{\eta i}(t)$. No period of time, T_{Ei}, is discernable which defines a periodic excitation pattern. In contrast, fig. 6.3.1(b) displays a periodic excitation. Up to an interval T_{Ei}, the pattern is as before, but subsequently the repetition sets in, yielding a marked deviation from the original non-periodic excitation. To overcome possible conceptual difficulties, we may use the device of selecting a very large T_{Ei}, thus postponing the onset of repetition. Hence the operation

$$T_{Ei} \to \infty \tag{6.3.1}$$

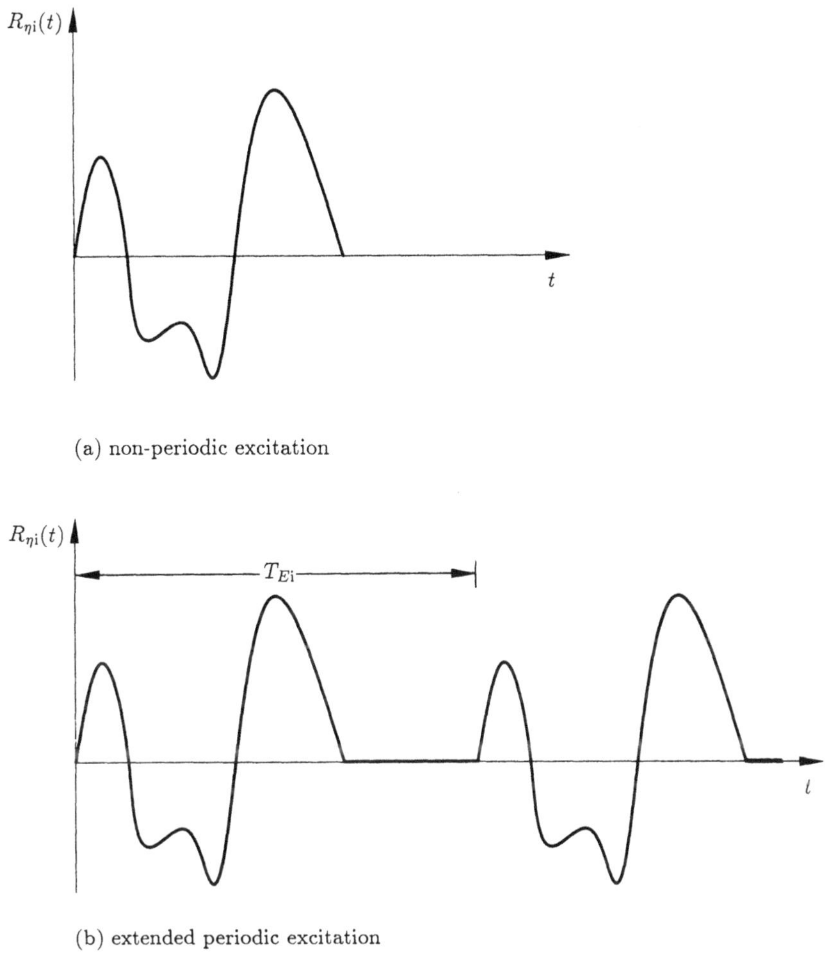

Fig. 6.3.1 Non-periodic and analogous periodic excitation

defines the transition from a periodic to a non-periodic excitation form. On the other hand, a very large period implies a very low circular frequency, Ω_0, and thus a finely stepped Fourier series. In the limit of an infinite period, the value Ω_0 becomes infinitesimal

$$\Omega_0 \to \mathrm{d}\Omega \tag{6.3.2}$$

Consequently, the sequence of the frequencies $k\Omega_0$ appearing in the terms of the series gradually tends to the limit

$$k\Omega_0 \to \Omega \tag{6.3.3}$$

Following equs. (6.2.34) and (6.2.35), we now write the Fourier series in a form suitable for the limes operation. In this way, we obtain

$$R_{\eta i}^{(F)}(t) = \sum_{k=-\infty}^{+\infty} \left(\frac{1}{2} \int_{-1}^{+1} R_{\eta i}(\tau)\, e^{-ik\pi\tau} d\tau \right) e^{ik\Omega_0 t}$$

$$= \sum_{k=-\infty}^{+\infty} \frac{1}{T_{Ei}} \int_{-T_{Ei}/2}^{+T_{Ei}/2} \left(R_{\eta i}(t)\, e^{-ik\Omega_0 t} dt \right) e^{ik\Omega_0 t}$$

$$= \sum_{k=-\infty}^{+\infty} \frac{\Omega_0}{2\pi} \int_{-T_{Ei}/2}^{+T_{Ei}/2} \left(R_{\eta i}(t)\, e^{-ik\Omega_0 t} dt \right) e^{ik\Omega_0 t}$$

Taking the limit, we find

$$R_{\eta i}^{(F)}(t) = \int_{-\infty}^{+\infty} \frac{1}{2\pi} \int_{-\infty}^{+\infty} \left(R_{\eta i}(t)\, e^{-i\Omega t} dt \right) e^{i\Omega t} d\Omega \tag{6.3.4}$$

In addition to the initial assumptions of equs. (6.3.1) to (6.3.3), we applied here the transition of the finely stepped series into an integral. In this way, we generate non-periodic excitation forms. We define

$$k_i(\Omega) = \frac{1}{2\pi} \int_{-\infty}^{+\infty} R_{\eta i}(t)\, e^{-i\Omega t} dt \tag{6.3.5}$$

as the frequency spectrum of the exitation. This replaces the discrete c_k-values of the original Fourier series. For those who dislike the operations with complex numbers, it is advantageous to split equ. (6.3.5) into two real integrals using Euler's formula

$$k_i^{(c)}(\Omega) = \frac{1}{2\pi} \int_{-\infty}^{+\infty} R_{\eta i}(t) \cos \Omega t \, dt$$

$$k_i^{(s)}(\Omega) = \frac{1}{2\pi} \int_{-\infty}^{+\infty} R_{\eta i}(t) \sin \Omega t \, dt \tag{6.3.6}$$

We note that

$$k_i(\Omega) = k_i^{(c)} - ik_i^{(s)} \tag{6.3.7}$$

The integral of equ. (6.3.4) or its more compact formulation

$$R_{\eta \mathrm{i}}^{(F)}(t) = \int_{-\infty}^{+\infty} k_{\mathrm{i}}(\Omega)\, \mathrm{e}^{i\Omega t} \mathrm{d}\Omega \tag{6.3.8}$$

is now termed the Fourier integral. Naturally, its representation in real variables is equally feasible as for the Fourier series

$$\begin{aligned} R_{\eta \mathrm{i}}^{(F)}(t) &= \int_{0}^{+\infty} \left(k_{\mathrm{i}}(-\Omega)\, \mathrm{e}^{-i\Omega t} + k_{\mathrm{i}}(\Omega)\, \mathrm{e}^{i\Omega t}\right) \mathrm{d}\Omega \\ &= \int_{0}^{+\infty} \left[\left(k_{\mathrm{i}}^{(c)} + i k_{\mathrm{i}}^{(s)}\right) \mathrm{e}^{-i\Omega t} + \left(k_{\mathrm{i}}^{(c)} - i k_{\mathrm{i}}^{(s)}\right) \mathrm{e}^{i\Omega t}\right] \mathrm{d}\Omega \\ &= \int_{0}^{+\infty} \left[k_{\mathrm{i}}^{(c)} \left(\mathrm{e}^{-i\Omega t} + \mathrm{e}^{i\Omega t}\right) + i k_{\mathrm{i}}^{(s)} \left(\mathrm{e}^{-i\Omega t} - \mathrm{e}^{i\Omega t}\right)\right] \mathrm{d}\Omega \\ &= 2 \int_{0}^{+\infty} \left(k_{\mathrm{i}}^{(c)} \cos \Omega t + k_{\mathrm{i}}^{(s)} \sin \Omega t\right) \mathrm{d}\Omega \end{aligned} \tag{6.3.9}$$

The Fourier integral of equs. (6.3.8) and (6.3.9) can thus be considered a harmonic excitation with an infinite sequence of forcing frequencies, Ω. The response to such an excitation is also known: each excitation component merely has to be multiplied by the frequency response characteristic and the entire particular solution then summed up or, more precisely, integrated. Consequently, we obtain

$$\eta_{p\mathrm{i}} = \int_{-\infty}^{+\infty} F_{\mathrm{i}}(\Omega) k_{\mathrm{i}}(\Omega)\, \mathrm{e}^{i\Omega t} \mathrm{d}\Omega \tag{6.3.10}$$

where

$$F_{\mathrm{i}}(\Omega) = \frac{1}{1 - (\Omega/\omega_{\mathrm{i}})^2} \tag{6.3.11}$$

is now a continuous function.

This integration can be split further into sub-critical and over-critical excitation regimes. In the sub-critical state, the phase shift of the response is zero whilst in the over-critical state it is 180°. Equation (6.3.10) can be written equally in a real form

$$\eta_{pi}^{(F)}(t) = \int_0^{+\infty} F_i(\Omega)\left(k_i(-\Omega)\,e^{-i\Omega t} + k_i(\Omega)\,e^{i\Omega t}\right)d\Omega$$

$$= \int_0^{\omega_i} V_i(\Omega)\left(k_i^{(c)}\cos\Omega t + k_i^{(s)}\sin\Omega t\right)d\Omega$$

$$+ \int_{\omega_i}^{\infty} V_i(\Omega)\left(k_i^{(c)}\cos(\Omega t+\pi) + k_i^{(s)}\sin(\Omega t+\pi)\right)d\Omega \qquad (6.3.12)$$

where

$$V_i(\Omega) = |F_i(\Omega)| \qquad (6.3.13)$$

A final and important point to be noted is that the convergence is only ensured if the area under $R_{\eta i}(t)$ is bounded

$$\int_{-\infty}^{+\infty} |R_{\eta i}(t)|dt < \infty \qquad (6.3.14)$$

Unfortunately, the aforementioned technique is not easily realised. Anticipating the discussion in the next section, we illustrate this by applying upon the structure a force histogram in the form of a rectangular block (see fig. 6.3.2) extending over a limited time interval. The functional representation of the load histogram is

$$R_{\eta i} = d_i\,, \qquad 0 \leqslant t \leqslant t_1$$

$$R_{\eta i} = 0\,, \qquad -\infty \leqslant t < 0\,, \quad t_1 < t \leqslant +\infty$$

The complex frequency spectrum of the excitation is correspondingly

$$k_i(\Omega) = \frac{1}{2\pi}\int_0^{t_1} d_i e^{-i\Omega t}dt = \frac{d_i}{2\pi\Omega i}(1 - e^{-i\Omega t_1})$$

$$= k_i^{(c)} - ik_i^{(s)}$$

with the real and imaginary terms

$$k_i^{(c)} = \frac{d_i t_1}{2\pi}\frac{\sin\Omega t_1}{\Omega t_1}$$

$$k_i^{(s)} = \frac{d_i t_1}{2\pi}\frac{1-\cos\Omega t_1}{\Omega t_1}$$

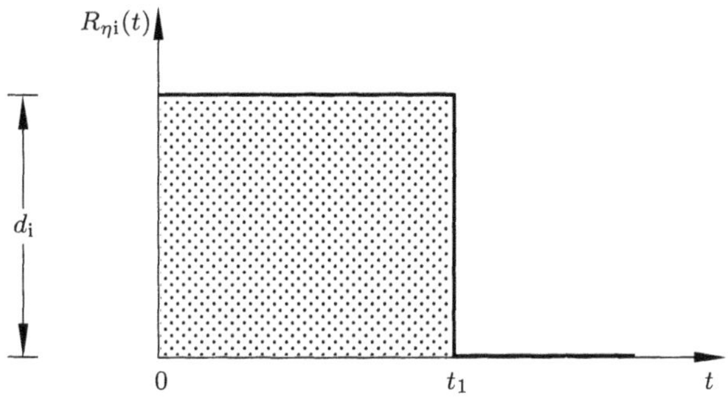

Fig. 6.3.2
Example of a non-periodic excitation (rectangular impulse of finite duration)

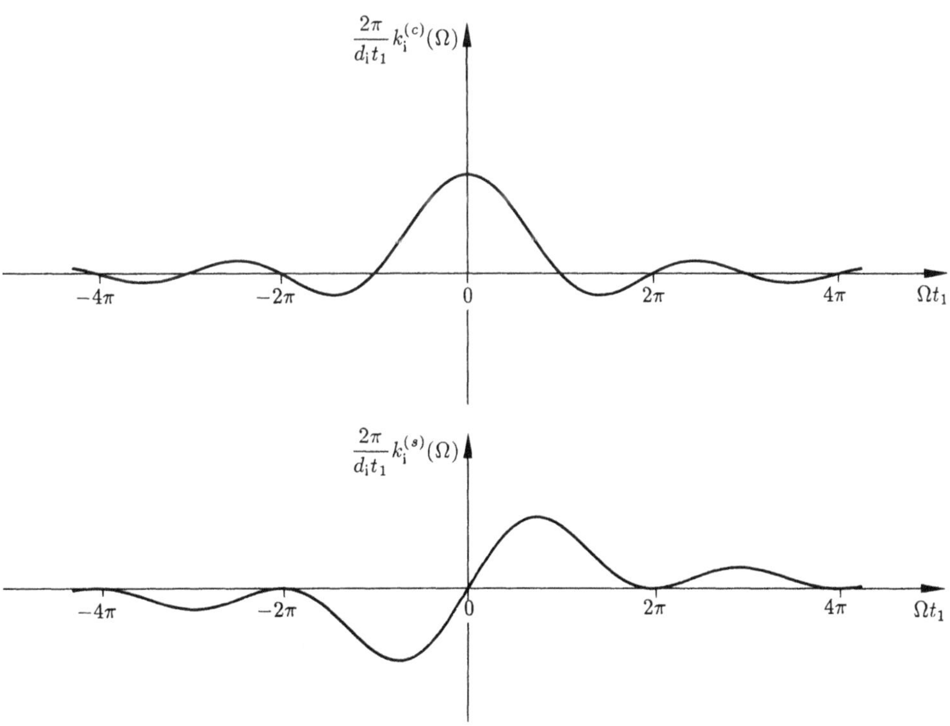

Fig. 6.3.3 Real part and imaginary part of the excitation frequency spectrum for a rectangular impulse of duration t_1

shown in fig. 6.3.3. The complex representation of the particular solution is simply

$$\eta_{\mathrm{pi}}^{(F)}(t) = \int\limits_{-\infty}^{+\infty} \frac{1}{1-(\Omega_{\mathrm{i}}/\omega_{\mathrm{i}})^2} \frac{d_{\mathrm{i}}}{2\pi\Omega i}(1-\mathrm{e}^{-i\Omega t_1})\,\mathrm{e}^{i\Omega t}\mathrm{d}\Omega$$

$$= \frac{d_{\mathrm{i}}\omega_{\mathrm{i}}^2}{2\pi i} \int\limits_{-\infty}^{+\infty} \frac{1}{\Omega(\omega_{\mathrm{i}}-\Omega)(\omega_{\mathrm{i}}+\Omega)} \left(\mathrm{e}^{i\Omega t} - \mathrm{e}^{i\Omega(t-t_1)}\right) \mathrm{d}\Omega$$

At first sight, this expression does not appear to allow a straighforward integration. However, developing the expression in the factors of the polynomial denominator and comparing the coefficients, we find

$$\frac{1}{\Omega(\omega_{\mathrm{i}}-\Omega)(\omega_{\mathrm{i}}+\Omega)} = \frac{1}{2\omega_{\mathrm{i}}^2}\left(\frac{2}{\Omega} + \frac{1}{\omega_{\mathrm{i}}-\Omega} - \frac{1}{\omega_{\mathrm{i}}+\Omega}\right)$$

In this way, we obtain integrals of the form

$$\int\limits_{-\infty}^{+\infty} \frac{\mathrm{e}^{iax}}{x}\mathrm{d}x = \int\limits_{-\infty}^{+\infty}\left(\frac{\cos ax}{x} + i\frac{\sin ax}{x}\right)\mathrm{d}x = 2i\int\limits_{0}^{+\infty}\frac{\sin ax}{x}\mathrm{d}x$$

$$= \begin{cases} i\pi, & \text{if } \ a>0 \\ -i\pi, & \text{if } \ a<0 \end{cases}$$

or

$$= \mathrm{sign}(a)\, i\pi$$

Hence, the overall first integral becomes

$$\int\limits_{-\infty}^{+\infty}\left(\frac{2}{\Omega} + \frac{1}{\omega_{\mathrm{i}}-\Omega} - \frac{1}{\omega_{\mathrm{i}}+\Omega}\right)\mathrm{e}^{i\Omega t}\mathrm{d}\Omega = \mathrm{sign}(t)\, i\pi\left(2 - \mathrm{e}^{i\omega_{\mathrm{i}}t} - \mathrm{e}^{-i\omega_{\mathrm{i}}t}\right)$$

and the second one reads

$$\int\limits_{-\infty}^{+\infty}\left(\frac{2}{\Omega} + \frac{1}{\omega_{\mathrm{i}}-\Omega} - \frac{1}{\omega_{\mathrm{i}}+\Omega}\right)\mathrm{e}^{i\Omega(t-t_1)}\mathrm{d}\Omega = \mathrm{sign}(t-t_1)\, i\pi\left(2 - \mathrm{e}^{i\omega_{\mathrm{i}}(t-t_1)} - \mathrm{e}^{-i\omega_{\mathrm{i}}(t-t_1)}\right)$$

The formulation of the particular response can then be tabulated conveniently for three time intervals

$$-\infty < t \leqslant 0: \qquad \eta_{\mathrm{pi}} = \tfrac{1}{2}d_{\mathrm{i}}\{\cos\omega_{\mathrm{i}}t - \cos\omega_{\mathrm{i}}(t-t_1)\}$$

$$0 \leqslant t \leqslant t_1: \qquad \eta_{\mathrm{pi}} = \tfrac{1}{2}d_{\mathrm{i}}\{2 - \cos\omega_{\mathrm{i}}t - \cos\omega_{\mathrm{i}}(t-t_1)\}$$

$$t_1 \leqslant t < \infty : \qquad \eta_{p\mathrm{i}} = \tfrac{1}{2} d_\mathrm{i} \{ -\cos \omega_\mathrm{i} t + \cos \omega_\mathrm{i} (t - t_1) \}$$

This particular solution may somewhat surprise the reader. The differential equation of equ. (6.1.7) and the special excitation function, $R_{\eta\mathrm{i}}$, (see the definition of the rectangular impulse above) suggest the following 'pure' particular solution

$$\begin{aligned} \eta_{p\mathrm{i}} &= 0 && \text{for} \quad t < 0 \\ \eta_{p\mathrm{i}} &= d_\mathrm{i} && \text{for} \quad 0 \leqslant t \leqslant t_1 \\ \eta_{p\mathrm{i}} &= 0 && \text{for} \quad t > t_1 \end{aligned}$$

with points of discontinuity at the instants $t = 0$ and $t = t_1$. In this particular solution, 'pure' means that the solution comprises no homogeneous part whatsoever. Our alternative form of the particular solution thus consists of the 'pure' solution plus homogeneous contributions (cosine and sine terms with the eigenfrequency, ω_i). In this way, the solution becomes continuous over the entire range $(-\infty, +\infty)$.

The complete response of the system due to the rectangular impulse can be obtained now by superposition of the free oscillations induced by the initial conditions appertaining to the aforementioned time intervals. For example, we can assume that

$$\begin{aligned} \eta_\mathrm{i}(t = 0) &= 0 \\ \dot{\eta}_\mathrm{i}(t = 0) &= 0 \end{aligned}$$

which specify a state of rest at the onset of the rectangular impulse. If we introduce these conditions into the complete solution of the first interval of time $(t \leqslant 0)$

$$\begin{aligned} \eta_\mathrm{i} &= \eta_{p\mathrm{i}} + \eta_{h\mathrm{i}} \\ &= \tfrac{1}{2} d_\mathrm{i} \Big(\cos \omega_\mathrm{i} t - \cos \omega_\mathrm{i} (t - t_1) \Big) + a \sin \omega_\mathrm{i} t + b \sin \omega_\mathrm{i} t \end{aligned}$$

we obtain a zero response

$$\eta_\mathrm{i}(t) = 0 \qquad \text{for } t \leqslant 0$$

Substituting the aforementioned initial conditions into the complete solution over the second time interval, $0 \leqslant t \leqslant t_1$, we find

$$\eta_\mathrm{i}(t) = d_\mathrm{i}(1 - \cos \omega_\mathrm{i} t) \qquad \text{if } 0 \leqslant t \leqslant t_1$$

yielding the specific response at time t_1

$$\begin{aligned} \eta_\mathrm{i}(t_1) &= d_\mathrm{i}(1 - \cos \omega_\mathrm{i} t_1) \\ \dot{\eta}_\mathrm{i}(t_1) &= d_\mathrm{i} \omega_\mathrm{i} \sin \omega_\mathrm{i} t \end{aligned}$$

These supply the initial conditions for the third time interval, $t \geqslant t_1$, for which we obtain the response

$$\eta_\mathrm{i}(t) = d_\mathrm{i} \Big(\cos \omega_\mathrm{i} (t - t_1) - \cos \omega_\mathrm{i} t \Big) \qquad \text{for } t \geqslant t_1$$

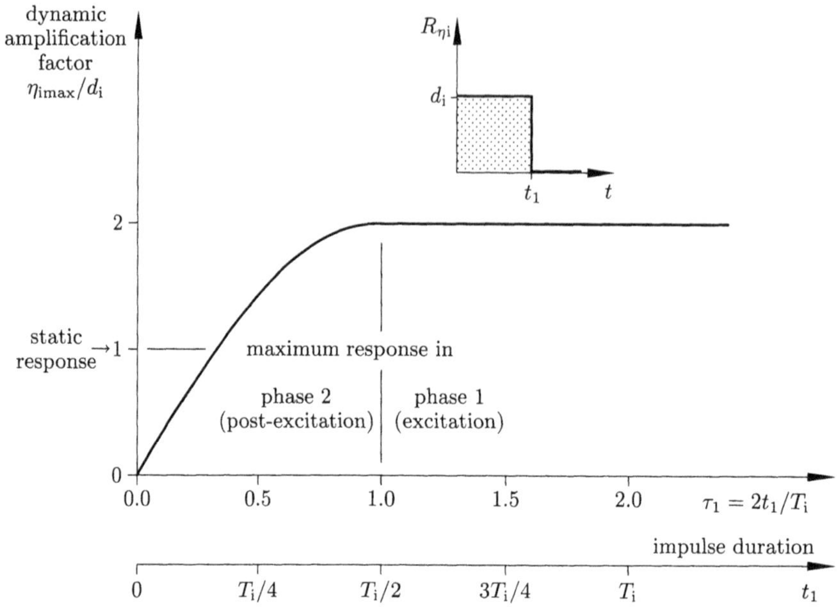

Fig. 6.3.4 Dynamic amplification factor for a rectangular impulse

Whilst the response until the instant t_1 – provided, of course, that t_1 is not too large and real damping is not too extreme – is relatively realistic, the subsequent free vibration is, in reality, damped out.

The determination of the maximum deflection is also a matter of interest. This value depends significantly upon the ratio

$$\tau_1 = 2\frac{t_1}{T_i}$$

Introducing also the corresponding dimensionless time τ

$$\tau = 2\frac{t}{T_i}$$

we obtain the response

$$\frac{\eta_i}{d_i} = 1 - \cos \pi\tau \qquad \text{for } 0 \leqslant \tau \leqslant \tau_1$$

$$\frac{\eta_i}{d_i} = \frac{\dot{\eta}_i(\tau_1)}{d_i\omega_i} \sin \pi(\tau - \tau_1) + \frac{\eta_i(\tau_1)}{d_i} \cos \pi(\tau - \tau_1) \qquad \text{for } \quad \tau \geqslant \tau_1$$

The amplitude is seen to increase steadily during the active application of the impulse until τ_1, as long as $\tau_1 \leqslant 1$ (i.e. $t_1 \leqslant T_i/2$). For an impulse of duration $\tau_1 = 1$ (and longer action), the maximum value of 2 for the non-dimensional amplitude is obtained in the first phase, otherwise this value is smaller

$$\left.\begin{array}{ll}\eta_{\mathrm{i\,max}}/d_{\mathrm{i}} = 2 & \text{for } \tau_1 \geqslant 1 \\ \eta_{\mathrm{i\,max}}/d_{\mathrm{i}} = 1 - \cos\pi\tau_1 & \text{for } \tau_1 < 1\end{array}\right\} \text{ and } 0 \leqslant \tau \leqslant \tau_1$$

In the subsequent free oscillatory phase, the maximum deflection is given by

$$\eta_{\mathrm{i\,max}}/d_{\mathrm{i}} = \left[\left(\frac{\dot{\eta}_{\mathrm{i}}(\tau_1)}{d_{\mathrm{i}}\omega_{\mathrm{i}}}\right)^2 + \left(\frac{\eta_{\mathrm{i}}(\tau_1)}{d_{\mathrm{i}}}\right)^2\right]^{1/2} = 2\sin\left(\frac{\pi}{2}\tau_1\right) \qquad \text{if } \tau_1 \leqslant \tau$$

Since $\eta_{\mathrm{i}} = d_{\mathrm{i}}$ represents the static deflection for a static force $R_{\eta\mathrm{i}} = d_{\mathrm{i}}$, the ratio $\eta_{\mathrm{i\,max}}/d_{\mathrm{i}}$ fulfils a function similar to that of the amplification factor for a harmonic excitation. This ratio is consequently often denoted the dynamic amplification factor. If the rectangular impulse extends beyond $t_1 = T_{\mathrm{i}}/2$, the amplification factor is always 2 and the maximum deflection occurs during the interval of active excitation (phase 1). If, on the other hand, the rectangular impulse is shorter than $t_1 = T_{\mathrm{i}}/2$, then the amplification factor lies between 0 and 2 and is achieved during the free oscillatory phase (fig. 6.3.4).

The above information is summarised once more in fig. 6.3.5, in which the middle diagram depicts the transition. The third diagram is of particular interest: when the impulse duration equals the period of the eigenoscillation, phase 2 remains completely motionless (state of rest) since both $\eta(\tau_1)$ and $\dot{\eta}(\tau_1)$ are zero at the initiation of this phase.

We referred previously to the complicatedness that may arise when applying Fourier integrals to such problems. Consequently, the reader may at this stage be interested to be informed of an alternative and faster solution to the problem in question (fig. 6.3.6). Here, the rectangular impulse block is simulated by superposition of two simple, time-shifted step functions. The first step function leads to the general solution

$$\eta_{\mathrm{i}}^{(1)}(t) = a_{\mathrm{i}}\sin\omega_{\mathrm{i}}t + b_{\mathrm{i}}\cos\omega_{\mathrm{i}}t + d_{\mathrm{i}} \qquad t \geqslant 0$$

or, if the system is motionless at $t = 0$, to the special solution

$$\eta_{\mathrm{i}}^{(1)}(t) = d_{\mathrm{i}}(1 - \cos\omega_{\mathrm{i}}t) \qquad t \geqslant 0$$

This expression represents the complete solution until the second step-function is initiated at the instant $t = t_1$. The special solution for the latter can immediately be written as

$$\eta_{\mathrm{i}}^{(2)}(t) = -d_{\mathrm{i}}\big(1 - \cos\omega_{\mathrm{i}}(t - t_1)\big) \qquad t \geqslant t_1$$

The overall solution is hence

$$\eta_{\mathrm{i}}(t) = d_{\mathrm{i}}\,(1 - \cos\omega_{\mathrm{i}}t) \qquad \text{in phase 1}, \qquad 0 \leqslant t \leqslant t_1$$

$$\eta_{\mathrm{i}}(t) = d_{\mathrm{i}}\big(\cos\omega_{\mathrm{i}}(t - t_1) - \cos\omega_{\mathrm{i}}t\big) \qquad \text{in phase 2}, \qquad t \geqslant t_1$$

which is in agreement with the result obtained by the Fourier integral method.

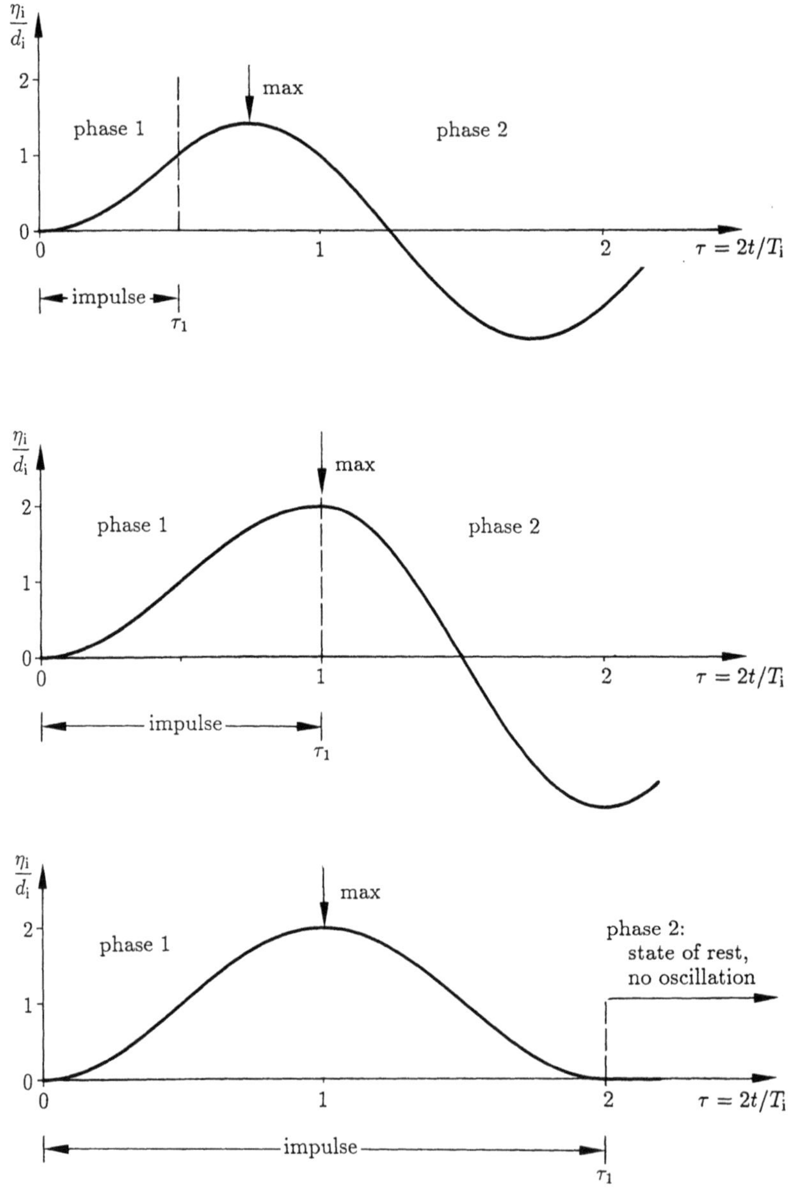

Fig. 6.3.5 Response due to rectangular impulses of varying duration

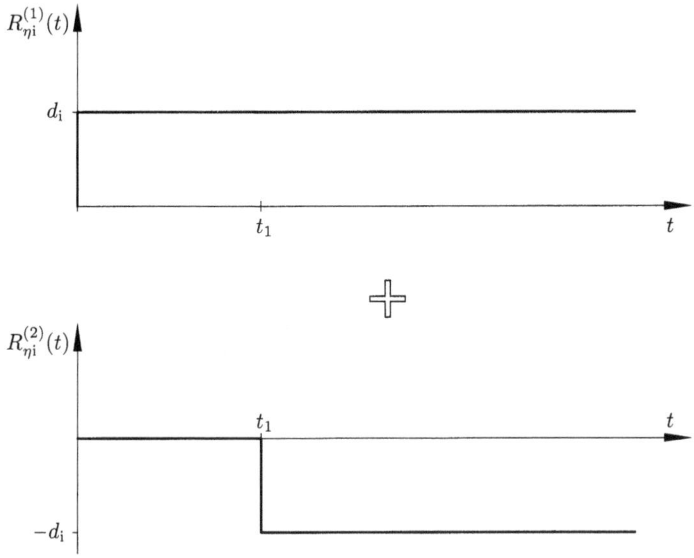

Fig. 6.3.6 Representation of a rectangular impulse by two time-shifted step functions

6.4 General excitation functions, Dirac impulse and Duhamel integral

Impulsive loadings on structures imply short-time applications of forces on the system. Generally, the effects of such forces, whilst being of only short duration, are intense. A car journey along a poor road or a flight in turbulent weather are two characteristic examples of the disagreeable nature of such dynamic effects. Damping plays a less dominant role here than it does in harmonic or periodic excitation where the total response to the system is comprised of both a particular and a homogeneous part. This total response is not really meaningful for practical evaluations, since the homogeneous part is significantly affected by damping (as we shall soon have opportunity to learn). However, the particular solution is of major relevance in engineering calculations, as long as the excitation is sufficiently remote from the point of resonance. These difficulties do not arise for impulsive loads in engineering applications, since the maximum deflection and hence the highest straining of the system is attained in a very short time and is hardly affected by damping effects. But obviously, damping has to be considered in long-range calculations.

The above conditions are clearly reflected in the example of a rectangular impulse analysed in section 6.3. Figure 6.3.5 demonstrates the rapid appearance of the maximum amplitude, namely for $t_1 \leqslant T_i/2$ in phase 2, i.e. in the free oscillatory state. It is evident here that the maximum deflection will have been realised on or before the time $t_1 + T_i/2$. Figure 6.3.5 also indicates a horizontal tangent of the response $\eta_i(t)$ at the start $t = 0$. Consequently, a very brief rectangular impulse can only generate a small deflection. On the other hand, this does not hold for the velocity

$$\dot{\eta}_i(t) = d_i \omega_i \sin \omega_i t \tag{6.4.1}$$

Its gradient at $t = 0$ is $d_i\omega_i^2$ and the velocity accordingly undergoes a significant change as a result of a very short rectangular impulse.

These theoretical considerations lead to a useful approximation. To this purpose, we define the impulse strength as

$$I_i = \int_0^{t_1} R_{\eta i}(t)\mathrm{d}t \,, \qquad \text{in general}$$

$$= d_i t_1 \,, \qquad \text{for a rectangular impulse} \tag{6.4.2}$$

Integration of the differential equation in phase 1 (excitation)

$$\frac{1}{\omega_i^2}\ddot{\eta}_i + \eta_i = d_i \,, \qquad 0 \leqslant t \leqslant t_1 \tag{6.4.3}$$

yields at the instant t_1

$$\Delta\dot{\eta}_i = \omega_i^2\left(d_i t_1 + \int_0^{t_1} \eta_i \mathrm{d}t\right)$$

$$\approx \omega_i^2 I_i \tag{6.4.4}$$

the system being assumed to be initially at rest.

Whilst the strength of the impulse even for very brief durations may become significant due to a large amplitude of the force, d_i, the area below the $\eta_i(t)$-curve is practically zero as a result of the vanishing gradient at the start. In the limiting case, when

$$I_i = \lim_{\substack{t_1 \to 0 \\ d_i \to \infty}} (d_i t_1) = \text{finite} \tag{6.4.5}$$

the force application reduces to the so-called Dirac impulse. Other short-duration impulses with arbitary shape may also be defined approximately by equ. (6.4.2). Once the strength of the impulse, I_i, is known, the detailed variation of $R_{\eta i}(t)$ is then no longer relevant from the practical point of view. We shall return to this point later on in the discussion.

One attractive feature of this theoretical development is that the excitation phase is specified adequately by equ. (6.4.4). We are aware that in any case the maximum displacement occurs in the free oscillatory phase. Applying now the initial conditions

$$\eta_i(t_1) = 0$$

$$\dot{\eta}_i(t_1) = \Delta\dot{\eta}_i = \omega_i^2 I_i \tag{6.4.6}$$

to this phase alone, the response of the system may immediately be expressed in the form

$$\eta_i(t) = \omega_i I_i \sin\omega_i(t - t_1) \,, \qquad t > t_1 \tag{6.4.7}$$

For a Dirac impulse ($t_1 \to 0$), this simplifies further to

$$\eta_i(t) = \omega_i I_i \sin \omega_i t \ , \qquad t > 0 \tag{6.4.8}$$

Note that equ. (6.4.8) is analytically exact whilst equ. (6.4.7) is merely an approximation for impulses of a finite but brief duration, t_1. As a limiting value for t_1, we may adopt $t_1 < T_i/4$. Thus, the first diagram in fig. 6.3.5 satisfies *in extremis* this condition. As a result, we reconsider this example and note that

$$I_i = d_i t_1 = \frac{1}{4} d_i T_i = \frac{\pi}{2} d_i \frac{1}{\omega_i}$$

As an approximation to the response, equ. (6.4.7) then offers the expression

$$\eta_i(t) = \frac{\pi}{2} d_i \sin \omega_i (t - t_1) \ , \qquad t > t_1$$

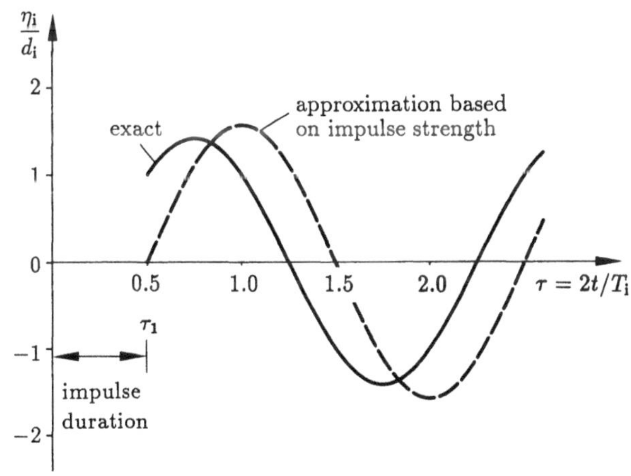

Fig. 6.4.1
Response due to a long-duration impulse

Figure 6.4.1 demonstrates a comparison between the exact and approximated motion. We observe a significant difference in the phase shift. In contrast, the maximum amplitude, which for technical purposes is of primary importance, does not appear to be affected greatly. Its deviation does not exceed 11%. We should also remember that we are operating here at the limit of the permissible range. We therefore investigate one more rectangular impulse of the same strength but of a shorter duration, $\tau_1 = 0.1$ ($t_1 = 0.05\, T_i$). In this case, we obtain the approximation

$$\begin{aligned}\eta_i(t) &= \frac{\pi}{2} d_i \sin \pi(\tau - 0.1) \\ &= 1.57 d_i \sin \pi(\tau - 0.1) \ , \qquad \tau \geqslant 0.1\end{aligned}$$

whilst the exact answer is

$$\begin{aligned}\eta_i(t) &= 5\sin(\pi\tau_1)d_i\left(\sin\pi(\tau-0.1)+\frac{1-\cos\pi\tau_1}{\sin\pi\tau_1}\cos\pi(\tau-0.1)\right)\\ &= 1.55\,d_i\Big(\sin\pi(\tau-0.1)+0.16\cos\pi(\tau-0.1)\Big)\\ &= 1.57\,d_i\sin\pi(\tau+0.06)\ , \qquad \tau\geqslant 0.1\end{aligned}$$

We observe – at least within the present accuracy of 2 decimal places – no difference in the amplitude. Only the phase shift deviates to an appreciable degree. It is evident that the approximation based on the strength of a Dirac impulse is a very useful tool for impulses of short duration.

The Dirac impulse and the corresponding response may be applied to the analysis of general excitation functions. This offers a useful alternative to the more elaborate technique of the Fourier integral method already discussed.

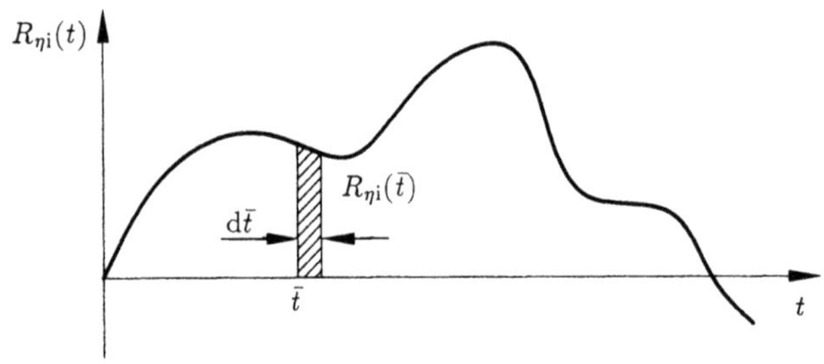

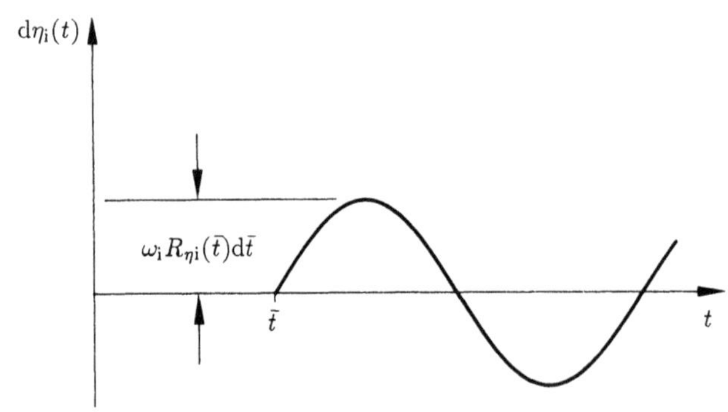

Fig. 6.4.2
General excitation, elemental impulse and associated response

Figure 6.4.2 shows a general excitation which we may interpret as an infinite sequence of impulses. In fact, at the instant $\bar{t}$, we note the instantaneous infinitesimal impulse

$$dI_i = R_{\eta i}(\bar{t})d\bar{t} \tag{6.4.9}$$

which induces the response

$$\begin{aligned} d\eta_i(t) &= \omega_i dI_i \sin\omega_i(t-\bar{t}) \\ &= \omega_i R_{\eta i}(\bar{t}) \sin\omega_i(t-\bar{t}) d\bar{t}\ , \qquad t>\bar{t} \end{aligned} \tag{6.4.10}$$

Proceeding from a state of rest, we obtain by integration over the sequence of all impulses the overall response of the system

$$\eta_i(t) = \omega_i \int_{\bar{t}=0}^{\bar{t}=t} R_{\eta i}(\bar{t}) \sin\omega_i(t-\bar{t}) d\bar{t} \tag{6.4.11}$$

which reproduces the well-known Duhamel integral. If the structure is not at a state of rest at $t = 0$, i.e. at the initiation of the excitation, it is necessary to superpose an additional free oscillation

$$\eta_i(t) = \frac{\dot{\eta}_i(0)}{\omega_i} \sin\omega_i t + \eta_i(0)\cos\omega_i t + \omega_i \int_0^t R_{\eta i}(\bar{t}) \sin\omega_i(t-\bar{t})\, d\bar{t} \tag{6.4.12}$$

We return to the simple expression of equ. (6.4.11) and reshape it for a numerical evaluation. In fact, application of the addition theorems for the trigonometric functions yields

$$\begin{aligned} \eta_i(t) &= \omega_i \int_{\bar{t}=0}^{t} R_{\eta i}(\bar{t})(\sin\omega_i t \cos\omega_i\bar{t} - \cos\omega_i t \sin\omega_i \bar{t})\, d\bar{t} \\ &= \omega_i \Big(\int_{\bar{t}=0}^{t} R_{\eta i}(\bar{t}) \cos\omega_i\bar{t}\, d\bar{t} \Big) \sin\omega_i t - \omega_i \Big(\int_{\bar{t}=0}^{t} R_{\eta i}(\bar{t}) \sin\omega_i\bar{t}\, d\bar{t} \Big) \cos\omega_i t \end{aligned} \tag{6.4.13}$$

To facilitate the calculations, we introduce a dimensionless time

$$\begin{aligned} \bar{\tau} &= \frac{2\bar{t}}{T_i} = \frac{1}{\pi}\omega_i \bar{t} \\ \tau &= \frac{2t}{T_i} = \frac{1}{\pi}\omega_i t \end{aligned} \tag{6.4.14}$$

and note the following integrals

$$\begin{aligned} \bar{a}(\tau) &= \pi \int_0^\tau R_{\eta i}(\bar{\tau}) \cos\pi\bar{\tau}\, d\bar{\tau} \\ \bar{b}(\tau) &= \pi \int_0^\tau R_{\eta i}(\bar{\tau}) \sin\pi\bar{\tau}\, d\bar{\tau} \end{aligned} \tag{6.4.15}$$

In this way, the response may be expressed as

$$\eta_{\mathrm{i}}(\tau) = \bar{a}(\tau)\sin\pi\tau - \bar{b}(\tau)\cos\pi\tau \tag{6.4.16}$$

Numerous procedures exist for the numerical integration of equs. (6.4.15), for example, the methods of Romberg, Gauss, Simpson and the trapezoidal rule, to mention but a few (see also section 1.4.5). In [6.1], the effectiveness of these methods has been examined for a few standard functions. The Gauss integration proved particularly advantageous here. At this point, the reader is referred to the one-dimensional integration scheme for finite elements (fig. 5.4.5 and equ. (5.4.24) in Vol. I, statics). In this connection, the time axis $\bar{\tau}$ is divided into finite ranges or intervals $\Delta\bar{\tau}_{\mathrm{g}}$ as is shown in fig. 6.4.3. The nodes of these 'flange elements' indicate the instants of time for which the response $\eta_{\mathrm{i}}(\tau)$ has to be evaluated. The partition into intervals or elements of time need not necessarily be based on a uniform scheme. A single element of time yields the contribution

$$\Delta\bar{a}_{\mathrm{g}} = \frac{\pi}{2}\sum_{\mathrm{k=1}}^{\mathrm{na}} w_{\mathrm{k}} f_c(\zeta_{\mathrm{k}})$$

$$\Delta\bar{b}_{\mathrm{g}} = \frac{\pi}{2}\sum_{\mathrm{k=1}}^{\mathrm{na}} w_{\mathrm{k}} f_s(\zeta_{\mathrm{k}}) \tag{6.4.17}$$

Here, w_{k} are the weighting factors and the sums run over the field or integration points of the Gauss scheme. Also,

$$f_c = R_{\eta\mathrm{i}}(\bar{\tau})\cos\pi\bar{\tau}$$

$$f_s = R_{\eta\mathrm{i}}(\bar{\tau})\sin\pi\bar{\tau} \tag{6.4.18}$$

are the so-called field values. The time variable within an element g is thereby given by

$$\bar{\tau}_{\mathrm{k}} = \sum_{\mathrm{j=1}}^{\mathrm{g-1}} \Delta\bar{\tau}_{\mathrm{j}} + \tfrac{1}{2}(1+\zeta_{\mathrm{k}})\Delta\tau_{\mathrm{g}} \tag{6.4.19}$$

Following h time steps, i.e. at the instant $\tau_{\mathrm{h}} = \sum_{\mathrm{g=1}}^{\mathrm{h}} \Delta\tau_{\mathrm{g}}$, the response is defined by

$$\bar{a}(\tau_{\mathrm{h}}) = \sum_{\mathrm{g=1}}^{\mathrm{h}} \Delta\bar{a}_{\mathrm{g}}$$

$$\bar{b}(\tau_{\mathrm{h}}) = \sum_{\mathrm{g=1}}^{\mathrm{h}} \Delta\bar{b}_{\mathrm{g}} \tag{6.4.20}$$

The numerical evaluation of the Duhamel integral hence poses fewer problems than that of the Fourier integral.

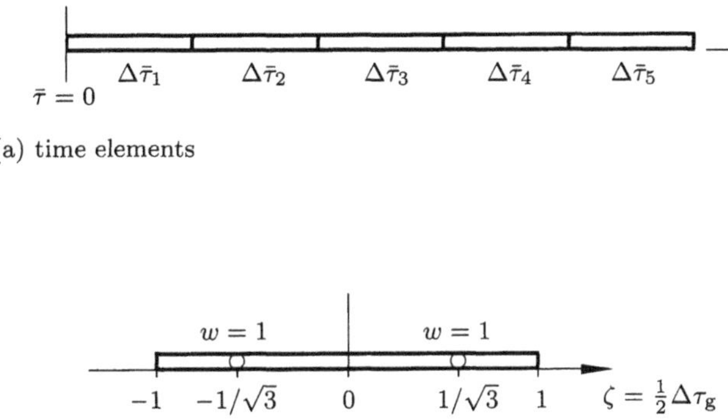

(a) time elements

(b) single time element, pivotal points and weighting factors for two-point integration

Fig. 6.4.3 Discretisation of the time-axis and Gauss integration over the time-element (example for 2 point rule)

The technique is illustrated by a small example, which can easily be integrated exactly. This example, displayed in fig. 6.3.2, shows a rectangular impulse of arbitrary duration. Introducing

$$R_{\eta i}(\bar{\tau}) = d_i \;, \qquad 0 \leqslant \bar{\tau} \leqslant \bar{\tau}_1$$
$$R_{\eta i}(\bar{\tau}) = 0 \;, \qquad \bar{\tau} > \bar{\tau}_1$$

we obtain

$$\left.\begin{aligned} \bar{a}(\tau) &= d_i \sin \pi\tau \;, \\ \bar{b}(\tau) &= d_i(1 - \cos \pi\tau) \;, \end{aligned}\right. \qquad 0 \leqslant \tau \leqslant \tau_1$$

and

$$\left.\begin{aligned} \bar{a}(\tau) &= d_i \sin \pi\tau_1 \;, \\ \bar{b}(\tau) &= d_i(1 - \cos \pi\tau_1) \;, \end{aligned}\right. \qquad \tau \geqslant \tau_1$$

The above data immediately yield the response

$$\eta_i(\tau) = d_i(1 - \cos \pi\tau) \qquad 0 \leqslant \tau \leqslant \tau_1$$
$$\eta_i(\tau) = d_i\big(\cos \pi(\tau - \tau_1) - \cos \pi\tau\big) \qquad \tau \geqslant \tau_1$$

These results have already been deduced before. At this stage, the reader may find it a useful exercise to carry out a numerical integration on a computer and examine the influence of the selected number of elements and the adopted order of integration upon the final result.

An additional example of numerical integration is given in fig. 6.4.4 which represents a trapezoidal impulse. The exact solution can easily be obtained through

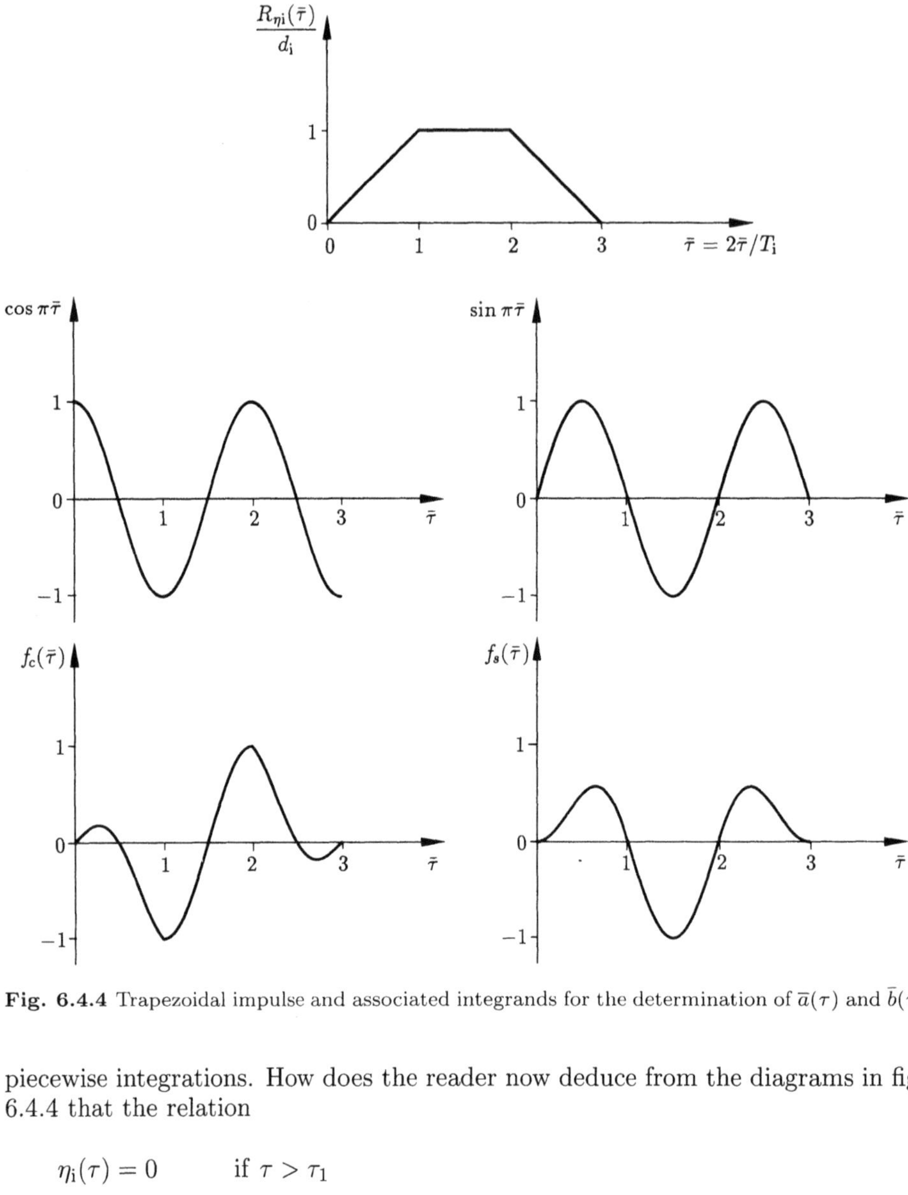

Fig. 6.4.4 Trapezoidal impulse and associated integrands for the determination of $\overline{a}(\tau)$ and $\overline{b}(\tau)$

piecewise integrations. How does the reader now deduce from the diagrams in fig. 6.4.4 that the relation

$$\eta_i(\tau) = 0 \qquad \text{if } \tau > \tau_1$$

holds?

Bibliography for Chapter 6

[6.1] *R.W. Clough, J. Penzien*; Dynamics of structures (McGraw-Hill, New York, 1975).

[6.2] *W.C. Hurty, M.F. Rubinstein*; Dynamics of structures (Prentice-Hall, Englewood Cliffs., 1964).

7 The nature of damping forces; modal damping

"The act of dying is also one of the acts of life"
Marcus Aurelius, Meditations, Bk. VI, sec. 2, (ca. 174 A.D.)

The reader will have observed in preceding chapters that significant discrepancies may arise between an ideal analytical response of an undamped body and its attempted experimental simulation, inevitably affected by damping. Damping in engineering structures is, however, not such an easily definable characteristic as the elastic and inertial forces of a body; the task of finding a suitable model and its associated damping forces is a continuous concern in engineering research.

The most commonly employed damping model is the viscous one. This applies to mechanical systems subject to slow motion in both liquids and gases, such as shock absorbers, hydraulic cylinders and sliding bearings, to name but a few examples. Figure 7.0.1 illustrates the effects of various damping or friction models; concentrating for the moment on a viscous damping model, we note that the corresponding damping force is taken to be proportional to the relative velocity.

In contrast, Coulomb damping or dry friction arises through the motion of a body along a dry surface. In this case, the friction which is otherwise constant depends to a large extent upon the contact pressure and the friction coefficients. This type of friction can also occur in compression joints (e.g. bolts and rivets).

It is the internal friction within the material itself which generates structural or hysteretic damping. Occasionally, we also include in this model the damping forces arising at the joints. The resulting forces depend in the main on the strains or displacements in the structure. The term 'hysteretic damping' stems from the consideration that these damping effects can be deduced experimentally from the hysteresis perceptible in a load-displacement diagram of a body subject to harmonic excitation. When the material is of the linear-elastic type, the damping forces are assumed to be proportional to the elastic forces and point in the opposite direction to the velocity.

If a structure oscillates, for example, in a flowing medium, forces arise which depend upon the motion of the system in the same way as damping forces do. If we also include the externally applied aerodynamic or hydrodynamic forces in this group of forces, we are led to the unusual concept of negative damping. In fact, a phenomenon can occur which appears strange at first sight: an oscillating structure may extract energy from the flow, in which case the oscillation is not damped, as in the presence of positive damping which consumes energy, but incited, as would arise from a so-called negative damping model. Well-known examples of the above are the vibration (and collapse) of open-air transmission lines, suspension bridges and cooling towers or the flutter of aeroplanes and car wheels. The latter phenomena, for which the name aeroelasticity has been coined, are discussed at length in the companion Vol. VI.

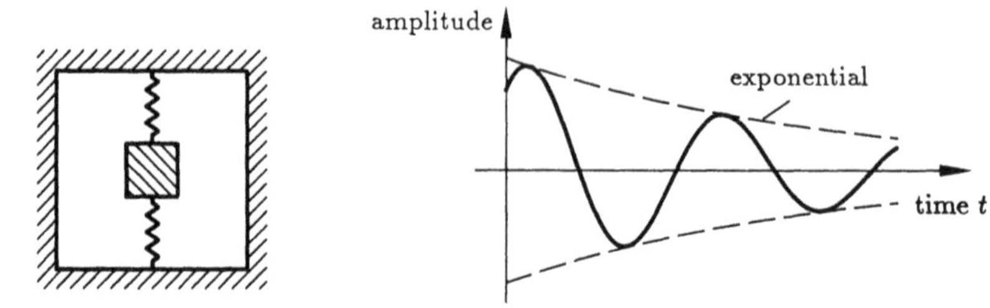

(a) viscous damping (fluid friction on the surface)

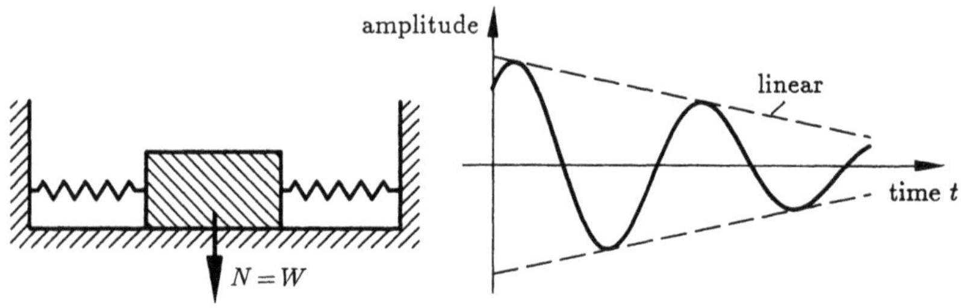

(b) Coulomb damping (solid friction on the surface)

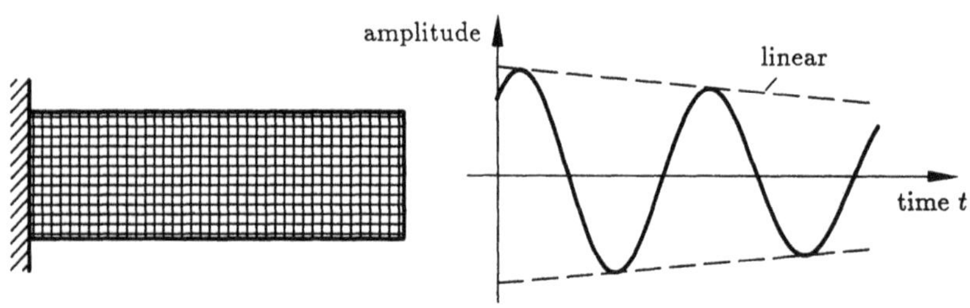

(c) structural or hysteretic damping (internal friction of the material)

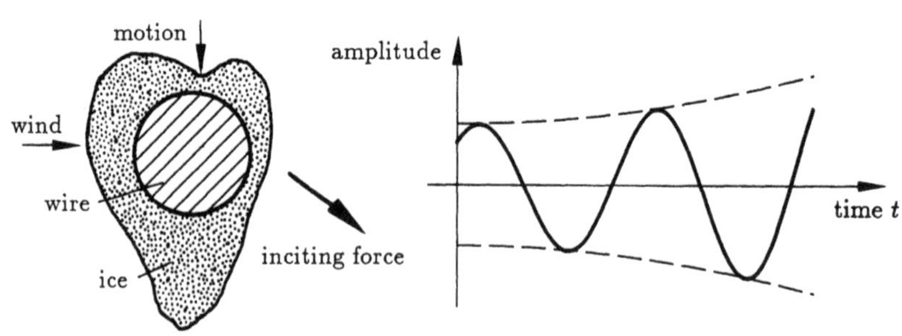

(d) negative damping (energy extracted from the fluid)

Fig. 7.0.1 Damping models commonly adopted in dynamics

None of the damping models mentioned so far offers an exact description of damping in engineering structures. On many occasions, we have to apply different models of damping simultaneously. This complexity induced engineers in the past to adopt models which provide simple solutions to the differential equations of motion. Consequently, researchers were led to substitute an equivalent viscous damping model in place of rather complex damping mechanisms. However, this model should ensure the same amount of energy dissipation as in the original damping mechanism.

7.1 Viscous damping matrix; modal and non-modal damping

The formal assembly of a damping matrix may be achieved as for a mass matrix. If we are to assume that all the elements of a body are subject to the same viscous damping effect or, otherwise expressed, that the viscous damping is uniformly distributed, we obtain in analogy to equ. (3.1.1)

$$\boldsymbol{p}_{VD} = -\mu(x, y, z)\begin{bmatrix} \dot{u} \\ \dot{v} \\ \dot{w} \end{bmatrix} = -\mu\dot{\boldsymbol{u}} \tag{7.1.1}$$

If the element now moves with the velocity $\dot{\boldsymbol{u}}$ in a direction, corresponding damping (volume) forces are applied to the element in the opposite direction. The kinematically equivalent nodal forces equ. (7.1.1) are hence (see equ. (5.1.87) and fig. 5.1.17 of Vol. I, statics)

$$\boldsymbol{R}_D = \sum_{\mathrm{g}=1}^{\mathrm{m}} \mathbf{a}_{\mathrm{g}}^{\mathrm{t}} \int_{V_{\mathrm{g}}} \boldsymbol{\omega}^{\mathrm{t}} \boldsymbol{p}_{VD} \mathrm{d}V \tag{7.1.2}$$

Assuming that the velocities are interpolated in the same way as the displacements

$$\dot{\boldsymbol{u}} = \boldsymbol{\omega}\dot{\boldsymbol{\rho}}_{\mathrm{g}} = \boldsymbol{\omega}\mathbf{a}_{\mathrm{g}}\dot{\boldsymbol{r}} \qquad \text{in region } V_{\mathrm{g}} \tag{7.1.3}$$

we obtain

$$\boldsymbol{R}_D = \left[\sum_{\mathrm{g}=1}^{\mathrm{m}} \mathbf{a}_{\mathrm{g}}^{\mathrm{t}} \left[\int_{V_{\mathrm{g}}} \mu\boldsymbol{\omega}^{\mathrm{t}}\boldsymbol{\omega}\mathrm{d}V\right] \mathbf{a}_{\mathrm{g}}\right] \dot{\boldsymbol{r}} \tag{7.1.4}$$

Thus, the element damping matrix is

$$\boldsymbol{c}_{\mathrm{g}} = \int_{V_{\mathrm{g}}} \mu\boldsymbol{\omega}^{\mathrm{t}}\boldsymbol{\omega}\mathrm{d}V \tag{7.1.5}$$

and the damping matrix of the complete structure becomes

$$\boldsymbol{C} = \sum_{\mathrm{g}=1}^{\mathrm{m}} \mathbf{a}_{\mathrm{g}}^{\mathrm{t}} \boldsymbol{c}_{\mathrm{g}} \mathbf{a}_{\mathrm{g}} \tag{7.1.6}$$

Fig. 7.1.1
A simple example of a viscous dash-pot applied at a discrete point of a beam

No special problem arises if damping forces are assumed to act at discrete points in addition to the distributed damping effects. In the most simple case, these forces are exerted directly on the structural nodes, the damping 'dash-pot' being attached to the foundation (fig. 7.1.1). In this case, the damping force becomes

$$R_\mathrm{i} = -c\dot{r}_\mathrm{i} \tag{7.1.7}$$

where the damping coefficient, c, is to be added to the diagonal element $\boldsymbol{C}(\mathrm{i},\mathrm{i})$ of the overall damping matrix. If a discrete damping force is applied within an element, it becomes necessary to establish its kinematically equivalent nodal forces. For example, assuming a damper is attached in the x-direction to a point $P(x_P, y_P, z_P)$ of an element g, the Cartesian damping force at the point P is

$$\begin{bmatrix} U \\ V \\ W \end{bmatrix}_P = -\lfloor c_{xP} \quad 0 \quad 0 \rfloor \dot{\boldsymbol{u}}_P \tag{7.1.8}$$

and the equivalent nodal forces are then

$$\begin{aligned} \boldsymbol{R}_D &= \mathbf{a}_\mathrm{g}^\mathrm{t} \boldsymbol{\omega}^\mathrm{t}(x_P, y_P, z_P) \begin{bmatrix} U \\ V \\ W \end{bmatrix}_P \\ &= -\mathbf{a}_\mathrm{g}^\mathrm{t} \boldsymbol{\omega}^\mathrm{t}(x_P, y_P, z_P) \lfloor c_{xP} \quad 0 \quad 0 \rfloor \boldsymbol{\omega}(x_P, y_P, z_P) \mathbf{a}_\mathrm{g} \dot{\boldsymbol{r}} \end{aligned} \tag{7.1.9}$$

Consequently, for the element g, additional damping terms have to be introduced into the damping matrix.

No difficulty arises if a discrete damper is placed between two nodal points of the oscillatory model. Let us assume, for example, that a damper is inserted along the line connecting two concurrent degrees of freedom, r_i and r_j, the associated damping constant being c. Here, the relative velocity $\dot{r}_\mathrm{i} - \dot{r}_\mathrm{j}$ determines the damping force. Thus, we have

$$\begin{aligned} R_\mathrm{i} &= -c(\dot{r}_\mathrm{i} - \dot{r}_\mathrm{j}) \\ R_\mathrm{j} &= +c(\dot{r}_\mathrm{i} - \dot{r}_\mathrm{j}) \end{aligned} \tag{7.1.10}$$

and the damping matrix, $\boldsymbol{C}$, contains the additional terms

	(i)	(j)
(i)	c	$-c$
(j)	$-c$	c

where it is tacitly assumed that r_i and r_j point in the same direction. The more complex case of a damper acting along a line connecting two arbitrary points, neither of which is a nodal point, may also be solved by the aforementioned technique. The overall idealisation determines a system of coupled differential equations

$$\boldsymbol{M}\ddot{\boldsymbol{r}} + \boldsymbol{C}\dot{\boldsymbol{r}} + \boldsymbol{K}\boldsymbol{r} = \boldsymbol{R}(t) \tag{7.1.11}$$

where the damping matrix is symmetric and sparsely populated. One decisive disadvantage of this type of viscous damping is that the diagonalisation of the system is somewhat problematic because $\boldsymbol{C}$ is generally a matrix which is unrelated to $\boldsymbol{M}$ and $\boldsymbol{K}$ (this is known as non-proportional damping). The decoupling of the differential equations, however, remains an important task in the solution of equ. (7.1.11). To resolve this difficulty, we shall provide a fast and simple method in Chapter 9.

The solution of the general system of equ. (7.1.11) is still an expensive undertaking since in the present case, the spectral analysis involves complex eigenvalues and eigenvectors. In an attempt to simplify the task, researchers have considered special forms of $\boldsymbol{C}$. One promising technique assumes a so-called proportional damping. In this approach, we set

$$\boldsymbol{C} = \alpha \boldsymbol{K} + \beta \boldsymbol{M} \tag{7.1.12}$$

One great advantage of this scheme arises from the fact that the eigenvectors of the undamped system diagonalise equ. (7.1.11); a disadvantage is that the damping model contains only two free parameters, α and β. The damping matrix resulting from diagonalisation with $\boldsymbol{K}$-orthonormal eigenvectors becomes

$$\begin{aligned} \boldsymbol{D} &= \boldsymbol{X}^t \boldsymbol{C} \boldsymbol{X} \\ &= \lceil d_1 \quad d_2 \quad d_3 \quad \dots \quad d_{nx} \rfloor \end{aligned} \tag{7.1.13}$$

where

$$d_i = \alpha + \frac{\beta}{\omega_i^2} \tag{7.1.14}$$

are denoted as 'modal' damping coefficients. It must be admitted, however, that the modal damping coefficients derived from equ. (7.1.14) yield only a poor simulation of the real damping characteristics of a structure. An improvement of the model of equ. (7.1.13) may nevertheless be achieved by ignoring the proportionality condition of equ. (7.1.12) and instead establishing directly the values of the diagonalised damping matrix, $\boldsymbol{D}$, of equ. (7.1.13). This multi-parameter model is much more flexible than the proportional damping model. In this approach, the nature of the original $\boldsymbol{C}$ matrix is of little significance. However, the actual expression for $\boldsymbol{C}$ may be deduced from an assumed $\boldsymbol{D}$-matrix by application of a complete set of linearly independent eigenvectors

$$\boldsymbol{C} = [\boldsymbol{X}^t]^{-1} \boldsymbol{D} \boldsymbol{X}^{-1} \tag{7.1.15}$$

The question as to the numerical values of proportional and modal damping is, of course, of practical importance. The relation

$$\zeta_i = \tfrac{1}{2} d_i \omega_i \tag{7.1.16}$$

offers a rough guideline for the evaluation of the dynamic behaviour of the structure (see also sections 1.3 and 7.4). If

$$\zeta_i < 1 \tag{7.1.17}$$

the damping is characterised as sub-critical. Following an excitation, the structure experiences a damped oscillation; when the critical point $\zeta_i = 1$ is reached, oscillation vanishes. For most real structures, damping is weak and ζ_i has to be assigned a value much smaller than 1.

A heuristic method of selecting the modal damping values may be obtained from

$$\boldsymbol{D} = \mathrm{diag}[\boldsymbol{X}^t \boldsymbol{C} \boldsymbol{X}] \tag{7.1.18}$$

A comparison calculation carried out on an example confirms that this modal model yields relatively satisfactory amplitudes; in contrast, the phase shift is in general unreliable. In principle, the models of equs. (7.1.12) and (7.1.13) are mathematical fictions. In fact, a physically consistent viscous damping model inevitably involves a complex spectral analysis.

7.2 A consideration of Coulomb damping forces

The single mass model of fig. 7.2.1(a) serves as a typical example for the exploration of Coulomb damping and is listed in most standard dynamic reference books. Inevitably, such simple examples circumvent the difficulties usually associated with more general formulations of the problem. When the mass moves along a straight horizontal path, the surface pressure, which in this case equals the dead weight, remains constant. On the other hand, when the mass slides along a cylindrical path, the significant components of the weight and the centrifugal force, Z, vary continuously, inducing a considerable complication in an oscillatory motion. In any case, we ought strictly to allow for different friction coefficients under gliding and stationary conditions since at the reversal point of oscillation, solid (or static) friction takes place. Very often, this fact is simply disregarded. A common feature of the one-dimensional problems in figs. 7.2.1(a) and 7.2.1(b) is that the direction of the damping force derives from the plus or minus sign of the velocity. If the mass begins to swing to the right, the damping force will point to the left but after the reversal point has been reached, the damping force will start pointing to the right. Consequently, the problem must be solved piecewise (see example in fig. 1.3.3). If, however, we consider the motion on a plane (see fig. 7.2.1(c)) in place of a linear movement, then we still initially know the absolute value of the damping force and its direction (as given by the initial conditions) but up to the point of reversal, the direction of the damping force will continuously change. A piecewise integration of the differential equations, as we demonstrated for a linear oscillator, is therefore hardly a practical proposition. Let us now consider the more

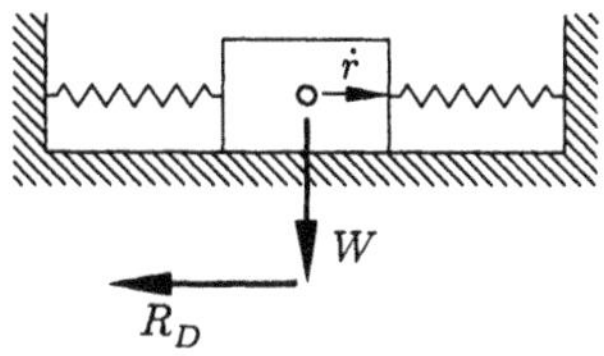

(a) one-dimensional, rectilinear

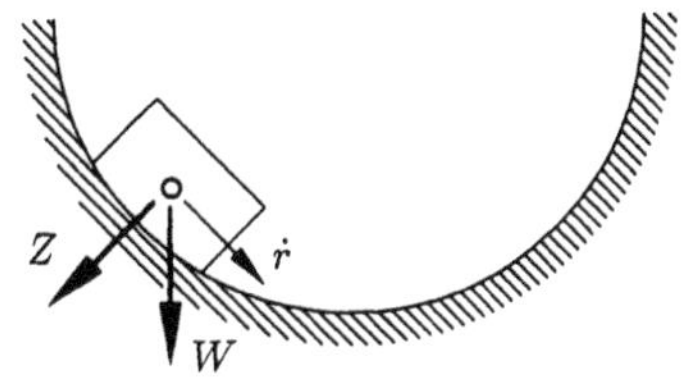

(b) one-dimensional, curved

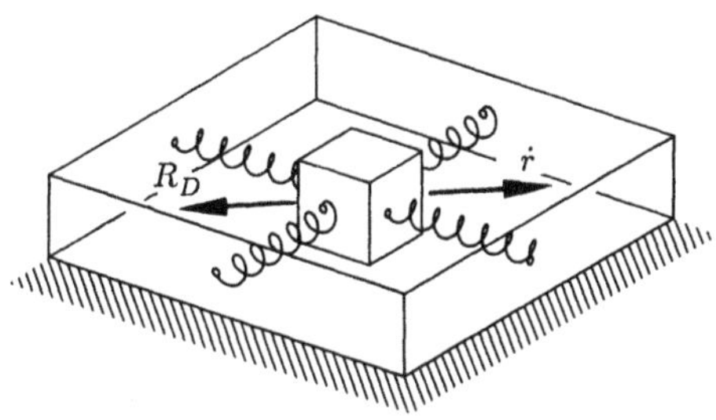

(c) two-dimensional, planar

Fig. 7.2.1
Coulomb friction for single-mass oscillators

general case of a multi-mass oscillator constrained to move on curved surfaces. A closed-form solution is now out of the question and only a step-by-step technique of the type to be demonstrated in Chapter 10 can be applied.

When an oscillator rubs against a surface at a point P_R with a surface pressure of p_n (N/m^2) and moves with a vectorial velocity $\dot{\boldsymbol{u}}$ and the coefficient of dry sliding friction is μ at the point P_R, it experiences per unit surface a vectorial force of

$$\boldsymbol{p}_D = -\mu p_n \frac{\dot{\boldsymbol{u}}}{|\dot{\boldsymbol{u}}|} \tag{7.2.1}$$

The kinematically equivalent nodal damping forces may be evaluated by the standard integration procedure using the current frictional force $\boldsymbol{p}_D$

$$\boldsymbol{R}_D = \sum_{\mathrm{g}=1}^{\mathrm{m}} \mathbf{a}_{\mathrm{g}}^{\mathrm{t}} \int\limits_{S_{\mathrm{g}R}} \boldsymbol{\omega}^{\mathrm{t}} \boldsymbol{p}_D \mathrm{d}S \tag{7.2.2}$$

Within the g-th element, the displacement is

$$\boldsymbol{u} = \boldsymbol{\omega}\boldsymbol{\rho}_{\mathrm{g}} = \boldsymbol{\omega}\mathbf{a}_{\mathrm{g}}\boldsymbol{r} \tag{7.2.3}$$

and hence the velocity is

$$\dot{\boldsymbol{u}} = \boldsymbol{\omega}\mathbf{a}_{\mathrm{g}}\dot{\boldsymbol{r}} \tag{7.2.4}$$

Consequently, the dynamic motion is described by the complicated integro-differential equation

$$\boldsymbol{M}\ddot{\boldsymbol{r}} + \boldsymbol{K}\boldsymbol{r} = \boldsymbol{R}(t) - \sum_{\mathrm{g}=1}^{\mathrm{m}} \mathbf{a}_{\mathrm{g}}^{\mathrm{t}} \int_{S_{\mathrm{g}R}} \mu p_n \boldsymbol{\omega}^{\mathrm{t}} \frac{\boldsymbol{\omega}\mathbf{a}_{\mathrm{g}}\dot{\boldsymbol{r}}}{|\boldsymbol{\omega}\mathbf{a}_{\mathrm{g}}\dot{\boldsymbol{r}}|} \mathrm{d}S \tag{7.2.5}$$

which does not even take into account the difference between gliding and static friction. Expression (7.2.5) generally proves too complicated for most practical applications.

7.3 A survey of structural (material) damping

When a solid body is harmonically excited, a steady-state oscillation with the excitation frequency Ω establishes itself once the transient oscillation has died down. The internal or material damping forces here are approximately independent of the excitation frequency. If we test this observation on the viscous damping model of equ. (7.1.11), we will not, however, confirm this assertion. Indeed, consider the excitation

$$\boldsymbol{R} = \boldsymbol{a} \sin \Omega t \tag{7.3.1}$$

The steady-state response follows the same frequency and is given by

$$\boldsymbol{r} = \boldsymbol{b} \sin(\Omega t - \varphi) \tag{7.3.2}$$

Consequently, the viscous damping forces become

$$\boldsymbol{R}_D = -\boldsymbol{C}\dot{\boldsymbol{r}} = -\boldsymbol{C}\boldsymbol{b}\Omega \cos(\Omega t - \varphi) \tag{7.3.3}$$

and are clearly seen to be proportional to the excitation frequency. This shortcoming of the viscous damping model when applied to structural damping can be overcome by the adoption of the hysteretic damping model defined by

$$\boldsymbol{R}_D = -\frac{1}{\Omega}\boldsymbol{C}_S\dot{\boldsymbol{r}} \tag{7.3.4}$$

This formulation, as the reader should not forget, is, however, closely tied to the special type of harmonic excitation and the steady-state response of equs. (7.3.1) and (7.3.2). The dynamic equation for hysteretic damping can now be put in the form

$$\boldsymbol{M}\ddot{\boldsymbol{r}} + \frac{1}{\Omega}\boldsymbol{C}_S\dot{\boldsymbol{r}} + \boldsymbol{K}\boldsymbol{r} = \boldsymbol{a}\sin\Omega t \tag{7.3.5}$$

The appearance of the excitation frequency on the left-hand side of equ. (7.3.4) can be circumvented if we adopt a complex formulation. Let the steady-state response of the system due to a sine excitation be given by $\boldsymbol{r}_s$. The dynamic motion for hysteretic damping is then described by

$$\boldsymbol{M}\ddot{\boldsymbol{r}}_s + \frac{1}{\Omega}\boldsymbol{C}_S\dot{\boldsymbol{r}}_s + \boldsymbol{K}\boldsymbol{r}_s = \boldsymbol{a}\sin\Omega t \tag{7.3.6}$$

Let also $\boldsymbol{r}_c$ be the steady-state response due to a cosine excitation (which merely induces a phase shift of 90^0 with respect to $\boldsymbol{r}_s$). We thus have

$$\boldsymbol{M}\ddot{\boldsymbol{r}}_c + \frac{1}{\Omega}\boldsymbol{C}_S\dot{\boldsymbol{r}}_c + \boldsymbol{K}\boldsymbol{r}_c = \boldsymbol{a}\cos\Omega t \tag{7.3.7}$$

Multiplication of the first equation by the imaginary unit, i, and addition of equs. (7.3.6) and (7.3.7) then yield with the notation

$$\widetilde{\boldsymbol{r}} = \boldsymbol{r}_c + i\boldsymbol{r}_s \tag{7.3.8}$$

the complex differential equation

$$\boldsymbol{M}\ddot{\widetilde{\boldsymbol{r}}} + \frac{1}{\Omega}\boldsymbol{C}_S\dot{\widetilde{\boldsymbol{r}}} + \boldsymbol{K}\widetilde{\boldsymbol{r}} = \boldsymbol{a}\,\mathrm{e}^{i\Omega t} = \widetilde{\boldsymbol{R}} \tag{7.3.9}$$

in which we take account of equ. (6.2.16). The real part of this equation is identical to equ. (7.3.7) and the imaginary part to equ. (7.3.6). For a steady-state condition, we express the solution as

$$\widetilde{\boldsymbol{r}} = \widetilde{\boldsymbol{b}}\,\mathrm{e}^{i\Omega t} \tag{7.3.10}$$

where $\widetilde{\boldsymbol{b}}$ can naturally be complex. Hence

$$\dot{\widetilde{\boldsymbol{r}}} = i\Omega\widetilde{\boldsymbol{b}}\,\mathrm{e}^{i\Omega t} = i\Omega\widetilde{\boldsymbol{r}} \tag{7.3.11}$$

It is now possible to express equ. (7.3.9) with a frequency-independent damping term. In fact, we have

$$\boldsymbol{M}\ddot{\widetilde{\boldsymbol{r}}} + \left[\boldsymbol{K} + i\boldsymbol{C}_S\right]\widetilde{\boldsymbol{r}} = \widetilde{\boldsymbol{R}}(t) \tag{7.3.12}$$

Note that equ. (7.3.5) now coincides with the imaginary part of this equation. For linearly elastic systems, the $\boldsymbol{C}_S$ matrix may be set approximately proportional to the stiffness matrix, $\boldsymbol{K}$. Denoting the factor of proportionality by g, the value of which is usually in the range

$$g \leqslant 0.05 \tag{7.3.13}$$

equ. (7.3.12) simplifies to

$$\boldsymbol{M}\ddot{\widetilde{\boldsymbol{r}}} + (1 + ig)\boldsymbol{K}\widetilde{\boldsymbol{r}} = \widetilde{\boldsymbol{R}}(t) \tag{7.3.14}$$

This formulation does not at first sight reflect the basic pre-supposition of the imposition of a harmonic excitation

$$\widetilde{\boldsymbol{R}} = \boldsymbol{a}\,(\cos \Omega t + i \sin \Omega t) \tag{7.3.15}$$

(for a real formulation, see equ. (7.3.1)), but the reader should keep this essential prerequisite of our theory in mind. At most, one may speculate whether periodic processes, which may be developed in Fourier series, can be analysed with equ. (7.3.14).

7.4 The response of modally damped systems

The present section, which serves as a continuation of Chapters 5 and 6, outlines a method for the determination of the response of an oscillatory system subject to modal-viscous damping. In this context, the term 'modal' applies to all damping models which permit a decoupling of the equations of motion by transformations with the eigenvectors of the undamped system. By this definition, it follows that the material damping as defined in equ. (7.3.14) is also 'modal'. In a less broad sense, the word 'modal' is usually applied in connection with viscous damping and should be understood here as such.

Thus, the computation of modally damped systems initially requires a spectral analysis of the undamped system. This means that all the techniques described in section 6.1 have now to be applied. Initiating our investigation with equ. (6.1.6), we may introduce directly the modal (diagonal) damping matrix $\boldsymbol{D}$

$$\boldsymbol{\Lambda}\ddot{\boldsymbol{\eta}} + \boldsymbol{D}\dot{\boldsymbol{\eta}} + \boldsymbol{\eta} = \boldsymbol{R}_\eta(t) \tag{7.4.1}$$

A typical scalar equation of this then takes the form

$$\ddot{\eta}_\mathrm{i} + 2\zeta_\mathrm{i}\,\omega_\mathrm{i}\,\dot{\eta}_\mathrm{i} + \omega_\mathrm{i}^2\eta_\mathrm{i} = \omega_\mathrm{i}^2 R_{\eta\mathrm{i}} \qquad \mathrm{i} = 1 \ldots \mathrm{nx} \tag{7.4.2}$$

where ζ_i represents the damping ratio

$$\zeta_\mathrm{i} = \tfrac{1}{2}\omega_\mathrm{i}\,d_\mathrm{i} \tag{7.4.3}$$

and ω_i is an eigenfrequency of the undamped system.

Considering first the free oscillations of a modally damped system for which $\boldsymbol{R}_\eta(t) = \boldsymbol{o}$, resp. $R_{\eta\mathrm{i}} = 0$, we have

$$\ddot{\eta}_\mathrm{i} + 2\zeta_\mathrm{i}\,\omega_\mathrm{i}\,\dot{\eta}_\mathrm{i} + \omega_\mathrm{i}^2\eta_\mathrm{i} = 0 \tag{7.4.4}$$

The differential equation of the i-th generalised coordinate (eigenvector $\boldsymbol{x}_\mathrm{i}$) is seen to be identical to the equation of motion for a damped single-mass oscillator. Consequently, the modal answer $\eta_\mathrm{i}(t)$ can be extracted directly from section 1.3 and we find

$$\eta_i(t) = e^{-\omega_i \zeta_i t}\Big(a_i \sin \omega_{Di} t + b_i \cos \omega_{Di} t\Big)$$

$$\text{if } \zeta_i < 1 \qquad \text{(sub-critical)} \tag{7.4.5}$$

$$\eta_i(t) = (a_i + b_i t)\, e^{-\omega_i t} \qquad \text{if } \zeta_i = 1 \qquad \text{(critical)} \tag{7.4.6}$$

$$\eta_i(t) = a_i e^{-\omega_i(\zeta_i - \sqrt{\zeta_i^2 - 1})t} + b_i e^{-\omega_i(\zeta_i + \sqrt{\zeta_i^2 - 1})t}$$

$$\text{if } \zeta_i > 1 \qquad \text{(over-critical)} \tag{7.4.7}$$

where we have introduced the abbreviation ω_{Di} for the damped angular frequency

$$\omega_{Di} = \omega_i \sqrt{1 - \zeta_i^2} \tag{7.4.8}$$

In analogy to our analysis of a free undamped system, we now turn to the solution of a sub-critically damped system for which the initial conditions are as in equ. (5.1.8). In this way, we obtain the constants

$$b_i = \eta_{0i}$$

$$a_i = \frac{\dot{\eta}_{0i} + \omega_i \zeta_i \eta_{0i}}{\omega_i \sqrt{1 - \zeta_i^2}} = \frac{\dot{\eta}_{0i} + \omega_i \zeta_i \eta_{0i}}{\omega_{Di}} \tag{7.4.9}$$

The special solution hence becomes

$$\begin{aligned} \eta_i &= e^{-\omega_i \zeta_i t}\left(\frac{\dot{\eta}_{0i} + \omega_i \zeta_i \eta_{0i}}{\omega_i \sqrt{1 - \zeta_i^2}} \sin \omega_i \sqrt{1 - \zeta_i^2}\, t + \eta_{0i} \cos \omega_i \sqrt{1 - \zeta_i^2}\, t\right) \\ &= e^{-\omega_i \zeta_i t}\left(\frac{\dot{\eta}_{0i} + \omega_i \zeta_i \eta_{0i}}{\omega_{Di}} \sin \omega_{Di} t + \eta_{0i} \cos \omega_{Di} t\right) \end{aligned} \tag{7.4.10}$$

The transformation of the initial conditions $\boldsymbol{r}_0$ respectively $\dot{\boldsymbol{r}}_0$ into $\boldsymbol{\eta}$ respectively $\dot{\boldsymbol{\eta}}$ has already been discussed in conjunction with equ. (5.1.13) and can be adopted here as it stands.

As demonstrated before for an undamped structure, we may impose a special excitation on a modally damped system, thus ensuring that a steady-state oscillation is immediately attained. For this purpose, the structure need only be deflected in the form of an eigenvector $\boldsymbol{x}_i$ and then let free. The corresponding initial conditions in this case would be $\eta_{0i} \neq 0$ and $\eta_{0j} = 0$ ($j \neq i$, $j = 1 \ldots$ to n) and all corresponding velocities zero. For such an excitation, the body undergoes an oscillation with the same deflection shape as that for zero damping. The eigenfrequency is modified, however, in accordance with equ. (1.3.16) and the amplitude decreases continuously. Figure 7.4.1 compares an undamped eigenoscillation to a modally damped one.

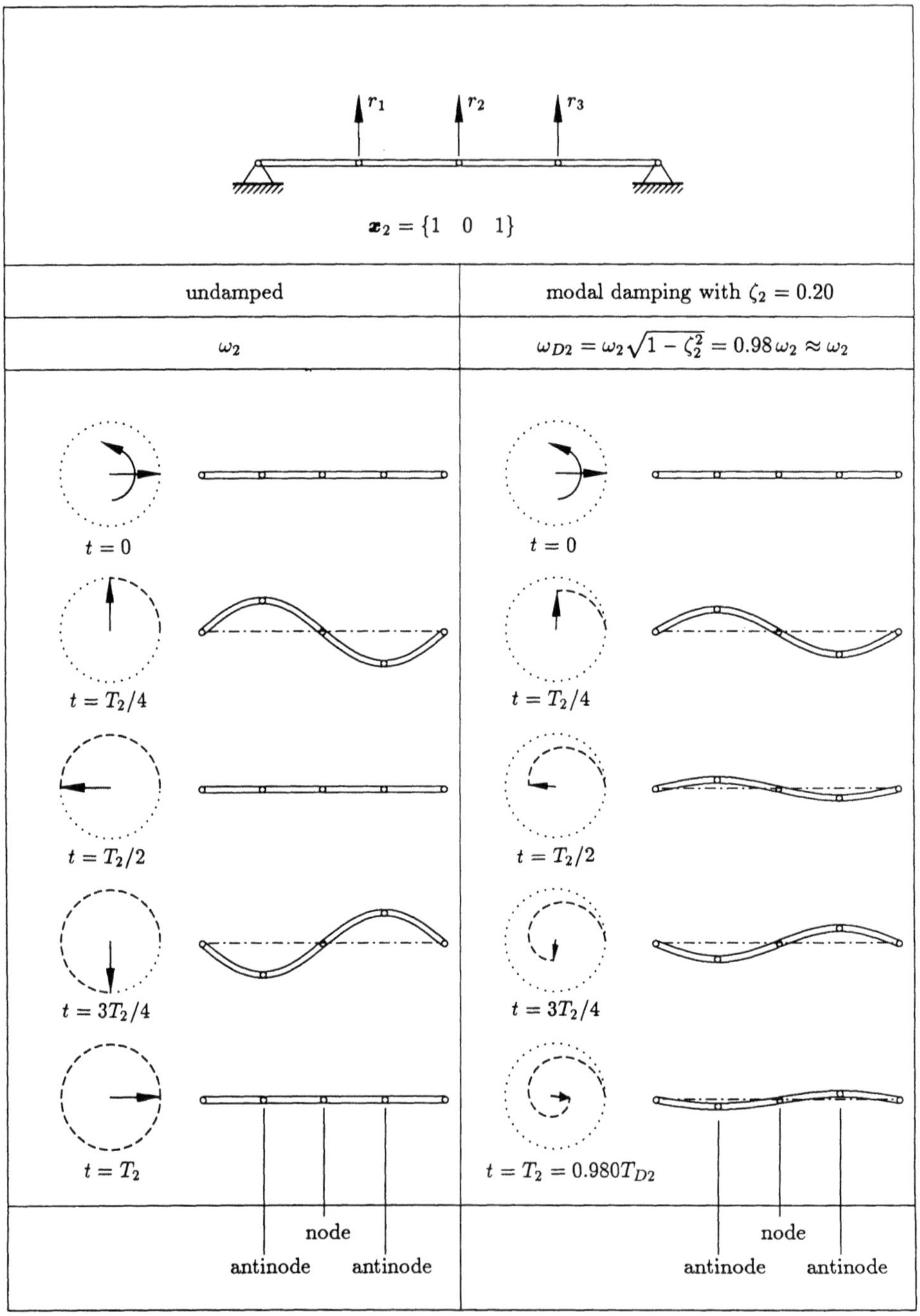

Fig. 7.4.1 Undamped and modally damped eigenoscillations for identical initial conditions recorded at the same instant t

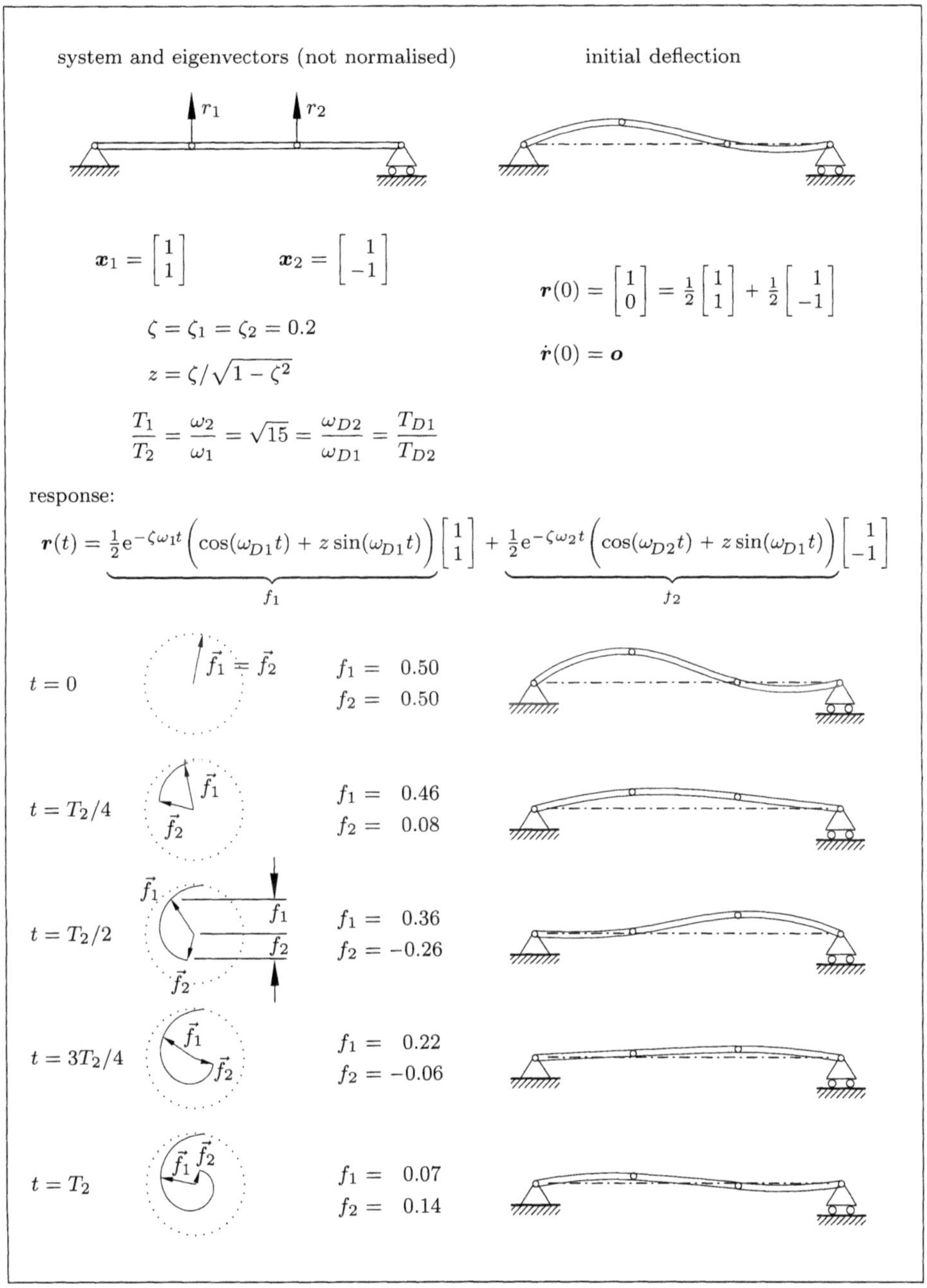

Fig. 7.4.2 Unsteady free oscillation of a modally damped structure, simulated by a superposition of eigenoscillations

For a general excitation, the response involves a combination of eigenoscillations. A steady-state oscillation with stationary nodes and antinodes can no longer be realised (fig. 7.4.2).

Let us now consider the forced oscillation of a modally damped structure. The solution to the typical differential equation (7.4.2) now consists of two parts: a homogeneous part, with which we are already familiar from equs. (7.4.5) to (7.4.7), and a particular solution η_{pi}, which depends upon the type of excitation. Thus,

$$\eta_i(t) = e^{-\omega_i \zeta_i t}(a_i \sin \omega_{Di} t + b_i \cos \omega_{Di} t) + \eta_{pi}(t) \tag{7.4.11}$$

and is given here for the most common type of sub-critical damping. As was the case for undamped structures, there are excitation functions $R_{\eta i}(t)$ that are integrated exactly and others that must be handled numerically. All considerations of the oscillation of a single-degree-of-freedom system given in section 1.4 apply here, too. A brief summary of the steady-state response to a harmonic excitation is also appended since the results are required in subsequent chapters.

The complex unit excitation of equ. (6.2.17)

$$R_{\eta i} = e^{i\Omega t} = \cos \Omega t + i \sin \Omega t$$

contains a pure cosine excitation in the real part and a pure sine excitation in the imaginary part. For the particular response, we once more apply equ. (6.2.26)

$$\eta_{pi} = F_i\, e^{i\Omega t}$$

and obtain the complex frequency response characteristic F_i as

$$F_i = \frac{1}{1 - (\Omega/\omega_i)^2 + i\, 2\zeta_i\, (\Omega/\omega_i)} \tag{7.4.12}$$

Here, the reader is reminded of the standard representation of complex numbers

$$\frac{1}{x + iy} = \frac{1}{r\,(\cos \varphi + i \sin \varphi)} = \frac{1}{r} e^{-i\varphi} \tag{7.4.13}$$

where

$$r = (x^2 + y^2)^{1/2}$$

$$\varphi = \tan^{-1} \frac{y}{x} \tag{7.4.14}$$

The value of $1/r$ leads us to the absolute value of the complex frequency characteristic F_i, i.e. to the amplification factor V_i

$$V_i = \left(\left[1 - \left(\frac{\Omega}{\omega_i} \right)^2 \right]^2 + \left[2\zeta_i \left(\frac{\Omega}{\omega_i} \right) \right]^2 \right)^{-1/2} \tag{7.4.15}$$

The angle φ_i defines the phase shift

$$\varphi_i = \tan^{-1} \frac{2\zeta_i \Omega/\omega_i}{1 - (\Omega/\omega_i)^2} \tag{7.4.16}$$

The particular solution thus becomes

$$\begin{aligned} \eta_{pi} &= V_i \, e^{-i\varphi_i} \, e^{i\Omega t} \\ &= V_i \, e^{i(\Omega t - \varphi_i)} \\ &= V_i \Big(\cos(\Omega t - \varphi_i) + i \sin(\Omega t - \varphi_i)\Big) \end{aligned} \tag{7.4.17}$$

As we already know, the homogeneous solution to modal damping is gradually damped away and the above particular solution represents the steady-state oscillation of the system. The real part yields the response to a cosine excitation whilst the imaginary part corresponds to a sinusoidal excitation. Damping affects the phase shift which is either 0 or π for an undamped body, but is now determined by equ. (7.4.16); moreover, the amplification factor is modified so that at $\Omega = \omega_i$, the singularity has vanished and V_i remains always finite. Thus, even for weak damping, the response is considerably affected in the neighbourhood of resonance and much depends here upon an exact measure of damping. Figure 1.4.2 gives a graphic representation of the amplification factor V_i and the phase shift φ_i. It also demonstrates the possibility of errors that may creep in if damping is not defined precisely.

Bibliography for Chapter 7

[7.1] *B.J. Lazan*; Damping of materials and members in structural mechanics (Pergamon Press, 1968).

[7.2] *T.K. Caughey*; Classical normal modes in damped linear dynamic systems, J. Appl. Mech., 27 (1960) 269–271.

[7.3] *E.L. Wilson, J. Penzien*; Evaluation of orthogonal damping matrices, Int. J. Num. Meth. Eng., 4 (1972) 5–10.

[7.4] *S.H. Crandall*; The role of damping in vibration theory, J. Sound Vibr., 11 (1970) 3–18.

[7.5] *W.T. Thomson, T. Calkins, P. Caravani*; A numerical study of damping, Earthqu. Eng. Structr. Dyn., 3 (1974) 97–103.

[7.6] *R.H. Scanlan*; Linear damping models and causality in vibrations, J. Sound Vibr., 13 (1970) 499–503.

[7.7] *N.O. Myklestad*; The concept of complex damping, J. Appl. Mech., 19 (1952) 284–286.

[7.8] *P. Lancaster*; Free vibration and hysteretic damping, J. Roy. Aer. Soc., 64 (1960) 229.

[7.9] *C.M. Harris, Ch.E. Crede*; Shock and vibration handbook (McGraw-Hill, New York, 1976).

[7.10] *L. Meirovitch*; Computational methods in structural dynamics (Sijthoff & Noordhoff, Alphen aan den Rijn, 1980).

8 Random vibrations of modally damped systems

"Chance is a dicer"
Thomas Fuller, Gnomologia, No. 1080 (1732)

The specific forms of excitation already discussed in Chapters 6 (undamped systems) and 7 (modally damped systems) are 'deterministic' in nature and the histogram of $\boldsymbol{R}(t)$ or $\boldsymbol{R}_\eta(t)$ is uniquely prescribed.

If we test the loading action on a specific vehicle as it travels along a surface, for example a rough country path or a motorway, and attune the testing programme to the future utilisation of the vehicle, we can use an instrumental array to measure the forces exerted upon the vehicle's axes. The result would probably resemble the diagram in fig. 8.1.1. It is alas not possible to express this load histogram in a simple functional formula and a complete deterministic analysis becomes far too complicated as well as being of little value. If two further vehicles were to be sent along the same path, different values at a certain point in time t_1 would inevitably be measured. Thus, there seems to be little point in searching for *the* specific excitation value $R_{\eta\mathrm{i}}(t_1)$ and *the* specific response $\eta_\mathrm{i}(t_1)$ for each type of vehicle and road. The statistical approach offers the only practical solution here.

In statistics, it is not the specific value of a function $x(t)$ at a point in time t_1 we search for, but instead we inquire after the probability of a function $x(t)$ being larger than a value x_0 at a point in time t_1, where x stands for the excitation $R_{\eta\mathrm{i}}(t)$ or the response $\eta_\mathrm{i}(t)$. A typical result of a stochastic excitation tells us, for example, that $\eta_\mathrm{i}(t_1)$ is with an 80% probability or likelihood of occurrence not larger than 0.15. This chapter attempts to formulate and solve such problems.

8.1 Survey of the nature of a single stochastic process

We now set out to define 'probability,' the most important technical term used in this chapter. Analogous to the series of measurements given in fig. 8.1.1, we now consider N series of measurements ($x_\mathrm{k}(t)$, $\mathrm{k} = 1, \mathrm{N}$) and thus, at a certain instant t_1, we also possess the functional values $x_\mathrm{k}(t_1)$. Let N_x be the number of measured values that are smaller than x_0. We define now the probability that 'x is at the instant t_1 smaller than x_0' by the mathematical statement

$$P(x_0, t_1) = \lim_{\mathrm{N}\to\infty} \frac{\mathrm{N}_x}{\mathrm{N}} \tag{8.1.1}$$

(providing that the limit exists). It is clear that in practical terms, we require an additional elaboration of equ. (8.1.1), for no manufacturer would be prepared to provide us with an infinite number of test vehicles. We shall return to this problem later. If we set x_0 as our variable here and attempt to discover how many values

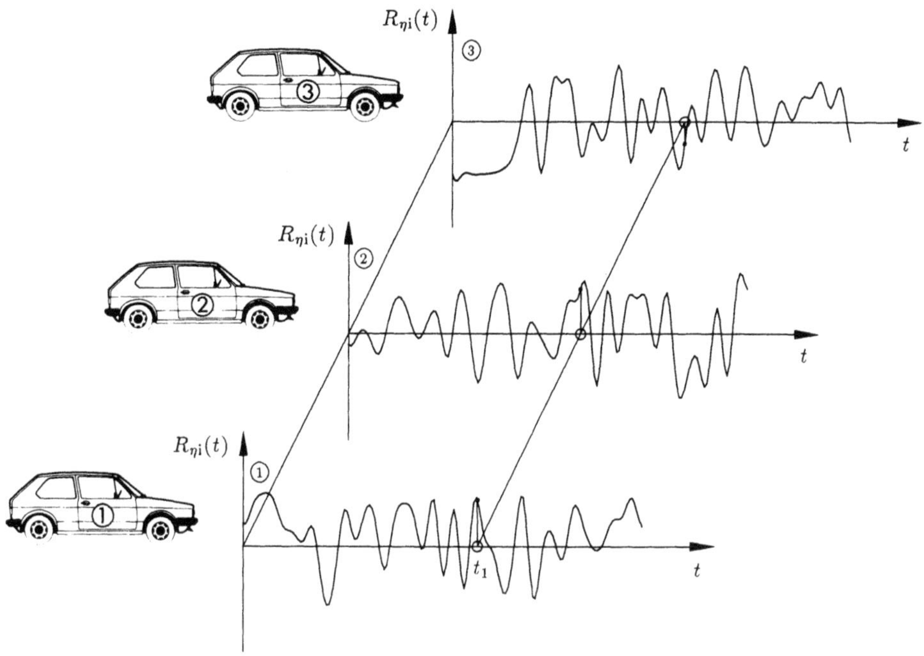

Fig. 8.1.1 Measured excitation histograms of three nominally identical vehicles

of $x(t_1)$ are smaller than $-\infty$, then the only possible answer is none. On the other hand, however, all measured values will be smaller than $+\infty$. This means that

$$\begin{aligned} P(-\infty, t_1) &= 0 \\ P(+\infty, t_1) &= 1 \end{aligned} \tag{8.1.2}$$

Between these two limiting values, $P(x, t_1)$ behaves as a monotonically ascending function. The greater the chosen value of x_0, the larger the number of measurements N_x that are smaller than x_0. If we now form the gradient of probability – known in what follows as the probability density –

$$p(x, t_1) = \frac{\mathrm{d}P(x, t_1)}{\mathrm{d}x} \tag{8.1.3}$$

we observe that it is either positive or equal to zero

$$p(x, t_1) \geqslant 0 \tag{8.1.4}$$

Since the probability density defines the increment of probability for an increasing value of x_0, we may conversely deduce from it the probability itself

$$P(x_0, t_1) = \int_{-\infty}^{x_0} p(x, t_1)\mathrm{d}x \tag{8.1.5}$$

Furthermore, we may determine without difficulty the probability of $x(t_1)$ possessing a value between x_a and x_b. In fact, this probability is

$$\int_{x_a}^{x_b} p(x,t_1)\,\mathrm{d}x \tag{8.1.6}$$

The preceding state of information is illustrated in fig. 8.1.2.

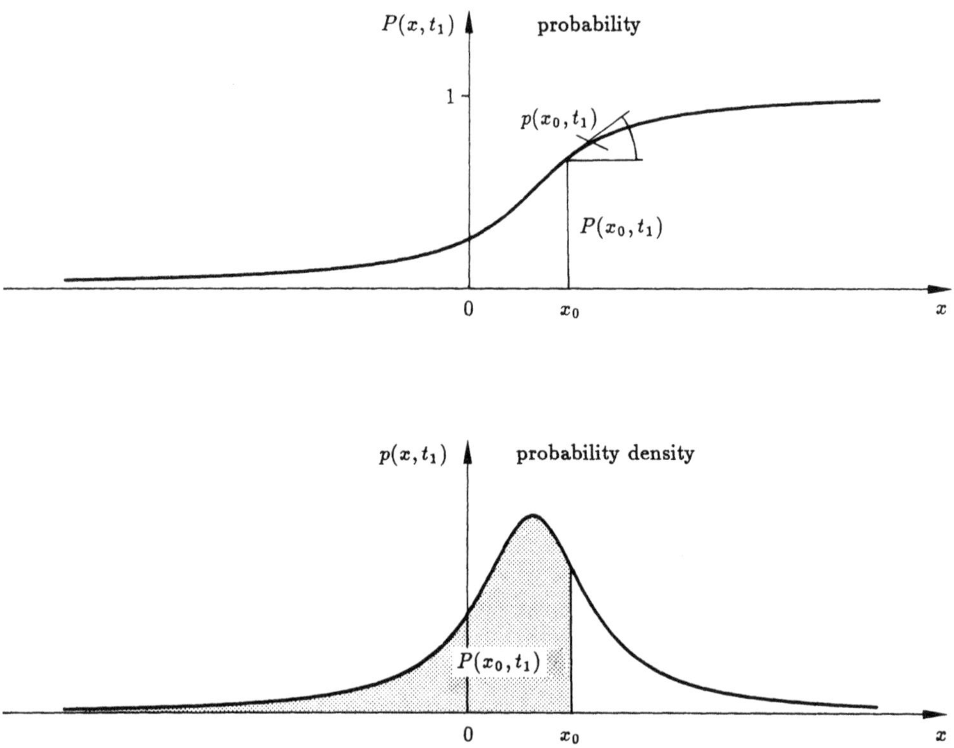

Fig. 8.1.2 Probability P and probability density p of a natural, stochastic process x at an instant t_1

For all natural stochastic processes, we may assume that extreme values only occur very rarely. For example, let us take x to represent the vertical gust velocity on a flight route from Frankfurt to Moscow. In practice, this value is limited and, with all probability, an aircraft on this route will never experience a down gust of $x = -100$ m/s or a rising gust of $x = +100$ m/s. The probability $P(x,t_1)$ is thus zero when $x \leqslant -100$ m/s and it is 1 when $x \geqslant 100$ m/s. On the other hand, small gust velocities will often occur. The increase of probability from 0 to 1 will take place in this range. The fact that no value at all is registered below -100 m/s or above 100 m/s means that the probability remains constant in these ranges and the probability density is zero. For low gust velocities, there is a rapid transition of the probability from 0 to 1, which is linked to high p gradients and consequently to high probability densities. The bell-shaped curve in fig. 8.1.2 is often observed in natural processes.

So far, we have only discussed the situation at an instant t_1 of our inspection of a whole series of measurements. If we now turn to another point of time, $t_1 + \Delta t$, other conditions may arise. For a special class of random processes, the so-called stationary ones, however, the following applies

$$P(x, t_1) = P(x, t_1 + \Delta t), \qquad \Delta t \text{ arbitrary} \tag{8.1.7}$$

The time t is then no longer of relevance and can be omitted in all definitions of the stationary processes. Throughout the following discussion, we presume the existence of a stationary process.

In addition to the probability that $x(t_1)$ does not exceed a certain value x_0 in all series of measurements, we are also interested in other quantities. They appertain to the category of the so-called expectation values. First we name the mean value

$$\overline{x} = E[x] = \int_{-\infty}^{+\infty} x p(x) \mathrm{d}x \tag{8.1.8}$$

as well as the mean square value

$$\overline{x^2} = E[x^2] = \int_{-\infty}^{+\infty} x^2 p(x) \mathrm{d}x \tag{8.1.9}$$

and the mean square deviation or variance

$$\sigma_x^2 = E[(x - \overline{x})^2] = \int_{-\infty}^{+\infty} (x - \overline{x})^2 p(x) \mathrm{d}x \tag{8.1.10}$$

The root of the variance, σ_x, is termed the standard deviation or dispersion. According to this scheme, countless other expectation values can be formulated and the more expectation values of the type

$$E[f(x)] = \int_{-\infty}^{+\infty} f(x) p(x) \mathrm{d}x \tag{8.1.11}$$

are available, the better the random process is described.

In connection with the expectation values, it should be noted that $p(x_0)\mathrm{d}x$ represents the increment in the probability when increasing the limit value from x_0 to $x_0 + \mathrm{d}x$ or, to be more precise, the probability of the variable x lying between x_0 and $x_0 + \mathrm{d}x$. This probability acts as the weighting factor for all expectation

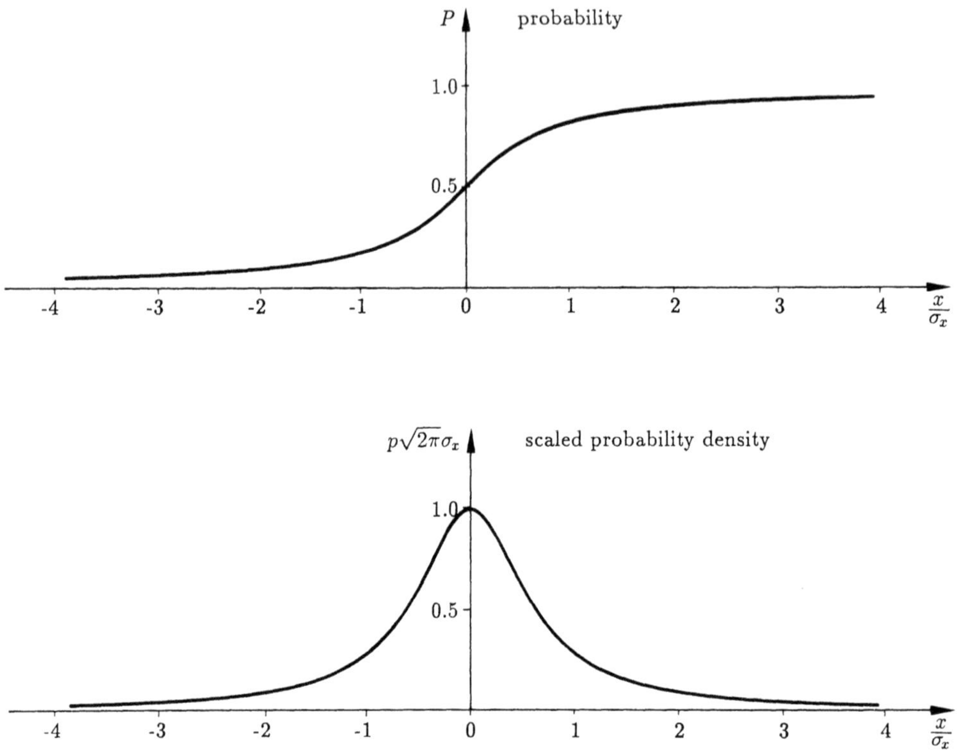

Fig. 8.1.3 Gaussian or normal distribution with a mean value of $\overline{x} = 0$: probability and probability density

values and, therefore, the most frequently occuring values are seen to enter in the summation with a reinforced weighting.

Gaussian distribution, also known as normal distribution, is the most frequently employed density function

$$p(x) = \frac{1}{\sqrt{2\pi}\sigma_x} \exp\left(-\frac{(x-\overline{x})^2}{2\sigma_x^2}\right) \tag{8.1.12}$$

which is obviously uniquely determined by the mean value and the variance. When applying such a density function to the excitation functions in dynamics, the condition

$$\overline{x} = \int\limits_{-\infty}^{+\infty} xp(x)\mathrm{d}x = 0 \tag{8.1.13}$$

can always be fulfilled since x can be shifted by $\overline{x}$, and the ensuing constant term can be evaluated deterministically. In such a case, normal distribution is determined by the variance alone. A graphic representation of this density function may be found in fig. 8.1.3. In what follows, we shall employ a mean value of zero; consequently, the Gaussian density function becomes symmetrical with respect to $x = 0$. Thus

$$\int_{-\infty}^{0} p(x)\mathrm{d}x = \int_{0}^{+\infty} p(x)\mathrm{d}x = \frac{1}{2} \tag{8.1.14}$$

The derivation of the probability from the probability density function leads to the so-called probability integral, the evaluation of which is given in tabular form in many standard mathematical handbooks

$$\begin{aligned} P(x) = \int_{-\infty}^{x} p(x)\mathrm{d}x &= \frac{1}{\sigma_x\sqrt{2\pi}} \int_{-\infty}^{x} \exp(-x^2/2\sigma_x^2)\mathrm{d}x \\ &= \frac{1}{\sqrt{2\pi}} \int_{\infty}^{x/\sigma_x} \mathrm{e}^{-\xi^2/2}\mathrm{d}\xi \qquad (8.1.15) \\ &= \frac{1}{\sqrt{2\pi}} \left(\frac{1}{2} \mp \frac{1}{\sqrt{2\pi}} \int_{0}^{x/\sigma_x} \mathrm{e}^{-\xi^2/2}\mathrm{d}\zeta\right), \quad \text{where} \quad \begin{matrix} -, \text{ if } x < 0 \\ +, \text{ if } x > 0 \end{matrix} \end{aligned}$$

The expectation values can also easily be obtained for normal or Gaussian distribution through corresponding definite integrals.

The evaluation of a series of measurements so far discussed involved at an instant t (for stationary processes, this point in time is arbitrary) the determination – across N series of measurements – of the number of instances for which a quantity x is smaller than a limit x_0 (see fig. 8.1.1). We have already pointed out that whilst this may be statistically correct, it is economically inconvenient. Which aeronautical consortium would, for example, be willing to provide us with 1000 pilots and machines to statistically evaluate manufacturing and operating tolerances? Moreover, the figure of 1000 quoted here is statistically an insignificant number. At the same time, we are forced in practice to resort to very few specimens, often to only a single one. Intuitively, we as engineers feel that even a single aeroplane, if built with sufficient production precision, can yield satisfactory statistical data. To this purpose, we presume that the standard operating data (such as flight-altitude, weather and pilot) are monitored continually over a longer period of time. The statistical evaluation of these measurements proceeds along the time-axis and no longer across the test objects (fig. 8.1.4). The statistical processes in which this procedure is permissible are termed ergodic. At the same time, ergodic processes are stationary in nature. In what follows, we take ergodicity for granted.

So far in the discussion, $p(x_0)\mathrm{d}x$ defined the probability of a measured value lying between the two points x_0 and $x_0 + \mathrm{d}x$ and the readings being carried out over all series of measurements at a fixed point in time. In the case of ergodic processes, it has to be ascertained how often this measured value lies between x_0 and $x_0 + \Delta x$.

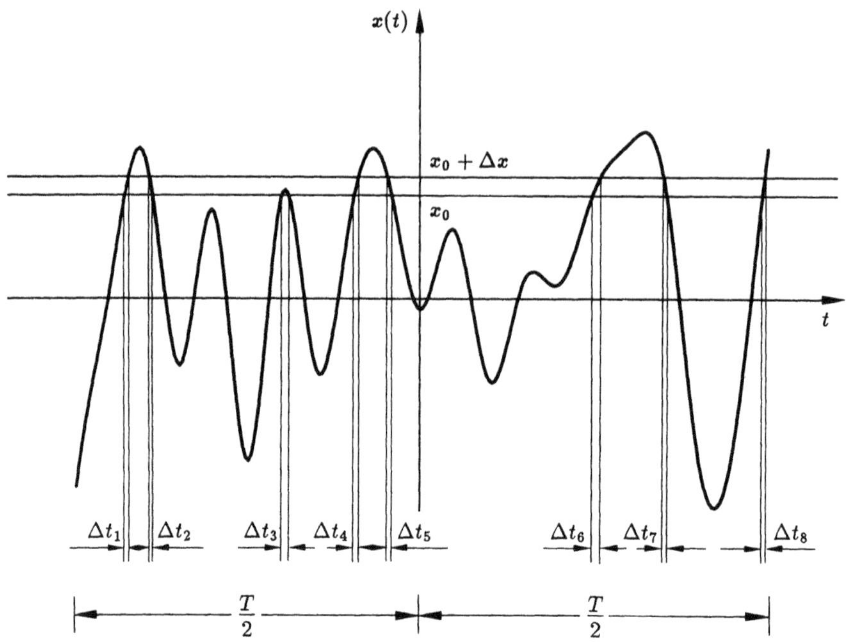

Fig. 8.1.4 On the calculation of the probability density of an ergodic stochastic process

This is illustrated in fig. 8.1.4. The ratio of this information to the complete period of measurements, T, then yields the probability. Thus, we have

$$p(x_0)\Delta x = \frac{\Delta t_1 + \Delta t_2 + \Delta t_3 + \dots}{T} = \frac{1}{T}\sum_{\mathrm{i}} \Delta t_{\mathrm{i}} \tag{8.1.16}$$

where, from the mathematical point of view, T should be as large as possible. A numerical evaluation of the probability density according to equ. (8.1.16) can be carried out on a computer by dividing the time range, T, into small, equal intervals Δt and considering the functional values $x(t)$ over the time period of each interval. We now let the total number of intervals be N (i.e. $T = \mathrm{N}\Delta t$) and count the number of intervals N_x at which $x(t)$ lies between x_0 and $x_0 + \Delta x$ where Δx is small. We then have

$$p(x_0)\Delta x = \frac{\mathrm{N}_x}{\mathrm{N}} \tag{8.1.17}$$

To obtain the expectation values for an ergodic process, we can adapt equs. (8.1.8), (8.1.9), (8.1.10) and (8.1.11), including equ. (8.1.16). This yields the following mean value

$$\overline{x} = \lim_{T\to\infty} \frac{1}{T} \int_{-T/2}^{+T/2} x(t)\mathrm{d}t \tag{8.1.18}$$

as well as the mean square value

$$\overline{x^2} = \lim_{T\to\infty} \frac{1}{T} \int_{-T/2}^{+T/2} x^2(t)\mathrm{d}t \tag{8.1.19}$$

and the variance

$$\sigma_x^2 = \lim_{T\to\infty} \frac{1}{T} \int_{-T/2}^{+T/2} \left(x(t) - \overline{x}\right)^2 \mathrm{d}t \tag{8.1.20}$$

All of these will, as a rule, rapidly converge within a finite time interval, T, to a stationary value. Thus, in practice, the limiting process $T \to \infty$ is no longer required. The mean values mentioned so far can also be applied to non-ergodic processes, in which case they are known as temporal mean values and differ from the mean values evaluated across the measured readings. This observation may provide a method of testing the ergodicity of a process.

In addition to the expectation values already mentioned (mean value, mean square value and variance), additional quantities which can easily be experimentally obtained are useful in the analysis of stochastic excitations in dynamics. These are the autocorrelation function and the spectral density. The general definition of the autocorrelation function presumes once more a parallel series of measurements $x_k(t)$ (fig. 8.1.5). Observations of the measured values at the instants t_1 and $t_1 + \Delta t$ yield the expectation value

$$\begin{aligned}\Phi_x(t_1, \Delta t) &= E[x(t_1)x(t_1 + \Delta t)] \\ &= \int_{-\infty}^{+\infty} x(t_1)x(t_1 + \Delta t)p(x)\mathrm{d}x\end{aligned} \tag{8.1.21}$$

We at once confirm the general validity of

$$\Phi_x(t_1, \Delta t = 0) = \overline{x^2} \tag{8.1.22}$$

For all processes for which the mean value is zero, the following is also true

$$\Phi_x(t_1, \Delta t = 0) = \overline{x^2} = \sigma_x^2 , \qquad \text{if} \quad \overline{x} = 0 \tag{8.1.23}$$

For stationary processes, Φ_x is thus no longer dependent upon the time t_1, but only upon the time-difference Δt in the series of measurements. Hence, t_1 can be ignored as a determining variable and instead of equ. (8.1.23), we have the simplified expression

$$\Phi_x(0) = \sigma_x^2 , \qquad \text{if} \quad \overline{x} = 0 \text{ and } x \text{ stationary} \tag{8.1.24}$$

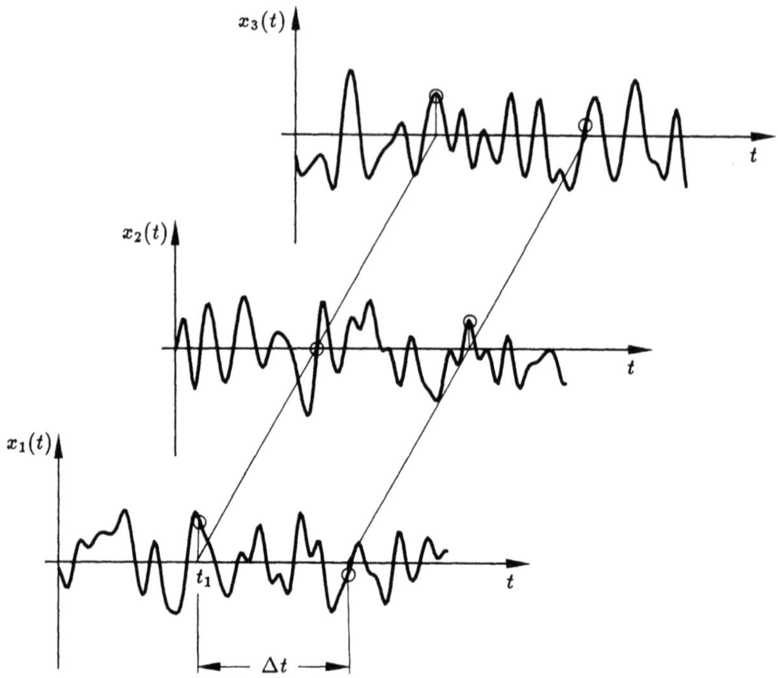

Fig. 8.1.5 On the general definition of the autocorrelation function

Furthermore, $\Phi_x(\Delta t)$ is a symmetrical function for stationary processes since the following applies

$$\begin{aligned}\Phi_x(-\Delta t) &= E[x(t_1)x(t_1 - \Delta t)] \\ &= E[x(t_1 - \Delta t)x(t_1 - \Delta t + \Delta t)] \\ &= \Phi_x(+\Delta t)\end{aligned} \tag{8.1.25}$$

By the skilful selection of a new expectation value, we can easily prove that $\Phi_x(\Delta t)$ is always smaller than $\Phi_x(0)$ in stationary processes. To this purpose, we define

$$\begin{aligned}E\Big[\big(x(t_1) \pm x(t_1+\Delta t)\big)^2\Big] &= \int_{-\infty}^{+\infty} \big(x(t_1) \pm x(t_1+\Delta t)\big)^2 p(x)\,\mathrm{d}x \\ &= \int_{-\infty}^{+\infty} \big(x^2(t_1) \pm 2x(t_1)x(t_1+\Delta t) + x^2(t_1+\Delta t)\big)\, p(x)\,\mathrm{d}x \\ &= \Phi_x(0) \pm 2\Phi_x(\Delta t) + \Phi_x(0) \geqslant 0\end{aligned} \tag{8.1.26}$$

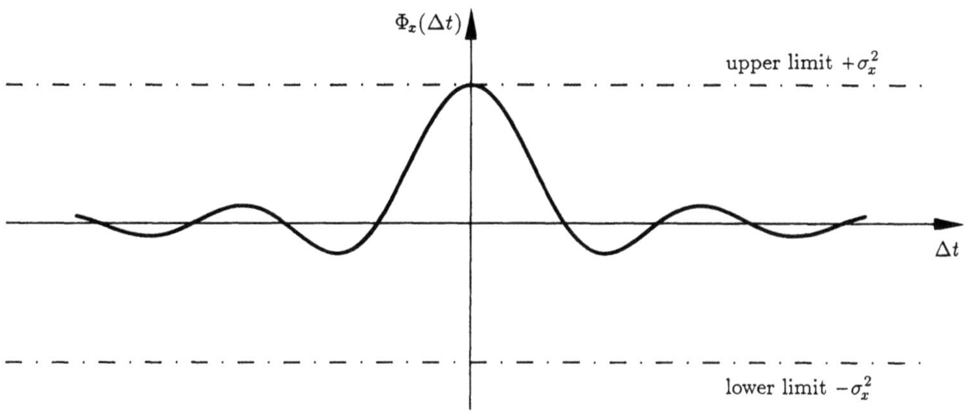

Fig. 8.1.6 Typical example of an autocorrelation function for a stationary process with a mean value of zero

The fact that the expectation function defined in this manner must be positive can be seen from the right-hand side of the first line of equ. (8.1.26). Since a quadratic term is always positive and $p(x)$ is also positive or zero, integration also yields a positive value. For a stationary process, $\Phi_x(t_1, 0)$ and $\Phi_x(t_1 + \Delta t, 0)$ are of equal magnitude. The last line of equ. (8.1.26) leads to

$$\Phi_x(0) \geqslant |\Phi_x(\Delta t)|, \qquad \text{if } x \text{ stationary} \tag{8.1.27}$$

No more general remarks can be made, except that for a great many stationary stochastic processes, the autocorrelation function Φ_x decreases rapidly with increasing Δt (see the qualitative course in fig. 8.1.6).

In an ergodic process, the derivation of the expectation value from the series of measurements can be replaced by a mean value deduced from the temporal evolution of one measurement point, as discussed before. Proceeding in this way, we obtain from equ. (8.1.21)

$$\Phi_x(\Delta t) = \lim_{T\to\infty} \frac{1}{T} \int\limits_{-T/2}^{+T/2} x(t)x(t + \Delta t)\mathrm{d}t, \qquad \text{if } x \text{ ergodic} \tag{8.1.28}$$

In addition to the autocorrelation function, the spectral density function S_x is of particular importance in the treatment of stochastic processes. A strict mathematical description of this function, however, lies outside the scope of this book and instead, the reader is advised to consult [3.1] and other standard works on the mathematics of stochastic processes. We have already mentioned (section 6.3) that a function $x(t)$ can only be represented without cautionary restrictions by a Fourier integral if the condition

$$\int\limits_{-\infty}^{+\infty} |x(t)|\mathrm{d}t < \infty \tag{8.1.29}$$

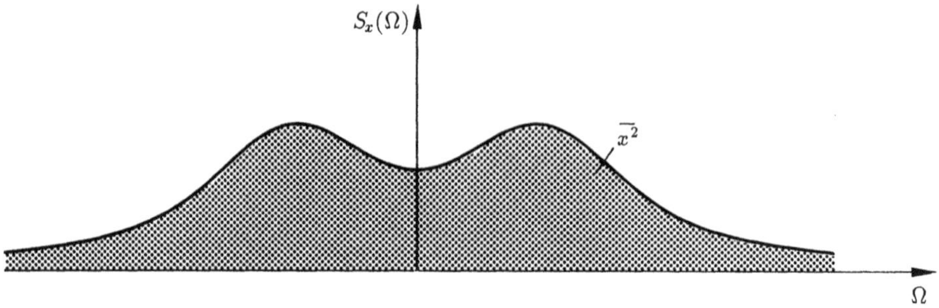

(a) spectral density and mean square value ($\bar{x} = 0$)

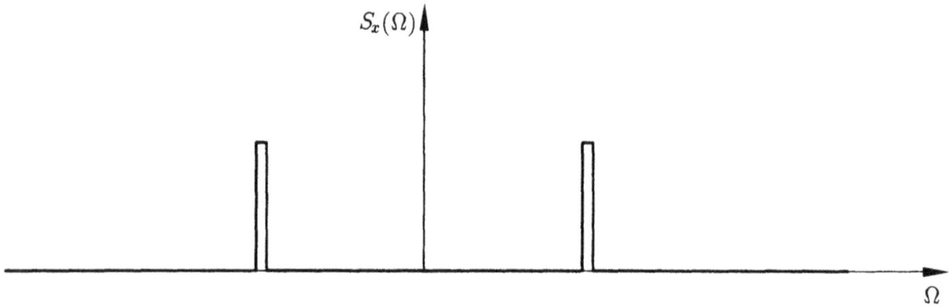

(b) spectral density of a stochastic process with a narrow frequency band

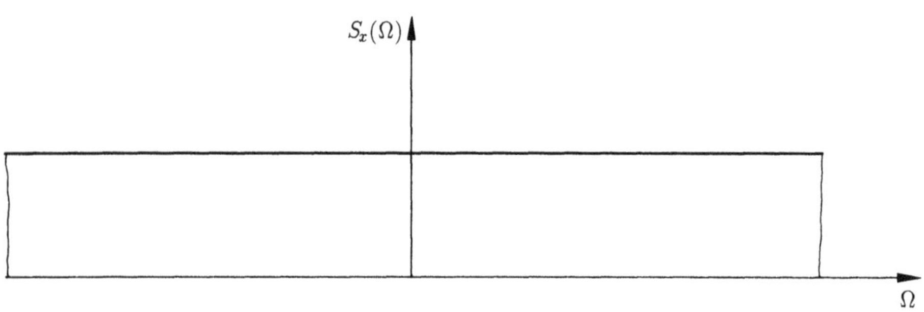

(c) spectral density of a stochastic process with a wide frequency band (white noise)

Fig. 8.1.7 Spectral density distributions of stochastic processes

applies. If $x(t)$ is here taken to represent the excitation, then this proves a most adverse illustration. Since a stochastic excitation (from the theoretical point of view, at least) can last indefinitely, equ. (8.1.29) need not hold even though the mean value $\overline{x}$ is zero. If, on the other hand, we turn our attention to the autocorrelation function $\Phi_x(\Delta t)$, the chance of the absolute area below the curve remaining finite increases. For small values of Δt, it follows from the continuity of the $x(t)$-curve that both $x(t)$ and $x(t + \Delta t)$ have the same sign and $\Phi_x(\Delta t)$ must thus grow along the time-axis. There is no predictable relationship in a random process between $x(t)$ and $x(t + \Delta t)$ for large values of Δt; occasionally, negative values may arise and at other times positive ones, which may cancel each other out. The value Φ_x approaches zero for large Δt. An exception arises for periodic processes $x(t)$ for which $\Phi_x(\Delta t)$ can also be periodic. Since the condition of equ. (8.1.29) can then no longer be fulfilled, the periodic components of $\Phi_x(\Delta t)$ have to be extracted and considered separately.

Let us now turn our attention to the standard case of an aperiodic autocorrelation $\Phi_x(\Delta t)$ which satisfies equ. (8.1.29). In order to represent this function by means of a Fourier integral, we may apply the procedure of section 6.3, where nonperiodic excitation functions were described in this way. From equs. (6.3.4) and (6.3.5), we deduce analogously

$$\Phi_x(\Delta t) = \int_{-\infty}^{+\infty} S_x(\Omega)\, \mathrm{e}^{i\Omega\Delta t} \mathrm{d}\Omega \tag{8.1.30}$$

where

$$S_x(\Omega) = \frac{1}{2\pi} \int_{-\infty}^{+\infty} \Phi_x(\Delta t)\, \mathrm{e}^{-i\Omega\Delta t} \mathrm{d}(\Delta t) \tag{8.1.30a}$$

is the aforementioned spectral density function. In addition, $S_x(\Omega)$ is the Fourier transform of] the function $\Phi_x(\Delta t)$ and $\Phi_x(\Delta t)$ is the inverse Fourier transform of the spectral density $S_x(\Omega)$. The continuous function $S_x(\Omega)$ obtained in this manner may, in addition, have to be superimposed by a series of impulses if $\Phi_x(\Delta t)$ or $x(t)$ contains periodic components.

An important characteristic of the $S_x(\Omega)$ function can be deduced from equ. (8.1.30) for $\Delta t = 0$

$$\Phi_x(0) = \int_{-\infty}^{+\infty} S_x(\Omega)\, \mathrm{d}\Omega = \overline{x^2} \tag{8.1.31}$$

Hence, the shaded area below the plot of the spectral density function (see fig. 8.1.7) represents the mean square value of the process. Using Euler's formula, the complex representation in equ. (8.1.30a) can, as has already been observed in section 6.3, be written in the alternative form

$$\begin{aligned} S_x(\Omega) &= \frac{1}{2\pi} \int_{-\infty}^{+\infty} \Phi_x(\Delta t) \cos(\Omega\Delta t)\, \mathrm{d}(\Delta t) - \frac{i}{2\pi} \int_{-\infty}^{+\infty} \Phi_x(\Delta t) \sin(\Omega\Delta t)\, \mathrm{d}(\Delta t) \\ &= \frac{1}{2\pi} \int_{-\infty}^{+\infty} \Phi_x(\Delta t) \cos(\Omega\Delta t)\, \mathrm{d}(\Delta t) \end{aligned} \tag{8.1.32}$$

the general random process

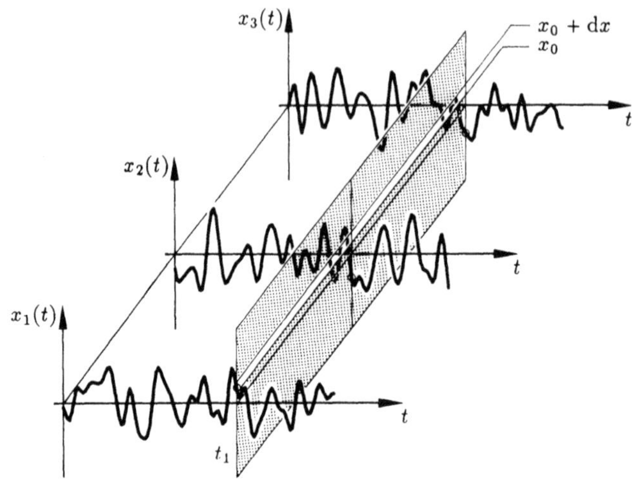

probability at x_0 and t_1

let $x < x_0$ in N_x of N series of measurements

$$P(x_0, t_1) = \lim_{N\to\infty} \frac{N_x}{N} \quad , \quad 0 \leqslant P \leqslant 1 \quad \text{monotonic}$$

probability density at x_0 and t_1

$$p(x_0, t_1) = \frac{\mathrm{d}P}{x}(x_0) \geqslant 0$$

general expectation value

$$E\left[f(x)\right] = \int_{-\infty}^{+\infty} f(x)\, p(x)\, \mathrm{d}x$$

special forms of expectation value

mean value of x

$$\overline{x}(t_1) = \int_{-\infty}^{+\infty} x(t_1)\, p(x)\, \mathrm{d}x$$

mean square value of x

$$\overline{x^2}(t_1) = \int_{-\infty}^{+\infty} x^2(t_1)\, p(x)\, \mathrm{d}x$$

Fig. 8.1.8 The general random process: the most important definitions and relations, *Part 1* variance

$$\sigma_x^2(t_1) = \int_{-\infty}^{+\infty} (x(t_1) - \overline{x}(t_1))^2 \, p(x) \, \mathrm{d}x$$

autocorrelation function

$$\Phi_x(t_1, \Delta t) = \int_{-\infty}^{+\infty} x(t_1) \, x(t_1 + \Delta t) \, p(x) \, \mathrm{d}x$$

relations

$$\Phi_x(t_1, 0) = \overline{x^2}(t_1)$$

$$\Phi_x(t_1, 0) = \sigma_x^2(t_1) \qquad \text{if} \quad \overline{x}(t_1) = 0$$

Note:
if the random process is stationary, all results become independent of the time t_1; consequently, t_1 can then be omitted.

Fig. 8.1.8 The general random process: the most important definitions and relations, *Part 2*

Here, the imaginary part must vanish since $\Phi_x(\Delta t)$ is symmetrical and sin $\Omega\Delta t$ is antisymmetrical with respect to $\Delta t = 0$. Consequently, the areas below the antisymmetrical integrand function cancel one another out. If we next examine the behaviour of the density function $S_x(\Omega)$ with respect to Ω, we immediately obtain

$$\begin{aligned} S_x(-\Omega) &= \frac{1}{2\pi} \int_{-\infty}^{+\infty} \Phi_x(\Delta t) \cos(-\Omega \Delta t) \, \mathrm{d}(\Delta t) \\ &= S_x(\Omega) \end{aligned} \tag{8.1.33}$$

The spectral density function is thus not only real, but also symmetrical with respect to $\Omega = 0$, as is the autocorrelation. Whilst Davenport [8.2] has proved $S_x(\Omega)$ to be positive, we are at least able to deduce from equ. (8.1.31) that the area below the $S_x(\Omega)$ curve must be positive, which is in any case only conceivable if S_x is dominantly positive. Figure 8.1.7 shows several typical spectral density distributions.

The authors are aware that in this chapter the reader has been introduced to many new definitions and this, moreover, in an almost unfamiliar area. To alleviate the burden on the reader, we summarise in figs. 8.1.8 and 8.1.9 the most important relationships so far discussed.

the ergodic random process

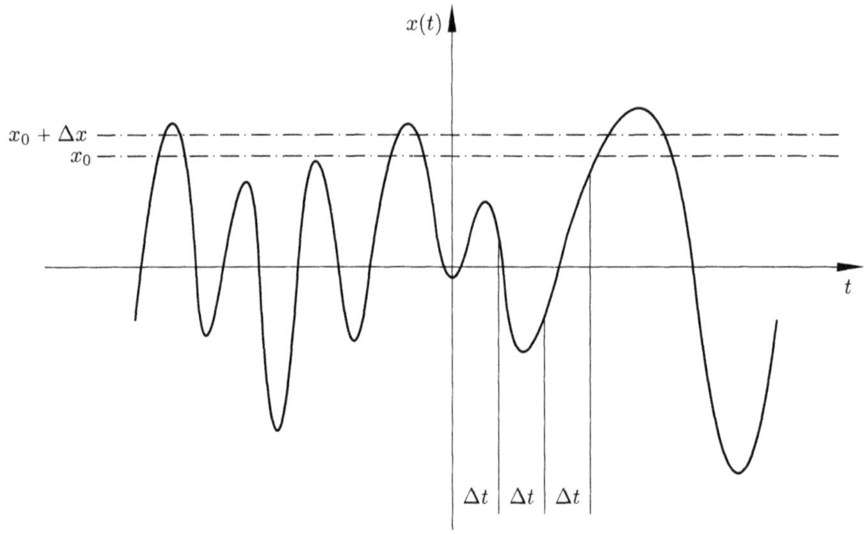

probability at x_0

in N_x of N intervals Δt, let $x < x_0$; then

$$P(x_0) = \lim_{\substack{\Delta t \to 0 \\ N \to \infty}} \frac{N_x}{N} \quad , \quad 0 \leqslant P \leqslant 1 \quad \text{monotonic}$$

probability density at x_0

let $x < x_0 + \Delta x$ in $N_{x+\Delta x}$ intervals Δt, then

$$p(x_0) = \lim_{\Delta x \to 0} \frac{P(x_0 + \Delta x) - P(x_0)}{\Delta x}$$

$$= \lim_{\substack{\Delta t \to 0 \\ N \to \infty \\ \Delta x \to 0}} \frac{N_{x+\Delta x} - N_x}{\Delta x N} \quad \geqslant 0$$

general expectation value

$$E\left[f(x)\right] = \lim_{T\to\infty} \frac{1}{T} \int_{-T/2}^{+T/2} f(x)\,\mathrm{d}t$$

special forms of expectation value

mean value of x

$$\overline{x} = \lim_{T\to\infty} \frac{1}{T} \int_{-T/2}^{+T/2} x\,\mathrm{d}t$$

Fig. 8.1.9 The ergodic random process: the most important definitions and relations, *Part 1* mean square value of x

$$\overline{x^2} = \lim_{T \to \infty} \frac{1}{T} \int_{-T/2}^{+T/2} x^2 \, dt$$

variance

$$\sigma_x^2 = \lim_{T \to \infty} \frac{1}{T} \int_{-T/2}^{+T/2} (x - \overline{x})^2 \, dt$$

autocorrelation function

$$\Phi_x(\Delta t) = \lim_{T \to \infty} \frac{1}{T} \int_{-T/2}^{+T/2} x(t) \, x(t + \Delta t) \, dt$$

spectral density function

$$S_x(\Omega) = \frac{1}{2\pi} \int_{-\infty}^{+\infty} \Phi_x(\Delta t) \, e^{-i\Omega\Delta t} \, d(\Delta t)$$

$$= \frac{1}{2\pi} \int_{-\infty}^{+\infty} \Phi_x(\Delta t) \, \cos(\Omega\Delta t) \, d(\Delta t)$$

relations

$$\Phi_x(0) = \overline{x^2}$$

$$\Phi_x(0) = \sigma_x^2 \qquad \text{if} \quad \overline{x} = 0$$

$$\Phi_x(\Delta t) = \int_{-\infty}^{+\infty} S_x(\Omega) \, e^{-i\Omega\Delta t} \, d\Omega$$

$$\Phi_x(0) = \int_{-\infty}^{+\infty} S_x(\Omega) \, d\Omega = \sigma_x^2 \qquad \text{if} \quad \overline{x} = 0$$

Fig. 8.1.9 The ergodic random process: the most important definitions and relations, *Part 2*

8.2 Random excitation of a modal degree of freedom in an uncoupled system (modal damping)

We begin with the dynamic analysis of a single random process already dealt with in section 8.1. For this purpose, we consider the following component equation of the diagonalised system

$$\ddot{\eta}_i + 2\zeta_i \omega_i \dot{\eta}_i + \omega_i^2 \eta_i = \omega_i^2 R_{\eta i}(t) \tag{8.2.1}$$

where $R_{\eta i}(t)$ is assumed to be an ergodic random force with a mean value of zero. In this way, we are in a position to discuss both systems with a single degree of freedom and multi-degree but uncoupled stochastic processes. In the next section, we shall extend our analysis to the more general case where each degree of freedom r_i is subjected to an independent, random excitation $R_i(t)$.

The Duhamel integral, with which the reader is already familiar, can now be used to compute the response η of the system (for convenience, the index i is omitted). For undamped systems, equ. (6.4.11) provides the solution

$$\eta(t) = \int\limits_{\bar{t}=0}^{\bar{t}=t} R_\eta(\bar{t})h(t-\bar{t})\,\mathrm{d}\bar{t} \tag{8.2.2}$$

in which

$$h(t-\bar{t}) = \omega \sin \omega(t-\bar{t}) \tag{8.2.3}$$

represents the response of the system due to a unit Dirac impulse (see equs. (6.4.9) and (6.4.10)). The response to a Dirac impulse in a modally damped system has not been detailed before but the reader should have no difficulties in verifying the expression himself. Indeed, proceeding from equ. (7.4.10) for a damped free vibration defined by the initial conditions

$$\begin{aligned} \eta_0 &= 0 \\ \dot{\eta}_0 &= \omega_\mathrm{i}^2 \int R_{\eta\mathrm{i}} \mathrm{d}t = \omega_\mathrm{i}^2 \cdot 1 \end{aligned} \tag{8.2.4}$$

and introducing the necessary time shift, we obtain, assuming weak damping, the solution

$$h(t-\bar{t}) = \frac{\omega}{\sqrt{1-\zeta^2}} \mathrm{e}^{-\omega\zeta(t-\bar{t})} \sin\left(\omega_D(t-\bar{t})\right) \tag{8.2.5}$$

with the damped angular frequency $\omega_D = \omega\sqrt{1-\zeta^2}$. This expression can be applied up to the critical damping value $\zeta = 1$. For $\zeta = 0$, the solution reduces to the familiar one in equ. (8.2.3). Both equs. (8.2.3) and (8.2.5) should be supplemented by

$$h(t-\bar{t}) = 0 \quad \text{for } t < \bar{t} \tag{8.2.6}$$

since no response can exist before the onset of the unit impulse. However, it should be noted that our random excitation $R_\eta(\bar{t})$ does not necessarily begin at $\bar{t}$ but could have been going on prior to $t = \bar{t}$ (at least from the theoretical point of view) for an indefinite period of time (fig. 8.2.1). Thus, equ. (8.2.2) for a random excitation can also be written

$$\begin{aligned} \eta(t) &= \int\limits_{\bar{t}=-\infty}^{\bar{t}=t} R_\eta(\bar{t})h(t-\bar{t})\,\mathrm{d}\bar{t} \\ &= \int\limits_{\bar{t}=-\infty}^{\bar{t}=+\infty} R_\eta(\bar{t})h(t-\bar{t})\,\mathrm{d}\bar{t} \end{aligned} \tag{8.2.7}$$

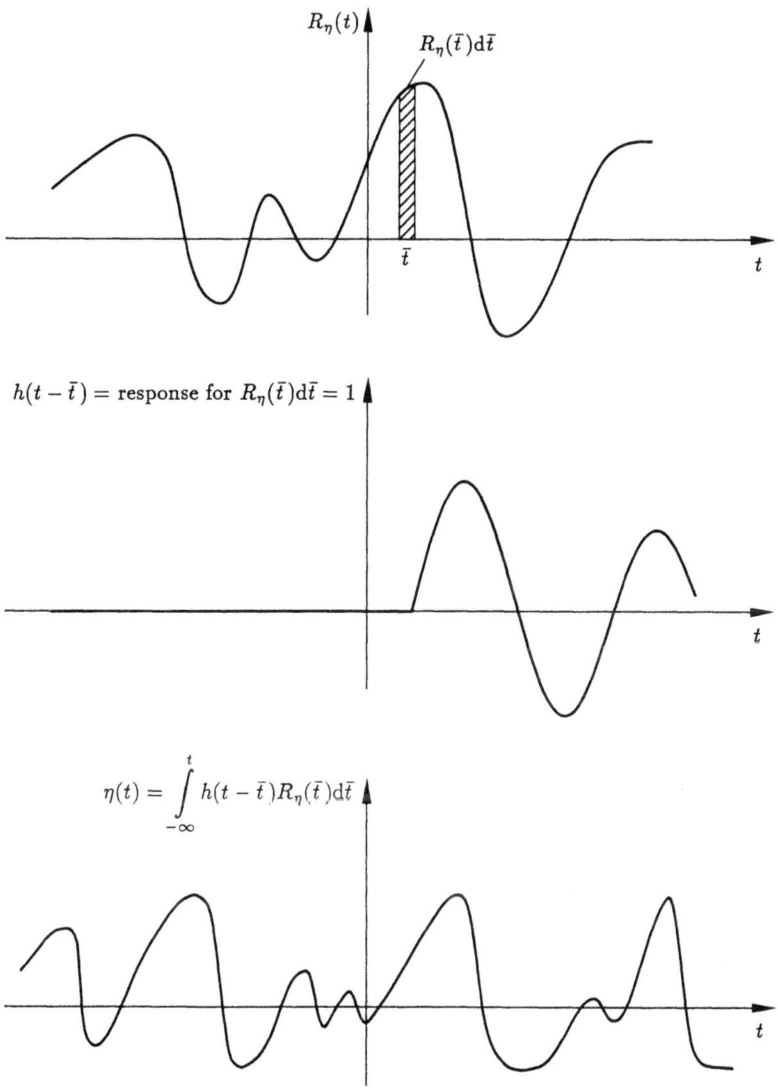

Fig. 8.2.1 Random excitation and infinitesimal impulse; unit response and superposed complete response

Introducing the new time-variable

$$\theta = t - \bar{t} \tag{8.2.8}$$

we find the alternative expression for the Duhamel integral

$$\eta(t) = \int_{-\infty}^{+\infty} h(\theta) R_\eta(t-\theta)\,\mathrm{d}\theta \tag{8.2.9}$$

Since $R_\eta(t)$ is presumed random, it follows that $\eta(t)$ is random, too. The mean value of the response being our first concern, we form the expression

$$\begin{aligned}\overline{\eta} &= E[\eta] \\ &= E\Big[\int_{-\infty}^{+\infty} h(\theta) R_\eta(t-\theta)\,\mathrm{d}\theta\Big] = \int_{-\infty}^{+\infty} h(\theta) E\Big[R_\eta(t-\theta)\Big]\,\mathrm{d}\theta \\ &= \int_{-\infty}^{+\infty} h(\theta)\overline{R}_\eta\,\mathrm{d}\theta \\ &= \overline{R}_\eta \int_{-\infty}^{+\infty} h(\theta)\,\mathrm{d}\theta \end{aligned} \tag{8.2.10}$$

It should be noted here that the formation of the mean value for an ergodic process $\eta(t)$ takes place over the time axis t; but since only $R_\eta(t-\theta)$ is dependent upon the time t, the mean value can be executed before integration. The excitation process R_η being also ergodic and therefore stationary, the mean value $\overline{R}_\eta$ must be time-independent. The mean value of the response must hence take the form given in the final line of equ. (8.2.10). As a most important consequence of the above we state: if the mean value of the excitation is zero, so, too, is the mean value of the response. The distinct consideration and deterministic analysis of a mean value $\overline{R}_\eta \neq 0$ has already been discussed.

The next stage is the discussion on the autocorrelation of the response. Once again, we are faced by an expectation value and have

$$\begin{aligned}\Phi_\eta(\Delta t) &= E\big[\eta(t)\eta(t+\Delta t)\big] \\ &= E\Big[\Big(\int_{-\infty}^{+\infty} h(\theta_1)R_\eta(t-\theta_1)\mathrm{d}\theta_1\Big)\Big(\int_{-\infty}^{+\infty} h(\theta_2)R_\eta(t+\Delta t-\theta_2)\mathrm{d}\theta_2\Big)\Big] \\ &= E\Big[\int_{-\infty}^{+\infty}\int_{-\infty}^{+\infty} h(\theta_1)h(\theta_2)R_\eta(t-\theta_1)R_\eta(t-\theta_2+\Delta t)\mathrm{d}\theta_1\mathrm{d}\theta_2\Big] \\ &= \int_{-\infty}^{+\infty}\int_{-\infty}^{+\infty} h(\theta_1)h(\theta_2)E\Big[R_\eta(t-\theta_1)R_\eta(t-\theta_2+\Delta t)\Big]\mathrm{d}\theta_1\mathrm{d}\theta_2 \\ &= \int_{-\infty}^{+\infty}\int_{-\infty}^{+\infty} h(\theta_1)h(\theta_2)\Phi_E(\Delta t+\theta_1-\theta_2)\mathrm{d}\theta_1\mathrm{d}\theta_2 \end{aligned} \tag{8.2.11}$$

Here, Φ_E is the autocorrelation of the excitation and equ. (8.2.11) determines the autocorrelation of the response. Unfortunately, however, the detailed evaluation is rather complicated in practice. We deduce incidentally from equations (8.2.10) and (8.2.11) that the response $\eta(t)$ is also stationary for a stationary excitation process $R_\eta(t)$ since neither the mean value $\overline{\eta}$ nor the autocorrelation $\Phi_\eta(\Delta t)$ are dependent upon time t.

The conversion of the spectral density S_E of the excitation into that of the response, S_η, can easily be calculated. For this purpose, we employ the definition of the spectral density given in equ. (8.1.30a) and apply equ. (8.2.11) to obtain the corresponding autocorrelation function

$$
\begin{aligned}
S_\eta(\Omega) &= \frac{1}{2\pi}\int\limits_{\Delta t=-\infty}^{+\infty} \Phi_\eta(\Delta t)\,\mathrm{e}^{-i\Omega\Delta t}\mathrm{d}(\Delta t) \\
&= \frac{1}{2\pi}\int\limits_{\Delta t=-\infty}^{+\infty}\int\limits_{\theta_1=-\infty}^{+\infty}\int\limits_{\theta_2=-\infty}^{+\infty} h(\theta_1)h(\theta_2)\Phi_E(\Delta t+\theta_1-\theta_2)\mathrm{d}\theta_2\mathrm{d}\theta_1\,\mathrm{e}^{-i\Omega\Delta t}\mathrm{d}(\Delta t) \\
&= \int\limits_{-\infty}^{+\infty}\int\limits_{-\infty}^{+\infty} h(\theta_1)h(\theta_2)\Big(\frac{1}{2\pi}\int\limits_{\Delta t=-\infty}^{+\infty}\Phi_E(\Delta t+\theta_1-\theta_2)\,\mathrm{e}^{-i\Omega\Delta t}\mathrm{d}(\Delta t)\Big)\mathrm{d}\theta_1\mathrm{d}\theta_2 \\
&= S_E(\Omega)\int\limits_{-\infty}^{+\infty}\int\limits_{-\infty}^{+\infty} h(\theta_1)h(\theta_2)\,\mathrm{e}^{-i\Omega(\theta_2-\theta_1)}\mathrm{d}\theta_1\mathrm{d}\theta_2
\end{aligned} \tag{8.2.12}
$$

The statistical measures obtained so far for the response still contain some integral expressions. These can be simplified, however, by means of the Fourier integral method, with some surprising results. In this context, the reader is reminded of the response given by equ. (6.3.10) where, for an undamped oscillator, the frequency characteristic of equ. (6.3.11) has to be inserted and, for a modally damped system, the complex frequency characteristic of equ. (7.4.12). Applying this Fourier integral method to an impulse excitation with the unit value at $\theta = 0$, we deduce the response $h(\theta)$ as

$$
h(\theta) = \int\limits_{-\infty}^{+\infty} F(\Omega)k(\Omega)\,\mathrm{e}^{i\Omega\theta}\mathrm{d}\Omega \tag{8.2.13}
$$

where the complex frequency response characteristic is

$$
F(\Omega) = \frac{1}{1-(\frac{\Omega}{\omega})^2 + i2\zeta\frac{\Omega}{\omega}} \tag{8.2.14}
$$

and the frequency spectrum of the impulse excitation becomes

$$\begin{aligned} k(\Omega) &= \frac{1}{2\pi}\int_{-\infty}^{+\infty} R_\eta e^{-i\Omega\theta} d\theta \\ &= \frac{1}{2\pi} \end{aligned} \tag{8.2.15}$$

This last result is evident because R_η operates only in the range of small values of θ where $e^{-i\Omega\theta} \approx 1$. The integral then yields the impulse and this is indeed 1. In conjunction with equ. (8.2.15), equ. (8.2.13) gives

$$h(\theta) = \frac{1}{2\pi}\int_{-\infty}^{+\infty} F(\Omega)\, e^{i\Omega\theta} d\Omega \tag{8.2.16}$$

A comparison with equs. (8.1.30) and (8.1.31) yields the expression $F(\Omega)/2\pi$ for the Fourier transform of the function $h(\theta)$. Analogous to equ. (8.1.31), we then have

$$F(\Omega) = \int_{-\infty}^{+\infty} h(\theta)\, e^{-i\Omega\theta} d\theta \tag{8.2.17}$$

where the complex frequency characteristic is defined by the simple equ. (8.2.14). Thus, the integrals with infinite limits are elegantly evaluated and, instead of equ. (8.2.10), we obtain for the mean value of the response when $\Omega = 0$

$$\begin{aligned} \overline{\eta} &= \overline{R}_\eta F(0) \\ &= \overline{R}_\eta \end{aligned} \tag{8.2.18}$$

For equ. (8.2.12), we first deduce

$$\begin{aligned} \int_{-\infty}^{+\infty}\int_{-\infty}^{+\infty} h(\theta_1)h(\theta_2)\, e^{-i\Omega(\theta_2-\theta_1)} d\theta_1 d\theta_2 &= \int_{-\infty}^{+\infty} h(\theta)\, e^{+i\Omega\theta} d\theta \int_{-\infty}^{+\infty} h(\theta)\, e^{-i\Omega\theta} d\theta \\ &= F^*(\Omega)F(\Omega) \\ &= V^2 \end{aligned} \tag{8.2.19}$$

Here, F^* is the complex conjugate value of F and V is the real amplification factor of equ. (7.4.15) which coincides with the modulus of the complex frequency characteristic. We have

$$V^2 = \left(\left[1 - \left(\frac{\Omega}{\omega}\right)^2\right]^2 + \left[2\varsigma\frac{\Omega}{\omega}\right]^2\right)^{-1} \tag{8.2.20}$$

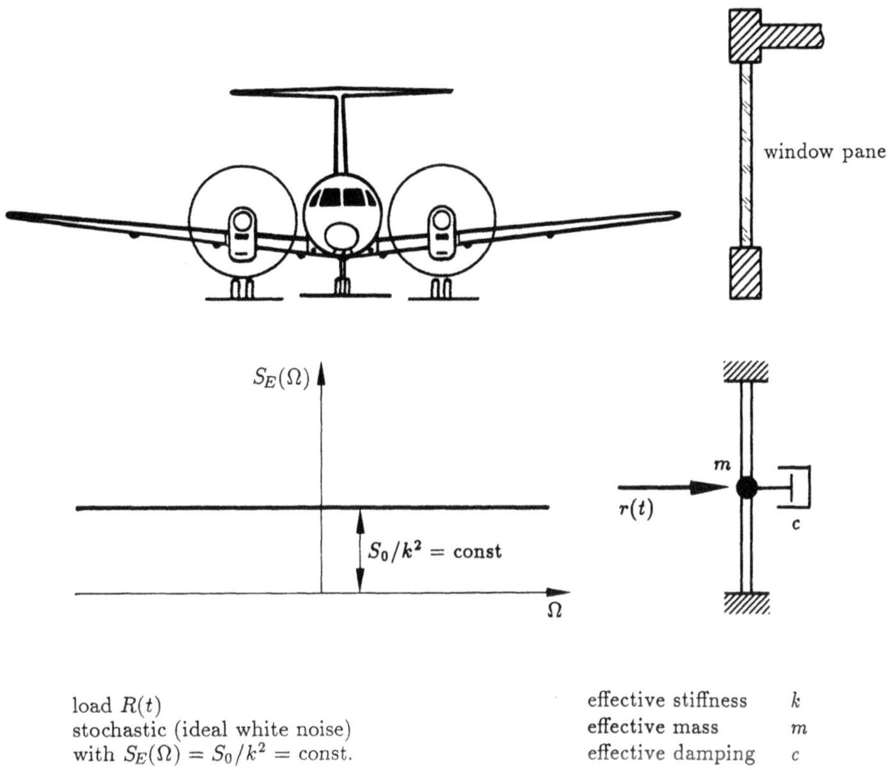

Fig. 8.2.2 Greatly simplified model of the vibration of a window pane due to acoustic excitation

Equation (8.2.12) can now be written in the simplified form

$$S_\eta(\Omega) = V^2 S_E(\Omega) \tag{8.2.21}$$

In order to illustrate the somewhat abstract concepts, we consider the simple example of a single-degree-of-freedom system of fig. 8.2.2. In this figure, the pressure impulses generated by the engine noise of a jet plane impinge upon a window. The ensuing excitation is stochastic and is idealised in terms of a constant frequency band (white noise). To establish the amplification function V, we require the eigenfrequency of the window pane

$$\omega = \left(\frac{k}{m}\right)^{1/2}$$

and the dimensionless damping coefficient

$$\begin{aligned}\zeta &= \tfrac{1}{2}\omega\frac{c}{k} \\ &= \tfrac{1}{2}c\,(km)^{-1/2}\end{aligned}$$

The spectral density of the response can then be obtained by application of the generalised force of excitation

$$R_\eta(t) = \frac{1}{k}R(t)$$

in conjunction with the spectral density of the force excitation

$$S_R = S_0 = \text{const.}$$

and the spectral density of the generalised force excitation

$$S_E(\Omega) = \frac{1}{k^2}S_0$$

in the form

$$\begin{aligned} S_r(\Omega) = S_\eta &= V^2\frac{S_0}{k^2} \\ &= \frac{S_0}{k^2}\frac{1}{\left(1-(\frac{\Omega}{\omega})^2\right)^2 + \left(2\zeta\frac{\Omega}{\omega}\right)^2} \\ &= \frac{S_0}{(k-m\Omega^2)^2 + (c\Omega)^2} \end{aligned}$$

Since the force function R_η enters in a quadratic form into the autocorrelation and since the spectral density corresponds to the Fourier transform of the autocorrelation, the factor $(1/k)^2$ can be absorbed into the expression of the spectral density of the generalised force excitation.

A qualitative representation of the process is reproduced in fig. 8.2.3. The largest contributions to the response arise in the range of the eigenfrequencies, as long as damping is not too pronounced.

The mean square value of the response can be evaluated from equ. (8.1.31)

$$\begin{aligned} \overline{r^2} &= \int_{-\infty}^{+\infty} S_r(\Omega)\mathrm{d}\Omega \\ &= \int_{-\infty}^{+\infty} \frac{S_0}{(k-m\Omega^2)^2 + c^2\Omega^2}\mathrm{d}\Omega \\ &= \frac{\pi S_0}{kc} \end{aligned}$$

The mean value of the amplitude square of the window pane is thus independent of the mass. Consequently, this value cannot be reduced by increasing the mass of the pane, but by increasing the stiffness and the damping.

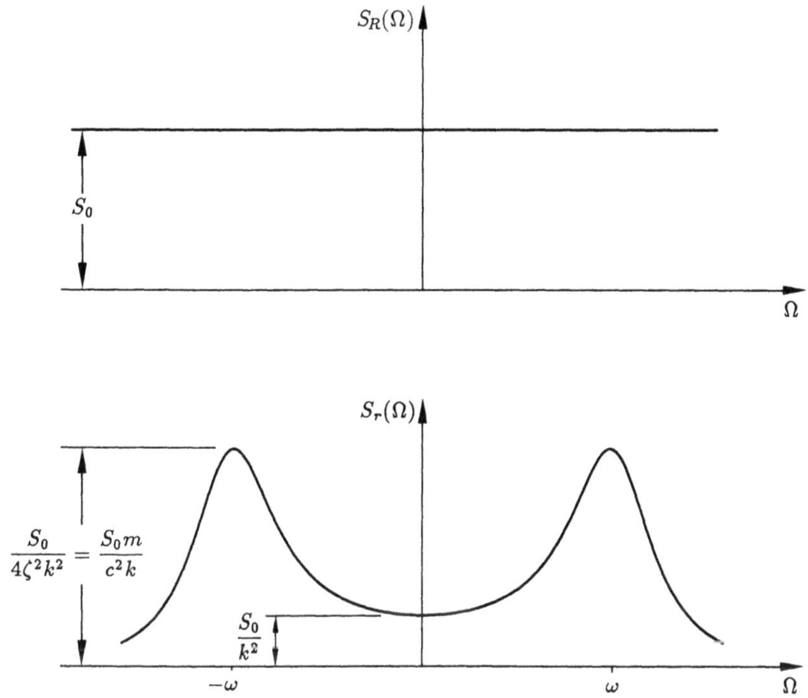

Fig. 8.2.3 Spectral density of input and response

In our preceding discussion of the response, we have referred only to the expectation values and have not mentioned the probability density and the probability itself of the response. For general distributions, there is no simple method of proceeding from the excitation probability to the response probability. However, a Gaussian excitation process can be shown to induce a Gaussian response process, provided that the system is linear [8.2]. In many cases, the response of the system can be approximated by Gaussian distribution, even when this approach is inapplicable for the excitation. Taking account of this fact, we can expand upon our example. If the mean value $\overline{r}$ of the response is zero, the probability distribution becomes as shown in fig. 8.1.3. We also note that the dispersion is given by

$$\sigma_x = \sqrt{\frac{\pi S_0}{kc}}$$

We deduce from the upper diagram in fig. 8.1.3

$$P\left(\frac{x}{\sigma_x} = 2\right) = 0.97725$$

and hence a probability of only 2.275% for the amplitude r to exceed the value σ_x by a factor of two, i.e.

$$P = 2.275\% \quad \text{for} \quad r(t) \geqslant 2\sqrt{\frac{\pi S_0}{kc}}$$

By special modifications of the design, it can be ensured that unduly high amplitudes are assigned to the range of the improbable, i.e. infrequent, occurrences. However, in contrast to the conditions in a deterministic approach, a failure damage can never be excluded completely here.

8.3 Random processes in multi-degree-of-freedom systems and the analysis of distributed random excitations according to the finite element concept

Let a structure discretised according to the finite element method possess n degrees of freedom, each of which may also be assigned discrete forces. In the case of random excitation, we must presume that the force $R_{\mathrm{i}}(t)$ at the node i undergoes a different random process than the force $R_{\mathrm{j}}(t)$ at the node j. This raises the question as to how such simultaneous random processes can be analysed. For this purpose, the definitions set out in section 8.2 have to be readjusted.

Let $x_{\mathrm{i}}(t)$ and $x_{\mathrm{j}}(t)$ be two stochastic processes for each of which we possess N series of measurements, ${x_{\mathrm{i}}}^{(\mathrm{k})}(t)$ and ${x_{\mathrm{j}}}^{(\mathrm{k})}(t)$. At a certain instant, t_1, we evaluate the number of measuring values, N_x, for which the conditions

$$x_{\mathrm{i}}^{(\mathrm{k})}(t_1) < x_{0\mathrm{i}}$$

and

$$x_{\mathrm{j}}^{(\mathrm{k})}(t_1) < x_{0\mathrm{j}} \tag{8.3.1}$$

hold. We thus establish the so-called cross probability

$$P_{\mathrm{ij}}(x_{0\mathrm{i}}, x_{0\mathrm{j}}, t_1) = \lim_{\mathrm{N}\to\infty} \frac{\mathrm{N}_x}{\mathrm{N}} \tag{8.3.2}$$

and obtain in this way for n degrees of freedom a quadratic matrix $\boldsymbol{P}$ which represents the cross probability for all intervening processes. Analogous to equ. (8.1.2), the following also applies

$$P_{\mathrm{ij}}(-\infty, -\infty, t) = 0$$

$$P_{\mathrm{ij}}(+\infty, +\infty, t) = 1 \tag{8.3.3}$$

Furthermore, we notice that for $x_{0\mathrm{i}} = +\infty$, every value of $x_{\mathrm{i}}^{(\mathrm{k})}$ satisfies the condition of equ. (8.3.1); therefore $x_{\mathrm{j}}^{(\mathrm{k})}$ remains as the only decisive factor determining the

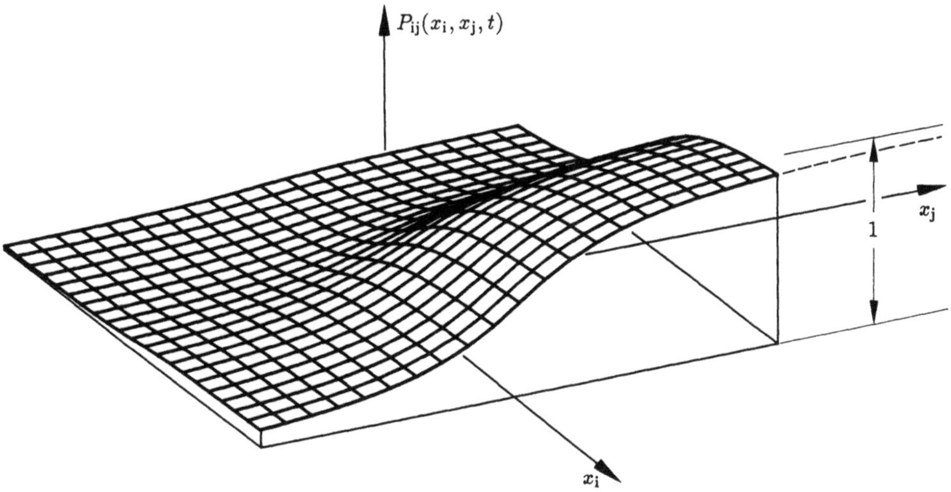

Fig. 8.3.1 Cross probability of two random processes x_i and x_j

number N_x and the probability. Consequently, we obtain the same probability as for a single process in section 8.1

$$P_{ij}(+\infty, x_{0j}, t) = P_j$$
$$P_{ij}(x_{0i}, +\infty, t) = P_i \tag{8.3.4}$$

If we consider the limiting values x_{0i} and x_{0j} as variables, then the cross probability P_{ij} represents a surface which rises over the x_{0i}, x_{0j}-plane and extends between 0 and 1. The 0 index is often omitted for economy of notation; one thus refers to the probability $P_{ij}(x_i, x_j, t)$ (fig. 8.3.1). The probability P_{ij} of x_i and x_j being located within a stipulated band can easily be determined. It is best evaluated from the difference between the probabilites $P_{ij}(x_{0i}, x_{0j}, t)$ and $P_{ij}(x_{0i} + \Delta x_i, x_{0j} + \Delta x_j, t)$. In fact, the probability that the conditions

$$x_{0i} \leqslant x_i(t) < x_{0i} + \Delta x_i \tag{8.3.5}$$

and

$$x_{0j} \leqslant x_j(t) < x_{0j} + \Delta x_j$$

are fulfilled is simply

$$\Delta P_{ij} = P_{ij}(x_{0i} + \Delta x_i, x_{0j} + \Delta x_j, t) - P_{ij}(x_{0i}, x_{0j}, t) \tag{8.3.6}$$

Alternatively, we may state: if the limiting values are incremented by Δx_i or Δx_j, the probability increases by ΔP_{ij}. Proceeding to infinitesimal increments, we thus deduce the cross probability density in the form

$$p_{ij}(x_i, x_j, t) = \frac{\partial^2 P_{ij}}{\partial x_i \partial x_j} \tag{8.3.7}$$

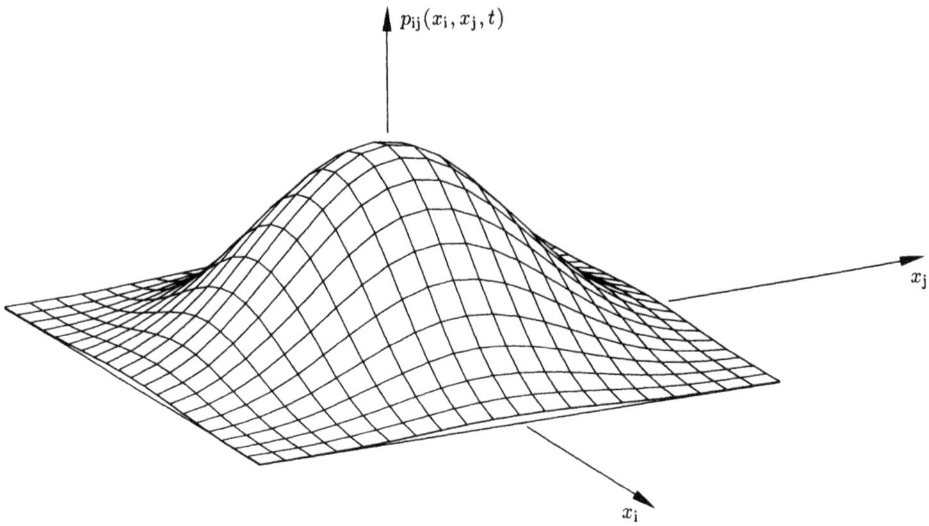

Fig. 8.3.2 Typical distribution of the cross probability density

Since the probability must increase or at least remain constant when the limits are raised, p has to be positive

$$p_{ij}(x_i, x_j, t) \geqslant 0 \tag{8.3.8}$$

The cross probability can be determined once more from the probability density

$$P_{ij}(x_{0i}, x_{0j}, t) = \int_{-\infty}^{x_{0j}} \int_{-\infty}^{x_{0i}} p_{ij}(x_i, x_j, t)\, dx_i dx_j \tag{8.3.9}$$

Since the statement in the second line of equ. (8.3.3) applies, we have

$$\int_{-\infty}^{+\infty} \int_{-\infty}^{+\infty} p_{ij}(x_i, x_j, t)\, dx_i dx_j = 1 \tag{8.3.10}$$

which expresses that the volume enclosed by the 'hilly terrain' of cross probability densities has the value of 1. Since extreme values for $x_i(t)$ and $x_j(t)$ should arise only rarely, the value p_{ij} remains small for large limiting values x_{0i} and x_{0j}. The hilly terrain p_{ij} possesses a peak located in the range of smaller values of x_{0i} and x_{0j} (fig. 8.3.2).

Elementary probability densities can easily be calculated from the cross probability density. For this purpose, we proceed from equ. (8.3.4) and, using equ. (8.3.9), obtain the alternative expression

$$P_i = \int_{-\infty}^{x_{0i}} \int_{-\infty}^{+\infty} p_{ij}(x_i, x_j, t)\, dx_j dx_i \tag{8.3.11}$$

Comparison with equ. (8.1.5) yields

$$p_{\mathrm{i}}(x_{\mathrm{i}},t) = \int\limits_{-\infty}^{+\infty} p_{\mathrm{ij}}(x_{\mathrm{i}},x_{\mathrm{j}},t)\,\mathrm{d}x_{\mathrm{j}} \tag{8.3.12}$$

and

$$p_{\mathrm{j}}(x_{\mathrm{j}},t) = \int\limits_{-\infty}^{+\infty} p_{\mathrm{ij}}(x_{\mathrm{i}},x_{\mathrm{j}},t)\,\mathrm{d}x_{\mathrm{i}} \tag{8.3.13}$$

Turning our attention next to the expectation values and using for typographical economy the symbol p in place of p_{ij}, we define the general second order expectation value as

$$E_{\mathrm{ij}}[f(x_{\mathrm{i}},x_{\mathrm{j}})] = \int\limits_{-\infty}^{+\infty}\int\limits_{-\infty}^{+\infty} f(x_{\mathrm{i}},x_{\mathrm{j}})\,p(x_{\mathrm{i}},x_{\mathrm{j}})\,\mathrm{d}x_{\mathrm{i}}\mathrm{d}x_{\mathrm{j}} \tag{8.3.14}$$

In this way, we establish the mean values

$$\begin{aligned}
\overline{x}_{\mathrm{i}} = E[x_{\mathrm{i}}] &= \int\limits_{-\infty}^{+\infty}\int\limits_{-\infty}^{+\infty} x_{\mathrm{i}}\,p(x_{\mathrm{i}},x_{\mathrm{j}})\,\mathrm{d}x_{\mathrm{i}}\mathrm{d}x_{\mathrm{j}} \\
\overline{x}_{\mathrm{j}} = E[x_{\mathrm{j}}] &= \int\limits_{-\infty}^{+\infty}\int\limits_{-\infty}^{+\infty} x_{\mathrm{j}}\,p(x_{\mathrm{i}},x_{\mathrm{j}})\,\mathrm{d}x_{\mathrm{i}}\mathrm{d}x_{\mathrm{j}}
\end{aligned} \tag{8.3.15}$$

In place of the mean square value, the more general mean values of products are evaluated

$$\begin{aligned}
\overline{x_{\mathrm{i}}^2} = E[x_{\mathrm{i}}^2] &= \int\limits_{-\infty}^{+\infty}\int\limits_{-\infty}^{+\infty} x_{\mathrm{i}}^2 p\,\mathrm{d}x_{\mathrm{i}}\mathrm{d}x_{\mathrm{j}} \\
\overline{x_{\mathrm{i}}x_{\mathrm{j}}} = E[x_{\mathrm{i}}x_{\mathrm{j}}] &= \int\limits_{-\infty}^{+\infty}\int\limits_{-\infty}^{+\infty} x_{\mathrm{i}}x_{\mathrm{j}} p\,\mathrm{d}x_{\mathrm{i}}\mathrm{d}x_{\mathrm{j}} \\
\overline{x_{\mathrm{j}}^2} = E[x_{\mathrm{j}}^2] &= \int\limits_{-\infty}^{+\infty}\int\limits_{-\infty}^{+\infty} x_{\mathrm{j}}^2 p\,\mathrm{d}x_{\mathrm{i}}\mathrm{d}x_{\mathrm{j}}
\end{aligned} \tag{8.3.16}$$

where $\overline{x_i x_j}$ represents the cross correlation. We also establish the more general expressions for the variance

$$\sigma_{ii} = \sigma_i^2 = E\big[(x_i - \overline{x}_i)^2\big] = \int\limits_{-\infty}^{+\infty}\int\limits_{-\infty}^{+\infty} (x_i - \overline{x}_i)^2 \, p \, dx_i dx_j$$

$$\sigma_{ij} = E\Big[(x_i - \overline{x}_i)(x_j - \overline{x}_j)\Big] = \int\limits_{-\infty}^{+\infty}\int\limits_{-\infty}^{+\infty} (x_i - \overline{x}_i)(x_j - \overline{x}_j) \, p \, dx_i dx_j$$

$$\sigma_{jj} = \sigma_j^2 = E\big[(x_j - \overline{x}_j)^2\big] = \int\limits_{-\infty}^{+\infty}\int\limits_{-\infty}^{+\infty} (x_j - \overline{x}_j)^2 \, p \, dx_i dx_j \tag{8.3.17}$$

where σ_{ij} is the covariance. Naturally, if $\overline{x}_i$ and $\overline{x}_j$ have zero mean values, we obtain

$$\sigma_i^2 = \overline{x_i^2}$$
$$\sigma_{ij} = \overline{x_i xj} \qquad \text{if} \quad \overline{x}_i = 0,\ \overline{x}_j = 0 \tag{8.3.18}$$
$$\sigma_j^2 = \overline{x_j^2}$$

The dimensionless factor

$$\rho_{ij} = \frac{\sigma_{ij}}{\sigma_i \sigma_j} \tag{8.3.19}$$

is known as the correlation coefficient or the normalised covariance and informs us about the correlation of the statistical processes x_i and x_j. In particular, the correlation coefficient is zero if the two processes are statistically independent; in contrast, if the two processes x_i and x_j are fully dependent, its value is 1. Figure 8.3.3 shows two examples. Let us now imagine a jet plane engaged in radio locations. Here, x_i could represent the measuring signal of an antenna pointing into outer space and x_j the stress at a measuring point of the wing root. Only in a Utopian world would the two processes be related. If we were to enter the various test values $x_i(t_1)$ and $x_j(t_1)$, measured at instant t_1 in a number of aeroplanes, in an x_i, x_j-diagram, we would obtain a very haphazard arrangement (fig. 8.3.3, left). The correlation of the two values here is clearly zero. At best, we can detect that the large values for x_i and x_j occur only rarely. Alternatively, we can imagine $x_i(t)$ to be the strain and $x_j(t)$ the stress at a measuring point of the wing root. Both of these processes are stochastic, but we know that they are – at least to a high precision – linearly related. Depending on the intensity of the superimposed interferences, we record from the measuring signals of the flying vehicles a diagram similar to that shown in fig. 8.3.3, right. We require no prior knowledge of the law of elasticity to observe that a definite correlation between x_i and x_j must exist.

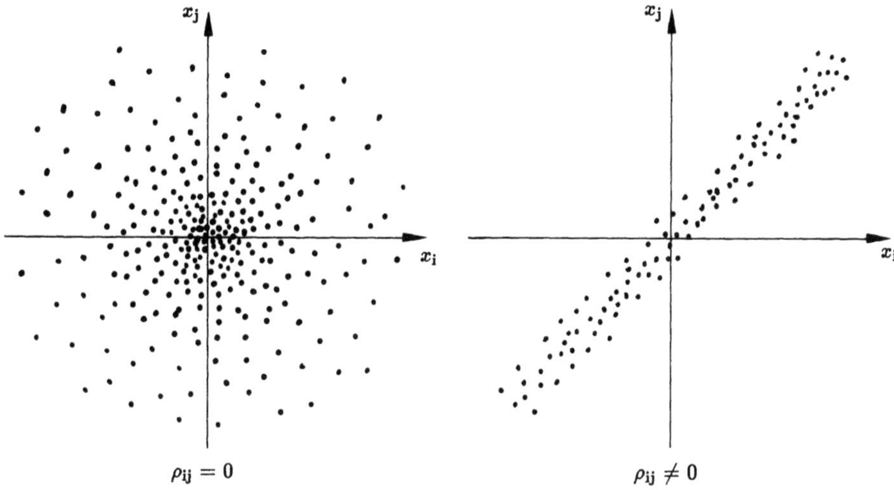

Fig. 8.3.3 Distribution of measured values at an instant t_1 and their correlation

Now assuming such a linear relation, we write

$$\widetilde{x}_j = a_0 + a_1 x_i \tag{8.3.20}$$

and evaluate the deviation of the linearly evaluated test values, $\widetilde{x}_j$, from the true ones, x_j. The expectation value of the square of these differences would then be

$$\begin{aligned} E[(x_j - \widetilde{x}_j)^2] &= E[x_j^2 - 2x_j\widetilde{x}_j + \widetilde{x}_j^2] \\ &= E[x_j^2] - 2E[a_0 x_j + a_1 x_i x_j] + E[a_0^2 + 2a_0 a_1 x_i + a_1^2 x_i^2] \end{aligned} \tag{8.3.21}$$

We can now determine the unknown coefficients a_0 and a_1 in such a way that the expectation value of the square of the difference becomes a minimum. From equ. (8.3.21), we deduce

$$\frac{\partial}{\partial a_0}\left(E[(x_j - \widetilde{x}_j)^2]\right) = 0 = -2\overline{x}_j + 2a_0 + 2a_1\overline{x}_i$$

$$\frac{\partial}{\partial a_1}\left(E[(x_j - \widetilde{x}_j)^2]\right) = 0 = -2\overline{x_i x_j} + 2a_0\overline{x}_i + 2a_1\overline{x_i^2} \tag{8.3.22}$$

Following some transformations, we obtain

$$\frac{\widetilde{x}_j - \overline{x}_j}{\sigma_j} = \frac{\sigma_{ij}}{\sigma_i\sigma_j}\frac{x_i - \overline{x}_i}{\sigma_i} = \rho_{ij}\frac{x_i - \overline{x}_i}{\sigma_i} \tag{8.3.23}$$

The linear relation represented by equ. (8.3.23) is denoted as a regression line. The gradient of this regression line is identical to the correlation coefficient. The correlation coefficient is independent of the sequential order of the indices ($\rho_{ij} = \rho_{ji}$) and an interchange of the indices yields the alternative expression

$$\frac{\widetilde{x}_i - \overline{x}_i}{\sigma_i} = \rho_{ij} \frac{x_j - \overline{x}_j}{\sigma_j} \tag{8.3.24}$$

The correlation can be deduced directly from the gradient of the regression line (fig. 8.3.4). If we plot the test values as given in equ. (8.3.23), they either form a haphazard pattern for independent processes or are approximated by a straight line for fully dependent processes; in the latter case, ρ_{ij} is the gradient of the straight line. This special plotting enables such a dependence to be verified at once.

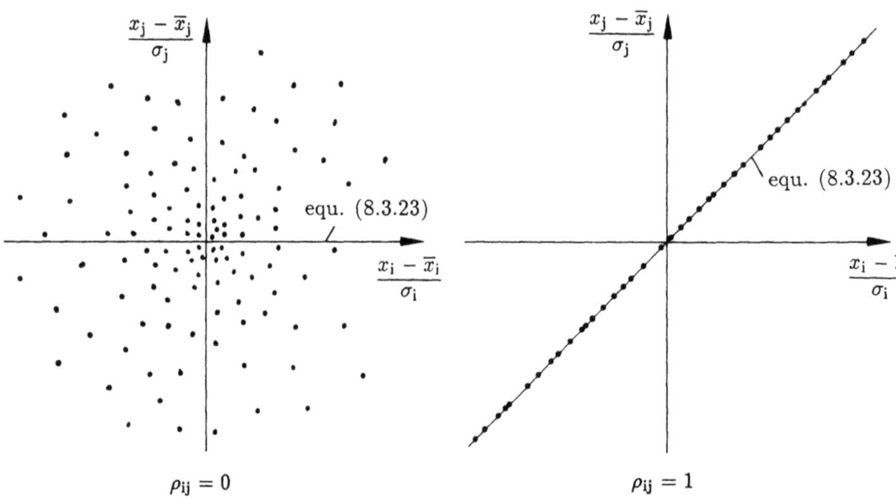

Fig. 8.3.4 Measured values and regression lines for independent (left) and totally dependent statistical processes (right)

In addition to the linear and quadratic expectation values dealt with so far, we had also defined the expectation value of the autocorrelation function in a single process. Here, it has to be replaced by the more general cross correlation function. In analogy to the autocorrelation function, we obtain

$$\begin{aligned}\Phi_{ij}(t, \Delta t) &= E[x_i(t) \cdot x_j(t + \Delta t)] \\ &= \int_{-\infty}^{+\infty} \int_{-\infty}^{+\infty} x_i(t)\, x_j(t + \Delta t)\, p_{ij} \mathrm{d}x_i \mathrm{d}x_j\end{aligned} \tag{8.3.25}$$

A cross correlation factor $\rho_{\mathrm{ij}}(\Delta t)$ corresponding to Φ_{ij} can now be established. For simplicity, we restrict the argument to stationary processes, for which we obtain

$$\rho_{\mathrm{ij}}(\Delta t)=\frac{E\big[(x_{\mathrm{i}}(t)-\overline{x}_{\mathrm{i}})(x_{\mathrm{j}}(t+\Delta t)-\overline{x}_{\mathrm{j}})\big]}{\sigma_{\mathrm{i}}\sigma_{\mathrm{j}}} \tag{8.3.26}$$

Whilst the correlation factor given in equ. (8.3.19) handles both x_{i} and x_{j} at the same instant t, two time-shifted processes are observed here. Clearly, the following applies

$$\rho_{\mathrm{ij}}=\rho_{\mathrm{ij}}(0) \tag{8.3.27}$$

The assertion of $\rho_{\mathrm{ij}}^2\leqslant 1$ has already been reported without proof, the same statement holds for the more general $\rho_{\mathrm{ij}}(\Delta t)$. We now offer a proof. For simplicity, we temporarily introduce the abbreviations

$$\begin{aligned} u&=x_{\mathrm{i}}(t)-\overline{x}_{\mathrm{i}}\\ v&=x_{\mathrm{j}}(t+\Delta t)-\overline{x}_{\mathrm{j}} \end{aligned} \tag{8.3.28}$$

For the stationary processes, we note

$$\rho_{\mathrm{ij}}^2(\Delta t)=\frac{E^2[u,v]}{E[u^2]E[v^2]} \tag{8.3.29}$$

By virtue of

$$E^2[u,v]\leqslant E[u^2]E[v^2] \tag{8.3.30}$$

we confirm that

$$\rho_{\mathrm{ij}}^2(\Delta t)\leqslant 1 \tag{8.3.31}$$

holds. In order to prove equ. (8.3.30), we evaluate the following expectation value which is positive for every constant c

$$\begin{aligned} E[(cu-v)^2]&=E[u^2]c^2-2E[u,v]c+E[v^2]\\ &=(c-c_1)(c-c_2)\geqslant 0 \end{aligned} \tag{8.3.32}$$

Here, c_1 and c_2 are solutions to a quadratic equation in c

$$E[u^2]c^2-2E[u,v]c+E[v^2]=0$$

and are defined by

$$c_{1,2}=\frac{E[u,v]\pm\left(E^2[u,v]-E[u^2]E[v^2]\right)^{1/2}}{E[u^2]} \tag{8.3.33}$$

Consequently, we obtain in conjunction with equ. (8.3.32)

$$\left\{E[u^2]c - E[u,v] - \left(E^2[u,v] - E[u^2]E[v^2]\right)^{\frac{1}{2}}\right\}\left\{E[u^2]c - E[u,v] + \left(E^2[u,v] - E[u^2]E[v^2]\right)^{\frac{1}{2}}\right\} \geqslant 0 \tag{8.3.34}$$

which applies for any value of c and therefore also for

$$c = \frac{E[u,v]}{E[u^2]}$$

for which value, equ. (8.3.34) becomes

$$-\left(E^2[u,v] - E[u^2]E[v^2]\right) \geqslant 0 \tag{8.3.35}$$

which proves the assertions of equs. (8.3.30) and (8.3.31). Using the cross correlation of the covariance, we deduce from equ. (8.3.26)

$$\begin{aligned} \sigma_i\sigma_j\rho_{ij}(\Delta t) &= E\left[x_i(t)x_j(t+\Delta t) - \overline{x}_i\overline{x}_j\right] \\ &= \Phi_{ij}(\Delta t) - \overline{x}_i\overline{x}_j \end{aligned} \tag{8.3.36a}$$

or

$$\Phi_{ij}(\Delta t) = \overline{x}_i\overline{x}_j + \rho_{ij}(\Delta t)\sigma_i\sigma_j \tag{8.3.36b}$$

Application of equ. (8.3.31) in equ. (8.3.36b) yields the limiting values

$$\overline{x}_i\overline{x}_j - \sigma_i\sigma_j \leqslant \Phi_{ij}(\Delta t) \leqslant \overline{x}_i\overline{x}_j + \sigma_i\sigma_j \tag{8.3.37}$$

In many statistical processes, there is no cross correlation between $x_i(t) - \overline{x}_i$ and $x_j(t+\Delta t) - \overline{x}_j$ when the time-shift Δt is very large. If we consider, for example, the statistical processes associated with stress and strain at the wing root of a jet plane, it is evident that they are fully correlated for a given limited Δt. However, a comparison of the stresses with the strains measured 1000 hours previously demonstrates the case of zero correlation. Thus, the following is often found to be valid

$$\Phi_{ij}(\infty) = \overline{x}_i\overline{x}_j \tag{8.3.38}$$

The cross correlation is not an even function in Δt since

$$\begin{aligned} \Phi_{ij}(-\Delta t) &= E[x_i(t)\,x_j(t-\Delta t)] \\ &= E[x_i(t+\Delta t)\,x_j(t)] \\ &= \Phi_{ji}(\Delta t) \\ &\neq \Phi_{ij}(\Delta t) \end{aligned} \tag{8.3.39}$$

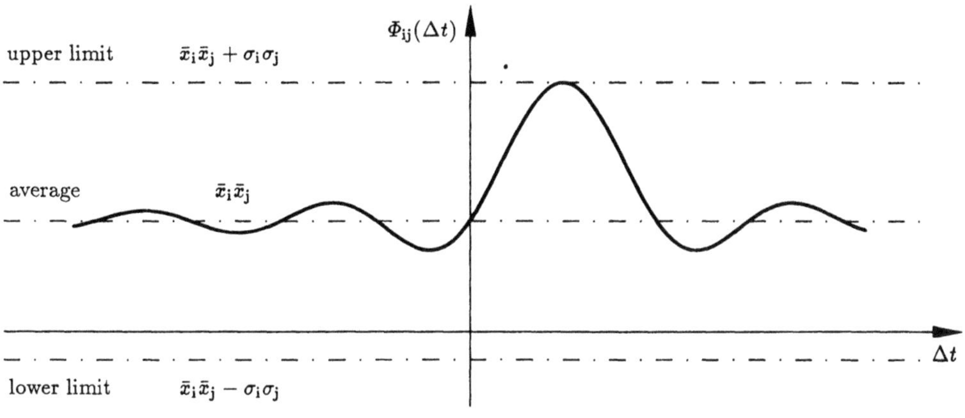

Fig. 8.3.5 Functional behaviour of the cross correlation

It should be noted that in the last expression, $x_i(t+\Delta t), x_i(t), x_j(t)$ and $x_j(t+\Delta t)$ generally represent different processes. In this connexion, the expectation values $\overline{x}_i, \sigma_i^2$ or $\overline{x}_j, \sigma_j^2$ can be time-independent. Therefore, the correlation matrix is unsymmetric. Figure 8.3.5 shows the qualitative course of a cross correlation function. In this example, the processes x_i and x_j possess maximum correlation for a positive Δt. For $i = j$, we obtain the familiar autocorrelation function. The cross correlation oscillates around the average $\overline{x}_i\overline{x}_j$ and is thus only suited for the adoption of a cross spectral density if the mean values $(\overline{x}_i, \overline{x}_j)$ are zero. In addition to the requirement that Φ_{ij} should fulfil the Dirichlet conditions (piecewise continuous and monotonic; limiting values on both sides at points of discontinuity), the following

$$\int_{-\infty}^{+\infty} |\Phi_{ij}(\Delta t)| \mathrm{d}(\Delta t) < \infty \tag{8.3.40}$$

should also be satisfied, but this is possible only for zero mean values of $\overline{x}_i$ and $\overline{x}_j$. If, on the other hand, the covariance of time-shifted processes is introduced, this difficulty is automatically circumvented for $x_i(t) - \overline{x}_i$ and $x_j(t+\Delta t) - \overline{x}_j$ both have mean values of zero

$$\begin{aligned} \sigma_{ij}(\Delta t) &= E\Big[\big(x_i(t) - \overline{x}_i\big)\big(x_j(t+\Delta t) - \overline{x}_j\big)\Big] \\ &= \int_{-\infty}^{+\infty}\int_{-\infty}^{+\infty} \big(x_i(t) - \overline{x}_i\big)\big(x_j(t+\Delta t) - \overline{x}_j\big) p_{ij} \mathrm{d}x_i \mathrm{d}x_j \\ &= \rho_{ij}\sigma_i\sigma_j \end{aligned} \tag{8.3.41}$$

Naturally, there holds

$$\sigma_{ij} = \Phi_{ij}, \qquad \text{if } \overline{x}_i = 0 \text{ and } \overline{x}_j = 0 \tag{8.3.42}$$

Also for the covariance of time-shifted processes, we generally have

$$\sigma_{ij}(\pm\infty) = 0 \tag{8.3.43}$$

which means that the processes $x_i(t) - \overline{x}_i$ and $x_j(t+\Delta t) - \overline{x}_j$ are not correlated for large Δt values in true stochastic processes. The cross spectral density can now easily be formed

$$S_{ij}(\Omega) = \frac{1}{2\pi}\int\limits_{-\infty}^{+\infty} \sigma_{ij}(\Delta t)\,\mathrm{e}^{-i\Omega\Delta t}\mathrm{d}(\Delta t) \tag{8.3.44}$$

the inverse transformation of which yields

$$\sigma_{ij}(\Delta t) = \int\limits_{-\infty}^{+\infty} S_{ij}(\Omega)\,\mathrm{e}^{i\Omega\Delta t}\mathrm{d}\Omega \tag{8.3.45}$$

Application of Euler's formula gives for the cross spectral density

$$S_{ij}(\Omega) = \frac{1}{2\pi}\int\limits_{-\infty}^{+\infty} \sigma_{ij}(\Delta t)(\cos\Omega\Delta t - i\sin\Omega\Delta t)\,\mathrm{d}(\Delta t) \tag{8.3.46}$$

Once more assuming stationary processes and considering the mirrored element, we deduce for

$$\begin{aligned}\sigma_{ij}(-\Delta t) &= E\Big[\big(x_i(t) - \overline{x}_i\big)\big(x_j(t-\Delta t) - \overline{x}_j\big)\Big]\\ &= E\Big[\big(x_i(t+\Delta t) - \overline{x}_i\big)\big(x_j(t) - \overline{x}_j\big)\Big]\\ &= \sigma_{ji}(\Delta t)\end{aligned} \tag{8.3.47}$$

in analogy to equ. (8.3.39) the expression for the cross correlation

$$\begin{aligned}S_{ji}(\Omega) &= \frac{1}{2\pi}\int\limits_{-\infty}^{+\infty} \sigma_{ji}(\Delta t)\,\mathrm{e}^{-i\Omega\Delta t}\mathrm{d}(\Delta t)\\ &= \frac{1}{2\pi}\int\limits_{-\infty}^{+\infty} \sigma_{ij}(-\Delta t)\,\mathrm{e}^{-i\Omega\Delta t}\mathrm{d}(\Delta t) = \frac{1}{2\pi}\int\limits_{-\infty}^{+\infty} \sigma_{ij}(\Delta t)\,\mathrm{e}^{i\Omega\Delta t}\mathrm{d}(\Delta t)\\ &= \frac{1}{2\pi}\int\limits_{-\infty}^{+\infty} \sigma_{ij}(\Delta t)(\cos\Omega\Delta t + i\sin\Omega\Delta t)\,\mathrm{d}(\Delta t)\end{aligned} \tag{8.3.48}$$

The spectral density matrix $\boldsymbol{S}$, the elements of which refer to pairs within n processes, possesses a symmetric real part and a skew-symmetric imaginary part. Thus, the S_{ij} element is conjugate complex to S_{ji}

$$\begin{aligned} S_{ij} &= S_{ji}^* \\ S_{ji} &= S_{ij}^* \end{aligned} \tag{8.3.49}$$

The diagonal elements of the spectral density matrix $\boldsymbol{S}$ are, of course, purely real and correspond to the spectral density of the single processes x_i $(i = 1 \ldots n)$.

Following these general definitions of the most important statistical quantities, we must once again return to the cross probability density. Here also, normal or Gaussian distribution is often observed in nature. Assuming zero mean values for $\overline{x}_i$ and $\overline{x}_j$, we obtain

$$p_{ij} = \frac{1}{2\pi\sigma_i\sigma_j\sqrt{1-\sigma_{ij}^2}} \exp\left(-\frac{1}{2(1-\sigma_{ij}^2)}\left[\left(\frac{x_i}{\sigma_i}\right)^2 + \left(\frac{x_j}{\sigma_j}\right)^2 - 2\rho_{ij}\frac{x_i x_j}{\sigma_i\sigma_j}\right]\right) \tag{8.3.50}$$

If $\rho_{ij} = 0$, the two processes x_i and x_j are statistically uncoupled and we obtain

$$p_{ij} = p_i p_j \,, \qquad \text{if } \rho_{ij} = 0 \tag{8.3.51}$$

Here, p_i and p_j are the Gaussian distributions of the individual processes. Often, the system response is assumed to be Gaussian.

Up until now, the statistical processes have been assumed to be stationary, the statistical quantities P, p and E being time-independent. For a statistical evaluation, we consequently require many parallel series of measurements for ${x_i}^{(k)}(t)$ $(i = 1, n)$, where k represents the serial index. A more convenient approach, however, would apply to an ergodic process where the formation of the mean values does not proceed over the series of measurements, k, but rather along the time axis t. The probability that

$$x_{0i} \leqslant x_i(t) \leqslant x_{0i} + \Delta x_i$$

and (8.3.52)

$$x_{0j} \leqslant x_j(t) \leqslant x_{0j} + \Delta x_j$$

hold amounts to $p(x_{0i}, x_{0j})\Delta x_i \Delta x_j$. In the case of ergodicity of the processes, we operate along the time axis and count the intervals of time, Δt_k, for which the condition in equ. (8.3.52) is satisfied, and then compare these with the entire period of time, T. Thus, once again, we obtain the cross probability density (fig. 8.3.6).

$$p_{ij}\Delta x_i \Delta x_j = \frac{1}{T}\sum_k \Delta t_k \tag{8.3.53}$$

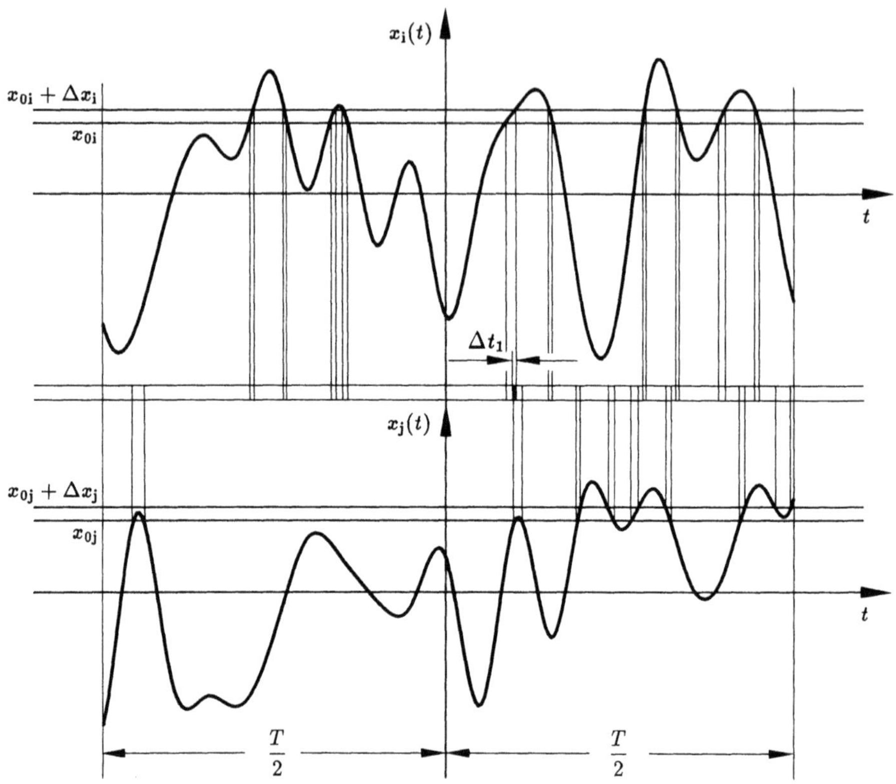

Fig. 8.3.6 Calculation of the cross probability density for ergodic random processes

A practical evaluation would once again divide the time axis into finite equal intervals Δt and determine the number of intervals, N_{ij}, for which the condition equ. (8.3.52) is fulfilled. With a total of N intervals Δt, we have

$$p_{ij}(x_{0i}, x_{0j})\Delta x_i \Delta x_j = \frac{N_{ij}}{N} \tag{8.3.54}$$

Using equ. (8.3.53), we can adapt all general formulae of the expectation values for the special case of ergodic processes. To this purpose, we note

$$E[f(x_i, x_j)] = \int_{-\infty}^{+\infty}\int_{-\infty}^{+\infty} f(x_i, x_j)\, p_{ij} \mathrm{d}x_i \mathrm{d}x_j$$

$$= \lim_{T\to\infty} \frac{1}{T} \int_{-T/2}^{T/2} f(x_i, x_j)\, \mathrm{d}t \tag{8.3.55}$$

Correspondingly, we obtain for the mean value

$$\overline{x}_{\mathrm{i}} = E[x_{\mathrm{i}}] = \lim_{T\to\infty} \frac{1}{T} \int\limits_{-T/2}^{T/2} x_{\mathrm{i}}(t)\,\mathrm{d}t \tag{8.3.56}$$

the cross mean value

$$\overline{x_{\mathrm{i}}x_{\mathrm{j}}} = E[x_{\mathrm{i}}x_{\mathrm{j}}] = \lim_{T\to\infty} \frac{1}{T} \int\limits_{-T/2}^{T/2} x_{\mathrm{i}}(t)x_{\mathrm{j}}(t)\,\mathrm{d}t \tag{8.3.57}$$

the variance (or σ_{i}= standard deviation)

$$\sigma_{\mathrm{ii}} = \sigma_{\mathrm{i}}^2 = E\big[(x_{\mathrm{i}} - \overline{x}_{\mathrm{i}})^2\big] = \lim_{T\to\infty} \frac{1}{T} \int\limits_{-T/2}^{T/2} \big(x_{\mathrm{i}}(t) - \overline{x}_{\mathrm{i}}\big)^2 \mathrm{d}t \tag{8.3.58}$$

the covariance

$$\sigma_{\mathrm{ij}} = E\big[(x_{\mathrm{i}} - \overline{x}_{\mathrm{i}})(x_{\mathrm{j}} - \overline{x}_{\mathrm{j}})\big] = \lim_{T\to\infty} \frac{1}{T} \int\limits_{-T/2}^{T/2} \big(x_{\mathrm{i}}(t) - \overline{x}_{\mathrm{i}}\big)\big(x_{\mathrm{j}}(t) - \overline{x}_{\mathrm{j}}\big)\mathrm{d}t \tag{8.3.59}$$

and the cross correlation

$$\Phi_{\mathrm{ij}}(\Delta t) = E\big[x_{\mathrm{i}}(t)x_{\mathrm{j}}(t+\Delta t)\big] = \lim_{T\to\infty} \frac{1}{T} \int\limits_{-T/2}^{T/2} x_{\mathrm{i}}(t)\,x_{\mathrm{j}}(t+\Delta t)\,\mathrm{d}t \tag{8.3.60}$$

We have so far restricted our discussion of n random processes $x_{\mathrm{i}}(t)$ to second order probabilities and expectation values, i.e. we have linked two processes $x_{\mathrm{i}}(t)$ and $x_{\mathrm{j}}(t)$. It is, of course, possible to set up definitions which combine all processes simultaneously. Thus,

$$\underset{(\mathrm{n}\times 1)}{p(\boldsymbol{x}_0)}\,\mathrm{d}x_1\mathrm{d}x_2\ldots\mathrm{d}x_{\mathrm{n}} \tag{8.3.61}$$

represents the probability that

$$\boldsymbol{x}_0 \leqslant \underset{(\mathrm{n}\times 1)}{\boldsymbol{x}(t)} \leqslant \boldsymbol{x}_0 + \mathrm{d}\boldsymbol{x} \tag{8.3.62}$$

is true. Here, $p(\boldsymbol{x}_0)$ is a cross-probability density of n-th order and the vector $\boldsymbol{x}$ includes all single processes

$$\boldsymbol{x} = \{x_1(t) \quad x_2(t) \quad \ldots \quad x_{\mathrm{n}}(t)\} \tag{8.3.63}$$

Each statistical mean can now easily be formulated in terms of the expectation function

$$E[f(\boldsymbol{x})] = \int_{-\infty}^{+\infty}\int_{-\infty}^{+\infty}\dots\int_{-\infty}^{+\infty} f(\boldsymbol{x})\,p(\boldsymbol{x})\,\mathrm{d}x_1\mathrm{d}x_2\dots\mathrm{d}x_\mathrm{n} \tag{8.3.64}$$

An example of n-th order cross probability density often employed in practice is normal or Gaussian distribution. In this case, we obtain in general terms

$$p(\boldsymbol{x}) = \left((2\pi)^\mathrm{n}\det\boldsymbol{\Sigma}\right)^{1/2}\exp\left(-\tfrac{1}{2}(\boldsymbol{x}-\overline{\boldsymbol{x}})^\mathrm{t}\boldsymbol{\Sigma}^{-1}(\boldsymbol{x}-\overline{\boldsymbol{x}})\right) \tag{8.3.65}$$

where $\overline{\boldsymbol{x}}$ is the vector of mean values

$$\begin{aligned}\overline{\boldsymbol{x}} &= \int_{-\infty}^{+\infty}\int_{-\infty}^{+\infty}\dots\int_{-\infty}^{+\infty} \boldsymbol{x}\,p(\boldsymbol{x})\,\mathrm{d}x_1\mathrm{d}x_2\dots\mathrm{d}x_\mathrm{n} \\ &= \int_{-\infty}^{+\infty}\{x_1p(x_1)\mathrm{d}x_1 \quad x_2p(x_2)\mathrm{d}x_2 \quad \dots \quad x_\mathrm{n}p(x_\mathrm{n})\mathrm{d}x_\mathrm{n}\}\end{aligned} \tag{8.3.66}$$

and $\boldsymbol{\Sigma}$ is the covariance matrix with which we are already familiar

$$\underset{(\mathrm{n}\times\mathrm{n})}{\boldsymbol{\Sigma}} = \int_{-\infty}^{+\infty}\int_{-\infty}^{+\infty}\dots\int_{-\infty}^{+\infty} \underset{(\mathrm{n}\times 1)}{[\boldsymbol{x}-\overline{\boldsymbol{x}}]}\,\underset{(1\times\mathrm{n})}{[\boldsymbol{x}-\overline{\boldsymbol{x}}]^\mathrm{t}}\,p(\boldsymbol{x})\,\mathrm{d}x_1\mathrm{d}x_2\dots\mathrm{d}x_\mathrm{n}$$

$$= \begin{bmatrix} \int_{-\infty}^{+\infty}(x_1-\overline{x}_1)^2p(x_1)\mathrm{d}x_1 & \int_{-\infty}^{+\infty}\int_{-\infty}^{+\infty}(x_1-\overline{x}_1)(x_2-\overline{x}_2)p(x_1,x_2)\mathrm{d}x_1\mathrm{d}x_2 & \dots \\ & \int_{-\infty}^{+\infty}(x_2-\overline{x}_2)^2p(x_2)\mathrm{d}x_2 & \dots \\ & & \ddots \\ \mathrm{sym} & & \end{bmatrix} \tag{8.3.67}$$

The time-shifted covariance matrix can be written as a more general expression of the matrix $\mathbf{\Sigma}$

$$\mathbf{\Sigma}(t,\Delta t) = \int\limits_{-\infty}^{+\infty}\int\limits_{-\infty}^{+\infty}\dots\int\limits_{-\infty}^{+\infty}[\boldsymbol{x}(t)-\overline{\boldsymbol{x}}(t)][\boldsymbol{x}(t+\Delta t)-\overline{\boldsymbol{x}}(t+\Delta t)]^{\mathrm{t}}p(\boldsymbol{x})\mathrm{d}\boldsymbol{x}$$

$$= \begin{bmatrix} \Sigma_{11} & \Sigma_{12} & \dots\dots \\ \Sigma_{21} & \Sigma_{22} & \dots\dots \\ \vdots & \vdots & \ddots \end{bmatrix} \tag{8.3.68}$$

with the matrix elements

$$\Sigma_{11} = \int\limits_{-\infty}^{+\infty}\Big(x_1(t)-\overline{x}_1(t)\Big)\Big(x_1(t+\Delta t)-\overline{x}_1(t+\Delta t)\Big)p(x_1)\,\mathrm{d}x_1$$

$$\Sigma_{12} = \int\limits_{-\infty}^{+\infty}\int\limits_{-\infty}^{+\infty}\Big(x_1(t)-\overline{x}_1(t)\Big)\Big(x_2(t+\Delta t)-\overline{x}_2(t+\Delta t)\Big)p(x_1,x_2)\,\mathrm{d}x_1\mathrm{d}x_2$$

$$\Sigma_{21} = \int\limits_{-\infty}^{+\infty}\int\limits_{-\infty}^{+\infty}\Big(x_2(t)-\overline{x}_2(t)\Big)\Big(x_1(t+\Delta t)-\overline{x}_1(t+\Delta t)\Big)p(x_2,x_1)\,\mathrm{d}x_2\mathrm{d}x_1$$

$$\Sigma_{22} = \int\limits_{-\infty}^{+\infty}\Big(x_2(t)-\overline{x}_2(t)\Big)\Big(x_2(t+\Delta t)-\overline{x}_2(t+\Delta t)\Big)p(x_2)\,\mathrm{d}x_2$$

(8.3.68a)

for which there holds

$$\mathbf{\Sigma}(t,0) = \mathbf{\Sigma}(t) \tag{8.3.69}$$

The covariance matrix and the correlation matrix are identical for zero mean values

$$\mathbf{\Sigma}(t,\Delta t) = \mathbf{\Phi}(t,\Delta t), \qquad \text{if } \overline{\boldsymbol{x}} = \boldsymbol{o} \tag{8.3.70}$$

Finally, we define the spectral density matrix $\boldsymbol{S}$ as the Fourier transform of the covariance with time-shift. In the most general case, both matrices would depend on time, but we reproduce them here for stationary processes

$$\boldsymbol{S}(\Omega) = \frac{1}{2\pi}\int\limits_{-\infty}^{+\infty}\mathbf{\Sigma}(\Delta t)\,\mathrm{e}^{-i\Omega\Delta t}\mathrm{d}(\Delta t) \tag{8.3.71}$$

the general random process for a multi-degree-of-freedom system

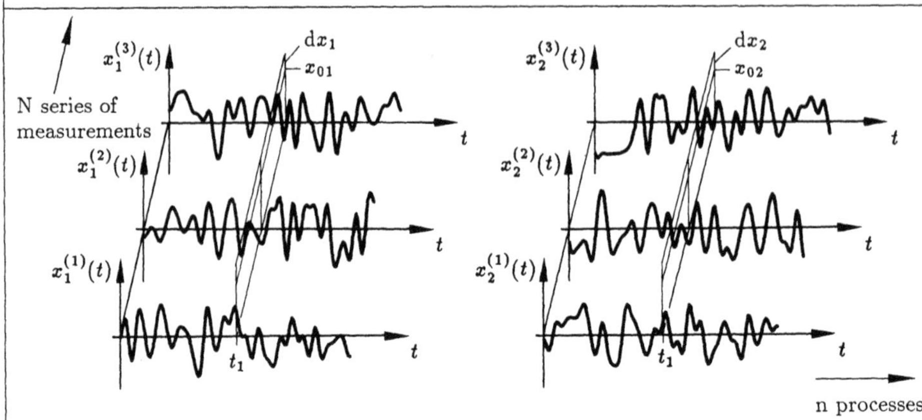

probability at $\boldsymbol{x}_0$ and t_1

let $\boldsymbol{x} < \boldsymbol{x}_0$ in N_x of N series of measurement

$$P(\boldsymbol{x}_0, t_1) = \lim_{N\to\infty} \frac{N_x}{N} \quad , \quad 0 \leqslant P \leqslant 1 \quad \text{monotonic}$$

probability density for $\boldsymbol{x}_0$ and t_1

$$p(\boldsymbol{x}_0, t_1) = \frac{\partial^n P}{\partial x_1 \partial x_2 \dots \partial x_n} \geqslant 0$$

general expectation value

$$E[f(\boldsymbol{x})] = \int_{-\infty}^{+\infty}\int_{-\infty}^{+\infty}\dots\int_{-\infty}^{+\infty} f(\boldsymbol{x})\, p(\boldsymbol{x})\, dx_1 dx_2 \dots dx_n$$

special expectation values

mean value for $\boldsymbol{x}$

$$\bar{\boldsymbol{x}}(t_1) = \int_{-\infty}^{+\infty}\int_{-\infty}^{+\infty}\dots\int_{-\infty}^{+\infty} \boldsymbol{x}\, p(\boldsymbol{x})\, dx_1 dx_2 \dots dx_n$$

correlation matrix

$$\boldsymbol{\Phi}(t_1) = \int_{-\infty}^{+\infty}\int_{-\infty}^{+\infty}\dots\int_{-\infty}^{+\infty} \boldsymbol{x}(t_1)\,[\boldsymbol{x}(t_1)]^t\, p(\boldsymbol{x})\, dx_1 dx_2 \dots dx_n$$

covariance matrix

$$\boldsymbol{\Sigma}(t_1) = \int_{-\infty}^{+\infty}\int_{-\infty}^{+\infty}\dots\int_{-\infty}^{+\infty} [\boldsymbol{x}(t_1) - \bar{\boldsymbol{x}}(t_1)][\boldsymbol{x}(t_1) - \bar{\boldsymbol{x}}(t_1)]^t\, p(\boldsymbol{x})\, dx_1 dx_2 \dots dx_n$$

Fig. 8.3.7 The most important statistical measures for general random processes in multi-degree-of-freedom systems, *Part 1* correlation matrix with time shift

$$\boldsymbol{\Phi}(t_1, \Delta t) = \int\limits_{-\infty}^{+\infty}\int\limits_{-\infty}^{+\infty}\cdots\int\limits_{-\infty}^{+\infty} \boldsymbol{x}(t_1)\,[\boldsymbol{x}(t_1+\Delta t)]^{\mathrm{t}}\, p(\boldsymbol{x})\,\mathrm{d}x_1\mathrm{d}x_2\ldots\mathrm{d}x_{\mathrm{n}}$$

covariance matrix with time shift

$$\boldsymbol{\Sigma}(t_1, \Delta t) = \int\limits_{-\infty}^{+\infty}\int\limits_{-\infty}^{+\infty}\cdots\int\limits_{-\infty}^{+\infty} [\boldsymbol{x}(t_1) - \overline{\boldsymbol{x}}(t_1)][\boldsymbol{x}(t_1+\Delta t) - \overline{\boldsymbol{x}}(t_1+\Delta t)]^{\mathrm{t}} p(\boldsymbol{x})\,\mathrm{d}x_1\mathrm{d}x_2\ldots\mathrm{d}x_{\mathrm{n}}$$

spectral density matrix

$$\boldsymbol{S}(t_1, \Omega) = \frac{1}{2\pi}\int\limits_{-\infty}^{+\infty} \boldsymbol{\Sigma}(t_1, \Delta t)\,\mathrm{e}^{-i\Omega\Delta t}\mathrm{d}(\Delta t)$$

special relations

$$\boldsymbol{\Sigma}(t_1) = \boldsymbol{\Phi}(t_1) \quad \text{if} \quad \overline{\boldsymbol{x}}(t_1) = \boldsymbol{o}$$

$$\boldsymbol{\Phi}(t_1, 0) = \boldsymbol{\Phi}(t_1)$$

$$\boldsymbol{\Sigma}(t_1, 0) = \boldsymbol{\Sigma}(t_1)$$

$$\boldsymbol{\Sigma}(t_1, 0) = \boldsymbol{\Phi}(t_1) \quad \text{if} \quad \overline{\boldsymbol{x}}(t_1) = \boldsymbol{o}$$

$$\boldsymbol{S}(t_1, 0) = \frac{1}{2\pi}\int\limits_{-\infty}^{+\infty} \boldsymbol{\Sigma}(t_1, \Delta t)\,\mathrm{d}(\Delta t)$$

$$\boldsymbol{S}(t_1, \Omega) = \frac{1}{2\pi}\int\limits_{-\infty}^{+\infty} \boldsymbol{\Phi}(t_1, \Delta t)\,\mathrm{e}^{-i\Omega\Delta t}\mathrm{d}(\Delta t) \quad \text{if} \quad \overline{\boldsymbol{x}}(t_1) = \boldsymbol{o}$$

Note:
if the random process is stationary, all results are independent of the time t_1; consequently, t_1 can be ignored.

Fig. 8.3.7 The most important statistical measures for general random processes in multi-degree-of-freedom systems, *Part 2*

Equation (8.3.49) indicates that the two elements ij and ji of the spectral density matrix are complex conjugate. Thus,

$$\boldsymbol{S}(\Omega) = \boldsymbol{A} + i\boldsymbol{B} \tag{8.3.72}$$

where $\boldsymbol{A}$ is real symmetric and $\boldsymbol{B}$ is real skew-symmetric. The formation of the mean value for ergodic processes proceeds over the time-axis, i.e. an arbitrary expectation value can be obtained from

$$E[f(\boldsymbol{x})] = \lim_{T\to\infty}\frac{1}{T}\int\limits_{-T/2}^{T/2} f(\boldsymbol{x})\mathrm{d}t \tag{8.3.73}$$

the ergodic random process for a multi-degree-of-freedom system

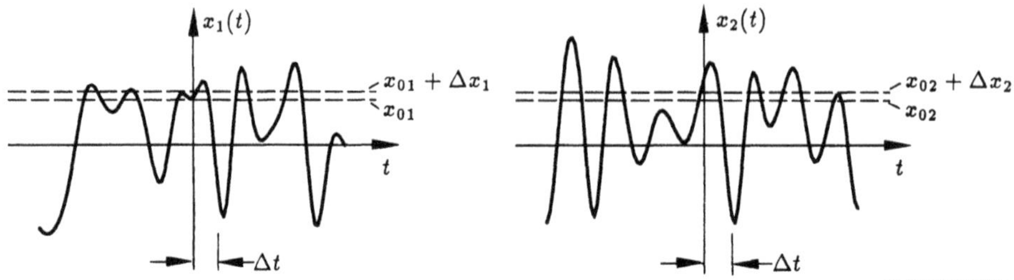

n processes

probability at $\boldsymbol{x}_0$

in N_x of N intervals Δt, let $\boldsymbol{x} < \boldsymbol{x}_0$; then

$$P(\boldsymbol{x}_0) = \lim_{\substack{\Delta t \to 0 \\ N \to \infty}} \frac{N_x}{N} \quad , \quad 0 \leqslant P \leqslant 1 \quad \text{monotonic}$$

probability density at $\boldsymbol{x}_0$

in $N_{x+\Delta x}$ of N intervals Δt, let $\boldsymbol{x} < \boldsymbol{x}_0 + \Delta\boldsymbol{x}$; then

$$p(\boldsymbol{x}_0) = \lim_{\Delta \boldsymbol{x} \to \boldsymbol{o}} \frac{P(\boldsymbol{x}_0 + \Delta\boldsymbol{x}) - P(\boldsymbol{x}_0)}{\Delta x_1 \Delta x_2 \ldots \Delta x_n}$$

$$= \lim_{\substack{\Delta t \to 0 \\ N \to \infty \\ \Delta \boldsymbol{x} \to \boldsymbol{o}}} \frac{N_{x+\Delta x} - N_x}{\Delta x_1 \Delta x_2 \ldots \Delta x_n N}$$

general expectation value

$$E[f(\boldsymbol{x})] = \lim_{T \to \infty} \frac{1}{T} \int_{-T/2}^{+T/2} f(\boldsymbol{x}) \, \mathrm{d}t$$

special forms of expectation value

mean value of $\boldsymbol{x}$

$$\overline{\boldsymbol{x}} = \lim_{T \to \infty} \frac{1}{T} \int_{-T/2}^{+T/2} \boldsymbol{x} \, \mathrm{d}t$$

correlation matrix

$$\boldsymbol{\Phi} = \lim_{T \to \infty} \frac{1}{T} \int_{-T/2}^{+T/2} \boldsymbol{x}\,\boldsymbol{x}^{\mathrm{t}} \, \mathrm{d}t$$

covariance matrix

$$\boldsymbol{\Sigma} = \lim_{T \to \infty} \frac{1}{T} \int_{-T/2}^{+T/2} [\boldsymbol{x} - \overline{\boldsymbol{x}}][\boldsymbol{x} - \overline{\boldsymbol{x}}]^{\mathrm{t}} \mathrm{d}t$$

Fig. 8.3.8 The most important statistical measures for ergodic processes in multi-degree-of-freedom systems, *Part 1* correlation matrix with time shift

$$\boldsymbol{\Phi}(\Delta t) = \lim_{T\to\infty} \frac{1}{T} \int_{-T/2}^{+T/2} \boldsymbol{x}(t)[\boldsymbol{x}(t+\Delta t)]^{\mathrm{t}} \mathrm{d}t$$

covariance matrix with time shift

$$\boldsymbol{\Sigma}(\Delta t) = \lim_{T\to\infty} \frac{1}{T} \int_{-T/2}^{+T/2} [\boldsymbol{x}(t) - \bar{\boldsymbol{x}}][\boldsymbol{x}(t+\Delta t) - \bar{\boldsymbol{x}}]^{\mathrm{t}} \mathrm{d}t$$

spectral density matrix

$$\boldsymbol{S}(\Omega) = \frac{1}{2\pi} \int_{-\infty}^{+\infty} \boldsymbol{\Sigma}(\Delta t)\, \mathrm{e}^{-\mathrm{i}\Omega\Delta t} \mathrm{d}(\Delta t)$$

special relations

$$\boldsymbol{\Sigma} = \boldsymbol{\Phi} \quad \text{if} \quad \bar{\boldsymbol{x}} = \boldsymbol{o}$$

$$\boldsymbol{\Phi}(0) = \boldsymbol{\Phi}$$

$$\boldsymbol{\Sigma}(0) = \boldsymbol{\Sigma}$$

$$\boldsymbol{\Sigma}(0) = \boldsymbol{\Phi} \quad \text{if} \quad \bar{\boldsymbol{x}} - \boldsymbol{o}$$

$$\boldsymbol{S}(0) = \frac{1}{2\pi} \int_{-\infty}^{+\infty} \boldsymbol{\Sigma}(\Delta t)\, \mathrm{d}(\Delta t)$$

$$\boldsymbol{S}(\Omega) = \frac{1}{2\pi} \int_{-\infty}^{+\infty} \boldsymbol{\Phi}(\Delta t)\, \mathrm{e}^{-\mathrm{i}\Omega\Delta t} \mathrm{d}(\Delta t) \quad \text{if} \quad \bar{\boldsymbol{x}} = \boldsymbol{o}$$

Fig. 8.3.8 The most important statistical measures for ergodic processes in multi-degree-of-freedom systems, *Part 2*

The reader should now be in a position to determine all special expectation values that may arise. They are summarised in fig. 8.3.8 whilst fig. 8.3.7 presents the more general relations for non-ergodic processes.

We have so far taken the n stochastic processes $x_{\mathrm{i}}(t)$ as given. If, however, we consider the application of excitation functions $R_{\eta\mathrm{i}}$ (i $= 1, \mathrm{nx}$) or $R_{\mathrm{i}}(t)$ (i $= 1, \mathrm{n}$) in a practical case, the availability of discrete load functions can no longer be taken for granted. In general, natural processes are subjected to distributed loads. The problem then arises as to how to discretise a continuous stochastic process $p(x, y, z, t)$ where x, y, z are space coordinates.

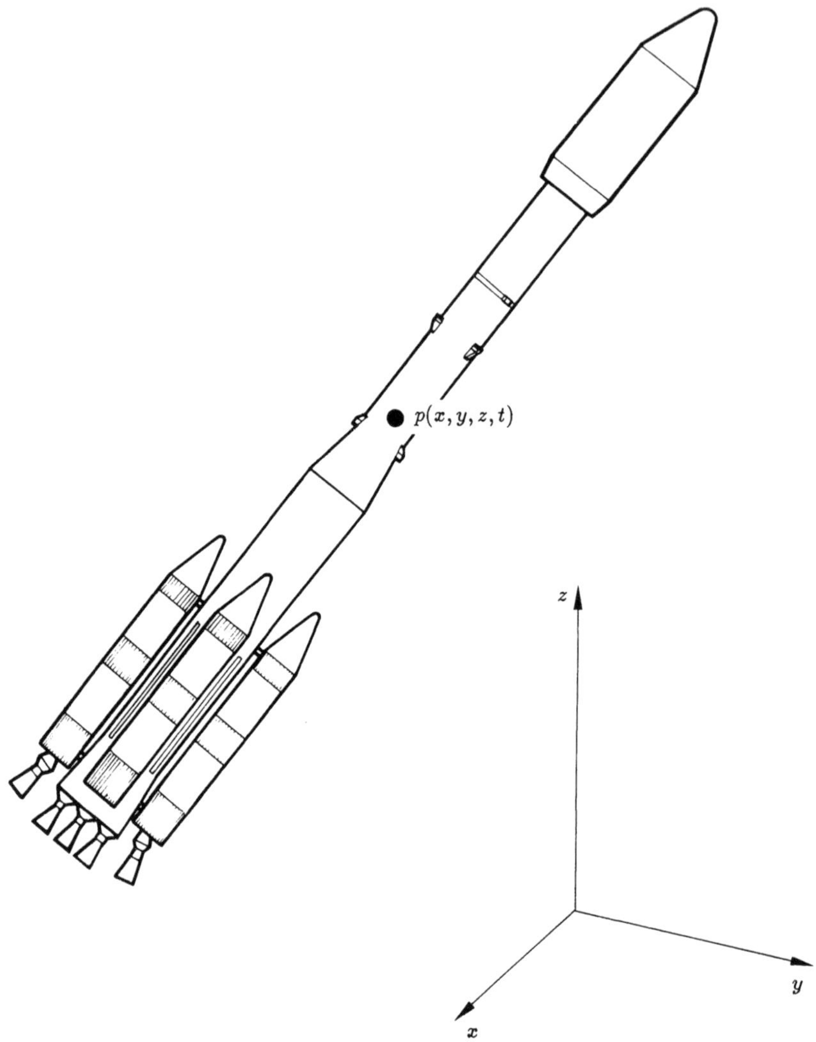

Fig. 8.3.9 Sound induced pressure fluctuation on the surface point of a rocket

A typical example of such processes is the sound generation due to jet engine noise. The stochastic fluctuation of pressure on the rocket's shell depends most clearly on the distance from the engine (fig. 8.3.9).

To begin with, we calculate kinematically equivalent nodal forces from the distributed pressures. Once a stochastic process contains volume forces $p_V(x, y, z, t)$

and surface forces $p_S(x, y, z, t)$, the vector for the discrete loads at the nodes becomes

$$\begin{aligned}\boldsymbol{R}_e(t) &= \mathbf{a}^{\mathrm{t}}\mathbf{P}_e(t)\\ &= \sum_{\mathrm{g}=1}^{\mathrm{m}} \mathbf{a}_{\mathrm{g}}^{\mathrm{t}}\mathbf{P}_{e\mathrm{g}}(t)\end{aligned} \tag{8.3.74}$$

with the individual element nodal forces

$$\mathbf{P}_{e\mathrm{g}}(t) = \int_{V_{\mathrm{g}}} \boldsymbol{\omega}^{\mathrm{t}}(x, y, z)\boldsymbol{p}_V(x, y, z, t)\,\mathrm{d}V + \int_{S_{\mathrm{g}}} \boldsymbol{\omega}^{\mathrm{t}}(x, y, z, t)\boldsymbol{p}_S(x, y, z, t)\,\mathrm{d}S \tag{8.3.75}$$

In order to be more specific in our discussion, we restrict ourselves at this point to surface loads. The most important expectation values, amongst them the mean value, can now be calculated

$$\begin{aligned}\overline{\boldsymbol{R}}_e = E[\boldsymbol{R}_e] &= E\Big[\sum_{\mathrm{g}=1}^{\mathrm{m}} \mathbf{a}_{\mathrm{g}}^{\mathrm{t}} \int_{S_{\mathrm{g}}} \boldsymbol{\omega}^{\mathrm{t}}\boldsymbol{p}_S \mathrm{d}S\Big]\\ &= \sum_{\mathrm{g}=1}^{\mathrm{m}} \mathbf{a}_{\mathrm{g}}^{\mathrm{t}} \int_{S_{\mathrm{g}}} \boldsymbol{\omega}^{\mathrm{t}} E[\boldsymbol{p}_S]\mathrm{d}S\end{aligned} \tag{8.3.76}$$

Note that the mean value $E[\boldsymbol{p}_S]$ of the surface load is a (3×1)-vector which depends on the location in space.

A further interesting quantity is the covariance matrix with time shift. In order to economise on our discussion, we presume zero mean values in which case the covariance and correlation matrices are identical

$$\begin{aligned}&\underset{(\mathrm{n}\times\mathrm{n})}{\boldsymbol{\Sigma}}(t, \Delta t) = \boldsymbol{\Phi}(t, \Delta t)\\ &= E\big[\boldsymbol{R}_e(t)\boldsymbol{R}_e^{\mathrm{t}}(t + \Delta t)\big] = E\Big[\Big(\sum_{\mathrm{g}=1}^{\mathrm{m}} \mathbf{a}_{\mathrm{g}}^{\mathrm{t}}\mathbf{P}_{e\mathrm{g}}(t)\Big)\Big(\sum_{\mathrm{h}=1}^{\mathrm{m}} \mathbf{P}_{e\mathrm{h}}^{\mathrm{t}}(t + \Delta t)\mathbf{a}_{\mathrm{h}}\Big)\Big]\\ &= \sum_{\mathrm{g}=1}^{\mathrm{m}}\sum_{\mathrm{h}=1}^{\mathrm{m}} \mathbf{a}_{\mathrm{g}}^{\mathrm{t}} E\big[\mathbf{P}_{e\mathrm{g}}(t)\mathbf{P}_{e\mathrm{h}}^{\mathrm{t}}(t + \Delta t)\big]\mathbf{a}_{\mathrm{h}}\\ &= \sum_{\mathrm{g}=1}^{\mathrm{m}}\sum_{\mathrm{h}=1}^{\mathrm{m}} \mathbf{a}_{\mathrm{g}}^{\mathrm{t}} E\Big[\int_{S_{\mathrm{g}}} \boldsymbol{\omega}^{\mathrm{t}}\boldsymbol{p}_S(t)\mathrm{d}S \int_{S_{\mathrm{h}}} \boldsymbol{p}_S^{\mathrm{t}}(t + \Delta t)\boldsymbol{\omega}\mathrm{d}S\Big]\mathbf{a}_{\mathrm{h}}\\ &= \sum_{\mathrm{g}=1}^{\mathrm{m}}\sum_{\mathrm{h}=1}^{\mathrm{m}} \mathbf{a}_{\mathrm{g}}^{\mathrm{t}}\Big[\int\!\!\int \boldsymbol{\omega}^{\mathrm{t}}(x_{\mathrm{g}}, y_{\mathrm{g}}, z_{\mathrm{g}}) E\big[\boldsymbol{p}_S(x_{\mathrm{g}}, y_{\mathrm{g}}, z_{\mathrm{g}}, t)\boldsymbol{p}_S^{\mathrm{t}}(x_{\mathrm{h}}, y_{\mathrm{h}}, z_{\mathrm{h}}, t+\Delta t)\big]\boldsymbol{\omega}(x_{\mathrm{h}}, y_{\mathrm{h}}, z_{\mathrm{h}})\mathrm{d}S_{\mathrm{g}}\mathrm{d}S_{\mathrm{h}}\Big]\mathbf{a}_{\mathrm{h}},\\ &\qquad \text{if } \overline{\boldsymbol{R}}_e = \boldsymbol{o}\end{aligned} \tag{8.3.77}$$

Here,

$$E[\boldsymbol{p}_S(x_{\mathrm{g}}, y_{\mathrm{g}}, z_{\mathrm{g}}, t)\boldsymbol{p}_S^{\mathrm{t}}(x_{\mathrm{h}}, y_{\mathrm{h}}, z_{\mathrm{h}}, t+\Delta t)] = \underset{(3\times 3)}{\boldsymbol{\Sigma}}(x_{\mathrm{g}}, y_{\mathrm{g}}, z_{\mathrm{g}}, x_{\mathrm{h}}, y_{\mathrm{h}}, z_{\mathrm{h}}, \Delta t)$$
$$= \boldsymbol{\Phi}(x_{\mathrm{g}}, y_{\mathrm{g}}, z_{\mathrm{g}}, x_{\mathrm{h}}, y_{\mathrm{h}}, z_{\mathrm{h}}, \Delta t)$$
$$\text{if } \overline{\boldsymbol{p}}_S = \boldsymbol{o} \tag{8.3.78}$$

denotes the covariance or correlation matrix of the continuous stochastic process $\boldsymbol{p}_S$ and $\boldsymbol{x}_{\mathrm{g}}$ defines a point in the g-element and $\boldsymbol{x}_{\mathrm{h}}$ a point in the h-element. Of course, we presume that the expectation functions of the continuous random process are known.

The final issue on discretisation to be considered concerns the transformation of the spectral densities. Applying a Fourier transformation to equ. (8.3.77), we obtain

$$\underset{(\mathrm{n}\times\mathrm{n})}{\boldsymbol{S}}(\Omega) = \sum_{\mathrm{g}=1}^{\mathrm{m}}\sum_{\mathrm{h}=1}^{\mathrm{m}} \mathbf{a}_{\mathrm{g}}^{\mathrm{t}} \iint \boldsymbol{\omega}^{\mathrm{t}}(\boldsymbol{x}_{\mathrm{g}})\boldsymbol{S}(\boldsymbol{x}_{\mathrm{g}}, \boldsymbol{x}_{\mathrm{h}}, \Omega)\,\boldsymbol{\omega}(\boldsymbol{x}_{\mathrm{h}})\,\mathrm{d}S_{\mathrm{g}}\mathrm{d}S_{\mathrm{h}}\mathbf{a}_{\mathrm{h}} \tag{8.3.79}$$

where

$$\underset{(3\times 3)}{\boldsymbol{S}}(\boldsymbol{x}_{\mathrm{g}}, \boldsymbol{x}_{\mathrm{h}}, \Omega) = \frac{1}{2\pi}\int_{-\infty}^{+\infty} \boldsymbol{\Sigma}(\boldsymbol{x}_{\mathrm{g}}, \boldsymbol{x}_{\mathrm{h}}, \Delta t)\,\mathrm{e}^{-i\Omega\Delta t}\mathrm{d}(\Delta t) \tag{8.3.80}$$

is the spectral density matrix of the continuous process $\boldsymbol{p}_S(\boldsymbol{x}, t)$.

Unfortunately, the covariance and the spectral density are generally fully populated matrices. Moreover, the fact that arbitrary elements are linked together makes their computation via the finite element method more difficult. It is thus often common practice to resort to rough approximate methods, analogous to the lumped-mass concept. Let

$$\boldsymbol{p}_S(\boldsymbol{x}_{\mathrm{i}}, t)$$

and

$$\boldsymbol{p}_S(\boldsymbol{x}_{\mathrm{j}}, t+\Delta t)$$

be the averaged surface pressures at the nodes i and j acting over their areas of influence, A_{i} and A_{j} respectively. The component $\boldsymbol{\Sigma}_{\mathrm{ij}}$ of the covariance matrix appertaining to the i and j nodes is then

$$\begin{aligned}\underset{(3\times 3)}{\boldsymbol{\Sigma}_{\mathrm{ij}}}(\Delta t) &= E[\boldsymbol{p}_S(\boldsymbol{x}_{\mathrm{i}}, t)\boldsymbol{p}_S(\boldsymbol{x}_{\mathrm{j}}, t+\Delta t)A_{\mathrm{i}}A_{\mathrm{j}}] \\ &= \underset{(3\times 3)}{\boldsymbol{\Sigma}}(\boldsymbol{x}_{\mathrm{i}}, \boldsymbol{x}_{\mathrm{j}}, \Delta t)A_{\mathrm{i}}A_{\mathrm{j}}\end{aligned} \tag{8.3.81}$$

and the corresponding spectral density term is

$$S_{ij}(\Omega) = \underset{(3\times3)}{S}(\boldsymbol{x}_i, \boldsymbol{x}_j, \Omega) A_i A_j \tag{8.3.82}$$

Both formulae are very simple but include only very globally the statistical characteristics which may cause large deviations. The matrices $\boldsymbol{\Sigma}$ and $\boldsymbol{S}$ remain fully populated but for special cases.

8.4 Random excitation and response of multi-degree-of-freedom systems with modal damping

This section begins where section 8.2 left off and applies the definitions formulated in section 8.3. The basis for our discussion is the coupled system

$$\boldsymbol{M}\ddot{\boldsymbol{r}} + \boldsymbol{C}\dot{\boldsymbol{r}} + \boldsymbol{K}\boldsymbol{r} = \boldsymbol{R}(t) \tag{8.4.1}$$

where $\boldsymbol{R}(t)$ is a vector whose elements represent random excitation processes. Here, we are interested in the random response $\boldsymbol{r}(t)$. In the case of modal damping, the system can be uncoupled using the eigenvectors $\boldsymbol{X}$. We set

$$\boldsymbol{r}(t) = \boldsymbol{X}\boldsymbol{\eta}(t) \tag{8.4.2}$$

and obtain a set of uncoupled equations (see also equ. (7.4.2))

$$\begin{aligned} \ddot{\boldsymbol{\eta}} + 2\boldsymbol{Z}\boldsymbol{\Omega}\dot{\boldsymbol{\eta}} + \boldsymbol{\Omega}^2\boldsymbol{\eta} &= \boldsymbol{\Omega}^2\boldsymbol{X}^t\boldsymbol{R}(t) \\ &= \boldsymbol{\Omega}^2\boldsymbol{R}_\eta(t) \end{aligned} \tag{8.4.3}$$

where the two diagonal matrices $\boldsymbol{Z}$ and $\boldsymbol{\Omega}$

$$\boldsymbol{Z} = \lceil \zeta_1 \quad \zeta_2 \quad \ldots\ldots \quad \zeta_n \rfloor \tag{8.4.4}$$

and

$$\boldsymbol{\Omega} = \lceil \omega_1 \quad \omega_2 \quad \ldots\ldots \quad \omega_n \rfloor \tag{8.4.5}$$

comprise the damping ratios and eigenfrequencies respectively. If the Duhamel integral of equ. (8.2.9) is extended to systems with multi degrees of freedom, the modal response becomes

$$\boldsymbol{\eta}(t) = \int_{-\infty}^{+\infty} \boldsymbol{h}(\theta)\boldsymbol{R}_\eta(t-\theta)\mathrm{d}\theta \tag{8.4.6}$$

where

$$\boldsymbol{h}(\theta) = \lceil h_1(\theta) \quad h_2(\theta) \quad \ldots\ldots \quad h_n(\theta) \rfloor \tag{8.4.7}$$

and, according to equ. (8.2.5),

$$h_{\mathrm{i}}(\theta) = (1-\zeta_{\mathrm{i}}^2)^{-1/2}\omega_{\mathrm{i}}\,\mathrm{e}^{-\omega_{\mathrm{i}}\zeta_{\mathrm{i}}\theta}\sin\left[\left(1-\zeta_{\mathrm{i}}^2\right)^{1/2}\omega_{\mathrm{i}}\theta_{\mathrm{i}}\right] \tag{8.4.8}$$

Next, we derive the stochastic measures for the modal response $\boldsymbol{\eta}$ from the known statistical measures for $\boldsymbol{R}(t)$. In this connection, it is interesting to discover which oscillatory eigenmodes are strongly excited. The mean value of the response for stationary processes is

$$\begin{aligned}\overline{\boldsymbol{\eta}} &= E[\boldsymbol{\eta}(t)]\\ &= E\Big[\int_{-\infty}^{+\infty}\boldsymbol{h}(\theta)\boldsymbol{X}^{\mathrm{t}}\boldsymbol{R}(t-\theta)\mathrm{d}\theta\Big] = \int_{-\infty}^{+\infty}\boldsymbol{h}(\theta)\boldsymbol{X}^{\mathrm{t}}E[\boldsymbol{R}(t-\theta)]\mathrm{d}\theta\\ &= \int_{-\infty}^{+\infty}\boldsymbol{h}(\theta)\mathrm{d}\theta\boldsymbol{X}^{\mathrm{t}}\overline{\boldsymbol{R}}\\ &= \boldsymbol{F}(0)\boldsymbol{X}^{\mathrm{t}}\overline{\boldsymbol{R}}\end{aligned} \tag{8.4.9}$$

where

$$\boldsymbol{F}(0) = \lceil F_1(0)\quad F_2(0)\quad\ldots\ldots\quad F_{\mathrm{n}}(0)\rfloor$$

represents the diagonal matrix of the complex frequency response characteristics with the individual entries

$$F_{\mathrm{i}}(\Omega) = \frac{1}{1-\left(\frac{\Omega}{\omega_{\mathrm{i}}}\right)^2 + i2\zeta_{\mathrm{i}}\left(\frac{\Omega}{\omega_{\mathrm{i}}}\right)} \tag{8.4.10}$$

and $\overline{\boldsymbol{R}}$ is the vector of the statistical mean values of the excitation. Following equ. (8.4.2), we then obtain for the mean values of the response

$$\begin{aligned}\overline{\boldsymbol{r}} &= E[\boldsymbol{r}(t)] = E[\boldsymbol{X}\boldsymbol{\eta}(t)] = \boldsymbol{X}\overline{\boldsymbol{\eta}}\\ &= \boldsymbol{X}\boldsymbol{F}(0)\boldsymbol{X}^{\mathrm{t}}\overline{\boldsymbol{R}}\\ &= \boldsymbol{X}\boldsymbol{X}^{\mathrm{t}}\overline{\boldsymbol{R}}\end{aligned} \tag{8.4.11}$$

since $\boldsymbol{F}(0)$ is the unit matrix (see equ. (8.4.10)).

In addition to the response, the statistical measures of the stresses may also be relevant. In the finite element concept, the stresses are generally deduced at certain points from the displacements. In general, the relationship is assumed to be linear, i.e.

$$\boldsymbol{\sigma} = \boldsymbol{D}_{\sigma}\boldsymbol{r} \tag{8.4.12}$$

$\boldsymbol{D}_\sigma$ can be considered as a matrix, the entries of which are already known in numerical terms. For $\boldsymbol{\sigma}$, we obtain the statistical mean values

$$\overline{\boldsymbol{\sigma}} = \boldsymbol{D}_\sigma \overline{\boldsymbol{r}} \tag{8.4.13}$$

which means that the input of the statistical mean values of the deflections is entered and the standard FEM-stress procedure then yields as output the mean values of the stresses.

Finally, we note that all mean values, i.e. $\overline{\boldsymbol{\eta}}$, $\overline{\boldsymbol{r}}$ and $\overline{\boldsymbol{\sigma}}$, vanish if the excitation $\boldsymbol{R}$ possesses zero mean values, $\overline{\boldsymbol{R}} = \boldsymbol{o}$. We refer incidentally to the possibility of deterministically dealing with finite mean values but presume in what follows that $\overline{\boldsymbol{R}} = \boldsymbol{o}$. In this case, we need not distinguish between the correlation matrix and covariance matrix for time-shifted processes. With this supposition, the covariance matrix of the modal response becomes

$$\begin{aligned}
\boldsymbol{\Sigma}_\eta(\Delta t) &= \boldsymbol{\Phi}_\eta(\Delta t) \\
&= E[\boldsymbol{\eta}(t)\boldsymbol{\eta}^{\mathrm{t}}(t+\Delta t)] \\
&= E\Big[\int_{-\infty}^{+\infty} \boldsymbol{h}(\theta)\boldsymbol{X}^{\mathrm{t}}\boldsymbol{R}(t-\theta)\mathrm{d}\theta \int_{-\infty}^{+\infty} \boldsymbol{R}^{\mathrm{t}}(t+\Delta t-\theta)\boldsymbol{X}\boldsymbol{h}(\theta)\mathrm{d}\theta\Big] \\
&= \int_{-\infty}^{+\infty}\int_{-\infty}^{+\infty} \boldsymbol{h}(\theta_1)\boldsymbol{X}^{\mathrm{t}} E\big[\boldsymbol{R}(t-\theta)\boldsymbol{R}^{\mathrm{t}}(t-\theta+\Delta t)\big]\, \boldsymbol{X}\boldsymbol{h}(\theta_2)\mathrm{d}\theta_1\mathrm{d}\theta_2 \\
&= \int_{-\infty}^{+\infty}\int_{-\infty}^{+\infty} \boldsymbol{h}(\theta_1)\boldsymbol{X}^{\mathrm{t}}\boldsymbol{\Sigma}_R\boldsymbol{X}\boldsymbol{h}(\theta_2)\mathrm{d}\theta_1\mathrm{d}\theta_2 \\
&= \int_{-\infty}^{+\infty} \boldsymbol{h}(\theta)\mathrm{e}^{i\Omega\theta}\mathrm{d}\theta\boldsymbol{X}^{\mathrm{t}}\boldsymbol{\Sigma}_R\boldsymbol{X}\int_{-\infty}^{+\infty} \boldsymbol{h}(\theta)\mathrm{e}^{-i\Omega\theta}\mathrm{d}\theta \\
&= \boldsymbol{F}^*(\Omega)\boldsymbol{X}^{\mathrm{t}}\boldsymbol{\Sigma}_R\boldsymbol{X}\boldsymbol{F}(\Omega)
\end{aligned} \tag{8.4.14}$$

The calculation of the covariance matrix of nodal point displacements now involves only a small effort. From equ. (8.4.2), we have

$$\begin{aligned}
\boldsymbol{\Sigma}_r(\Delta t) &= E[\boldsymbol{r}(t)\boldsymbol{r}^{\mathrm{t}}(t+\Delta t)] = E[\boldsymbol{X}\boldsymbol{\eta}(t)\boldsymbol{\eta}^{\mathrm{t}}(t+\Delta t)\boldsymbol{X}^{\mathrm{t}}] \\
&= \boldsymbol{X}\boldsymbol{\Sigma}_\eta\boldsymbol{X}^{\mathrm{t}} \\
&= \boldsymbol{X}\boldsymbol{F}^*(\Omega)\boldsymbol{\Sigma}_E\boldsymbol{F}(\Omega)\boldsymbol{X}^{\mathrm{t}} \\
&= \boldsymbol{X}\boldsymbol{F}^*(\Omega)\boldsymbol{X}^{\mathrm{t}}\boldsymbol{\Sigma}_R\boldsymbol{X}\boldsymbol{F}(\Omega)\boldsymbol{X}^{\mathrm{t}}
\end{aligned} \tag{8.4.15}$$

Finally, we discuss the covariance matrix of the stresses. With equ. (8.4.13), we obtain

$$\begin{aligned}\boldsymbol{\Sigma}_\sigma(\Delta t) &= E[\boldsymbol{\sigma}(t)\boldsymbol{\sigma}^{\mathrm{t}}(t+\Delta t)] = E[\boldsymbol{D}_\sigma \boldsymbol{r}(t)\boldsymbol{r}^{\mathrm{t}}(t+\Delta t)\boldsymbol{D}_\sigma^{\mathrm{t}}] \\ &= \boldsymbol{D}_\sigma \boldsymbol{\Sigma}_r \boldsymbol{D}_\sigma^{\mathrm{t}} \\ &= \boldsymbol{D}_\sigma [\boldsymbol{D}_\sigma \boldsymbol{\Sigma}_r^{\mathrm{t}}]^{\mathrm{t}}\end{aligned} \tag{8.4.16}$$

The final formulation of the covariance matrix indicates that with the n rows of $\boldsymbol{\Sigma}_r$ considered as displacement entries, we run through the standard stress processor $\boldsymbol{D}_\sigma$ and obtain in this way the $n\sigma \times n$ matrix $\boldsymbol{D}_\sigma \boldsymbol{\Sigma}_r^{\mathrm{t}}$. Using next the $n\sigma$ rows of the last matrix as displacement entries, we once again work through the stress processor and ultimately obtain $\boldsymbol{\Sigma}_\sigma$. It should be noted that equs. (8.4.14) to (8.4.16) are only valid when the mean value $\overline{\boldsymbol{R}}$ is zero and the $\boldsymbol{R}(t)$ processes are stationary. Since the calculation of the expectation values proceeds over the time axis and considers only separate single oscillators, ergodicity also has to be postulated.

The spectral density matrix $\boldsymbol{S}(\Omega)$ is the final quantity to be discussed in the present analysis of random excitations. Its derivation from the covariant matrix $\boldsymbol{\Sigma}(\Delta t)$ has already been given in equ. (8.3.71). Applying this transformation to equs. (8.4.14) to (8.4.16), we deduce the spectral density matrix $\boldsymbol{S}_\eta(\Omega)$ of the modal response from the spectral density matrix $\boldsymbol{S}_R(\Omega)$ of the excitation in the form

$$\begin{aligned}\boldsymbol{S}_\eta(\Omega) &= \boldsymbol{F}^*(\Omega)\boldsymbol{S}_E(\Omega)\boldsymbol{F}(\Omega) \\ &= \boldsymbol{F}^*(\Omega)\boldsymbol{X}^{\mathrm{t}}\boldsymbol{S}_R(\Omega)\boldsymbol{X}\boldsymbol{F}(\Omega)\end{aligned} \tag{8.4.17}$$

Furthermore, we can express the spectral density matrix of the nodal point displacements as

$$\boldsymbol{S}_r(\Omega) = \boldsymbol{X}\boldsymbol{S}_\eta(\Omega)\boldsymbol{X}^{\mathrm{t}} \tag{8.4.18}$$

and the spectral density matrix of the nodal point stresses as

$$\boldsymbol{S}_\sigma(\Omega) = \boldsymbol{D}_\sigma \boldsymbol{S}_r(\Omega)\boldsymbol{D}_\sigma^{\mathrm{t}} \tag{8.4.19}$$

Using equ. (8.3.45), the spectral densities can be converted into the covariance matrix. In fact, we have

$$\boldsymbol{\Sigma}(\Delta t) = \int_{-\infty}^{+\infty} \boldsymbol{S}(\Omega)\, \mathrm{e}^{i\Omega\Delta t}\mathrm{d}\Omega \tag{8.4.20}$$

The probability density for a Gaussian distribution (equ. 8.3.65) and zero mean values is simply given by $\boldsymbol{\Sigma}(0)$. We then find

$$\boldsymbol{\Sigma} = \boldsymbol{\Sigma}(0) = \int_{-\infty}^{+\infty} \boldsymbol{S}(\Omega)\mathrm{d}\Omega \tag{8.4.21}$$

which means that the covariance matrix with a zero time-shift can be deduced from the spectral density matrix by integration over the frequency. Experience has already taught us that a structure oscillates particularly violently in the vicinity of its eigenfrequencies. Thus, it seems hardly surprising that the entries $S_{r\mathrm{kk}}(\Omega)$ possess characteristic peaks at $\Omega = \omega_\mathrm{i}$. For this reason, numerical integration of equ. (8.4.21) demands some care. Thus, $\Omega = \omega_\mathrm{i}$ itself should be an integration point and in addition, the vicinity of an eigenfrequency should be assigned a denser net of integration pivotal points. Furthermore, it should be noted that in general the $\boldsymbol{\Sigma}$ matrix is real; in this case, it is sufficient to integrate the real part of $\boldsymbol{S}(\Omega)$ in equ. (8.4.21), which is symmetric.

In most instances, it is not the complete cross covariances that are of primary interest but rather the variances of the response listed on the diagonal. The spectral density matrix of the modal response $\boldsymbol{S}_\eta$ is, in many cases, diagonally dominant. One typical element of this spectral density matrix is

$$S_{\eta\mathrm{ij}} = \boldsymbol{x}_\mathrm{i}^\mathrm{t} \boldsymbol{S}_R \boldsymbol{x}_\mathrm{j} V_\mathrm{i} V_\mathrm{j} \big(\cos(\varphi_\mathrm{i} - \varphi_\mathrm{j}) + i \sin(\varphi_\mathrm{i} - \varphi_\mathrm{j}) \big) \tag{8.4.22}$$

where the amplification function V_i is the modulus of the complex frequency response characteristic F_i; see equ. (8.2.20),

$$V_\mathrm{i} = \frac{1}{\left[\left(1 - (\Omega/\omega_\mathrm{i})^2\right)^2 + \left(2\zeta_\mathrm{i}\Omega/\omega_\mathrm{i}\right)^2 \right]^{1/2}} \tag{8.4.23}$$

The angle φ_i represents the phase shift

$$\varphi_\mathrm{i} = \tan^{-1} \frac{2\zeta_\mathrm{i}(\Omega/\omega_\mathrm{i})}{1 - (\Omega/\omega_\mathrm{i})^2} \tag{8.4.24}$$

The two functions V_i and V_j possess a peak at $\Omega \approx \omega_\mathrm{i}$ and $\Omega \approx \omega_\mathrm{j}$, i.e. at the locations of the eigenfrequencies (see fig. 1.4.2). Providing that the eigenfrequencies are well separated and $\mathrm{i} \neq \mathrm{j}$, one function displays a peak value whilst the other only has a very small value. In contrast, if $\mathrm{i} = \mathrm{j}$, the peak amplitudes are multiplied and in addition the function $\cos(\varphi_\mathrm{i} - \varphi_\mathrm{j})$ assumes a value of 1. Therefore, it often suffices to work with the diagonal values alone and assume

$$\boldsymbol{S}_\eta(\Omega) \approx \lceil S_{\eta 11} \quad S_{\eta 22} \quad \cdots\cdots \quad S_{\eta\mathrm{nn}} \rfloor \tag{8.4.25}$$

where

$$S_{\eta\mathrm{ii}}(\Omega) = V_\mathrm{i}^2(\Omega) \boldsymbol{x}_\mathrm{i}^\mathrm{t} \boldsymbol{S}_R(\Omega) \boldsymbol{x}_\mathrm{i} \tag{8.4.26}$$

With this approximation, the spectral density of the displacements can be obtained from equ. (8.4.18) as

$$\begin{aligned} \boldsymbol{S}_r(\Omega) &= \sum_{\mathrm{i}=1}^{\mathrm{n}} S_{\eta\mathrm{ii}} \boldsymbol{x}_\mathrm{i} \boldsymbol{x}_\mathrm{i}^\mathrm{t} \\ &= \sum_{\mathrm{i}=1}^{\mathrm{n}} V_\mathrm{i}^2(\Omega) \big(\boldsymbol{x}_\mathrm{i}^\mathrm{t} \boldsymbol{S}_R(\Omega) \boldsymbol{x}_\mathrm{i} \big) \boldsymbol{x}_\mathrm{i} \boldsymbol{x}_\mathrm{i}^\mathrm{t} \end{aligned} \tag{8.4.27}$$

The diagonal elements of the spectral density of the displacements, which we require for the variance, are thus

$$
\begin{aligned}
S_{r\mathrm{kk}}(\Omega) &= \sum_{\mathrm{i}=1}^{\mathrm{n}} V_{\mathrm{i}}^2(\Omega)\left(\boldsymbol{x}_{\mathrm{i}}^{\mathrm{t}} \boldsymbol{S}_R(\Omega)\boldsymbol{x}_{\mathrm{i}}\right)x_{\mathrm{ik}}^2 \\
&\approx \sum_{\mathrm{i}=1}^{\mathrm{n}} c_{\mathrm{ik}} V_{\mathrm{i}}^2(\Omega)
\end{aligned}
\qquad (8.4.28)
$$

In this argument, we assume that the influence of $V_{\mathrm{i}}^2(\Omega)$ is dominant as against that of $\boldsymbol{S}_{\mathrm{R}}(\Omega)$, as is the case for white noise excitation. The variance can easily be determined from equ. (8.4.28) and integration over Ω yields

$$
\begin{aligned}
\sigma_{r\mathrm{kk}} &= 2\int_0^{\infty} S_{r\mathrm{kk}}(\Omega)\,\mathrm{d}\Omega \\
&= \sum_{\mathrm{i}=1}^{\mathrm{n}} c_{\mathrm{ik}} 2\zeta_{\mathrm{i}}\omega_{\mathrm{i}}\pi
\end{aligned}
\qquad (8.4.29)
$$

This formula defines a so-called quasi-analytical integration. This technique can also be extended to the stresses.

We close this somewhat abstract chapter with the simple example of fig. 8.4.1, for which a solution is given in [8.3] and [8.4]. Here, a multiply supported beam experiences the noise radiation of a jet-plane. Let the continuous cross-spectral density of the pressure be prescribed by

$$
S = \mathrm{e}^{-i\Omega(x_2 - x_1)/6}
$$

The discretisation of the spectral density proceeds in [8.3] in terms of kinematically equivalent loads, whilst [8.4] applies a simple engineering method. We apply the latter procedure here and note that the area of influence for the pressure at a nodal point is simply 1/4. For the degrees of freedom w at nodal points i and j, we then obtain the cross-spectral density of the excitation forces as

$$
\begin{aligned}
S_{\mathrm{ij}} &= \left(\frac{1}{4}\right)^2 \mathrm{e}^{-i\Omega(\mathrm{j}-\mathrm{i})/24} \\
&= \frac{1}{16}\left(\cos\frac{\mathrm{j}-\mathrm{i}}{24}\Omega - i\sin\frac{\mathrm{j}-\mathrm{i}}{24}\Omega\right), \qquad \mathrm{j} \geqslant \mathrm{i}
\end{aligned}
$$

Here, $(\mathrm{j}-\mathrm{i})/4$ is the distance of $x_2 - x_1$ connecting the nodes i and j. The rotations at the nodal points are assigned no moments in our simple model. The cross spectral density matrix can thus be represented by 19 standard functions where

$$
\mathrm{j} - \mathrm{i} = 0, 1, 2, \; \ldots \; 18
$$

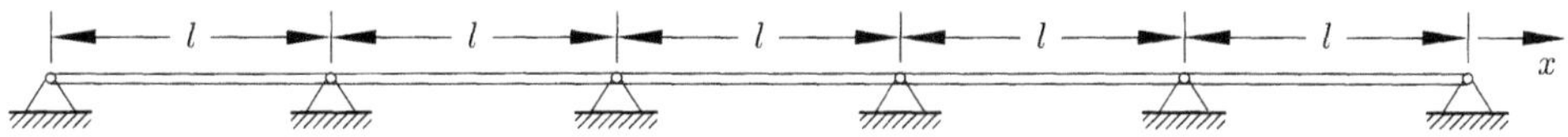

problem in plane

data: (coherent system of units)

cross-section area	$A = 1.0$
span	$l = 1.0$
moment of inertia	$I = 1.0$
Young's modulus	$E = 1.0$
modal damping	$\zeta_i = 0.01$, i=1…n

load: cross-spectral density of the pressures $p(x_1)$ and $p(x_2)$ due to jet noise,

$S = \mathrm{e}^{-i\Omega(x_2 - x_1)/6}$

idealisation

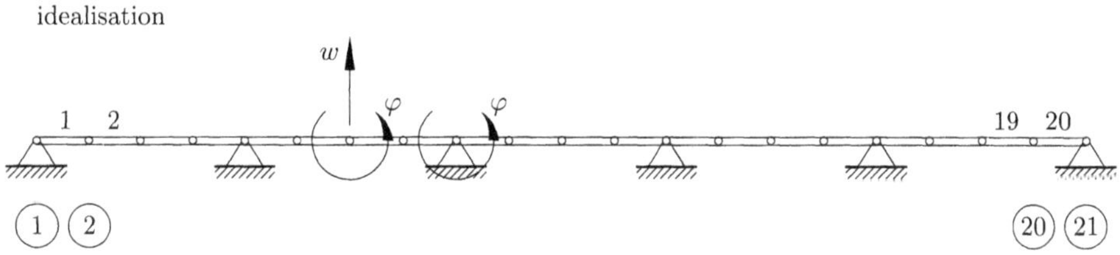

20 BECOS elements

21 nodal points

36 degrees of freedom : 21φ and $15\,w$

Fig. 8.4.1 Supported beam subject to the noise radiation of a jet-plane: original structure and system idealised by the finite element method

and the number 18 defines the distance between the free nodes 2 and 20. Since the spectral density matrix is dependent upon the frequency Ω, the 19 standard functions should be entered into the computer in a tabular form. We obtain the following extract from the computational results for the diagonal elements of the spectral density matrix for the nodal displacements w_3 and φ_1

$\Omega = \omega_i$	direct spectral density for	
	φ_1	w_3
9.87	0.1585	0.0157
10.95	1.4295	0.1344
13.70	0.2374	0.0179
17.26	0.0009	0.0001
20.73	0.0004	0.0000
39.63	0.0426	0.0000

The variances of the displacements φ_1 and w_3 can now be deduced by integration over the frequency. The root of the variances then yields the standard deviation

$$\sigma(\varphi_1) = 1.1807$$
$$\sigma(w_3) = 0.3433$$

These standard deviations will be assigned here a Gaussian distribution. Considering this to hold, we obtain for

$$P\Big(\frac{x}{\sigma_x} = 2\Big) = 0.9775$$

the following statements, which apply with a probability of 2.275%

$$\varphi_1 \geqslant 2.3614$$

or

$$w_3 \geqslant 0.6866$$

For readers who are interested in enriching their experience on random dynamics, we append a list of some more advanced treatises. In particular, we draw attention to the books of Newland [8.5], Lin [8.6] and Crandall's selection [8.7]. These publications also illuminate the subject from the experimental point of view in much greater detail than is possible in our treatise.

Bibliography for Chapter 8

[8.1] *A. Papoulis*; Probability, random variables and stochastic processes (McGraw-Hill, New York, 1965).

[8.2] *W.B. Davenport, W.L. Root*; An introduction to the theory of random signals and noise (McGraw-Hill, New York, 1958).

[8.3] *M.D. Olson*; A consistent finite element method for random response problems, Computers and Structures, 2 (1972) 163-180.

[8.4] *T.L. Johnsen, S.S. Dey*; ASKA Part II – Linear dynamic analysis, ASKA UM 218, Stuttgart (1978)

[8.5] *D.E. Newland*; An introduction to random vibrations and spectral analysis (Longman, London, 1975)

[8.6] *Y.K. Lin*; Probabilistic theory of structural dynamics (McGraw-Hill, New York, 1967).

[8.7] *St.H. Crandall*, ed.; Random vibrations (MIT and John Wiley, New York, 1958).

[8.8] *W. Heinrich, K. Hennig*; Zufallsschwingungen mechanischer Systeme (Vieweg, Braunschweig, 1978).

9 Dynamic analysis of structures with arbitrary viscous damping

"One should absorb the colour of life, but
one should never remember its details.
Details are always vulgar."
Oscar Wilde, The Picture of Dorian Gray (1891)

In Chapter 7, we proposed a solution to the eigenvalue problem of damped elastic systems by assigning a modal damping model to the damping mechanism of the structure. Since the eigenvectors of the proportionally damped system may be used to transform the damping matrix into a diagonal form as we demonstrated before for the stiffness and mass matrices, the system of equations can easily be uncoupled and the solution greatly simplified. The disadvantage of this method, however, is that it is difficult to conceive a realistic model of the damping mechanism with the two parameters allowed for in the proportional damping approach.

We have so far restricted ourselves to proportional damping but we now remove this restriction and propose to incorporate an arbitrary viscous damping into our calculation model. Nevertheless, we should be aware that it is much more difficult and computer-intensive to solve the eigenvalue problem when the solution is no longer real, but complex. Simultaneous vector iteration and the Eberlein method are two possible solution techniques which we shall discuss in the present chapter.

9.1 The special condition for non-modal damping

In Chapter 4, the undamped eigenoscillations of the system

$$\boldsymbol{M}\ddot{\boldsymbol{r}} + \boldsymbol{K}\boldsymbol{r} = \boldsymbol{o} \tag{9.1.1}$$

were discussed in detail. We also stated that the trial solution

$$\boldsymbol{r} = \boldsymbol{x}\,\mathrm{e}^{pt} \tag{9.1.2}$$

leads to 2n solutions for p. Each pair of solutions

$$\begin{aligned} p_{\mathrm{i}-} &= -i\omega_{\mathrm{i}} \\ p_{\mathrm{i}+} &= +i\omega_{\mathrm{i}} \end{aligned} \tag{9.1.3}$$

possesses the same solution vector $\boldsymbol{x}_{\mathrm{i}}$ (=eigenvector). As a result, the general solution to equ. (9.1.1) could be expressed in Chapter 4 by n contributions of the n eigenvectors $\boldsymbol{x}_{\mathrm{i}}$. An additional characteristic of the n eigenvectors $\boldsymbol{x}_{\mathrm{i}}$ is that they diagonalise the mass matrix $\boldsymbol{M}$ and the stiffness matrix $\boldsymbol{K}$ (see, for example, equs. (4.1.31b) and (4.1.32)). This property was applied in Chapter 6 to uncouple the forced, undamped oscillations

$$\boldsymbol{M}\ddot{\boldsymbol{r}} + \boldsymbol{K}\boldsymbol{r} = \boldsymbol{R}(t) \tag{9.1.4}$$

This was achieved by substitution of the trial solution

$$\boldsymbol{r}(t) = \boldsymbol{X}\boldsymbol{\eta}(t) \tag{9.1.5}$$

and pre-multiplication with $\boldsymbol{X}^{\mathrm{t}}$. In this way, the system is reduced to n independent equations, each with a single degree of freedom.

This special technique could also be applied to damped oscillations for a special type of damping. Thus, we showed in Chapter 8 that

$$\boldsymbol{M}\ddot{\boldsymbol{r}} + \boldsymbol{C}\dot{\boldsymbol{r}} + \boldsymbol{K}\boldsymbol{r} = \boldsymbol{R}(t) \tag{9.1.6}$$

can be uncoupled, if, for example, the damping matrix can be expressed as

$$\boldsymbol{C} = \alpha\boldsymbol{K} + \beta\boldsymbol{M} \tag{9.1.7}$$

This formulation is denoted as proportional damping . Even more general is the so-called modal damping

$$\boldsymbol{C} = \boldsymbol{X}^{-\mathrm{t}}\boldsymbol{D}\boldsymbol{X}^{-1} \tag{9.1.8}$$

Here, equ. (9.1.8) is only of formal significance and the damping coefficients $D_{\mathrm{ii}}\,(D_{\mathrm{ij}} = 0, \mathrm{i} \neq \mathrm{j})$ have to be specified directly for the uncoupled system.

The very simple academic example of fig. 9.1.1 indicates, however, that the special form given for the damping matrix in equs. (9.1.7) or (9.1.8) cannot be presumed, in general, to hold. In the present case, the mass matrix is diagonal

$$\boldsymbol{M} = \lceil m_1 \quad m_2 \rfloor = m\lceil 1 \quad 2 \rfloor$$

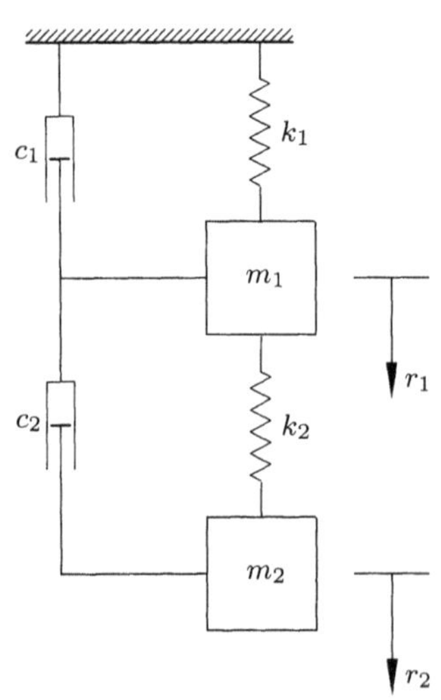

$m_1 = m$

$m_2 = 2m$

$c_1 = c_2 = 0.2m\,\omega_0$

$k_1 = m\,\omega_0^2$

$k_2 = 4m\,\omega_0^2$

Fig. 9.1.1

A simple example for general viscous damping [9.1]

whilst the stiffness matrix takes the form

$$\boldsymbol{K} = \begin{bmatrix} k_1 + k_2 & -k_2 \\ -k_2 & k_2 \end{bmatrix} = m\omega_0^2 \begin{bmatrix} 5 & -4 \\ -4 & 4 \end{bmatrix}$$

The associated viscous damping matrix

$$\boldsymbol{C} = \begin{bmatrix} c_1 + c_2 & -c_2 \\ -c_2 & c_2 \end{bmatrix} = 0.2m\omega_0 \begin{bmatrix} 2 & -1 \\ -1 & 1 \end{bmatrix}$$

is closely related to the formation of the stiffness matrix (dampers and springs are both connected in series), but $\boldsymbol{C}$ is at present not a linear combination of $\boldsymbol{M}$ and $\boldsymbol{K}$. This has repercussions upon the uncoupling of the system in equ. (9.1.6). Following the methodology outlined in Chapter 4, the reader should as a matter of fact have no difficulty in evaluating the eigenvalues and eigenvectors of the undamped system in fig. 9.1.1. Putting

$$\overline{\omega}_{\mathrm{i}} = -p_{\mathrm{i}}^2/\omega_0^2$$

we obtain

$$\begin{aligned} \overline{\boldsymbol{\Omega}} &= \lceil \overline{\omega}_{11} \;\; \overline{\omega}_{22} \rfloor \\ &= \lceil (7 - \sqrt{41})/2 \quad (7 + \sqrt{41})/2 \rfloor \end{aligned}$$

with the corresponding eigenvectors (ignoring any specific normalisation)

$$\boldsymbol{X} = [\boldsymbol{x}_1 \;\; \boldsymbol{x}_2] = \begin{bmatrix} -3 + \sqrt{41} & -3 - \sqrt{41} \\ 4 & 4 \end{bmatrix}$$

When we now substitute equ. (9.1.5) into equ. (9.1.6) and pre-multiply by $\boldsymbol{X}^{\mathrm{t}}$, the transformed mass matrix reduces to the diagonal form

$$\boldsymbol{X}^{\mathrm{t}}\boldsymbol{M}\boldsymbol{X} = m\lceil (82 - 6\sqrt{41}) \quad (82 + 6\sqrt{41}) \rfloor$$

Similarly, the transformed stiffness matrix becomes

$$\boldsymbol{X}^{\mathrm{t}}\boldsymbol{K}\boldsymbol{X} = m\omega_0^2 \lceil (410 - 62\sqrt{41}) \quad (410 + 62\sqrt{41}) \rfloor$$

But the transformed damping matrix does not attain a diagonal form

$$\boldsymbol{D} = \boldsymbol{X}^{\mathrm{t}}\boldsymbol{C}\boldsymbol{X} = 0.2m\omega_0 \begin{bmatrix} 140 - 20\sqrt{41} & -24 \\ -24 & 140 + 20\sqrt{41} \end{bmatrix}$$

Indeed, the transformed damping matrix does not even possess a dominant diagonal which casts doubt on the standard proposal of selecting the diagonal of the $\boldsymbol{D}$ matrix as the measure for modal damping. It is thus evident that the eigenvectors of the undamped system are of no use in the uncoupling process of equ. (9.1.6) when the damping matrix $\boldsymbol{C}$ is of a general nature. It follows that there must be a fundamental difference between a system defined by equ. (9.1.6) and the undamped and modally damped structures. Let us now inquire into the characteristics of its solution. Like equs. (9.1.1) and (4.1.1), equ. (9.1.6) represents a system of coupled, linear differential equations. Thus, the trial solution of equ. (9.1.2) or (4.1.2) for the homogeneous solution is still legitimate. Analogous to equ. (4.1.3), we obtain for the evaluation of the homogeneous part of the solution a homogeneous algebraic system

$$[p^2\boldsymbol{M} + p\boldsymbol{C} + \boldsymbol{K}]\boldsymbol{x} = \boldsymbol{o} \tag{9.1.9}$$

The trivial solution $\boldsymbol{x} = \boldsymbol{o}$ is of no interest to us. Non-trivial solutions only exist if the system coefficient matrix is singular, i.e. the determinant of this matrix vanishes. As an extension of equ. (4.1.4a), we obtain here

$$\det[p^2\boldsymbol{M} + p\boldsymbol{C} + \boldsymbol{K}] = 0 \tag{9.1.10}$$

This yields the pertinent equation for the exponents p of equ. (9.1.2). Its special characteristics now become apparent. The expansion of the determinant no longer leads to an n-th order polynomial in p^2 as in equ. (4.1.4c) whose n solutions for p^2 also determine the eigenvectors. Here, we are forced by the damping term $p\boldsymbol{C}$ to operate with p itself and thus, we obtain as a characteristic polynomial in place of equ. (4.1.4c) an equation of order 2n in p

$$b_{2n}p^{2n} + b_{2n-1}p^{2n-1} + \ldots + b_0 = 0 \tag{9.1.11}$$

It can easily be seen that the two relevant statements expressed by the equs. (4.1.5) and (4.1.6) are still true in the present case; consequently, we have

$$b_{2n} = \det[\boldsymbol{M}] \tag{9.1.12}$$

$$b_0 = \det[\boldsymbol{K}] \tag{9.1.13}$$

Elementary mathematics teaches us that the polynomial in equ. (9.1.11) generally yields 2n solutions for p_i which can also be complex. Consequently, we must also expect complex solution vectors $\boldsymbol{x}_i$ which may be found by substitution of p_i into equ. (9.1.9) and application of an auxiliary condition (note the singularity of the system). In principle, the same also applies to undamped systems, since 2n solutions to p_i, namely p_{i+} and p_{i-}, generally correspond to the n solutions in p_i^2. Each of these solutions, however, possesses the same eigenvector $\boldsymbol{x}_i$, for only p_i^2 enters the matrix of the system coefficients. This special condition no longer holds in the present case. We shall now attempt to explore the nature of the solution in analogy to Chapter 4. To this purpose, we write equ. (9.1.9) for a specific solution p_i and $\boldsymbol{x}_i$ and pre-multiply it with the complex conjugate transposed vector $\boldsymbol{x}_i^*$, see equ. (4.1.8). Analogous to equ. (4.1.9), we obtain the scalar expression

$$p_i^2\boldsymbol{x}_i^*\boldsymbol{M}\boldsymbol{x}_i + p_i\boldsymbol{x}_i^*\boldsymbol{C}\boldsymbol{x}_i + \boldsymbol{x}_i^*\boldsymbol{K}\boldsymbol{x}_i = 0 \tag{9.1.14}$$

In the case of a non-rotating reference system and a conservative excitation, the matrices $\boldsymbol{C}$, $\boldsymbol{K}$ and $\boldsymbol{M}$ are all symmetric. Consequently, as already observed for undamped systems in equ. (4.1.10), the three expressions

$$\begin{aligned} \overline{m}_{\mathrm{i}} &= \boldsymbol{x}_{\mathrm{i}}^{*}\boldsymbol{M}\boldsymbol{x}_{\mathrm{i}} \\ \overline{c}_{\mathrm{i}} &= \boldsymbol{x}_{\mathrm{i}}^{*}\boldsymbol{C}\boldsymbol{x}_{\mathrm{i}} \\ \overline{k}_{\mathrm{i}} &= \boldsymbol{x}_{\mathrm{i}}^{*}\boldsymbol{K}\boldsymbol{x}_{\mathrm{i}} \end{aligned} \tag{9.1.15}$$

remain real numbers even if the vector $\boldsymbol{x}_{\mathrm{i}}$ is complex. Normally, $\boldsymbol{M}$ is positive definite (every conceivable motion is linked to inertial effects); the same applies to $\boldsymbol{C}$ (mechanical energy is transformed into thermal energy for all motions) and $\boldsymbol{K}$ (each motion results in elastic strains). From equ. (4.1.10), we can then deduce that the values $\overline{m}_{\mathrm{i}}$, $\overline{c}_{\mathrm{i}}$ and $\overline{k}_{\mathrm{i}}$ are not only real, but also positive. Following our analysis of an oscillator with a single degree of freedom, see equ. (1.3.3) and (1.3.4), we can now define

$$\omega_{\mathrm{i}}\zeta_{\mathrm{i}} = \overline{c}_{\mathrm{i}}/(2\overline{m}_{\mathrm{i}}) > 0 \tag{9.1.16}$$

$$\omega_{\mathrm{i}} = (\overline{k}_{\mathrm{i}}/\overline{m}_{\mathrm{i}})^{1/2} > 0 \tag{9.1.17}$$

Hence, equ. (9.1.14) becomes

$$p_{\mathrm{i}}^2 + 2\zeta_{\mathrm{i}}\omega_{\mathrm{i}}p_{\mathrm{i}} + \omega_{\mathrm{i}}^2 = 0 \tag{9.1.18}$$

with the solutions

$$p_{\mathrm{i}} = -\omega_{\mathrm{i}}\left[\zeta_{\mathrm{i}} \pm \sqrt{\zeta_{\mathrm{i}}^2 - 1}\right] \tag{9.1.19}$$

One of the two solutions from equ. (9.1.19) defines the exponent p_{i}, corresponding to $\boldsymbol{x}_{\mathrm{i}}$. As was the case for the simple single-degree-of-freedom system, we distinguish between two cases. For weak damping, $\overline{c}_{\mathrm{i}}$, and thus ζ_{i}, is small. In particular, $\zeta_{\mathrm{i}} < 1$ (sub-critical damping) may apply. In this case, the solution p_{i} is complex

$$\begin{aligned} p_{\mathrm{i}} &= -\omega_{\mathrm{i}}\left[\zeta_{\mathrm{i}} \pm i\sqrt{1 - \zeta_{\mathrm{i}}^2}\right] \\ &= -\zeta_{\mathrm{i}}\omega_{\mathrm{i}} \pm i\omega_{D\mathrm{i}} \end{aligned} \tag{9.1.20}$$

with a negative real part. It follows that the i-th contribution to the solution

$$\boldsymbol{x}_{\mathrm{i}}\mathrm{e}^{p_{\mathrm{i}}t} = \boldsymbol{x}_{\mathrm{i}}\mathrm{e}^{-\omega_{\mathrm{i}}\zeta_{\mathrm{i}}t}\mathrm{e}^{\pm i\omega_{D\mathrm{i}}t} \tag{9.1.21}$$

comprises an exponentially decaying factor.

If the complex solution of equ. (9.1.20) is substituted into equ. (9.1.9), we obtain an algebraic system with complex coefficients, from which $\boldsymbol{x}_{\mathrm{i}}$ can be determined using a secondary condition (e.g. $\boldsymbol{x}_{\mathrm{i}}$ is assigned a length 'one'). In general, $\boldsymbol{x}_{\mathrm{i}}$ is complex as we have already guessed correctly. We may evaluate it only up to a numerical factor (observe the secondary condition). In contrast to the undamped case, however, a purely real representation of each component given by equ. (9.1.21) proves impossible.

The attentive reader will doubtless have been puzzled since equ. (9.1.19) yielded two values of p_i, of which only one can be allocated to the complex vector $\boldsymbol{x}_i$. What, however, is the significance of the other solution for p_i, which does not appertain to $\boldsymbol{x}_i$ and is clearly complex conjugate to the other root? If we inspect equ. (9.1.14) in the light of these reflections, we realise that the two vectors $\boldsymbol{x}_i$ and $(\boldsymbol{x}_i^*)^t$ are interchangeable without any changes arising in equ. (9.1.15) and in our subsequent discussion. Could $(\boldsymbol{x}_i^*)^t$ thus be the eigenvector for the second solution of p_i? We put this hunch to test. Let

$$p_{i+-} = -\omega_i\zeta_i \pm i\omega_{Di} \tag{9.1.22}$$

be two solutions which appertain to the eigenvectors

$$\boldsymbol{x}_{i+-} = \boldsymbol{a}_i \pm i\boldsymbol{b}_i \tag{9.1.23}$$

Substituting these expressions into equ. (9.1.9), we obtain

$$\begin{aligned}
&\left[(-\zeta_i\omega_i \pm i\omega_{Di})^2\boldsymbol{M} + (-\zeta_i\omega_i \pm i\omega_{Di})\boldsymbol{C} + \boldsymbol{K}\right]\left[\boldsymbol{a}_i \pm i\boldsymbol{b}_i\right] \\
&= \left[(\zeta_i^2\omega_i^2 - \omega_{Di}^2)\boldsymbol{M} - \zeta_i\omega_i\boldsymbol{C} + \boldsymbol{K} \mp 2i\zeta_i\omega_i\omega_{Di}\boldsymbol{M} \pm i\omega_{Di}\boldsymbol{C}\right]\left[\boldsymbol{a}_i \pm i\boldsymbol{b}_i\right] \\
&= \left[(\zeta_i^2\omega_i^2 - \omega_{Di}^2)\boldsymbol{M} - \zeta_i\omega_i\boldsymbol{C} + \boldsymbol{K}\right]\boldsymbol{a}_i + \omega_{Di}\left[2\zeta_i\omega_i\boldsymbol{M} - \boldsymbol{C}\right]\boldsymbol{b}_i \\
&\quad \pm i\left[(\zeta_i^2\omega_i^2 - \omega_{Di}^2)\boldsymbol{M} - \zeta_i\omega_i\boldsymbol{C} + \boldsymbol{K}\right]\boldsymbol{b}_i \mp i\omega_{Di}\left[2\zeta_i\omega_i\boldsymbol{M} - \boldsymbol{C}\right]\boldsymbol{a}_i \\
&= \boldsymbol{o}
\end{aligned} \tag{9.1.24}$$

If the upper sign in equs. (9.1.22) and (9.1.23) represents a solution to equ. (9.1.9), then the corresponding real and imaginary parts in equ. (9.1.24) must be zero. Hence, the imaginary part, associated with the lower sign, must also be zero. Consequently, the lower sign in equ. (9.1.22) or (9.1.23) also represents a solution. We conclude: if the system in equ. (9.1.9) possesses a complex solution p_i with the complex eigenvector $\boldsymbol{x}_i$, then the complex conjugate root p_i^* and the complex conjugate eigenvector $(\boldsymbol{x}_i^*)^t$ define a solution, too. Analogous to the undamped case, we thus obtain a pair of solutions, but with two different eigenvectors. This set of solutions yields the following contribution to the homogeneous solution (see also equ. (4.1.17))

$$\begin{aligned}
&c_{i+}[\boldsymbol{a}_i + i\boldsymbol{b}_i]e^{(-\omega_i\zeta_i + i\omega_{Di})t} + c_{i-}[\boldsymbol{a}_i - i\boldsymbol{b}_i]e^{(-\omega_i\zeta_i - i\omega_{Di})t} \\
&= e^{-\omega_i\zeta_i t}\Big[\boldsymbol{a}_i[(c_{i+} + c_{i-})(e^{i\omega_{Di}t} + e^{-i\omega_{Di}t})/2 + (c_{i+} - c_{i-})(e^{i\omega_{Di}t} - e^{-i\omega_{Di}t})/2] \\
&\qquad + i\boldsymbol{b}_i[(c_{i+} - c_{i-})(e^{i\omega_{Di}t} + e^{-i\omega_{Di}t})/2 + (c_{i+} + c_{i-})(e^{i\omega_{Di}t} - e^{-i\omega_{Di}t})/2]\Big] \\
&= e^{-\omega_i\zeta_i t}[\bar{\boldsymbol{a}}_i \sin\omega_{Di}t + \bar{\boldsymbol{b}}_i \cos\omega_{Di}t] \\
&= e^{-\omega_i\zeta_i t}\{\bar{c}_{i1}\sin(\omega_{Di}t + \varphi_{i1}) \quad \bar{c}_{i2}\sin(\omega_{Di}t + \varphi_{i2}) \quad \ldots \quad \bar{c}_{in}\sin(\omega_{Di}t + \varphi_{in})\}
\end{aligned} \tag{9.1.25}$$

i.e. an oscillatory motion in real terms where the two coefficients $\overline{\boldsymbol{a}}_i$ and $\overline{\boldsymbol{b}}_i$ are given by

$$\begin{aligned}\overline{\boldsymbol{a}}_i &= i(c_{i+} - c_{i-})\boldsymbol{a}_i - (c_{i+} + c_{i-})\boldsymbol{b}_i = \alpha_i \boldsymbol{a}_i + \beta_i \boldsymbol{b}_i \\ \overline{\boldsymbol{b}}_i &= (c_{i+} + c_{i-})\boldsymbol{a}_i + i(c_{i+} - c_{i-})\boldsymbol{b}_i = -\beta_i \boldsymbol{a}_i + \alpha_i \boldsymbol{b}_i\end{aligned} \tag{9.1.26}$$

For the corresponding component in the undamped case, we deduce from equ. (4.1.17)

$$\boldsymbol{x}_i(a_i \sin\omega_i t + b_i \cos\omega_i t) = \boldsymbol{x}_i c_i \sin(\omega_i t + \psi_i) \tag{9.1.27}$$

and for the modally damped system from equ. (7.4.5)

$$\boldsymbol{x}_i \mathrm{e}^{-\omega_i \zeta_i t}(a_i \sin\omega_{Di} t + b_i \cos\omega_{Di} t) = \boldsymbol{x}_i \mathrm{e}^{-\omega_i \zeta_i t} c_i \sin(\omega_{Di} t + \psi_{Di}) \tag{9.1.28}$$

We observe that in the case of modal damping, the phase lag for each specific mode remains the same for all degrees of freedom. The characteristic shape of the motion is represented totally by $\boldsymbol{x}_i$, and the time-dependence is merely expressed by a scaling factor. For general viscous damping, we have the same decay rate and the same beat for all degrees of freedom of the pair of solutions, but the phase relationship differs for each degree of freedom. Thus, in general, no steady-state oscillations can arise having fixed positions of nodes and antinodes. This fundamental difference is illustrated in the simple example of fig. 9.1.2.

Our solution in equ. (9.1.19) contains another type of solution which has not been mentioned so far. In the case of strong damping, $\zeta_i \geqslant 1$ (critical or over-critical damping) could arise. We note that

$$(\zeta_i^2 - 1)^{1/2} < \zeta_i \tag{9.1.29}$$

holds in all conditions and thus, the solutions p_i for $\zeta_i \geqslant 1$ are always negative and real

$$p_i < 0, \qquad \text{if} \quad \zeta_i \geqslant 1 \tag{9.1.30}$$

Analogous to modal damping, we now observe no oscillating motion, but an aperiodic fading-out of the deflection instead. If this real solution p_i is substituted into equ. (9.1.9), we obtain a real, linear system for the determination of the corresponding eigenvector $\boldsymbol{x}_i$. In this way, we can establish a real vector $\boldsymbol{x}_i$ if we adopt an appropriate auxiliary constraint. Since the complex solutions occur in pairs, an even number of real eigenvalues and corresponding real eigenvectors also arises. In this case, however, the speculation that the second root of equ. (9.1.19), which does not appertain to $\boldsymbol{x}_i$, defines the second real eigenvalue is inappropriate. This is hardly surprising since for real values of $\boldsymbol{x}_i$, we have $(\boldsymbol{x}_i^*)^t = \boldsymbol{x}_i$ and consequently, the determining equation can only be associated with a single vector.

Following the detailed discussion of the homogeneous solution to equ. (9.1.6), we can assemble all component solutions in the form

$$\underset{(n\times 1)}{\boldsymbol{r}(t)} = \underset{(n\times 1)}{[\,\boldsymbol{x}_1 \quad \boldsymbol{x}_2} \ \dots \ \boldsymbol{x}_{2n}]\lceil e^{p_1 t} \quad e^{p_2 t} \ \dots \ e^{p_{2n} t}\rfloor\{c_1 \quad c_2 \ \dots \ c_{2n}\}$$

$$= \underset{(n\times 2n)}{\boldsymbol{X}} \ \underset{(2n\times 2n)}{\boldsymbol{p}(t)} \ \underset{(2n\times 1)}{\boldsymbol{c}} \tag{9.1.31}$$

Here, the matrix $\boldsymbol{X}$ contains eigenvectors $\boldsymbol{x}_i$, some of which may be real and others complex. A pair of associated complex contributions can be combined into a real expression which contains two real vectors in this alternative formulation – see also equ. (9.1.25). Furthermore,

$$\boldsymbol{p}(t) = \lceil e^{p_1 t} \quad e^{p_2 t} \ \dots \ e^{p_{2n} t}\rfloor \tag{9.1.32}$$

is the diagonal matrix of the time functions which comprise some real and some complex exponents with negative real parts. Finally, the vector

$$\boldsymbol{c} = \{c_1 \quad c_2 \ \dots \ c_{2n}\} \tag{9.1.33}$$

defines the factors which may be assigned at liberty to a general homogeneous solution. Note that $\boldsymbol{c}$ can also contain partly real, partly complex numbers. When only real eigenvalues exist, we may reformulate equ. (9.1.31) into the real expression

$$\boldsymbol{r}(t) = \sum_{i=1}^{nr} c_i \boldsymbol{x}_i e^{p_i t} + \sum_{j=1}^{n-nr/2} e^{-\omega_j \zeta_j t}[\overline{\boldsymbol{a}}_j \sin \omega_{Dj} t + \overline{\boldsymbol{b}}_j \cos \omega_{Dj} t] \tag{9.1.34}$$

Here, the nr real negative exponents p_i are given by equ. (9.1.19) and ω_j, ζ_j, ω_{Dj} are defined by equs. (9.1.16), (9.1.17) and (9.1.20). The column vectors $\overline{\boldsymbol{b}}_j$ and $\overline{\boldsymbol{a}}_j$ are identical with the imaginary and real parts of the corresponding complex pair of eigenvectors (see also equ. (9.1.25)).

Combining the trigonometrical functions according to the standard addition formulae as in equ. (4.1.17), we obtain the solution

$$\boldsymbol{r}(t) = \sum_{i=1}^{nr} c_i \boldsymbol{x}_i e^{p_i t} + \sum_{j=1}^{n-nr/2} e^{-\omega_j \zeta_j t} \boldsymbol{c}_j \sin(\omega_{Dj} t \boldsymbol{e}_n + \boldsymbol{\varphi}_j) \tag{9.1.35}$$

This representation differs quite considerably from that for modal damping. The components of the modal contributions are obviously phase shifted to each other, whilst for modal damping, the components are always in phase. In the latter case, φ_j is scalar, but for general damping, $\boldsymbol{\varphi}_j$ is a vector. For modal damping, each component – a so-called eigenoscillation – describes a steady-state motion and oscillatory nodes and antinodes exist. In contrast, for arbitrary viscous damping, there is no longer a steady-state pattern, even when the response $\boldsymbol{r}(t)$ contains a single modal component only. The deflection shape repeats itself with a period

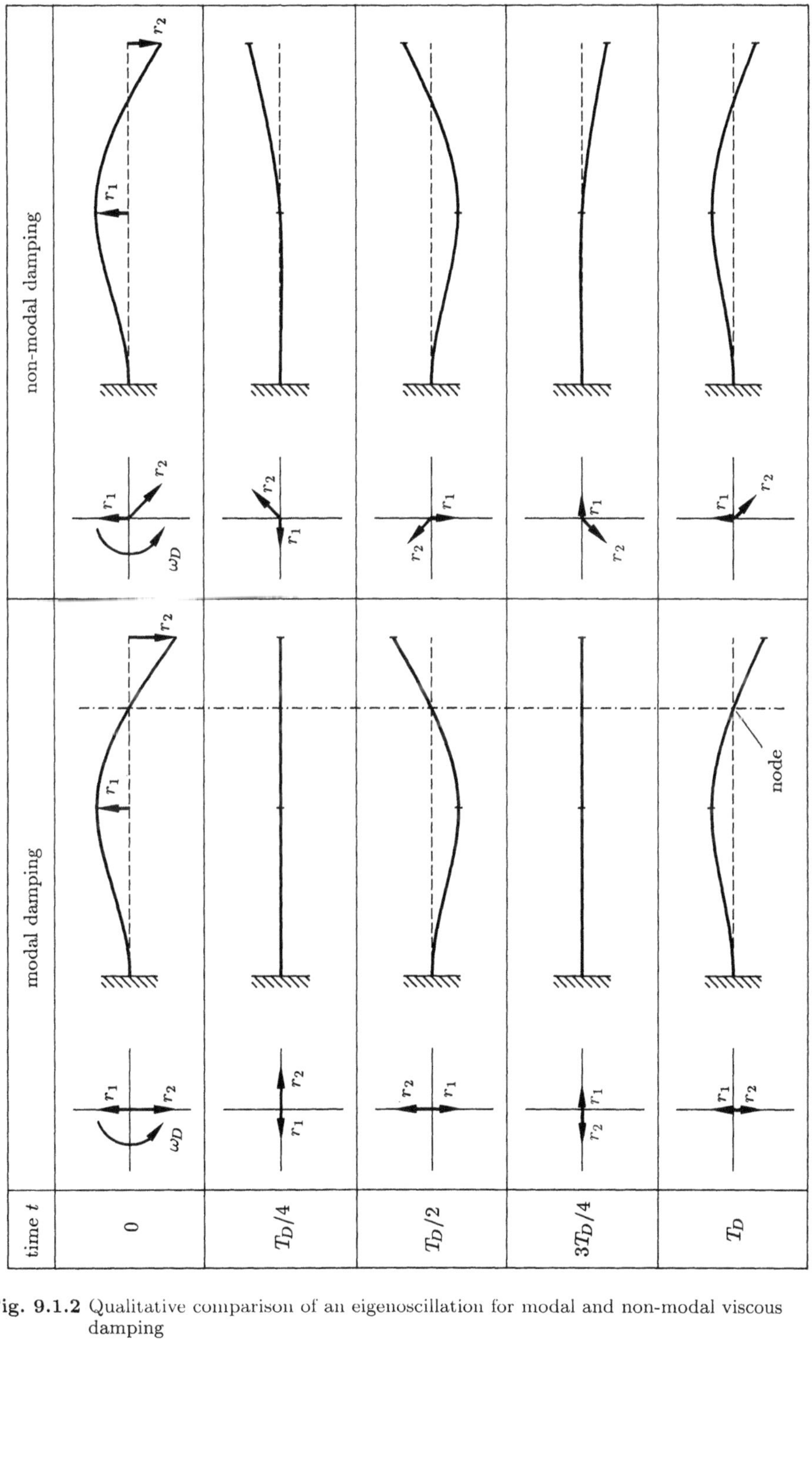

Fig. 9.1.2 Qualitative comparison of an eigenoscillation for modal and non-modal viscous damping

$T_{D\mathrm{j}} = 2\pi/\omega_{D\mathrm{j}}$ and with a decreasing amplitude but nodes and antinodes of oscillation are no longer existent. This information is illustrated in the vector diagram of fig. 9.1.2.

Both equ. (9.1.30) and the real formulation of equ. (9.1.34) are expressed in terms of 2n vectors, whilst the solution for modal damping only contains n vectors. In the latter case, an uncoupling of the system of equ. (9.1.6) is obtained by the modal transformation

$$\underset{(\mathrm{n}\times 1)}{\boldsymbol{r}(t)} = \underset{(\mathrm{n}\times\mathrm{n})}{\boldsymbol{X}}\;\underset{(\mathrm{n}\times 1)}{\boldsymbol{\eta}(t)} \tag{9.1.36}$$

thus enabling a convenient analysis of the inhomogeneous system. An analogous formulation would also be formally possible for viscous damping. Thus, in accordance with equ. (9.1.31), we have

$$\underset{(\mathrm{n}\times 1)}{\boldsymbol{r}(t)} = \underset{(\mathrm{n}\times 2\mathrm{n})}{\boldsymbol{X}}\;\underset{(2\mathrm{n}\times 1)}{\boldsymbol{\eta}(t)} \tag{9.1.37}$$

It is at once apparent, however, that this cannot yield the required solution because it contains 2n vectors with which n equations should be transformed.

Before exploring this matter further from the theoretical point of view, we rehearse our present state of knowledge on the simple example of fig. 9.1.1. The system matrices $\boldsymbol{M}$, $\boldsymbol{C}$ and $\boldsymbol{K}$ have already been set up, so that the algebraic system in equ. (9.1.9) can immediately be written as

$$\begin{bmatrix} \psi^2 + 0.4\psi + 5 & -0.2\psi - 4 \\ -0.2\psi - 4 & 2\psi^2 + 0.2\psi + 4 \end{bmatrix} \begin{bmatrix} x_1 \\ x_2 \end{bmatrix} = \begin{bmatrix} 0 \\ 0 \end{bmatrix}$$

where we use the abbreviation

$$\psi^2 = p_{\mathrm{i}}^2/\omega_0^2$$

Evaluating the determinant, we obtain the characteristic equation

$$\psi^4 + 0.50\psi^3 + 7.02\psi^2 + 0.50\psi + 2 = 0$$

Since all coefficients of this expression are positive, the roots can only be negative real or complex. The solutions are, in fact,

$$\begin{aligned} \psi_{1,2} &= -0.0274 \pm 0.5458\,i = (-\omega_1\zeta_1 \pm i\omega_{D1})/\omega_0 \\ \psi_{3,4} &= -0.2226 \pm 2.5783\,i = (-\omega_2\zeta_2 \pm i\omega_{D2})/\omega_0 \end{aligned}$$

We observe two complex pairs of solutions, since in our case damping is too weak for the generation of aperiodic contributions. The next step involves the calculation of the eigenvectors appertaining to $\psi_{1,2}$. To this purpose, we substitute the solution $\psi_{1,2}$ into the above algebraic system. From the first equation, we deduce

$$\boldsymbol{x}_{1,2} = \left[\begin{bmatrix} 1.0000 \\ 1.1750 \end{bmatrix} \pm i \begin{bmatrix} 0 \\ 0.0151 \end{bmatrix}\right] = \boldsymbol{a}_1 \pm i\boldsymbol{b}_1$$

Once again, this confirms that the two eigenvectors associated with a pair of roots or eigenvalues are complex conjugate. Since the system matrix is singular, it would also have been possible to use the second constituent equation to calculate $\boldsymbol{x}_{1,2}$. Here, we apply this second equation to compute the other two eigenvectors associated with $\psi_{3,4}$ and find

$$\boldsymbol{x}_{3,4} = \left[\begin{bmatrix} 1.0000 \\ -0.4232 \end{bmatrix} \pm i \begin{bmatrix} 0 \\ 0.0257 \end{bmatrix}\right] = \boldsymbol{a}_2 \pm i\boldsymbol{b}_2$$

The free motion of the system in fig. 9.1.1 may now be obtained by superposition of the four solution components using four arbitrary factors. We thus deduce from equ. (9.1.31)

$$\begin{aligned} \boldsymbol{r}(t) = \; & [\boldsymbol{a}_1 + i\boldsymbol{b}_1]c_1 \mathrm{e}^{(-\omega_1\zeta_1 + i\omega_{D1})t} + [\boldsymbol{a}_1 - i\boldsymbol{b}_1]c_2 \mathrm{e}^{(-\omega_1\zeta_1 - i\omega_{D1})t} \\ & + [\boldsymbol{a}_2 + i\boldsymbol{b}_2]c_3 \mathrm{e}^{(-\omega_2\zeta_2 + i\omega_{D2})t} + [\boldsymbol{a}_2 - i\boldsymbol{b}_2]c_4 \mathrm{e}^{(-\omega_2\zeta_2 - i\omega_{D2})t} \end{aligned}$$

where the complex constants c_1 to c_4 can be calculated from the initial conditions. If, for example, $\boldsymbol{r}(0)$ and $\dot{\boldsymbol{r}}(0)$ are given, equ. (9.1.31) yields

$$\dot{\boldsymbol{r}}(t) = \boldsymbol{X}\dot{\boldsymbol{p}}(t)\boldsymbol{c} \tag{9.1.38}$$

where

$$\dot{\boldsymbol{p}}(t) = \boldsymbol{p}(t)\boldsymbol{d}_p \tag{9.1.39}$$

and

$$\boldsymbol{d}_p = \lceil p_1 \quad p_2 \;\dots\; p_{2n} \rfloor \tag{9.1.40}$$

is the diagonal matrix of the solution exponents. Now, however, the following always applies

$$\boldsymbol{p}(0) = \boldsymbol{I}_{2n} \tag{9.1.41}$$

Consequently, a complex algebraic system can be set up with relative ease to determine the four $\boldsymbol{c}$ constants. We find

$$\begin{bmatrix} \underset{(n\times 2n)}{\boldsymbol{X}} \\ \underset{(n\times 2n)(2n\times 2n)}{\boldsymbol{X} \quad \boldsymbol{d}_p} \end{bmatrix} \underset{(2n\times 1)}{\boldsymbol{c}} = \begin{bmatrix} \underset{(n\times 1)}{\boldsymbol{r}(0)} \\ \underset{(n\times 1)}{\dot{\boldsymbol{r}}(0)} \end{bmatrix} \tag{9.1.42}$$

The right-hand side of this equation is real. If, for example, the mass m_1 in fig. 9.1.1 is displaced statically by '1', whilst the mass m_2 is kept at rest and then the system is released, we obtain

$$\{\boldsymbol{r}(0) \quad \dot{\boldsymbol{r}}(0)\} = \{1 \quad 0 \quad 0 \quad 0\}$$

The coefficients of the system in equ. (9.1.42) are complex and consequently the $\boldsymbol{c}$ constants are complex also. To simplify matters, we set in our example

$$\omega_0 = 1$$

The following solution for $\boldsymbol{c}$ is then obtained for the aforementioned initial static deflection

$$\boldsymbol{c} = \begin{bmatrix} 0.1316 & -0.0360\,i \\ 0.1316 & +0.0360\,i \\ 0.3684 & -0.0256\,i \\ 0.3684 & +0.0256\,i \end{bmatrix}$$

Substitution of these values confirms that the system's response $\boldsymbol{r}(t)$ is indeed real. Naturally, the preceding explanations also permit the selection of a purely real representation of the response, but it is admittedly not as simple and as easy to handle as the complex one. Combining the complex pairs of solutions in equ. (9.1.25) gives the following useful expression

$$\boldsymbol{r}(t) = \mathrm{e}^{-\omega_1\zeta_1 t}[\overline{\boldsymbol{a}}_1 \sin\omega_{D1}t + \overline{\boldsymbol{b}}_1 \cos\omega_{D1}t] + \mathrm{e}^{-\omega_2\zeta_2 t}[\overline{\boldsymbol{a}}_2 \sin\omega_{D2}t + \overline{\boldsymbol{b}}_2 \cos\omega_{D2}t]$$

where in accordance with equ. (9.1.26), the amplitude vectors are expressed as

$$\begin{aligned} \overline{\boldsymbol{a}}_\mathrm{i} &= \alpha_\mathrm{i}\boldsymbol{a}_\mathrm{i} + \beta_\mathrm{i}\boldsymbol{b}_\mathrm{i} \\ \overline{\boldsymbol{b}}_\mathrm{i} &= -\beta_\mathrm{i}\boldsymbol{a}_\mathrm{i} + \alpha_\mathrm{i}\boldsymbol{b}_\mathrm{i}, \qquad \mathrm{i} = 1,2 \end{aligned}$$

and the four real constants α_1, β_1, α_2, β_2 are once again obtained from the initial conditions, e.g. $\boldsymbol{r}(0)$ and $\dot{\boldsymbol{r}}(0)$. In fact, we deduce from the general solution

$$\begin{aligned} \boldsymbol{r}(0) &= \overline{\boldsymbol{b}}_1 + \overline{\boldsymbol{b}}_2 \\ \dot{\boldsymbol{r}}(0) &= -\omega_1\zeta_1\overline{\boldsymbol{b}}_1 + \omega_{D1}\overline{\boldsymbol{a}}_1 - \omega_2\zeta_2\overline{\boldsymbol{b}}_2 + \omega_{D2}\overline{\boldsymbol{a}}_2 \end{aligned}$$

Substitution of the preceding relation between $\overline{\boldsymbol{a}}_\mathrm{i}$, $\overline{\boldsymbol{b}}_\mathrm{i}$ and $\boldsymbol{a}_\mathrm{i}$, $\boldsymbol{b}_\mathrm{i}$ into the last two equations yields a real algebraic system for the determination of α_i, β_i. The reader may also deduce that these solutions are real from equ. (9.1.26) and the aforementioned solution for $\boldsymbol{c}$. For example, we find

$$\begin{aligned} \alpha_1 &= i\,(c_1 - c_2) = 0.0720 \\ \beta_1 &= -(c_1 + c_2) = -0.2632 \end{aligned}$$

and hence ultimately

$$\begin{aligned} \boldsymbol{r}(t) = \; & \mathrm{e}^{-0.0274t}\left[\begin{bmatrix} 0.0720 \\ 0.0807 \end{bmatrix} \sin 0.5458\,t + \begin{bmatrix} 0.2633 \\ 0.3104 \end{bmatrix} \cos 0.5458\,t\right] \\ & + \mathrm{e}^{-0.2226t}\left[\begin{bmatrix} 0.0512 \\ -0.0406 \end{bmatrix} \sin 2.5783\,t + \begin{bmatrix} 0.7367 \\ -0.3104 \end{bmatrix} \cos 2.5783\,t\right] \end{aligned}$$

or alternatively

$$\boldsymbol{r}(t) = \mathrm{e}^{-0.0274t}\left[\begin{bmatrix}0.2730\\0.3208\end{bmatrix}\sin\left(0.5458\,t + \begin{bmatrix}1.3037\\1.3165\end{bmatrix}\right)\right]$$

$$+\,\mathrm{e}^{-0.2226t}\left[\begin{bmatrix}0.7385\\0.3131\end{bmatrix}\sin\left(2.5783\,t + \begin{bmatrix}1.5015\\1.4408\end{bmatrix}\right)\right]$$

The first real formulation demonstrates dominant cosine terms, but the high frequency contribution is rapidly damped away (fig. 9.1.3). The second indicates as a result of small sine values only minimal differences in the phase of r_1 and r_2 for both low and high frequency contributions. This suggests that the present conditions fit well into the modal damping model, although the constitution of the transformed damping matrix $\boldsymbol{D}$, as derived at the beginning of the chapter, really suggests the contrary. Applying the enforced diagonalisation

$$\boldsymbol{d} = \mathrm{diag}\boldsymbol{D} = 0.2m\omega_0\lceil(140 - 20\sqrt{41}) \quad (140 + 20\sqrt{41})\rfloor$$

we obtain the solution

$$\psi_{1,2} = -\,0.0274 + 0.5436\,i$$
$$\psi_{3,4} = -\,0.2226 + 2.5791\,i$$

$$\boldsymbol{r}(t) = \mathrm{e}^{-0.0274t}\left[\begin{bmatrix}0.0133\\0.0157\end{bmatrix}\sin 0.5458t + \begin{bmatrix}0.2657\\0.3123\end{bmatrix}\cos 0.5458\,t\right]$$

$$+\,\mathrm{e}^{-0.2226t}\left[\begin{bmatrix}0.0634\\-0.0270\end{bmatrix}\sin 2.5791\,t + \begin{bmatrix}0.7343\\-0.3123\end{bmatrix}\cos 2.5791t\right]$$

$$= \mathrm{e}^{-0.0274t}\begin{bmatrix}0.2661\\0.3127\end{bmatrix}\sin(0.5458\,t + 1.5206)$$

$$+\,\mathrm{e}^{-0.2226t}\begin{bmatrix}0.7370\\0.3135\end{bmatrix}\sin(2.5791\,t + 1.4847)$$

which on the whole agrees surprisingly well with the preceding solution for arbitrary viscous damping. This is also confirmed by fig. 9.1.3 where the response for general viscous damping is set against the results of modal damping. This agreement appears so satisfactory that it may induce our reader to accept the effectiveness of the simplifying rule 'diagonal matrix of the modally transformed damping = modal damping matrix'. To check this, the same example has been analysed (see figs. 9.1.4 and 9.1.5) with modified damping values c_1 and c_2. Substantial discrepancies are seen to occur between exact analysis and modal approximation. The considerable additional effort required in the computations for arbitrary viscous damping is thus not only of theoretical interest, but also generally guarantees better results. Let us now close the discussion of this example and return once more to the general theory.

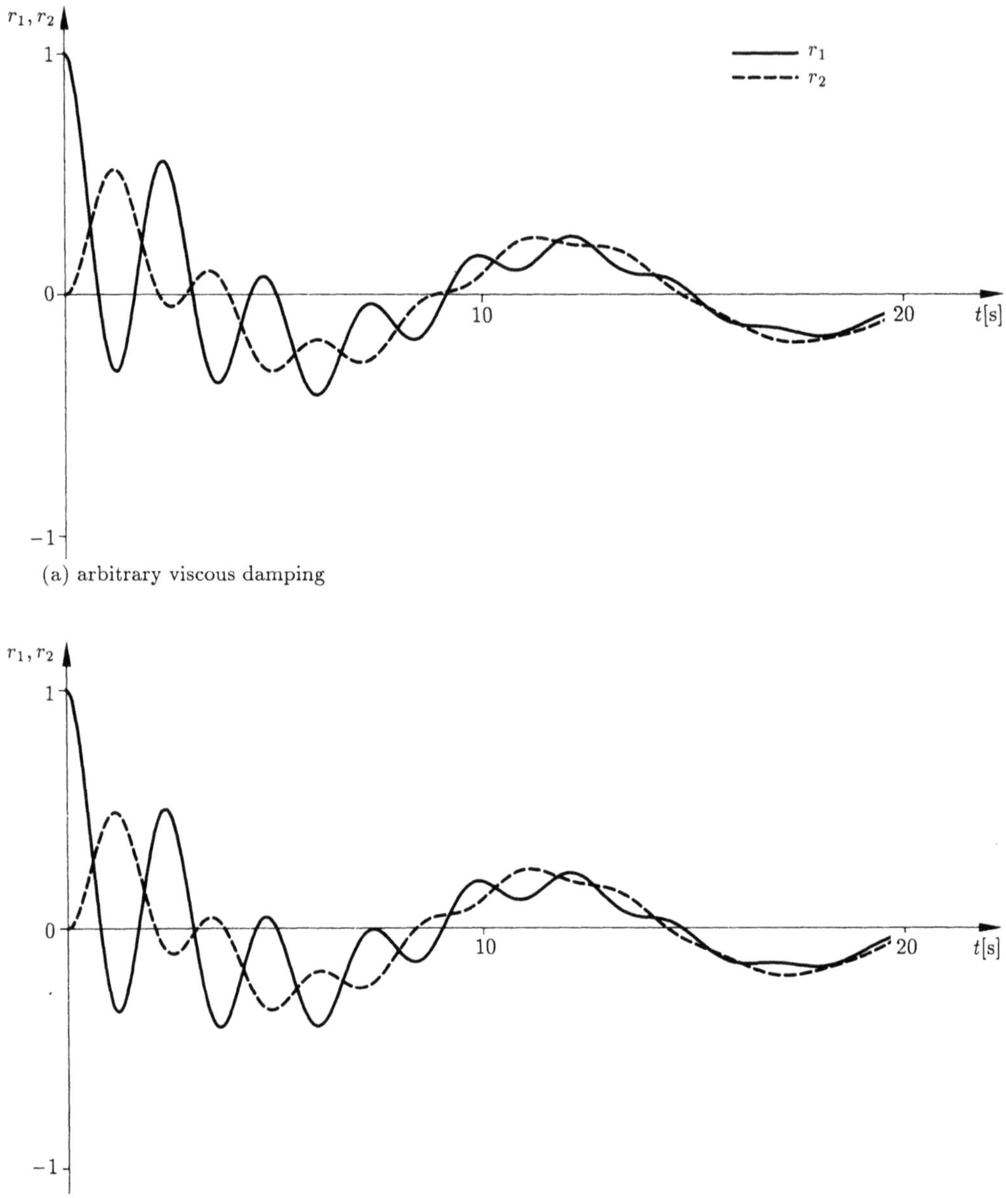

(b) approximated modal damping

Fig. 9.1.3 Free damped oscillation of the example shown in fig. 9.1.1 for $c_1 = c_2 = 0.2m\omega_0$ $(\omega_0 = 1)$

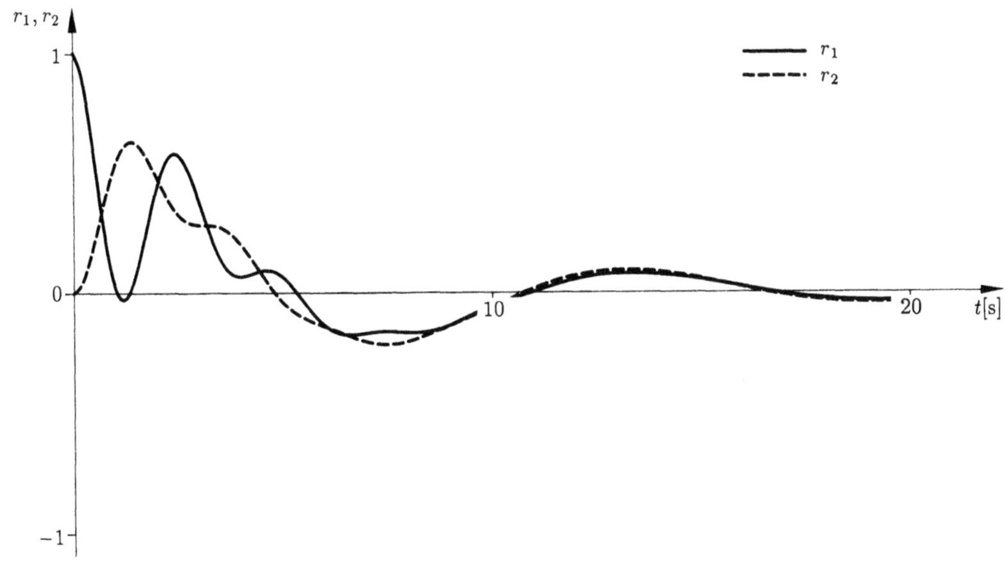

(a) arbitrary viscous damping

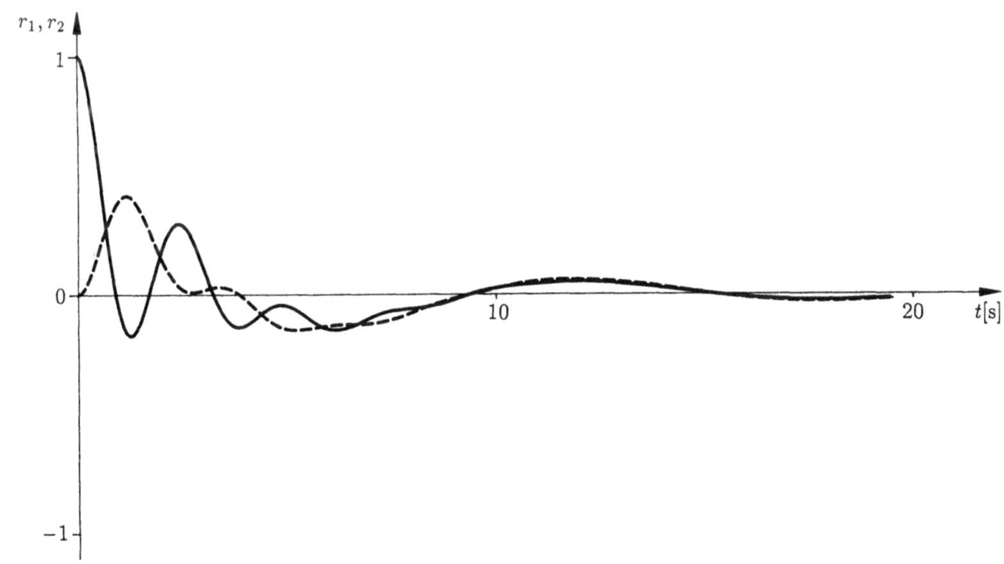

(b) approximated modal damping

Fig. 9.1.4 Free damped oscillation of the example in fig. 9.1.1 for $c_1 = m\omega_0$ and $c_2 = 0.2m\omega_0$ ($\omega_0 = 1$)

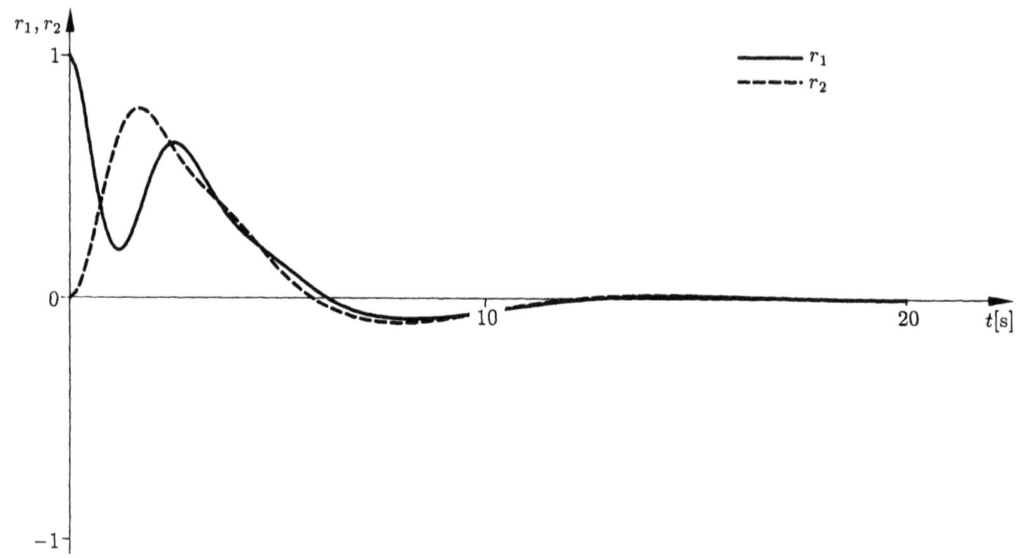

(a) arbitrary viscous damping

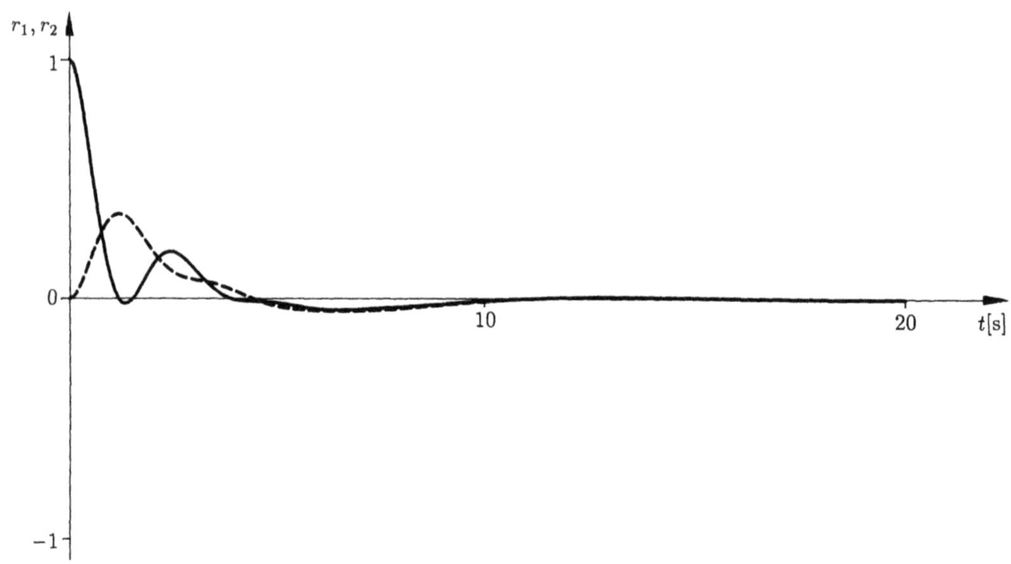

(b) approximated modal damping

Fig. 9.1.5 Free damped oscillation of the example in fig. 9.1.1 for $c_1 = 2m\omega_0$ and $c_2 = 0.2m\omega_0$ $(\omega_0 = 1)$

Equation (9.1.37) indicated that equ. (9.1.6) defining the motion of an arbitrarily damped system cannot be uncoupled by its 2n eigenvectors. In fact, there are insufficient equations – only n as against 2n eigenvectors. An extension of the system of equations into 2n suggests, however, an elegant technique of circumventing the difficulties. In fact, this can be realised by introducing the trivial relationship

$$\boldsymbol{M}\dot{\boldsymbol{r}} - \boldsymbol{M}\dot{\boldsymbol{r}} = \boldsymbol{o} \tag{9.1.43}$$

This system together with that of equ. (9.1.6) enables us to describe the motion by the set

$$\begin{bmatrix} \boldsymbol{O} & \boldsymbol{M} \\ \boldsymbol{M} & \boldsymbol{C} \end{bmatrix} \begin{bmatrix} \ddot{\boldsymbol{r}} \\ \dot{\boldsymbol{r}} \end{bmatrix} + \begin{bmatrix} -\boldsymbol{M} & \boldsymbol{O} \\ \boldsymbol{O}^{\mathrm{t}} & \boldsymbol{K} \end{bmatrix} \begin{bmatrix} \dot{\boldsymbol{r}} \\ \boldsymbol{r} \end{bmatrix} = \begin{bmatrix} \boldsymbol{o} \\ \boldsymbol{R}(t) \end{bmatrix} \tag{9.1.44}$$

This (2n×2n)-system represents a first order differential equation in $\{\boldsymbol{r} \quad \dot{\boldsymbol{r}}\}$, whilst the original system of motion (9.1.6) is a second order differential equation system. Expansion to doubled size thus reduces the order by one. The present technique is, of course, standard practice in mathematics. Both system matrices

$$\boldsymbol{A} = \begin{bmatrix} \boldsymbol{O} & \boldsymbol{M} \\ \boldsymbol{M} & \boldsymbol{C} \end{bmatrix} \tag{9.1.45}$$

and

$$\boldsymbol{B} = \begin{bmatrix} -\boldsymbol{M} & \boldsymbol{O} \\ \boldsymbol{O}^{\mathrm{t}} & \boldsymbol{K} \end{bmatrix} \tag{9.1.46}$$

of the order 2n are real and symmetric (the latter applies only to the case of symmetric damping $\boldsymbol{C}$ and stiffness $\boldsymbol{K}$, as is true for non-rotating structures). This suggests that the working technique for the solution of such a system ought to be similar to that for an undamped oscillator, which is described by two symmetric matrices $\boldsymbol{M}$ and $\boldsymbol{K}$. Whilst we could presume in the latter case at least one of the two system matrices to be positive definite – which enabled us to introduce a real symmetric dynamic matrix $\boldsymbol{A}$ in equ. (4.1.39) in place of the two system matrices – this technique proves impossible here. Neither $\boldsymbol{A}$ nor $\boldsymbol{B}$ are positive definite even if $\boldsymbol{M}$, $\boldsymbol{C}$ and $\boldsymbol{K}$ were. Therefore, it proves impossible to replace the two matrices $\boldsymbol{A}$ and $\boldsymbol{B}$ by one real symmetric matrix. We thus have to choose between an unsymmetric real or a symmetric complex dynamic matrix of the problem.

In the first variant, we could proceed by pre-multiplying the system in equ. (9.1.44) by $\boldsymbol{B}^{-1}$. This presumes, of course, that $\boldsymbol{M}$ and $\boldsymbol{K}$ are non-singular. However, even such a difficulty can be circumvented by a simple trick. To this purpose, the system in equ. (9.1.6) is expanded with the identity

$$\dot{\boldsymbol{r}} - \dot{\boldsymbol{r}} = \boldsymbol{o} \tag{9.1.47}$$

and equ. (9.1.6) is pre-multiplied with $\boldsymbol{K}^{-1}$ to yield

$$\boldsymbol{K}^{-1}\boldsymbol{M}\ddot{\boldsymbol{r}} + \boldsymbol{K}^{-1}\boldsymbol{C}\dot{\boldsymbol{r}} + \boldsymbol{r} = \boldsymbol{K}^{-1}\boldsymbol{R}(t) \tag{9.1.48}$$

$\boldsymbol{K}$ is now the only matrix to be presumed non-singular and we thus obtain

$$\widetilde{\boldsymbol{A}}\dot{\widetilde{\boldsymbol{r}}} + \widetilde{\boldsymbol{r}} = \widetilde{\boldsymbol{R}} \tag{9.1.49}$$

with

$$\widetilde{\boldsymbol{A}} = \begin{bmatrix} \boldsymbol{O} & -\boldsymbol{I} \\ \boldsymbol{K}^{-1}\boldsymbol{M} & \boldsymbol{K}^{-1}\boldsymbol{C} \end{bmatrix} \tag{9.1.50}$$

and

$$\widetilde{\boldsymbol{r}} = \{\dot{\boldsymbol{r}} \quad \boldsymbol{r}\} \tag{9.1.51}$$

as well as

$$\widetilde{\boldsymbol{R}} = \{\boldsymbol{O} \quad \boldsymbol{K}^{-1}\boldsymbol{R}(t)\} \tag{9.1.52}$$

As a trial solution for the homogeneous part of this first order system of differential equations, we set once more

$$\widetilde{\boldsymbol{r}} = \widetilde{\boldsymbol{x}}\, \mathrm{e}^{\widetilde{p}t} \tag{9.1.53}$$

and obtain the real unsymmetric algebraic system

$$[\widetilde{p}\widetilde{\boldsymbol{A}} + \boldsymbol{I}_{2\mathrm{n}}]\widetilde{\boldsymbol{x}} = \boldsymbol{o} \tag{9.1.54}$$

which again possesses non-trivial solutions only if

$$\det[\widetilde{p}\widetilde{\boldsymbol{A}} + \boldsymbol{I}_{2\mathrm{n}}] = 0 \tag{9.1.55}$$

applies. From equ. (9.1.55), we deduce a real characteristic polynomial of order 2n having in general 2n complex solutions for $\widetilde{p}$. In accordance with equ. (9.1.54), each of these complex solutions $\widetilde{p}_\mathrm{i}$ is associated with one eigenvector $\widetilde{\boldsymbol{x}}_\mathrm{i}$ which is, in general, complex. If the negative inverse value is defined as the eigenvalue $\widetilde{\lambda}_\mathrm{i}$

$$\widetilde{\lambda}_\mathrm{i} = -1/\widetilde{p}_\mathrm{i} \tag{9.1.56}$$

and all solutions of $\widetilde{\lambda}_\mathrm{i}$ are assembled in the diagonal matrix

$$\widetilde{\boldsymbol{\Lambda}} = \lceil \widetilde{\lambda}_1 \quad \widetilde{\lambda}_2 \ \ldots \ \widetilde{\lambda}_{2\mathrm{n}} \rfloor \tag{9.1.57}$$

and all eigenvectors in the array

$$\widetilde{\boldsymbol{X}} = [\widetilde{\boldsymbol{x}}_1 \quad \widetilde{\boldsymbol{x}}_2 \ \ldots \ \widetilde{\boldsymbol{x}}_{2\mathrm{n}}] \tag{9.1.58}$$

we obtain the complex form of the eigenvalue problem with which we are already familiar

$$\widetilde{\boldsymbol{A}}\widetilde{\boldsymbol{X}} = \widetilde{\boldsymbol{X}}\widetilde{\boldsymbol{\Lambda}} \tag{9.1.59}$$

A number of eigenvalue techniques, such as the vector iteration process or Eberlein's method, exists for the solution of this eigenvalue problem.

The eigenvectors of a system with arbitrary viscous damping possess, as in the undamped case, the corresponding orthogonality characteristics and this is also reflected in the mathematical deduction. Let there be two solutions $\widetilde{p}_i$, $\widetilde{\boldsymbol{x}}_i$ and $\widetilde{p}_j$, $\widetilde{\boldsymbol{x}}_j$, both of which satisfy the homogeneous part of equ. (9.1.44)

$$\widetilde{p}_i \boldsymbol{A}\widetilde{\boldsymbol{x}}_i + \boldsymbol{B}\widetilde{\boldsymbol{x}}_i = \boldsymbol{o} \tag{9.1.60}$$

$$\widetilde{p}_j \boldsymbol{A}\widetilde{\boldsymbol{x}}_j + \boldsymbol{B}\widetilde{\boldsymbol{x}}_j = \boldsymbol{o} \tag{9.1.61}$$

Pre-multiplying the first equation by $\widetilde{\boldsymbol{x}}_j^t$ and the second by $\widetilde{\boldsymbol{x}}_i^t$, transposing the first equation, subtracting one from the other and noting that $\boldsymbol{A}$ and $\boldsymbol{B}$ are symmetric, we obtain

$$(\widetilde{p}_i - \widetilde{p}_j)\widetilde{\boldsymbol{x}}_i^t \boldsymbol{A}\widetilde{\boldsymbol{x}}_j = (\widetilde{p}_i - \widetilde{p}_j)\overline{a}_{ij} = 0 \tag{9.1.62}$$

Assuming now that the two solution exponents $\widetilde{p}_i$ and $\widetilde{p}_j$ are different, we find

$$\overline{a}_{ij} = \widetilde{\boldsymbol{x}}_i^t \boldsymbol{A}\widetilde{\boldsymbol{x}}_j = 0 \tag{9.1.63}$$

and hence also

$$\overline{b}_{ij} = \widetilde{\boldsymbol{x}}_i^t \boldsymbol{B}\widetilde{\boldsymbol{x}}_j = 0 \tag{9.1.64}$$

This proves that the eigenvectors of the inflated system simultaneously diagonalise the matrices $\boldsymbol{A}$ and $\boldsymbol{B}$, i.e. that the vectors are orthogonal with respect to $\boldsymbol{A}$ and $\boldsymbol{B}$. This is analogous to the undamped case where $\boldsymbol{M}$ and $\boldsymbol{K}$ were simultaneously diagonalised. Thus, for all eigenvectors (with distinct eigenvalues) we have

$$\widetilde{\boldsymbol{X}}^t \boldsymbol{A}\widetilde{\boldsymbol{X}} = \boldsymbol{d}_A = \lceil \overline{a}_{11} \quad \overline{a}_{22} \ \ldots \ \overline{a}_{2n2n} \rfloor \tag{9.1.65}$$

$$\widetilde{\boldsymbol{X}}^t \boldsymbol{B}\widetilde{\boldsymbol{X}} = \boldsymbol{d}_B = \lceil \overline{b}_{11} \quad \overline{b}_{22} \ \ldots \ \overline{b}_{2n2n} \rfloor \tag{9.1.66}$$

Once again, it is possible to normalise the columns of $\widetilde{\boldsymbol{X}}$ and reduce the eigenvectors to a $\boldsymbol{B}$-orthonormal state

$$\widetilde{\boldsymbol{X}}^t \boldsymbol{B}\widetilde{\boldsymbol{X}} = \boldsymbol{I}_{2n} \tag{9.1.67}$$

$$\widetilde{\boldsymbol{X}}^t \boldsymbol{A}\widetilde{\boldsymbol{X}} = \widetilde{\boldsymbol{\Lambda}} \tag{9.1.68}$$

Here, it should be remembered that $\widetilde{\boldsymbol{\Lambda}}$ contains the negative inverse values of the solution exponents. Since we are aware that the eigenvectors for arbitrary viscous damping can be complex, it may be of interest to examine the orthogonality properties with respect to the real and imaginary parts of the vectors [9.2]. Considering one pair of solutions

$$\widetilde{\boldsymbol{x}}_i = \widetilde{\boldsymbol{\xi}}_i + i\widetilde{\boldsymbol{\eta}}_i$$

$$\tilde{\boldsymbol{x}}_j = \tilde{\boldsymbol{\xi}}_j + i\tilde{\boldsymbol{\eta}}_j$$

we obtain from equ. (9.1.63)

$$\tilde{\boldsymbol{\xi}}_i^t \boldsymbol{A}\tilde{\boldsymbol{\xi}}_j - \tilde{\boldsymbol{\eta}}_i^t \boldsymbol{A}\tilde{\boldsymbol{\eta}}_j + i(\tilde{\boldsymbol{\xi}}_i^t \boldsymbol{A}\tilde{\boldsymbol{\eta}}_j + \tilde{\boldsymbol{\eta}}_i^t \boldsymbol{A}\tilde{\boldsymbol{\xi}}_j) = 0 \tag{9.1.69}$$

Here, the real and the imaginary parts must separately be equal to zero. We know that if the solution $\tilde{\boldsymbol{\xi}}_i + i\tilde{\boldsymbol{\eta}}_i$ exists then $\tilde{\boldsymbol{\xi}}_i - i\tilde{\boldsymbol{\eta}}_i$ is also a solution vector. If we now substitute this second solution vector instead of the solution $\tilde{\boldsymbol{x}}_i$ in equ. (9.1.63), we find

$$\tilde{\boldsymbol{\xi}}_i^t \boldsymbol{A}\tilde{\boldsymbol{\xi}}_j + \tilde{\boldsymbol{\eta}}_i^t \boldsymbol{A}\tilde{\boldsymbol{\eta}}_j + i(\tilde{\boldsymbol{\xi}}_i^t \boldsymbol{A}\tilde{\boldsymbol{\eta}}_j - \tilde{\boldsymbol{\eta}}_i^t \boldsymbol{A}\tilde{\boldsymbol{\xi}}_j) = 0 \tag{9.1.70}$$

Summation of the real parts of equs. (9.1.69) and (9.1.70) at once yields

$$\tilde{\boldsymbol{\xi}}_i^t \boldsymbol{A}\tilde{\boldsymbol{\xi}}_j = 0 \tag{9.1.71}$$

which means, in effect, that the real parts of the solution vectors also diagonalise the $\boldsymbol{A}$ matrix. Subtraction of the real parts of equ. (9.1.69) and (9.1.70) gives

$$\tilde{\boldsymbol{\eta}}_i^t \boldsymbol{A}\tilde{\boldsymbol{\eta}}_j = 0 \tag{9.1.72}$$

which means that the imaginary parts of the solution vectors diagonalise the $\boldsymbol{A}$ matrix, too.

Applying next the same procedure to the imaginary parts of equs. (9.1.69) and (9.1.70) yields the expressions

$$\tilde{\boldsymbol{\xi}}_i^t \boldsymbol{A}\tilde{\boldsymbol{\eta}}_j = 0 \tag{9.1.73}$$

$$\tilde{\boldsymbol{\eta}}_i^t \boldsymbol{A}\tilde{\boldsymbol{\xi}}_j = 0 \tag{9.1.74}$$

Since the system matrices $\boldsymbol{A}$ and $\boldsymbol{B}$ are interchangeable – see also equs. (9.1.63) and (9.1.64) – the above assertions are also valid for the $\boldsymbol{B}$ matrix. A further condition is obtained for $i = j$ in equ. (9.1.70) (here, we are dealing with the pair of solutions $\tilde{\boldsymbol{\xi}}_i \pm i\tilde{\boldsymbol{\eta}}_i$). Since, as before, the real part must be zero, the following holds

$$\tilde{\boldsymbol{\xi}}_i^t \boldsymbol{A}\tilde{\boldsymbol{\xi}}_i = \tilde{\boldsymbol{\eta}}_i^t \boldsymbol{A}\tilde{\boldsymbol{\eta}}_i \tag{9.1.75}$$

So far, the orthogonality relation has been discussed for a general eigenvalue problem, defined by the two system matrices $\boldsymbol{A}$ and $\boldsymbol{B}$. We now turn to a consideration of the special eigenvalue problem of equ. (9.1.50) with only one real but unsymmetric matrix $\boldsymbol{A}$

$$\tilde{\boldsymbol{A}}\tilde{\boldsymbol{x}} = \tilde{\lambda}\tilde{\boldsymbol{x}} \tag{9.1.76}$$

Formally, the eigenvalues $\tilde{\lambda}$ of this problem are derived from

$$\det[\tilde{\boldsymbol{A}} - \tilde{\lambda}\boldsymbol{I}] = 0 \tag{9.1.77}$$

Apart from an arbitrary scaling factor, the eigenvalue $\widetilde{\boldsymbol{x}}$ is then deduced from equ. (9.1.76). Our next step is to determine the eigenvalues and the eigenvectors of the $\widetilde{\boldsymbol{A}}^{\mathrm{t}}$ matrix. Since the dynamic matrix for undamped problems is expressed in a real symmetric form, the new problem with an unsymmetric matrix will be unfamiliar to the reader. Obviously, the value of the determinant in equ. (9.1.77) is independent of whether we work with $\widetilde{\boldsymbol{A}}$ or $\widetilde{\boldsymbol{A}}^{\mathrm{t}}$. It is thus at once possible to predict that the eigenvalues of $\widetilde{\boldsymbol{A}}^{\mathrm{t}}$ agree with those of $\widetilde{\boldsymbol{A}}$. If, however, a specific value λ_{i} is substituted, then the formal calculation of the eigenvector $\boldsymbol{x}_{\mathrm{i}}$ is seen to depend decisively on whether $\widetilde{\boldsymbol{A}}^{\mathrm{t}}$ or $\widetilde{\boldsymbol{A}}$ is entered in equ. (9.1.76). The corresponding eigenvectors thus differ and are denoted by $\widetilde{\boldsymbol{y}}$ for a dynamic matrix $\widetilde{\boldsymbol{A}}^{\mathrm{t}}$, i.e.

$$\widetilde{\boldsymbol{A}}^{\mathrm{t}}\widetilde{\boldsymbol{y}} = \widetilde{\lambda}\widetilde{\boldsymbol{y}} \tag{9.1.78}$$

In place of equ. (9.1.78), we can also write

$$\widetilde{\boldsymbol{y}}^{\mathrm{t}}\widetilde{\boldsymbol{A}} = \widetilde{\lambda}\widetilde{\boldsymbol{y}}^{\mathrm{t}} \tag{9.1.79}$$

We thus define $\widetilde{\boldsymbol{y}}$ as the left-hand eigenvector of the matrix $\widetilde{\boldsymbol{A}}$, whilst $\widetilde{\boldsymbol{x}}$ is the right-hand eigenvector.

We now consider two distinct eigenvalues $\widetilde{\lambda}_{\mathrm{i}}$ and $\widetilde{\lambda}_{\mathrm{j}}$ and write

$$\widetilde{\boldsymbol{A}}\widetilde{\boldsymbol{x}}_{\mathrm{i}} = \widetilde{\lambda}_{\mathrm{i}}\widetilde{\boldsymbol{x}}_{\mathrm{i}} \tag{9.1.80}$$

$$\widetilde{\boldsymbol{y}}_{\mathrm{j}}^{\mathrm{t}}\widetilde{\boldsymbol{A}} = \widetilde{\lambda}_{\mathrm{j}}\widetilde{\boldsymbol{y}}_{\mathrm{j}}^{\mathrm{t}} \tag{9.1.81}$$

We next pre-multiply the first equation by $\widetilde{\boldsymbol{y}}_{\mathrm{j}}^{\mathrm{t}}$ and post-multiply the second by $\widetilde{\boldsymbol{x}}_{\mathrm{i}}$ and subtract the resulting new expressions. We find

$$(\widetilde{\lambda}_{\mathrm{i}} - \widetilde{\lambda}_{\mathrm{j}})\widetilde{\boldsymbol{y}}_{\mathrm{j}}^{\mathrm{t}}\widetilde{\boldsymbol{x}}_{\mathrm{i}} = 0 \tag{9.1.82}$$

For distinct eigenvalues we deduce

$$\widetilde{\boldsymbol{y}}_{\mathrm{j}}^{\mathrm{t}}\widetilde{\boldsymbol{x}}_{\mathrm{i}} = 0\,, \qquad \widetilde{\lambda}_{\mathrm{i}} \neq \widetilde{\lambda}_{\mathrm{j}} \tag{9.1.83}$$

The left-hand and right-hand eigenvectors of the $\widetilde{\boldsymbol{A}}$ matrix are thus orthogonal if the eigenvalues are different, but, in contrast to the undamped case, the eigenvectors $\widetilde{\boldsymbol{X}}$ are in themselves not orthogonal. The eigenvectors can now be normalised so that

$$\widetilde{\boldsymbol{Y}}^{\mathrm{t}}\widetilde{\boldsymbol{X}} = \boldsymbol{I}_{2\mathrm{n}} \tag{9.1.84}$$

applies, where all $\widetilde{\boldsymbol{y}}_{\mathrm{i}}$ values are collected in $\widetilde{\boldsymbol{Y}}$

$$\widetilde{\boldsymbol{Y}} = [\widetilde{\boldsymbol{y}}_1 \quad \widetilde{\boldsymbol{y}}_2 \quad \ldots \quad \widetilde{\boldsymbol{y}}_{2\mathrm{n}}] \tag{9.1.85}$$

If all eigenvalues of the $\widetilde{\boldsymbol{A}}$ matrix are distinct, then each eigenvector is determined from a system with a rank deficiency of 1, from which we deduce that all eigenvectors are linearly independent. Hence, we infer in conjunction with equ. (9.1.82) for the left-hand eigenvector

$$\widetilde{\boldsymbol{Y}} = \widetilde{\boldsymbol{X}}^{-\mathrm{t}} \tag{9.1.86}$$

Following equ. (9.1.59) and using equ. (9.1.84), we then obtain

$$\begin{aligned}\widetilde{\boldsymbol{Y}}^{\mathrm{t}}\widetilde{\boldsymbol{A}}\widetilde{\boldsymbol{X}} &= \widetilde{\boldsymbol{Y}}^{\mathrm{t}}\widetilde{\boldsymbol{X}}\widetilde{\boldsymbol{\Lambda}} \\ &= \widetilde{\boldsymbol{X}}^{-1}\widetilde{\boldsymbol{X}}\widetilde{\boldsymbol{\Lambda}} \\ &= \widetilde{\boldsymbol{\Lambda}}\end{aligned} \tag{9.1.87}$$

Thus, for the special eigenvalue problem, we conclude that the left-hand and right-hand eigenvectors simultaneously diagonalise or respectively do not destroy the diagonal form of two matrices, namely $\widetilde{\boldsymbol{A}}$ and $\boldsymbol{I}_{2\mathrm{n}}$. Alternatively, this also means that these vectors are orthonormal and diagonalise $\widetilde{\boldsymbol{A}}$.

9.2 Vector iteration and simultaneous vector iteration for a real unsymmetric dynamic matrix

As was the case for undamped vibrations, vector iteration is the simplest eigenvalue method available for the calculation of the largest eigenvalue λ_1 of an elastic system with arbitrary viscous damping. In fact, the present case does not entail any difference to the power step technique applied in the undamped case

$$\widetilde{\boldsymbol{x}}^{(\mathrm{k}+1)} = \widetilde{\boldsymbol{A}}\widetilde{\boldsymbol{x}}^{(\mathrm{k})} \tag{9.2.1}$$

As a result of the texture of the real unsymmetric iteration matrix $\widetilde{\boldsymbol{A}}$ (see equ. (9.1.50)), the power step technique is applied in a series of sub-steps, each of which comprises only operations with sparsely populated matrices. The iteration vector is subdivided in conformity with the form of $\widetilde{\boldsymbol{A}}$

$$\widetilde{\boldsymbol{x}} = \{\widetilde{\boldsymbol{x}}_{\mathrm{I}} \quad \widetilde{\boldsymbol{x}}_{\mathrm{II}}\} \tag{9.2.2}$$

In this way, we obtain the following sequence of operations

$$\begin{aligned}\widetilde{\boldsymbol{x}}_{\mathrm{I}}^{(\mathrm{k}+1)} &= -\widetilde{\boldsymbol{x}}_{\mathrm{II}}^{(\mathrm{k})} \\ \boldsymbol{K}\widetilde{\boldsymbol{x}}_{\mathrm{II}}^{(\mathrm{k}+1)} &= [\boldsymbol{M} \quad \boldsymbol{C}]\widetilde{\boldsymbol{x}}^{(\mathrm{k})}\end{aligned} \tag{9.2.3}$$

We are already informed that the first eigenvector can be both real and complex. It is at first somewhat surprising that the arithmetic operations in equ. (9.2.3) can be purely real and that this still yields the correct result. We note in this respect that for a real starting vector $\widetilde{\boldsymbol{x}}^{(0)}$, each ensuing result must also be real. Furthermore, we presume the total assembly of the eigenvectors of $\widetilde{\boldsymbol{A}}$, namely $\widetilde{\boldsymbol{X}}$, to be linearly independent. In this way, each starting vector can be expressed as a linear combination of these eigenvectors

$$\widetilde{\boldsymbol{x}}^{(0)} = \widetilde{\boldsymbol{X}}\boldsymbol{c} = \sum_{i=1}^{2n} c_i \widetilde{\boldsymbol{x}}_i \tag{9.2.4}$$

Substituting this expression into equ. (9.2.1), we obtain in conjunction with equ. (9.1.59)

$$\begin{aligned} \widetilde{\boldsymbol{x}}^{(1)} &= \widetilde{\boldsymbol{A}}\widetilde{\boldsymbol{x}}^{(0)} \\ &= \widetilde{\boldsymbol{A}}\widetilde{\boldsymbol{X}}\boldsymbol{c} \\ &= \widetilde{\boldsymbol{X}}\widetilde{\boldsymbol{\Lambda}}\boldsymbol{c} \\ &= \sum_{i=1}^{2n} \widetilde{\lambda}_i c_i \widetilde{\boldsymbol{x}}_i \end{aligned} \tag{9.2.5}$$

which proves that the coefficients of the expansion are amplified by the eigenvalues $\widetilde{\lambda}_i$. Applying the polar representation

$$\widetilde{\lambda}_i^{(1)} = \rho_i e^{i\varphi_i} \tag{9.2.6}$$

we find after k power steps

$$\widetilde{\boldsymbol{x}}^{(k)} = \sum_{i=1}^{2n} \rho_i^k e^{ik\varphi_i} c_i \widetilde{\boldsymbol{x}}_i \tag{9.2.7}$$

This result demonstrates that the largest eigenvalue is being amplified. For a complex eigenvalue (respectively an eigenvector), a rotation $e^{ik\varphi_i}$ also takes place. Thus, we can predict the existence of a real maximum eigenvalue if the vector components converge monotonically. If, on the other hand, these components fluctuate severely from power step to power step (leading even to a change of sign), we have to expect a complex eigenvalue with the largest modulus. We are already informed that in the case of a complex eigenvalue, the solution occurs in complex conjugate pairs, i.e.

$$\begin{aligned} \widetilde{\lambda}_1 &= \mu_1 + i\nu_1 = \rho_1 e^{i\varphi_1} \\ \widetilde{\lambda}_2 &= \mu_1 - i\nu_1 = \rho_1 e^{-i\varphi_1} \end{aligned} \tag{9.2.8}$$

with the associated eigenvectors

$$\begin{aligned} \widetilde{\boldsymbol{x}}_1 &= \widetilde{\boldsymbol{\xi}}_1 + i\widetilde{\boldsymbol{\eta}}_1 \\ \widetilde{\boldsymbol{x}}_2 &= \widetilde{\boldsymbol{\xi}}_1 - i\widetilde{\boldsymbol{\eta}}_1 \end{aligned} \tag{9.2.9}$$

In such a situation, the first and second eigenvectors are amplified equally in accordance with equ. (9.2.7)

$$\widetilde{\boldsymbol{x}}^{(k)} = \rho_1^k\Big[c_1 e^{ik\varphi_1}(\widetilde{\boldsymbol{\xi}}_1 + i\widetilde{\boldsymbol{\eta}}_1) + c_2 e^{-ik\varphi_1}(\widetilde{\boldsymbol{\xi}}_1 - i\widetilde{\boldsymbol{\eta}}_1)\Big] + \sum_{i=3}^{2n} \rho_i^k e^{ik\varphi_i}\widetilde{\boldsymbol{x}}_i \qquad (9.2.10)$$

In the case of a real starting vector $\widetilde{\boldsymbol{x}}_1^{(0)}$, the factors c_1 and c_2 must be conjugate complex, for otherwise, the imaginary part of equ. (9.2.4) could not vanish. Writing

$$\begin{aligned} c_1 &= \sigma_1 e^{i\psi_1} \\ c_2 &= \sigma_1 e^{-i\psi_1} \end{aligned} \qquad (9.2.11)$$

we obtain from equ. (9.2.10)

$$\begin{aligned} \widetilde{\boldsymbol{x}}^{(k)} &= \sigma_1\rho_1^k\Big[e^{i(\psi_1+k\varphi_1)}(\widetilde{\boldsymbol{\xi}}_1 + i\widetilde{\boldsymbol{\eta}}_1) + e^{-i(\psi_1+k\varphi_1)}(\widetilde{\boldsymbol{\xi}}_1 - i\widetilde{\boldsymbol{\eta}}_1)\Big] + \ldots \\ &= 2\sigma_1\rho_1^k\Big[\cos(\psi_1 + k\varphi_1)\widetilde{\boldsymbol{\xi}}_1 - \sin(\psi_1 + k\varphi_1)\widetilde{\boldsymbol{\eta}}_1\Big] + \ldots \end{aligned} \qquad (9.2.12)$$

Once again, we observe that not only the starting vector $\widetilde{\boldsymbol{x}}_1^{(0)}$, but all following iteration vectors $\widetilde{\boldsymbol{x}}_1^{(k)}$ are real and can, in theory, contain contributions of the real and imaginary parts of the pair of solutions. The question now is in which way these real and imaginary parts of the solution can be calculated. In additon, the pair of eigenvalues $\widetilde{\lambda}_1$ and $\widetilde{\lambda}_2$ should also be determined. In principle, there are two solution techniques. In conformity with the two eigenvectors, we may use two vectors for the iteration, a process contained in principle in the simultaneous vector iteration to be discussed later on. Alternatively, we may start with the simple vector iteration and observe in the course of the iteration process that the eigenvalues with the largest modulus must be complex conjugate. In such a case, it proves more advantageous if all relevant information is deduced from this one vector. This is possible if the results produced by successive power steps are registered [9.3].

For this purpose, we assume the process to have converged in q steps and we introduce the abbreviation

$$\alpha_{q+1} = \psi_1 + (q+1)\varphi_1 \qquad (9.2.13)$$

From equ. (9.2.12), we obtain for a succession of three iterations

$$\begin{aligned} \widetilde{\boldsymbol{x}}^{(q)} &= 2\sigma_1\rho_1^q\,[\cos(\alpha_{q+1} - \varphi_1)\,\widetilde{\boldsymbol{\xi}}_1 - \sin(\alpha_{q+1} - \varphi_1)\,\widetilde{\boldsymbol{\eta}}_1] \\ \widetilde{\boldsymbol{x}}^{(q+1)} &= 2\sigma_1\rho_1^{q+1}[\cos(\alpha_{q+1})\,\widetilde{\boldsymbol{\xi}}_1 - \sin(\alpha_{q+1})\,\widetilde{\boldsymbol{\eta}}_1] \\ \widetilde{\boldsymbol{x}}^{(q+2)} &= 2\sigma_1\rho_1^{q+2}[\cos(\alpha_{q+1} + \varphi_1)\,\widetilde{\boldsymbol{\xi}}_1 - \sin(\alpha_{q+1} + \varphi_1)\,\widetilde{\boldsymbol{\eta}}_1] \end{aligned} \qquad (9.2.14)$$

where the contributions of the higher eigenvectors in equs. (9.2.10) and (9.2.12) can be ignored since for an assumed convergence $\widetilde{\boldsymbol{x}} \rightarrow \widetilde{\boldsymbol{x}}_{1,2}$, they must tend to zero.

By virtue of the trigonometrical addition theorems, we deduce from the first and last lines of equ. (9.2.14)

$$\rho_1^2 \widetilde{\boldsymbol{x}}^{(q)} + \widetilde{\boldsymbol{x}}^{(q+2)} = 2\rho_1 \cos\varphi_1 \widetilde{\boldsymbol{x}}^{(q+1)} \tag{9.2.15}$$

where ρ_1 is the modulus and $\rho_1 \cos\varphi_1$ the real part of the first as also of the second eigenvalue of the problem. Since equ. (9.2.15) holds for all values of q, we also have

$$\rho_1^2 \widetilde{\boldsymbol{x}}^{(q+1)} + \widetilde{\boldsymbol{x}}^{(q+3)} = 2\rho_1 \cos\varphi_1 \widetilde{\boldsymbol{x}}^{(q+2)} \tag{9.2.16}$$

Pre-multiplication of equ. (9.2.15) by $(\widetilde{\boldsymbol{x}}^{(q+1)})^t$ and equ. (9.2.16) by $(\widetilde{\boldsymbol{x}}^{(q)})^t$ yields scalar equations which are then subtracted. With the abbreviations

$$\begin{aligned} a_{12} &= (\widetilde{\boldsymbol{x}}^{(q+1)})^t \widetilde{\boldsymbol{x}}^{(q+2)} \\ a_{11} &= (\widetilde{\boldsymbol{x}}^{(q+1)})^t \widetilde{\boldsymbol{x}}^{(q+1)} \\ a_{03} &= (\widetilde{\boldsymbol{x}}^{(q)})^t \widetilde{\boldsymbol{x}}^{(q+3)} \\ a_{02} &= (\widetilde{\boldsymbol{x}}^{(q)})^t \widetilde{\boldsymbol{x}}^{(q+2)} \end{aligned} \tag{9.2.17}$$

we obtain, without much complication, the real part of the first and second eigenvalues

$$\mu_1 = \rho_1 \cos\varphi_1 = (a_{12} - a_{03})/[2(a_{11} - a_{02})] \tag{9.2.18}$$

Applying the additional abbreviation

$$a_{10} = (\widetilde{\boldsymbol{x}}^{(q+1)})^t \widetilde{\boldsymbol{x}}^{(q)} \tag{9.2.19}$$

and using equ. (9.2.15), we establish the modulus of the first and second eigenvalue

$$\rho_1^2 = (2\rho_1 \cos\varphi_1 a_{11} - a_{12})/a_{10} \tag{9.2.20}$$

The trivial relation

$$\nu_1 = \rho_1 \sin\varphi_1 = (\rho_1^2 - \rho_1^2 \cos^2\varphi_1)^{1/2} \tag{9.2.21}$$

finally yields the imaginary part of the first and second eigenvalue. Thus, the complex eigenvalues have been determined by means of a skilful evaluation of four real iteration steps

$$\lambda_{1,2} = \mu_1 \pm i\nu_1 = \rho_1 e^{\pm i\varphi_1} \tag{9.2.22}$$

Our next step is to determine the associated complex eigenvectors. Following equ. (9.2.12), we obtain for the iteration steps q+1 and q+2

$$\begin{aligned}\tilde{\boldsymbol{x}}^{(q+1)} &= \sigma_1\rho_1^{q+1}[\quad e^{i\alpha_{q+1}}\tilde{\boldsymbol{x}}_1 + \quad e^{-i\alpha_{q+1}}\tilde{\boldsymbol{x}}_2]\\ \tilde{\boldsymbol{x}}^{(q+2)} &= \sigma_1\rho_1^{q+2}[e^{i(\alpha_{q+1}+\varphi_1)}\tilde{\boldsymbol{x}}_1 + e^{-i(\alpha_{q+1}+\varphi_1)}\tilde{\boldsymbol{x}}_2]\end{aligned} \tag{9.2.23}$$

and in this connection, we refer the reader to the remarks on equ. (9.2.14).

We first multiply the first line of equ. (9.2.23) by λ_2 of equ. (9.2.22) and then form

$$\tilde{\boldsymbol{x}}^{(q+2)} - \lambda_2\tilde{\boldsymbol{x}}^{(q+1)} = \sigma_1\rho_1^{q+2}(e^{i(\alpha_{q+1}+\varphi_1)} - e^{i(\alpha_{q+1}-\varphi_1)})\tilde{\boldsymbol{x}}_1 \tag{9.2.24}$$

Since the eigenvectors $\tilde{\boldsymbol{x}}_i$ can only be determined up to a complex factor, the preceding process directly supplies the first eigenvector. At the same time, we can also deduce

$$\tilde{\boldsymbol{x}}^{(q+2)} - \lambda_1\tilde{\boldsymbol{x}}^{(q+1)} = \sigma_1\rho_1^{q+2}(e^{-i(\alpha_{q+1}+\varphi_1)} - e^{-i(\alpha_{q+1}-\varphi_1)})\tilde{\boldsymbol{x}}_2 \tag{9.2.25}$$

which is actually superfluous since, in this case, the first and second eigenvectors are complex conjugate to each other. The real and imaginary parts of these vectors can also be calculated directly from equ. (9.2.24). Using equ. (9.2.22), we obtain

$$\tilde{\boldsymbol{\xi}}_1 = \tilde{\boldsymbol{x}}^{(q+2)} - \mu_1\tilde{\boldsymbol{x}}^{(q+1)} \tag{9.2.26}$$

$$\tilde{\boldsymbol{\eta}}_1 = \nu_1\,\tilde{\boldsymbol{x}}^{(q+1)} \tag{9.2.27}$$

and thus

$$\tilde{\boldsymbol{x}}_{1,2} = \tilde{\boldsymbol{\xi}}_1 \pm i\tilde{\boldsymbol{\eta}}_1 \tag{9.2.28}$$

Once the first two complex eigenvalues are determined, the complex eigenvectors can thus easily be established with two successive real iteration steps.

We put this method to the test on our standard example in fig. 9.1.1, the eigenvalues and eigenvectors of which have already been calculated. In fig. 9.2.1, we show the flow chart of a small demonstration programme. The iteration process is extremely simple. The somewhat complex branching arises from the necessity of having to ascertain and take into account the real or complex solution character. Following three power steps, we possess sufficient iteration vectors $\tilde{\boldsymbol{x}}$ to compute – even for a complex eigenvalue λ_i – the associated complex eigenvector. We hence examine at k = 3 the sequence of Rayleigh-Ritz factors

$$r = \tilde{\boldsymbol{x}}^t\tilde{\boldsymbol{A}}\tilde{\boldsymbol{x}} = \tilde{\boldsymbol{x}}^t\tilde{\boldsymbol{y}} \tag{9.2.29}$$

arising from the preceding three power steps. If λ_1 is real, a monotonic convergence behaviour $r \to \lambda_1$ takes place. If, on the other hand, we observe an oscillatory behaviour of r (even with a sign change of r), then λ_1 is complex. Depending on this particular situation, the ensuing computation proceeds in real terms and is very simple or involves an additional secondary determination of the real and imaginary parts of λ_1, p_1 and $\tilde{\boldsymbol{x}}_1$ (more time-consuming). In the first case, the break-off or termination criterion is based upon a real error vector

$$\boldsymbol{d} = \tilde{\boldsymbol{A}}\tilde{\boldsymbol{x}} - r\tilde{\boldsymbol{x}} = \tilde{\boldsymbol{y}} - r\tilde{\boldsymbol{x}} \tag{9.2.30}$$

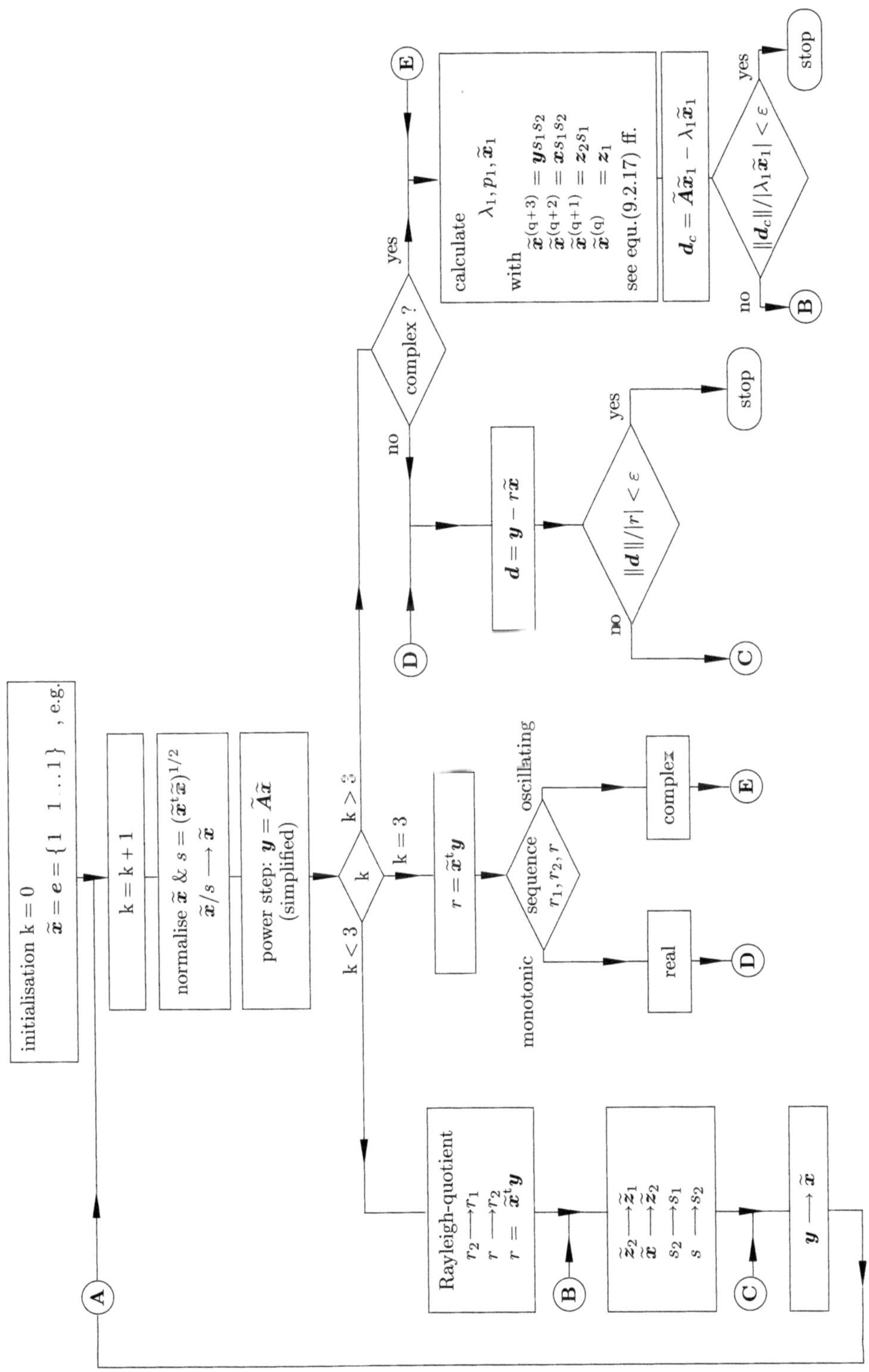

Fig. 9.2.1 Flow chart of simple vector iteration for non-modal viscous damping

whilst for a complex solution the error vector

$$\boldsymbol{d}_c = \widetilde{\boldsymbol{A}}\widetilde{\boldsymbol{x}}_1 - \lambda_1\widetilde{\boldsymbol{x}}_1 \tag{9.2.31}$$

becomes complex, but the break-off criterion is once again real.

In fig. 9.2.2, this process is applied to the example in fig. 9.1.1. The starting vector possesses throughout equal entries and is normalised to a length of 1, whereby $s^{(k)}$ represents the normalisation factor

$$\widetilde{\boldsymbol{y}} = s^{(k)}\widetilde{\boldsymbol{x}}^{(k)} \tag{9.2.32}$$

For k + 1 = 4 (k=3), the behaviour of the Rayleigh-Ritz quotients indicates that the first eigenvalue λ_1 must be complex. Accordingly, the calculation of the real and imaginary parts of the eigenvalue λ_1, of the solution exponent p_1 and of the eigenvector $\widetilde{\boldsymbol{x}}_1$ must be initiated for the first time. For easy comparison with the results of section 9.1, the eigenvector is divided by the third complex element $\widetilde{x}_{13}$ of $\widetilde{\boldsymbol{x}}_1$. The second group of $\widetilde{\boldsymbol{x}}_1$ contains the eigenvector $\boldsymbol{x}_1$ of the uninflated problem, as already calculated. The first group, on the other hand, contains $p_1\boldsymbol{x}_1$, i.e. effectively the eigenvector $\boldsymbol{x}_1$ once more. Since this vector has been normalised so that the first entry equals 'one', we obtain in the converged state, as the first entry of this group, the complex value p_1. In addition, we can observe for k+1=4 that there is already a practically acceptable approximation to the ultimate solution. From k+1=8 onwards, the eigenvalue is exact to four digits, from k+1=9 the same also applies to the complex eigenvector.

Our next step is to generalise the algorithm in order to obtain not only the eigenvector $\boldsymbol{x}_1$ and the largest eigenvalue λ_1, but also simultaneously the whole group appertaining to the eigenvalues with largest moduli. The fundamental principles of simultaneous vector iteration for unsymmetric real matrices $\boldsymbol{A}$ are identical to those for symmetric matrices (section 4.3.2). To simplify the discussion, we neglect all subtleties and discuss only some fundamental aspects as in [9.4] and [9.5]. In fact, we make no distinction as to whether a solution is real or complex and we iterate with a real set of vectors which yields the right-hand eigenvectors. Starting with a set of nx linearly independent real vectors $\boldsymbol{W}^{(k)}(2\mathrm{n} \times \mathrm{nx})$, we proceed, first of all, to an orthonormalisation applying, for example, the Cholesky process

$$\underset{(\mathrm{nx}\times\mathrm{nx})}{\boldsymbol{H}} = \boldsymbol{W}^{(k)^t}\boldsymbol{W}^{(k)} \tag{9.2.33}$$

$$\boldsymbol{H} = \boldsymbol{U}_H^t\boldsymbol{U}_H \tag{9.2.34}$$

$$\boldsymbol{V} = \boldsymbol{W}^{(k)}\boldsymbol{U}_H^{-1} \tag{9.2.35}$$

Operating next the power step

$$\underset{(2\mathrm{n}\times\mathrm{nx})}{\boldsymbol{Y}} = \underset{(2\mathrm{n}\times 2\mathrm{n})}{\widetilde{\boldsymbol{A}}}\;\underset{(2\mathrm{n}\times\mathrm{nx})}{\boldsymbol{V}} \tag{9.2.36}$$

k+1		1	2	3	4	5	6	7	8	9
$\widetilde{x}^{(k)}$		0.5000	−0.1964	−0.5223	0.1354	0.5718	−0.0978	−0.6144	0.0619	0.6414
		0.5000	−0.1964	−0.6039	0.1627	0.6685	−0.1201	−0.7198	0.0780	0.7524
		0.5000	0.6284	−0.3852	−0.6352	0.3003	0.6414	−0.2009	−0.6455	0.0884
		0.5000	0.7266	−0.4627	−0.7427	0.3687	0.7514	−0.2532	−0.7572	0.1216
$s^{(k)}$		2.0000	2.5461	1.2032	2.8448	1.1110	3.0697	1.0440	3.2441	1.0065
r		0.0000	1.2250	−0.4297	1.1932	−0.3094	1.0235	−0.1802	0.7371	−0.0510
λ_1		0.0000	0.0000	0.0000	0.0958	0.1026	0.0918	0.0922	0.0918	0.0918
		0.0000	0.0000	0.0000	1.8026	1.8209	1.8264	1.8274	1.8275	1.8276
p_1		0.0000	0.0000	0.0000	−0.0294	−0.0308	−0.0275	−0.0275	−0.0274	−0.0274
		0.0000	0.0000	0.0000	0.5532	0.5474	0.5462	0.5458	0.5458	0.5458
$\mathrm{Re}(\widetilde{x}_1)$	$p_1 x_1$	0.0000	0.0000	0.0000	−0.0323	−0.0250	−0.0276	−0.0273	−0.0274	−0.0274
		0.0000	0.0000	0.0000	−0.0458	−0.0354	−0.0406	−0.0402	−0.0404	−0.0404
	x_1	0.0000	0.0000	0.0000	1.0000	1.0000	1.0000	1.0000	1.0000	1.0000
		0.0000	0.0000	0.0000	1.1729	1.1744	1.1749	1.1750	1.1750	1.1750
$\mathrm{Im}(\widetilde{x}_1)$	$p_1 x_1$	0.0000	0.0000	0.0000	0.5445	0.5458	0.5457	0.5458	0.5458	0.5458
		0.0000	0.0000	0.0000	0.6329	0.6399	0.6404	0.6409	0.6409	0.6409
	x_1	0.0000	0.0000	0.0000	0.0000	0.0000	0.0000	0.0000	0.0000	0.0000
		0.0000	0.0000	0.0000	0.0111	0.0151	0.0149	0.0151	0.0150	0.0151

Fig. 9.2.2 Simple vector iteration for non-modal damping and real arithmetics for the example of fig. 9.1.1

which incorporates here, of course, the sparsely populated nature of the matrices $\boldsymbol{M}$, $\boldsymbol{C}$ and $\boldsymbol{K}$ contained in $\widetilde{\boldsymbol{A}}$, we have

$$\underset{(2\mathrm{n}\times\mathrm{nx})}{\boldsymbol{V}} = \{ \underset{(\mathrm{n}\times\mathrm{nx})}{\boldsymbol{V}_\mathrm{I}} \quad \underset{(\mathrm{n}\times\mathrm{nx})}{\boldsymbol{V}_\mathrm{II}} \} \tag{9.2.37}$$

$$\boldsymbol{Y}_\mathrm{I} = -\boldsymbol{V}_\mathrm{II} \tag{9.2.38}$$

$$\begin{aligned} \boldsymbol{K}\boldsymbol{Y}_\mathrm{II} &= [\boldsymbol{M} \quad \boldsymbol{C}]\boldsymbol{V} \\ &= \boldsymbol{M}\boldsymbol{V}_\mathrm{I} + \boldsymbol{C}\boldsymbol{V}_\mathrm{II} \end{aligned} \tag{9.2.39}$$

Thus, in principle, our discussion may be concluded since $\boldsymbol{Y}$ can be employed as a new starting set $\boldsymbol{W}^{(\mathrm{k}+1)}$. However, a Rayleigh-Ritz step also proves useful here to achieve an uncoupling of the vectors. To this end, we calculate the system matrix of the sub-space appertaining to $\boldsymbol{V}$

$$\underset{(\mathrm{nx}\times\mathrm{nx})}{\widetilde{\boldsymbol{B}}} = \boldsymbol{V}^\mathrm{t}\boldsymbol{Y} = \boldsymbol{V}^\mathrm{t}\widetilde{\boldsymbol{A}}\boldsymbol{V} \tag{9.2.40}$$

For a symmetric matrix $\boldsymbol{A}$ and real eigenvectors, this step leads in due course to a diagonal matrix. In the present more complex case, we are certainly aware that $\widetilde{\boldsymbol{A}}$ can only be diagonalised with the joint assistance of the left-hand and right-hand eigenvectors, see equ. (9.1.87). As a result, $\widetilde{\boldsymbol{B}}$ is not in general diagonal here but simply real. The matrix $\widetilde{\boldsymbol{B}}$, generated in this manner, is unsymmetric and an eigenvalue analysis, e.g. the Eberlein method described in the next section, is carried out on it

$$\widetilde{\boldsymbol{B}}\widetilde{\boldsymbol{X}}_B = \widetilde{\boldsymbol{X}}_B\boldsymbol{\Lambda}_B \tag{9.2.41}$$

Such an eigenvalue analysis yields the complete right-hand eigenvectors of the subspace together with the corresponding eigenvalues, which are in general complex. The left-hand eigenvectors are given by $\widetilde{\boldsymbol{X}}_B^{-t}$ as has already been demonstrated in equ. (9.1.86). Analogous to equ. (9.1.87), we then have

$$\begin{aligned}\widetilde{\boldsymbol{X}}_B^{-1}\widetilde{\boldsymbol{B}}\widetilde{\boldsymbol{X}}_B &= \boldsymbol{\Lambda}_B \\ &= \widetilde{\boldsymbol{X}}_B^{-1}\boldsymbol{V}^{\mathrm{t}}\widetilde{\boldsymbol{A}}\boldsymbol{V}\widetilde{\boldsymbol{X}}_B\end{aligned} \tag{9.2.42}$$

In conjunction with the corresponding left-hand eigenvectors

$$\begin{aligned}[\boldsymbol{V}\widetilde{\boldsymbol{X}}_B]^{-1} &= \widetilde{\boldsymbol{X}}_B^{-1}\boldsymbol{V}^{-1} \\ &= \widetilde{\boldsymbol{X}}_B^{-1}\boldsymbol{V}^{\mathrm{t}}\end{aligned} \tag{9.2.43}$$

the vector set $\boldsymbol{V}\widetilde{\boldsymbol{X}}_B$, in general complex, diagonalises the unsymmetric matrix $\widetilde{\boldsymbol{A}}$. We thus obtain an uncoupling of the subspace specified by nx. As convergence proceeds, we then have

$$\boldsymbol{V}\widetilde{\boldsymbol{X}}_B \rightarrow \underset{(2\mathrm{n}\times\mathrm{nx})}{\widetilde{\boldsymbol{X}}} \tag{9.2.44}$$

which contains the true and, in general, complex eigenvectors whilst $\boldsymbol{\Lambda}_B$ contains the corresponding eigenvalues. The aforementioned eigenvalue analysis of the $\widetilde{\boldsymbol{B}}$ matrix can, for example, be executed with the Eberlein algorithm as will be discussed next. This procedure is analogous to the Jacobi algorithm and diagonalises iteratively $\widetilde{\boldsymbol{B}}$. The arithmetics are real and complex solutions appear as (2×2) blocks in the diagonal. The eigenvectors $\widetilde{\boldsymbol{X}}_B$ can be stored and used to initiate the vector iteration in the following loop. The quality of results of the vector iteration can be assessed by the complex residue

$$\begin{aligned}\boldsymbol{E}_{\mathrm{c}} &= \widetilde{\boldsymbol{A}}\boldsymbol{V}\widetilde{\boldsymbol{X}}_B - \boldsymbol{V}\widetilde{\boldsymbol{X}}_B\boldsymbol{\Lambda}_B \\ &= \boldsymbol{Y} - \boldsymbol{V}\widetilde{\boldsymbol{B}}\widetilde{\boldsymbol{X}}_B\end{aligned} \tag{9.2.45}$$

which according to equ. (9.1.59) must tend to zero. For example, [9.5] proposes that the column norm $\boldsymbol{E}_{\mathrm{cj}}$ be normalised with the modulus of λ_1, ρ_1, and of λ_{j}, ρ_{j} and be tested with a break-off criterion

$$\|\boldsymbol{E}_{\mathrm{cj}}\|_E \,/\, (|\lambda_{B1}|\,|\lambda_{B\mathrm{j}}|)^{1/2} \leqslant \epsilon = 10^{-\mathrm{THRESH}} \tag{9.2.46}$$

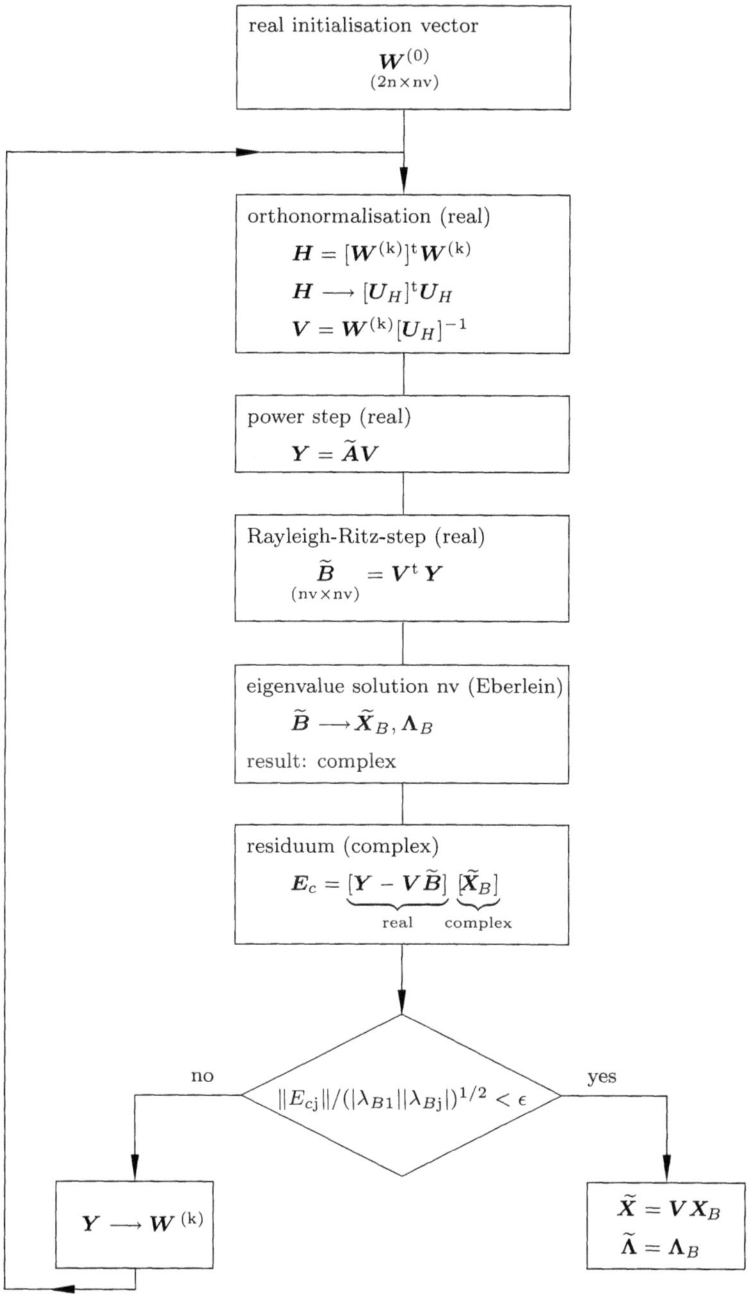

Fig. 9.2.3 Simultaneous vector iteration for the unsymmetric real matrix $\widetilde{A}$ using real arithmetics

k+1	1		2		3		4		5	
$\boldsymbol{V}^{(k)}$	0.7071	0.0000	−0.5155	0.7136	−0.6060	0.6840	0.1336	0.0635	0.5983	−0.3043
	0.0000	0.7071	0.0000	−0.5035	−0.5807	−0.2332	0.1340	0.9739	0.6925	−0.2690
	0.7071	0.0000	0.6186	0.1507	−0.3838	−0.0345	−0.6419	0.2179	0.2547	0.5758
	0.0000	0.7071	0.5929	0.4633	−0.3850	−0.6903	−0.7430	−0.0012	0.3124	0.7096
$\lambda^{(k)}$	1.2675	−0.3925	0.2318	−0.5336	0.4433	0.4433	0.0965	0.0965	0.1174	0.1174
	0.0000	0.0000	0.0000	0.0000	0.7090	−0.7090	1.5957	−1.5957	1.8145	−1.8145
$p^{(k)}$	−0.7890	2.5477	−4.3138	1.8741	−0.6340	−0.6340	−0.0378	−0.0378	−0.0355	−0.0355
	0.0000	0.0000	0.0000	0.0000	1.0140	−1.0140	0.6244	−0.6244	0.5488	−0.5488
$\varepsilon^{(k)}$	2.7500	1.7253	1.6717	2.8686	3.0290	3.0290	0.3969	0.3969	0.1025	0.1025
$\mathrm{Re}(\tilde{\boldsymbol{x}}^{(k)})$	1.0000	1.0000	−2.6708	0.2478	−0.7299	−0.7299	−0.1787	−0.1787	−0.0961	−0.0961
	1.1675	−0.4925	1.1025	−0.6487	2.0788	2.0788	0.0663	0.0663	0.0116	0.0116
	1.0000	1.0000	1.0000	1.0000	1.0000	1.0000	1.0000	1.0000	1.0000	1.0000
	1.1675	−0.4925	0.2601	1.3692	3.0530	3.0530	1.0891	1.0891	1.2316	1.2316
$\mathrm{Im}(\tilde{\boldsymbol{x}}^{(k)})$	0.0000	0.0000	0.0000	0.0000	2.0877	−2.0877	0.0762	−0.0762	0.5828	−0.5828
	0.0000	0.0000	0.0000	0.0000	−0.5115	0.5115	0.7135	−0.7135	0.6453	−0.6453
	0.0000	0.0000	0.0000	0.0000	0.0000	0.0000	0.0000	0.0000	0.0000	0.0000
	0.0000	0.0000	0.0000	0.0000	−1.8533	1.8533	−0.1774	0.1774	−0.0012	0.0012

k+1	6		7		8		9		10	
$\boldsymbol{V}^{(k)}$	−0.0806	−0.6251	−0.6303	0.1549	0.0443	0.6461	0.6473	−0.0360	−0.0089	−0.6482
	−0.0989	−0.7700	−0.7385	0.1823	0.0574	0.7597	0.7598	−0.0474	−0.0157	−0.7612
	0.6439	−0.0814	−0.1465	−0.6305	−0.6468	0.0385	0.0297	0.6472	0.6479	−0.0027
	0.7544	−0.0982	−0.1895	−0.7384	−0.7592	0.0623	0.0528	0.7600	0.7615	−0.0210
$\lambda^{(k)}$	0.0937	0.0937	0.0931	0.0931	0.0918	0.0918	0.0918	0.0918	0.0918	0.0918
	1.8338	−1.8338	1.8272	−1.8272	1.8278	−1.8278	1.8276	−1.8276	1.8276	−1.8276
$p^{(k)}$	−0.0278	−0.0278	−0.0278	−0.0278	−0.0274	−0.0274	−0.0274	−0.0274	−0.0274	−0.0274
	0.5439	−0.5439	0.5459	−0.5459	0.5457	−0.5457	0.5458	−0.5458	0.5458	−0.5458
$\varepsilon^{(k)}$	0.0144	0.0144	0.0056	0.0056	0.0006	0.0006	0.0003	0.0003	0.0000	0.0000
$\mathrm{Re}(\tilde{\boldsymbol{x}}^{(k)})$	−0.0279	−0.0279	−0.0312	−0.0312	−0.0273	−0.0273	−0.0276	−0.0276	−0.0274	−0.0274
	−0.0338	−0.0338	−0.0379	−0.0379	−0.0402	−0.0402	−0.0403	−0.0403	−0.0404	−0.0404
	1.0000	1.0000	1.0000	1.0000	1.0000	1.0000	1.0000	1.0000	1.0000	1.0000
	1.1721	1.1721	1.1770	1.1770	1.1749	1.1749	1.1751	1.1751	1.1750	1.1750
$\mathrm{Im}(\tilde{\boldsymbol{x}}^{(k)})$	0.5280	−0.5280	0.5470	−0.5470	0.5452	−0.5452	0.5458	−0.5458	0.5458	−0.5458
	0.6503	−0.6503	0.6411	−0.6411	0.6413	−0.6413	0.6409	−0.6409	0.6409	−0.6409
	0.0000	0.0000	0.0000	0.0000	0.0000	0.0000	0.0000	0.0000	0.0000	0.0000
	0.0024	−0.0024	0.0147	−0.0147	0.0144	−0.0144	0.0151	−0.0151	0.0150	−0.0150

Fig. 9.2.4 Simultaneous vector iteration of the example in fig. 9.1.1 using two vectors

The process described is summarised in fig. 9.2.3 in the form of a flow chart. We apply this to our standard example in fig. 9.1.1 which we have already used to test the simple method of vector iteration. In the latter case, the adoption of a refined calculating technique yielded a complex pair of solutions using only one iteration vector. In this case, the contributions of the first and second eigenvectors to the starting vectors are equally amplified. On the other hand, the calculation technique applied here requires two starting vectors for a complex solution λ_1, since λ_2 has an equal modulus. In return, however, we need not store a set of four iteration vectors. Just as it is essential for the convergence of the simple vector iteration process to have distinctly different moduli, $\lambda_1 > \lambda_2$ (respectively $\lambda_1 = \lambda_2 > \lambda_3$ for a complex solution), so we require a jump in the eigenvalue moduli at the end of the spectrum in question when using the simultaneous vector iteration. In general, one would start with more vectors than are actually necessary (for example, one may double the amount) and then, following the emergence of the eigenvalue spectrum, one could select a suitable range of eigenvalues.

If we now return to our example, we note that the simultaneous vector iteration with two linearly independent starting vectors can yield the complete set of solutions λ_1, λ_2 and the corresponding eigenvectors $\boldsymbol{x}_1$, $\boldsymbol{x}_2$. This is illustrated in fig. 9.2.4. As the reader may not yet be familiar with the Eberlein method, he can make do by solving the (2×2)-eigenvalue problem directly 'by hand'

$$\begin{bmatrix} \widetilde{b}_{11} - \lambda_B & \widetilde{b}_{12} \\ \widetilde{b}_{21} & \widetilde{b}_{22} - \lambda_B \end{bmatrix} \begin{bmatrix} \widetilde{x}_{B1} \\ \widetilde{x}_{B2} \end{bmatrix} = \begin{bmatrix} 0 \\ 0 \end{bmatrix}$$

The characteristic equation is a quadratic one and gives the two solutions $\lambda_{B1,2}$. From this we deduce, for example,

$$\widetilde{\boldsymbol{x}}_B = \left\{ 1.0 \quad \frac{\lambda_B - \widetilde{b}_{11}}{\widetilde{b}_{12}} \right\}$$

for the corresponding eigenvector. The first 10 loops of the procedure are reproduced in fig. 9.2.4. In the present case, we used a sequence of unit matrices as a pair of starting vectors. We note that to start with, the approximation is highly deficient and that at first, real solutions for $\boldsymbol{x}_1$ and $\boldsymbol{x}_2$ are emerging. However, once the complex conjugate solution has appeared, we obtain a relatively fast and, as far as the error criterion is concerned, monotonic convergence to the true solution. At the same time, the eigenvectors are normalised so that they can immediately be compared to the preceding solution.

9.3 The Eberlein process for the calculation of the eigenspectrum of a real unsymmetric dynamic matrix

For the simultaneous vector iteration, it proved necessary in section 9.2 to use a procedure yielding all the eigenvalues and eigenvectors of a real unsymmetric matrix. For this reason, but also for other practical considerations, it seems advisable to discuss at least one such process. Here, our concern is not so much efficiency as the didactic aspects. For symmetric dynamic matrices, we applied the Jacobi process for the calculation of all eigenvalues and eigenvectors. A consequent extension of this framework to unsymmetric dynamic matrices leads us to the Eberlein process [9.4, 9.5, 9.6, 9.7] which also permits a hypermatrix formulation as the Jacobi method does.

In analogy to the Jacobi method, a series of two-dimensional transformations reduces the dynamic matrix $\widetilde{\boldsymbol{A}}$ to a diagonal or block-diagonal form. We are already aware that an unsymmetric matrix possesses left-hand eigenvectors $\boldsymbol{Y} = \boldsymbol{X}^{-1}$ and right-hand eigenvectors $\boldsymbol{X}$. Correspondingly, the total transformation can be expressed as

$$\widetilde{\boldsymbol{A}}_T = \boldsymbol{T}^{-1}\widetilde{\boldsymbol{A}}\boldsymbol{T} \tag{9.3.1}$$

where $\boldsymbol{T}^{-1}$ contains the left-hand eigenvectors, $\boldsymbol{T}$ the right-hand eigenvectors and $\widetilde{\boldsymbol{A}}_T$ the eigenvalues. The complete transformation is built up in individual steps

$$\widetilde{\boldsymbol{A}}^{(\mathrm{k}+1)} = \boldsymbol{T}_{\mathrm{k}}^{-1}\widetilde{\boldsymbol{A}}^{(\mathrm{k})}\boldsymbol{T}_{\mathrm{k}} \tag{9.3.2}$$

Consequently, we obtain

$$\boldsymbol{T} = \boldsymbol{T}_1\boldsymbol{T}_2 \quad \ldots \quad \boldsymbol{T}_{\mathrm{k}} \quad \ldots \tag{9.3.3}$$

The sequence of the $\widetilde{\boldsymbol{A}}^{(\mathrm{k})}$ matrices converges against $\widetilde{\boldsymbol{A}}_T$, and the purpose of each transformation is to amplify the norm of the diagonal elements as against that of the off-diagonal elements.

A specific transformation matrix $\boldsymbol{T}_{\mathrm{k}}$ takes the form

$$\boldsymbol{T}_{\mathrm{k}} = \begin{bmatrix} 1 & & & & & & & & \\ & \ddots & & & & & & & \\ & & 1 & & & & & & \\ & & & t_{\mathrm{pp}} & \cdots & & t_{\mathrm{pq}} & & \\ & & & \vdots & 1 & & \vdots & & \\ & & & & & \ddots & & & \\ & & & \vdots & & 1 & \vdots & & \\ & & & t_{\mathrm{qp}} & \cdots & & t_{\mathrm{qq}} & & \\ & & & & & & & 1 & \\ & & & & & & & & \ddots \\ & & & & & & & & & 1 \end{bmatrix} \begin{matrix} \\ \\ \\ \mathrm{p} \\ \\ \\ \\ \mathrm{q} \\ \\ \\ \\ \end{matrix} \tag{9.3.4}$$

(columns p and q)

and comprises a rotation $\boldsymbol{R}$, as in the Jacobi method, and an additional shear $\boldsymbol{S}$

$$\boldsymbol{T}_{\mathrm{k}} = \boldsymbol{R}\boldsymbol{S} \tag{9.3.5}$$

In principle, the matrices $\boldsymbol{R}$ and $\boldsymbol{S}$ have the same form as $\boldsymbol{T}_{\mathrm{k}}$

$$\boldsymbol{R} = \begin{bmatrix} 1 & & & & & & & & \\ & \ddots & & & & & & & \\ & & 1 & & & & & & \\ & & & r_{\mathrm{pp}} & \cdots & & r_{\mathrm{pq}} & & \\ & & & \vdots & 1 & & \vdots & & \\ & & & & & \ddots & & & \\ & & & \vdots & & 1 & \vdots & & \\ & & & r_{\mathrm{qp}} & \cdots & & r_{\mathrm{qq}} & & \\ & & & & & & & 1 & \\ & & & & & & & & \ddots \\ & & & & & & & & & 1 \end{bmatrix} \begin{matrix} \\ \\ \\ \mathrm{p} \\ \\ \\ \\ \mathrm{q} \\ \\ \\ \\ \end{matrix} \tag{9.3.6}$$

(columns p and q)

$$
\boldsymbol{S} = \left[\begin{array}{ccccccccc}
1 & & & & & & & & \\
 & \ddots & & & & & & & \\
 & & 1 & & & & & & \\
 & & & s_{pp} & \cdots & s_{pq} & & & \\
 & & & & 1 & & & & \\
 & & & \vdots & \ddots & \vdots & & & \\
 & & & & 1 & & & & \\
 & & & s_{qp} & \cdots & s_{qq} & & & \\
 & & & & & & 1 & & \\
 & & & & & & & \ddots & \\
 & & & & & & & & 1
\end{array}\right] \begin{array}{c} \\ \\ \\ p \\ \\ \\ \\ q \\ \\ \\ \\ \end{array} \tag{9.3.7}
$$

We already know that an unsymmetric real matrix $\widetilde{\boldsymbol{A}}$ possibly possesses complex eigenvalues and eigenvectors. A similarity transformation like that given in equ. (9.3.1) can generally only guarantee a diagonal matrix $\widetilde{\boldsymbol{A}}_T$ if $\boldsymbol{T}_k$ itself is allowed to be complex. Hence, complex values must also be assigned to the unspecified elements in $\boldsymbol{R}$ and $\boldsymbol{S}$ in order to achieve diagonality. In the present context, the following scheme is selected

$$
\begin{aligned}
r_{pp} = \quad r_{qq} &= \cos x \\
r_{pq} = \qquad\quad &= -e^{i\alpha} \sin x \\
r_{qp} = -r^*_{pq} &= e^{-i\alpha} \sin x
\end{aligned} \tag{9.3.8}
$$

The matrix $\boldsymbol{R}$ thus construed is unitary – as is the rotation matrix $\boldsymbol{Q}_k$ in the Jacobi method – which means that the inverse of $\boldsymbol{R}$ is simply its conjugate transposed (= transposed in the real case)

$$
\boldsymbol{R}^{-1} = \boldsymbol{R}^* \tag{9.3.9}
$$

Such a unitary transformation or rotation does not modify either the eigenvalues or the Euclidean norm of the matrix $\widetilde{\boldsymbol{A}}^{(k)}$

$$
\|\boldsymbol{R}^*\widetilde{\boldsymbol{A}}^{(k)}\boldsymbol{R}\| = \|\widetilde{\boldsymbol{A}}^{(k)}\| = \Big(\sum_{i,j} |a_{ij}^{(k)}|^2\Big)^{1/2} \tag{9.3.10}
$$

$$
\boldsymbol{\Lambda}(\boldsymbol{R}^*\widetilde{\boldsymbol{A}}^{(k)}\boldsymbol{R}) = \boldsymbol{\Lambda}(\widetilde{\boldsymbol{A}}^{(k)}) = \boldsymbol{\Lambda}(\widetilde{\boldsymbol{A}}) \tag{9.3.11}
$$

The round brackets indicate that $\boldsymbol{\Lambda}$ encompasses the eigenvalues of the matrix expression in brackets.

The adopted transformation aims (as in the Jacobi method) at the reduction of the norm of the off-diagonal elements

$$
\tau = \Big(\sum_{i \neq j} |a_{ij}^{(k)}|^2\Big)^{1/2} \tag{9.3.12}
$$

The specification of the two parameters α and x is somewhat complicated and is relegated to the appendix. It is based upon the known elements of the $\widetilde{\boldsymbol{A}}^{(k)}$ matrix.

The shear matrix $\boldsymbol{S}$ is defined by

$$\begin{aligned} s_{pp} = s_{qq} &= \operatorname{ch} y \\ s_{pq} &= -ie^{i\beta}\operatorname{sh} y \\ s_{qp} = s^*_{pq} &= ie^{-i\beta}\operatorname{sh} y \end{aligned} \tag{9.3.13}$$

where the abbreviations cosh = ch and sinh = sh are adopted for typographical simplicity. This matrix is Hermitian and represents the complex generalisation of a real symmetric matrix. A Hermitian matrix contains a symmetric real part and a skew-symmetric imaginary part and, in analogy to a real symmetric matrix, possesses the property

$$\boldsymbol{S}^* = \boldsymbol{S} \tag{9.3.14}$$

The inverse of the complex matrix $\boldsymbol{S}$ is simply specified by the pivot elements

$$\begin{bmatrix} \operatorname{ch} y & ie^{i\beta}\operatorname{sh} y \\ -ie^{-i\beta}\operatorname{sh} y & \operatorname{ch} y \end{bmatrix} \tag{9.3.15}$$

from which it follows that the transformation with $\boldsymbol{S}$ represents merely a similarity transformation and not a unitary congruent transformation as the rotation $\boldsymbol{R}$. A similarity transformation does not alter the eigenvalues of the matrix $\boldsymbol{A}^{(k)}$ either (see the comments on equ. (9.3.11))

$$\Lambda(\boldsymbol{S}^{-1}\widetilde{\boldsymbol{A}}^{(k)}\boldsymbol{S}) = \Lambda(\widetilde{\boldsymbol{A}}^{(k)}) = \Lambda(\widetilde{\boldsymbol{A}}) \tag{9.3.16}$$

but the shear modifies the Euclidean norm of $\widetilde{\boldsymbol{A}}^{(k)}$, and this intentionally.

In our discussion on the calculation of the eigenvalues of $\widetilde{\boldsymbol{A}}$, we move into the class of diagonalisable or nondefective matrices. These can be diagonalised by a similarity transformation even when the eigenvalues are complex. Until now, this fact has simply been taken for granted. A special sub-class of diagonalisable matrices are the so-called normal matrices $\widetilde{\boldsymbol{A}}$ with the property

$$\widetilde{\boldsymbol{A}}^*\widetilde{\boldsymbol{A}} = \widetilde{\boldsymbol{A}}\widetilde{\boldsymbol{A}}^* \tag{9.3.17}$$

Furthermore, for a normal matrix $\widetilde{\boldsymbol{A}}$, the following relation applies

$$\|\widetilde{\boldsymbol{A}}\| = (\textstyle\sum |\lambda_i|^2)^{1/2} \tag{9.3.18}$$

which means that if the eigenvalues remain unaltered, the Euclidean matrix norm also does not change. Furthermore, we note that a normal matrix $\widetilde{\boldsymbol{A}}$ can be diagonalised by means of a unitary transformation, e.g. $\boldsymbol{R}$. From this, we conclude that the handling of normal matrices is a degree simpler than that of non-normal matrices.

In general, however, $\widetilde{\boldsymbol{A}}$ is not normal, which means that the rule of equ. (9.3.17) does not hold

$$\widetilde{\boldsymbol{B}} = \widetilde{\boldsymbol{A}}^* \widetilde{\boldsymbol{A}} - \widetilde{\boldsymbol{A}} \widetilde{\boldsymbol{A}}^* \neq \boldsymbol{O} \tag{9.3.19}$$

In this case, equ. (9.3.18) no longer applies either, but can be replaced by

$$\|\widetilde{\boldsymbol{A}}\| > (\sum |\lambda_i|^2)^{1/2} \tag{9.3.20}$$

which means that the Euclidean norm is larger than the root of the sum of the squares of the eigenvalue moduli. For a non-normal matrix $\widetilde{\boldsymbol{A}}$, though, a series of unitary congruent transformations with $\boldsymbol{R}$ may not yield a diagonalisation. Hence, one attempts to give a more normal form to the matrix by a shear operation with $\boldsymbol{S}$. This involves imposing a reduction of the Euclidean norm in such a way that the rotation yields the required effect. Proceeding from $\widetilde{\boldsymbol{A}}^{(1)} = \widetilde{\boldsymbol{A}}$, a rotation $\boldsymbol{R}_k^* \widetilde{\boldsymbol{A}}^{(k)} \boldsymbol{R}_k$ does not, of course, alter the Euclidean norm or the eigenvalues. The subsequent operation with the shear $\boldsymbol{S}_k^{-1} \boldsymbol{R}_k^* \widetilde{\boldsymbol{A}}^{(k)} \boldsymbol{R}_k \boldsymbol{S}_k$ reduces the Euclidean norm but not the eigenvalues. This entails a progressive and simultaneous normalisation and diagonalisation of the matrix $\widetilde{\boldsymbol{A}}^{(k+1)}$. Consequently, the calculation of the shearing parameters in equ. (9.3.13), still to be determined, has to be based upon the matrix

$$\widetilde{\boldsymbol{B}} = [\widetilde{\boldsymbol{A}}^{(k)}]^* \widetilde{\boldsymbol{A}}^{(k)} - \widetilde{\boldsymbol{A}}^{(k)} [\widetilde{\boldsymbol{A}}^{(k)}]^* \tag{9.3.21}$$

and once $\widetilde{\boldsymbol{A}}^{(k)}$ is normal, $\widetilde{\boldsymbol{B}}$ is empty and thus $\boldsymbol{S}$ is transformed into the unit matrix $\boldsymbol{I}$ ($\beta = y = 0$). Shear can then be omitted and diagonalisation is achieved simply by means of rotations as in the Jacobi method. The details for the evaluation of β and y are given in section *9.3, an appendix to this chapter.

The procedure mentioned so far [9.6, 9.7] entails a small shortcoming: following the initial step, we have to face onerous complex arithmetics even for a real starting matrix $\widetilde{\boldsymbol{A}}$. At the same time, this has the advantage that $\widetilde{\boldsymbol{A}}$ need not necessarily be real; our procedure is also applicable to a complex matrix $\widetilde{\boldsymbol{A}}$. Nevertheless, one is inevitably led to wonder if it would not be possible to adopt some simplifications and reduce the operations to real arithmetics when applying a real starting matrix. At the same time, it is evident that no complex diagonals can be achieved by means of real similarity transformations. However, we can attain a block diagonalisation by means of real arithmetics. The real eigenvalues appear on the diagonal and the complex conjugate pairs of eigenvalues are placed as (2×2)-diagonal blocks, e.g.

$$\widetilde{\boldsymbol{A}}_T = \begin{bmatrix} \lambda_1 & \lambda_2 & \begin{bmatrix} \lambda_{3r} & \lambda_{3i} \\ -\lambda_{3i} & \lambda_{3r} \end{bmatrix} & \cdots \end{bmatrix} \tag{9.3.22}$$

where λ_{3r} represents the real part and λ_{3i} the imaginary part of the third eigenvalue. The eigenvectors, i.e. their real and imaginary parts, occur as columns in the transformation matrix $\boldsymbol{T}$

$$\boldsymbol{T} = [\boldsymbol{x}_1 \quad \boldsymbol{x}_2 \quad \boldsymbol{\xi}_3 \quad \boldsymbol{\eta}_3 \quad \ldots] \tag{9.3.23}$$

Applying a slight modification of the complex algorithm described earlier, we obtain in place of equ. (9.3.8) the real operation

$$\begin{aligned} r_{pp} &= r_{qq} = \cos x \\ r_{pq} &= -r_{qp} = -\sin x \end{aligned} \tag{9.3.24}$$

as also in place of equ. (9.3.13)

$$\begin{aligned} s_{pp} &= s_{qq} = \operatorname{ch} y \\ s_{pq} &= s_{qp} = -\operatorname{sh} y \end{aligned} \tag{9.3.25}$$

As before, $\boldsymbol{R}$ is unitary and $\boldsymbol{S}$ is symmetric (generally: Hermitian). In order to determine the pivot elements, only two (previously: four) parameters, x and y, are required. These are described in more detail in the aforementioned appendix.

Although the real algorithm seems to be ideally suited to our needs as compared to the generally complex diagonalisation, it also entails some disadvantages which should not remain unmentioned: the existence of multiple complex roots considerably slows down the convergence, and in the case of defective roots, the convergence becomes even slower. Such difficulties do not arise, however, for the complex algorithm.

We return once again to the example of fig. 9.1.1 and use fig. 9.3.1 to trace the development of the sequence of $\widetilde{\boldsymbol{A}}^{(k)}$ matrices and apply in this case the real algorithm. The Euclidean norm of the initial matrix is 3.78, but it soon converges to the ultimate value of 2.64. We can see that the initial matrix $\widetilde{\boldsymbol{A}}$ is not normal but is rapidly becoming normalised. Interesting information is further obtained from the Euclidean norm $\|\widetilde{\boldsymbol{A}}_{HD}\|$ of the (2×2)-hyperdiagonal, which contains the complex eigenvalues, i.e. its real and imaginary parts, and the real eigenvalues collected in pairs as also from the Euclidean norm $\|\widetilde{\boldsymbol{A}}_{HND}\|$ of the remaining non-diagonal elements. The first measure converges to the Euclidean norm of $\widetilde{\boldsymbol{A}}_T$, whilst the second measure tends to zero. The termination criterion is satisfied following 21 loops, at the same time the (2×2)-blocks of the complex eigenvalues have been formed. In addition, we quote in the figure the eigenvectors (= matrix $\boldsymbol{T}$) as also the special eigenvectors normalised with the condition $x_{3j} = 1$ whereby the real and the imaginary parts appear alternately in the columns.

Supplement to 9.3

*9.3 Determination of transformation parameters in the Eberlein process

We first discuss the complex iteration in some detail. Proceeding from the complex iterated $\widetilde{\boldsymbol{A}}^{(k)}$ ($\widetilde{\boldsymbol{A}}^{(0)} = \widetilde{\boldsymbol{A}} =$ real), we form with the pivot elements of $\widetilde{\boldsymbol{A}}^{(k)}$

$$\begin{aligned} n_1 &= |\widetilde{a}_{pq} + \widetilde{a}^*_{qp}| \\ z_1 &= \operatorname{Re}(\widetilde{a}_{pp} - \widetilde{a}_{qq}) \end{aligned} \tag{*9.3.1}$$

$$\widetilde{\boldsymbol{A}}^{(0)} = \widetilde{\boldsymbol{A}} = \begin{bmatrix} \boldsymbol{O} & -\boldsymbol{I} \\ \boldsymbol{K}^{-1}\boldsymbol{M} & \boldsymbol{K}^{-1}\boldsymbol{C} \end{bmatrix} \qquad \epsilon = 0.37835170 \cdot 10^{-6}$$

start $\quad \|\widetilde{\boldsymbol{A}}^{(0)}\| = 3.7835169 \quad \|\widetilde{\boldsymbol{A}}^{(0)}_{HD}\| = 0.2549510 \quad \|\widetilde{\boldsymbol{A}}^{(0)}_{HND}\| = 3.7749171$

$$\widetilde{\boldsymbol{A}}^{(0)} = \begin{bmatrix} 0.0000000E+00 & 0.0000000E+00 & -0.1000000E+01 & 0.0000000E+00 \\ 0.0000000E+00 & 0.0000000E+00 & 0.0000000E+00 & -0.1000000E+01 \\ 0.1000000E+01 & 0.2000000E+01 & 0.2000000E+00 & 0.0000000E+00 \\ 0.1000000E+01 & 0.2500000E+01 & 0.1500000E+00 & 0.5000000E-01 \end{bmatrix}$$

loop 5 $\quad \|\widetilde{\boldsymbol{A}}^{(5)}\| = 2.6609998 \quad \|\widetilde{\boldsymbol{A}}^{(5)}_{HD}\| = 2.0147474 \quad \|\widetilde{\boldsymbol{A}}^{(5)}_{HND}\| = 1.7383075$

$$\widetilde{\boldsymbol{A}}^{(5)} = \begin{bmatrix} 0.1495421E+00 & 0.1410194E+01 & 0.9207101E+00 & 0.1555172E-01 \\ -0.1417580E+01 & 0.1779273E-03 & -0.1175916E-02 & 0.8135234E+00 \\ -0.9185054E+00 & 0.3745297E-01 & 0.1778729E+00 & 0.2243089E-01 \\ -0.6024274E-01 & -0.8144066E+00 & -0.2243089E-01 & -0.7759311E-01 \end{bmatrix}$$

loop 10 $\quad \|\widetilde{\boldsymbol{A}}^{(10)}\| = 2.6449447 \quad \|\widetilde{\boldsymbol{A}}^{(10)}_{HD}\| = 2.5903456 \quad \|\widetilde{\boldsymbol{A}}^{(10)}_{HND}\| = 0.5346420$

$$\widetilde{\boldsymbol{A}}^{(10)} = \begin{bmatrix} 0.9451319E-01 & -0.1786790E+01 & -0.9560172E-02 & 0.1312621E-01 \\ 0.1786774E+01 & 0.8778678E-01 & -0.3777098E+00 & 0.1978039E-02 \\ -0.9388644E-02 & 0.3776913E+00 & 0.3460078E-01 & 0.3910117E+00 \\ -0.1306203E-01 & 0.4618093E-03 & -0.3910116E+00 & 0.3309909E-01 \end{bmatrix}$$

loop 15 $\quad \|\widetilde{\boldsymbol{A}}^{(15)}\| = 2.6449211 \quad \|\widetilde{\boldsymbol{A}}^{(15)}_{HD}\| = 2.6449196 \quad \|\widetilde{\boldsymbol{A}}^{(15)}_{HND}\| = 0.0027805$

$$\widetilde{\boldsymbol{A}}^{(15)} = \begin{bmatrix} 0.9176150E-01 & -0.1827582E+01 & 0.6552515E-04 & -0.3446300E-06 \\ 0.1827583E+01 & 0.9176133E-01 & 0.1964975E-02 & -0.1225765E-04 \\ 0.5497156E-04 & -0.1964975E-02 & 0.3323849E-01 & 0.3849896E+00 \\ -0.1252595E-05 & 0.3763846E-04 & -0.3849896E+00 & 0.3323855E-01 \end{bmatrix}$$

loop 21 $\quad \|\widetilde{\boldsymbol{A}}^{(21)}\| = 2.6449215 \quad \|\widetilde{\boldsymbol{A}}^{(21)}_{HD}\| = 2.6449215 \quad \|\widetilde{\boldsymbol{A}}^{(21)}_{HND}\| = 0.0000080$

$$\widetilde{\boldsymbol{A}}^{(21)} = \begin{bmatrix} 0.9176147E-01 & -0.1827584E+01 & -0.1028425E-06 & -0.4112120E-07 \\ 0.1827584E+01 & 0.9176142E-01 & 0.5498335E-05 & 0.1324036E-05 \\ 0.4400172E-06 & -0.5498336E-05 & 0.3323850E-01 & 0.3849893E+00 \\ -0.4139998E-07 & -0.1253013E-05 & -0.3849895E+00 & 0.3323849E-01 \end{bmatrix}$$

final result

eigenvectors

$$\boldsymbol{X} = [\widetilde{\boldsymbol{\xi}}_1 \ \widetilde{\boldsymbol{\eta}}_1 \ \widetilde{\boldsymbol{\xi}}_2 \ \widetilde{\boldsymbol{\eta}}_2] = \begin{bmatrix} 0.7607260E-01 & 0.3867622E+00 & -0.1367175E+01 & -0.1309176E+01 \\ 0.9520023E-01 & 0.4532996E+00 & 0.6121286E+00 & 0.5187986E+00 \\ -0.7138203E+00 & 0.1035301E+00 & -0.4585786E+00 & 0.5698619E+00 \\ -0.8371783E+00 & 0.1323943E+00 & 0.1793835E+00 & -0.2529074E+00 \end{bmatrix}$$

normalised eigenvectors

$$\boldsymbol{X}_n = \left[\begin{array}{cc|cc} * & * & * & * \\ \boldsymbol{\xi}_1 & \boldsymbol{\eta}_1 & \boldsymbol{\xi}_2 & \boldsymbol{\eta}_2 \end{array}\right] = \begin{bmatrix} -0.2741072E-01 & -0.5457956E+00 & -0.2225876E+00 & 0.2578254E+01 \\ -0.4041402E-01 & -0.6408947E+00 & 0.2791304E-01 & -0.1096632E+01 \\ 0.1000000E+01 & 0.0000000E+00 & 0.1000000E+01 & 0.0000000E+00 \\ 0.1174997E+01 & -0.1505527E-01 & -0.4231180E+00 & 0.2570676E-01 \end{bmatrix}$$

Fig. 9.3.1 The real Eberlein method for the example of fig. 9.1.1

respectively

$$n_2 = |\widetilde{a}_{pq} - \widetilde{a}^*_{qp}|$$
$$z_2 = \mathrm{Im}(\widetilde{a}_{pp} - \widetilde{a}_{qq}) \tag{*9.3.2}$$

where, as is also the case in the following expressions, the superior iteration index k is not appended here to the elements $\widetilde{a}_{ij}$. Furthermore, let $\widetilde{a}^*_{qp}$ be the conjugate complex value of $\widetilde{a}_{qp}$. If now

$$n_1^2 + z_1^2 \geqslant n_2^2 + z_2^2 \tag{*9.3.3}$$

and

$$n_1 > 0 \tag{*9.3.4}$$

the following applies

$$\begin{aligned} \cos\alpha &= \mathrm{Re}\,(\widetilde{a}_{pq} + \widetilde{a}_{qp})/n_1 \\ \sin\alpha &= \mathrm{Im}\,(\widetilde{a}_{pq} - \widetilde{a}_{qp})/n_1 \\ \cot 2x &= z_1/n_1 \end{aligned} \tag{*9.3.5}$$

If, however,

$$n_1^2 + z_1^2 < n_2^2 + z_2^2 \tag{*9.3.6}$$

and

$$n_2 > 0 \tag{*9.3.7}$$

then we have

$$\begin{aligned} \cos\alpha &= \mathrm{Im}\,(\widetilde{a}_{pq} + \widetilde{a}_{qp})/n_2 \\ \sin\alpha &= -\mathrm{Re}\,(\widetilde{a}_{pq} - \widetilde{a}_{qp})/n_2 \\ \cot 2x &= z_2/n_2 \end{aligned} \tag{*9.3.8}$$

On the other hand, if we have either

$$n_1 = 0 \tag{*9.3.9}$$

or

$$n_2 = 0 \tag{*9.3.10}$$

we set

$$\begin{aligned} \cos x &= \cos \alpha = 1 \\ \sin x &= \sin \alpha \ = 0 \end{aligned} \qquad (*9.3.11)$$

which entails zero rotation. In this way, the first two parameters α and x, which are required for the specification of the rotation, can now be determined and we can next consider the shear which is governed by β and y. The shear transformation aims at an increase of the normality and thus the operation has to be based on a matrix which measures the degree of normality. This is ensured, for example, by the matrix $\widetilde{\boldsymbol{B}}$ given in equ. (9.3.19). However, since a rotation does not alter the normality, it is here more convenient to use

$$\boldsymbol{G} = \boldsymbol{R}^{*}\widetilde{\boldsymbol{A}}^{(\mathrm{k})}\boldsymbol{R}^{*} \qquad (*9.3.12)$$

and thus to form the matrix

$$\boldsymbol{H} = \boldsymbol{G}^{*}\boldsymbol{G} - \boldsymbol{G}\boldsymbol{G}^{*} \qquad (*9.3.13)$$

If $\widetilde{\boldsymbol{A}}^{(\mathrm{k})}$ is normal, $\boldsymbol{H}$ is an empty matrix. If now

$$n_3 = |h_{\mathrm{pq}}| > 0 \qquad (*9.3.14)$$

we set

$$\begin{aligned} \cos \beta &= -\mathrm{Im}\,(h_{\mathrm{pq}})/n_3 \\ \sin \beta &= \ \ \mathrm{Re}\,(h_{\mathrm{pq}})/n_3 \\ \coth x &= \ \ z_3/n_3 \end{aligned} \qquad (*9.3.15)$$

where

$$z_3 = \sum_{\mathrm{i} \neq \mathrm{p,q}} \left(|g_{\mathrm{pi}}|^2 + |g_{\mathrm{ip}}|^2 + |g_{\mathrm{qi}}|^2 + |g_{\mathrm{iq}}|^2\right) + 2\left(|g_{\mathrm{pp}} - g_{\mathrm{qq}}|^2 + |g_{\mathrm{pq}}\mathrm{e}^{-i\beta} + g_{\mathrm{qp}}\mathrm{e}^{i\beta}|^2\right) \qquad (*9.3.16)$$

In case, though,

$$n_3 = 0$$

we obtain $\beta = y = 0$ and no shear is entailed. It is self-evident that this procedure does not necessitate the formation of the $\boldsymbol{G}$ and $\boldsymbol{H}$ matrices since effectively only the p and q rows and columns of the $\boldsymbol{G}$ matrix, respectively the h_{pq} element of the $\boldsymbol{H}$ matrix, are required.

At this point, we turn our attention to the alternative real algorithm. Analogous to equ. (*9.3.1), we form

$$\begin{aligned} n_1 &= \widetilde{a}_{pq} + \widetilde{a}_{qp} \\ z_1 &= \widetilde{a}_{pp} - \widetilde{a}_{qq} \end{aligned} \tag{*9.3.17}$$

and obtain for x (α is not applicable here)

$$\cot 2x = z_1/n_1 \tag{*9.3.18}$$

as also in equ. (*9.3.5). The determination of the shear parameter y is somewhat more complicated. We find

$$\coth y = z_3/n_3$$

with

$$n_3 = ab - c/2 \tag{*9.3.19}$$

$$z_3 = d + 2(a^2 + b^2) \tag{*9.3.20}$$

and

$$a = \widetilde{a}_{pq} - \widetilde{a}_{qp}$$

$$b = \cos 2x\,(\widetilde{a}_{pp} - \widetilde{a}_{qq}) + \sin 2x\,(\widetilde{a}_{pq} + \widetilde{a}_{qp})$$

$$c = \cos 2x\Big(2\sum_{i\neq p,q}(\widetilde{a}_{pi}\widetilde{a}_{qi} - \widetilde{a}_{ip}\widetilde{a}_{iq})\Big) - \sin 2x\Big(\sum_{i\neq p,q}(\widetilde{a}^2_{pi} + \widetilde{a}^2_{iq} - \widetilde{a}^2_{ip} - \widetilde{a}^2_{iq})\Big)$$

$$d = \sum_{i\neq p,q}(\widetilde{a}^2_{pi} + \widetilde{a}^2_{ip} + \widetilde{a}^2_{iq} + \widetilde{a}^2_{qi})$$

Bibliography for Chapter 9

[9.1] *L. Meirovitch*; Analytical methods in vibrations (Macmillan, New York, 1967).

[9.2] *W.C. Hurty, M.F. Rubinstein*; Dynamics of structures (Prentice-Hall, Englewood Cliffs, 1964).

[9.3] *D.K. Faddeev, W.N. Faddeeva*; Computational methods of linear algebra, W.H. Freeman & Co., San Francisco (1963).

[9.4] *K.A. Braun*; Spektralanalyse großer, linear elastischer, allgemein gedämpfter Schwingungsprobleme, Dissertation, Universität Stuttgart (1979).

[9.5] *E. Schelkle*; LARSTRAN 80, Theoretical background, User's reference manual, ISD-Report No. 290, Stuttgart (1981).

[9.6] *J.H. Wilkinson, C. Reinsch*; eds.; Handbook for automatic computation, Linear Algebra; Contribution II/12, P.J. Eberlein and J. Boothroyd; Solution to the eigenproblem by a norm reducing Jacobi type method, Springer, Berlin (1971) 327-338.

[9.7] *P.J. Eberlein*; A Jacobi-like method for the automatic computation of eigenvectors of an arbitrary matrix, J. Soc. Indust. Appl. Math. SIAM, 10, March (1962) 74-87.

[9.8] *L. Meirovitch*; Computational methods in structural dynamics (Sijthoff & Noordhoff, Alphen aan den Rijn, 1980).

10 Direct integration methods for the equation of dynamic equilibrium

"The journey of a thousand miles begins
with one step"
Lao-Tsze, Tao-teh-king, Maxim 64 (ca. 550 B.C.)

10.1 Introduction and basic principles

Our main concern in Vols. I to IV has been to establish approximate (but engineeringwise reliable) expressions for the displacements, $\boldsymbol{u}$, or stresses, $\boldsymbol{\sigma}$, at each point $\boldsymbol{x}$ of a body or structure under static loading conditions. Within the displacement method, we adopted for each restricted volume of a body – in fact the finite element – an assumed displacement pattern. If the structure is now subjected to a time-dependent or dynamic loading, the variables will not only encompass the coordinates x, y and z, but also the time t; moreover, the finite elements will not only be identified with specific volumes of the body but will also extend over a specific time interval. A real, technical structure covers only a limited space and can hence be discretised by a limited number of elements. In principle, the same idealisation technique also applies to the time domain. Although the realm of time is unlimited, a real technical problem will necessitate an analysis only within a finite stretch of time. This concept does not change even when we seek so-called stationary solutions, since these are realised in practice after a finite interval of time. Anyhow, the direct integration techniques presently to be discussed are usually applied only to the calculation of the short-term response of a structure. Thus, in technical problems, the discretisation in space and time involves a finite number of four-dimensional elements.

The simultaneous discretisation in space and time, developed, amongst others, by Nickel and Sackman [10.1], Oden [10.2], Fried [10.3] and also by the senior author [10.21, 10.22], is demonstrated in fig. 10.1.1 on the longitudinal motion $u(x,t)$ in a rod placed along the x-axis, whereby the TRIM3 concept is applied to the x and t variables (linear interpolation). There is evidently one serious disadvantage of this procedure: a static calculation for such an example (FLA2-chain) would require five unknown u-values; the dynamic analysis with nt=4 time points increases the number of unknown variables to 20, i.e. by a factor of nt. This is especially unwelcome in three-dimensional problems. For this reason, the field of discretisation in space and time was soon abandoned by scientists.

A further look at fig. 10.1.1 suggests another approach to our problem. If we imagine that a number of t_i =const cuts are placed through the surface, we observe distributions which remind us of the well-known one-dimensional FLA2 concept. When proceeding from the instant t_1 to t_2, only the nodal values change. Thus, $u(x,t)$ can be represented as a product of a spatial *ansatz* in x with pivotal point

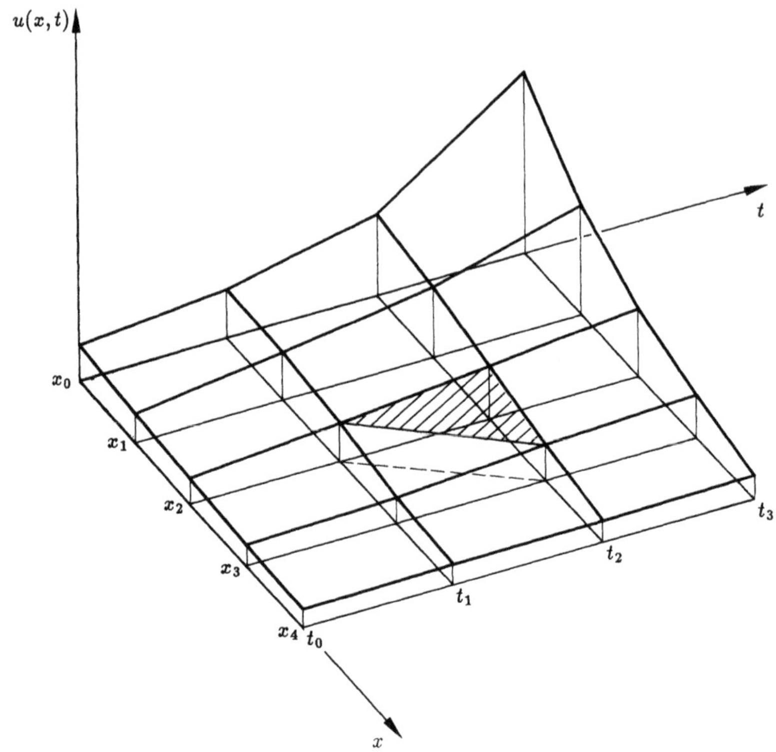

Fig. 10.1.1 Finite elements in space (x) and time (t): the concept of TRIM3 (linear interpolation) applied to the displacement $u(x,t)$

values dependent only upon the time t. This leads us to the generalised product formulation for the displacement function in space and time within the g-th element

$$\begin{aligned} \boldsymbol{u} &= \boldsymbol{\omega}_{\mathrm{g}}(x,y,z)\boldsymbol{\rho}(t) \\ &= \boldsymbol{\omega}_{\mathrm{g}}(x,y,z)\mathbf{a}_{\mathrm{g}}\boldsymbol{r}(t) \qquad \text{in } \mathrm{V_g} \end{aligned} \tag{10.1.1}$$

where $\mathbf{a}_{\mathrm{g}}$ is the well-known indicial Boolean matrix. The mathematical justification of this procedure and the discussion of its limitations is ignored so as to avoid complexities at this initial stage. We only wish to point out that equ. (10.1.1) may be considered a special case of the Kantorovich method and may be denoted as a semi-discretisation. Here, 'semi' refers to the space idealisation for which we adopt the familiar static displacement approximation. On the other hand, the functional dependence on time, i.e. $\boldsymbol{\rho}(t)$ or $\boldsymbol{r}(t)$, still remains open and does not involve a discretisation at this stage. If the product formulation (10.1.1) is now substituted into the partial differential equation of motion as a trial solution, one obtains a system of ordinary differential equations in time (cf. statics: algebraic equations). This formulation can in fact be realised using element modal functions already known to us from statics. Having established this system of ordinary differential equations in time, we can proceed to a new discretisation over a finite time interval.This problem is, however, one-dimensional and decoupled from the spatial idealisation. The partitioning of this time-span into time-elements is generally uniform; we thus select ns time-steps of the length Δt as finite time elements (fig. 10.1.2). In fact, we do not possess in most cases the necessary information for a non-uniform partitioning.

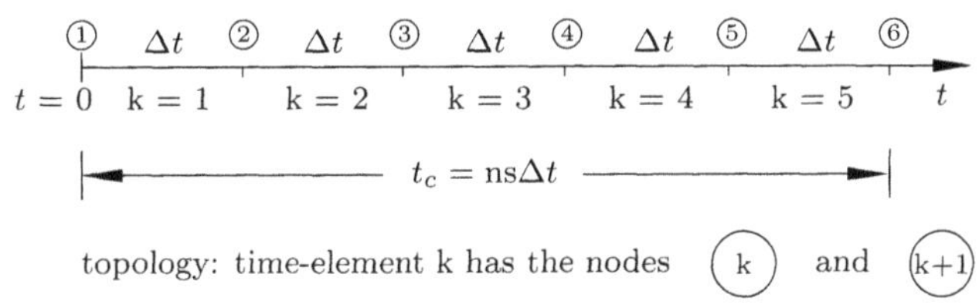

Fig. 10.1.2 Discretisation of the time axis with line elements and topology of the time space

All step-by-step calculations, whether linear or non-linear, are based upon the fundamental equation which already formed the basis of our preceding dynamic calculations. In fact, this entails d'Alembert's principle which expresses the equilibrium of inertial forces $\boldsymbol{R}_I$, damping forces $\boldsymbol{R}_D$ and elastic forces $\boldsymbol{R}_e$ with the external forces $\boldsymbol{R}$ at a certain instant t. We have

$$\boldsymbol{R}_I(t) + \boldsymbol{R}_D(t) + \boldsymbol{R}_e(t) = \boldsymbol{R}(t) \tag{10.1.2}$$

To establish a step-by-step algorithm, we have to deduce an incremental equilibrium relationship from equ. (10.1.2). Thus, equilibrium must not only hold at the beginning of the time-step (t_k) but also at its end ($t_{\mathrm{k}+1} = t_\mathrm{k} + \Delta t$). We note therefore

$$\boldsymbol{R}_I(t_\mathrm{k}) + \boldsymbol{R}_D(t_\mathrm{k}) + \boldsymbol{R}_e(t_\mathrm{k}) = \boldsymbol{R}(t_\mathrm{k}) \tag{10.1.3a}$$

$$\boldsymbol{R}_I(t_\mathrm{k}+\Delta t) + \boldsymbol{R}_D(t_\mathrm{k}+\Delta t) + \boldsymbol{R}_e(t_\mathrm{k}+\Delta t) = \boldsymbol{R}(t_\mathrm{k}+\Delta t) \tag{10.1.3b}$$

and form the difference

$$\Delta\boldsymbol{R}_I(t_\mathrm{k}) + \Delta\boldsymbol{R}_D(t_\mathrm{k}) + \Delta\boldsymbol{R}_e(t_\mathrm{k}) = \Delta\boldsymbol{R}(t_\mathrm{k}) \tag{10.1.4}$$

In this expression, the individual force increments are given by

$$\begin{aligned}
\Delta\boldsymbol{R}_I(t_\mathrm{k}) &= \boldsymbol{R}_I(t_\mathrm{k}+\Delta t) - \boldsymbol{R}_I(t_\mathrm{k})\\
\Delta\boldsymbol{R}_D(t_\mathrm{k}) &= \boldsymbol{R}_D(t_\mathrm{k}+\Delta t) - \boldsymbol{R}_D(t_\mathrm{k})\\
\Delta\boldsymbol{R}_e(t_\mathrm{k}) &= \boldsymbol{R}_e(t_\mathrm{k}+\Delta t) - \boldsymbol{R}_e(t_\mathrm{k})\\
\Delta\boldsymbol{R}(t_\mathrm{k}) &= \boldsymbol{R}(t_\mathrm{k}+\Delta t) - \boldsymbol{R}(t_\mathrm{k})
\end{aligned} \tag{10.1.5}$$

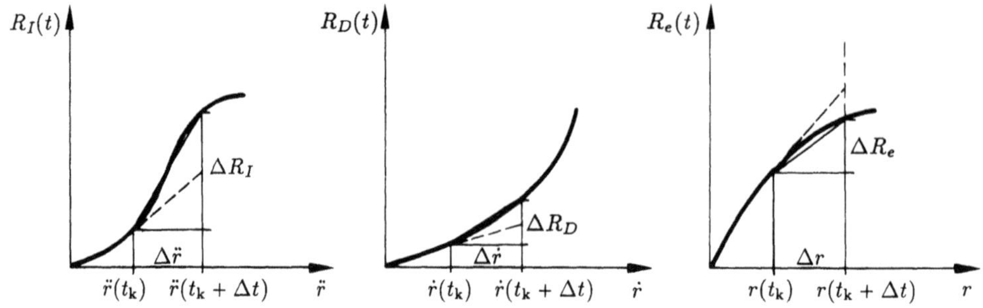

Fig. 10.1.3 Exact reproduction of force increments by a secant relation: the tangential progress (broken line) is a commonly accepted approximation

Only the increment of the external forces $\Delta\boldsymbol{R}(t_k)$ is known initially. In addition, we know that the inertial forces $\boldsymbol{R}_I$ depend solely upon the acceleration $\ddot{\boldsymbol{r}}$ of the system, whilst the elastic forces derive from the displacements $\boldsymbol{r}$. The most difficult point concerns damping. Here, in order to secure some conceptual advance, it is necessary to restrict ourselves to a specific type of damping, for example, to the frequently adopted viscous damping, which is proportional to the velocity. Using this information, the force increments in equ. (10.1.5) are expressed as

$$\begin{aligned}\Delta\boldsymbol{R}_I(t_k) &= \boldsymbol{M}_s(t_k)\Delta\ddot{\boldsymbol{r}}(t_k)\\ \Delta\boldsymbol{R}_D(t_k) &= \boldsymbol{C}_s(t_k)\Delta\dot{\boldsymbol{r}}(t_k)\\ \Delta\boldsymbol{R}_e(t_k) &= \boldsymbol{K}_s(t_k)\Delta\boldsymbol{r}(t_k)\end{aligned} \tag{10.1.6}$$

It should be noted that equ. (10.1.6) is generally also applicable to non-linear problems, in which even the mass may be variable. Furthermore, we should not forget that – providing equ. (10.1.6) reproduces an exact relationship – $\boldsymbol{M}_s(t_k), \boldsymbol{C}_s(t_k)$ and $\boldsymbol{K}_s(t_k)$ represent secantial properties. This is most easily seen to hold for a one-degree-of-freedom system (fig. 10.1.3). To apply a step-by-step procedure, we have to rely on our knowledge of the initial state from which we cannot deduce, however, the secantial characteristics of the relevant matrices. Therefore, the exact relationship (equ. 10.1.6) can be only iteratively approximated. Consequently, we generally adopt the first term of a Taylor expansion

$$\begin{aligned}\Delta\boldsymbol{R}_I(t_k) &\approx \frac{\partial\boldsymbol{R}_I}{\partial\ddot{\boldsymbol{r}}}(t_k)\Delta\ddot{\boldsymbol{r}} = \boldsymbol{M}_t(t_k)\Delta\ddot{\boldsymbol{r}}\\ \Delta\boldsymbol{R}_D(t_k) &\approx \frac{\partial\boldsymbol{R}_D}{\partial\dot{\boldsymbol{r}}}(t_k)\Delta\dot{\boldsymbol{r}} = \boldsymbol{C}_t(t_k)\Delta\dot{\boldsymbol{r}}\\ \Delta\boldsymbol{R}_e(t_k) &\approx \frac{\partial\boldsymbol{R}_e}{\partial\boldsymbol{r}}(t_k)\Delta\boldsymbol{r} = \boldsymbol{K}_t(t_k)\Delta\boldsymbol{r}\end{aligned} \tag{10.1.7}$$

Here, the mass matrix $\boldsymbol{M}_t(t_k)$, the damping matrix $\boldsymbol{C}_t(t_k)$ and the stiffness matrix $\boldsymbol{K}_t(t_k)$ now represent the tangential characteristics at the beginning of the time-step. With this approximation, we obtain an incremental equilibrium relationship which does not differ formally from the fundamental equation of linear dynamics. It is valid even if all possible non-linearities are allowed for

$$\boldsymbol{M}_t(t_\mathrm{k})\Delta\ddot{\boldsymbol{r}} + \boldsymbol{C}_t(t_\mathrm{k})\Delta\dot{\boldsymbol{r}} + \boldsymbol{K}_t(t_\mathrm{k})\Delta\boldsymbol{r} = \Delta\boldsymbol{R}(t_\mathrm{k}) \tag{10.1.8}$$

The index t is omitted for linear dynamic problems in which case $\boldsymbol{M}$, $\boldsymbol{C}$ and $\boldsymbol{K}$ are reduced to constant matrices.

All significant step-by-step calculation procedures approximate the acceleration $\ddot{\boldsymbol{r}}$ within a time-step. Having this information, we deduce the velocity $\dot{\boldsymbol{r}}$ and subsequently the deflection $\boldsymbol{r}$ applying a temporal integration. For this purpose, it is advantageous to introduce the dimensionless time τ

$$\tau = \frac{t - t_\mathrm{k}}{\Delta t} \tag{10.1.9}$$

which forms the instantaneous variable in our time element. Assuming a knowledge of the acceleration $\ddot{\boldsymbol{r}}(\tau)$ within the time-step $(0 \leqslant \tau \leqslant 1)$, we obtain by temporal integration the velocity $\dot{\boldsymbol{r}}$ at an instant τ within the time-step k

$$\dot{\boldsymbol{r}}(\tau) = \dot{\boldsymbol{r}}_\mathrm{k} + \Delta t \int_0^\tau \ddot{\boldsymbol{r}}(\tau)\mathrm{d}\tau \,, \qquad 0 \leqslant \tau \leqslant 1 \tag{10.1.10}$$

where the velocity at the beginning of the time-step is assumed to be given

$$\dot{\boldsymbol{r}}_\mathrm{k} = \dot{\boldsymbol{r}}(t = t_\mathrm{k}) = \dot{\boldsymbol{r}}(\tau = 0) \tag{10.1.11}$$

Further integration of equ. (10.1.10) yields the displacement histogram

$$\boldsymbol{r}(\tau) = \boldsymbol{r}_\mathrm{k} + \tau\Delta t\,\dot{\boldsymbol{r}}_\mathrm{k} + (\Delta t)^2 \int_0^\tau\!\!\int_0^\tau \ddot{\boldsymbol{r}}(\tau)\mathrm{d}\tau\mathrm{d}\tau \,, \qquad 0 \leqslant \tau \leqslant 1 \tag{10.1.12}$$

Here again, the deflection at the beginning of the step is presumed known

$$\boldsymbol{r}_\mathrm{k} = \boldsymbol{r}(t = t_\mathrm{k}) = \boldsymbol{r}(\tau = 0) \tag{10.1.13}$$

Those readers who find double temporal integration in equ. (10.1.12) too cumbersome can apply a transformation to the corresponding integral by means of an inverse partial integration and in this way, obtain a formulation which serves *inter alia* as a basis to the Newmark method [10.5]. Thus, we write

$$\underbrace{\int_0^\tau \left(\int_0^\tau \ddot{\boldsymbol{r}}(\tau)\mathrm{d}\tau \right) \mathrm{d}\tau} = \tau \int_0^\tau \ddot{\boldsymbol{r}}(\tau)\mathrm{d}\tau - \int_0^\tau \tau\ddot{\boldsymbol{r}}(\tau)\mathrm{d}\tau$$

which reproduces the classic operation

$$\int_a^b v \quad \cdot \mathrm{d}u = [u \cdot v]_a^b \quad - \int_a^b u\mathrm{d}v \tag{10.1.14}$$

Applying this transformation and noting that the increments of displacement and velocity can easily be derived from equs. (10.1.10) and (10.1.12) for the time interval $\tau = 1$, we obtain

$$\Delta \boldsymbol{r} = \Delta t \dot{\boldsymbol{r}}_{\mathrm{k}} + (\Delta t)^2 \int_0^1 (1-\tau)\, \ddot{\boldsymbol{r}}(\tau)\, \mathrm{d}\tau \tag{10.1.15}$$

and

$$\Delta \dot{\boldsymbol{r}} = \Delta t \int_0^1 \ddot{\boldsymbol{r}}(\tau)\, \mathrm{d}\tau \tag{10.1.16}$$

Thus, once the acceleration response $\ddot{\boldsymbol{r}}(\tau)$ is available, the corresponding displacement and velocity increments can easily be calculated. For a given initial information on $\boldsymbol{r}_{\mathrm{k}}$ and $\dot{\boldsymbol{r}}_{\mathrm{k}}$, we establish in this way the input data for the next step, $\boldsymbol{r}_{\mathrm{k}+1}$ and $\dot{\boldsymbol{r}}_{\mathrm{k}+1}$. The formulation of a recurrence formula hence suggests itself.

Within a time-step or a finite time element, we now introduce an *ansatz* for the time-dependence of the acceleration, analogous to the discretisation in space

$$\ddot{\boldsymbol{r}}(\tau) = \boldsymbol{\omega}_A(\tau) \boldsymbol{\rho}_A \tag{10.1.17}$$

Here, $\boldsymbol{\omega}_A(\tau)$ defines the assumed variation of the acceleration within the time element and $\boldsymbol{\rho}_A$ contains the corresponding values at the interpolation sampling points. Substitution of equ. (10.1.17) in equs. (10.1.15), (10.1.16) yields at the end of the time-step k

$$\Delta \boldsymbol{r} = \boldsymbol{r}_{\mathrm{k}+1} - \boldsymbol{r}_{\mathrm{k}} = \Delta t \dot{\boldsymbol{r}}_{\mathrm{k}} + (\Delta t)^2 \Big(\int_0^1 (1-\tau)\, \boldsymbol{\omega}_A(\tau)\, \mathrm{d}\tau \Big) \boldsymbol{\rho}_A \tag{10.1.18}$$

for the displacement and

$$\Delta \dot{\boldsymbol{r}} = \dot{\boldsymbol{r}}_{\mathrm{k}+1} - \dot{\boldsymbol{r}}_{\mathrm{k}} = \Delta t \Big(\int_0^1 \boldsymbol{\omega}_A(\tau)\, \mathrm{d}\tau \Big) \boldsymbol{\rho}_A \tag{10.1.19}$$

for the velocity increment. Since knowledge of $\boldsymbol{\omega}_A$ is pre-supposed and is taken to be the same for all time elements, integration can be applied *a priori*, and need be executed only once for each specific type of time element. We denote these (in general rectangular) influence matrices by

$$\boldsymbol{C}_{DA} = \int_0^1 (1-\tau)\, \boldsymbol{\omega}_A(\tau)\, \mathrm{d}\tau \tag{10.1.20}$$

(where 'D' stands for *displacement*) and

$$C_{VA} = \int_0^1 \omega_A(\tau)\,d\tau \tag{10.1.21}$$

(where 'V' stands for *velocity*).

Equations (10.1.18) and (10.1.19) do not suffice for the calculation of the increments Δr and $\Delta \dot{r}$ since ρ_A can still contain unknown values. For example, in the family of Newmark methods, the vector ρ_A takes the form

$$\rho_A = \{\ddot{r}_k \quad \ddot{r}_{k+1}\} = \{\ddot{r}_k \quad (\ddot{r}_k + \Delta\ddot{r})\} \tag{10.1.22}$$

in which $\ddot{r}_{k+1}$ respectively $\Delta\ddot{r}$ are unknown. Hence, to specify ρ_A as defined in equ. (10.1.22), we require a set of n additional equations in order to determine the unknown values r_{k+1} and $\dot{r}_{k+1}$. For instance, we may use the incremental equilibrium expression given by equ. (10.1.8). This would put us in a position to establish for the calculation of increments Δr and $\Delta\dot{r}$ a linear algebraic system of the type

$$D_1 \begin{bmatrix} \Delta r \\ \Delta \dot{r} \end{bmatrix} = b \tag{10.1.23}$$

where b contains only known information. Unfortunately, this calculation of increments generally involves the solution of a system of equations

$$\begin{bmatrix} \Delta r \\ \Delta \dot{r} \end{bmatrix} = D_1^{-1} b \tag{10.1.24}$$

All procedures in which this inversion cannot be avoided are termed implicit. The explicit algorithms, on the other hand, are characterised by the following direct representation

$$\begin{bmatrix} \Delta r \\ \Delta \dot{r} \end{bmatrix} = b \tag{10.1.25}$$

which circumvents the inversion. Obviously, explicit procedures yield a very fast solution in each step. Newmark's method with $\beta = 0$, $\gamma = 1/2$ [10.5], which is to be discussed in more detail in the next section, is a simple example of this technique. In this connection, we shall be informed, however, that the explicitness is tied to certain initial pre-conditions which can be realised only for simple approximation models. This reflection also shows that explicit processes are certainly fast and cheap for each step, but require small time-steps to ensure numerical stability and accuracy. The advantage achieved within each step may thus be ultimately lost. In addition to the Newmark process mentioned above, a number of further explicit

explicit methods	implicit methods
$$\begin{bmatrix} \boldsymbol{r} \\ \dot{\boldsymbol{r}} \end{bmatrix}_{k+1} = \boldsymbol{A} \begin{bmatrix} \boldsymbol{r} \\ \dot{\boldsymbol{r}} \end{bmatrix}_{k} + \boldsymbol{L}$$	$$\boldsymbol{D}_1 \begin{bmatrix} \boldsymbol{r} \\ \dot{\boldsymbol{r}} \end{bmatrix}_{k+1} = \boldsymbol{D}_0 \begin{bmatrix} \boldsymbol{r} \\ \dot{\boldsymbol{r}} \end{bmatrix}_{k} + \boldsymbol{F}$$
$\boldsymbol{A}$,$\boldsymbol{L}$ determinable without inversion	$\left.\begin{array}{l} \boldsymbol{A} = \boldsymbol{D}_1^{-1}\boldsymbol{D}_0 \\ \boldsymbol{L} = \boldsymbol{D}_1^{-1}\boldsymbol{F} \end{array}\right\}$ solution of equations required
fast operation within time step; small time steps required	slow operation within time step; numerically stable, as a rule
example: Newmark, $\beta = 0 \quad \gamma = \frac{1}{2}$	example: Newmark, average acceleration $\beta = \frac{1}{4} \quad \gamma = \frac{1}{2}$

type of interpolation		
unfixed	Lagrangian polynomials	Hermitian polynomials
Newmark $\beta = 0$ $\gamma = \frac{1}{2}$ explicit [10.5]	linear 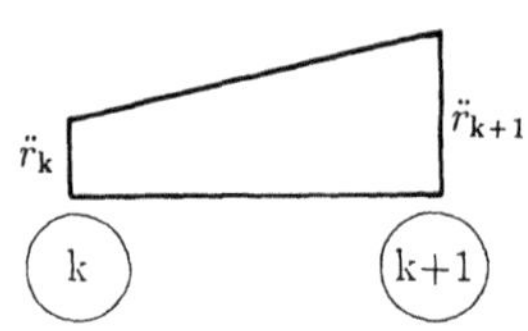 (FLA2) implicit	average acceleration 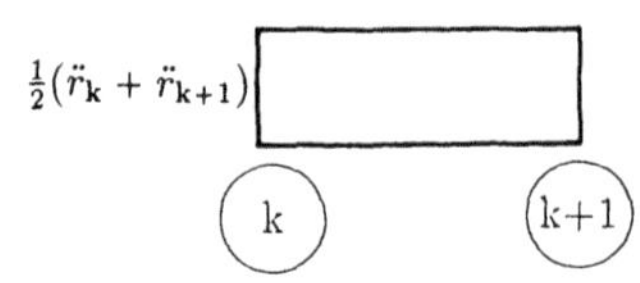 Newmark implicit [10.5]
	quadratic 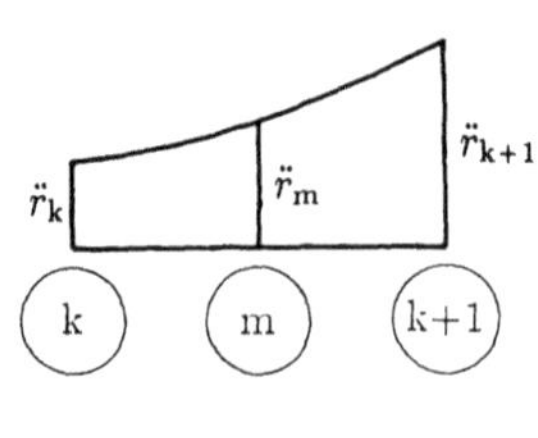 (FLA3) implicit Zienkiewicz [10.9]	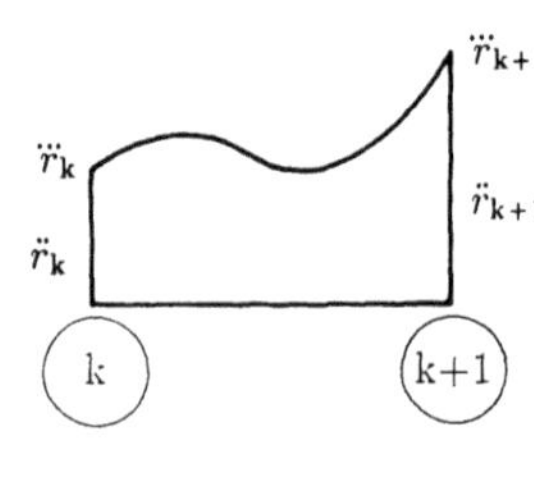 (planar beam) implicit Argyris *et al.* [10.4]

Fig. 10.1.4 Integration procedures classified according to the adopted step solution and the interpolation scheme

procedures exist, such as the method of central differences [10.6], Fu's algorithm [10.7] and Trujillo's method [10.8].

In contrast, implicit processes can be set up for all possible interpolation models and though they may be costly in each step, they are in return numerically stable. Simple examples are Newmark's method of average accelerations [10.5], the method of linear accelerations, the quadratic acceleration model [10.9] and the Hermitian algorithms (cubic or higher) of the senior author *et al.* [10.4] and [10.13].

As we shall observe, however, the specification of the interpolation function $\boldsymbol{\omega}_A(\tau)$ is by no means always necessary. In such cases where this takes place, however, we may adopt another classification of the procedures, which depends upon the type of the temporal interpolation. We are already well familiar with Lagrangian polynomials as applied to the discretisation with respect to space. Thus, we can extract the linear acceleration scheme from the flange element FLA2, whilst FLA3 may serve as the prototype of a quadratic acceleration model. On the other hand, we are also familiar with the Hermitian polynomials from the beam and plate models and these also yield their own element class in the time domain. For example, the plane beam contains a cubic interpolation model. Figure 10.1.4 summarises the possible classification of integration methods. A further grouping of the procedures will be possible following the discussion on stability in section 10.3.

10.2 Elaboration of the simplest procedures

The family of Newmark algorithms

The algorithms of the Newmark family employ as pivotal entries

$$\boldsymbol{\rho}_A = \{\ddot{\boldsymbol{r}}_{\mathrm{k}} \quad \ddot{\boldsymbol{r}}_{\mathrm{k+1}}\} = \{\ddot{\boldsymbol{r}}_{\mathrm{k}} \quad (\ddot{\boldsymbol{r}}_{\mathrm{k}} + \Delta\ddot{\boldsymbol{r}})\} \tag{10.2.1}$$

Without prior specification of the interpolation function, we may express $\boldsymbol{\omega}_A$ in the form

$$\boldsymbol{\omega}_A(\tau) = \left[\omega_1(\tau)\boldsymbol{I}_{\mathrm{n}} \quad \omega_2(\tau)\boldsymbol{I}_{\mathrm{n}}\right] \tag{10.2.2}$$

Analogous to the discretisation procedure in space, we must ensure in the temporal idealisation that a constant acceleration behaviour can be represented by equ. (10.2.2). This is indeed a prerequisite for an assumed infinitesimally small time-step. Obviously, this condition is fulfilled only if

$$\omega_1(\tau) + \omega_2(\tau) = 1 \qquad 0 \leqslant \tau \leqslant 1 \tag{10.2.3}$$

Using equ. (10.2.3), we can also write in place of equ. (10.2.2)

$$\boldsymbol{\omega}_A(\tau) = \left[\omega_1(\tau)\boldsymbol{I}_{\mathrm{n}} \quad (1 - \omega_1(\tau))\boldsymbol{I}_{\mathrm{n}}\right] \tag{10.2.4}$$

where $\omega_1(\tau)$ is a scalar function of time which must fulfil the following conditions

$$\omega_1(0) = 1 \quad , \qquad \omega_1(1) = 0 \tag{10.2.5}$$

for a continuous acceleration variation along the time axis. Applying next equ. (10.2.4) for the determination of the influence matrices $\boldsymbol{C}_{DA}$ and $\boldsymbol{C}_{VA}$, we obtain the two integrals

$$\int_0^1 (1-\tau)\,\omega_1(\tau)\,\mathrm{d}\tau = \tfrac{1}{2} - \beta \tag{10.2.6}$$

and

$$\int_0^1 \omega_1(\tau)\,\mathrm{d}\tau \qquad = 1 - \gamma \tag{10.2.7}$$

The definitions of the constants β and γ correspond to those of Newmark [10.5]. In this way, we find

$$\boldsymbol{C}_{DA} = \left[\left(\tfrac{1}{2} - \beta\right) \boldsymbol{I}_{\mathrm{n}} \quad \beta \boldsymbol{I}_{\mathrm{n}} \right] \tag{10.2.8}$$

and

$$\boldsymbol{C}_{VA} = \left[(1-\gamma)\boldsymbol{I}_{\mathrm{n}} \quad \gamma \boldsymbol{I}_{\mathrm{n}} \right] \tag{10.2.9}$$

Hence, the incremental expressions for the displacement and velocity vectors become

$$\begin{aligned} \Delta\boldsymbol{r} &= \Delta t \dot{\boldsymbol{r}}_{\mathrm{k}} + (\Delta t)^2 (\tfrac{1}{2} - \beta) \ddot{\boldsymbol{r}}_{\mathrm{k}} + (\Delta t)^2 \beta \ddot{\boldsymbol{r}}_{\mathrm{k}+1} \\ &= \Delta t \dot{\boldsymbol{r}}_{\mathrm{k}} + \tfrac{1}{2} (\Delta t)^2 \ddot{\boldsymbol{r}}_{\mathrm{k}} + (\Delta t)^2 \beta \Delta \ddot{\boldsymbol{r}} \end{aligned} \tag{10.2.10}$$

and

$$\begin{aligned} \Delta\dot{\boldsymbol{r}} &= \Delta t (1-\gamma) \ddot{\boldsymbol{r}}_{\mathrm{k}} + \Delta t \gamma \ddot{\boldsymbol{r}}_{\mathrm{k}+1} \\ &= \Delta t \ddot{\boldsymbol{r}}_{\mathrm{k}} + \Delta t \gamma \Delta \ddot{\boldsymbol{r}} \end{aligned} \tag{10.2.11}$$

To evaluate these increments, we require the acceleration increment $\Delta\ddot{\boldsymbol{r}}$ respectively the total acceleration $\ddot{\boldsymbol{r}}_{\mathrm{k}+1}$ at the end of the step. It should also be noted that for non-linear applications, the condition of incremental equilibrium represents only an approximate relationship if tangential stiffness, damping and mass matrices are employed. In this case, the expression

$$\Delta\ddot{\boldsymbol{r}} = \boldsymbol{M}_t^{-1} [\Delta \boldsymbol{R}(t_{\mathrm{k}}) - \boldsymbol{C}_t(t_{\mathrm{k}}) \Delta \dot{\boldsymbol{r}} - \boldsymbol{K}_t(t_{\mathrm{k}}) \Delta \boldsymbol{r}] \tag{10.2.12}$$

may lead to an accumulation of errors due to the tangential approximation. To circumvent this, we require the secant expressions. If the mass matrix is constant, the total acceleration can be calculated from the overall equilibrium at the time $(t_k + \Delta t)$

$$\ddot{\boldsymbol{r}}_{k+1} = \boldsymbol{M}^{-1}\left[\boldsymbol{R}(t_{k+1}) - \boldsymbol{R}_D(t_{k+1}) - \boldsymbol{R}_e(t_{k+1})\right] \tag{10.2.13}$$

This avoids an accumulation of errors. For linear dynamic problems, which are our main concern here, we at once obtain

$$\ddot{\boldsymbol{r}}_{k+1} = \boldsymbol{M}^{-1}\left[\boldsymbol{R}(t_{k+1}) - \boldsymbol{C}\dot{\boldsymbol{r}}_{k+1} - \boldsymbol{K}\boldsymbol{r}_{k+1}\right] \tag{10.2.14}$$

and the incremental relationship of equ. (10.2.12) holds exactly. We next substitute equ. (10.2.12) into equs. (10.2.10) and (10.2.11) and find

$$\begin{bmatrix} \boldsymbol{M}_t + \Delta t^2\beta\boldsymbol{K}_t & \Delta t^2\beta\boldsymbol{C}_t \\ \Delta t\gamma\boldsymbol{K}_t & \boldsymbol{M}_t + \Delta t\gamma\boldsymbol{C}_t \end{bmatrix}\begin{bmatrix} \Delta\boldsymbol{r} \\ \Delta\dot{\boldsymbol{r}} \end{bmatrix} = \begin{bmatrix} \Delta t\boldsymbol{M}_t\dot{\boldsymbol{r}}_k + \frac{1}{2}\Delta t^2\boldsymbol{M}_t\ddot{\boldsymbol{r}}_k + \Delta t^2\beta\Delta\boldsymbol{R} \\ \Delta t\boldsymbol{M}_t\ddot{\boldsymbol{r}}_k + \Delta t\gamma\Delta\boldsymbol{R} \end{bmatrix} \tag{10.2.15}$$

The hypermatrix on the left-hand side of this expression is identical to $\boldsymbol{D}_1$ in equ. (10.1.23). It is obvious that the procedure is implicit for arbitrary values of β and γ. Thus, a system of order (2n × 2n) with an unsymmetric coefficient matrix $\boldsymbol{D}_1$ must be solved at each step. It should be noted that the matrices $\boldsymbol{M}_t$, $\boldsymbol{C}_t$, $\boldsymbol{K}_t$ vary from step to step when the problem is non-linear. The procedure is clearly highly costly in such a case.

An iterative solution within each step offers a possible alternative. The rules for such an iteration are very simple and are based upon the three fundamental equs. (10.2.10), (10.2.11) and (10.2.12)

initial values: $\boldsymbol{r}_k,\ \dot{\boldsymbol{r}}_k,\ \ddot{\boldsymbol{r}}_k,\ \Delta\ddot{\boldsymbol{r}}^{(0)} = \boldsymbol{o}$

$$\mathrm{i} = 0, \mathrm{ni}\quad \begin{aligned} \Delta\boldsymbol{r}^{(i+1)} &= \Delta t\dot{\boldsymbol{r}}_k + \tfrac{1}{2}\Delta t^2\ddot{\boldsymbol{r}}_k + \Delta t^2\beta\Delta\ddot{\boldsymbol{r}}^{(i)} \\ \Delta\dot{\boldsymbol{r}}^{(i+1)} &= \Delta t\ddot{\boldsymbol{r}}_k + \Delta t\gamma\Delta\ddot{\boldsymbol{r}}^{(i)} \\ \Delta\ddot{\boldsymbol{r}}^{(i+1)} &= \boldsymbol{M}_t^{-1}\left[\Delta\boldsymbol{R} - \boldsymbol{C}_t\Delta\dot{\boldsymbol{r}}^{(i+1)} - \boldsymbol{K}_t\Delta\boldsymbol{r}^{(i+1)}\right] \end{aligned} \tag{10.2.16}$$

In our experience, this procedure has the advantage of converging very rapidly and is hence preferable to the direct solution of equ. (10.2.15). Moreover, the programming of equ. (10.2.16) is straightforward whilst equ. (10.2.15) requires a special solution procedure for large unsymmetric matrices. A simple realisation of equ. (10.2.16) presumes $\boldsymbol{M}_t$ to be diagonal. Even when this is not the case, there are, however, a large number of non-linear problems for which the mass matrix may be assumed constant. For such non-diagonal mass matrices, Cholesky decomposition need only be carried out once in the step-by-step procedure.

For the special case when $\beta = 0$ and $\boldsymbol{M}_t$, $\boldsymbol{C}_t$ are diagonal, we obtain an explicit procedure to be discussed subsequently with specific reference to linear problems. A suitable transformation of equ. (10.2.15) yields

$$\begin{bmatrix} \Delta \boldsymbol{r} \\ \Delta \dot{\boldsymbol{r}} \end{bmatrix} = \begin{bmatrix} \Delta t \dot{\boldsymbol{r}}_{\mathrm{k}} + \frac{1}{2}\Delta t^2 \ddot{\boldsymbol{r}}_{\mathrm{k}} \\ [\boldsymbol{M}_t + \Delta t \gamma \boldsymbol{C}_t]^{-1}[-\Delta t^2 \gamma \boldsymbol{K}_t \dot{\boldsymbol{r}}_{\mathrm{k}} + (\Delta t \boldsymbol{M}_t - \Delta t^3 \gamma \boldsymbol{K}_t)\ddot{\boldsymbol{r}}_{\mathrm{k}} + \Delta t \gamma \Delta \boldsymbol{R}] \end{bmatrix} \tag{10.2.17}$$

For linear problems, the total acceleration $\ddot{\boldsymbol{r}}_{\mathrm{k}}$ on the right-hand side of equ. (10.2.15) can be eliminated using the overall equilibrium condition at time t_{k}. In this way, we deduce the standard formulation

$$\boldsymbol{D}_1 \begin{bmatrix} \boldsymbol{r} \\ \dot{\boldsymbol{r}} \end{bmatrix}_{\mathrm{k}+1} = \boldsymbol{D}_0 \begin{bmatrix} \boldsymbol{r} \\ \dot{\boldsymbol{r}} \end{bmatrix}_{\mathrm{k}} + \boldsymbol{F} \tag{10.2.18}$$

By inversion of $\boldsymbol{D}_1$ in the case of implicit methods and directly in explicit variants, the alternative expression

$$\begin{bmatrix} \boldsymbol{r} \\ \dot{\boldsymbol{r}} \end{bmatrix}_{\mathrm{k}+1} = \boldsymbol{A} \begin{bmatrix} \boldsymbol{r} \\ \dot{\boldsymbol{r}} \end{bmatrix}_{\mathrm{k}} + \boldsymbol{L} \tag{10.2.19}$$

is obtained which is important for our subsequent stability analysis. For implicit procedures, we find

$$\boldsymbol{A} = \boldsymbol{D}_1^{-1} \boldsymbol{D}_0 \tag{10.2.20}$$

$$\boldsymbol{L} = \boldsymbol{D}_1^{-1} \boldsymbol{F} \tag{10.2.21}$$

We now substitute

$$\boldsymbol{M} \ddot{\boldsymbol{r}}_{\mathrm{k}} = \boldsymbol{R}_{\mathrm{k}} - \boldsymbol{C} \dot{\boldsymbol{r}}_{\mathrm{k}} - \boldsymbol{K} \boldsymbol{r}_{\mathrm{k}} \tag{10.2.22}$$

into equ. (10.2.15) and deduce directly from the resulting equation the following expression valid for the Newmark family in linear problems

$$\boldsymbol{D}_1 = \begin{bmatrix} \boldsymbol{M} + \Delta t^2 \beta \boldsymbol{K} & \Delta t^2 \beta \boldsymbol{C} \\ \Delta t \gamma \boldsymbol{K} & \boldsymbol{M} + \Delta t \gamma \boldsymbol{C} \end{bmatrix} \tag{10.2.23}$$

as also the matrices

$$\boldsymbol{D}_0 = \begin{bmatrix} \boldsymbol{M} - (\frac{1}{2} - \beta)\Delta t^2 \boldsymbol{K} & \Delta t \boldsymbol{M} - (\frac{1}{2} - \beta)\Delta t^2 \boldsymbol{C} \\ -\Delta t(1 - \gamma)\boldsymbol{K} & \boldsymbol{M} - (1 - \gamma)\Delta t \boldsymbol{C} \end{bmatrix} \tag{10.2.24}$$

and

$$\boldsymbol{F} = \begin{bmatrix} \Delta t^2\big[(\tfrac{1}{2} - \beta)\boldsymbol{R}_\mathrm{k} + \beta\boldsymbol{R}_{\mathrm{k+1}}\big] \\ \Delta t\big[(\tfrac{1}{2} - \gamma)\boldsymbol{R}_\mathrm{k} + \gamma\boldsymbol{R}_{\mathrm{k+1}}\big] \end{bmatrix} \tag{10.2.25}$$

In what follows, we discuss a few special cases. Let us begin with

Newmark $\beta = 0$, $\gamma = \frac{1}{2}$;
Method of central differences, explicit for diagonal $\boldsymbol{M}$ and $\boldsymbol{C}$

We have already seen that $\beta = 0$ yields an explicit procedure on certain preconditions. In fact, we prove in the following section that γ has to be at least 1/2, in order to ensure a conditionally stable process.

It would, of course, be ideal if such reflections were to provide the two values β and γ. On the other hand, however, these values depend on the modal form $\omega_1(\tau)$ as shown in equs. (10.2.6) and (10.2.7). Such considerations are circumvented here by prescribing the integral values in equs. (10.2.6) and (10.2.7) instead of $\omega_1(\tau)$ itself. The procedure is permissible if we do not seek any information within the time-step. To understand the technique, we first illustrate the significance of the two factors β and γ and hence draw conclusions on the underlying modal shape. From equ. (10.2.4), we deduce in conjunction with equ. (10.2.1) the acceleration

$$\ddot{\boldsymbol{r}}(\tau) = \ddot{\boldsymbol{r}}_\mathrm{k} + (1 - \omega_1(\tau))\Delta\ddot{\boldsymbol{r}} \tag{10.2.26}$$

The functional variation consists of a constant term $\ddot{\boldsymbol{r}}_\mathrm{k}$ and a time-dependent variation (fig. 10.2.1). A first integration of this equation yields the velocity and if the integration is extended over the full time-step, we obtain once more equ. (10.2.11)

$$\begin{aligned} \Delta\dot{\boldsymbol{r}} &= \Delta t\int_0^1 \ddot{\boldsymbol{r}}(\tau)\mathrm{d}\tau \\ &= \Delta t\,\ddot{\boldsymbol{r}}_\mathrm{k} + \Delta t\Big(\int_0^1 (1 - \omega_1(\tau))\,\mathrm{d}\tau\Big)\Delta\ddot{\boldsymbol{r}} \\ &= \Delta t\,\ddot{\boldsymbol{r}}_\mathrm{k} + \Delta t\,\gamma\Delta\ddot{\boldsymbol{r}} \end{aligned} \tag{10.2.27}$$

But for a factor Δt, the first term in equ. (10.2.26) represents the area beneath the dominant acceleration term $\ddot{\boldsymbol{r}}_\mathrm{k}$= const, whilst the second term defines the additional area which is due to the variation of the acceleration. This additional area can be approximated by $\gamma = 1/2$ (triangle) to a first order and this is independent of the special form of $\ddot{\boldsymbol{r}}\,(\tau)$.

Proceeding to equ. (10.1.15) and integrating over $(1-\tau)\ddot{\boldsymbol{r}}(\tau)$, we obtain the displacement increment and thereby confirm equ. (10.2.10)

$$\begin{aligned}\Delta\boldsymbol{r} &= \Delta t\dot{\boldsymbol{r}}_{\mathrm{k}} + \Delta t^2\int_0^1(1-\tau)\ddot{\boldsymbol{r}}(\tau)\mathrm{d}\tau \\ &= \Delta t\dot{\boldsymbol{r}}_{\mathrm{k}} + \tfrac{1}{2}\Delta t^2\ddot{\boldsymbol{r}}_{\mathrm{k}} + \Delta t^2\beta\Delta\ddot{\boldsymbol{r}} \end{aligned} \tag{10.2.28}$$

But for the factor Δt^2, the second term represents the area below the curve $(1-\tau)\ddot{\boldsymbol{r}}_{\mathrm{k}}$, whilst the third term $\beta\Delta\ddot{\boldsymbol{r}}$ represents the incremental area between $(1-\tau)\ddot{\boldsymbol{r}}(\tau)$ and the triangular area $\frac{1}{2}\ddot{\boldsymbol{r}}_{\mathrm{k}}$. To a first approximation, this additional area is zero from which we deduce $\beta=0$ as a useful estimate. To satisfy the condition $\beta=0$ exactly, we must assume in addition a variation of the acceleration as shown in fig. 10.2.1(c). The course of this curve leads, amongst other things, to the values $\beta=0$ and $\gamma=1/2$ and explains the origin of the Newmark algorithm with $\beta=0, \gamma=1/2$.

If we presume now the known initial conditions $\boldsymbol{r}_{\mathrm{k}}$ and $\dot{\boldsymbol{r}}_{\mathrm{k}}$ at the beginning of the step, we first determine the corresponding acceleration from

$$\ddot{\boldsymbol{r}}_{\mathrm{k}} = \boldsymbol{M}^{-1}[\boldsymbol{R}_{\mathrm{k}} - \boldsymbol{C}\dot{\boldsymbol{r}}_{\mathrm{k}} - \boldsymbol{K}\boldsymbol{r}_{\mathrm{k}}] \tag{10.2.29}$$

and then, using equ. (10.2.10), find the following expression for the displacement at the end of the step

$$\boldsymbol{r}_{\mathrm{k+1}} = \boldsymbol{r}_{\mathrm{k}} + \Delta t\dot{\boldsymbol{r}}_{\mathrm{k}} + \tfrac{1}{2}(\Delta t)^2\ddot{\boldsymbol{r}}_{\mathrm{k}} \tag{10.2.30}$$

Since the acceleration increment is required in order to calculate the velocity at the end of the step in accordance with equ. (10.2.11), it is also useful to consider the incremental equilibrium of equ. (10.2.12) and combine both equations into a single expression. Thus, we find

$$\dot{\boldsymbol{r}}_{\mathrm{k+1}} = \dot{\boldsymbol{r}}_{\mathrm{k}} + [\boldsymbol{M} + \Delta t\gamma\boldsymbol{C}]^{-1}\Delta t\big[\boldsymbol{M}\ddot{\boldsymbol{r}}_{\mathrm{k}} + \gamma(\boldsymbol{R}_{\mathrm{k+1}} - \boldsymbol{R}_{\mathrm{k}}) + \gamma\boldsymbol{K}(\boldsymbol{r}_{\mathrm{k+1}} - \boldsymbol{r}_{\mathrm{k}})\big] \tag{10.2.31}$$

Naturally, the algorithm can also be arranged differently and the following computational scheme applied [10.7]

$$\begin{aligned}\boldsymbol{r}_{\mathrm{k+1}} &= \boldsymbol{r}_{\mathrm{k}} + \Delta t\dot{\boldsymbol{r}}_{\mathrm{k}} + \tfrac{1}{2}(\Delta t)^2\ddot{\boldsymbol{r}}_{\mathrm{k}} \\ \ddot{\boldsymbol{r}}_{\mathrm{k+1}} &= [\boldsymbol{M} + \Delta t\gamma\boldsymbol{C}]^{-1}\big[\boldsymbol{R}_{\mathrm{k+1}} - \boldsymbol{C}\dot{\boldsymbol{r}}_{\mathrm{k}} - \Delta t(1-\gamma)\boldsymbol{C}\ddot{\boldsymbol{r}}_{\mathrm{k}} - \boldsymbol{K}\boldsymbol{r}_{\mathrm{k+1}}\big] \\ \dot{\boldsymbol{r}}_{\mathrm{k+1}} &= \dot{\boldsymbol{r}}_{\mathrm{k}} + \Delta t\left[(1-\gamma)\ddot{\boldsymbol{r}}_{\mathrm{k}} + \gamma\ddot{\boldsymbol{r}}_{\mathrm{k+1}}\right]\end{aligned} \tag{10.2.32}$$

Newmark $\beta=\frac{1}{4}, \gamma=\frac{1}{2}$; average accelerations, implicit

Here, the acceleration within a time element is assumed to be constant. Then its amplitude is given by the average of the nodal point accelerations. Compatability in accordance with equ. (10.2.5) is no longer possible and we thus obtain

$$\omega_1(\tau) = \tfrac{1}{2} \tag{10.2.33}$$

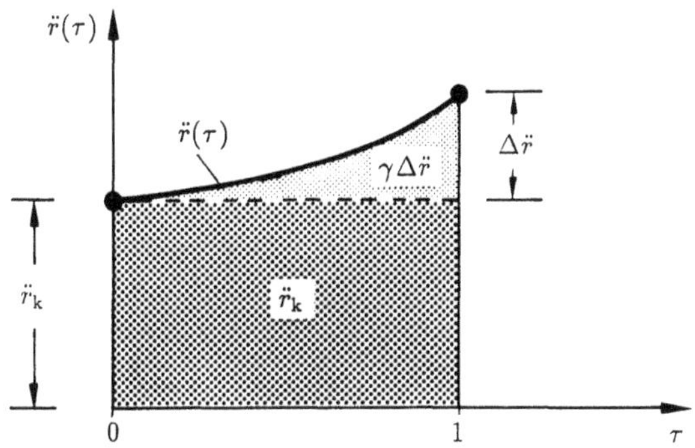

(a) Significance of the integration factor γ (Newmark)

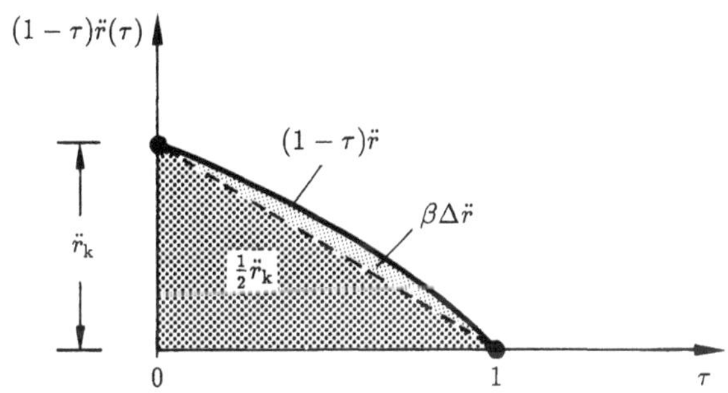

(b) Significance of the integration factor β (Newmark)

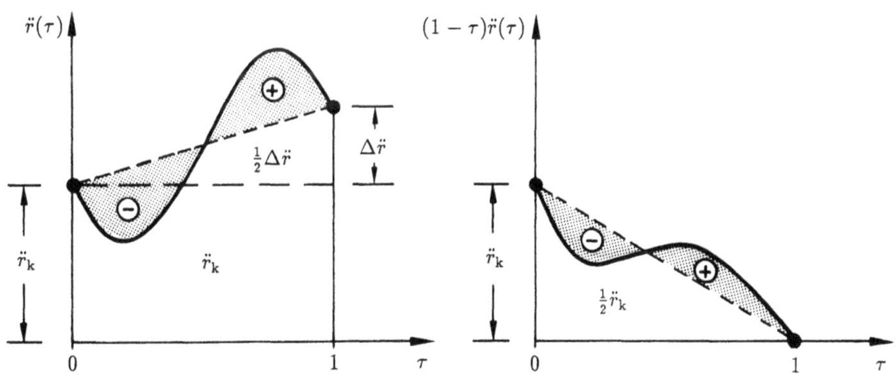

(c) Approximated course of $\ddot{r}(\tau)$, when $\beta = 0$, $\gamma = \frac{1}{2}$ are exactly fulfilled

Fig. 10.2.1 Interpolation of the acceleration and interpretation of the Newmark factors β and γ

Using equs. (10.2.6) and (10.2.7), we immediately deduce

$$\begin{aligned} \beta &= \tfrac{1}{4} \\ \gamma &= \tfrac{1}{2} \end{aligned} \tag{10.2.34}$$

Since the procedure here is implicit, we have the choice only between an iterative and a direct solution within the step. For linear problems, we can also rewrite the iteration procedure of equ. (10.2.16) as follows

$$\ddot{\boldsymbol{r}}^{(0)}_{\mathrm{k}+1} = \ddot{\boldsymbol{r}}_{\mathrm{k}}$$

$$\mathrm{i} = 1, \mathrm{ni} \left| \begin{aligned} \boldsymbol{r}^{(\mathrm{i}+1)}_{\mathrm{k}+1} &= \boldsymbol{r}_{\mathrm{k}} + \Delta t \dot{\boldsymbol{r}}_{\mathrm{k}} + (\Delta t)^2 (\tfrac{1}{2} - \beta) \ddot{\boldsymbol{r}}_{\mathrm{k}} + (\Delta t)^2 \beta \ddot{\boldsymbol{r}}^{(\mathrm{i})}_{\mathrm{k}+1} \\ \dot{\boldsymbol{r}}^{(\mathrm{i}+1)}_{\mathrm{k}+1} &= \dot{\boldsymbol{r}}_{\mathrm{k}} + \Delta t (1 - \gamma) \ddot{\boldsymbol{r}}_{\mathrm{k}} + \Delta t \gamma \ddot{\boldsymbol{r}}^{(\mathrm{i})}_{\mathrm{k}+1} \\ \ddot{\boldsymbol{r}}^{(\mathrm{i}+1)}_{\mathrm{k}+1} &= \boldsymbol{M}^{-1} \Big[\boldsymbol{R}_{\mathrm{k}+1} - \boldsymbol{C} \dot{\boldsymbol{r}}^{(\mathrm{i}+1)}_{\mathrm{k}+1} - \boldsymbol{K} \boldsymbol{r}^{(\mathrm{i}+1)}_{\mathrm{k}+1} \Big] \end{aligned} \right. \tag{10.2.35}$$

To realise the direct solution, we note that the standard system of equ. (10.2.18) may be reduced from an order (2n×2n) to (n×n) without difficulty. The most simple of possible reduction techniques is obtained by substitution of equs. (10.2.10) and (10.2.11) into equ. (10.2.14). Rearranging, we find

$$[\boldsymbol{M} + \Delta t \gamma \boldsymbol{C} + \Delta t^2 \beta \boldsymbol{K}] \ddot{\boldsymbol{r}}_{\mathrm{k}+1} = \boldsymbol{R}_{\mathrm{k}+1} - \boldsymbol{C}[\dot{\boldsymbol{r}}_{\mathrm{k}} + \Delta t (1-\gamma) \ddot{\boldsymbol{r}}_{\mathrm{k}}] - \boldsymbol{K}[\boldsymbol{r}_{\mathrm{k}} + \Delta t \dot{\boldsymbol{r}}_{\mathrm{k}} + \Delta t^2 (\tfrac{1}{2} - \beta) \ddot{\boldsymbol{r}}_{\mathrm{k}}] \tag{10.2.36}$$

Solving with respect to $\ddot{\boldsymbol{r}}_{\mathrm{k}+1}$, we find

$$\boldsymbol{r}_{\mathrm{k}+1} = \boldsymbol{r}_{\mathrm{k}} + \Delta t \dot{\boldsymbol{r}}_{\mathrm{k}} + \Delta t^2 (\tfrac{1}{2} - \beta) \ddot{\boldsymbol{r}}_{\mathrm{k}} + \Delta t^2 \beta \ddot{\boldsymbol{r}}_{\mathrm{k}+1} \tag{10.2.37a}$$

$$\dot{\boldsymbol{r}}_{\mathrm{k}+1} = \dot{\boldsymbol{r}}_{\mathrm{k}} + \Delta t (1 - \gamma) \ddot{\boldsymbol{r}}_{\mathrm{k}} + \Delta t^2 \gamma \ddot{\boldsymbol{r}}_{\mathrm{k}+1} \tag{10.2.37b}$$

To solve equ. (10.2.36), we have to operate with a modified mass matrix. Generally, this matrix will be symmetric if $\boldsymbol{C}$ does not contain unsymmetric components, e.g. due to Coriolis forces. For a constant time-step, Cholesky decomposition has to be carried out only once.

It is also possible to select $\boldsymbol{r}$ as a primary variable and to work with a modified stiffness matrix. For this purpose, we calculate the acceleration from equ. (10.2.10), substitute this into equ. (10.2.11) and then both expressions into equ. (10.2.14). The result is

$$\begin{aligned} &\left[\boldsymbol{K} + \frac{\gamma}{\Delta t \beta} \boldsymbol{C} + \frac{1}{\Delta t^2 \beta} \boldsymbol{M} \right] \boldsymbol{r}_{\mathrm{k}+1} \\ &\quad = \boldsymbol{R}_{\mathrm{k}} + \boldsymbol{C} \left[\frac{\gamma}{\Delta t \beta} \boldsymbol{r}_{\mathrm{k}} - (1 - \frac{\gamma}{\beta}) \dot{\boldsymbol{r}}_{\mathrm{k}} - \Delta t (1 - \tfrac{1}{2}\gamma + \gamma\beta) \ddot{\boldsymbol{r}}_{\mathrm{k}} \right] \\ &\qquad + \boldsymbol{K} \big[\boldsymbol{r}_{\mathrm{k}} + \Delta t \dot{\boldsymbol{r}}_{\mathrm{k}} + \Delta t^2 (\tfrac{1}{2} - \beta) \ddot{\boldsymbol{r}}_{\mathrm{k}} \big] \end{aligned} \tag{10.2.38}$$

Having calculated the deflection, the following computational steps remain

$$\dot{\boldsymbol{r}}_{\mathrm{k}+1} = \frac{\gamma}{\Delta t \beta}(\boldsymbol{r}_{\mathrm{k}+1} - \boldsymbol{r}_{\mathrm{k}}) + (1 - \frac{\gamma}{\beta})\dot{\boldsymbol{r}}_{\mathrm{k}} + \Delta t(1 - \tfrac{1}{2}\gamma + \gamma\beta)\ddot{\boldsymbol{r}}_{\mathrm{k}} \qquad (10.2.39\mathrm{a})$$

$$\ddot{\boldsymbol{r}}_{\mathrm{k}+1} = \frac{1}{\Delta t^2 \beta}(\boldsymbol{r}_{\mathrm{k}+1} - \boldsymbol{r}_{\mathrm{k}}) - \frac{1}{\Delta t \beta}\dot{\boldsymbol{r}}_{\mathrm{k}} - \frac{\frac{1}{2} - \beta}{\beta}\ddot{\boldsymbol{r}}_{\mathrm{k}} \qquad (10.2.39\mathrm{b})$$

Newmark $\beta = \frac{1}{6}, \gamma = \frac{1}{2}$; *linear acceleration, implicit*

Here, the FLA2-concept is employed and the linear interpolation yields

$$\omega_1(\tau) = 1 - \tau \qquad (10.2.40)$$

The acceleration over the time-axis is continuous since equ. (10.2.5) is satisfied. From equs. (10.2.6) and (10.2.7), we can easily deduce

$$\beta = \tfrac{1}{6}$$
$$\gamma = \tfrac{1}{2} \qquad (10.2.41)$$

Since we developed in the preceding section ($\beta = 1/4, \gamma = 1/2$) all computational steps for arbitrary values of β and γ, these can also be applied here. A further discussion appears hardly necessary.

Newmark $\beta = \frac{1}{12}, \gamma = \frac{1}{2}$; *Fox-Goodwin method, implicit*

In [10.10] – an excellent summary, which we recommend to readers wishing to deepen their understanding of the subject – Hughes refers to a further variant of the Newmark process, also known as the 'royal road' method. This technique is implicit and only conditionally stable as will be seen in our subsequent discussion on stability. These characteristics make it unlikely that this method can compete either in general or in particular with the method of average acceleration, for instance, when analysing large systems. For undamped systems, however, it possesses an accuracy of fourth order.

Cubic Hermitian interpolation of accelerations [10.4]

In analogy to the plane beam model, we adopt the following pivotal values

$$\boldsymbol{\rho}_A = \left\{ \ddot{\boldsymbol{r}}_{\mathrm{k}} \quad \left(\frac{\mathrm{d}\ddot{\boldsymbol{r}}}{\mathrm{d}\tau}\right)_{\mathrm{k}} \quad \ddot{\boldsymbol{r}}_{\mathrm{k}+1} \quad \left(\frac{\mathrm{d}\ddot{\boldsymbol{r}}}{\mathrm{d}\tau}\right)_{\mathrm{k}+1} \right\} \qquad (10.2.42)$$

and define in this way a cubic polynomial in time

$$\boldsymbol{\omega}_A = \left[(1 - 3\tau^2 + 2\tau^3)\boldsymbol{I}_{\mathrm{n}} \quad (\tau - 2\tau^2 + \tau^3)\boldsymbol{I}_{\mathrm{n}} \quad (3\tau^2 - 2\tau^3)\boldsymbol{I}_{\mathrm{n}} \quad (-\tau^2 + \tau^3)\boldsymbol{I}_{\mathrm{n}}\right] \qquad (10.2.43)$$

We note here one particular characteristic of this approach which concerns the pivotal incorporation of derivatives of the acceleration. The approximated acceleration thus displays neither jumps (Newmark $\beta = 0$, Newmark averaged) nor kinks (Newmark linear). Using equ. (10.2.28), we easily find the influence matrices

$$\boldsymbol{C}_{DA} = \frac{1}{60}\begin{bmatrix} 21\boldsymbol{I}_n & 3\boldsymbol{I}_n & 9\boldsymbol{I}_n & -2\boldsymbol{I}_n \end{bmatrix} \tag{10.2.44}$$

and

$$\boldsymbol{C}_{VA} = \frac{1}{12}\begin{bmatrix} 6\boldsymbol{I}_n & \boldsymbol{I}_n & 6\boldsymbol{I}_n & -\boldsymbol{I}_n \end{bmatrix} \tag{10.2.45}$$

In addition, we note

$$\frac{d\ddot{\boldsymbol{r}}}{d\tau} = \frac{d\ddot{\boldsymbol{r}}}{dt}\frac{dt}{d\tau} = \Delta t\dddot{\boldsymbol{r}} \tag{10.2.46}$$

and can thus easily establish the increments of displacement and velocity

$$\begin{aligned} \Delta\boldsymbol{r} &= \Delta t\dot{\boldsymbol{r}}_k + \tfrac{1}{60}\Delta t^2(21\ddot{\boldsymbol{r}}_k + 3\Delta t\dddot{\boldsymbol{r}}_k + 9\ddot{\boldsymbol{r}}_{k+1} - 2\Delta t\dddot{\boldsymbol{r}}_{k+1}) \\ &= \Delta t\dot{\boldsymbol{r}}_k + \tfrac{1}{2}\Delta t^2\ddot{\boldsymbol{r}}_k + \tfrac{1}{60}\Delta t^3\dddot{\boldsymbol{r}}_k + \tfrac{3}{20}\Delta t^2\Delta\ddot{\boldsymbol{r}} - \tfrac{1}{30}\Delta t^3\Delta\dddot{\boldsymbol{r}} \end{aligned} \tag{10.2.47}$$

$$\begin{aligned} \Delta\dot{\boldsymbol{r}} &= \tfrac{1}{12}\Delta t(6\ddot{\boldsymbol{r}}_k + \Delta t\dddot{\boldsymbol{r}}_k + 6\ddot{\boldsymbol{r}}_{k+1} - \Delta t\dddot{\boldsymbol{r}}_{k+1}) \\ &= \Delta t\ddot{\boldsymbol{r}}_k + \tfrac{1}{2}\Delta t\Delta\ddot{\boldsymbol{r}} - \tfrac{1}{12}\Delta t^2\Delta\dddot{\boldsymbol{r}} \end{aligned} \tag{10.2.48}$$

In these expressions, not only the acceleration increment, but also the increment of its time derivative remains unknown. Thus, we require two additional systems of equations in order to establish the unknown vectors. One such system is given by the incremental equilibrium, equ. (10.1.8), and an additional one can be obtained from the temporal differentiation of the overall force equilibrium, equ. (10.1.2). For non-linear problems, it is often advantageous to work directly with inertial forces and their derivatives with respect to time instead of the accelerations employed here; see also [10.22]. At this point, however, we restrict ourselves to linear dynamics and assume constant mass, damping and stiffness matrices. In this case, differentiation of equ. (10.1.8) is straightforward and the two additional incremental equations become

$$\boldsymbol{M}\Delta\ddot{\boldsymbol{r}} + \boldsymbol{C}\Delta\dot{\boldsymbol{r}} + \boldsymbol{K}\Delta\boldsymbol{r} = \Delta\boldsymbol{R} \tag{10.2.49}$$

$$\boldsymbol{M}\Delta\dddot{\boldsymbol{r}} + \boldsymbol{C}\Delta\ddot{\boldsymbol{r}} + \boldsymbol{K}\Delta\dot{\boldsymbol{r}} = \Delta\dot{\boldsymbol{R}} \tag{10.2.50}$$

or – if we are working with total displacements –

$$\boldsymbol{M}\ddot{\boldsymbol{r}}_{k+1} + \boldsymbol{C}\dot{\boldsymbol{r}}_{k+1} + \boldsymbol{K}\boldsymbol{r}_{k+1} = \boldsymbol{R}_{k+1} \tag{10.2.51}$$

$$\boldsymbol{M}\dddot{\boldsymbol{r}}_{k+1} + \boldsymbol{C}\ddot{\boldsymbol{r}}_{k+1} + \boldsymbol{K}\dot{\boldsymbol{r}}_{k+1} = \dot{\boldsymbol{R}}_{k+1} \tag{10.2.52}$$

Here, we must note critically that the time derivative of the external excitation also enters into the calculation. This causes no difficulty when the excitation function is defined mathematically, but this requirement may be disadvantageous in the case of experimentally recorded discrete excitation values.

The system of equs. (10.2.47) to (10.2.52) can be processed once again iteratively. We obtain

$$\dddot{\boldsymbol{r}}_{k+1}^{(0)} = \dddot{\boldsymbol{r}}_k$$

$$\ddot{\boldsymbol{r}}_{k+1}^{(0)} = \ddot{\boldsymbol{r}}_k + \Delta t \dddot{\boldsymbol{r}}_k$$

$$i = 0, ni \quad \left| \begin{aligned} \boldsymbol{r}_{k+1}^{(i+1)} &= \boldsymbol{r}_k + \Delta t \dot{\boldsymbol{r}}_k + \tfrac{1}{60}\Delta t^2 (21\ddot{\boldsymbol{r}}_k + 3\Delta t \dddot{\boldsymbol{r}}_k + 9\ddot{\boldsymbol{r}}_{k+1}^{(i)} - 2\Delta t \dddot{\boldsymbol{r}}_{k+1}^{(i)}) \\ \dot{\boldsymbol{r}}_{k+1}^{(i+1)} &= \dot{\boldsymbol{r}}_k + \tfrac{1}{12}\Delta t (6\ddot{\boldsymbol{r}}_k + \Delta t \dddot{\boldsymbol{r}}_k + 6\ddot{\boldsymbol{r}}_{k+1}^{(i)} - \Delta t \dddot{\boldsymbol{r}}_{k+1}^{(i)}) \\ \ddot{\boldsymbol{r}}_{k+1}^{(i+1)} &= \boldsymbol{M}^{-1}\left[\boldsymbol{R}_{k+1} - \boldsymbol{C}\dot{\boldsymbol{r}}_{k+1}^{(i+1)} - \boldsymbol{K}\boldsymbol{r}_{k+1}^{(i+1)}\right] \\ \dddot{\boldsymbol{r}}_{k+1}^{(i+1)} &= \boldsymbol{M}^{-1}\left[\dot{\boldsymbol{R}}_{k+1} - \boldsymbol{C}\ddot{\boldsymbol{r}}_{k+1}^{(i+1)} - \boldsymbol{K}\dot{\boldsymbol{r}}_{k+1}^{(i+1)}\right] \end{aligned} \right. \tag{10.2.53}$$

A direct solution procedure within a step can also be easily established by mere substitution of equs. (10.2.47) and (10.2.48) into equs. (10.2.51) and (10.2.52). The result is an algebraic system of order (2n × 2n)

$$\begin{bmatrix} \boldsymbol{M} + \frac{1}{2}\Delta t \boldsymbol{C} + \frac{3}{20}\Delta t^2 \boldsymbol{K} & -\frac{1}{12}\Delta t^2 \boldsymbol{C} - \frac{1}{30}\Delta t^3 \boldsymbol{K} \\ \boldsymbol{C} + \frac{1}{2}\Delta t \boldsymbol{K} & \boldsymbol{M} - \frac{1}{12}\Delta t^2 \boldsymbol{K} \end{bmatrix} \begin{bmatrix} \ddot{\boldsymbol{r}} \\ \dddot{\boldsymbol{r}} \end{bmatrix}_{k+1}$$

$$= \begin{bmatrix} \boldsymbol{R}_{k+1} - \boldsymbol{C}(\dot{\boldsymbol{r}}_k + \frac{1}{2}\Delta t \ddot{\boldsymbol{r}}_k + \frac{1}{12}\Delta t^2 \dddot{\boldsymbol{r}}_k) - \boldsymbol{K}(\boldsymbol{r}_k + \Delta t \dot{\boldsymbol{r}}_k + \frac{7}{20}\Delta t^2 \ddot{\boldsymbol{r}}_k + \frac{1}{20}\Delta t^3 \dddot{\boldsymbol{r}}_k) \\ \dot{\boldsymbol{R}}_{k+1} - \boldsymbol{K}(\dot{\boldsymbol{r}}_k + \frac{1}{2}\Delta t \ddot{\boldsymbol{r}}_k + \frac{1}{12}\Delta t^2 \dddot{\boldsymbol{r}}_k) \end{bmatrix} \tag{10.2.54}$$

Unfortunately, however, this system is unsymmetric and very large. Even when the Gaussian decomposition has to be performed only once, as is the case for a constant time-step Δt, and only one forward and back substitution has to be carried out in each step, many disadvantages are associated with this procedure. Thus, in general, the iterative solution is preferable.

Finally, the reader is advised to consult [10.4] in which higher order Hermitian algorithms are developed; quintic interpolation in particular has been sucessfully employed in celestial mechanics (see also section 11.4.4).

10.3 Review of the simplest methods: stability and accuracy

All the procedures discussed in section 10.2 are recursive in nature. Proceeding from displacement $\boldsymbol{r}_{\mathrm{k}}$ and velocity $\dot{\boldsymbol{r}}_{\mathrm{k}}$ at the beginning of the step k, $\boldsymbol{r}_{\mathrm{k}+1}$ and $\dot{\boldsymbol{r}}_{\mathrm{k}+1}$ are calculated and serve as the new initial conditions for the next step. If we examine a free oscillation with the initial conditions $\boldsymbol{r}_1$ and $\dot{\boldsymbol{r}}_1$, our experience with the modal solution procedure leads us to expect that the responses $\boldsymbol{r}(t)$ and $\dot{\boldsymbol{r}}(t)$ will be bounded as long as damping is non-negative. The deflections and velocities calculated by means of numerical integration should in no circumstances grow unlimitedly, provided the procedure itself is stable. Thus, a distinction has to be made between those procedures that are conditionally stable and others that are unconditionally stable. We are still to prove that conditionally stable procedures yield a bounded response only below a certain time-step limit, whilst unconditionally stable algorithms ensure a limited response even for large time-steps. In order to confirm this, we set the excitation forces $\boldsymbol{R}(t)$ equal to zero and calculate the response of the free oscillations using the standard form of equs. (10.2.18) and (10.2.19). Multiple application of the recurrence formula in equ. (10.2.19) then yields the following expression for the response after the k-th step

$$\begin{bmatrix} \boldsymbol{r} \\ \dot{\boldsymbol{r}} \end{bmatrix}_{\mathrm{k}+1} = \boldsymbol{A}^{\mathrm{k}} \begin{bmatrix} \boldsymbol{r} \\ \dot{\boldsymbol{r}} \end{bmatrix}_{1} \quad , \qquad \mathrm{k} = 1, 2, \ldots \tag{10.3.1}$$

where $\{\boldsymbol{r} \quad \dot{\boldsymbol{r}}\}_1$ represents the initial conditions at the time node 1 ($t = 0$).

Whether the answer obtained from these initial conditions remains bounded depends upon the powers of the $\boldsymbol{A}$ matrix; only if these remain limited is the stability ensured. We now enquire as to whether there are specific vectors $\boldsymbol{x}_A$ for which

$$\boldsymbol{A}\boldsymbol{x}_A = \boldsymbol{x}_A \lambda_A \tag{10.3.2}$$

applies. The matrix $\boldsymbol{A}$ is real here and of order 2n but may be unsymmetric. Non-trivial solutions of equ. (10.3.2) are possible only if

$$\det\left[\boldsymbol{A} - \lambda_A \boldsymbol{I}\right] = 0 \tag{10.3.3}$$

which corresponds to an algebraic equation of order 2n for the eigenvalues λ_A (characteristic or secular equation). In this way, we obtain 2n solutions λ_A, with the proviso that repeated roots are to be counted according to their multiplicity. To each of the solutions $\lambda_{A\mathrm{i}}$ appertains an eigenvector $\boldsymbol{x}_{A\mathrm{i}}$; having $\lambda_{A\mathrm{i}}$, it can be determined from equ. (10.3.2), whereby the eigenvectors associated with distinct eigenvalues are linearly independent. All this holds without special assumptions on the structure of the $\boldsymbol{A}$ matrix. In order not to complicate the discussion at this point, we assume that $\boldsymbol{A}$ has 2n linearly independent eigenvectors. If, for example, a multiple root $\lambda_{A\mathrm{i}}$ were to exist with a multiplicity of p_{i}, then it is presumed that the matrix $\boldsymbol{A} - \lambda_{A\mathrm{i}}\boldsymbol{I}$ also has a rank deficiency of p_{i} (this, it should be noted, is not self-evident since the rank deficiency can in some cases be smaller).

Following this brief digression into the eigenvalue analysis of a general matrix $\boldsymbol{A}$, we return to our starting point and, to begin with, generalise equ. (10.3.2) for all 2n eigenvectors

$$\boldsymbol{A}\boldsymbol{X}_A = \boldsymbol{X}_A\boldsymbol{\Lambda}_A \tag{10.3.4}$$

Here, $\boldsymbol{\Lambda}_A$ is a diagonal matrix of the eigenvalues λ_A. Since $\boldsymbol{X}_A$ contains 2n linearly independent column vectors, the vector $\{\boldsymbol{r} \quad \dot{\boldsymbol{r}}\}_1$ can be represented as a linear combination of these eigenvectors

$$\begin{bmatrix} \boldsymbol{r} \\ \dot{\boldsymbol{r}} \end{bmatrix}_1 = \boldsymbol{X}_A\boldsymbol{c}_1 \tag{10.3.5}$$

The operation of equ. (10.3.1) can be executed once again and yields in conjunction with equ. (10.3.4)

$$\begin{bmatrix} \boldsymbol{r} \\ \dot{\boldsymbol{r}} \end{bmatrix}_{k+1} = \boldsymbol{X}_A\boldsymbol{\Lambda}_A^{k}\boldsymbol{c}_1 \tag{10.3.6}$$

It thus emerges that the original coefficients $\boldsymbol{c}_1$ are reinforced by the powers of the eigenvalues, $\boldsymbol{\Lambda}_A^{k}$. A boundedness of the solution can only be expected in this case if the modulus of the maximum eigenvalue does not exceed 1. In general, the eigenvalues of the $\boldsymbol{A}$ matrix are complex numbers and they can be thought of as being dispersed in the complex plane. Their distance from the origin corresponds to the modulus. Our requirement amounts to all the eigenvalues being contained within a certain circle with a radius of 1. In fact, mathematical theory states that for stability, the spectral radius ρ_A of the matrix $\boldsymbol{A}$ must not exceed the value of 1

$$\rho_A = \max|\lambda_{Ai}| \leqslant 1 \tag{10.3.7}$$

If we are now to assume this condition is fulfilled, the response in equ. (10.3.6) is bounded. This, however, does not necessarily ensure a satisfactory behaviour of the algorithm. For this purpose, we consider an arbitrary complex eigenvalue

$$\begin{aligned} \lambda_{Aj} &= a_j + ib_j \\ &= \sqrt{a_j^2 + b_j^2}\, e^{i\tan^{-1}(b_j/a_j)} \\ &= \rho_j e^{i\varphi_j} \end{aligned} \tag{10.3.8}$$

Involution of this eigenvalue yields

$$\begin{aligned} \lambda_{Aj}^{k} &= \rho_j^{k}\, e^{i\,k\varphi_j} \\ &= \rho_j^{k}\left(\cos k\varphi_j + i\sin k\varphi_j\right) \end{aligned} \tag{10.3.9}$$

We thus deduce

- positive real eigenvalues do not lead to a change in sign;
- negative real eigenvalues lead to a change in sign at each step;
- complex eigenvalues *may* lead to changes in sign, this depending on each individual step;
- if the modulus of the eigenvalue is smaller than 1, the involution converges towards zero.

Our algorithm, equ. (10.3.6), should, in fact, be capable of reproducing a free, undamped oscillation. The reversal of direction of this oscillatory motion should not be dependent on the step-size of the integration and thus, at least a few complex eigenvalues must exist. If the modulus of all eigenvalues is indeed smaller than 1, stability is ensured, but our algorithm would simulate a damped motion, a phenomenon often referred to as algorithmic damping. Thus, to achieve a satisfactory algorithmic performance, we must demand the presence of complex eigenvalues with a modulus of 1. All component factors $\boldsymbol{c}_1$ of the free motion appertaining to $\rho_\mathrm{j} < 1$ are algorithmically damped away anyhow.

Following these general remarks, we review critically the algorithms already discussed.

The Newmark procedures

The reader is reminded at this point of the standard formulation of equ. (10.2.18) for $\boldsymbol{F} = \boldsymbol{o}$ as well as of the matrices $\boldsymbol{D}_1$, equ. (10.2.23), and $\boldsymbol{D}_0$, equ. (10.2.24). In the present discussion on the stability behaviour, we follow a contribution by Goudreau and Taylor [10.11] and may consequently restrict ourselves to undamped systems with one degree of freedom. A system with one degree of freedom has the eigenfrequency

$$\omega^2 = \frac{k}{m} \tag{10.3.10}$$

Also introducing the abbreviation

$$\vartheta = \Delta t\, \omega = \frac{2\pi\Delta t}{T} \tag{10.3.11}$$

defining a dimensionless step-size, we obtain the following system

$$\begin{bmatrix} 1+\beta\vartheta^2 & 0 \\ \frac{\gamma}{\Delta t}\vartheta^2 & 1 \end{bmatrix} \begin{bmatrix} \boldsymbol{r} \\ \dot{\boldsymbol{r}} \end{bmatrix}_{\mathrm{k}+1} = \begin{bmatrix} 1-(\frac{1}{2}-\beta)\vartheta^2 & \Delta t \\ -\frac{1-\gamma}{\Delta t}\vartheta^2 & 1 \end{bmatrix} \begin{bmatrix} \boldsymbol{r} \\ \dot{\boldsymbol{r}} \end{bmatrix}_{\mathrm{k}} \tag{10.3.12}$$

The first sub-equation yields with relative ease the transition to the alternative standard form in equ. (10.2.19) with $\boldsymbol{L} = \boldsymbol{o}$. Thus, we obtain

$$\boldsymbol{A} = \frac{1}{1+\beta\vartheta^2} \begin{bmatrix} 1-(\frac{1}{2}-\beta)\vartheta^2 & \Delta t \\ \frac{\vartheta^2}{\Delta t}((\frac{1}{2}\gamma-\beta)\vartheta^2-1) & 1+(\beta-\gamma)\vartheta^2 \end{bmatrix} \tag{10.3.13}$$

The characteristic equation for this (2 × 2) matrix is of the form

$$\lambda^2 - (a_{11} + a_{22})\lambda + a_{11}a_{22} - a_{12}a_{21} = 0 \tag{10.3.14}$$

and yields the eigenvalues. For the specific coefficients of equ. (10.3.13), we find

$$a_{11} + a_{22} = 2 - \frac{\vartheta^2}{1 + \beta\vartheta^2}(\gamma + \tfrac{1}{2}) \tag{10.3.15}$$

and

$$a_{11}a_{22} - a_{12}a_{21} = \det \boldsymbol{A} = 1 - \frac{(\gamma - \frac{1}{2})\vartheta^2}{1 + \beta\vartheta^2} \tag{10.3.16}$$

Introducing for typographic simplicity the abbreviation

$$\alpha = \frac{\vartheta^2}{1 + \beta\vartheta^2} \tag{10.3.17}$$

the characteristic equ. (10.3.14) reduces to

$$\lambda^2 - \left(2 - \alpha(\gamma + \tfrac{1}{2})\right)\lambda + 1 - \alpha(\gamma - \tfrac{1}{2}) = 0 \tag{10.3.18}$$

and has the roots

$$\lambda = 1 - \tfrac{1}{2}\alpha(\gamma + \tfrac{1}{2}) \pm \left(\tfrac{1}{4}\alpha^2(\gamma + \tfrac{1}{2})^2 - \alpha\right)^{1/2} \tag{10.3.19}$$

If the radicand is positive, two real solutions exist which cannot, of course, reproduce an oscillatory solution. Therefore, the radicand has to be negative so that we obtain a pair of conjugate complex solutions. This applies if

$$\alpha(\gamma + \tfrac{1}{2})^{1/2} < 4 \tag{10.3.20}$$

or

$$\vartheta^2\left(\tfrac{1}{4}(\gamma + \tfrac{1}{2})^2 - \beta\right) < 1 \tag{10.3.21}$$

This requirement holds independently of the dimensionless step-size if

$$\beta \geqslant \tfrac{1}{4}(\gamma + \tfrac{1}{2})^2 \tag{10.3.22}$$

applies. For $\gamma = 1/2$, $\beta \geqslant 1/4$ must hence be ensured. Newmark's average acceleration method fulfils this requirement ($\gamma = 1/2, \beta = 1/4$) whilst the ($\beta = 0$)

procedure and the linear acceleration method ($\beta = 1/6$) do not. In the latter cases, the step-width is restricted to

$$\vartheta = \Delta t\omega < \left(\frac{4}{(\gamma+\frac{1}{2})^2 - 4\beta}\right)^{1/2} \tag{10.3.23}$$

and we obtain the following bounds

$$\vartheta = \Delta t\,\omega < 2 \qquad \text{for } \beta = 0\,,\quad \gamma = \tfrac{1}{2} \tag{10.3.24}$$

$$\vartheta = \Delta t\,\omega < 2\sqrt{3} \qquad \text{for } \beta = \tfrac{1}{6}\,,\quad \gamma = \tfrac{1}{2} \qquad \text{(linear acceleration)} \tag{10.3.25}$$

All that remains is to check the spectral radius. Assuming equ. (10.3.21) to be satisfied, we find for complex roots

$$\begin{aligned}\rho^2 = \rho_1^2 = \rho_2^2 &= \left(1 - \tfrac{1}{2}\alpha(\gamma+\tfrac{1}{2})\right)^2 + \left(\alpha - \tfrac{1}{4}\alpha^2(\gamma+\tfrac{1}{2})^2\right)\\ &= 1 - \alpha(\gamma - \tfrac{1}{2})\\ &= 1 - \frac{(\gamma-\frac{1}{2})\vartheta^2}{1+\beta\vartheta^2}\end{aligned} \tag{10.3.26}$$

Obviously, the spectral radius for $\gamma = 1/2$ is equal to 1 – irrespective of the value of β and the dimensionless step-size. Equation (10.3.26) furthermore reveals that for $\gamma < 1/2$, the spectral radius becomes larger than 1 since ϑ is finite. Consequently, instability takes place. In contrast, however, the algorithm for $\gamma > 1/2$ possesses a spectral radius of less than 1 and therefore generates algorithmic damping. Thus, $\gamma = 1/2$, the value adopted in all three Newmark procedures, appears ideal.

Cubic Hermitian interpolation of accelerations

In order to attain the standard form, we begin with equs. (10.2.47) and (10.2.48) and substitute equs. (10.2.49) to (10.2.52) into them. Following some matricial operations, we obtain

$$\boldsymbol{D}_1 = \begin{bmatrix} \boldsymbol{M} + \frac{3}{20}\Delta t^2\boldsymbol{K} + \frac{1}{30}\Delta t^3\boldsymbol{C}\boldsymbol{M}^{-1}\boldsymbol{K} & \frac{3}{20}\Delta t^2\boldsymbol{C} - \frac{1}{30}\Delta t^3\left[\boldsymbol{K} - \boldsymbol{C}\boldsymbol{M}^{-1}\boldsymbol{C}\right] \\ \frac{1}{2}\Delta t\boldsymbol{K} + \frac{1}{12}\Delta t^2\boldsymbol{C}\boldsymbol{M}^{-1}\boldsymbol{K} & \boldsymbol{M} + \frac{1}{2}\Delta t\boldsymbol{C} - \frac{1}{12}\Delta t^2\left[\boldsymbol{K} - \boldsymbol{C}\boldsymbol{M}^{-1}\boldsymbol{C}\right] \end{bmatrix} \tag{10.3.27}$$

and

$$\boldsymbol{D}_0 = \begin{bmatrix} \boldsymbol{M} - \frac{7}{20}\Delta t^2\boldsymbol{K} + \frac{1}{20}\Delta t^3\boldsymbol{C}\boldsymbol{M}^{-1}\boldsymbol{K} & \Delta t\boldsymbol{M} - \frac{7}{20}\Delta t^2\boldsymbol{C} - \frac{1}{20}\Delta t^3\left[\boldsymbol{K} - \boldsymbol{C}\boldsymbol{M}^{-1}\boldsymbol{C}\right] \\ -\frac{1}{2}\Delta t\boldsymbol{K} + \frac{1}{12}\Delta t^2\boldsymbol{C}\boldsymbol{M}^{-1}\boldsymbol{K} & \boldsymbol{M} - \frac{1}{2}\Delta t\boldsymbol{C} - \frac{1}{12}\Delta t^2\left[\boldsymbol{K} - \boldsymbol{C}\boldsymbol{M}^{-1}\boldsymbol{C}\right] \end{bmatrix} \tag{10.3.28}$$

which appertain to the standard formulation of equ. (10.2.18) for $\boldsymbol{F} = \boldsymbol{o}$. It suffices to restrict ourselves in what follows to a single degree of freedom and zero damping, $\boldsymbol{C} = \boldsymbol{O}$. As before, the two equs. (10.2.10) and (10.3.11) apply, and the specific recurrence formula becomes

$$\begin{bmatrix} 1+\frac{3}{20}\vartheta^2 & -\frac{1}{30}\Delta t\vartheta^2 \\ \frac{1}{2}\frac{\vartheta^2}{\Delta t} & 1-\frac{1}{12}\vartheta^2 \end{bmatrix}\begin{bmatrix} \boldsymbol{r} \\ \dot{\boldsymbol{r}} \end{bmatrix}_{k+1} = \begin{bmatrix} 1-\frac{7}{20}\vartheta^2 & \Delta t(1-\frac{1}{20}\vartheta^2) \\ -\frac{1}{2}\frac{\vartheta^2}{\Delta t} & 1-\frac{1}{12}\vartheta^2 \end{bmatrix}\begin{bmatrix} \boldsymbol{r} \\ \dot{\boldsymbol{r}} \end{bmatrix}_{k} \tag{10.3.29}$$

Analogous to the stability investigation in the Newmark procedures, this equation can be transformed into the standard form of equ. (10.1.19) with $\boldsymbol{L} = \boldsymbol{o}$. However, this is not as easy to achieve here as it was for equ. (10.3.12). Following some algebraic manipulations, the $\boldsymbol{A}$ matrix is found to be

$$\boldsymbol{A} = \frac{1}{1+\frac{1}{15}\vartheta^2+\frac{1}{240}\vartheta^4}\begin{bmatrix} 1-\frac{13}{30}\vartheta^2+\frac{1}{80}\vartheta^4 & \Delta t(1-\frac{1}{10}\vartheta^2+\frac{1}{720}\vartheta^4) \\ -\frac{\vartheta^2}{\Delta t}(1-\frac{1}{10}\vartheta^2) & 1-\frac{13}{30}\vartheta^2+\frac{1}{80}\vartheta^4 \end{bmatrix} \tag{10.3.30}$$

Formally, the characteristic equation for the calculation of the eigenvalues is identical to equ. (10.3.14). In addition, $a_{11} = a_{22}$ also applies. Thus, we may easily obtain the following expression for the eigenvalues

$$\lambda_{A\,1,2} = a_{11} \pm \sqrt{a_{12}a_{21}} \tag{10.3.31}$$

In order to realise an oscillatory motion, it is necessary for the eigenvalues to be complex which means that

$$a_{12}a_{21} < 0 \tag{10.3.32}$$

Then, the spectral radius becomes

$$\rho^2 = \rho_1^2 = \rho_2^2 = a_{11}^2 - a_{12}a_{21} = \det(\boldsymbol{A}) \tag{10.3.33}$$

From equ. (10.3.32), we deduce immediately in conjunction with equ. (10.3.30)

$$\frac{-\vartheta^2}{(1+\frac{1}{15}\vartheta^2+\frac{1}{240}\vartheta^4)^2}(1-\frac{1}{10}\vartheta^2)(1-\frac{1}{10}\vartheta^2+\frac{1}{720}\vartheta^4) < 0 \tag{10.3.34}$$

As is at once evident, this condition can only be satisfied if

$$\vartheta^2 < 10 \tag{10.3.35a}$$

or

$$\vartheta = \omega\Delta t < \sqrt{10} \tag{10.3.35b}$$

It thus follows that the aforementioned cubic Hermitian algorithm is merely *conditionally stable.* The spectral radius of equ. (10.3.33) becomes

$$\rho^2 = \frac{\left(1 - \frac{13}{30}\vartheta^2 + \frac{1}{80}\vartheta^4\right)^2 + \vartheta^2\left(1 - \frac{1}{10}\vartheta^2\right)\left(1 - \frac{1}{10}\vartheta^2 + \frac{1}{720}\vartheta^4\right)}{\left(1 + \frac{1}{15}\vartheta^2 + \frac{1}{240}\vartheta^4\right)^2}$$

$$= 1 \tag{10.3.36}$$

as required.

So far, we have limited our discussion to an examination of the stability and the algorithmic damping of the recurrence formula. We now wish to supplement the discussion by considering additional error sources. Once more, we confine our analysis to an undamped system with a single degree of freedom. We consider a free oscillation and write the exact response in analogy with equ. (5.1.9) as

$$r(t) = r(0)\cos\omega t + \frac{1}{\omega}\dot{r}(0)\sin\omega t \tag{10.3.37}$$

from which we immediately deduce the exact formula

$$\begin{bmatrix} r(t) \\ \dot{r}(t) \end{bmatrix} = \begin{bmatrix} \cos\omega t & \frac{1}{\omega}\sin\omega t \\ -\omega\sin\omega t & \cos\omega t \end{bmatrix} \begin{bmatrix} r(0) \\ \dot{r}(0) \end{bmatrix} \tag{10.3.38}$$

On the other hand, we obtain for a step-by-step integration

$$\begin{aligned} \begin{bmatrix} r \\ \dot{r} \end{bmatrix}_{k+1} &= \boldsymbol{A}^k \begin{bmatrix} r(0) \\ \dot{r}(0) \end{bmatrix} \\ &= \left[\boldsymbol{X}_A \boldsymbol{\Lambda}_A \boldsymbol{X}_A^{-1}\right]^k \begin{bmatrix} r(0) \\ \dot{r}(0) \end{bmatrix} \\ &= \boldsymbol{X}_A \boldsymbol{\Lambda}_A^k \boldsymbol{X}_A^{-1} \begin{bmatrix} r(0) \\ \dot{r}(0) \end{bmatrix} \end{aligned} \tag{10.3.39}$$

We now assume that step-size limitations, if existent, are observed. For all algorithms discussed so far, the eigenvalues then have to be conjugate complex and have, in addition, a modulus of 1

$$\begin{aligned} \boldsymbol{\Lambda}_A &= \lceil a_1 + ib_1 \quad a_1 - ib_1 \rfloor \\ &= \lceil e^{i\Phi} \quad e^{-i\Phi} \rfloor \\ &= \lceil \cos\Phi + i\sin\Phi \quad \cos\Phi - i\sin\Phi \rfloor \end{aligned} \tag{10.3.40}$$

where

$$\tan\Phi = \frac{b_1}{a_1} \tag{10.3.41}$$

applies. For the powers of the $\mathbf{\Lambda}_A$ matrix, we then simply obtain

$$\begin{aligned}\mathbf{\Lambda}_A^{k} &= \lceil e^{ik\Phi} \quad e^{-ik\Phi} \rfloor \\ &= \lceil \cos(k\Phi) + i\sin(k\Phi) \quad \cos(k\Phi) - i\sin(k\Phi) \rfloor\end{aligned} \tag{10.3.42}$$

Even without a prior knowledge of the eigenvectors $\boldsymbol{X}_A$, it is possible to conclude that the response

$$\begin{bmatrix} r(k\Delta t) \\ \dot{r}(k\Delta t) \end{bmatrix} = \boldsymbol{X}_A \lceil \cos(k\Phi) + i\sin(k\Phi) \quad \cos(k\Phi) - i\sin(k\Phi) \rfloor \boldsymbol{X}_A^{-1} \begin{bmatrix} r(0) \\ \dot{r}(0) \end{bmatrix} \tag{10.3.43}$$

must be real and that $\boldsymbol{X}_A$ has to be independent of the number of steps, k, and thus also of the time $t = k\Delta t$. Comparison with the exact solution in equ. (10.3.37) demonstrates that the exact period of the response is modified due to numerical integration. In fact, we now observe in equ. (10.3.42) the argument $k\Phi$ in place of $\omega t = \omega k \Delta t = k\vartheta$. The approximate period is denoted by $\widetilde{T}$, whilst the exact period is given by $T = 2\pi/\omega$. Their ratio is

$$\frac{\widetilde{T}}{T} = \frac{\omega}{\widetilde{\omega}} = \frac{\omega t}{\widetilde{\omega} t} = \frac{k\vartheta}{k\Phi} = \frac{\vartheta}{\Phi} \tag{10.3.44}$$

Our next step is to examine the period deviation or anomaly for the different algorithms.

Newmark's procedures

The eigenvalues have already been given in equ. (10.3.19) where the radicand has to be negative. Thus, we obtain

$$\tan\Phi = \frac{\left(\alpha - \frac{1}{4}\alpha^2 (\gamma + \frac{1}{2})^2\right)^{1/2}}{1 - \frac{1}{2}\alpha(\gamma + \frac{1}{2})} \tag{10.3.45}$$

Substituting equ. (10.3.17), we immediately find

$$\tan\Phi = \vartheta \frac{\left(1 + \vartheta^2 \left[\beta - \frac{1}{4}(\gamma + \frac{1}{2})^2\right]\right)^{1/2}}{1 + \vartheta^2 \left[\beta - \frac{1}{2}(\gamma + \frac{1}{2})\right]} \tag{10.3.46}$$

The algorithmically undamped case is represented by $\gamma = 1/2$

$$\tan\Phi = \vartheta \frac{\left(1 - (\frac{1}{4} - \beta)\vartheta^2\right)^{1/2}}{1 - (\frac{1}{2} - \beta)\vartheta^2} \tag{10.3.47}$$

At this point, we note the convergence of this function towards ϑ for small step-sizes ϑ, thus yielding in the limit $\Phi = \vartheta$. We conclude from this result that for small step-sizes, no period alteration takes place independently of β , and thus $\widetilde{T}/T = 1$ applies. Furthermore, we note that equ. (10.3.47) for $\beta \geqslant 1/4$, i.e. in the case of unconditional stability, always yields a solution. On the other hand, for conditional stability, i.e. $\beta < 1/4$, the range of equ. (10.3.46) is limited through the stability bounds of equs. (10.3.23) to (10.3.25)

$$\vartheta_{\text{lim}} = (\tfrac{1}{4} - \beta)^{-1/2} \qquad \text{if} \quad \beta < \tfrac{1}{4}$$

$$\tan\Phi(\vartheta_{\text{lim}}) = 0$$

$$\frac{\widetilde{T}}{T}(\vartheta_{\text{lim}}) = \frac{\vartheta_{\text{lim}}}{\Phi(\vartheta_{\text{lim}})} = \frac{1}{\pi(\frac{1}{4} - \beta)^{1/2}} \tag{10.3.48}$$

Furthermore, the function on the right-hand side of equ. (10.3.47) possesses a vertical asymptote for $\beta < 1/2$. At such a point, we have

$$\vartheta_{\text{As}} = (\tfrac{1}{2} - \beta)^{-1/2} \qquad \text{if} \quad \beta < \tfrac{1}{2}$$

$$\tan\Phi(\vartheta_{\text{As}}) = \pm\infty$$

$$\frac{\widetilde{T}}{T}(\vartheta_{\text{As}}) = \frac{\vartheta_{\text{As}}}{\Phi(\vartheta_{\text{As}})} = \frac{1}{\pi(\frac{1}{2} - \beta)^{1/2}} \tag{10.3.49}$$

For the Newmark algorithms discussed so far, our results are summarised in the following table ($\gamma = 1/2$)

β	ϑ_{As}	$\frac{\widetilde{T}}{T}(\vartheta_{\text{As}})$	ϑ_{lim}	$\frac{\widetilde{T}}{T}(\vartheta_{\text{lim}})$
0 (explicit)	$\sqrt{2}$	$\frac{2\sqrt{2}}{\pi} = 0.90$	2	$\frac{2}{\pi} = 0.64$
$\frac{1}{6}$ (linear)	$\sqrt{3}$	$\frac{2\sqrt{3}}{\pi} = 1.10$	$2\sqrt{3}$	$\frac{2\sqrt{3}}{\pi} = 1.10$
$\frac{1}{4}$ (average)	2	$\frac{4}{\pi} = 1.27$	∞	∞

(10.3.50)

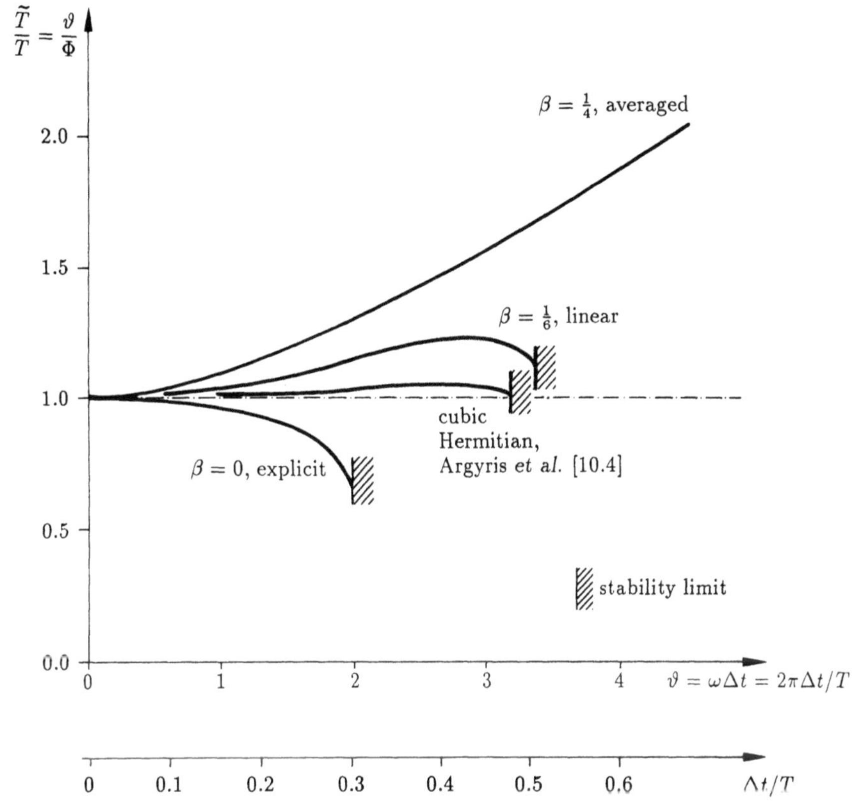

Fig. 10.3.1 Period elongation due to numerical integration for various step-sizes: Newmark algorithm and cubic Hermitian interpolation [2]

The family of curves for the period deviation is plotted in fig. 10.3.1 against the nondimensional step ϑ. The Newmark procedures discussed so far unfortunately deviate from the ideal relationship $\widetilde{T}/T = 1$. The unconditionally stable procedure of average accelerations ($\beta = 1/4$), in particular, cannot be applied for arbitrarily large step-sizes since this would result in extremely large period anomalies. Close observation of the family of curves as against the parameter β leads us to suspect that there is a β_{opt}-value with an ideal relationship of $\widetilde{T}/T(\vartheta_{\lim}) = 1$ and a maximum step-size. Since $\beta_{\text{opt}} < 1/6$ evidently applies here, this method can only be conditionally stable. From the third expression in equ. (10.3.48), we easily find

$$\beta_{\text{opt}} = \frac{1}{4} - \frac{1}{\pi^2} = 0.1487 \tag{10.3.51}$$

and from equ. (10.3.47), we deduce the maximum step-size

$$\begin{aligned} \vartheta(\beta_{\text{opt}})_{\lim} &= \pi \\ \frac{\Delta t}{T}(\beta_{\text{opt}})_{\lim} &= 0.5 \end{aligned} \tag{10.3.52}$$

At this stage, some reference should be made to the restrictions on this algorithm with respect to multi-degree-of-freedom systems: the β_{opt} in accordance with Newmark, and value $\gamma = 1/2$ with step-size of $\vartheta = \pi$ will integrate a *single-degree-of-freedom* system without algorithmic damping and without period elongation. Before we proceed to multi-degree-of-freedom systems, we ought first to discuss the Hermitian acceleration interpolation according to Argyris *et al.* [10.4] from the point of view of period elongation.

Hermitian interpolation

We now return to the eigenvalues in equ. (10.3.31) and also consider equ. (10.3.30). Hence, we obtain

$$\tan\Phi = \frac{(-a_{12}a_{21})^{1/2}}{a_{11}} = \vartheta\frac{\left((1-\frac{1}{10}\vartheta^2)(1-\frac{1}{10}\vartheta^2+\frac{1}{720}\vartheta^4)\right)^{1/2}}{1-\frac{13}{30}\vartheta^2+\frac{1}{80}\vartheta^4} \qquad (10.3.53)$$

The validity of this formula terminates at the aforementioned stability bound of

$$\vartheta_{\text{lim}} = \sqrt{10}$$

with

$$\tan\Phi(\vartheta_{\text{lim}}) = 0$$

$$\frac{\widetilde{T}}{T}(\vartheta_{\text{lim}}) = \frac{\vartheta_{\text{lim}}}{\Phi(\vartheta_{\text{lim}})} = \frac{\sqrt{10}}{\pi} = 1.0066 \qquad (10.3.54)$$

As before, the function on the right-hand side of equ. (10.3.53) possesses a vertical asymptote within the admissible range at

$$\vartheta_{\text{As}} = \left(\tfrac{1}{3}(52-\sqrt{1984}\,)\right)^{1/2} = 1.5767$$

$$\tan\Phi(\vartheta_{\text{As}}) = \pm\infty$$

$$\frac{\widetilde{T}}{T}(\vartheta_{\text{As}}) = \frac{\vartheta_{\text{As}}}{\Phi(\vartheta_{\text{As}})} = \frac{\vartheta_{\text{As}}}{\pi/2} = 1.0038 \qquad (10.3.55)$$

Both these values demonstrate the extraordinarily good performance of the cubic Hermitian interpolation. Figure 10.3.1 reproduces the complete curve. The maximum deviation within the permissible range amounts to about 3%, and the quintic Hermitian interpolation yields even more precise results [10.4].

Whilst we have so far discussed only single-degree-of-freedom systems, such cases are, however, trivial. Our main aim is indeed to achieve the direct integration of the coupled differential equation system with numerous degrees of freedom. In order to obtain some useful criteria for the behaviour of such systems, one takes the hypothetical point of view that they are decoupled. For n degrees of freedom, one would then have n differential equations, representing single-degree-of-freedom systems which have to be integrated simultaneously with a common step-size Δt.

Since each of the decoupled equations has a different eigenfrequency $\omega_\mathrm{i}(\mathrm{i}=1,\mathrm{n})$, the dimensionless step-length $\vartheta_\mathrm{i}=\omega_\mathrm{i}\Delta t$ or $\Delta t/T_\mathrm{i}$ is different for each equation. The equation with the largest circular eigenfrequency (smallest period) then possesses the maximum allowable step-length. For conditionally stable algorithms, the following global bound thus applies

$$\Delta t < \frac{\vartheta_{\mathrm{lim}}}{\omega_{\mathrm{max}}} = \frac{1}{2\pi}\vartheta_{\mathrm{lim}}T_{\mathrm{min}} \tag{10.3.56}$$

for the step-size ensuring a stable integration of all constituent equations. At this point, we stress our aim to secure not only a stable but also an exact integration. Since each equation possesses a different dimensionless step-size ϑ_i, the integration of the contributing modal forms involves a different distortion in each case. Lower eigenfrequencies ω_i lead to small step-sizes ϑ_i and thus (see fig. 10.3.1) to nearly exact results for all the algorithms so far discussed. On the other hand, if we are to apply the unconditionally stable method of average accelerations and large step-widths ϑ_i, which is permissible from the point of view of stability, very large errors for the high eigenfrequencies possibly arise. In such a case, pseudo-oscillations can occur with a period of $2\Delta t$. This may be conjectured from the observation that the $\boldsymbol{A}$ matrix is reduced for $\gamma=1/2$ and $\beta=1/4$ and an unlimited step-size to

$$\lim_{\Delta t\to\infty}\boldsymbol{A} = \lceil -1 \quad -1 \rfloor \tag{10.3.57}$$

Unconditionally stable algorithms are hence not automatically better working tools and the unpleasant nature of such spurious high frequency oscillations has led researchers to reconsider artificial algorithmic damping once again. One opts for the lesser of two evils and attempts to suppress false oscillatory components with high eigenfrequencies by introducing large algorithmic damping. One such very simple method is Wilson's θ-method [10.12]. Unfortunately, however, this procedure not only removes spurious high-frequency oscillations but also correct ones. Whilst oscillations involving only lower eigenfrequencies can be integrated correctly, a short-time response cannot be determined by this means. A decisive yardstick in assessing the reliability of a method for general applications is the value of $\tilde{T}/T$ which should be 1, independently of ϑ. Figure 10.3.1 illustrates how the cubic Hermitian algorithm developed by Argyris *et al.* [10.4] proves the most effective. The optimum value for the Newmark procedures in equ. (10.3.51) is correct only for a single-degree-of- freedom system. Below the maximum step-length, $\tilde{T}/T$ will deviate from 1 which is also demonstrated by the set of curves for the Newmark procedures shown in fig. 10.3.2. From a global point of view, smaller β-values are the more promising choices if the maximum admissible step-size is not fully exploited. In this way, the period errors quoted in table 10.3.1 are obtained. From this summary, it is clear that a well-designed Newmark algorithm can produce results of the same quality as a cubic Hermitian algorithm. For large step-widths (e.g. $\vartheta=2.7$), however, the cubic Hermitian interpolation remains superior. Table 10.3.2 once again summarises all the algorithms mentioned so far from the point of view of stability and accuracy. In addition to the step-size limit ϑ_{lim} arising from stability considerations, the step-size corresponding to a maximum error of 1% or 3% for any participating period is also quoted. The extreme cases are the first to catch our attention. The most simple explicit algorithm $\beta=0$ can only

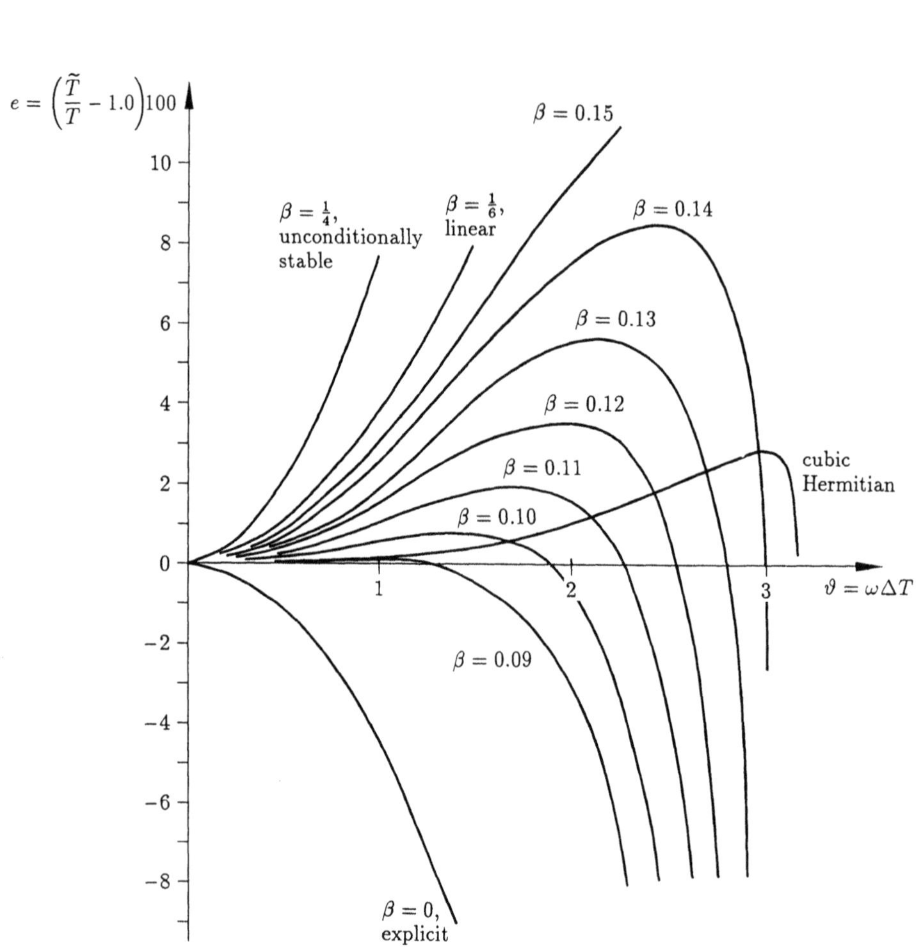

Fig. 10.3.2 Percentage distortion of the period due to numerical integration: cubic Hermitian interpolation, favourable range of Newmark algorithms with explicit as well as unconditionally stable variant

operate exactly with a very sharp step-size reduction. The same applies to the unconditionally stable algorithm $\beta = 1/4$. Both methods are little suited to an exact analysis of oscillations involving high eigenfrequencies. Limiting the error to 3%, the cubic Hermitian algorithm permits the largest step-size; the same applies for the error limit of 1% where only the Newmark procedure $\beta = 0.10$ is nearly as good. Of course, the point of view alters when dealing with the integration of oscillatory processes with low eigenfrequencies only. In this case, a procedure with a large step-length of $\vartheta_{\lim}$ could be advantageous when provisions are made against spurious high-frequency components (e.g. $\gamma > 1/2, \beta \geqslant (0.5 + \gamma)^2/4$). Since, however, the engineer can only possess a limited precognition of the oscillatory behaviour of a system, there is little point in pursuing this special case here.

Table 10.3.1 Period error e with the cubic Hermitian and the well-designed Newmark algorithm

type	$\vartheta_{\max}$	$e = (\frac{\widetilde{T}}{T} - 1)100$
$\beta = 0.09\,,\ \gamma = 0.5$	≈ 1.4	$< 0.2\%$
$\beta = 0.10\,,\ \gamma = 0.5$	≈ 2.0	$< 0.8\%$
$\beta = 0.11\,,\ \gamma = 0.5$	≈ 2.4	$< 2.0\%$
$\beta = 0.12\,,\ \gamma = 0.5$	≈ 2.7	$< 3.7\%$
cubic Hermitian	≈ 1.4	$< 0.3\%$
cubic Hermitian	≈ 2.0	$< 0.9\%$
cubic Hermitian	≈ 2.7	$< 2.3\%$
cubic Hermitian	≈ 3.0	$< 3.0\%$

Table 10.3.2 Some Newmark parameters and the cubic Hermitian interpolation: step-size limits due to stability ($\vartheta_{\lim}$) and period preservation ($\vartheta_{1\%}, \vartheta_{3\%}$)

type	$\vartheta_{\lim} = \omega_{\max}\Delta t$	$\vartheta_{1\%}$ $(e < 1\%)$	$\vartheta_{3\%}$ $(e < 3\%)$
$\beta = 0$, $\gamma = 0.5$ explicit	2.0	0.48	0.83
$\beta = 0.09$, $\gamma = 0.5$ explicit	2.5	1.68	2.00
$\beta = 0.10$, $\gamma = 0.5$ explicit	2.58	2.05	2.25
$\beta = 0.11$, $\gamma = 0.5$ explicit	2.67	0.96	2.47
$\beta = 0.12$, $\gamma = 0.5$ explicit	2.77	0.80	1.50
$\beta = 0.13$, $\gamma = 0.5$ explicit	2.89	0.75	1.22
$\beta = 0.14$, $\gamma = 0.5$ explicit	3.02	0.61	1.07
$\beta = 0.15$, $\gamma = 0.5$ explicit	3.16	0.55	0.98
$\beta = 1/6$, $\gamma = 0.5$ linear	$2\sqrt{3} = 3.46$	0.48	0.87
$\beta = 1/4$, $\gamma = 0.5$ average	∞	0.35	0.60
cubic Hermitian	$\sqrt{10} = 3.16$	2.07	3.16

The step-size plays a decisive role in the evaluation of the stability and accuracy of numerical integration and thus a little more should be mentioned about the manner in which the maximum circular frequency $\omega_{\max}$ is obtained. Up to now, our error discussion was restricted to undamped systems and we now continue our discussion, retaining this assumption. It should be noted, however, that the step-sizes mentioned so far have to be reduced in the presence of damping.

The required frequency is obtained from the eigenvalue problem

$$\left[\omega^2 \boldsymbol{M} - \boldsymbol{K}\right]\boldsymbol{x} = \boldsymbol{o} \tag{10.3.58}$$

for example with the aid of the vector iteration

$$\boldsymbol{M}\boldsymbol{x}^{(\mathrm{k}+1)} = \boldsymbol{K}\boldsymbol{x}^{(\mathrm{k})} \tag{10.3.59}$$

and Schwarz's constants

$$\omega_{\max}^{(k)} = \left(\frac{\boldsymbol{x}^{(k)t} \boldsymbol{K} \boldsymbol{x}^{(k)}}{\boldsymbol{x}^{(k)t} \boldsymbol{M} \boldsymbol{x}^{(k)}} \right)^{1/2} = \left(\frac{\boldsymbol{x}^{(k)t} \boldsymbol{M} \boldsymbol{x}^{(k+1)}}{\boldsymbol{x}^{(k)t} \boldsymbol{M} \boldsymbol{x}^{(k)}} \right)^{1/2} \tag{10.3.60}$$

Solving equ. (10.3.59) where $\boldsymbol{M}$ is a coefficient matrix is not a serious task since $\boldsymbol{M}^{-1}$ occurs anyhow in all integration procedures and, in general, one operates with a diagonal $\boldsymbol{M}$. Applying $\boldsymbol{U}_M$, the Cholesky factor of $\boldsymbol{M}$ which is required anyway, the vector iteration can also be symmetrically formulated. Writing

$$\boldsymbol{M} = \boldsymbol{U}_M^t \boldsymbol{U}_M \tag{10.3.61}$$

and introducing the transformed eigenvector

$$\overline{\boldsymbol{x}} = \boldsymbol{U}_M \boldsymbol{x} \tag{10.3.62}$$

we form the corresponding dynamic matrix

$$\boldsymbol{A} = \boldsymbol{U}_M^{-t} \boldsymbol{K} \boldsymbol{U}_M^{-1} \tag{10.3.63}$$

which, of course, does not have to be construed explicitly but is to be operated upon in its product form. In place of equ. (10.3.59), the following rule applies

$$\overline{\boldsymbol{x}}^{(k+1)} = \boldsymbol{A} \overline{\boldsymbol{x}}^{(k)} \tag{10.3.64}$$

and in place of equ. (10.3.60), we obtain

$$\omega_{\max}^{(k)} = \left(\overline{\boldsymbol{x}}^{(k+1)t} \, \overline{\boldsymbol{x}}^{(k)} \right)^{1/2} \tag{10.3.65}$$

if

$$\overline{\boldsymbol{x}}^{(k)t} \, \overline{\boldsymbol{x}}^{(k)} = 1 \tag{10.3.66}$$

holds. This procedure generally converges slowly since the high eigenfrequencies are not well separated in an FE-idealisation; as a guide, we quote 10 to 20 iteration loops. In contrast, useful approximations for the smallest eigenfrequency can generally be achieved with a third of this effort.

Those who consider the above process too laborious are advised to switch over to norms of the $\boldsymbol{A}$ matrix, but should realise that this involves a reduced accuracy for $\omega_{\max}$. Of course, in the present case, it is advisable to restrict ourselves to a diagonal $\boldsymbol{M}$ (thus $\boldsymbol{U}_M^{-1} = \boldsymbol{M}^{-1/2}$). It is known that any matrix norm $\|\boldsymbol{A}\|$ yields an upper bound for the maximum eigenvalue of the matrix

$$\|\boldsymbol{A}\| \geqslant \omega_{\max}^2 \tag{10.3.67}$$

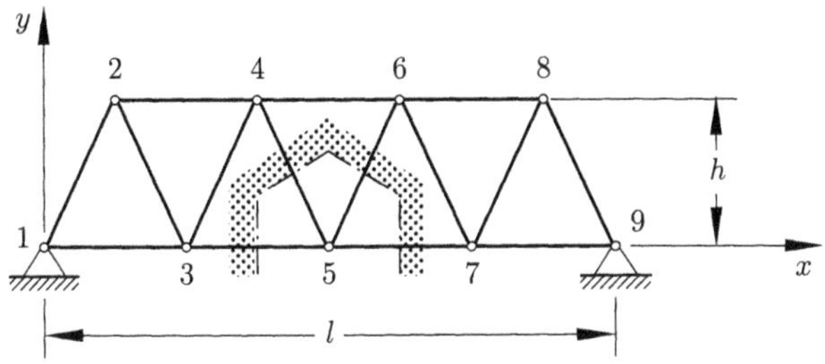

data:

span	$l = 8$ m
height	$h = 2$ m
cross-section area	$A = 10^{-3}$ m^2
density	$\rho = 8000$ Ns^2/m^4
Young's modulus	$E = 2.1 \cdot 10^{11}$ N/m^2

Fig. 10.3.3 Evaluation of the highest eigenfrequency generated by the horizontal motion of nodal point 5

In order to keep the calculation efforts within reasonable limits, we can choose the Euclidean norm which has already been introduced

$$\|\boldsymbol{A}\|_E = \left(\sum_{\mathrm{i,j=1}}^{\mathrm{n}} a_{\mathrm{ij}}^2 \right)^{1/2} \geqslant \omega_{\mathrm{max}}^2 \qquad (10.3.68)$$

or the column norm (Hölder 1-norm)

$$\|\boldsymbol{A}\|_1 = \max_j \sum_{\mathrm{i=1}}^{\mathrm{n}} |a_{\mathrm{ij}}| \geqslant \omega_{\mathrm{max}}^2 \qquad (10.3.69)$$

or the row norm (Hölder ∞-norm)

$$\|\boldsymbol{A}\|_\infty = \max_i \sum_{\mathrm{j=1}}^{\mathrm{n}} |a_{\mathrm{ij}}| \geqslant \omega_{\mathrm{max}}^2 \qquad (10.3.70)$$

The practically inclined reader will possibly prefer to opt for a more intuitive determination of the bounds. We know from experience that the high-frequency eigenoscillations involve numerous oscillatory nodal lines or nodal surfaces. The highest eigenfrequency of a continuum is infinite, the highest eigenfrequency of a discretised structure, on the other hand, is finite. It is to be expected that the corresponding oscillation pictogram will contain a very high number of nodal lines, so distributed that each nodal point of the discretisation is separated from the neighbouring ones by oscillatory nodal lines. To obtain a rough estimate, we may consider a nodal point with a high ratio of stiffness to mass as being isolated. As an example, we view the framework discussed in [4.1] (see fig. 10.3.3). Obviously, the horizontal motion of the nodes 3, 4, 5, 6, 7 is associated with the largest ratio of stiffness to mass. Applying the unit displacement method to node 5 and assuming an effective length of half for the flange (note that the nodal line resides in the middle of the bar!), we find

$$k_5 = \sum_{\mathrm{g}} \left(\frac{EA}{l_{\mathrm{eff}}}\right)_{\mathrm{g}} (\Delta l_{\mathrm{g}})^2 = 2EA\left(1 + \frac{2}{5\sqrt{5}}\right) = 2.36EA$$

$$m_5 = \sum_{\mathrm{g}} \tfrac{1}{2}(\rho A l_{\mathrm{eff}})_{\mathrm{g}} = \rho A\left(1 + \frac{\sqrt{5}}{2}\right) = 2.12\rho A$$

$$\omega_{\mathrm{max}} \approx \left(\frac{k}{m}\right)^{1/2} = \left(1.11\,\frac{E}{\rho}\right)^{1/2} = 5406\ \mathrm{s}^{-1}$$

$$\nu_{\mathrm{max}} = \frac{\omega_{\mathrm{max}}}{2\pi} \approx 860\ \ \mathrm{Hz}$$

The exact maximum eigenfrequency, on the other hand, is

$$\nu_{14} = 785.0\ \mathrm{Hz}$$

As was to be expected, we observe an overestimate of the frequency by 9,6%. Bearing our elementary argument in mind, though, this is not excessive. Anyhow, we are on the safe side as far as the specification of the step-width is concerned.

10.4 Unconditionally stable algorithms based upon the Hermitian interpolation model [10.13]

Analogous to the development of the Newmark procedures, we proceed from the Hermitian concept of a pivotal specification but do not explicitly prescribe the interpolation forms. In place of equ. (10.2.43), we now consider the more general relation

$$\ddot{\boldsymbol{r}} = \omega_1(\tau)\ddot{\boldsymbol{r}}_{\mathrm{k}} + \omega_2(\tau)\ddot{\boldsymbol{r}}_{\mathrm{k+1}} + \omega_3(\tau)\Delta t\dddot{\boldsymbol{r}}_{\mathrm{k}} + \omega_4(\tau)\Delta t\dddot{\boldsymbol{r}}_{\mathrm{k+1}} \tag{10.4.1}$$

for the acceleration within a Δt $(0 \leqslant \tau \leqslant 1)$ time-step. The unknown interpolation functions have to satisfy, however, certain shape requirements, as is illustrated diagramatically in fig. 10.4.1. In fact, regardless of the analytical specification of the function, the following conditions must hold

$$
\begin{aligned}
\omega_1(0) &= 1 & \omega_2(0) &= 0 \\
\frac{\partial \omega_1}{\partial \tau}(0) &= 0 & \frac{\partial \omega_2}{\partial \tau}(0) &= 0 \\
\omega_1(1) &= 0 & \omega_2(1) &= 1 \\
\frac{\partial \omega_1}{\partial \tau}(1) &= 0 & \frac{\partial \omega_2}{\partial \tau}(1) &= 0 \\
\omega_3(0) &= 0 & \omega_4(0) &= 0 \\
\frac{\partial \omega_3}{\partial \tau}(0) &= 1 & \frac{\partial \omega_4}{\partial \tau}(0) &= 0 \\
\omega_3(1) &= 0 & \omega_4(1) &= 0 \\
\frac{\partial \omega_3}{\partial \tau}(1) &= 0 & \frac{\partial \omega_4}{\partial \tau}(1) &= 1
\end{aligned}
\qquad (10.4.2)
$$

which ensure a conforming course of acceleration at the junction of the time-steps. In addition, we require that $\ddot{\boldsymbol{r}}_{\mathrm{k}}$ and $\ddot{\boldsymbol{r}}_{\mathrm{k+1}}$ establish an analogous continuity of the acceleration within the element; the same holds for $\dddot{\boldsymbol{r}}_{\mathrm{k}}$ and $\dddot{\boldsymbol{r}}_{\mathrm{k+1}}$. As a result, we have to prescribe

$$
\begin{aligned}
\omega_1(\tau) &= \omega_2(1-\tau) \\
\omega_3(\tau) &= -\omega_4(1-\tau)
\end{aligned}
\qquad (10.4.3)
$$

If we now deduce the velocity by integration of equ. (10.4.1) and note the formulation of equ. (10.1.19), we find

$$
\dot{\boldsymbol{r}}_{\mathrm{k+1}} = \dot{\boldsymbol{r}}_{\mathrm{k}} + \Delta t \ddot{\boldsymbol{r}}_{\mathrm{k}} \int_0^1 \omega_1(\tau) \mathrm{d}\tau + \Delta t \ddot{\boldsymbol{r}}_{\mathrm{k+1}} \int_0^1 \omega_2(\tau) \mathrm{d}\tau + \Delta t^2 \dddot{\boldsymbol{r}}_{\mathrm{k}} \int_0^1 \omega_3(\tau) \mathrm{d}\tau + \Delta t^2 \dddot{\boldsymbol{r}}_{\mathrm{k+1}} \int_0^1 \omega_4(\tau) \mathrm{d}\tau
\qquad (10.4.4)
$$

The conditions of equ. (10.4.3) indicate, however, that the areas below ω_1 and ω_2 are equal and those below ω_3 and ω_4 are equal but opposite. This immediately yields the more general relation

$$
\dot{\boldsymbol{r}}_{\mathrm{k+1}} = \dot{\boldsymbol{r}}_{\mathrm{k}} + a_0 \Delta t (\ddot{\boldsymbol{r}}_{\mathrm{k}} + \ddot{\boldsymbol{r}}_{\mathrm{k+1}}) + a_1 \Delta t^2 (\dddot{\boldsymbol{r}}_{\mathrm{k}} - \dddot{\boldsymbol{r}}_{\mathrm{k+1}}) \qquad (10.4.5)
$$

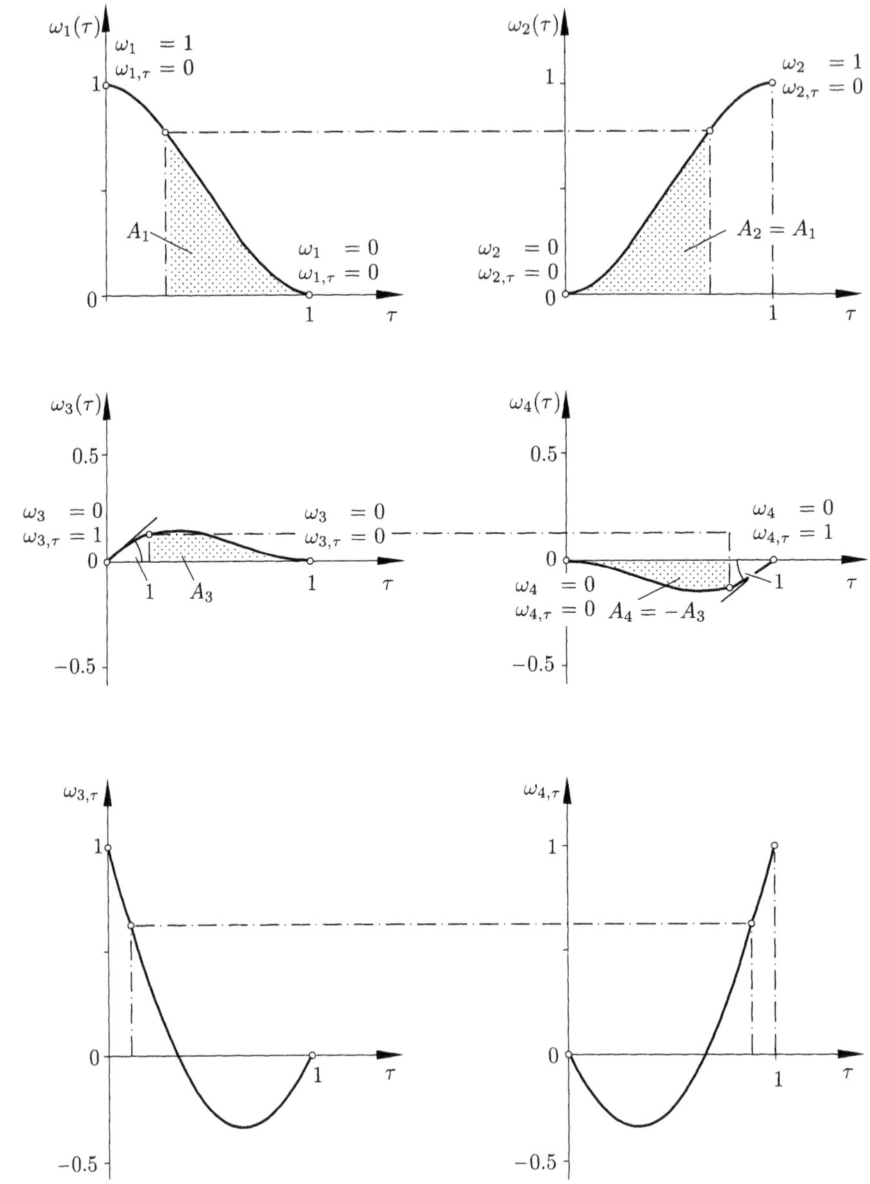

notation: $\dfrac{\partial \omega_i}{\partial \tau} = \omega_{i,\tau}$

Fig. 10.4.1 Essential conditions for the modified Hermitian interpolation scheme

To derive the displacements, we apply equ. (10.1.18)

$$\boldsymbol{r}_{k+1} = \boldsymbol{r}_k + \Delta t\dot{\boldsymbol{r}}_k + \Delta t^2\ddot{\boldsymbol{r}}_k\int_0^1(1-\tau)\,\omega_1(\tau)\,d\tau + \Delta t^2\ddot{\boldsymbol{r}}_{k+1}\int_0^1(1-\tau)\,\omega_2(\tau)\,d\tau$$
$$+ \Delta t^3\dddot{\boldsymbol{r}}_k\int_0^1(1-\tau)\,\omega_3(\tau)\,d\tau + \Delta t^3\dddot{\boldsymbol{r}}_{k+1}\int_0^1(1-\tau)\,\omega_4(\tau)\,d\tau \tag{10.4.6}$$

Noting that $f_1(\tau) = (1-\tau)\,\omega_1(\tau)$ is a mirror image of $\tau\omega_2(\tau)$ since

$$\begin{aligned} f_1(1-\tau) &= (1-1+\tau)\,\omega_1(1-\tau) \\ &= \tau\,\omega_2(\tau) \end{aligned} \tag{10.4.7}$$

we establish the relations

$$\int_0^1(1-\tau)\,\omega_1(\tau)\,d\tau = b_0 = \int_0^1\tau\,\omega_2(\tau)\,d\tau \tag{10.4.8}$$

and

$$\begin{aligned} \int_0^1(1-\tau)\,\omega_2(\tau)\,d\tau = c_0 &= \int_0^1\omega_2(\tau)\,d\tau - \int_0^1\tau\,\omega_2(\tau)\,d\tau \\ &= a_0 - b_0 \end{aligned} \tag{10.4.9}$$

Analogous to the above, we observe that $f_3(\tau) = (1-\tau)\,\omega_3(\tau)$ is a double mirror image of $\tau\omega_4(\tau)$ since

$$\begin{aligned} f_3(1-\tau) &= (1-1+\tau)\,\omega_3(1-\tau) \\ &= -\tau\,\omega_4(\tau) \end{aligned} \tag{10.4.10}$$

Hence

$$\int_0^1(1-\tau)\,\omega_3(\tau)\,d\tau = b_1 = -\int_0^1\tau\,\omega_4(\tau)\,d\tau \tag{10.4.11}$$

also holds, as does

$$\begin{aligned} \int_0^1(1-\tau)\,\omega_4(\tau)\,d\tau = c_1 &= \int_0^1\omega_4(\tau)\,d\tau - \int_0^1\tau\,\omega_4(\tau)\,d\tau \\ &= -a_1 + b_1 \end{aligned} \tag{10.4.12}$$

Thus, the ultimate expression for the displacement becomes

$$\boldsymbol{r}_{k+1} = \boldsymbol{r}_k + \Delta t\dot{\boldsymbol{r}}_k + \Delta t^2(b_0\ddot{\boldsymbol{r}}_k + c_0\ddot{\boldsymbol{r}}_{k+1}) + \Delta t^3(b_1\dddot{\boldsymbol{r}}_k + c_1\dddot{\boldsymbol{r}}_{k+1}) \tag{10.4.13}$$

in which connection we note the relations

$$b_0 + c_0 = a_0$$

$$b_1 - c_1 = a_1 \tag{10.4.14}$$

In order to guarantee convergence, it is necessary to ensure the exact representation of at least a constant acceleration. This means that the pivotal values

$$\ddot{\boldsymbol{r}}_k = \ddot{\boldsymbol{r}}_{k+1}$$

$$\dddot{\boldsymbol{r}}_k = \dddot{\boldsymbol{r}}_{k+1} = \boldsymbol{o} \tag{10.4.15}$$

should yield

$$\ddot{\boldsymbol{r}}(\tau) = \ddot{\boldsymbol{r}}_k = \text{const} \tag{10.4.16}$$

This, however, is only possible if the sum of the two mutually mirror-inverted functions ω_1 and ω_2 has a value of 1 for any τ

$$\omega_1(\tau) + \omega_2(\tau) = 1 \tag{10.4.17}$$

Equation (10.4.4) respectively (10.4.5) then determines the constant a_o as

$$a_0 = \tfrac{1}{2} \tag{10.4.18}$$

Noting equ. (10.4.14), we observe three disposable constants for the modified cubic algorithm and these can in principle be fixed arbitrarily without an explicit specification of the interpolation function as long as we renounce the condition of a consistent local acceleration. The cubic Hermitian polynomial approach thus forms a special case in the present, more general argument.

In addition to the two equs. (10.4.5) and (10.4.13) just derived, the equs. (10.2.51) and (10.2.52) played an important role in the development of the iteration formulae; in particular, we demanded that the differential equation of the problem and its first derivative be fulfilled at each pivotal point and thus also at t_{k+1}. In searching for the reason for the conditionality of the stability, we go on to test these four basic equations for very large time-steps in the case of $\boldsymbol{R}(t) = \boldsymbol{o}$, i.e. for free oscillations. Equation (10.4.5) leads to the following statement

$$\lim_{\Delta t \to \infty}\left(\frac{1}{\Delta t^2}\dot{\boldsymbol{r}}_{k+1}\right) = a_1\left[\dddot{\boldsymbol{r}}_k - \dddot{\boldsymbol{r}}_{k+1}\right] \tag{10.4.19}$$

Furthermore, equ. (10.4.13) yields

$$\lim_{\Delta t \to \infty}\left(\frac{1}{\Delta t^3}\boldsymbol{r}_{k+1}\right) = b_1\dddot{\boldsymbol{r}}_k + c_1\dddot{\boldsymbol{r}}_{k+1} \tag{10.4.20}$$

Hence, we deduce from equ. (10.2.52) in conjunction with $\dot{\boldsymbol{R}}_{k+1} = \boldsymbol{o}$

$$\lim_{\Delta t \to \infty} \left(\frac{1}{\Delta t^2} \left[\boldsymbol{M}\ddot{\boldsymbol{r}}_{k+1} + \boldsymbol{C}\ddot{\boldsymbol{r}}_{k+1} + \boldsymbol{K}\dot{\boldsymbol{r}}_{k+1} \right] \right) = \boldsymbol{K} a_1 \left[\dddot{\boldsymbol{r}}_k - \dddot{\boldsymbol{r}}_{k+1} \right] \overset{!}{=} \boldsymbol{o} \tag{10.4.21}$$

For very large time-steps Δt, the differential equ. (10.2.52) can only be satisfied if

$$\dddot{\boldsymbol{r}}_k - \dddot{\boldsymbol{r}}_{k+1} = \boldsymbol{o} \quad , \qquad \Delta t \to \infty \tag{10.4.22}$$

Otherwise, $\boldsymbol{K}$ must be singular, which we will exclude, or a_1 must be zero, which causes the disappearance of the forms ω_3 and ω_4. We now turn to equ. (10.2.51) and find

$$\begin{aligned} \lim_{\Delta t \to \infty} \left(\frac{1}{\Delta t^3} \left[\boldsymbol{M}\ddot{\boldsymbol{r}}_{k+1} + \boldsymbol{C}\dot{\boldsymbol{r}}_{k+1} + \boldsymbol{K}\boldsymbol{r}_{k+1} \right] \right) &= \boldsymbol{K} \left[b_1 \dddot{\boldsymbol{r}}_k + c_1 \dddot{\boldsymbol{r}}_{k+1} \right] \\ &= \boldsymbol{K} \left[b_1 + c_1 \right] \dddot{\boldsymbol{r}}_{k+1} \\ &\overset{!}{=} \boldsymbol{o} \end{aligned} \tag{10.4.23}$$

In order to satisfy this condition

$$b_1 + c_1 = 0 \tag{10.4.24}$$

must apply. From equ. (10.4.14), there follows

$$b_1 = -c_1 = \frac{a_1}{2} \tag{10.4.25}$$

The conditionally stable algorithm with cubic interpolation possesses the following constants [10.4]

$$a_1 = \tfrac{1}{12} \quad , \quad b_1 = \tfrac{1}{20} \quad , \quad c_1 = -\tfrac{1}{30} \tag{10.4.26}$$

and thus cannot satisfy equ. (10.4.23).

We deduce from equ. (10.4.24) that two disposable constants, e.g. a_1 and b_0, remain for the design of an unconditionally stable algorithm. To determine the constant a_1, we demand additionally that a linear acceleration is reproduced exactly, e.g.

$$\ddot{\boldsymbol{r}}(\tau) = \tau \ddot{\boldsymbol{r}}_0 \tag{10.4.27}$$

with the initial conditions

$$\begin{aligned} \boldsymbol{r}(0) &= \boldsymbol{o} \\ \dot{\boldsymbol{r}}(0) &= \boldsymbol{o} \end{aligned} \tag{10.4.28}$$

In accordance with equ. (10.4.27), we obtain by direct integration

$$\dot{\boldsymbol{r}}_{k+1} = \dot{\boldsymbol{r}}(\tau = 1) = \tfrac{1}{2}\Delta t\ddot{\boldsymbol{r}}_0 \tag{10.4.29}$$

and

$$\boldsymbol{r}_{k+1} = \boldsymbol{r}(\tau = 1) = \tfrac{1}{6}\Delta t^2\ddot{\boldsymbol{r}}_0 \tag{10.4.30}$$

for the aforementioned initial conditions. On the other hand, we deduce from equ. (10.4.27) the pivotal values

$$\begin{aligned} \ddot{\boldsymbol{r}}_k &= \boldsymbol{o} \\ \ddot{\boldsymbol{r}}_{k+1} &= \ddot{\boldsymbol{r}}_0 \\ \dddot{\boldsymbol{r}}_k &= \frac{1}{\Delta t}\ddot{\boldsymbol{r}}_0 = \dddot{\boldsymbol{r}}_{k+1} \end{aligned} \tag{10.4.31}$$

which lead in conjunction with equs. (10.4.5) and (10.4.28) to

$$\dot{\boldsymbol{r}}_{k+1} = a_0\Delta t\ddot{\boldsymbol{r}}_0 \tag{10.4.32}$$

and also from equ. (10.4.13) to

$$\boldsymbol{r}_{k+1} = \Delta t^2 c_0\ddot{\boldsymbol{r}}_0 + \Delta t^2(b_1 + c_1)\ddot{\boldsymbol{r}}_0 \tag{10.4.33}$$

For an exact reproduction of a linear polynomial in time, equs. (10.4.29) and (10.4.32) must be identical. Reference to equ. (10.4.18) shows this to be the case. Furthermore, equs. (10.4.30) and (10.4.33) must also be equal. If we observe equ. (10.4.24), we obtain

$$\begin{aligned} c_0 &= \tfrac{1}{6} \\ b_0 &= a_0 - c_0 = \tfrac{1}{2} - \tfrac{1}{6} = \tfrac{1}{3} \end{aligned} \tag{10.4.34}$$

Although a linear polynomial is thus exactly reproduced, the a_1 constant is still disposable. For this purpose, we go a step further and also require that a quadratic polynomial be realised exactly

$$\ddot{\boldsymbol{r}}(\tau) = \tau^2\ddot{\boldsymbol{r}}_0 \tag{10.4.35}$$

with the same initial conditions as in equ. (10.4.28). Equation (10.4.35) can once more be integrated directly and used for the determination of the pivotal values. This leads to the conditions

$$\dot{\boldsymbol{r}}_{k+1} = \tfrac{1}{3}\Delta t\ddot{\boldsymbol{r}}_0 \stackrel{!}{=} a_0\Delta t\ddot{\boldsymbol{r}}_0 - 2a_1\Delta t\ddot{\boldsymbol{r}}_0 \tag{10.4.36}$$

and

$$\boldsymbol{r}_{k+1} = \tfrac{1}{12}\Delta t^2\ddot{\boldsymbol{r}}_0 \stackrel{!}{=} \Delta t^2 c_0\ddot{\boldsymbol{r}}_0 + 2\Delta t^2 c_1\ddot{\boldsymbol{r}}_0 \tag{10.4.37}$$

From equ. (10.4.36), we deduce the constant a_1, and hence also c_1 and b_1

$$\begin{aligned} a_1 &= \tfrac{1}{2}a_0 - \tfrac{1}{6} = \tfrac{1}{12} \\ c_1 &= -\tfrac{1}{2}a_1 = -\tfrac{1}{24} = -b_1 \end{aligned} \tag{10.4.38}$$

In this argument, equ. (10.4.37) could not be used since all values have already been determined. However, since

$$c_0 + 2c_1 = \tfrac{1}{6} - \tfrac{1}{12} = \tfrac{1}{12} \tag{10.4.39}$$

is true, we can assert that our algorithm reproduces quadratic accelerations exactly with regard to velocity and displacement.

The resulting algorithmic expressions are

$$\dot{\boldsymbol{r}}_{k+1} = \dot{\boldsymbol{r}}_k + \tfrac{1}{2}\Delta t(\ddot{\boldsymbol{r}}_k + \ddot{\boldsymbol{r}}_{k+1}) + \tfrac{1}{12}\Delta t^2(\dddot{\boldsymbol{r}}_k - \dddot{\boldsymbol{r}}_{k+1}) \tag{10.4.40}$$

and

$$\boldsymbol{r}_{k+1} = \boldsymbol{r}_k + \Delta t\dot{\boldsymbol{r}}_k + \Delta t^2(\tfrac{1}{3}\ddot{\boldsymbol{r}}_k + \tfrac{1}{6}\ddot{\boldsymbol{r}}_{k+1}) + \tfrac{1}{24}\Delta t^3(\dddot{\boldsymbol{r}}_k - \dddot{\boldsymbol{r}}_{k+1}) \tag{10.4.41}$$

Using these equations as well as the ensuing equs. (10.4.51) and (10.4.52), we can construct alternative algorithmic variants with different coefficients. In order to check the stability and assess the accuracy, we recommend the following algorithmic form ($\boldsymbol{R} = \boldsymbol{o}$)

$$\boldsymbol{D}_1 \begin{bmatrix} \boldsymbol{r} \\ \dot{\boldsymbol{r}} \end{bmatrix}_{k+1} = \boldsymbol{D}_0 \begin{bmatrix} \boldsymbol{r} \\ \dot{\boldsymbol{r}} \end{bmatrix}_k \tag{10.4.42}$$

which should serve as a basis for the calculation of free oscillations. With a little effort, we may now deduce from the aforementioned four basic equations

$$\boldsymbol{D}_1 = \begin{bmatrix} \boldsymbol{M} + \tfrac{1}{6}\Delta t^2\boldsymbol{K} + \tfrac{1}{24}\Delta t^3\boldsymbol{C}\boldsymbol{M}^{-1}\boldsymbol{K} & \tfrac{1}{6}\Delta t^2\boldsymbol{C} - \tfrac{1}{24}\Delta t^3[\boldsymbol{K} - \boldsymbol{C}\boldsymbol{M}^{-1}\boldsymbol{C}] \\ \tfrac{1}{2}\Delta t\boldsymbol{K} + \tfrac{1}{12}\Delta t^2\boldsymbol{C}\boldsymbol{M}^{-1}\boldsymbol{K} & \boldsymbol{M} + \tfrac{1}{2}\Delta t\boldsymbol{C} - \tfrac{1}{12}\Delta t^2[\boldsymbol{K} - \boldsymbol{C}\boldsymbol{M}^{-1}\boldsymbol{C}] \end{bmatrix} \tag{10.4.43}$$

which is analogous to equ. (10.3.27), and

$$\boldsymbol{D}_0 = \begin{bmatrix} \boldsymbol{M} - \tfrac{1}{3}\Delta t^2\boldsymbol{K} + \tfrac{1}{24}\Delta t^3\boldsymbol{C}\boldsymbol{M}^{-1}\boldsymbol{K} & \Delta t\boldsymbol{M} - \tfrac{1}{3}\Delta t^2\boldsymbol{C} - \tfrac{1}{24}\Delta t^3[\boldsymbol{K} - \boldsymbol{C}\boldsymbol{M}^{-1}\boldsymbol{C}] \\ -\tfrac{1}{2}\Delta t\boldsymbol{K} + \tfrac{1}{12}\Delta t^2\boldsymbol{C}\boldsymbol{M}^{-1}\boldsymbol{K} & \boldsymbol{M} - \tfrac{1}{2}\Delta t\boldsymbol{C} - \tfrac{1}{12}\Delta t^2[\boldsymbol{K} - \boldsymbol{C}\boldsymbol{M}^{-1}\boldsymbol{C}] \end{bmatrix} \tag{10.4.44}$$

which is, in turn, analogous to equ. (10.3.28). To simplify the argument, we restrict our subsequent discussion to undamped systems with a single degree of freedom. Analogous to equ. (10.3.29), we obtain

$$\begin{bmatrix} 1+\frac{1}{6}\vartheta^2 & -\frac{1}{24}\Delta t\vartheta^2 \\ \frac{1}{2}\frac{\vartheta^2}{\Delta t} & 1-\frac{1}{12}\vartheta^2 \end{bmatrix} \begin{bmatrix} r \\ \dot{r} \end{bmatrix}_{k+1} = \begin{bmatrix} 1-\frac{1}{3}\vartheta^2 & \Delta t(1-\frac{1}{24}\vartheta^2) \\ -\frac{1}{2}\frac{\vartheta^2}{\Delta t} & 1-\frac{1}{12}\vartheta^2 \end{bmatrix} \begin{bmatrix} r \\ \dot{r} \end{bmatrix}_{k} \tag{10.4.45}$$

or applying the standard form of equ. (10.2.19)

$$\begin{bmatrix} r \\ \dot{r} \end{bmatrix}_{k+1} = \frac{1}{1+\frac{1}{12}\vartheta^2+\frac{1}{144}\vartheta^4} \begin{bmatrix} 1-\frac{5}{12}\vartheta^2+\frac{1}{144}\vartheta^4 & \Delta t(1-\frac{1}{12}\vartheta^2) \\ -\frac{\vartheta^2}{\Delta t}(1-\frac{1}{12}\vartheta^2) & 1-\frac{5}{12}\vartheta^2+\frac{1}{144}\vartheta^4 \end{bmatrix} \begin{bmatrix} r \\ \dot{r} \end{bmatrix}_{k}$$

$$= \boldsymbol{A}\{r \quad \dot{r}\}_k \tag{10.4.46}$$

In order to reproduce an oscillatory motion, inequality (10.3.32) must be satisfied once more. Hence

$$a_{12}a_{21} = \frac{-\vartheta^2\left(1-\frac{1}{12}\vartheta^2\right)^2}{\left(1+\frac{1}{12}\vartheta^2+\frac{1}{144}\vartheta^4\right)^2} < 0 \tag{10.4.47}$$

This requirement is obviously fulfilled for arbitrary step-sizes ϑ. Following equ. (10.3.33), we next calculate the spectral radius and obtain

$$\rho^2 = \frac{\left(1-\frac{5}{12}\vartheta^2+\frac{1}{144}\vartheta^4\right)^2+\vartheta^2\left(1-\frac{1}{12}\vartheta^2\right)^2}{\left(1+\frac{1}{12}\vartheta^2+\frac{1}{144}\vartheta^4\right)^2} \tag{10.4.48}$$

$$= 1$$

Clearly, no problems arise in this respect. As a final point, we should inquire into the accuracy and discuss, as before, the period distortion for a free oscillation. This is determined in general by equ. (10.3.44). Applying the first quotient of equ. (10.3.53), we obtain

$$\tan\Phi = \frac{\left(-a_{12}a_{21}\right)^{1/2}}{a_{11}} = \frac{\vartheta(1-\frac{1}{12}\vartheta^2)}{1-\frac{5}{12}\vartheta^2+\frac{1}{144}\vartheta^4} \tag{10.4.49}$$

Equation (10.4.49) hence possesses vertical asymptotes at

$$\vartheta_{\mathrm{As}}^2 = 30 \pm \sqrt{756} \tag{10.4.50}$$

for which

$$\tan\Phi(\vartheta_{\mathrm{As}}) = \pm\infty \tag{10.4.51}$$

Thus, we find

$$\frac{\widetilde{T}}{T}(\vartheta_{\mathrm{As}} = 1.58) = \frac{\vartheta_{\mathrm{As}}}{\pi/2} = 1.0075$$

and

$$\frac{\widetilde{T}}{T}(\vartheta_{\mathrm{As}} = 7.58) = \frac{\vartheta_{\mathrm{As}}}{3\pi/2} = 1.6091$$

With reference to the relative period elongation of 60%, we should note that the conditionally stable algorithm only works up to $\vartheta = 3.16$. As was the case for the unconditionally stable Newmark method, ($\beta = 1/4$), stability is achieved at the expense of accuracy. This also can be deduced from the error curves in fig. 10.4.2. In comparison to Wilson's unconditionally stable method ($\theta = 1.5$) and the Newmark procedure with $\beta = 1/4$, the modified cubic Hermitian algorithm wins hands down.

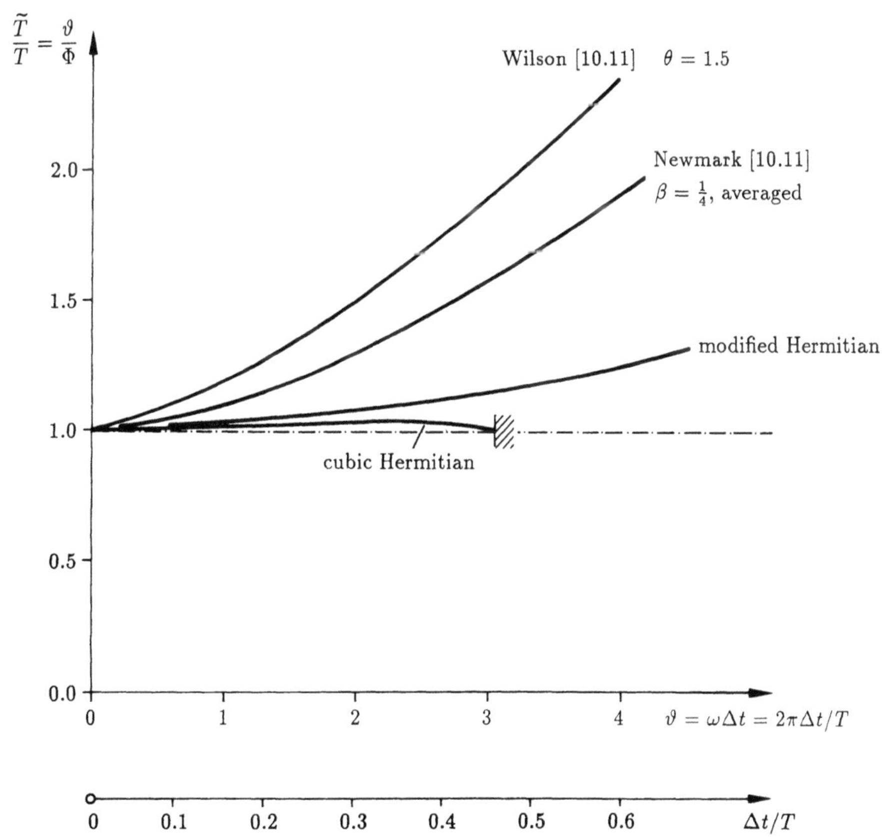

Fig. 10.4.2 Period elongation in dependence of the step-size for some unconditionally stable procedures and the conditionally stable cubic Hermitian algorithm

This chapter would be incomplete without a short discussion of how the modification can be extended to the whole family of Hermitian basic algorithms. In our presentation, we follow [10.13]. At the instants k and k+1, the complete group of Hermitian interpolations possesses the following pivotal values in a sequential order

$$
\begin{array}{llcl}
 & & \text{order} & \text{of interpolation} \\
1. & \ddot{\boldsymbol{r}} & \rightarrow & \text{quadratic (Newmark)} \\
2. & \ddot{\boldsymbol{r}}\,,\ \dddot{\boldsymbol{r}} & \rightarrow & \text{cubic} \\
3. & \ddot{\boldsymbol{r}}\,,\ \dddot{\boldsymbol{r}}\,,\ \ddddot{\boldsymbol{r}} & \rightarrow & \text{quartic} \\
\vdots & & \vdots & \\
\text{n.} & \ddot{\boldsymbol{r}}\,,\ \dots\quad \boldsymbol{r}^{(n+1)} = \ddot{\boldsymbol{r}}^{(n-1)} & \rightarrow & (n+1)\text{-th}
\end{array}
\tag{10.4.52}
$$

(here, the n-th time derivative of $\boldsymbol{r}$ is denoted by $\boldsymbol{r}^{(n)}$ or $\ddot{\boldsymbol{r}}^{(n-2)}$).

It is interesting to note that the first algorithm coincides with the Newmark family. The n-th group encompasses time-derivatives of order up to n−1 of the accelerations, i.e. up to order n+1 for displacements, as pivotal values. The interpolation forms appertaining to the steps k and k+1 are denoted by ω_{kj} (pivotal point k, pivotal value $\boldsymbol{r}_k^{(j+2)}$ or $\ddot{\boldsymbol{r}}_k^{(j)}$) or $\omega_{k+1,j}$ respectively. Consequently, we can express the course of acceleration for the n-th family as

$$\ddot{\boldsymbol{r}}(\tau) = \sum_{j=0}^{n-1} (\Delta t)^j \left[\omega_{kj}(\tau)\ddot{\boldsymbol{r}}_k^{(j)} + \omega_{k+1,j}(\tau)\ddot{\boldsymbol{r}}_{k+1}^{(j)}\right] \tag{10.4.53}$$

As before, it is necessary for $\ddot{\boldsymbol{r}}_k^{(j)}$ and $\ddot{\boldsymbol{r}}_{k+1}^{(j)}$ to have a balanced influence upon the acceleration within the element. This leads us to prescribe

$$\omega_{kj}(\tau) = (-1)^j \omega_{k+1,j}(1-\tau) \tag{10.4.54}$$

which corresponds to equ. (10.4.3). Applying this requirement, we obtain from equ. (10.4.53) by integration over the complete time-interval

$$\dot{\boldsymbol{r}}_{k+1} = \boldsymbol{r}_k + \sum_{j=0}^{n-1} a_j (\Delta t)^{j+1} \left[\ddot{\boldsymbol{r}}_k^{(j)} + (-1)^j \ddot{\boldsymbol{r}}_{k+1}^{(j)}\right] \tag{10.4.55}$$

for the velocity and

$$\boldsymbol{r}_{k+1} = \boldsymbol{r}_k + \Delta t \dot{\boldsymbol{r}}_k + \sum_{j=1}^{n} (\Delta t)^{j+2} \left[b_j \ddot{\boldsymbol{r}}_k^{(j)} + c_j \ddot{\boldsymbol{r}}_{k+1}^{(j)}\right] \tag{10.4.56}$$

for the displacement with

$$b_j + (-1)^j c_j = a_j \tag{10.4.57}$$

In order to establish the entire iteration procedure, equs. (10.4.55) and (10.4.56) must be supplemented by the basic equilibrium conditions and their derivatives. For the point in time (k+1), we obtain

$$
\begin{aligned}
\boldsymbol{M}\ddot{\boldsymbol{r}}_{k+1} &+ \boldsymbol{C}\dot{\boldsymbol{r}}_{k+1} + \boldsymbol{K}\boldsymbol{r}_{k+1} = \boldsymbol{R}_{k+1}\\
\boldsymbol{M}\ddot{\boldsymbol{r}}^{(1)}_{k+1} &+ \boldsymbol{C}\dot{\boldsymbol{r}}^{(1)}_{k+1} + \boldsymbol{K}\boldsymbol{r}^{(1)}_{k+1} = \boldsymbol{R}^{(1)}_{k+1}\\
&\vdots\\
\boldsymbol{M}\ddot{\boldsymbol{r}}^{(n-1)}_{k+1} &+ \boldsymbol{C}\dot{\boldsymbol{r}}^{(n-1)}_{k+1} + \boldsymbol{K}\boldsymbol{r}^{(n-1)}_{k+1} = \boldsymbol{R}^{(n-1)}_{k+1}
\end{aligned}
\tag{10.4.58}
$$

Proceeding to the limit, we deduce from equ. (10.4.55)

$$
\lim_{\Delta t\to\infty}\left(\frac{1}{(\Delta t)^{n}}\dot{\boldsymbol{r}}_{k+1}\right) = a_{n-1}\left[\ddot{\boldsymbol{r}}^{(n-1)}_{k} + (-1)^{n-1}\ddot{\boldsymbol{r}}^{(n-1)}_{k+1}\right]
\tag{10.4.59}
$$

and analogous to this, we have from equ. (10.4.56)

$$
\lim_{\Delta t\to\infty}\left(\frac{1}{(\Delta t)^{n+1}}\boldsymbol{r}_{k+1}\right) = b_{n-1}\ddot{\boldsymbol{r}}^{(n-1)}_{k} + c_{n-1}\ddot{\boldsymbol{r}}^{(n-1)}_{k+1}
\tag{10.4.60}
$$

In conjunction with equ. (10.4.59), the second equation in the set (10.4.58) yields the expression

$$
\lim_{\Delta t\to\infty}\frac{1}{(\Delta t)^{n}}\left[\boldsymbol{M}\dddot{\boldsymbol{r}}_{k+1} + \boldsymbol{C}\ddot{\boldsymbol{r}}_{k+1} + \boldsymbol{K}\dot{\boldsymbol{r}}_{k+1}\right] = \boldsymbol{K}a_{n-1}\left[\ddot{\boldsymbol{r}}^{(n-1)}_{k} + (-1)^{n-1}\ddot{\boldsymbol{r}}^{(n-1)}_{k+1}\right] \overset{!}{=} \boldsymbol{o}
\tag{10.4.61}
$$

for the case $\boldsymbol{R} = \boldsymbol{o}$ (free oscillations). This can only be satisfied in a non-trivial manner if

$$
\ddot{\boldsymbol{r}}^{(n-1)}_{k} + (-1)^{n-1}\ddot{\boldsymbol{r}}^{(n-1)}_{k+1} = \boldsymbol{o}, \qquad \text{for} \qquad \Delta t \to \infty
\tag{10.4.62}
$$

Following equs. (10.4.60) and (10.4.62), the first equation of (10.4.58) yields the following expression

$$
\begin{aligned}
\lim_{\Delta t\to\infty}\frac{1}{(\Delta t)^{n+1}}\left[\boldsymbol{M}\ddot{\boldsymbol{r}}_{k+1} + \boldsymbol{C}\dot{\boldsymbol{r}}_{k+1} + \boldsymbol{K}\boldsymbol{r}\right] &= b_{n-1}\ddot{\boldsymbol{r}}^{(n-1)}_{k} + c_{n-1}\ddot{\boldsymbol{r}}^{(n-1)}_{k+1}\\
&= \left(b_{n-1} - (-1)^{n-1}c_{n-1}\right)\ddot{\boldsymbol{r}}^{(n-1)}_{k+1}\\
&\overset{!}{=} \boldsymbol{o}
\end{aligned}
\tag{10.4.63}
$$

This requirement can be fulfilled only if

$$
b_{n-1} - (-1)^{n-1}c_{n+1} = 0
\tag{10.4.64}
$$

Observing the condition of equ. (10.4.57), we note that (2n−1) constants are still undetermined. To fix them, we demand that polynomials up to a certain order can be reproduced exactly. Considering the polynomial *ansatz*

$$\ddot{\boldsymbol{r}} = \tau^{\mathrm{m}}\ddot{\boldsymbol{r}}_0 \tag{10.4.65}$$

we find for $\tau = 1$ the velocity

$$\dot{\boldsymbol{r}}_{\mathrm{k}+1} = \frac{1}{\mathrm{m}+1}\Delta t\ddot{\boldsymbol{r}}_0 \tag{10.4.66}$$

and the displacement

$$\boldsymbol{r}_{\mathrm{k}+1} = \frac{1}{(\mathrm{m}+1)(\mathrm{m}+2)}\Delta t^2\ddot{\boldsymbol{r}}_0 \tag{10.4.67}$$

by direct integration. On the other hand, differentiation of equ. (10.4.65) furnishes the basis for the pivotal values at the points in time k ($\tau = 0$) and k + 1 ($\tau = 1$)

$$\begin{aligned}
\ddot{\boldsymbol{r}}^{(0)} &= \tau^{\mathrm{m}}\ddot{\boldsymbol{r}}_0 \\
\ddot{\boldsymbol{r}}^{(1)} &= \mathrm{m}\tau^{\mathrm{m}-1}\ddot{\boldsymbol{r}}_0\Delta t^{-1} \\
\ddot{\boldsymbol{r}}^{(2)} &= \mathrm{m}(\mathrm{m}-1)\tau^{\mathrm{m}-2}\ddot{\boldsymbol{r}}_0\Delta t^{-2} \\
&\vdots \\
\ddot{\boldsymbol{r}}^{(\mathrm{m})} &= \mathrm{m}(\mathrm{m}-1)(\mathrm{m}-2)\ldots 1\ddot{\boldsymbol{r}}_0\Delta t^{-\mathrm{m}} \\
\ddot{\boldsymbol{r}}^{(\mathrm{m}+1)} &= \boldsymbol{o}
\end{aligned} \tag{10.4.68}$$

There follows

$$\begin{aligned}
\ddot{\boldsymbol{r}}_{\mathrm{k}}^{(\mathrm{j})} &= \boldsymbol{o} && \text{if} \quad \mathrm{j} \neq \mathrm{m} \\
\ddot{\boldsymbol{r}}_{\mathrm{k}}^{(\mathrm{j})} &= \mathrm{m}!\ddot{\boldsymbol{r}}_0\Delta t^{-\mathrm{j}} && \text{if} \quad \mathrm{j} = \mathrm{m}
\end{aligned} \tag{10.4.69}$$

and furthermore

$$\begin{aligned}
\ddot{\boldsymbol{r}}_{\mathrm{k}+1}^{(\mathrm{j})} &= \frac{\mathrm{m}!}{(\mathrm{m}-\mathrm{j})!}\Delta t^{-\mathrm{j}}\ddot{\boldsymbol{r}}_0 && \text{if} \quad \mathrm{j} \leqslant \mathrm{m} \\
\ddot{\boldsymbol{r}} &= 0 && \text{if} \quad \mathrm{j} > \mathrm{m}
\end{aligned} \tag{10.4.70}$$

We next require that the direct integration of the τ^{m} polynomial (noting the result of equ. (10.4.66)) must coincide with the solution in equ. (10.4.55) based on the pivotal values

$$\begin{aligned}
\frac{1}{\mathrm{m}+1} &= \mathrm{m}!a_{\mathrm{m}} + \sum_{\mathrm{j}=0}^{\mathrm{m}} \frac{\mathrm{m}!}{(\mathrm{m}-\mathrm{j})!}(-1)^{\mathrm{j}}a_{\mathrm{j}} && \text{if} \quad \mathrm{m} \leqslant \mathrm{n}-1 \\
&= \sum_{\mathrm{j}=0}^{\mathrm{n}-1} \frac{\mathrm{m}!}{(\mathrm{m}-\mathrm{j})!}(-1)^{\mathrm{j}}a_{\mathrm{j}} && \text{if} \quad \mathrm{m} > \mathrm{n}-1
\end{aligned} \tag{10.4.71}$$

We select the n=2 (cubic) family as a test case and deduce from equ. (10.4.71) the equations for m=0 to m=3

$$\begin{aligned}
1 &= a_0 + a_0 \\
\tfrac{1}{2} &= a_1 + a_0 - a_1 \\
\tfrac{1}{3} &= \quad a_0 - 2a_1 \\
\tfrac{1}{4} &= \quad a_0 - 3a_1
\end{aligned} \tag{10.4.72}$$

We immediately observe that the system possesses supernumerary equations. However, in the derivation of the modified cubic Hermitian algorithm, we did not require the m=3 polynomial either. Nevertheless, the system (10.4.72) contains no contradiction, since the first two equations are linearly dependent on the final two. Hence, the modified cubic Hermitian algorithm exactly reproduces a polynomial with a power of up to 3 as far as the velocity is concerned.

For the solution of equ. (10.4.71), we must select n truly linear independent equations. The example n=2 demonstrates the unsuitability of the first group for m $\leqslant$ n$-$1, since the first and second equations are identical. On the other hand, sufficient linearly independent equations remain in the m > n$-$1 group. The system for the calculation of the a_j coefficients is thus

$$\sum_{j=0}^{n-1} \frac{m!}{(m-j)!}(-1)^j a_j = \frac{1}{m+1} \quad , \qquad m = n \quad \text{to} \quad 2n-1 \tag{10.4.73}$$

and can also be written in matrix form

$$\begin{bmatrix} 1 & \frac{n!}{(n-1)!} & \frac{n!}{(n-2)!} & \cdots & n! \\ 1 & \frac{(n+1)!}{n!} & \frac{(n+1)!}{(n-1!)} & & \frac{(n+1)!}{2!} \\ \vdots & & & & \vdots \\ 1 & \frac{(2n-1)!}{(2n-2)!} & \frac{(2n-1)!}{(2n-3)!} & \cdots & \frac{(2n-1)!}{n!} \end{bmatrix} \begin{bmatrix} a_0 \\ -a_1 \\ \vdots \\ (-1)^{n-1} a_{n-1} \end{bmatrix} = \begin{bmatrix} \frac{1}{n+1} \\ \frac{1}{n+2} \\ \vdots \\ \frac{1}{2n} \end{bmatrix} \tag{10.4.74}$$

For example, we obtain the following system for n=4, i.e. for the 4th member of the Hermitian family (quintic interpolation) which works with third derivatives of the acceleration as pivotal values

$$\begin{bmatrix} 1 & 4 & 12 & 24 \\ 1 & 5 & 20 & 60 \\ 1 & 6 & 30 & 120 \\ 1 & 7 & 42 & 210 \end{bmatrix} \begin{bmatrix} a_0 \\ -a_1 \\ a_2 \\ -a_3 \end{bmatrix} = \begin{bmatrix} \frac{1}{5} \\ \frac{1}{6} \\ \frac{1}{7} \\ \frac{1}{8} \end{bmatrix} \tag{10.4.75a}$$

the solution of which is

$$a_0 = \tfrac{1}{2}, \quad a_1 = \tfrac{3}{28}, \quad a_2 = \tfrac{1}{84}, \quad a_3 = \tfrac{1}{1680} \tag{10.4.75b}$$

The group of equations not considered in this argument

$$\begin{aligned} 1 &= a_0 + a_0 \\ \tfrac{1}{2} &= a_1 + a_0 - a_1 \\ \tfrac{1}{3} &= 2a_2 + a_0 - 2a_1 + 2a_2 \\ \tfrac{1}{4} &= 6a_3 + a_0 - 3a_1 + 6a_2 - 3a_3 \end{aligned} \tag{10.4.75c}$$

is obviously automatically fulfilled, which can, of course, also be shown to be true in general terms.

Let us now turn to the displacements. Direct integration of a time polynomial τ^{m} leads to equ. (10.4.67) whilst the solution deduced from the pivotal values in equs. (10.4.69) and (10.4.70) is obtained from equ. (10.4.56). Assuming the initial conditions $\boldsymbol{r}_{\mathrm{k}} = \boldsymbol{o}$ and $\dot{\boldsymbol{r}}_{\mathrm{k}} = \boldsymbol{o}$, we find

$$\begin{aligned} \frac{1}{(\mathrm{m}+1)(\mathrm{m}+2)} &= \mathrm{m}!\, b_{\mathrm{m}} + \sum_{\mathrm{j}=0}^{\mathrm{m}} \frac{\mathrm{m}!}{(\mathrm{m}-\mathrm{j})!} c_{\mathrm{j}} \qquad \text{if} \quad \mathrm{m} \leqslant \mathrm{n}-1 \\ &= \sum_{\mathrm{j}=0}^{\mathrm{n}-1} \frac{\mathrm{m}!}{(\mathrm{m}-\mathrm{j})!} c_{\mathrm{j}} \qquad \text{if} \quad \mathrm{m} > \mathrm{n}-1 \end{aligned} \tag{10.4.76}$$

The coefficients of this system are, in absolute terms, identical to those used in the calculation of a_{j}, but the right-hand sides of the equations are different. When establishing the system for the determination of the c_{j}, it should be noted that, following the calculation of the a_{j}, only (n−1) of the n values of c_{j} are deducable from equ. (10.4.76), since the condition of equ. (10.4.64) must be maintained. Therefore, we employ the polynomial up to a degree of 2n−2. The n=2 family (cubic) is considered again as a test case for which we obtain the equations

$$\begin{aligned} \mathrm{m}=0: \quad \tfrac{1}{2} &= b_0 + c_0 \\ \mathrm{m}=1: \quad \tfrac{1}{6} &= b_1 + c_0 + c_1 \\ \mathrm{m}=2: \quad \tfrac{1}{12} &= c_0 + 2c_1 \end{aligned} \tag{10.4.77}$$

where, additionally,

$$b_1 = -c_1 = \tfrac{1}{2} a_1 = \tfrac{1}{24} \tag{10.4.78}$$

must also hold. The only remaining unknown is c_0, and, once again, more equations are available than necessary. However, two equations are superfluous and thus no contradiction arises. In fact, the first equation confirms a_0 which is already known, whilst the second and third agree with the equations determining a_1, which has already been evaluated. In order to determine c_{j}, we utilise as before the equations of the second group in equ. (10.4.76) which guarantees linear independence. In this way, we obtain

$$\sum_{\mathrm{j}=0}^{\mathrm{n}-2} \frac{\mathrm{m}!}{(\mathrm{m}-\mathrm{j})!} c_{\mathrm{j}} = \frac{1}{(\mathrm{m}+1)(\mathrm{m}+2)} - \frac{\mathrm{m}!}{(\mathrm{m}-\mathrm{n}+1)!} \frac{a_1}{2}, \qquad \mathrm{m}=\mathrm{n} \text{ to } 2\mathrm{n}-2$$

(10.4.79)

or in matrix form

$$\begin{bmatrix} 1 & \frac{\mathrm{n}!}{(\mathrm{n}-1)!} & \frac{\mathrm{n}!}{(\mathrm{n}-2)!} & \cdots & \frac{\mathrm{n}!}{2!} \\ 1 & \frac{(\mathrm{n}+1)!}{\mathrm{n}!} & \frac{(\mathrm{n}+1)!}{(\mathrm{n}-1!)} & & \frac{(\mathrm{n}+1)!}{3!} \\ \vdots & & & & \vdots \\ 1 & \frac{(2\mathrm{n}-2)!}{(2\mathrm{n}-3)!} & \frac{(2\mathrm{n}-2)!}{(2\mathrm{n}-4)!} & \cdots & \frac{(2\mathrm{n}-2)!}{\mathrm{n}!} \end{bmatrix} \begin{bmatrix} c_0 \\ c_1 \\ \vdots \\ c_{\mathrm{n}-2} \end{bmatrix} = \begin{bmatrix} \frac{1}{2} \\ \frac{1}{6} \\ \vdots \\ \frac{1}{2\mathrm{n}(2\mathrm{n}-1)} \end{bmatrix} - \begin{bmatrix} \mathrm{n}! \\ \frac{(\mathrm{n}+1)!}{2!} \\ \vdots \\ \frac{(2\mathrm{n}-2)!}{(\mathrm{n}-1)} \end{bmatrix} c_{\mathrm{n}-1}$$

(10.4.80)

where the relation

$$b_{\mathrm{n}-1} = (-1)^{\mathrm{n}-1} c_{\mathrm{n}-1} = \frac{a_{\mathrm{n}-1}}{2} \qquad (10.4.81)$$

holds. The 4th member of the modified Hermitian family becomes (see also equ. (10.4.75a))

$$\begin{bmatrix} 1 & 4 & 12 \\ 1 & 5 & 20 \\ 1 & 6 & 30 \end{bmatrix} \begin{bmatrix} c_0 \\ c_1 \\ c_2 \end{bmatrix} = \begin{bmatrix} \frac{1}{2} \\ \frac{1}{6} \\ \frac{1}{20} \end{bmatrix} + \frac{1}{3360} \begin{bmatrix} 24 \\ 60 \\ 120 \end{bmatrix} \qquad (10.4.82a)$$

with

$$b_3 = -c_3 = \tfrac{1}{2} a_3 = \tfrac{1}{3360} \qquad (10.4.82b)$$

The solution of equ. (10.4.82a) then yields

$$c_0 = \tfrac{1}{7}, \quad c_1 = -\tfrac{1}{24}, \quad c_2 = \tfrac{3}{560} \qquad (10.4.82c)$$

from which, noting equ. (10.4.57), we obtain

$$b_0 = \tfrac{5}{14}, \quad b_1 = \tfrac{11}{168}, \quad b_2 = \tfrac{11}{1680} \qquad (10.4.82d)$$

The complete algorithmic family from n=1 to n=4 is summarised in fig. 10.4.3. In practice, we can ignore the range n>4 and in most cases, n=2 proves sufficient. We must also remember that for higher algorithms, we require the corresponding derivatives of the excitation force.

n		$\ddot{\boldsymbol{r}}_{k}$	$\ddot{\boldsymbol{r}}_{k+1}$	$\dddot{\boldsymbol{r}}_{k}$	$\dddot{\boldsymbol{r}}_{k+1}$	$\ddddot{\boldsymbol{r}}_{k}$	$\ddddot{\boldsymbol{r}}_{k+1}$	$\overset{\cdot\cdot\cdot\cdot\cdot}{\boldsymbol{r}}_{k}$	$\overset{\cdot\cdot\cdot\cdot\cdot}{\boldsymbol{r}}_{k+1}$
1	$\dot{\boldsymbol{r}}_{k+1} = \dot{\boldsymbol{r}}_{k} +$ $\boldsymbol{r}_{k+1} = \boldsymbol{r}_{k} + \Delta t \dot{\boldsymbol{r}}_{k} +$	$\frac{1}{2}\Delta t$ $\frac{1}{4}\Delta t^2$	$\frac{1}{2}\Delta t$ $\frac{1}{4}\Delta t^2$						
2	$\dot{\boldsymbol{r}}_{k+1} = \dot{\boldsymbol{r}}_{k} +$ $\boldsymbol{r}_{k+1} = \boldsymbol{r}_{k} + \Delta t \dot{\boldsymbol{r}}_{k} +$	$\frac{1}{2}\Delta t$ $\frac{1}{3}\Delta t^2$	$\frac{1}{2}\Delta t$ $\frac{1}{6}\Delta t^2$	$\frac{1}{12}\Delta t^2$ $\frac{1}{24}\Delta t^3$	$-\frac{1}{12}\Delta t^2$ $-\frac{1}{24}\Delta t^3$				
3	$\dot{\boldsymbol{r}}_{k+1} = \dot{\boldsymbol{r}}_{k} +$ $\boldsymbol{r}_{k+1} = \boldsymbol{r}_{k} + \Delta t \dot{\boldsymbol{r}}_{k} +$	$\frac{1}{2}\Delta t$ $\frac{7}{20}\Delta t^2$	$\frac{1}{2}\Delta t$ $\frac{3}{20}\Delta t^2$	$\frac{1}{10}\Delta t^2$ $\frac{7}{120}\Delta t^3$	$-\frac{1}{10}\Delta t^2$ $-\frac{1}{24}\Delta t^3$	$\frac{1}{120}\Delta t^3$ $\frac{1}{240}\Delta t^4$	$\frac{1}{120}\Delta t^3$ $\frac{1}{240}\Delta t^4$		
4	$\dot{\boldsymbol{r}}_{k+1} = \dot{\boldsymbol{r}}_{k} +$ $\boldsymbol{r}_{k+1} = \boldsymbol{r}_{k} + \Delta t \dot{\boldsymbol{r}}_{k} +$	$\frac{1}{2}\Delta t$ $\frac{5}{14}\Delta t^2$	$\frac{1}{2}\Delta t$ $\frac{1}{7}\Delta t^2$	$\frac{3}{28}\Delta t^2$ $\frac{11}{168}\Delta t^3$	$-\frac{3}{28}\Delta t^2$ $-\frac{1}{24}\Delta t^3$	$\frac{1}{84}\Delta t^3$ $\frac{11}{1680}\Delta t^4$	$\frac{1}{84}\Delta t^3$ $\frac{3}{560}\Delta t^4$	$\frac{1}{1680}\Delta t^4$ $\frac{1}{3360}\Delta t^5$	$-\frac{1}{1680}\Delta t^4$ $-\frac{1}{3360}\Delta t^5$

Fig. 10.4.3
The first four members in the family of unconditionally stable modified Hermitian algorithms

10.5 A short introduction into the linear multi-step methods (LMS) and the α-method in structural dynamics

All the methods discussed so far were concerned with the transition from a given state at the beginning of the step k ($\boldsymbol{r}_k, \dot{\boldsymbol{r}}_k, \ddot{\boldsymbol{r}}_k$) to a new state at the end of the k-th step ($\boldsymbol{r}_{k+1}, \dot{\boldsymbol{r}}_{k+1}, \ddot{\boldsymbol{r}}_{k+1}$), thereby using information at the beginning and at the end of this step only. These algorithms thus belong to the class of so-called single-step methods . If we are to seek the inclusion of pivotal values from several time-steps within a temporal integration procedure, we are led to the integration of the equations of motion via the so-called multi-step methods. Within this methodology, the linear multi-step or LMS methods, the step-by-step deflections $\boldsymbol{r}_i$(i=0,k) are linked in a linear expression (*sic nomen*) to the velocities $\dot{\boldsymbol{r}}_i$ (k-step procedure of the 1st order) and possibly also to some higher derivatives, e.g. to accelerations $\ddot{\boldsymbol{r}}_i$ (k-step method of the 2nd order). It should be noted that 'linear' does not refer to the differential equation to be integrated; this equation may be non-linear. In what follows, we present an *aperçu* of some LMS-methods.

Houbolt's method

The method [10.19] is mainly of historical significance, but it was one of the first procedures developed for the computer analysis of aircraft dynamics. It is based upon the following system of equations

$$\boldsymbol{M}\ddot{\boldsymbol{r}}_{k+1} + \boldsymbol{C}\dot{\boldsymbol{r}}_{k+1} + \boldsymbol{K}\boldsymbol{r}_{k+1} = \boldsymbol{R}_{k+1} \qquad (10.5.1)$$

$$\ddot{\boldsymbol{r}}_{k+1} = (2\boldsymbol{r}_{k+1} - 5\boldsymbol{r}_k + 4\boldsymbol{r}_{k-1} - \boldsymbol{r}_{k-2})/\Delta t^2 \qquad (10.5.2)$$

$$\dot{\boldsymbol{r}}_{k+1} = (11\boldsymbol{r}_{k+1} - 18\boldsymbol{r}_k + 9\boldsymbol{r}_{k-1} - 2\boldsymbol{r}_{k-2})/(6\Delta t) \qquad (10.5.3)$$

If we substitute equs. (10.5.2) and (10.5.3) into equ. (10.5.1), we obtain an equation determining $\boldsymbol{r}_{k+1}$ in terms of $\boldsymbol{r}_k, \boldsymbol{r}_{k-1}, \boldsymbol{r}_{k-2}$. This algorithm describes a 3-step method of 2nd order. As is the case for all multi-step procedures, the Houbolt method requires an initialisation since at the starting-point in time t_0, only $\boldsymbol{r}_0, \dot{\boldsymbol{r}}_0$ and $\ddot{\boldsymbol{r}}_0$ are known (from the equation of motion). For a consistent application of equs. (10.5.2) and (10.5.1), however, we have to know $\boldsymbol{r}_{-1}$ and $\boldsymbol{r}_{-2}$. Thus, one should begin either with a single-step procedure or introduce some auxiliary equations. Houbolt suggests the following equations

$$\ddot{\boldsymbol{r}}_0 = \boldsymbol{M}^{-1}[\boldsymbol{R}(t_0) - \boldsymbol{C}\dot{\boldsymbol{r}}_0 - \boldsymbol{K}\boldsymbol{r}_0] \qquad (10.5.4)$$

$$\boldsymbol{r}_{-1} = \Delta t^2 \ddot{\boldsymbol{r}}_0 + 2\boldsymbol{r}_0 - \boldsymbol{r}_1 \qquad (10.5.5)$$

$$\boldsymbol{r}_{-2} = 6\Delta t \dot{\boldsymbol{r}}_0 + 6\boldsymbol{r}_{-1} - 3\boldsymbol{r}_0 - 2\boldsymbol{r}_1 \qquad (10.5.6)$$

Following their substitution into equs. (10.5.2) and (10.5.3), these equations clearly determine the unknown vectors $\boldsymbol{r}_{-1}$ and $\boldsymbol{r}_{-2}$ in terms of the known $\boldsymbol{r}_0$ and $\dot{\boldsymbol{r}}_0$ The Houbolt method is unconditionally stable, exact to the 2nd order and obviously implicit. The algorithm possesses strong algorithmic damping which produces a stabilising effect in some cases, But can also falsify the solution. By present day knowledge, the following Park method is preferable due to its higher accuracy.

Park's method

The Park procedure [10.20] is based upon the relationship

$$\dot{\widetilde{\boldsymbol{r}}}_{\mathrm{k}+1} = \left[10\widetilde{\boldsymbol{r}}_{\mathrm{k}+1} - 15\widetilde{\boldsymbol{r}}_{\mathrm{k}} + 6\widetilde{\boldsymbol{r}}_{\mathrm{k}-1} - \widetilde{\boldsymbol{r}}_{\mathrm{k}-2}\right]/(6\Delta t) \tag{10.5.7}$$

and thus represents a 3-step procedure of 1st order. To apply equ. (10.5.7), the equation of motion first has to be reformulated as a system of 1st order. This is straightforward if we recollect the technique developed for arbitrary viscous damping (see equ. (9.1.44))

$$\boldsymbol{A}\dot{\widetilde{\boldsymbol{r}}} + \boldsymbol{B}\widetilde{\boldsymbol{r}} = \widetilde{\boldsymbol{R}} \tag{10.5.8}$$

and

$$\boldsymbol{A} = \begin{bmatrix} \boldsymbol{C} & \boldsymbol{M} \\ \boldsymbol{M} & \boldsymbol{O} \end{bmatrix} \tag{10.5.9}$$

$$\boldsymbol{B} = \begin{bmatrix} \boldsymbol{K} & \boldsymbol{O} \\ \boldsymbol{O} & -\boldsymbol{M} \end{bmatrix} \tag{10.5.10}$$

$$\widetilde{\boldsymbol{R}} = \{\boldsymbol{R} \quad \boldsymbol{o}\} \tag{10.5.11}$$

$$\widetilde{\boldsymbol{r}} = \{\boldsymbol{r} \quad \dot{\boldsymbol{r}}\} \tag{10.5.12}$$

Substitution of equ. (10.5.7) into equ. (10.5.8) yields

$$\left[\frac{10}{6\Delta t}\boldsymbol{A} + \boldsymbol{B}\right]\widetilde{\boldsymbol{r}}_{\mathrm{k}+1} = \widetilde{\boldsymbol{R}} + \frac{1}{6\Delta t}\boldsymbol{A}\left[15\widetilde{\boldsymbol{r}}_{\mathrm{k}} - 6\widetilde{\boldsymbol{r}}_{\mathrm{k}-1} + \widetilde{\boldsymbol{r}}_{\mathrm{k}-2}\right] \tag{10.5.13}$$

where the right-hand side of the system may be assumed known. In principle, $\widetilde{\boldsymbol{r}}_{\mathrm{k}+1}$ is thus determined. However, we prefer at this point to reduce the present system of order 2n × 2n (n = number of degrees of freedom in $\boldsymbol{r}$) to two simple equations of order n × n. To this purpose, we rewrite equ. (10.5.13) in the form

$$\frac{10}{6\Delta t}\boldsymbol{M}\dot{\boldsymbol{r}}_{\mathrm{k}+1} + \left[\frac{10}{6\Delta t}\boldsymbol{C} + \boldsymbol{K}\right]\boldsymbol{r}_{\mathrm{k}+1} = \boldsymbol{R}_1 \tag{10.5.14}$$

$$\frac{10}{6\Delta t}\boldsymbol{M}\boldsymbol{r}_{\mathrm{k}+1} - \boldsymbol{M}\dot{\boldsymbol{r}}_{\mathrm{k}+1} = \boldsymbol{R}_2 \tag{10.5.15}$$

where

$$\boldsymbol{R}_1 = \boldsymbol{R} + \frac{1}{6\Delta t}\boldsymbol{C}\left[15\boldsymbol{r}_{\mathrm{k}} - 6\boldsymbol{r}_{\mathrm{k}-1} + \boldsymbol{r}_{\mathrm{k}-2}\right] + \frac{1}{6\Delta t}\boldsymbol{M}\left[15\dot{\boldsymbol{r}}_{\mathrm{k}} - 6\dot{\boldsymbol{r}}_{\mathrm{k}-1} + \dot{\boldsymbol{r}}_{\mathrm{k}-2}\right] \tag{10.5.16}$$

and

$$\boldsymbol{R}_2 = \frac{1}{6\Delta t}\boldsymbol{M}\left[15\boldsymbol{r}_{\mathrm{k}} - 6\boldsymbol{r}_{\mathrm{k}-1} + \boldsymbol{r}_{\mathrm{k}-2}\right] \tag{10.5.17}$$

apply. The second equation (10.5.15) finally yields

$$\boldsymbol{M}\dot{\boldsymbol{r}}_{\mathrm{k}+1} = \frac{10}{6\Delta t}\boldsymbol{M}\boldsymbol{r}_{\mathrm{k}+1} - \boldsymbol{R}_2 \tag{10.5.18}$$

Substituting this result into the first equation, we ultimately obtain

$$\left[\left(\frac{10}{6\Delta t}\right)^2\boldsymbol{M} + \frac{10}{6\Delta t}\boldsymbol{C} + \boldsymbol{K}\right]\boldsymbol{r}_{\mathrm{k}+1} = \boldsymbol{R}_1 + \frac{10}{6\Delta t}\boldsymbol{R}_2 \tag{10.5.19}$$

Thus, we once again have a system of order n for the unknown displacements $\boldsymbol{r}_{\mathrm{k}+1}$ at the end of step k. The Park method is unconditionally stable, obviously implicit and generally gives better results than the Houbolt procedure, although the truncation error and the degree of accuracy are the same. The Park method also possesses algorithmic damping.

Collocation methods

The dissipative character of the Houbolt and Park methods cannot be influenced since all parameters are fixed. At this point, the collocation methods referred to in detail in [10.14] are brought into play. They combine elements of both the Newmark formulae and Wilson's θ-method. Here, θ is a collocation parameter determining the point in time k+ θ at which the dynamical force equilibrium is established. The value $\theta = 1$ represents the end of the k-th time-step, but Wilson often applied $\theta = 1.4$, i.e. a later point in time

$$\boldsymbol{M}\ddot{\boldsymbol{r}}_{\mathrm{k}+\theta} + \boldsymbol{C}\dot{\boldsymbol{r}}_{\mathrm{k}+\theta} + \boldsymbol{K}\boldsymbol{r}_{\mathrm{k}+\theta} = \boldsymbol{R}_{\mathrm{k}+\theta} \tag{10.5.20}$$

Following Wilson, accelerations and excitation forces at the point in time k+θ are deduced from pivotal values at k and k+1 by linear extrapolation

$$\ddot{\boldsymbol{r}}_{\mathrm{k}+\theta} = (1-\theta)\ddot{\boldsymbol{r}}_{\mathrm{k}} + \theta\ddot{\boldsymbol{r}}_{\mathrm{k}+1} \tag{10.5.21}$$

$$\boldsymbol{R}_{\mathrm{k}+\theta} = (1-\theta)\boldsymbol{R}_{\mathrm{k}} + \theta\boldsymbol{R}_{\mathrm{k}+1} \tag{10.5.22}$$

Displacements and velocities at this instant in time are established in analogy to the Newmark procedure

$$\boldsymbol{r}_{\mathrm{k}+\theta} = \boldsymbol{r}_{\mathrm{k}} + \theta\Delta t\dot{\boldsymbol{r}}_{\mathrm{k}} + \frac{(\theta\Delta t)^2}{2}\left[(1-2\beta)\ddot{\boldsymbol{r}}_{\mathrm{k}} + 2\beta\ddot{\boldsymbol{r}}_{\mathrm{k}+\theta}\right] \tag{10.5.23}$$

$$\dot{\boldsymbol{r}}_{\mathrm{k}+\theta} = \dot{\boldsymbol{r}}_{\mathrm{k}} + \theta\Delta t\left[(1-\gamma)\ddot{\boldsymbol{r}}_{\mathrm{k}} + \gamma\ddot{\boldsymbol{r}}_{\mathrm{k}+\theta}\right] \tag{10.5.24}$$

Clearly, equs. (10.5.23) and (10.5.24) can be substituted into equ. (10.5.20) which in its turn yields the remaining unknown accelerations $\ddot{\boldsymbol{r}}_{\mathrm{k}+\theta}$. From equ. (10.5.21) we then deduce the accelerations $\ddot{\boldsymbol{r}}_{k+1}$, and hence with the standard Newmark formulae the displacement and velocity at the end of the k-th time-step

$$\boldsymbol{r}_{k+1} = \boldsymbol{r}_k + \Delta t \dot{\boldsymbol{r}}_k + \frac{\Delta t^2}{2}\left[(1-2\beta)\ddot{\boldsymbol{r}}_k + 2\beta\ddot{\boldsymbol{r}}_{k+1}\right] \tag{10.5.25}$$

$$\dot{\boldsymbol{r}}_{k+1} = \dot{\boldsymbol{r}}_k + \Delta t\left[(1-\gamma)\ddot{\boldsymbol{r}}_k + \gamma\ddot{\boldsymbol{r}}_{k+1}\right] \tag{10.5.26}$$

For $\theta = 1$, the procedure reduces to the Newmark method, and for $\beta = 1/6$ and $\gamma = 1/2$ to the Wilson θ-method. It can be demonstrated that the above procedure corresponds to a 3-step procedure of 2nd order. For $\gamma = 1/2$, there is an accuracy of order 2 (truncation error $O(\Delta t^3)$. The accuracy can be additionally increased by an adjustment of β to reach the order 3 at $\gamma = 1/2$ with

$$\beta = \frac{1}{12} - \frac{1}{2}\theta(\theta - 1) \tag{10.5.27}$$

(see [10.14]). Unconditional stability together with a 2nd order accuracy can be achieved for

$$\gamma = \tfrac{1}{2}$$

$$\theta \geqslant 1$$

$$\frac{\theta}{2(\theta+1)} \geqslant \beta \geqslant \frac{2\theta^2 - 1}{4(2\theta^3 - 1)} \tag{10.5.28}$$

In addition, [10.14] proposes a special adaptation of the parametric pair (β, θ) which yields the so-called optimal collocation methods . This designation is applied to unconditionally stable algorithms that combine a maximum algorithmic damping with a minimum period distortion. The optimality is thus related to such problems in structural dynamics where the dynamical behaviour is described by a relatively small number of low eigenvibrations.

The α-method of Hilber, Hughes, Taylor [10.15]

Houbolt's method possesses an undesirably strong algorithmic damping also in the response characteristics of low frequency modes.The Park method has some positive characteristics from this point of view, but is nevertheless penalised by a larger period error rate than the collocation methods. The collocation methods demonstrate, on the other hand, a phenomenon that does not appear in the Houbolt and Park procedures, so-called "overshoot". This describes a tendency to short-time heavy oscillations for certain initial conditions or load jumps, an occurrence that bears no relation to the stability and accuracy characteristics of the algorithms discussed so far. The α-method is said not to possess this tendency.

If it is considered essential – especially in the dynamic analysis of low frequency processes – to rapidly damp away high freqency components in the response, the original requirement of amplitude fidelity, i.e. of a spectral radius $\rho = 1$ for arbitrary frequencies, may be dropped. Thus, according to equ. (10.3.26), the Newmark method with $\gamma = 1/2, \beta = 1/4$ (average accelerations) demonstrates the

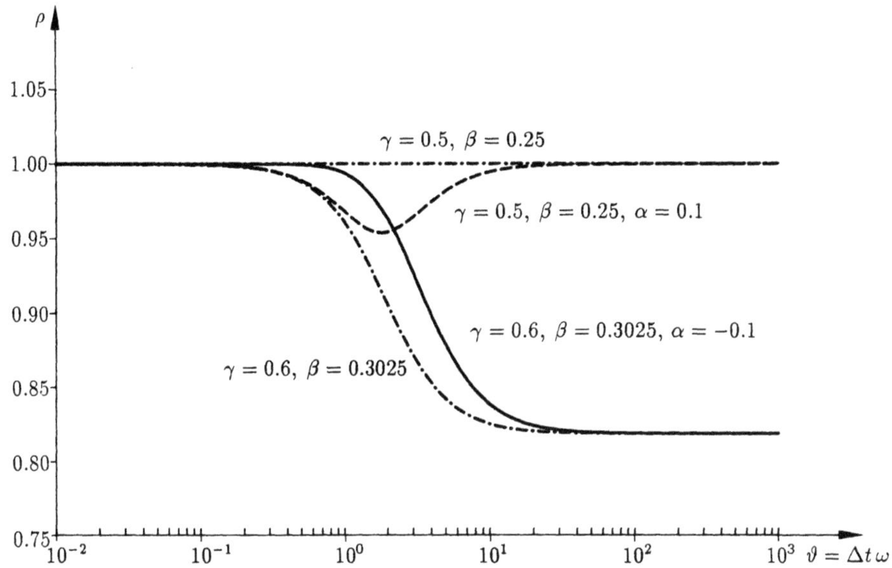

Fig. 10.5.1 Spectral radius for various Newmark procedures and the α-method

spectral radius $\rho = 1$ over the complete frequency range; this may be thought to be undesirable as far as high frequencies are concerned (see also section 10.5.1). One possibility exists in increasing γ, thereby introducing algorithmic damping. For example, for $\gamma = 0.6$, $\beta = 0.3025$, we obtain a procedure which satisfies the condition of equ. (10.3.22), is hence unconditionally stable and damps higher frequencies but not lower ones (see fig. 10.5.1). In principle, this characteristic property appears suitable for practical purposes. If alternatives to the increase of γ are examined, a modification of the original problem could be thought of in such a way that a pseudo- damping is introduced, with the property of tending to zero for small $\vartheta = \Delta t\omega$. This experiment has been carried out in [10.16]. One of the variants discussed in this publication operates with the modified equilibrium relationship

$$\boldsymbol{M}\ddot{\boldsymbol{r}}_{\mathrm{k}+1} + \boldsymbol{C}\dot{\boldsymbol{r}}_{\mathrm{k}+1} + \boldsymbol{K}\boldsymbol{r}_{\mathrm{k}+1} + \alpha\boldsymbol{K}[\boldsymbol{r}_{\mathrm{k}+1} - \boldsymbol{r}_{\mathrm{k}}] = \boldsymbol{R}_{\mathrm{k}+1} \tag{10.5.29}$$

Here, the additional term

$$\alpha\boldsymbol{K}[\boldsymbol{r}_{\mathrm{k}+1} - \boldsymbol{r}_{\mathrm{k}}] = \alpha\Delta t\boldsymbol{K}\frac{\boldsymbol{r}_{\mathrm{k}+1} - \boldsymbol{r}_{\mathrm{k}}}{\Delta t} \tag{10.5.30}$$

can be considered a $\boldsymbol{K}$-proportional damping which tends to zero for $\Delta t \to 0$. Its magnitude can be controlled by α. Relying on this scheme alone, however, turns out to be unsatisfactory since the procedure $\alpha = 0.1$, $\beta = 0.25$, $\gamma = 0.5$ possessing a viscous $\boldsymbol{K}$-proportional additional damping responds in the lower frequency range in a similar manner to the Newmark method for $\gamma = 0.6$, but possesses a spectral radius of 1 for higher frequencies and thus no damping (fig. 10.5.1). Consideration of the diagram reveals that the Newmark procedure for $\gamma > 1/2$ may be improved by negative α-damping. This is in fact true as shown in fig. 10.5.1. In [10.10], Hughes points out a further modification

$$\boldsymbol{M}\ddot{\boldsymbol{r}}_{k+1}+(1+\alpha)\boldsymbol{C}\dot{\boldsymbol{r}}_{k+1}-\alpha\boldsymbol{C}\dot{\boldsymbol{r}}_{k}+(1+\alpha)\boldsymbol{K}\boldsymbol{r}_{k+1}-\alpha\boldsymbol{K}\boldsymbol{r}_{k}=(1+\alpha)\boldsymbol{R}_{k+1}-\alpha\boldsymbol{R}_{k} \tag{10.5.31}$$

Comparative results to those obtained from equ. (10.5.29), however, have not yet been published. Finally, it still remains to be said that equs. (10.5.29) and (10.5.31) have to be complemented by the normal Newmark formulae (see e.g. equ. (10.5.23) and (10.5.24) for $\theta = 1$), which, since self-evident, have not been mentioned so far.

10.6 Examples of direct integration of the dynamic equations

Triangular pulse in a free-free bar [10.11, 10.13]

Our first example is taken from [10.11]. The triangular pulse load imposed upon a free-free bar runs along to the other free end of the bar, is reflected there and returns etc. When there is no damping, this process continues for an indefinite length of time and is thus very well suited to an appraisal of the numerical integration methods. Following fig. 10.6.1, the problem has been calculated with two unconditionally stable algorithms (Wilson and Newmark ($\beta = 1/4$); the latter is identical to the modified Hermitian algorithm with n=1) and the conditionally stable cubic Hermitian algorithm n=2 of [10.13].

The Wilson θ-method, it should be noted, possesses very strong artificial damping. Although undesired high-frequency response components are damped out in this manner, such excessive damping does not seem to be acceptable. The simple method of average accelerations according to Newmark (or modified Hermitian, n=1) shows, as expected, no artificial damping. Thus, much higher harmonic vibrations become clearly visible. As a result, Goudreau and Taylor [13.11] propose the introduction of slight artificial damping, e.g. $\gamma = 0.55$ and $\beta = 0.25\ (1.05)^2$. In this way, the higher harmonies are indeed damped away but, unfortunately, the amplitudes of the pulse are also decreased. Since engineering structures possess damping, the effect of the artificial damping may be insignificant. Since Newmark's method involves 10% fewer operations than Wilson's method, it appears to be the most recommendable variant of all the simpler algorithms.

The most exact observance of the triangular pulse is given by the cubic Hermitian algorithm of [10.13]. This procedure is only conditionally stable, but small time-steps are necessary in any case for a reliable representation of the force and also for reasons of accuracy – even when using unconditionally stable methods. Apart from the appearance of higher harmonics, the solution is perfect and justifies the higher expense of the calculations. If, however, the problem admits a large time-step, the stability limits of the cubic Hermitian algorithm possibly become a disadvantage. In this case, use of the unconditionally stable modified Hermitian algorithm n=2 is recommended.

Excited cantilever bar in resonance

In the example in fig. 10.6.2, a cantilever bar is sinusoidally excited with its smallest eigenfrequency (largest period). Although the Wilson procedure with

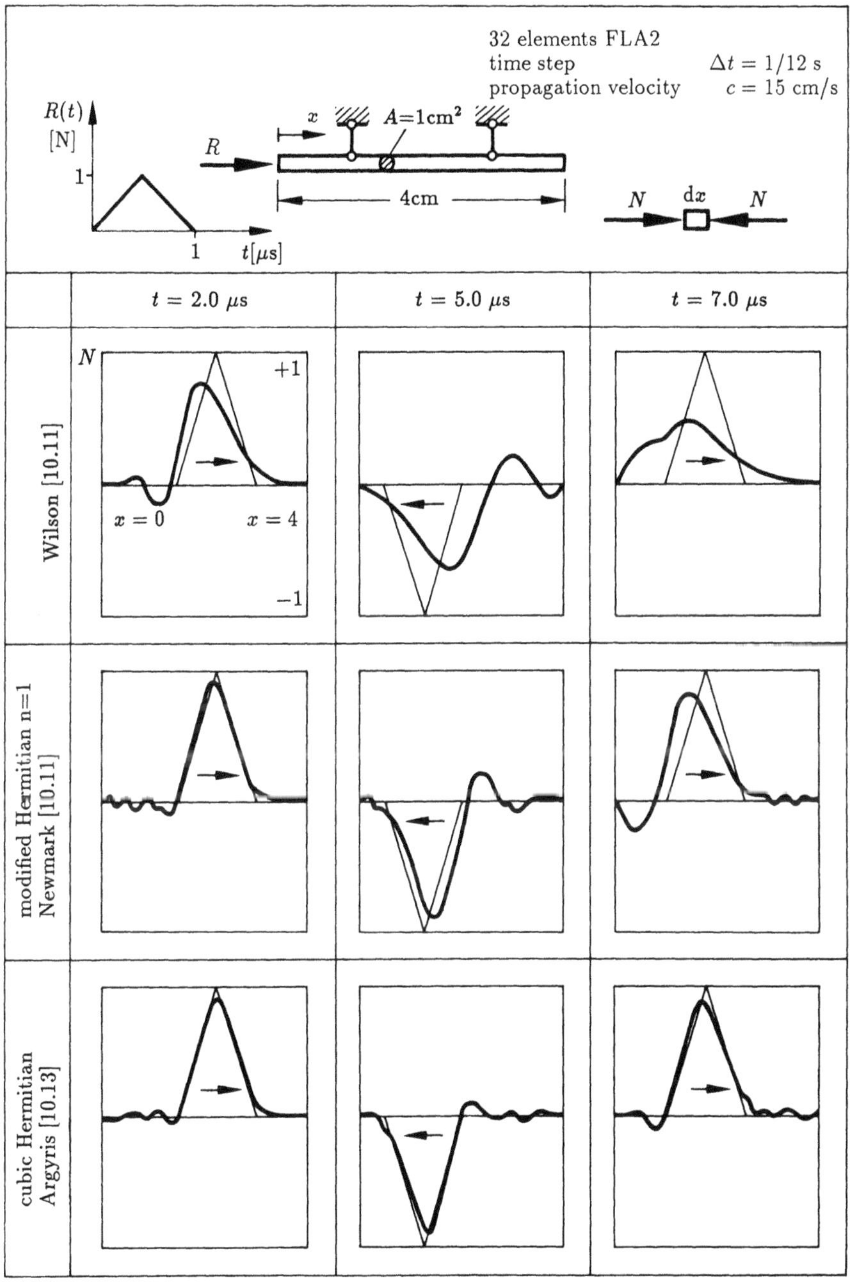

Fig. 10.6.1 Triangular longitudinal impulse in a free-free bar; normal force distributions resulting from three different integration procedures

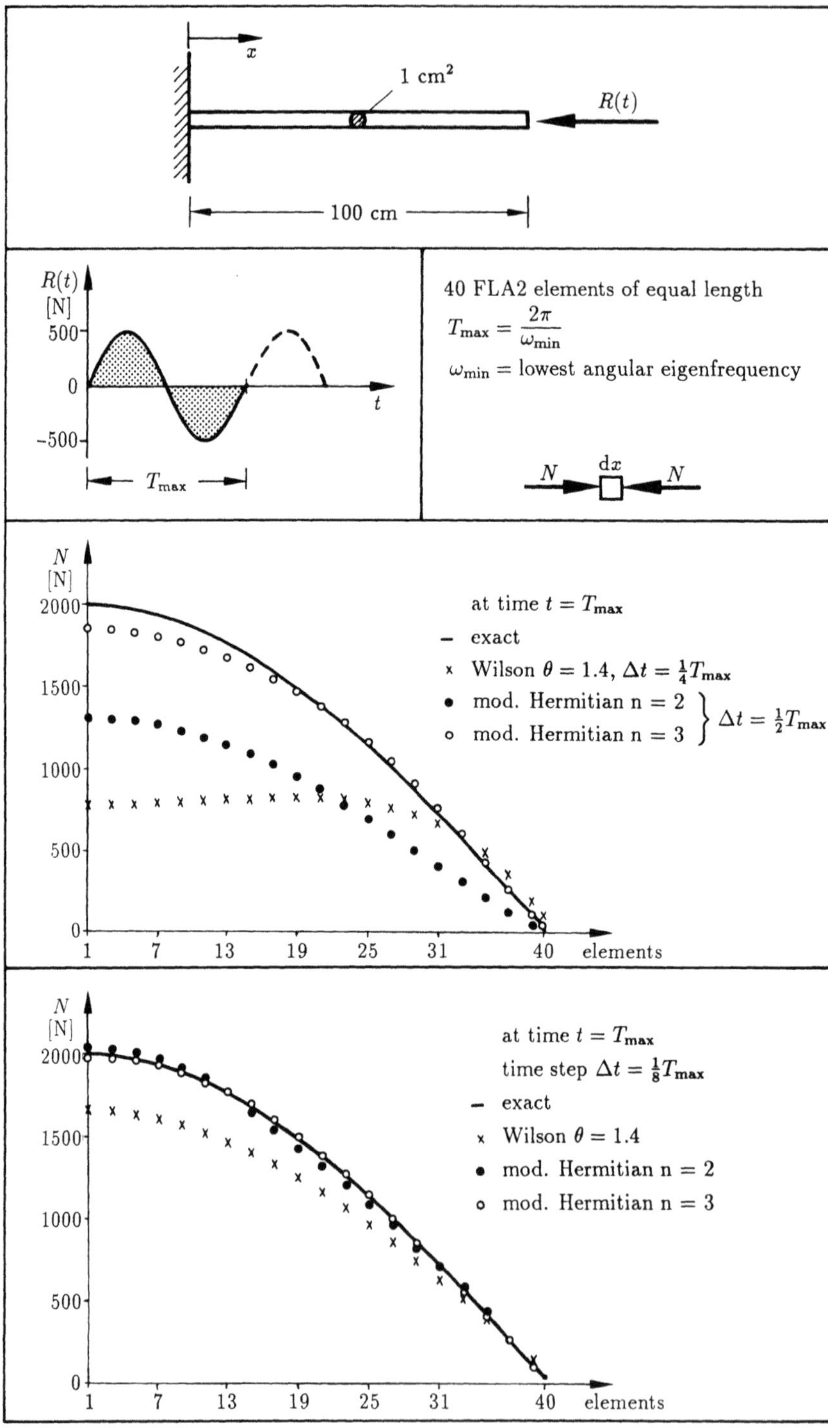

Fig. 10.6.2 Cantilever bar in resonance [10.13]; comparison of unconditionally stable algorithms

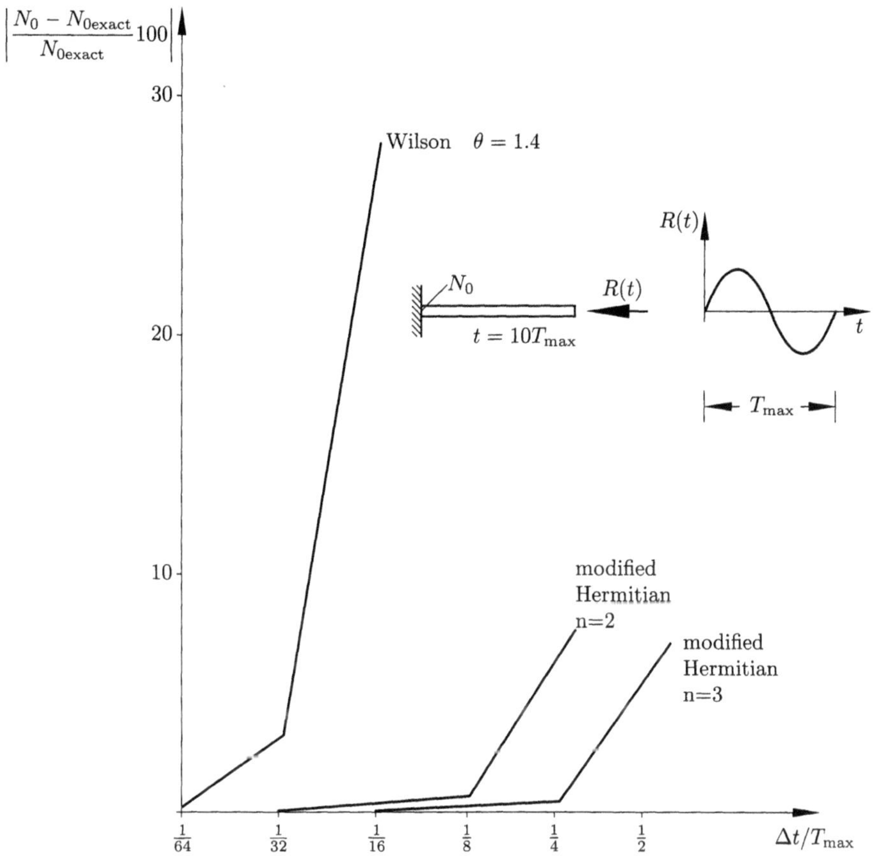

	N_0 ($N_{0exact} = 20000$)		
$\frac{\Delta t}{T_{max}}$	Wilson	n = 2	n = 3
1/2	—	—	18753
1/4	—	18673	20100
1/8	—	19894	20044
1/16	14393	20038	19986
1/32	19375	19987	20002
1/64	19932	20007	19995

Fig. 10.6.3 Cantilever bar in resonance [10.13]; normal force at the support after 10 load cycles

$\Delta t = 0.25T_{\max}$ uses only half the step-length in comparison to the modified Hermitian algorithms n=2 and n=3, very grave errors arise. The n=3 procedure, on the other hand, yields "engineering precision". For a reduced time-step $\Delta t = 1/8T_{\max}$, the Wilson method still remains very inaccurate, whilst the algorithms n=2 and n=3 according to [10.13] are practically exact. The improvement in accuracy with an increasing order n is remarkable. It is underlined by an additional investigation in which we apply the pulsating force over 10 periods and then determine the amplitude of the normal force at the support. Theoretically, this amplitude should become 40 times as large as the amplitude of the excitation force, i.e. 20 000N. In fig. 10.6.3, the results of the various algorithms are plotted against the step-size. In order to achieve the same accuracy with the Wilson scheme as for n=2 and $\Delta t/T_{\max} = 1/4$, the step-size would have to be reduced to $\Delta t/T_{\max} \approx 1/32$.

Proportionally damped free oscillation of a beam

Figure 10.6.4 illustrates a simply supported beam which, at time t=0, is subjected to a static deflection by a central load of 500N. Hereafter, the load is removed and a free damped oscillation is initiated. The amplitude error at a certain given instant is examined, in dependence of the step-size. The relative error of the modified Hermitian procedure n=2 is about 2 decimal powers smaller than that of the Wilson algorithm.

Non-modally damped pin-jointed framework [10.17]

The bridge-like structure shown in fig. 10.6.5 contains two dashpots and possesses the non-modal damping matrix

$$
\begin{array}{cc}
 & \begin{array}{cccccccccc} u_2 & v_2 & \dots & u_4 & v_4 & u_5 & v_5 & u_6 & \dots & v_8 \end{array} \\
\boldsymbol{C} = & \left[\begin{array}{cccccccccc}
0 & 0 & \dots & 0 & & \dots & & 0 & \dots & 0 \\
0 & 0 & & & & & & & & \\
\vdots & \vdots & & \vdots & & & & \vdots & & \vdots \\
0 & 0 & \dots & c_{46} & 0 & 0 & 0 & -c_{46} & \dots & 0 \\
\cdot & \cdot & & 0 & 0 & 0 & 0 & 0 & & 0 \\
\cdot & \cdot & & 0 & 0 & 0 & 0 & 0 & & 0 \\
\cdot & \cdot & & 0 & 0 & 0 & c_5 & 0 & \dots & 0 \\
0 & 0 & \dots & -c_{46} & 0 & 0 & 0 & c_{46} & \dots & 0 \\
\vdots & \vdots & & \vdots & & & & \vdots & & \vdots \\
0 & 0 & \dots & 0 & & \dots & & 0 & \dots & 0
\end{array}\right]
\begin{array}{c} u_2 \\ v_2 \\ \vdots \\ u_4 \\ v_4 \\ u_5 \\ v_5 \\ u_6 \\ \vdots \\ v_8 \end{array}
\end{array}
$$

which, as we already know, complicates the modal analysis in a dynamic response calculation. In contrast, the application of direct integration methods involves no difficulties. As a result, the direct integration method is preferable for this type of problem in dynamics. The response is dominated by the first eigensolution and consequently, the response components associated with higher eigenfrequencies can be neglected. For this reason, our choice of the time-step may be based on the smallest eigenfrequency (largest period $T_{\max}$). For the structure in question, we have $T_{\max} = 69.5$ ms. In accordance with fig. 10.3.2, the accuracy at $\Delta t/T = 0.1$ for the Newmark algorithm with $\beta = 1/4, \gamma = 1/2$ is about 3% with respect to

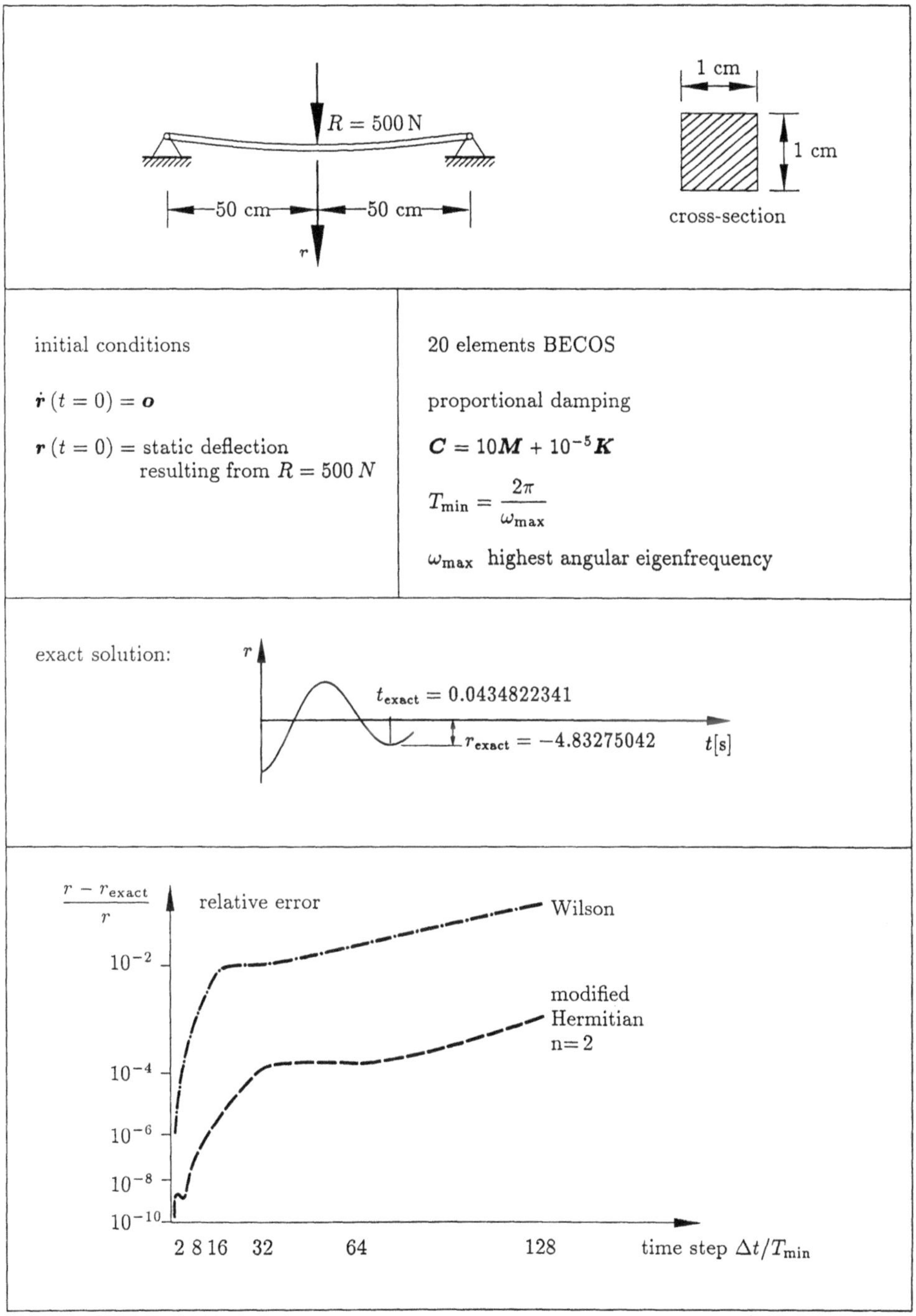

Fig. 10.6.4 Two unconditionally stable algorithms applied to a damped free oscillation

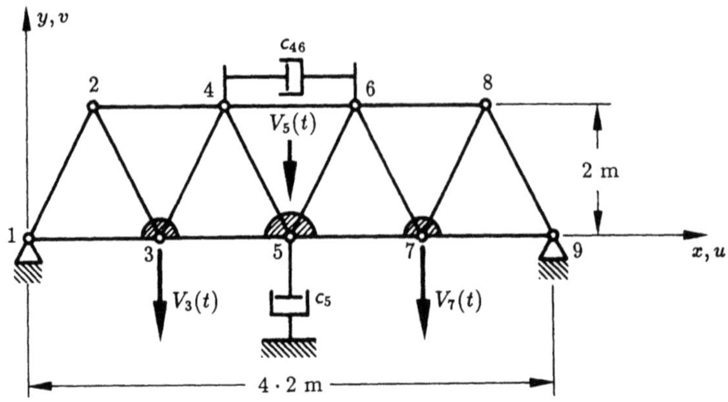

data:

cross section	$A = 10^{-3}\ \mathrm{m}^2$
Young's modulus	$E = 2.1 \cdot 10^{11}\ \mathrm{N/m}^2$
density	$\rho = 8000\ \mathrm{Ns}^2/\mathrm{m}^4$
additional masses	$\Delta m_3 = 1000$ kg
	$\Delta m_5 = 2000$ kg
	$\Delta m_7 = 1000$ kg
characteristic damping values	$c_5 = 30000$ Ns/m
	$c_{46} = 500$ Ns/m

excitation forces:

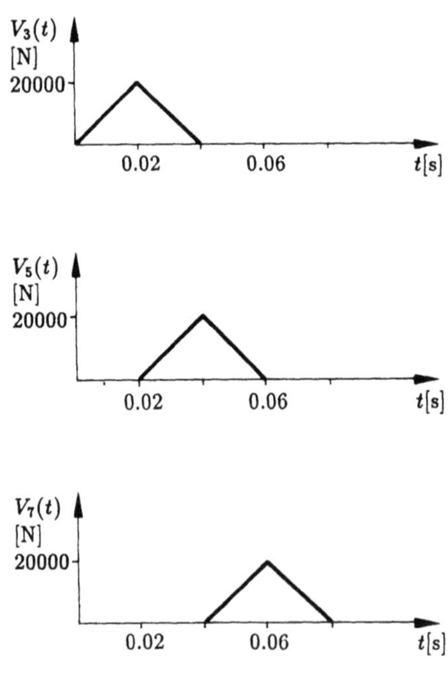

Fig. 10.6.5 Non-modally damped bridge-like structure: problem description and data [10.17]

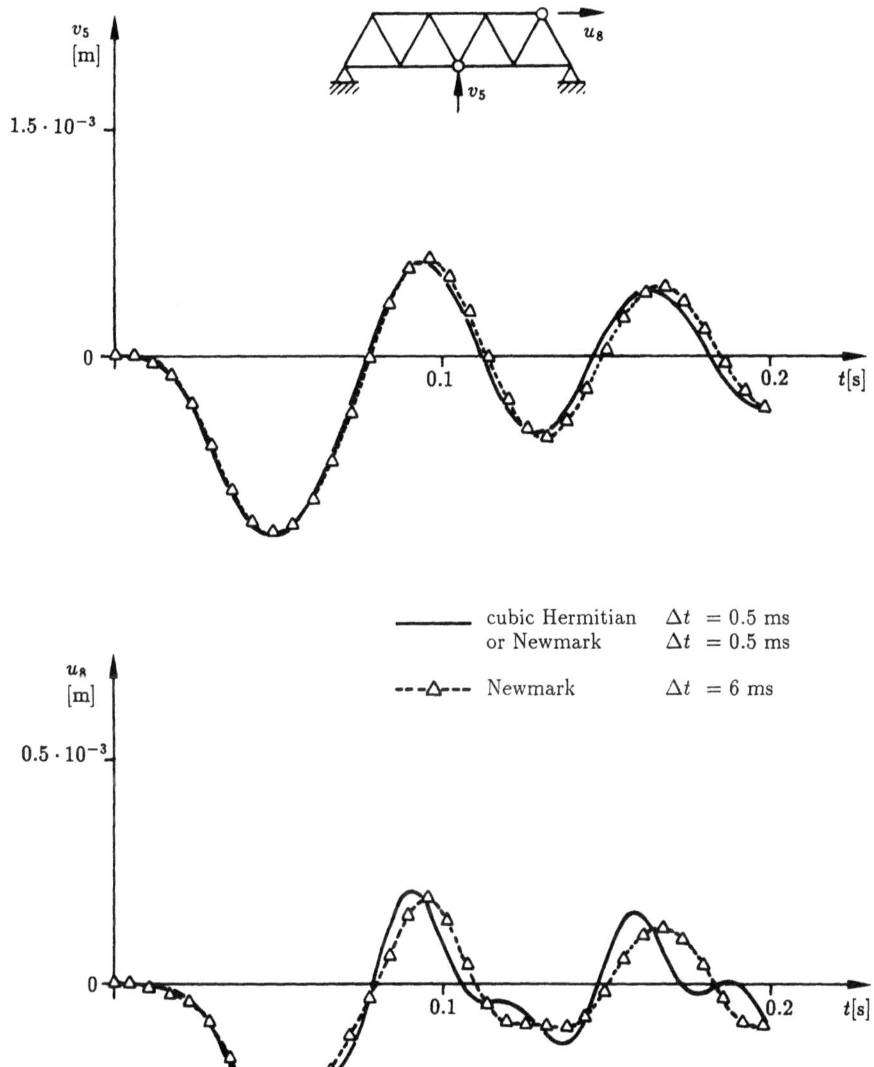

Fig. 10.6.6 Non-modally damped bridge-like structure: displacements v_5 and u_8 evaluated by the Newmark algorithm ($\beta = 1/4$) and the conditionally stable cubic Hermitian algorithm [10.17]

the period elongation. Hence, $\Delta t = 6$ms should ensure sufficient accuracy when applying the Newmark procedure.

If, on the other hand, the conditionally stable Hermitian algorithm is employed, we observe a loss of stability at $\Delta t/T \approx 0.55$. Since we have to avoid the explosion of high-frequency components in all circumstances, the fact that the response comprises mainly low-frequency components becomes irrelevant. Thus, for T, we have to set the minimum period of the structure at $T_{\min} = 1.8$ ms (undamped).

Since damping tends to lower this limiting value somewhat, $\Delta t = 0.5$ ms is selected for safety. Newmark's method of average accelerations is thus once again applied with this step-size. As may be seen from fig. 10.6.6, the Newmark process and the cubic Hermitian algorithm display identical results for a short time-step. For a long time-step, however, which can only be used for the Newmark algorithm, we obtain a good agreement for v_5, but only a moderate one for u_8. The lowest eigenoscillation is clearly not sufficient to yield a good representation of the u_8 displacement. Correspondingly, the time-step should not be based upon $T_{\max}$ but made smaller instead. A comparison of the computation time required proves interesting. We obtain the ratios

$$\begin{bmatrix}\text{cubic Hermitian}\\ \Delta t = 0.5\text{ms}\end{bmatrix} : \begin{bmatrix}\text{Newmark}\\ \Delta t = 0.5\text{ms}\end{bmatrix} : \begin{bmatrix}\text{Newmark}\\ \Delta t = 6\text{ms}\end{bmatrix}$$

with regard to the central processor time

$$30 \quad : \quad 10 \quad : \quad 1$$

and with regard to the transfer of data between central processor and background memory

$$6 \quad : \quad 8 \quad : \quad 1$$

Provided we are aware of important characteristics of the response and can correspondingly select a large time-step, the Newmark method proves very economical. Note, however, that the modified cubic Hermitian algorithm n=2 also permits large time-steps but is not taken into account in this discussion.

Proportionally damped piping system in a power plant [10.18]

The series of examples is concluded with a more significant application. The piping system in question, investigated within a reactor safety programme, is shown in fig. 10.6.7. The largest undamped period of the system is

$$T_1 = T_{\max} = 0.112\,\text{s}$$

and the smallest is

$$T_{150} = T_{\min} = 0.000053\,\text{s}$$

The excitation period is

$$T_E = 0.0785\,\text{s}$$

and hence lies between the second and third eigenperiod. It is thus not acceptable to be guided by $T_{\max}$ since the response will no doubt include the five lowest eigenforms. On the other hand, however, it would be uneconomical to be guided

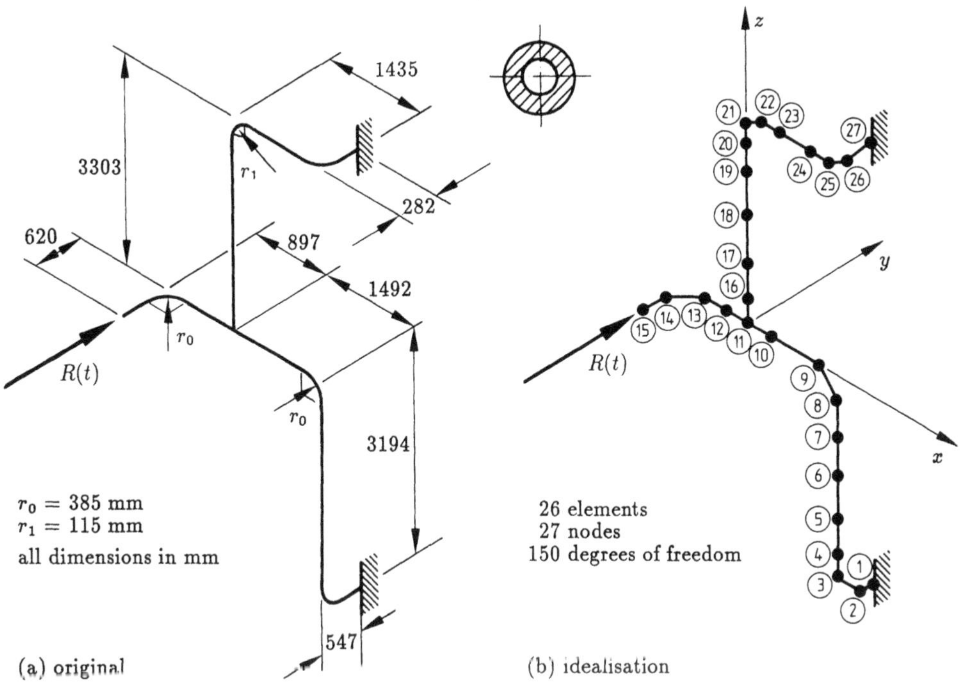

further data:

Young's modulus	$E = 2 \cdot 10^{11}$ N/m^2
Poisson's ratio	$\nu = 0.3$
density	$\rho = 7900$ Ns^2/m^4

element ⟶		① - ② to ⑭ - ⑮	⑪ - ⑯ to ㉖ - ㉗
cross-section area	A [m^2]	$1.2920 \cdot 10^{-2}$	$1.4650 \cdot 10^{-3}$
moment of inertia	I [m^4]	$1.0707 \cdot 10^{-4}$	$1.2769 \cdot 10^{-6}$
St.Venant torsion constant	J [m^4]	$2.1414 \cdot 10^{-4}$	$2.5538 \cdot 10^{-6}$
equivalent shear area	A_s [m^2]	$6.4599 \cdot 10^{-2}$	$7.3249 \cdot 10^{-3}$

damping: $\boldsymbol{C} = 3\boldsymbol{M} + 3 \cdot 10^{-4}\boldsymbol{K}$ [Ns/m]

excitation: $R(t) = 10000 \sin 80t$ [N]

Fig. 10.6.7 Piping system: data and discretisation

in a selection by the smallest period T_{min}, since the associated responses are immediately damped away. A useful compromise appears to be 0,001 s (1 ms) for the time-step, which corresponds to $0.1T_8$ and thus guarantees the exact participation of the lower eight modal components. In fig. 10.6.8, we observe that the modal analysis and the Newmark method yield identical results. If the time-step is increased fivefold, the result still appears to be acceptable. We operate here at a level of $0.1T_3$, ensuring the exact representation of the first three modal forms, which should be sufficient as may also be conjectured from an inspection of the

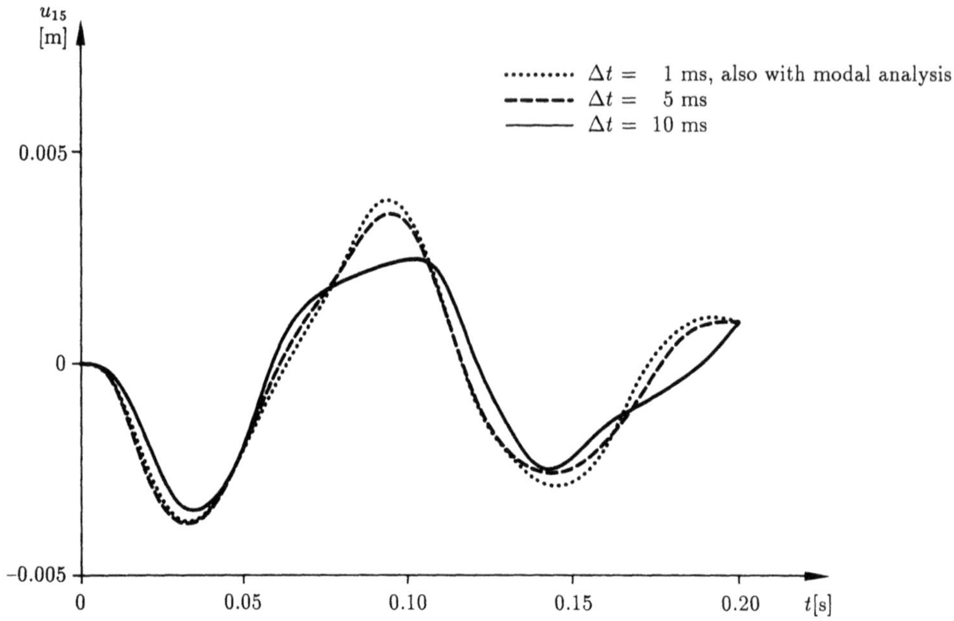

Fig. 10.6.8 Piping system: horizontal x-displacement at node 15 evaluated by the Newmark procedure ($\beta = 1/4, \gamma = 1/2$) for varying step-sizes

excitation frequency. The quality of the result, however, declines for a step-size of 10 ms, which corresponds to about 0.1 T_2.

Obviously, the response component associated with the third eigenform is important and its representation becomes deficient when $\Delta t/T_3 = 0.2$. Finally, let us compare the computation times. We have for

$$\begin{bmatrix} \text{Newmark} \\ \Delta t = 1\text{ms} \end{bmatrix} : \begin{bmatrix} \text{Newmark} \\ \Delta t = 5\text{ms} \end{bmatrix} : \begin{bmatrix} \text{Newmark} \\ \Delta t = 10\text{ms} \end{bmatrix}$$

the relations

$$10 \quad : \quad 2 \quad : \quad 1$$

with regard to the central processor time as well as for the data transfer. A prior knowledge of the prevailing conditions is hence valuable, provided, of course, that we possess the necessary information – a common uncertainty in all direct integration methods.

Bibliography for Chapter 10

[10.1] *R.E. Nickel, J.J. Sackmann*; Approximate solutions in linear coupled thermoelasticity, J. Appl. Mech. ser. E, 35,2 (1968) 255–266.

[10.2] *J.T. Oden*; A general theory of finite elements, part 2 (applications), Int. J. Num. Meth. Eng., 1,3 (1969) 247-259.

[10.3] *I. Fried*; Finite element analysis of time-dependent phenomena, AIAA J., 7,6 (1969) 1170-1172.

[10.4] *J. Argyris, P.C. Dunne, T. Angelopoulos*; Non-linear oscillations using the finite element technique, Comp. Meth. Appl. Mech. Eng., 2 (1973) 203-250.

[10.5] *N.M. Newmark*; A method of computation for structural dynamics, Journal of the Engineering Mechanics Division, ASCE 85, EM3 (1959) 67-94.

[10.6] *R.W. Clough*; Numerical integration of the equations of motion, Lectures on finite element methods in continuum mechanics, University of Alabama, Huntsville (1973) 525-533.

[10.7] *C.C. Fu*; A method for the numerical integration of the equations of motion arising from a finite element analysis, Transactions of the ASME, Journal of Applied Mechanics, 37 (1970) 599-605.

[10.8] *D.M. Trujillo*; An unconditionally stable explicit algorithm for structural dynamics, Int. J. Num. Meth. Eng., 11 (1977) 1579-1592.

[10.9] *O.C. Zienkiewicz*; The finite element method, 3rd ed. (McGraw-Hill Book Company, London, 1977).

[10.10] *T.J.R. Hughes*; Analysis of transient algorithms with particular reference to stability behaviour, first series: Computational methods in mechanics, ed. by T. Belytschko and T.J.R. Hughes (North-Holland, Amsterdam, 1983).

[10.11] *G.L. Goudreau, R.L. Taylor*; Evaluation of numerical integration methods in elastodynamics, Comp. Meth. Appl. Mech. Eng. 2 (1973) 69-97.

[10.12] *K.J. Bathe, E.L. Wilson*; Stability and accuracy analysis of direct integration methods, Earthqu. Eng. and Struct. Dyn., 1 (1973) 283-291.

[10.13] *J. Argyris, P.C. Dunne, T. Angelopoulos*; Dynamic response by large step integration, Earthqu. Eng. and Struct. Dyn., 2 (1973) 185-203.

[10.14] *H.M. Hilber, T.J.R. Hughes*; Collocation, dissipation and 'overshoot' for time integration schemes in structural dynamics, Earthqu. Eng. and Struct. Dyn., 6 (1978) 99-118.

[10.15] *H.M. Hilber, T.J.R. Hughes, R.L. Taylor*; Improved numerical dissipation for time integration algorithms in structural dynamics, Earthqu. Eng. and Struct. Dyn., 5 (1977) 283-292.

[10.16] *H.M. Hilber*; Analysis and design of numerical integration methods in structural dynamics, EECR report No. 76-29, Earthquake Engineering Research Center, University of California, Berkeley, CA (November 1976).

[10.17] *K. Straub, G. Vallianos, R. Walther*; ASKA Part II - Linear dynamic analysis, Direct integration of generally damped systems, ASKA UM 220, Stuttgart (1980).

[10.18] *R. Walther*; ASKA Part II - Linear dynamic analysis, Direct integration using the Newmark method, ASKA UM 228, Stuttgart (1980).

[10.19] *J.C. Houbolt*; A recurrence matrix solution for the dynamic analysis of aircraft, Journal of the Aeronautical Sciences, 17 (1950) 540-550.

[10.20] *K.C. Park*; Evaluating time integration methods for non-linear dynamic analysis, in: Finite element analysis of transient non-linear behaviour, eds T. Belytschko, J.R. Osias and P.V. Marcal, Applied Mechanics Symposia Series, ASME, New York (1975).

[10.21] *J. Argyris, D. Scharpf*; Finite elements in time and space, The Aeronautical Journal of the Royal Aeronautical Society, 73 (1969) 1041-1044.

[10.22] *J. Argyris, A.S.L. Chan*; Applications of finite elements in time and space, Ingenieur Archiv, 41 (1972) 235-257.

11 Aspects of non-linear structural dynamics

"Complicated monsters, head and tail,
Scorpion and asp, and Amphisbaena dire,
Cerases horned, Hydrus, and Ellops drear."

John Milton, Paradise lost

As demonstrated in the previous chapters, a wide range of mildly non-linear problems can successfully be solved using some kind of linearised formulation. In this way, a satisfactory degree of engineering accuracy can be achieved without exceeding the limits of an acceptable economic expenditure. The mathematical framework developed for the solution of these problems may nowadays be considered complete, so that reliable computations may be performed as standard engineering practice. In addition to numerous problems which can be solved approximately by linearisation techniques, there remain, however, major problems that cannot adequately be simplified in such a manner. The fascination of the study of these non-linear problems is highlighted by the appearance of new, unexpected phenomena, not even the merest trace of which can be discerned in a linearised presentation. Here, in contrast to the linearised approach, it is generally no longer possible to establish an exact analytical solution. Non-linear dynamics is thus *a priori* a domain of approximation techniques.

An initial insight into the astounding variety of physical manifestations observed in the analysis of non-linear dynamic processes will be illustrated here by a detailed discussion of a very simple – at least in a formal sense – non-linear equation of motion, the so-called Duffing equation. In addition to semi-analytical techniques, we shall demonstrate the significance of numerical procedures as evolved for modern computers. Much space is devoted to the interpretation of the results, up to the onset of irregular, 'chaotic' behaviour of non-linear dynamical systems which will be studied extensively in Vol. VII of this series [11.59].

The second main theme of this chapter is concerned with a critical appraisal of algorithms for the time integration of non-linear equations of motion. The essential foundations for this investigation were laid in Chapter 10, in which the most efficient procedures and methods were described. Here, the supplementations and modifications necessary for a successful non-linear application of the algorithms are developed.

The chapter is rounded off with an elaboration on a few characteristic features of non-linearities in structural mechanics. On the one hand, these concern special types of loading, dependent upon the motion itself (displacement, velocity) which lead to so-called load-correction matrices. On the other hand, they refer to the pronounced non-linear response of structures due to large displacements and/or plastic behaviour of the material. The 'crash' problems as arising in the automotive industry serve as examples for the computer analysis of such complex phenomena. However interesting structural dynamics may be, we also grant ourselves the luxury of enquiring into computational problems of celestial mechanics. Such unexpected excursions should intrigue the curious minds of students.

11.1 The Duffing equation as an illustrator of non-linear phenomena

11.1.1 Semi-analytical solutions

Within the framework of our introduction to dynamics, we first present a brief insight into the world of non-linear systems by reference to a simple example, namely the Duffing equation. To this end, we ignore in the present account the qualitative analysis of such phenomena, e.g. the discussion of the stability, and restrict ourselves to the quantitative analysis; in this connection, we apply only one of the numerous semi-analytical solution techniques; this corresponds, however, best to the style of this work. The technique selected here is the Ritz-Galerkin procedure, also recommended by Klotter [11.1]. Those readers interested in broadening their knowledge of this subject are advised to consult [11.2, 11.3, 11.4, 11.5 etc.].

All semi-analytical solution techniques have one characteristic feature in common: they provide an approximate description of non-linear phenomena of a certain differential equation or class of differential equations; however, they are not suited to the automatic computer analysis of general structures. To the last class of problems, however, we address this treatise. Practically only one tool is available in such cases, namely the direct integration methods which we have already come across. The realisation of these methods in the field of non-linear dynamics is the main concern of this chapter. But before proceeding to the large-scale engineering aspects, we consider the analysis of the Duffing equation by means of the Ritz-Galerkin procedure.

With this intention, we now enter *medias res* into the non-linear domain and consider the highly simplified mechanical system shown in fig. 11.1.1 which depicts a string under tension, with a point mass m in the middle. Let the initial pretension force be N_0, the inertial force R_I and the damping force R_D being identified by the same measures as in the linear solution. In fact, the restoring elastic force $R_e(t)$ contains the only non-linear contribution. Thus, the equation of motion becomes

$$m\ddot{r} + c\dot{r} + R_e(r,t) = R(t) \tag{11.1.1}$$

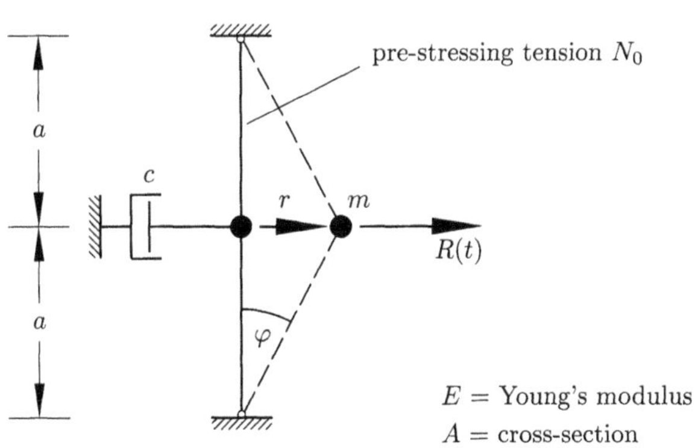

Fig. 11.1.1

Transverse motion of a point mass m on a string under tension

Due to a deflection r, the string is stretched and consequently, the pre-tension force is raised beyond N_0. Retaining the standard small strain definition, the additional elastic strain is obtained from

$$\begin{aligned} \varepsilon &= \frac{1}{a}\big[(a^2 + r^2)^{1/2} - a\big] \\ &= \sqrt{1 + (r/a)^2} - 1 \end{aligned} \tag{11.1.2}$$

Hence, the instantaneous tensional force, N, becomes

$$\begin{aligned} N &= N_0 + \varepsilon EA \\ &= N_0 + EA\{\sqrt{1 + (r/a)^2} - 1\} \end{aligned} \tag{11.1.3}$$

where E is Young's modulus and A is the cross-sectional area of the string. The restoring force R_e is then deduced from the standard parallelogram of forces

$$\begin{aligned} R_e(r) &= 2N \sin\varphi \\ &= 2N\frac{r}{(a^2 + r^2)^{1/2}} \\ &= 2N_0\frac{r}{a}\{1 + (r/a)^2\}^{-1/2} + 2EA\frac{r}{a}\{1 - \big(1 + (r/a)^2\big)^{-1/2}\} \end{aligned} \tag{11.1.4}$$

An exact solution of the resulting differential equation for this elementary oscillator is, *a priori*, illusory. Hence, it appears more advisable for us within our present analytical framework to exclude very large deflections. Consequently, we expand the inverse root terms in equ. (11.1.4) in a binomial series and find

$$\begin{aligned} R_e(r) &= 2N_0(r/a)\{1 - \tfrac{1}{2}(r/a)^2 + \ldots\} + 2EA(r/a)\{\tfrac{1}{2}(r/a)^2 - \ldots\} \\ &= 2N_0(r/a) + (EA - N_0)(r/a)^3 + O\big((r/a)^5\big) \end{aligned} \tag{11.1.5}$$

If terms of the order $(r/a)^5$ are ignored, equ. (11.1.1) yields for a harmonic excitation force the apparently harmless looking non-linear differential equation

$$\ddot{r} + 2\zeta\omega_0\dot{r} + \omega_0^2(r + \delta r^3) = \omega_0^2 r_0 \sin\omega_E t \tag{11.1.6}$$

where, in analogy to the linear formulation of equ. (1.4.5), the constants are defined as

$$k_0 = \frac{2N_0}{a} \qquad \text{linearised stiffness} \tag{11.1.7}$$

$$\delta = \frac{1}{2a^2}\left(\frac{EA}{N_0} - 1\right) \qquad \text{non-linear stiffness contribution} \tag{11.1.8}$$

$$\omega_0^2 = k_0/m \qquad \text{linearised eigenfrequency} \tag{11.1.9}$$

$$\zeta = c/(2m\omega_0) \qquad \text{damping ratio} \tag{11.1.10}$$

$$r_0 = R_0/k_0 \qquad \text{linearised static deflection} \tag{11.1.11}$$

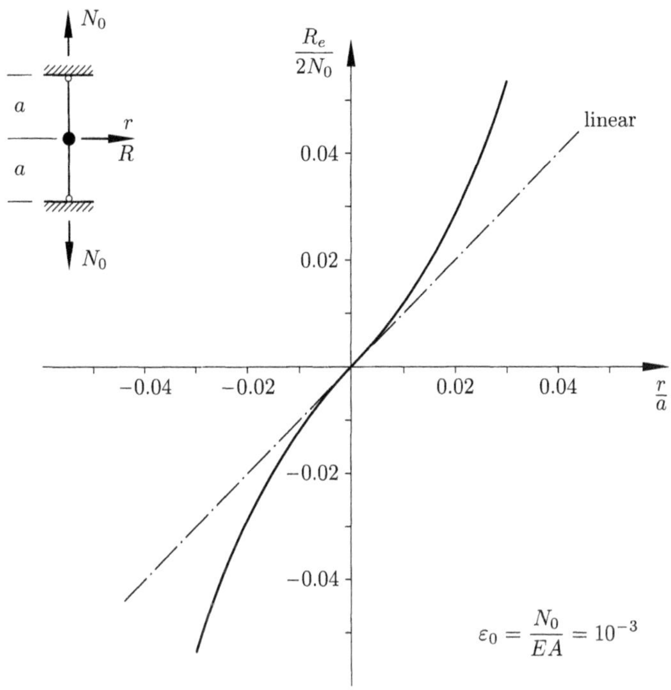

Fig. 11.1.2 Non-linear hardening spring characteristic of the oscillator in fig. 11.1.1 ($a \approx 100$ corresponds to $\delta = 0.05$)

In most cases, EA (hypothetical tension force of the string for a unit strain) is much larger than the initial pre-tension force N_0. Therefore,

$$\delta \approx \frac{EA}{2a^2 N_0} \tag{11.1.12}$$

represents a good approximation and we are justified in stating that the elastic response of the system is described by a hardening spring characteristic (fig. 11.1.2).

The deviation from linearity is given by the term

$$\begin{aligned}\delta r^2 &= \frac{1}{2}\left(\frac{r}{a}\right)^2\left(\frac{EA}{N_0} - 1\right) \\ &\approx \frac{1}{2\varepsilon_0}\left(\frac{r}{a}\right)^2\end{aligned} \tag{11.1.13}$$

and should be compared with 'one' which measures the linear effect. Here, ϵ_0 defines the initial strain of the string and is deduced from the initial pre-tension N_0

$$\varepsilon_0 = \frac{N_0}{EA} \tag{11.1.14}$$

We now observe that the non-linear influence is dependent on the magnitude of the displacement in relation to the pre-tension. In the case of small pre-tension (initial strain), small amplitudes already produce a significant influence (e.g. for ε_0 $= 0.001$ and $r/a = 0.1$, we find $\delta r^2 = 5$ which is by no means negligible as against 'one'). So far, our restriction to moderate amplitudes r/a does not necessarily reveal a measure of the non-linerarity of the system. We can only conjecture that a linear analysis is justifiable for high initial pre-tension and small amplitudes.

Differential equations of the type shown in equ. (11.1.6) also arise for pendulums, electrical resonant circuits and other problems. The equation is known as Duffing's equation and is referred to as such in literature. Despite its inconspicuous form, a general solution cannot be established. As a result of this, numerous approximation methods have been developed for the analysis of such non-linear systems; the Duffing and Rauscher methods are examples of such techniques, as are also the perturbation method and the procedure of Krylov and Bogoljubov. As for the linear oscillator discussed in section 1.4, we may inquire whether the non-linear system of equ. (11.1.6) can execute stationary periodic motions. In order to answer this question, we have to take recourse to a procedure well-suited to the style of this work and recommended by Klotter [11.1] as being the best method for this situation. This refers to the Ritz-Galerkin method or the method of weighted residuals (see also Vol. II of this series). Proceeding from the differential equation

$$D(r) = \ddot{r} + 2\zeta\omega_0\dot{r} + \omega_0^2(r + \delta r^3) - \omega_0^2 r_0 \sin\omega_E t = 0 \tag{11.1.15}$$

we attempt the following solution *ansatz*

$$\begin{aligned} r &= \sum_{\mathrm{k}=1}^{\mathrm{n}} \omega_{\mathrm{k}}(t) r_{\mathrm{k}} \\ &= \underset{(1\times \mathrm{n})}{\boldsymbol{\omega}} \underset{(\mathrm{n}\times 1)}{\boldsymbol{r}} \end{aligned} \tag{11.1.16}$$

In general, this trial solution does not satisfy equ. (11.1.15) for all times t, even if the coefficients are locally determinable. Thus, the operation $D(r)$ is not reduced to zero but produces a residuum. This residuum is weighted with the $\boldsymbol{\omega}$-functions and is integrated over a period of the assumed stationary motion and set to zero at least in a weighted mean. Thus, we have

$$\int_{t_1}^{t_1+T} \boldsymbol{\omega}^{\mathrm{t}} D(\boldsymbol{\omega}\boldsymbol{r})\mathrm{d}t = \boldsymbol{o} \tag{11.1.17}$$

Equation (11.1.17) represents a non-linear algebraic system of order n for the calculation of the n coefficients of the *ansatz*. This greatly simplified representation should suffice here and we proceed to the specific task.

Considering first the linear case of section 1.4, we note that a harmonic excitation $R_0 \sin \omega_E t$ of a linear oscillator produces a steady-state response $r_0 V \sin(\omega_E t - \varphi)$ with the same frequency but a different phase relationship (see fig. 1.4.15). Thus, in dynamics, the static amplitude r_0 is magnified by the amplification function V. Assuming a weak non-linear case, the same *ansatz* could be utilised at least as an initial approximation in order to ascertain a non-linear periodic condition

$$\begin{aligned} r &= [\sin \omega_E t \quad -\cos \omega_E t]\{r_1 \quad r_2\} \\ &= \bar{r} \sin(\omega_E t - \varphi) \end{aligned} \tag{11.1.18}$$

We note

$$\begin{aligned} r_1 &= \bar{r} \cos \varphi \\ r_2 &= \bar{r} \sin \varphi \end{aligned} \tag{11.1.19}$$

and correspondingly

$$\bar{r} = (r_1^2 + r_2^2)^{1/2} \tag{11.1.20}$$

$$\varphi = \tan^{-1} \frac{r_2}{r_1} \tag{11.1.21}$$

In our case, we thus have

$$\boldsymbol{\omega} = [\sin \omega_E t \quad -\cos \omega_E t] \tag{11.1.22}$$

and the differential operator D can be split into a linear part D_L and a non-linear part D_{NL}

$$D(r) = D_L(r) + D_{NL}(r) \tag{11.1.23}$$

Then

$$\begin{aligned} D_L(r) &= \ddot{r} + 2\zeta\omega_0\dot{r} + \omega_0^2 r - \omega_0^2 r_0 \sin \omega_E t \\ &= \omega_0^2\left[(1-\beta^2)\boldsymbol{\omega} + 2\zeta\frac{1}{\omega_0}\dot{\boldsymbol{\omega}}\right]\boldsymbol{r} - \omega_0^2 r_0 \sin \omega_E t \end{aligned} \tag{11.1.24}$$

applies where, in analogy to the linear case, we introduce the frequency ratio

$$\beta = \omega_E/\omega_0 \tag{11.1.25}$$

Furthermore, the non-linear part is

$$\begin{aligned} D_{NL}(r) &= \omega_0^2 \delta r^3 \\ &= \omega_0^2 \delta (\boldsymbol{\omega} \boldsymbol{r})^3 \end{aligned} \tag{11.1.26}$$

Assuming the period of the motion to be T_E, the weighted residual of $D(r)$ is set to zero

$$\begin{aligned}\int_{t_1}^{t_1+T_E} \boldsymbol{\omega}^{\mathrm{t}} D(r)\mathrm{d}t &= \int_{t_1}^{t_1+T_E} [\boldsymbol{\omega}^{\mathrm{t}} D_L(r) + \boldsymbol{\omega}^{\mathrm{t}} D_{NL}(r)]\mathrm{d}t \\ &= \omega_0^2 \int_{t_1}^{t_1+T_E} \left[(1-\beta^2)\boldsymbol{\omega}^{\mathrm{t}}\boldsymbol{\omega} + 2\zeta\frac{1}{\omega_0}\boldsymbol{\omega}^{\mathrm{t}}\dot{\boldsymbol{\omega}}\right] \mathrm{d}t\, \boldsymbol{r} - \omega_0^2 r_0 \int_{t_1}^{t_1+T_E} \boldsymbol{\omega}^{\mathrm{t}} \sin\omega_E t\, \mathrm{d}t \\ &\qquad + \omega_0^2\delta \int_{t_1}^{t_1+T_E} \boldsymbol{\omega}^{\mathrm{t}}(\boldsymbol{\omega r})^3 \mathrm{d}t \\ &= \boldsymbol{o}\end{aligned} \tag{11.1.27}$$

The first two integral expressions contain only integrals of the type

$$\int_{t_1}^{t_1+T_E} \sin\omega_E t \cos\omega_E t\, \mathrm{d}t = 0 \tag{11.1.28}$$

respectively

$$\int_{t_1}^{t_1+T_E} \sin^2\omega_E t\, \mathrm{d}t = \int_{t_1}^{t_1+T_E} \cos^2\omega_E t\, \mathrm{d}t = \frac{\pi}{\omega_E} \tag{11.1.29}$$

The evaluation of the third integral is somewhat more complex and leads to the following integral expressions

$$\int_{t_1}^{t_1+T_E} \sin^4\omega_E t\, \mathrm{d}t = \int_{t_1}^{t_1+T_E} \cos^4\omega_E t\, \mathrm{d}t = \frac{3}{4}\frac{\pi}{\omega_E} \tag{11.1.30}$$

$$\int_{t_1}^{t_1+T_E} \sin^3\omega_E t \cos\omega_E t\, \mathrm{d}t = \int_{t_1}^{t_1+T_E} \sin\omega_E t \cos^3\omega_E t\, \mathrm{d}t = 0 \tag{11.1.31}$$

$$\int_{t_1}^{t_1+T_E} \sin^2\omega_E t \cos^2\omega_E t\, \mathrm{d}t = \frac{1}{4}\frac{\pi}{\omega_E} \tag{11.1.32}$$

Correspondingly, we deduce from equ. (11.1.27)

$$\begin{bmatrix} 1-\beta^2 & 2\zeta\beta \\ -2\zeta\beta & 1-\beta^2 \end{bmatrix} \begin{bmatrix} r_1 \\ r_2 \end{bmatrix} + \frac{3}{4}\delta \begin{bmatrix} r_1^3 + r_1 r_2^2 \\ r_1^2 r_2 + r_2^3 \end{bmatrix} = \begin{bmatrix} r_0 \\ 0 \end{bmatrix} \tag{11.1.33}$$

Now considering equ. (11.1.20) and defining

$$\begin{aligned}\gamma^2 &= 1 + \tfrac{3}{4}\delta(r_1^2 + r_2^2)\\ &= 1 + \tfrac{3}{4}\delta\bar{r}^2\end{aligned} \tag{11.1.34}$$

we obtain in place of equ. (11.1.33)

$$\begin{bmatrix}\gamma^2-\beta^2 & 2\zeta\beta\\ -2\zeta\beta & \gamma^2-\beta^2\end{bmatrix}\begin{bmatrix}r_1\\ r_2\end{bmatrix} = \begin{bmatrix}r_0\\ 0\end{bmatrix} \tag{11.1.35}$$

This matrix relation can now be used to establish formally the amplitudes (γ^2, however, still contains $\bar{r}^2$) just as easily as in the linear case

$$\begin{aligned}r_1 &= r_0\frac{\gamma^2-\beta^2}{(\gamma^2-\beta^2)^2+(2\zeta\beta)^2}\\ r_2 &= r_0\frac{2\zeta\beta}{(\gamma^2-\beta^2)^2+(2\zeta\beta)^2}\end{aligned} \tag{11.1.36}$$

Using equ. (11.1.20), we find for the amplitude $\bar{r}$ of the response

$$\left(\frac{\bar{r}}{r_0}\right)^2\left[(\gamma^2-\beta^2)^2+(2\zeta\beta)^2\right] = 1 \tag{11.1.37}$$

and for the phase-relationship, we derive from equ. (11.1.21)

$$\tan\varphi = \frac{2\zeta\beta}{\gamma^2-\beta^2} \tag{11.1.38}$$

Whilst equ. (10.1.37) yields directly the amplification factor V for the linear case, this equation reflects in the present context a highly non-linear relationship for the calculation of the $\bar{r}$-β- diagram, since $\bar{r}$ is still contained implicitly in γ^2. If $\bar{r}$ is now explicitly introduced via equ. (11.1.34), we obtain a third order equation in $\bar{r}^2$. From this, we obtain $\bar{r}(\beta)$. Nevertheless, it is easier to analyse the inverse relation $\beta(\bar{r})$, since this relation is only of the second order in β^2. In fact, we have

$$\beta^4 - 2(1+\tfrac{3}{4}\delta\bar{r}^2-2\zeta^2)\beta^2 + \left[(1+\tfrac{3}{4}\delta\bar{r}^2)^2-(r_0/\bar{r})^2\right] = 0 \tag{11.1.39}$$

The following solutions for the frequency ratio β^2 as a function of the amplitude $\bar{r}$ are now obtained

$$\beta_{1,2}^2 = (1+\tfrac{3}{4}\delta\bar{r}^2-2\zeta^2) \pm \left[(r_0/\bar{r})^2-4\zeta^2(1+\tfrac{3}{4}\delta\bar{r}^2-\zeta^2)\right]^{1/2} \tag{11.1.40}$$

Proceeding on the rational assumption of positive β-values, equ. (11.1.40) yields two, one or even no positive β-value. It should also be noted that in contrast to linear systems, β is not a function of the amplification factor $\bar{r}/r_0$ alone but depends, in addition, upon r_0 itself. Naturally, equ. (11.1.40) for $\delta = 0$ also includes the special case of linear oscillations and thus also the expression of equ. (1.4.13) for the amplification factor which is illustrated in fig. 1.4.2. In the non-linear case too, the peaks of the resonance curves can be established relatively easily, since the curve there possesses for a given $\bar{r}$ value ($\bar{r}_S$) only a single β-solution (β_S). Thus, at the peak point, the root term in equ. (11.1.40) must vanish. This leads to the condition

$$3\zeta^2\delta\bar{r}_S^4 + 4\zeta^2(1-\zeta^2)\bar{r}_S^2 - r_0^2 = 0 \tag{11.1.41}$$

In the linear case ($\delta = 0$), we obtain a linear equation in $\bar{r}_S^2$ with the solution

$$\bar{r}_S^2(\delta = 0) = \frac{r_0^2}{4\zeta^2(1-\zeta^2)} \tag{11.1.42}$$

whilst in the non-linear case ($\delta \neq 0$), the relevant equation is quadratic in $\bar{r}_S^2$ with the positive solution

$$\bar{r}_S^2 = \frac{2(1-\zeta^2)}{3\delta}\left[\left(1 + \frac{3\delta r_0^2}{4\zeta^2(1-\zeta^2)^2}\right)^{1/2} - 1\right] \tag{11.1.43}$$

(an additional, negative solution is of no interest and is omitted). Equation (11.1.41) now predicts that the peak point for given values of ζ and r_0 is lower in the non-linear case than in the linear one. (Solve equ. (11.1.41) formally for $\bar{r}_S^2$ and consider the $\bar{r}_S^4$ term as a corrective increment).

Substituting equ. (11.1.43) into equ. (11.1.40), we obtain

$$\begin{aligned}\beta_S^2 &= 1 - 2\zeta^2 + \tfrac{3}{4}\delta\bar{r}_S^2 \\ &= 1 - 2\zeta^2 + \frac{(1-\zeta^2)}{2}\left[\left(1 + \frac{3\delta r_0^2}{4\zeta^2(1-\zeta^2)^2}\right)^{1/2} - 1\right]\end{aligned} \tag{11.1.44}$$

This solution also comprises the linear case and demonstrates through its first line that β_S is displaced to the right as a result of non-linear influences.

Thus, we observe the tendency of the resonance curve to bend to the right for a positive cubic non-linear spring characteristic. This situation is illustrated in fig. 11.1.3 for the specific case of a cubic spring coefficient $\delta = 0.05$ and a static amplitude $r_0 = 1$, and is compared with the linearised response. Essential differences can be recognised, especially for low damping in the range $\beta > 0.80$. The maximum amplitudes are highly amplified in the non-linear case and the resonance curves demonstrate a tendency towards 'overhanging'. The consequences of this behaviour will be discussed later. The attentive reader will certainly notice that the resonance curves do not begin in the static case $\beta = 0$ at $\bar{r} = 1.0$ but at a slightly smaller value. This is due to the hardening of the spring, i.e. to the additional cubic term in the stiffness. In fact, equ. (11.1.39) yields for $\beta = 0$

$$\bar{r}(\beta = 0) = \frac{r_0}{1 + \frac{3}{4}\delta\bar{r}^2} \tag{11.1.45}$$

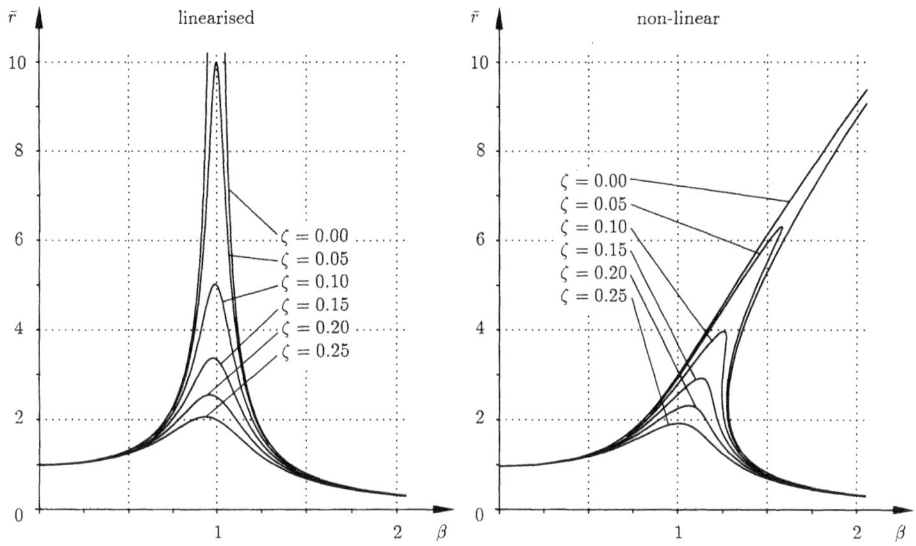

Fig. 11.1.3 Response amplitude $\bar{r}$ as a function of frequency ratio $\beta = \omega_E/\omega_0$ for the oscillator of fig. 11.1.1 under a harmonic excitation ($\delta = 0.05, r_0 = 1$)

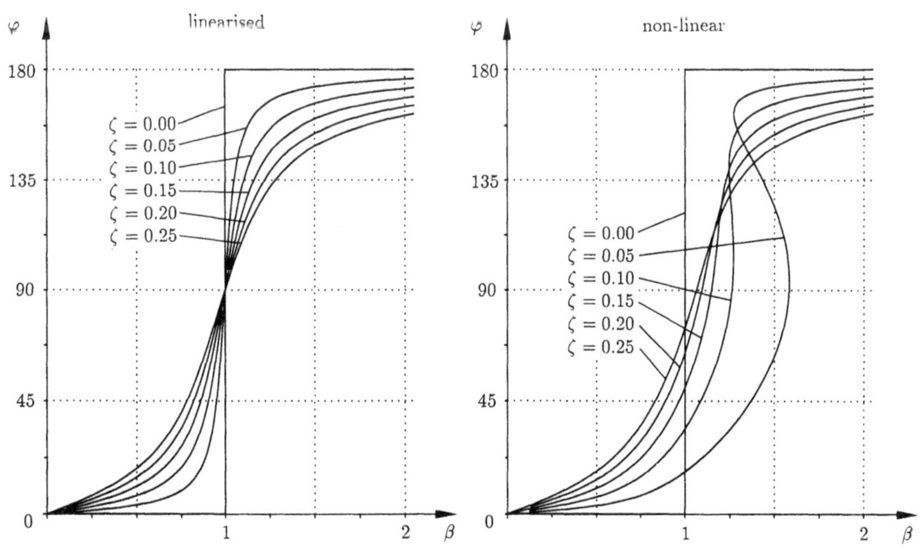

Fig. 11.1.4 Response phase angle φ as a function of frequency ratio $\beta = \omega_E/\omega_0$ for the oscillator of fig. 11.1.1 under a harmonic excitation ($\delta = 0.05, r_0 = 1$)

which proves that $\bar{r} < r_0$. Equation (11.1.45) is a third order expression for $\bar{r}$ and can be solved most simply in a few iteration steps, using $\bar{r} = r_0$ as a starting estimate (for $r_0 = 1, \delta = 0.05$, we have $\bar{r} = 0.9662$).

Let us now consider the phase shift between excitation and response. Equations (11.1.38) and (11.1.34) yield for known values of β and $\bar{r}$ (see also fig. 11.1.4)

$$\tan\varphi = \frac{2\zeta\beta}{1 - \beta^2 + \frac{3}{4}\delta\bar{r}^2} \tag{11.1.46}$$

For the undamped oscillator, no difference arises to the linearised case, i.e. the phase difference is either 0^0 or 180^0. Whilst in the linear, damped case, however, the characteristic attribute of $\varphi(\beta = 1) = 90°$ is observed – independently of the damping ratio ζ – and serves as a most convenient guide for the experimental determination of the resonance point, the corresponding curves in the non-linear case are bent to the right, and this characteristic property for ζ is lost.

Much still remains to be said on the remarkable behaviour of non-linear systems. If we scrutinise the 'overhanging' resonance curves in fig. 11.1.3, we observe that for a given value of β, up to three different amplitudes can arise. Figure 11.1.5 demonstrates for a specific non-linear oscillator the interesting jumps in the amplitudes. In this connection, we carry out a thought experiment as is often adopted in practice as an acceptance test. To this purpose, we excite our structure harmonically and thereby sweep over a certain frequency range. Starting from the static condition ($\beta = 0$), we first observe an increase of amplitude analogous to that in the linear case. If the excitation frequency ratio β is increased beyond the value 1, the amplitude abruptly decreases somewhere in the range of 1 to 2 (jump effect). Hereafter, analogous to the linear case, it tends towards zero. If the frequency is now reduced, the response remains from the 2nd to the 3rd point on the lower branch of the curve until an amplitude jump takes place to the point 4 on the upper branch of the response curve. From this, it follows that branch 1 - 3 must be unstable. Theoretically, this behaviour could be applied for switching purposes since it demonstrates the characteristics of a switch spring.

Until this point – having proceeded from linear conditions to limited non-linearities – we have assumed that a harmonic excitation with frequency ω_E induces a phase-shifted stationary response with the same frequency. This can only serve as an approximation of the 1st order and does not preclude the existence of other solutions.

Let us now adopt a more general approach to the evaluation of the steady-state response of the harmonically excited oscillator of equ. (11.1.6). We follow thereby Hayashi [11.3]. We first assume the presence of light damping ($\zeta \ll 1$) for which (according to fig. 11.1.4) the phase angle between excitation and response is small. The fundamental component of the response is again sinusoidal and synchronous with the excitation. We now supplement the response discussed so far by an additional term with a different frequency whereby, firstly, we must examine the conditions under which the response may contain oscillatory components not keeping beat with the frequency of the excitation. To this purpose, we express the frequency of excitation as an integral multiple of a basic frequency Ω

$$\omega_E = \nu\Omega \,, \qquad \nu = 2, 3, 4, \ldots \tag{11.1.47}$$

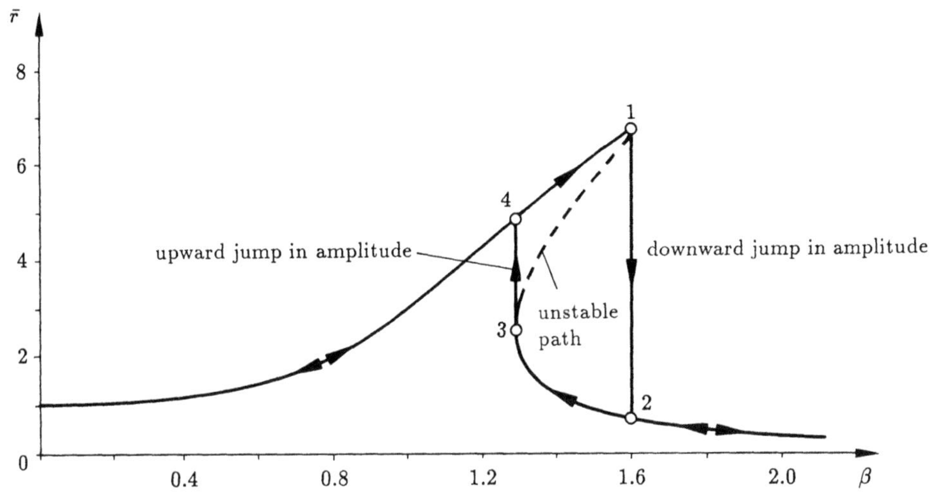

Fig. 11.1.5 Jump effect when sweeping over the frequencies of a non-linear oscillator ($\delta = 0.005, r_0 = 1, \zeta = 0.05$; see fig. 11.1.1)

and write the excitation force as

$$R(t) = R_0 \sin \omega_E t = R_0 \sin \nu\Omega t \tag{11.1.48}$$

The response is now assumed to be

$$r(t) = r_1 \sin \Omega t + r_2 \cos \Omega t + r_3 \sin \nu\Omega t \ , \qquad \nu = 2, 3, 4, \ldots \tag{11.1.49}$$

In addition to the term $r_3 \sin \nu\Omega t$ of the response expected for light damping, a phase-shifted low frequency oscillation is also permitted. In order to ascertain whether the existence of such sub-harmonic oscillation components is possible, we substitute this trial solution into the differential equation

$$\frac{1}{\omega_0^2}\ddot{r} + \frac{2\zeta}{\omega_0}\dot{r} + r - r_0 \sin \nu\Omega t + \delta r^3 = 0 \tag{11.1.50}$$

and compare the coefficients of the terms $\sin \Omega t$ and $\cos \Omega t$ (i.e. the sub-harmonic components). Introducing the abbreviations

$$\begin{aligned} s &= \sin \Omega t \\ c &= \cos \Omega t \end{aligned} \tag{11.1.51}$$

and

$$\begin{aligned} s_\nu &= \sin \nu\Omega t \\ c_\nu &= \cos \nu\Omega t \end{aligned} \tag{11.1.52}$$

as well as

$$\beta_s = \Omega/\omega_0$$

we obtain

$$\begin{aligned}&\big[(1-\beta_s^2)r_1 - 2\zeta\beta_s r_2\big]s + \big[2\zeta\beta_s r_1 + (1-\beta_s^2)r_2\big]c \\ &+ \big[(1-\nu^2\beta_s^2)r_3 - r_0\big]s_\nu + 2\zeta\nu\beta_s r_3 c_\nu + \delta(r_1 s + r_2 c + r_3 s_\nu)^3 = 0 \qquad (11.1.53)\end{aligned}$$

The evaluation and transformation of the non-linear cubic part causes the most effort, but the reader may benefit from a consultation of any of the numerous mathematical compendia of formulae containing powers and products of trigonometric functions. The solution for r^3 contains, in addition to the explicit terms in s and c, trigonometric functions of the arguments (frequencies)

$$(\nu-2)\Omega\,, \quad (\nu+2)\Omega\,, \quad (2\nu-1)\Omega\,, \quad (2\nu+1)\Omega\,, \quad 3\nu\Omega$$

Applying the notation of equ. (11.1.52), we find

$$\begin{aligned}4r^3 = \;& 3r_1^3 s - r_1^3 s_3 + 3r_1^2 r_2 c - 3r_1^2 r_2 c_3 \\ &+ 3r_1 r_2^2 s + 3r_1 r_2^2 s_3 + 3r_2^3 c + r_2^3 c_3 \\ &+ 6r_1^2 r_3 s_\nu - 3r_1^2 r_3 s_{\nu+2} - 3r_1^2 r_3 s_{\nu-2} \\ &+ 6r_1 r_2 r_3 c_{\nu-2} - 6r_1 r_2 r_3 c_{\nu+2} + 6r_2^2 r_3 s_\nu \\ &+ 3r_2^2 r_3 s_{\nu+2} + 3r_2^2 r_3 s_{\nu-2} + 6r_1 r_3^2 s \qquad (11.1.54) \\ &- 3r_1 r_3^2 s_{2\nu+1} + 3r_1 r_3^2 s_{2\nu-1} + 6r_2 r_3^2 c \\ &- 3r_2 r_3^2 c_{2\nu-1} - 3r_2 r_3^2 c_{2\nu+1} + 3r_3^3 s_\nu \\ &- r_3^3 s_{3\nu}\end{aligned}$$

In accordance with the tabular reference below, only one specific contribution to s and c, namely for $\nu = 3$ at $\nu - 2$, can be expected:

ν	$\nu-2$	$\nu+2$	$2\nu-1$	$2\nu+1$	3ν
2	0	4	3	5	6
3	①	5	5	7	9
4	2	6	7	9	12
5	3	7	9	11	15
⋮					

By comparing the coefficients with respect to s, we note for $\nu = 2, 4, 5, ...$

$$(1 - \beta_s^2)r_1 - 2\zeta\beta_s r_2 + \tfrac{3}{4}\delta(r_1^3 + r_1 r_2^2 + 2r_1 r_3^2) = 0 \tag{11.1.55}$$

Similarly, for the coefficients in c, we have

$$2\zeta\beta_s r_1 + (1 - \beta_s^2)r_2 + \tfrac{3}{4}\delta(r_1^2 r_2 + r_2^3 + 2r_2 r_3^2) = 0 \tag{11.1.56}$$

Applying a slight rearrangement, we obtain

$$\begin{aligned} &\left[1 - \beta_s^2 + \tfrac{3}{4}\delta(r_1^2 + r_2^2) + \tfrac{3}{2}\delta r_3^2\right]r_1 - 2\zeta\beta_s r_2 = 0 \\ &\left[1 - \beta_s^2 + \tfrac{3}{4}\delta(r_1^2 + r_2^2) + \tfrac{3}{2}\delta r_3^2\right]r_2 + 2\zeta\beta_s r_1 = 0 \end{aligned} \tag{11.1.57}$$

We now multiply the first equation by r_2, the second by r_1 and form their difference and find

$$2\zeta\beta_s(r_1^2 + r_2^2) = 0 \tag{11.1.58}$$

where $r_1^2 + r_2^2$ represents the amplitude of the sub-harmonic oscillation. Thus, for a damped oscillator, no sub-harmonic oscillations of the order 1/2, 1/4, 1/5, 1/6... (as against the excitation frequency ω_E) exist, i.e. oscillations with frequency ω_E/ν do not appear for $\nu = 2, 4, 5, 6...$.

Let us now turn to the special case $\nu = 3$. From the contribution due to $\nu - 2$ in equ. (11.1.54), additional terms arise which are placed on the right-hand side of the extended equ. (11.1.57)

$$\begin{aligned} \left[1 - \beta_s^2 + \tfrac{3}{4}\delta(r_1^2 + r_2^2) + \tfrac{3}{2}\delta r_3^2\right] r_1 - 2\zeta\beta_s r_2 &= \tfrac{3}{4}\delta r_3(r_1^2 - r_2^2) \\ \left[1 - \beta_s^2 + \tfrac{3}{4}\delta(r_1^2 + r_2^2) + \tfrac{3}{2}\delta r_3^2\right] r_2 + 2\zeta\beta_s r_1 &= -\tfrac{3}{4}\delta r_3 2r_1 r_2 \end{aligned} \tag{11.1.59}$$

For the sake of simplicity, we consider the undamped case $\zeta = 0$ and once more multiply the first equation by r_2 and the second by r_1 and form their difference. For $r_3 \neq 0$, we deduce the simple equation

$$(3r_1^2 - r_2^2)r_2 = 0 \tag{11.1.60}$$

with the solutions

$$r_2 = 0 \tag{11.1.61}$$

or respectively

$$r_2 = \pm\sqrt{3}\, r_1 \tag{11.1.62}$$

Equation (11.1.61) yields the sub-harmonic solution component

$$r_1 \sin \Omega t = r_1 \sin \frac{\omega_E t}{3}$$

whilst equ. (11.1.62) contributes the additional sub-harmonic solution components

$$r_1(\sin\Omega t + \sqrt{3}\cos\Omega t) = r_1 \sin\frac{\omega_E t + 2\pi}{3}$$

$$r_1(\sin\Omega t - \sqrt{3}\cos\Omega t) = r_1 \sin\frac{\omega_E t + 4\pi}{3}$$

The last two solution components can be defined as sub-harmonic oscillations due to a force excitation shifted by 2π respectively 4π. This, however, does not produce an essential new condition and it suffices to consider the case for $r_2 = 0$. In the undamped case, the motion can thus be expressed as follows

$$r = r_1 \sin\frac{\omega_E t}{3} + r_3 \sin\omega_E t \tag{11.1.63}$$

Having obtained this result, we now abandon the technique of comparing the coefficients. Apart from the computational difficulties, the procedure is in principle very simple but involves so-called secular terms in some non-linear cases. When considered in isolation, these tend towards infinity with time, but their overall contribution in fact remains finite when contained in a series representation of the global solution. The discussion of these technicalities is too involved for the present purpose and we prefer to return in conjunction with equ. (11.1.63) to the method of weighted residuals. Our interpolation matrix $\boldsymbol{\omega}$ is

$$\begin{aligned}\boldsymbol{\omega} &= [\sin\tfrac{1}{3}\omega_E t \qquad \sin\omega_E t] \\ &= [s_{1/3} \qquad s]\end{aligned} \tag{11.1.64}$$

and in place of equ. (11.1.27), we now have the somewhat more general expression

$$\int_{t_1}^{t_1+3T_E}\left[\frac{1}{\omega_0^2}\boldsymbol{\omega}^{\mathrm t}\ddot{\boldsymbol{\omega}} + \frac{2\zeta}{\omega_0}\boldsymbol{\omega}^{\mathrm t}\dot{\boldsymbol{\omega}} + \boldsymbol{\omega}^{\mathrm t}\boldsymbol{\omega}\right]\mathrm{d}t\,\boldsymbol{r} - r_0\int_{t_1}^{t_1+3T_E}\boldsymbol{\omega}^{\mathrm t}\sin\omega_E t\,\mathrm{d}t + \delta\int_{t_1}^{t_1+3T_E}\boldsymbol{\omega}^{\mathrm t}(\boldsymbol{\omega}\boldsymbol{r})^3\mathrm{d}t = \boldsymbol{o} \tag{11.1.65}$$

The range of integration extends over the longer of the two periods of the trial solution. Substitution of equ. (11.1.64) yields for $\zeta = 0$

$$\int_{t_1}^{t_1+3T_E}\left(\begin{bmatrix}(1-\beta^2/9)s_{1/3}^2 & (1-\beta^2)s_{1/3}s \\ (1-\beta^2/9)ss_{1/3} & (1-\beta^2)s^2\end{bmatrix}\begin{bmatrix}r_1 \\ r_3\end{bmatrix} - r_0\begin{bmatrix}s_{1/3}s \\ s^2\end{bmatrix}\right)\mathrm{d}t + \delta\int_{t_1}^{t_1+3T_E}\begin{bmatrix}s_{1/3}(s_{1/3}r_1 + sr_3)^3 \\ s(s_{1/3}r_1 + sr_3)^3\end{bmatrix}\mathrm{d}t = \boldsymbol{o} \tag{11.1.66}$$

where the following integrals

$$
\begin{aligned}
\int_{t_1}^{t_1+3T_E} s_{1/3}^2 \mathrm{d}t &= \frac{3}{\omega_E}\int_{\tau_1}^{\tau_1+2\pi} \sin^2\tau \mathrm{d}\tau &= \frac{3\pi}{\omega_E} \\
\int_{t_1}^{t_1+3T_E} s^2 \mathrm{d}t &= \frac{1}{\omega_E}\int_{\tau_1}^{\tau_1+6\pi} \sin^2\tau \mathrm{d}\tau &= \frac{3\pi}{\omega_E} \\
\int_{t_1}^{t_1+3T_E} s_{1/3} s \mathrm{d}t &= \frac{3}{\omega_E}\int_{\tau_1}^{\tau_1+2\pi} \sin\tau \sin 3\tau \mathrm{d}\tau = 0
\end{aligned}
\tag{11.1.67}
$$

determine the first and simpler expression in equ. (11.1.66). The evaluation of the non-linear cubic part of the equation is more laborious

$$
\int_{t_1}^{t_1+3T_E} \begin{bmatrix} s_{1/3} r^3 \\ s r^3 \end{bmatrix} dt = \int_{t_1}^{t_1+3T_E} \begin{bmatrix} s_{1/3}^4 r_1^3 + 3s_{1/3}^3 s r_1^2 r_3 + 3s_{1/3}^2 s^2 r_1 r_3^2 + s_{1/3} s^3 r_3^3 \\ s s_{1/3}^3 r_1^3 + 3s^2 s_{1/3}^2 r_1^2 r_3 + 3s_{1/3} s^3 r_1 r_3^2 + s^4 r_3^3 \end{bmatrix} \mathrm{d}t
\tag{11.1.68}
$$

The integrals arising here are

$$
\begin{aligned}
\int_{t_1}^{t_1+3T_E} s_{1/3}^4 \,\mathrm{d}t &= \frac{3}{\omega_E}\int_{\tau_1}^{\tau_1+2\pi} \sin^4\tau \,\mathrm{d}\tau &= \frac{3}{4}\frac{3\pi}{\omega_E} \\
\int_{t_1}^{t_1+3T_E} s_{1/3}^3 s \,\mathrm{d}t &= \frac{3}{\omega_E}\int_{\tau_1}^{\tau_1+2\pi} \sin^3\tau \sin 3\tau \,\mathrm{d}\tau &= -\frac{1}{4}\frac{3\pi}{\omega_E} \\
\int_{t_1}^{t_1+3T_E} s_{1/3}^2 s^2 \mathrm{d}t &= \frac{3}{\omega_E}\int_{\tau_1}^{\tau_1+2\pi} \sin^2\tau \sin^2 3\tau \,\mathrm{d}\tau &= \frac{1}{2}\frac{3\pi}{\omega_E} \\
\int_{t_1}^{t_1+3T_E} s_{1/3} s^3 \mathrm{d}t &= \frac{3}{\omega_E}\int_{\tau_1}^{\tau_1+2\pi} \sin\tau \sin^3 3\tau \,\mathrm{d}\tau &= 0 \\
\int_{t_1}^{t_1+3T_E} s^4 \mathrm{d}t &= \frac{1}{\omega_E}\int_{\tau_1}^{\tau_1+6\pi} \sin^4\tau \,\mathrm{d}\tau &= \frac{3}{4}\frac{3\pi}{\omega_E}
\end{aligned}
\tag{11.1.69}
$$

We are now in a position to rewrite the system of equ. (11.1.66) and obtain

$$
\begin{aligned}
&(1-\beta^2/9) r_1 + \tfrac{3}{4}\delta(r_1^3 - r_1^2 r_3 + 2r_1 r_3^2) = 0 \\
&(1-\beta^2) r_3 + \tfrac{1}{4}\delta(-r_1^3 + 6r_1^2 r_3 + 3r_3^3) = r_0
\end{aligned}
\tag{11.1.70}
$$

The same expressions could have been found by a comparison of the coefficients, which confirms the effectiveness of the method of weighted residuals. In fact, the first equation in (11.1.70) follows directly from the first expression in equ. (11.1.59) by substitution of $r_2 = 0$, $\zeta = 0$ and $\beta_s^2 = \beta^2/9$.

The return journey to a steady-state oscillation having only one frequency component is also contained in this presentation. Evidently, the first expression in equ. (11.1.70) is satisfied by

$$r_1 = 0$$

and this, when entered into the second expression, yields

$$(1 - \beta^2 + \tfrac{3}{4}\delta r_3^2)r_3 = r_0 \tag{11.1.71}$$

The same result can also be deduced from equ. (11.1.36) by setting $\zeta = 0$ and by noting equ. (11.1.34). In this case, r_3 is identical to $\overline{r}$.

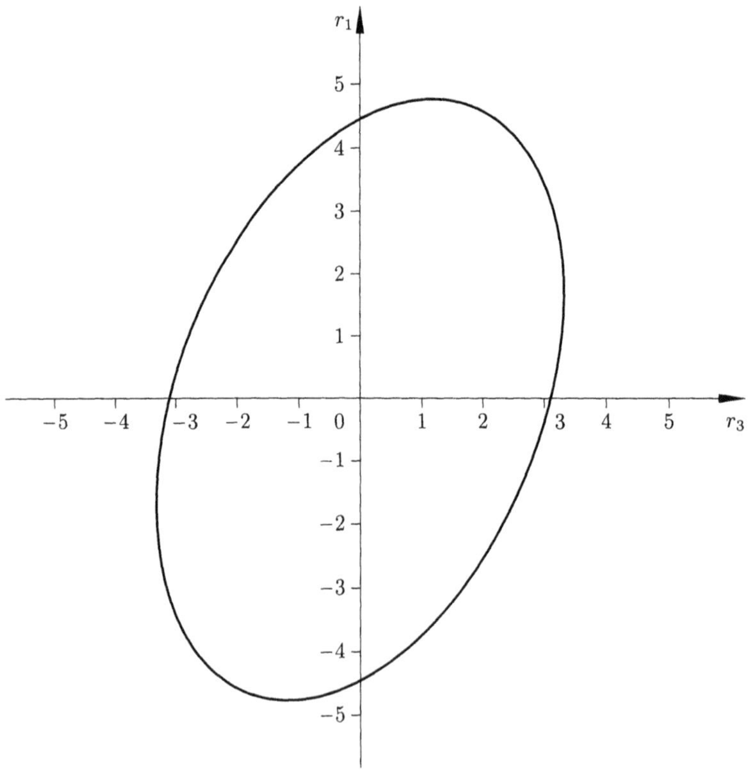

Fig. 11.1.6 Relationship between harmonic and sub-harmonic amplitudes with frequency ratio $\beta = \omega_E/\omega_0 = 4$ and cubic coefficient $\delta = 0.05$

If we now proceed to the more general case $r_1 \neq 0$, the first expression in equ. (11.1.70) at once yields

$$r_1^2 - r_3 r_1 + 2r_3^2 - \frac{4}{27\delta}(\beta^2 - 9) = 0 \tag{11.1.72}$$

This relationship between the amplitudes r_1 and r_3 represents a conic section in the r_1, r_3-plane. If we now examine the invariants of this curve of second order, we observe for $\beta > 3$ a real ellipse with the mid-point at $r_1 = 0, r_3 = 0$ (fig. 11.1.6). For $\beta = 3$, the ellipse degenerates into an imaginary pair of straight lines and for $\beta < 3$, the ellipse becomes imaginary. From this simple geometrical discussion, we are able to conclude that sub-harmonic oscillations can only be generated in the over-critical range $\beta > 3$. This is also confirmed by the formal solution of equ. (11.1.72) for r_1

$$r_1 = 0.5 r_3 \pm \sqrt{\frac{4}{27\delta}(\beta^2 - 9) - \frac{7}{4} r_3^2} \tag{11.1.73}$$

This equation, in addition, demonstrates explicitly for which amplitudes r_3 and a fixed frequency ratio β real solutions r_1 can exist. Once the radicand vanishes, the limiting amplitude is achieved for

$$r_3^2 = \frac{4}{7}\frac{4}{27\delta}(\beta^2 - 9) \tag{11.1.74}$$

This limit reproduces the maximum extension of the elliptical r_1, r_3-area in the r_3-direction. From a certain amplitude r_3 onwards, no sub-harmonic component can arise. If equ. (11.1.72) is formally solved for r_3 and the discussion is repeated, the following expression is obtained

$$r_1^2 \leqslant \frac{8}{7}\frac{4}{27\delta}(\beta^2 - 9) \tag{11.1.75}$$

i.e. the amplitude r_1 cannot exceed this limiting value. This corresponds to the extension of the ellipse in the r_1-direction. Finally, it should be noted that there are two solutions r_1 for every amplitude r_3. Here, we observe two stationary conditions corresponding to two static states of equilibrium, but are not in a position to assess the stability of these situations. Hayashi [11.3] showed, however, that the lower branch of the ellipse defines only unstable states.

Let us now consider the general solution of the system described by equ. (11.1.70). In a practical situation, this solution can most simply be determined iteratively. In particular, the second expression of equ. (11.1.70) can be selected to yield an iterative calculation of the amplitude r_3

$$r_3 = [r_0 - \tfrac{1}{4}\delta(-r_1^3 + 6r_1^2 r_3 + 3r_3^3)]/(1 - \beta^2) \tag{11.1.76}$$

which can be initiated with $r_1 = r_3 = 0$. Applying this approximate value of r_3, we can solve the first expression for r_1 in equ. (11.1.73) exactly. This yields two solutions for r_1, both of which lead to two corresponding r_3 values in the next iteration loop. In the subsequent iteration procedure, the convergence progress of these pairs is observed. This usually happens rapidly. We demonstrate the procedure on an example with $\beta = \omega_E/\omega_0 = 4, r_0 = 1$ and $\delta = 0.05$:

Iteration	solution branch $+\sqrt{}$			solution branch $-\sqrt{}$		
	r_1	r_3	$e_2[\%]$	r_1	r_3	$e_2[\%]$
1	4.520013	-0.066667	-125.65	-4.586679	-0.066667	110.10
2	4.474634	-0.150433	-8.94	-4.550826	0.006731	8.76
3	4.471301	-0.156396	-0.61	-4.547885	0.012570	0.68
4	4.471072	-0.156804	-0.04	-4.547657	0.013021	0.05
5	4.471056	-0.156832	0.00	-4.547640	0.013056	0.00

The quoted error is generated by the second expression in equ. (11.1.70) (the first one is satisfied exactly) and is normalised with reference to r_0 as follows

$$e_2 = 100\left[(1-\beta^2)r_3 + \tfrac{1}{4}\delta(-r_1^3 + 6r_1^2 r_3 + 3r_3^3) - r_0\right]/r_0 \tag{11.1.77}$$

This measure is used as the break-off criterion in the iteration of fig. 11.1.7. The process converges extraordinarily rapidly. If we now examine the resonance curve in fig. 11.1.7, we observe that no sub-harmonic oscillation can arise for $\beta < 3$, as was also demonstrated before. Thus, in this range, we note the same response as for the simple harmonic solution *ansatz*. For $\beta > 3$, there is the possibility of a sub-harmonic motion; a keen eye will note that the corresponding branching point (see r_1, β-diagram in (a)) lies very close to $\beta = 3$ but is nevertheless placed somewhat above it. This is also confirmed by equ. (11.1.73) since no real solutions are possible for $\beta = 3$ and $r_3^2 > 0$. The corresponding transition point is deduced from equ. (11.1.70) for an arbitrary small value of $r_1 = \varepsilon$ in the form

$$r_{3T} = \frac{-r_0}{8 + 51\delta r_{3T}^2/4} \tag{11.1.78}$$

from which r_3 is obtained iteratively. There follows the frequency ratio β_T

$$\beta_T = 3(1 + 3\delta r_{3T}^2/2)^{1/2} \tag{11.1.79}$$

For $r_0 = 1$ and $\delta = 0.05$ we have, for example,

$$r_{3T} = -0.1248$$

$$\beta_T = 3.0018$$

For $r_3 = r_{3T}$, the rapidly growing sub-harmonic response component emerges and generates for each given frequency ratio two amplitudes r_1 and three values of r_3. Our calculation does not assess which of these stationary states are stable and which unstable. At a first glance, the two branches of r_1 appear identical but for the sign. This, however, is not the case (see fig. 11.1.6) and the misleading impression arises from the small (absolute) value of r_3. Considering the three solution branches of $|r_3|$, we note that two of them appertain to the two r_1 solutions

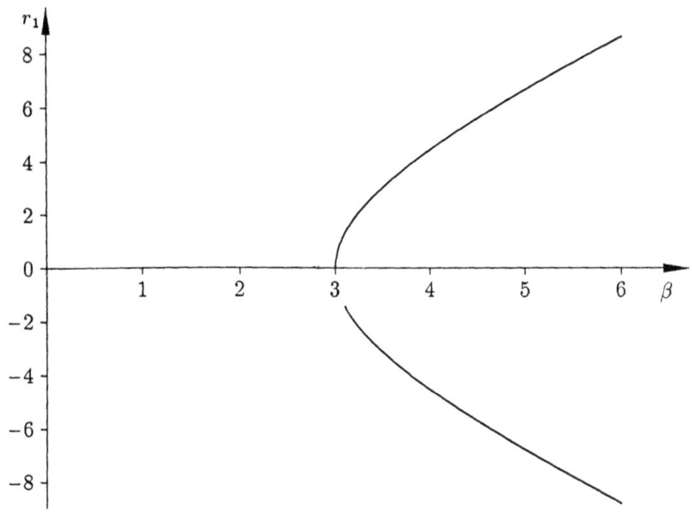

(a) sub-harmonic amplitude

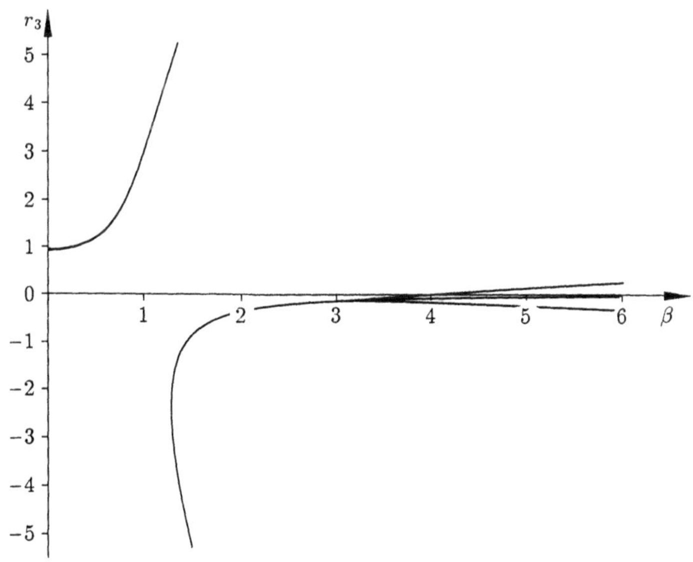

(b) harmonic amplitude

Fig. 11.1.7 Resonance behaviour of a cubically non-linear oscillator in conjunction with sub-harmonic oscillations ($\zeta = 0, r_0 = 1, \delta = 0.05$)

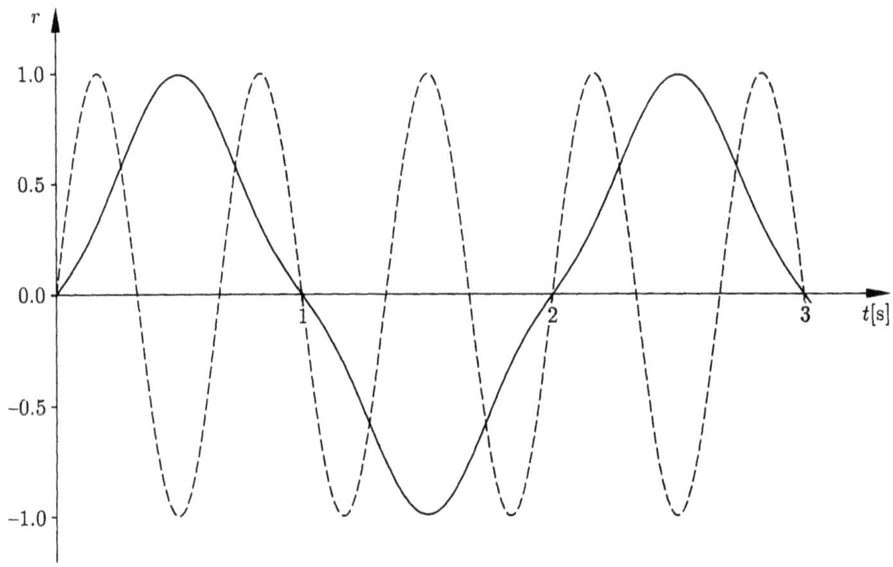

Fig. 11.1.8 Sub-harmonic motion with $T_0 = 1\,\text{s}, \beta = 3.05, \zeta = 0, r_0 = 1, \delta = 0.05$; the broken curve represents the linearised steady-state motion $r_0 \sin \omega_E t = (R_0/k) \sin \omega_E t$

in the diagram above and the third one belongs to the solution $r_1 = 0$ which can also represent a 'state of equilibrium'.

As an example of a stationary sub-harmonic motion, we have selected in fig. 11.1.8 the case $\beta = 3.05$ for $T_0 = 1\text{s}$. All other parameters are equal to those in the previous examples ($\zeta = 0$, $\delta = 0.05$, $r_0 = 1$). Here, we have deliberately exceeded the value of $\beta = 3$ only slightly in order to ensure the visualisation of the two solution contributions $\omega_E t/3$ and $\omega_E t$. For higher frequency ratios, the large sub-harmonic component overshadows all other effects. Even in the present case, the sub-harmonic part with $r_1 = 0.87$ is much more dominant than the harmonic part $r_3 = -0.12$.

So far, we have restricted ourselves to the special case of an undamped oscillator, so as not to complicate this introduction. A more thorough discussion reveals that the sub-harmonic oscillations are extremely sensitive to the presence of damping. Relatively low damping values already completely suppress the appearance of sub-harmonic oscillations. In [11.1], Klotter mentions that for

$$\sqrt{\tfrac{3}{4}\delta}\, r_0 = 1$$

(our example possesses a value of 0.19), the critical damping ratio which suppresses sub-harmonic oscillations is merely $\zeta = 0.0244$.

If a linear oscillator is harmonically excited with a circular frequency ω_E, it displays harmonic steady-state response in beat with ω_E. This is the only steady response we can observe experimentally as a result of damping which is inevitably present in a real oscillator. For non-linear systems, however, the situation is totally different and the additional, apparently inconspicuous cubic term in the Duffing equation leads to solutions containing sub-harmonic oscillations. The amplitudes of such sub-oscillatory component motion can well be dominant. In practice, the sub-harmonic oscillations have also been observed and are used technically in order to achieve a frequency reduction of quartz and atomic chronometers. Our formulation in equ. (11.1.63) can be understood only as a first attempt to comprehend the nature of sub-harmonic oscillations. Further components with frequencies of, e.g., $5\,\omega_E/3$, $7\omega_E/3$ (super-sub-harmonics) are imaginable but their manifestation is controlled by more severe limits on damping than for the basic component $\omega_E/3$.

Let us now return once again to the discussion of the simple harmonic trial solution and consider matters on the other side of the fence, so to speak. For this purpose, we restrict the argument to the special case of an undamped oscillator ($\zeta = 0$) and, in place of equ. (11.1.18), write the trial solution as

$$r = r_1 \sin \omega_E t = r_1 s \tag{11.1.80}$$

This time, however, we substitute this *ansatz* into the differential equation (11.1.6) in analogy to the procedure for sub-harmonic oscillations, and obtain

$$(1 - \beta^2) r_1 s + \delta r_1^3 s^3 - r_0 s = 0 \tag{11.1.81}$$

where β has already been defined in equ. (11.1.25). Using standard trigonometric transformations for the cubic term, we find

$$s^3 = \sin^3 \omega_E t = \tfrac{1}{4}(3 \sin \omega_E t - \sin 3\omega_E t) \tag{11.1.82}$$

and observe that the assumed solution of equ. (11.1.80) cannot be correct. It may be justified as a first approximation in the neighbourhood of $\delta = 0$, but equ. (11.1.82) indicates that the solution must contain super-harmonic components, here of the frequency $3\omega_E$. If we return once again to the differential equation, using such an extended *ansatz*, still higher frequencies are observed. Within this introduction, however, we prefer to limit ourselves to the first step in an iterative improvement and write

$$\begin{aligned} r &= r_1 \sin \omega_E t + r_3 \sin 3\omega_E t \\ &= [s \quad s_3]\{r_1 \quad r_3\} \end{aligned} \tag{11.1.83}$$

We now apply once again the method of weighted residues. We begin our investigation with equ. (11.1.65) and once more extend the integration over the longer one of the two periods, in this case T_E. With $\zeta = 0$ (otherwise the trial equation (11.1.83) would also have to be modified), we obtain

$$\int_{t_1}^{t_1+3T_E} \left(\begin{bmatrix} (1-\beta^2)s^2 & (1-9\beta^2)ss_3 \\ (1-\beta^2)ss_3 & (1-9\beta^2)s_3^2 \end{bmatrix} \begin{bmatrix} r_1 \\ r_3 \end{bmatrix} - r_0 \begin{bmatrix} s^2 \\ ss_3 \end{bmatrix} \right) \mathrm{d}t + \delta \int_{t_1}^{t_1+3T_E} \begin{bmatrix} s(r_1 s + r_3 s_3)^3 \\ s_3(r_1 s + r_3 s_3)^3 \end{bmatrix} \mathrm{d}t = \boldsymbol{o} \tag{11.1.84}$$

Here, the integrals

$$\int_{t_1}^{t_1+T_E} s^2 \mathrm{d}t = \frac{\pi}{\omega_E}$$

$$\int_{t_1}^{t_1+T_E} s_3^2 \mathrm{d}t = \frac{\pi}{\omega_E} \tag{11.1.85}$$

$$\int_{t_1}^{t_1+T_E} s s_3 \mathrm{d}t = 0$$

facilitate the numerical evaluation of the first, more simple part of equ. (11.1.84). Analogous to equ. (11.1.68), we obtain the following expression for the additional cubic term

$$\int_{t_1}^{t_1+T_E} \begin{bmatrix} s r^3 \\ s_3 r^3 \end{bmatrix} \mathrm{d}t = \int_{t_1}^{t_1+T_E} \begin{bmatrix} r_1^3 s^4 + 3r_1^2 r_3 s^3 s_3 + 3r_1 r_3^2 s^2 s_3^2 + r_3^3 s s_3^3 \\ r_1^3 s_3 s^3 + 3r_1^2 r_3 s^2 s_3^2 + 3r_1 r_3^2 s s_3^3 + r_3^3 s_3^4 \end{bmatrix} \mathrm{d}t \tag{11.1.86}$$

the evaluation of which is achieved with the integrals

$$\int_{t_1}^{t_1+T_E} s^4 \mathrm{d}t = \frac{3}{4}\frac{\pi}{\omega_E}$$

$$\int_{t_1}^{t_1+T_E} s^3 s_3 \mathrm{d}t = -\frac{1}{4}\frac{\pi}{\omega_E}$$

$$\int_{t_1}^{t_1+T_E} s^2 s_3^2 \mathrm{d}t = \frac{1}{2}\frac{\pi}{\omega_E}$$

$$\int_{t_1}^{t_1+T_E} s s_3^3 \mathrm{d}t = 0$$

$$\int_{t_1}^{t_1+T_E} s_3^4 \mathrm{d}t = \frac{3}{4}\frac{\pi}{\omega_E} \tag{11.1.87}$$

We are now in a position to write the system of equ. (11.1.84) in an explicit form and obtain

$$(1-\beta^2)r_1 + \tfrac{3}{4}\delta(r_1^3 - r_1^2 r_3 + 2r_1 r_3^2) = r_0$$

$$(1-9\beta^2)r_3 + \tfrac{1}{4}\delta(-r_1^3 + 6r_1^2 r_3 + 3r_3^3) = 0 \tag{11.1.88}$$

Although this system is very similar to the sub-harmonic formulation in equ. (11.1.70), it is much more difficult to solve as a consequence of the inhomogeneous form of the first equation. Theoretically, a linearisation in r_3 would be possible providing that the super-harmonic component remains small. However, our knowledge of sub-harmonic oscillations tells us of the risk of such pre-suppositions since such components could also become dominant. We thus settle for the lesser of the two evils and attempt to solve equ. (11.1.88) iteratively in a numerical way. To this purpose, we introduce the ratio

$$\alpha = \frac{r_3}{r_1} \tag{11.1.89}$$

and solve formally the first equation for β^2

$$\beta^2 = 1 - \frac{r_0}{r_1} + \frac{3}{4}\delta r_1^2(1 - \alpha + 2\alpha^2) \tag{11.1.90}$$

We then substitute this result into the second equ. (11.1.88) and obtain for $\alpha \neq 0$

$$24\frac{r_1}{r_0} + \frac{3}{4}\delta\frac{r_1^3}{r_0}\left(\frac{1}{\alpha} + 21 - 27\alpha + 51\alpha^2\right) - 27 = 0 \tag{11.1.91}$$

For given values of

$$\alpha = \pm c \tag{11.1.92}$$

the left-hand side of equ. (11.1.91) can be expressed as an error function $e(r_1)$ which, of course, does not vanish for arbitrary values of r_1. Applying Newton's method, the zeros of $e(r_1)$ can approximately be located through

$$\Delta r_1 = -e(r_1)/\frac{\partial e}{\partial r_1}(r_1) \tag{11.1.93}$$

In this way, we obtain the recurrence formula

$$\left(\frac{r_1}{r_0}\right)_{\nu+1} = \left(\frac{r_1}{r_0}\right)_{\nu} + \frac{27 - 24(r_1/r_0)_\nu - \frac{3}{4}\delta r_{1\nu}^2(\frac{1}{\alpha} + 21 - 27\alpha + 51\alpha^2)}{24 + 3\cdot\frac{3}{4}\delta r_{1\nu}^2(\frac{1}{\alpha} + 21 - 27\alpha + 51\alpha^2)} \tag{11.1.94}$$

For any given α, we apply an iteration until the convergence of r_1 is achieved, and then deduce r_3 from equ. (11.1.89). Subsequently, the corresponding frequency ratio is calculated from equ. (11.1.90). Depending upon the sign of α in equ. (11.1.92), we observe two solution branches. The convergence is achieved with extreme rapidity as is illustrated by the following example with $\alpha = \pm 2.5$:

ν	$r_1(+2.5)$	$r_1(-2.5)$
1	1.1250000	1.1250000
2	0.8932649	0.8598520
3	0.8576620	0.8021817
4	0.8569261	0.7997877
5	0.8569257	0.7997838
6	0.8569257	0.7997838

The corresponding solutions are found to be

α	β	r_1	r_3
+2.5	0.368708	0.856926	2.142314
−2.5	0.365314	0.799784	−1.999460

and the residual errors of equ. (11.1.88) are seen to be effectively zero.

Figure 11.1.9 reproduces the resonance curves calculated for a frequency range of practical interest and the standard parameter values used before. In a wide range, the super-harmonic component appears to be diminishingly small, although it is interesting to note that it does not vanish for $\beta = 0$. Obviously, at $\beta \approx 1/3$, a new, super-harmonic resonance emerges. The resonance peak curve is bent towards the right, analogous to the harmonic resonance $\beta = 1$ in the non-linear case. Thus, we observe once again that up to three amplitudes r_3 are associated with a given frequency ratio β. If, in addition, we carry out a stability analysis, we realise in analogy to fig. 11.1.5 that super-harmonic jump-effects can occur and that the branch between the upper and the lower solution is unstable. It should be added that r_3 itself reaches very large values ($\zeta = 0!$) and that the peak in the graph is here cut off at $r_3 = 2$, but continues as a matter of fact its upward trend indefinitely. The associated behaviour of the harmonic component is also extremely strange. With the onset of the super-harmonic resonance, r_1 decreases considerably. Thus, in the case of low damping, the super-harmonic component can be substantially larger than the harmonic one. The non-linearity of the system induces a transfer of energy into the higher harmonic oscillations. With regard to r_1, it is to be noted once more that a given value of β can be associated with up to three amplitude values. A stability analysis demonstrates that the bottommost solution branch is unstable, up to the point of intersection. The solution must contain super-harmonic components, whereas the sub-harmonic ones will only appear under certain conditions. If we had inserted terms with frequencies $5\,\omega, 7\omega$... into the trial solution, then the behaviour just demonstrated would repeat itself for the values of $\beta = 1/5, 1/7...$. Thus, we are faced with extremely complex manifestations in real situations.

As an example of a super-harmonic steady-state motion, we consider in fig. 11.1.10 the case of $\alpha = 0.85$ and $\beta = 0.3492623$. Both solution components are equally important here and it emerges clearly that non-linear oscillators can carry out periodic motions which are, however, unharmonic as a general rule. The high frequency components significantly alter the pattern of motion within the period.

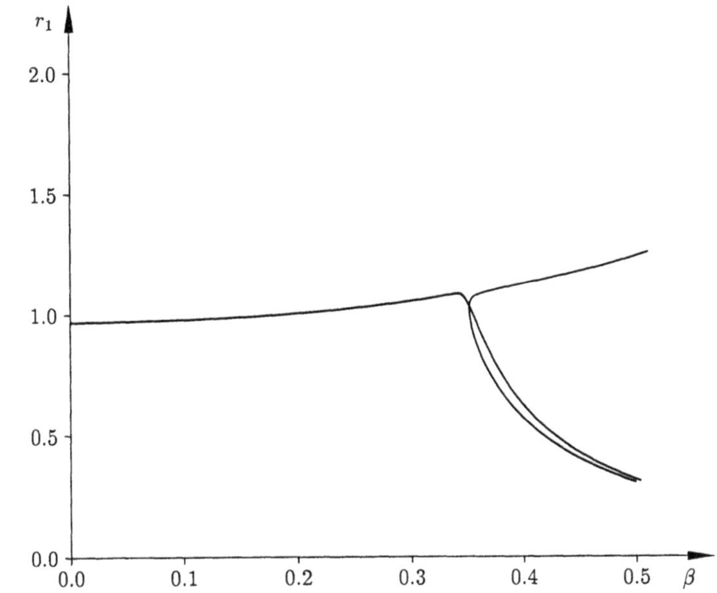

(a) harmonic component of solution

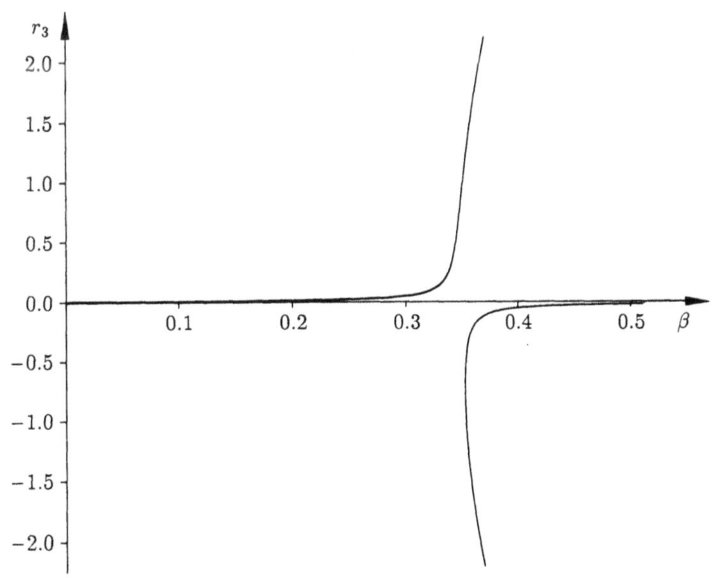

(b) super-harmonic component of solution

Fig. 11.1.9 Resonance behaviour of a cubically non-linear oscillator containing super-harmonic oscillations ($\zeta = 0, r_0 = 1, \delta = 0.05$)

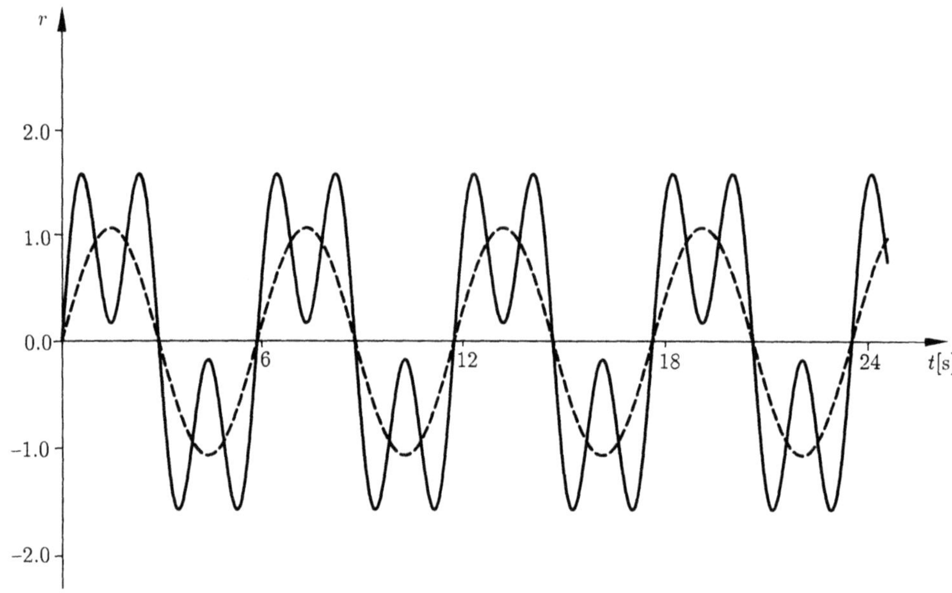

Fig. 11.1.10 Super-harmonic motion with $T_0 = 1\,\mathrm{s}, \beta = 0.3492623, \zeta = 0, r_0 = 1, \delta = 0.05$; the broken curve reproduces the linearised steady-state motion $r_0 \sin \omega_E t = (R_0/k) \sin \omega_E t$

11.1.2 Direct integration methods and experiments in the 'computer laboratory'

With the aid of a few very simple trial solutions, we were able to obtain some interesting, but unfortunately not comprehensive insights into the behaviour of non-linear oscillators. A generalisation of these techniques to multi-degree realistic non-linear problems does not appear to be attractive and closed analytical solutions are not possible in any case. Thus, the only remaining alternative is the application of the direct integration procedures for the solution of the non-linear equilibrium equation. This yields the response of the system for any given initial conditions with relative ease, as exemplified for linear systems in Chapter 10. We have to consider with some care, however, the question of numerical accuracy and also of computational expenditure.

The fact that classical linear systems admit only one solution derived, say, by the deceptively simple application of the direct integration methods (selection of the parameters and initial conditions leads to a more or less automatic solution) may generate a false sense of security as to the reliability of numerical solutions to non-linear problems. We may believe we have obtained a relevant computation of a physical process especially if it happens to represent a steady-state response to a sinusoidal excitation, but may forget too often the possible significant influence of the imprecisely definable initial conditions in a realistic problem or the possibility of a gradual alteration of the system parameters which can in turn induce serious consequences. With somewhat intensive effort, we may attempt to overcome the numerical difficulties of a problem and obtain seemingly meaningless, irregular, non-periodic responses, but may not realise that the strange result is not false, but represents the true response.

Thus, for a complete investigation of dynamical systems, we require not only a computer and the direct integration methods. These provide no more than an ideal computer laboratory in which an arbitrary number of experiments can be performed, yielding an immense data flow. We require, in addition, certain principles according to which the data may be evaluated and displayed, thus giving an insight into the astonishing variety of responses of dynamical systems. The authors hope to familiarise the reader with this new field of research by selecting the Duffing equation as an illustrator of such phenomena. Understandably, we can provide only an initial insight here, but refer the reader to the more general, but, from the engineer's point of view, still elementary description of the theoretical edifice of chaotic dynamics presented in Volume VII. Those readers interested in pursuing this fascinating field of research further are recommended to consult the ever-expanding literature on the topic which has appeared in recent years (e.g. [11.6], [11.7], [11.8], [11.59]).

Let us begin our elementary discussion with the example of a linear, damped, free oscillator as defined in equ. (1.3.5). Here, the damping ratio of $\zeta = 0.05$ and the eigenperiod of $T_0 = 1\mathrm{s}$ are selected as the system parameters. For given initial conditions, for example $r(0) = 1$ and $\dot{r}(0) = 0$, the solution may be deduced from equs. (1.3.19) and (1.3.21). Since this solution technique, however, is not appropriate for the subsequent transition to the non-linear Duffing equation, we do not apply it here, but employ the computer and a numerical integration procedure instead. Simplicity and numerical accuracy are the two important selection criterea here. A brief glance at table 10.3.2 indicates that the Newmark procedure with $\beta = 0$ and $\gamma = 0.50$ qualifies for this case. The algorithmic damping is zero and the period elongation can be kept small by means of a small step-size. Here, we are not concerned with the computational effort involved and select also in the following demonstrations $\Delta t = 0.005$ as the time-step. The computational scheme for this integration step is based upon the procedure of equ. (10.2.35). However, we already adjust it here with a view to its subsequent application to the Duffing equation and obtain

$$
\begin{aligned}
&\text{given} && r(0), \dot{r}(0) \\
&\text{calculate} && \ddot{r}(0) = \omega_0 r_0 \sin \omega_E t - 2\zeta\omega_0 \dot{r}(0) - \omega_0^2 \left(r(0) + \delta[r(0)]^3 \right) \\
& && t = 0 \\
&\text{loop } \Delta t: \\
&\quad \ddot{r}(t+\Delta t) = \ddot{r}(t) \\
&\quad \text{loop convergence } \ddot{r}(t+\Delta t): \\
&\qquad r(t+\Delta t) = r(t) + 0.4\Delta t^2 \ddot{r}(t) + 0.1\Delta t^2 \ddot{r}(t+\Delta t) \\
&\qquad \dot{r}(t+\Delta t) = \dot{r}(t) + 0.5\Delta t \ddot{r}(t) + 0.5\Delta t \ddot{r}(t+\Delta t) \\
&\qquad \ddot{r}(t+\Delta t) = \omega_0^2 r_0 \sin \omega_E (t+\Delta t) - 2\zeta\omega_0 \dot{r}(t+\Delta t) \\
&\qquad\qquad - \omega_0^2 \left(r(t+\Delta t) + \delta[r(t+\Delta t)]^3 \right) \\
&\quad t = t + \Delta t
\end{aligned}
\tag{11.1.95}
$$

Selecting as parameters the static excitation amplitude $r_0 = 0$ and the coefficient of the non-linear term $\delta = 0$, we now proceed to the solution of our initial problem. The 'laboratory' computer yields the well-known exponentially decaying eigenoscillation shown in fig. 11.1.11(a). In place of the time-history response diagram, it is also possible to demonstrate the motion in the phase space, but this has not yet been discussed in this treatise; see, however, the account in Volume VII. In the present case, the phase space is limited to a two-dimensional plane referred to a Cartesian system of displacement and velocity, r and $\dot{r}$ (free oscillator, 1 degree of freedom), two measures which direct integration supplies anyhow. In the phase space, the motion takes the form of a spiral, which approaches asymptotically the point of rest of the system , $r = 0, \dot{r} = 0$. At this point, we may forgo the effort required to repeat the procedure for various initial conditions. Each distinct initial condition obviously yields a motion in the form of a non-crossing spiral converging towards the origin of the phase space which represents the stationary equilibrium point in all cases. An almost magic significance might be attributed to the origin of the phase space since it inexorably attracts all individual motions; it is thus referred to as a point attractor. Evidently, this point attractor expresses a significant and fundamental concept, defining the motion in an overall sense.

The Fourier analysis of the time-history response provides a further possible interesting technique of interpretation often used in the experimental field. In section 1.4.2, we discussed the Fourier expansion of a periodic loading $R(t)$ and in section 1.4.3, we extended this procedure to the representation of more general excitation processes. We now apply these preceding considerations to the response $r(t)$. Let the information for $r(t)$ be registered at N equal time-steps Δt, i.e. within a time interval of $T_r = N\Delta t$. This is executed by a numerical integration. The Fourier expansion of the $r(t)$-function is now

$$r_F(t) = \tfrac{1}{2}a_0 + \sum_{\mathrm{n}=1}^{\infty}(a_\mathrm{n} \cos \mathrm{n}\omega_r t + b_\mathrm{n} \sin \mathrm{n}\omega_r t) \tag{11.1.96}$$

where

$$\begin{aligned} a_\mathrm{n} &= \frac{2}{T_r}\int_0^{T_r} r(t) \cos \mathrm{n}\omega_r t \, \mathrm{d}t \\ b_\mathrm{n} &= \frac{2}{T_r}\int_0^{T_r} r(t) \sin \mathrm{n}\omega_r t \, \mathrm{d}t \end{aligned} \tag{11.1.97}$$

This representation implicitly assumes a periodicity with a period T_r of the function $r(t)$. Generally, however, this is not the case and thus, components with a period larger than T_r are generally excluded. Thus, the smallest circular frequency it is possible to reproduce is given by

$$\omega_r = \frac{2\pi}{T_r} = \frac{2\pi}{\mathrm{N}\Delta t} = \Delta\omega \tag{11.1.98}$$

Here, ω_r also defines the step-up distance of the frequency sequence which can be reduced optionally by elongation of T_r. If we consider at the same time the

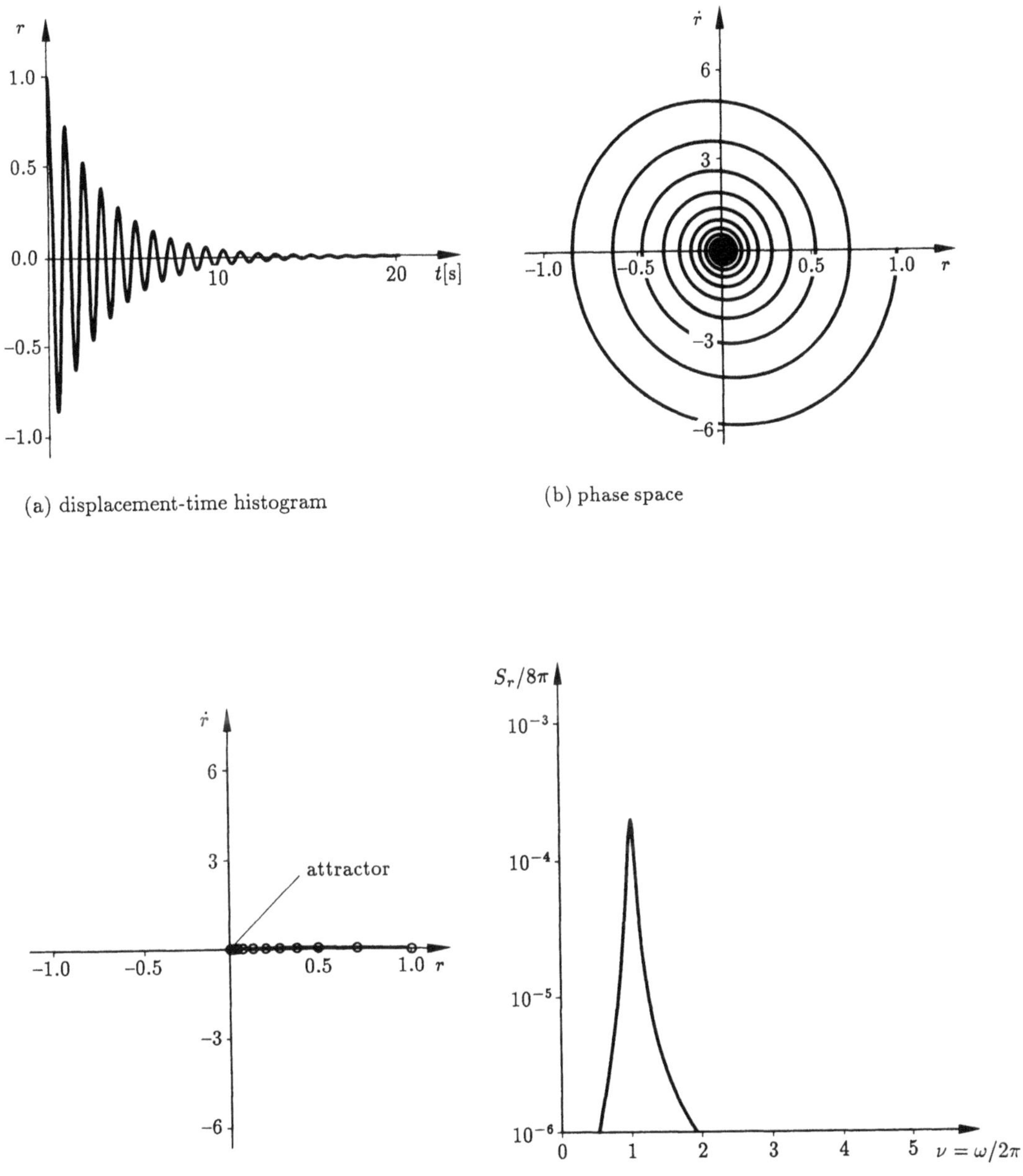

(a) displacement-time histogram

(b) phase space

(c) point attractor

(d) power spectral density

Fig. 11.1.11 Graphical representations of the damped free oscillations of a linear oscillator fact that we possess no continuous information on the function $r(t)$, but only a sequence of discrete values $r(t_{\mathrm{m}})$, $\mathrm{m} = 0, 1, 2, \ldots$ with

$$t_{\mathrm{m}} = \mathrm{m}\Delta t \tag{11.1.99}$$

and that we are well advised within a meaningful numerical analysis to set an upper limit to the frequency components, e.g. $\mathrm{n_{max}} = \mathrm{N} - 1$, we obtain from equs. (11.1.96) and (11.1.97) the fundamental formulae for a discrete Fourier transformation

$$r_F(t) = \tfrac{1}{2}a_0 + \sum_{\mathrm{n}=1}^{\mathrm{N}-1} \left(\bar{a}_{\mathrm{n}} \cos(\mathrm{n}\Delta\omega t) + \bar{b}_{\mathrm{n}} \sin(\mathrm{n}\Delta\omega t)\right) \Delta\omega \tag{11.1.100}$$

where

$$\begin{aligned} \bar{a}_{\mathrm{n}} &= \frac{\Delta t}{\pi} \sum_{\mathrm{m}=0}^{\mathrm{N}-1} r(t_{\mathrm{m}}) \cos 2\pi \frac{\mathrm{nm}}{\mathrm{N}} \\ \bar{b}_{\mathrm{n}} &= \frac{\Delta t}{\pi} \sum_{\mathrm{m}=0}^{\mathrm{N}-1} r(t_{\mathrm{m}}) \sin 2\pi \frac{\mathrm{nm}}{\mathrm{N}} \end{aligned} \tag{11.1.101}$$

applies. Putting

$$\begin{aligned} \Delta\omega &\rightarrow \mathrm{d}\omega \\ \mathrm{n}\Delta\omega &\rightarrow \omega \\ \Delta t &\rightarrow \mathrm{d}t \end{aligned} \tag{11.1.102}$$

we can incorporate in this representation the Fourier integrals briefly discussed in section 1.4.3. The numerical evaluation of equs. (11.1.100) and (11.1.101) can effectively be achieved by means of the fast Fourier transform which, however, requires

$$\mathrm{N} = 2^{\mathrm{p}} \quad \text{steps}, \qquad \mathrm{p} = \text{positive integer} \tag{11.1.103}$$

This possibility was allowed for in the development of the ASKA II system [11.9]. Equation (11.1.100) sums up the $\mathrm{n}\omega_r$-harmonics over the given frequency range at each of the instants determined by m. In order to evaluate the intensity of the individual frequency contributions over the entire time range, we require a suitable criterion. To this purpose, we may apply a statistic expectation value, i.e. the mean square value of Chapter 8. Thus, from equ. (8.1.19), we deduce

$$\overline{r_F^2} = \frac{1}{T_r} \int_0^{T_r} r_F(t)^2 \mathrm{d}t \tag{11.1.104}$$

From a physical point of view, the mean square value can be interpreted here as being proportional to the average power dissipated in the time interval T_r by a linear damper c. From this background, the name power spectral density, used in the following, appears appropriate. In a harmonic case, it is readily observed that this power is proportional to the square of the amplitude (see also equ. (1.4.22a). If two frequencies partake in this motion, the power will be proportional to the sum of the individual amplitude squares. The same also applies to the $\mathrm{N}-1$ frequency contributions, as can be demonstrated by elementary integrations (substitute equ. (11.1.100) into equ. (11.1.104))

$$\overline{r_F^2} = \tfrac{1}{4}a_0^2 + \tfrac{1}{2}\sum_{\mathrm{n}=1}^{\mathrm{N}-1}(\bar{a}_{\mathrm{n}}^2 + \bar{b}_{\mathrm{n}}^2) \tag{11.1.105}$$

The increment of the mean square value when progressing from $\mathrm{n}\Delta\omega$ to $(\mathrm{n}+1)\Delta\omega$ is given by

$$\Delta(\overline{r_F^2}) = \tfrac{1}{2}(\bar{a}_{\mathrm{n}}^2 + \bar{b}_{\mathrm{n}}^2) \tag{11.1.106}$$

from which we deduce the rate of increase or the power spectral density

$$S_r(\omega_{\mathrm{n}}) = \tfrac{1}{2}\frac{\Delta(\overline{r_F^2})}{\Delta\omega} = \frac{1}{4\Delta\omega}(\bar{a}_{\mathrm{n}}^2 + \bar{b}_{\mathrm{n}}^2) \tag{11.1.107}$$

Care should be taken when comparing our formulation to the power spectral density quoted in the literature. The definition employed here agrees with that of Clough and Penzien [11.14]; Meirovitch [11.11] and Ueda [11.12] use an expression which is larger by a factor of 2π. Finally, Hurty [11.13] defines the spectral density with an additional factor of 4π, whilst Harris and Crede [11.10] utilise the double value.

In analogy to its experimental determination, the spectral density can be obtained computationally by processing the $r(t)$-distribution through a frequency analyser. In this way, for any given circular frequency ω, only the frequency band between ω and $\omega + \Delta\omega$ is filtered through. Thus, the procedure evaluates the mean square value from the processed $r(t)$-values and then refers it to $\Delta\omega$. At the same time, the area element $S_r(\omega)\Delta\omega$ represents the contribution of the frequency ω to the mean square value $\Delta\overline{r^2}/2$ whilst the entire area below the curve $S_r(\omega)$ determines the mean square value of $\overline{r^2}/2$ itself (fig. 11.1.12). The power spectral density thus illustrates how the energy of the motion is distributed over the frequency spectrum. If a particular motion, e.g. a free or a steady-state oscillation as shown in fig. 11.1.11, contains only one circular frequency ω_0, then for an appropriate step-up of the frequency band, a single Fourier coefficient will be dominant, namely that appertaining to $\mathrm{n}\omega_r = \omega_0$. In the absence of damping, a Dirac function emerges at ω_0 in the limiting case. If damping is present, a transient process arises in a free motion and the spectral density function displays a peak at ω_0 which possesses a more or less wide basis depending on the cut-off error. We obtain a similar scenario for a motion involving a number of discrete frequencies, as for example for oscillations with sub-harmonic or super-harmonic components. The 'needles' in the S_r-diagram demonstrate the participation of distinct frequencies. In contrast to this, a band-like trend of S_r indicates the presence of numerous frequencies and thus signalises a stochastic character of the motion. The 'white noise ', which is

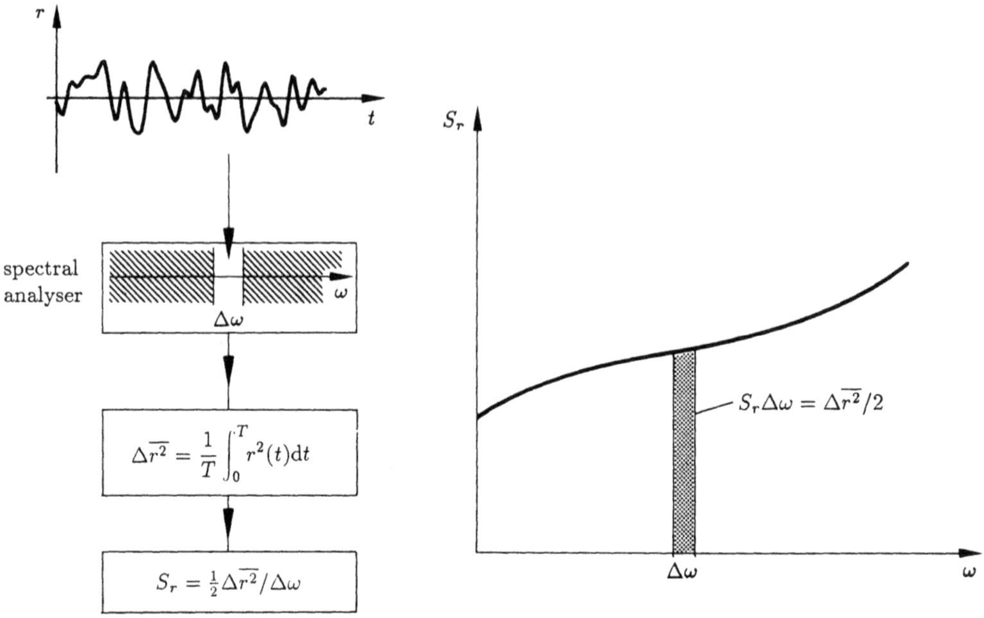

Fig. 11.1.12 Derivation and definition of the power spectral density

represented by a nearly constant $S_r(\omega)$ -function, is a typical example of this (see also fig. 8.1.7).

The S_r-diagram of fig. 11.1.11 should thus have received sufficient attention. The motion consists of a damped eigenoscillation with an approximate circular frequency of $\omega_0 = 2\pi/1$ (light damping) and the S_r-diagram displays a single peak which appears as a more or less pronounced needle according to the adopted step-up of $\Delta\omega$. In his own numerical experiments, the attentive reader should remember that the aforementioned procedure implies a periodicity in $r(t)$ with the observation interval T_r as the period. Since T_r is finite, a considerable 'jump' can accidentally occur between the initial and final value of the $r(t)$-record, thus causing a marked reduction in the quality of the S_r-distribution. In this case, we recommend an extension of the operational interval.

Let us now proceed one step further in our discussion and consider a damped linear oscillator with a harmonic excitation, thereby approaching our ultimate Duffing problem. This simple problem has already been discussed in section 1.4, see, for example, equ. (1.4.10). The motion of the system is composed of a transient part with the (damped) eigenfrequency of the system and a steady-state part in beat with the excitation frequency. The transient part is eventually damped out, and ultimately, only the steady-state part remains. Let us now discuss this course of motion from the point of view of our new concepts (fig. 11.1.13). The system parameters are similar to those of the aforementioned free oscillator ($\zeta = 0.05, T_0 =$ 1 s) and the excitation is defined by $R_0/k = r_0 = 1$ and $T_E = 0.25$ s. The initial conditions are taken for the present as $r(0) = 0$ and $\dot{r}(0) = 0$. The motion shown in fig. 11.1.13(a) is obtained by numerical integration and is practically exact; it has

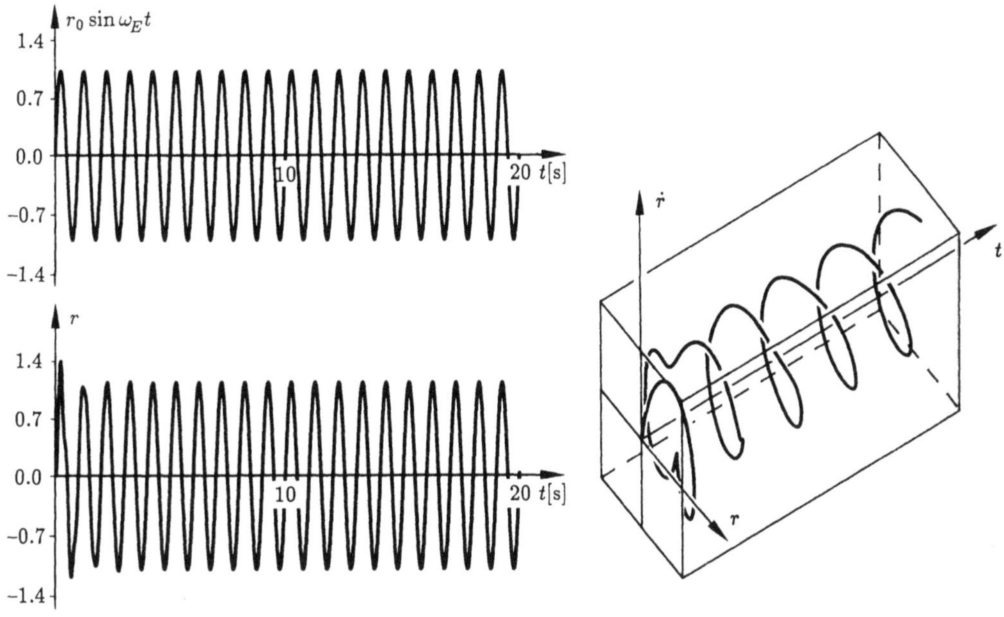

(a) excitation and response histogram

(b) phase space

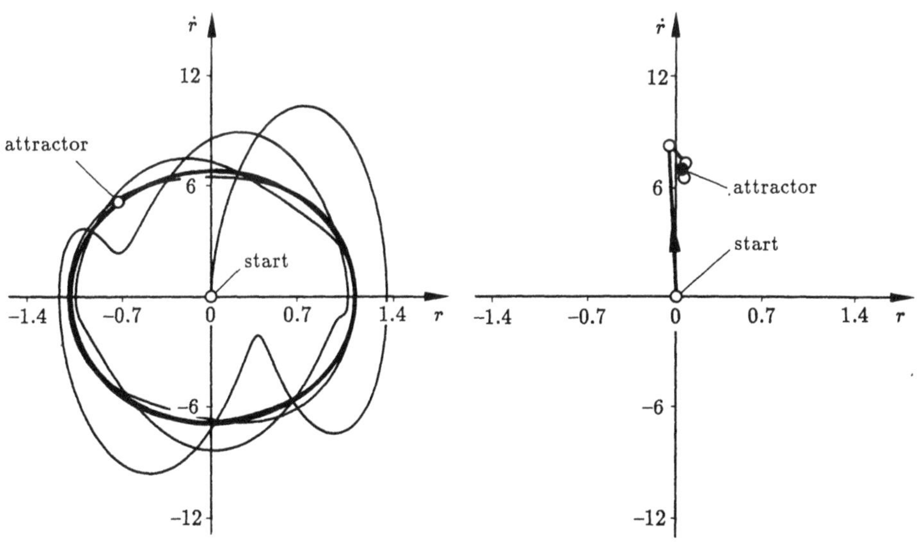

(c) projection of the phase space onto the $r, \dot{r}$ - plane

(d) periodic attractor

Fig. 11.1.13 Interpretation of the response of a damped linear oscillator subject to harmonic excitation, *Part 1* been referred to in the introduction. The present force-excited system possesses a three-dimensional, so-called extended phase space (see [11.59]), determined by the coordinates $r, \dot{r}$ and the time t (fig. 11.1.13). In this extended phase space, the motion is described initially as a wavy spiral free of any crossing-points. This space curve is often designated as an integral curve or trajectory. It is, of course, more convenient to work in a plane and thus, use is made of the projection of the extended phase space onto the $r, \dot{r}$-plane (we refer to this as the phase portrait); in the latter, the cross-free property of the trajectories is lost. For the selected initial conditions, the motion starts at the origin and, following the decaying of the transient part of the solution, proceeds to an elliptical orbit in the phase portrait which remains invariant. Such an invariant orbit is known as a periodic attractor or limit cycle.

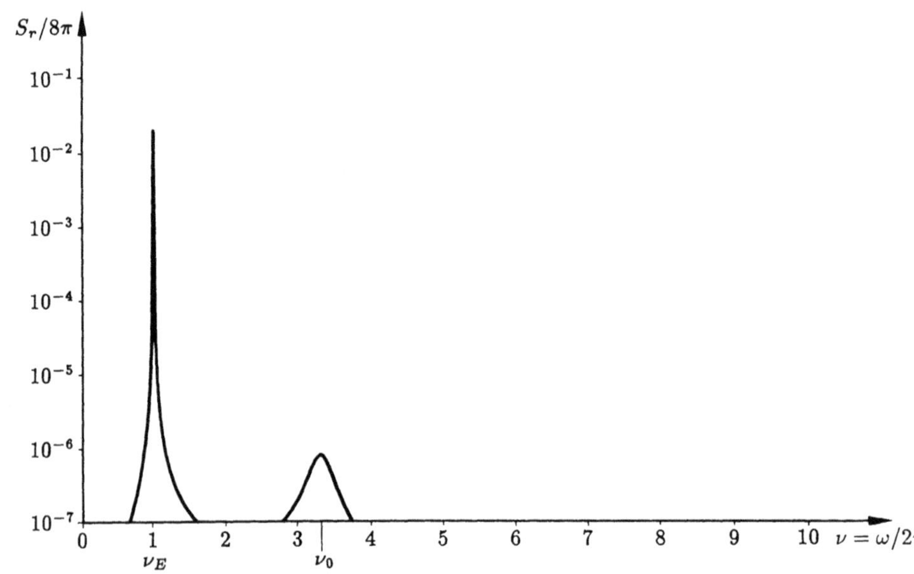

(e) power spectral density

Fig. 11.1.13 Interpretation of the response of a damped linear oscillator subject to harmonic excitation, *Part 2*

We thus have now gained some first characteristic information on the complex motion. If the motion is generated by different initial conditions, as for example in fig. 11.1.14(a) for $r(0) = 1$, $\dot{r}(0) = 0$ and in fig. 11.1.14(b) for $r(0) = -1, \dot{r}(0) = 0$, we will ultimately obtain the same elliptic orbit which seems to magically attract all motions, i.e. the orbit functions as an attractor. Since the magnitude of the response amplitude is proportional to the excitation amplitude, the motion converges for $t \to \infty$ and different excitation amplitudes to a family of ellipses as phase portraits (fig. 11.1.15). This information can be further compressed when points in phase space, separated from each other by the excitation period, are marked by a small circle (fig. 11.1.13(d)). In doing so, we obviously execute planar intersections

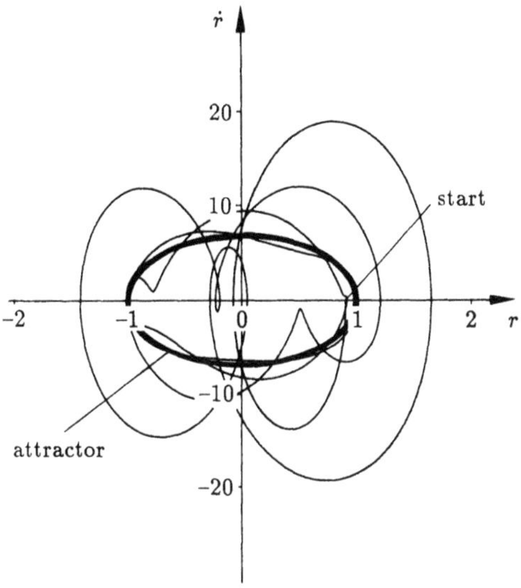

(a) $r(0) = 1.0$
$\dot{r}(0) = 0.0$

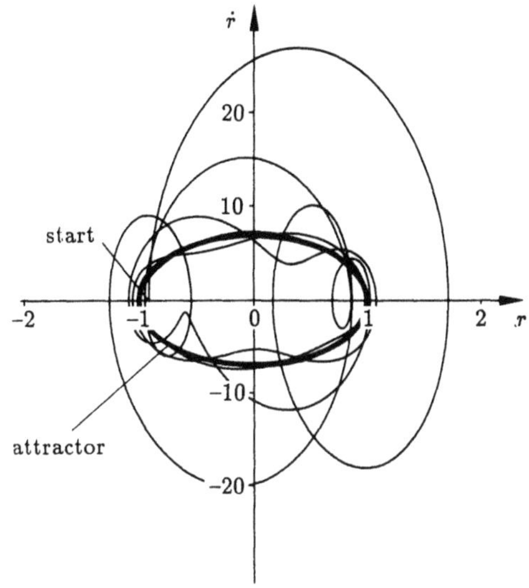

(b) $r(0) = -1.0$
$\dot{r}(0) = 0.0$

Fig. 11.1.14 Phase portrait of a damped linear oscillator harmonically excited for different initial conditions

of the trajectory at a distance T_E (stroboscopic method; fig. 11.1.16). The points of intersection with the integral curve are thereby identified and transferred ultimately onto an $r, \dot{r}$-phase portrait. In our case, these points always proceed from the individual starting point to the same end point (fig. 11.1.17). The periodic attractor or limit cycle is mapped on a single point by the stroboscopic method. The spiral approach to the attractor signalises the stable nature of the limit cycle.

A final point to be considered is the power spectral density (fig. 11.1.13(e)). In accordance with the components of the present motion, we observe two peaks in the S_r-diagram, one at the excitation frequency for the steady-state component and the other at the damped eigenfrequency for the transient part. The more the observed time interval is extended, the less important the transient peak becomes. If the motion is first inspected after a certain lead time in the course of which the transient motion has faded away, practically only the steady-state peak is observed.

We now return to the main issue of our discussion and in particular to the Duffing equation. The following data are selected as system parameters

$\zeta = 0.05$	damping
$\delta = 0.05$	cubic stiffness term
$T_0 = 1$ s	linearised eigenperiod
$T_E = 1$ s	excitation period
$R_0/k = r_0 = 1$	linearised static amplitude

and as initial conditions, we choose the state of rest $r(0) = 0$, $\dot{r}(0) = 0$.

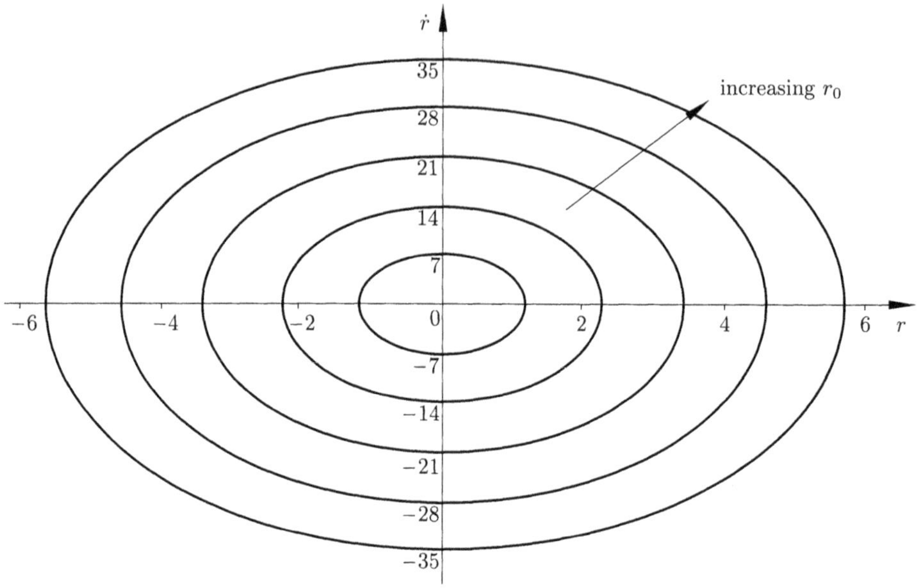

Fig. 11.1.15 Attractors in the phase portrait of a linear oscillator with varying excitation amplitudes

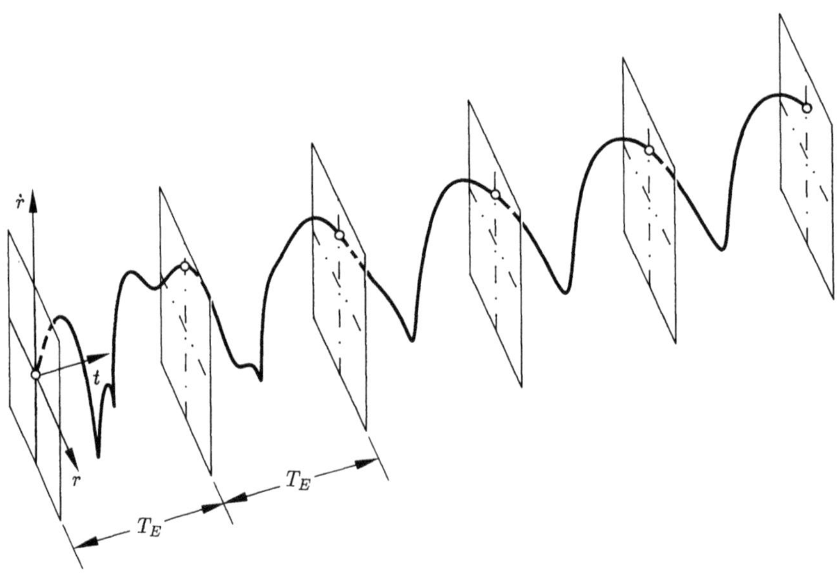

Fig. 11.1.16 Stroboscopic method or Poincaré's sections in the phase space; definition of the periodic attractor

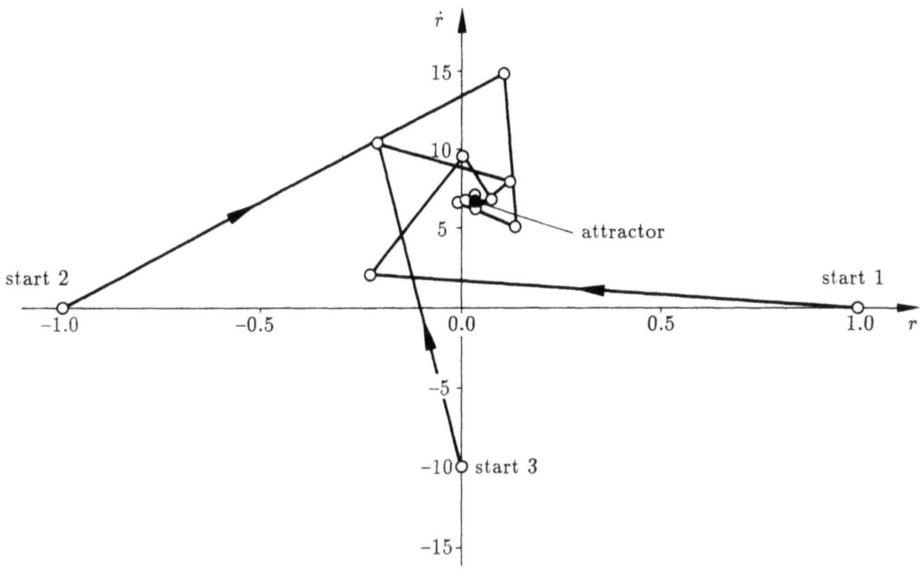

Fig. 11.1.17 Poincaré's sections for different initial conditions: the periodic attractor attracts all motions

The transient building-up oscillations are only of secondary interest here and a lead time of 4000 time steps, i.e. of 20 s = 20 load cycles, leads to a decay of the transient effects. In fig. 11.1.18, the excitation and the deflection for the follow-on 4000 time-steps are shown and the corresponding phase portrait is plotted. Deflection histogram and phase portrait point to a stationary periodic oscillation. This is underlined by the fact that the periodic attractor is reduced to a point in our representation. We confirm from the spectral density diagram that the periodic motion is essentially a harmonic one with an excitation frequency ω_E, but a small but sharp super-harmonic component $3\omega_E$ can also be detected. If, however, we observe the logarithmic scale, then the component $3\omega_E$ proves to be insignificant. It should still be noted that the spectral density diagram (as also the periodic attractor) was construed using $2^{13} = 8192$ time-steps in the fast Fourier transformation. The behaviour of the non-linear oscillator shown in the figure appears to be utterly harmless and confirms the legitimacy of a trial solution with a single frequency, namely the excitation frequency. The fact that the deflection amplitude proves to be essentially smaller than in the linear case without a cubic term is remarkable.

In the case of a linear oscillator, an increase of the excitation amplitude leads to a proportional amplification of the displacement amplitude. This is demonstrated in fig. 11.1.15 by the family of similar ellipses or limit cycles. We are already aware through our preceding experiments that the proportionality is lost in the non-linear case. Consequently, we employ our computer laboratory to determine the influence of the excitation amplitude upon the long-term behaviour of the oscillator. All other conditions remain unaltered.

Amplifying the excitation amplitude to $r_0 = 5$ appears for the present to have no significant effects (fig. 11.1.19). Once again, we obtain a stationary periodic solution and a phase portrait, pointing to a 'slightly dented ellipse-like picture'. The periodic attractor appears as a point and the power spectral density points to a more dominant, but nevertheless small super-harmonic component with a frequency $3\omega_E$. Here, an additional small super-harmonic component $5\omega_E$ is already visible but its size is insignificant. This trend is intensified for an excitation amplitude of $r_0 = 10$ (fig. 11.1.20). Furthermore, the deflection amplitude has become smaller than the excitation amplitude. For an excitation amplitude of $r_0 = 12.5$, the super-harmonic components, and in particular and very suddenly the frequency $2\,\omega_E$, have grown so strong that clear traces are visible in the time-history response and in the phase portrait (fig. 11.1.21). The time-history remains stationary and periodic in T_E, although it deviates distinctly from the sinusoidal form within the period. The phase portrait points to a stationary orbital curve somewhat unusual in shape; the orbital crossings (not present in the extended phase space $r, \dot{r}, t$!) are clearly visible. The power spectral density points to the presence of super-harmonics with all integral multiples of ω_E from 1 to 7. It should be noted, however, that these super-harmonics cannot show up in the periodic attractor with increment T_E. As before, the attractor appears as a single point, i.e. the integral curve maintains for all T_E intersections the same $r, \dot{r}$-position. Next raising the amplitude to $r_0 = 15$ causes the interlacing in the phase portrait to move to the left and indicates the possibility of its dissolution (fig. 11.1.22). The steady-state periodic response has altered markedly in comparison to $r_0 = 12.5$ and a super-harmonic $8\omega_E$ component emerges for the first time in the power spectral density diagram. In a final figure, we consider the behaviour for $r_0 = 30$ (fig. 11.1.23). The interlacing in the phase portrait is now removed and in contrast to $r_0 = 15$, where the super-harmonic $2\omega_E$ was dominant, $3\omega_E$ is now the largest super-harmonic component.

The variety of the stationary responses discussed so far illustrates the rich behaviour of the Duffing equation. We should not fail to mention that these calculations of the most elementary kind were stimulated by Ueda's publication [11.15], which, for different system parameters, displays similar stationary phase portraits. The reader should by now be in a position to generate such images independently, using the most simple desk computer (a PC should be sufficient).

The series of calculations illustrated in figs. 11.1.18 to 11.1.23 induces in us a belief that the present oscillator possesses deterministic stationary solutions – see the presence of a periodic attractor. However, we should not forget that we applied only a single initial condition. For a linear oscillator, harmonically excited, we could demonstrate that for any initial condition, the same definite attractor is attained. This statement cannot, however, be confirmed by our series of non-linear computations. This should not appear surprising, bearing in mind the amplitude frequency diagrams in figs. 11.1.5 and 11.1.3, right. Here, frequency ranges arise for which three distinct amplitudes are associated with a single frequency. To elucidate this behaviour, we use our computer laboratory to perform calculations in just such a frequency range. Thereby, the excitation amplitude is kept constant ($r_0 = 1$), but the initial conditions are varied. The system parameters ($\zeta = 0.05$, $\delta = 0.05$) remain unaltered as does the excitation period $T_E = 1$ s, whilst the linear eigenperiod is fixed at $T_0 = 1.4 T_E$, in order to operate at the frequency ratio of $\beta = \omega_E/\omega_0 = 1.40$ within the 'jump range' (see fig. 11.1.3). From fig. 11.1.24, we deduce the possibility of steady-state solutions, but clearly, the application of nearly equal initial conditions of $r(0) = 8$, $\dot{r}(0) = 0$ and $r(0) = 8.5$, $\dot{r}(0) = 0$ leads

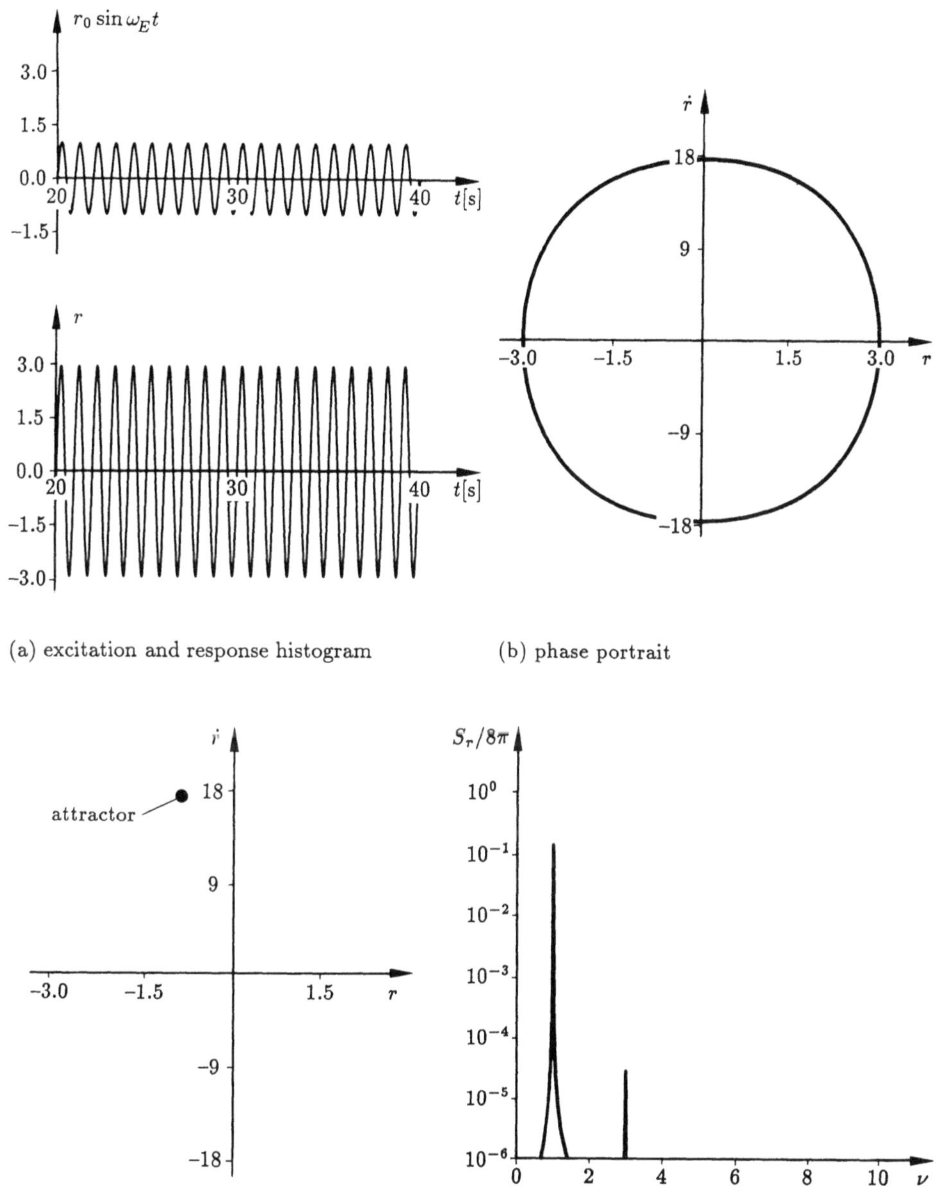

(a) excitation and response histogram

(b) phase portrait

(c) periodic attractor

(d) power spectral density

Fig. 11.1.18 Response of the linear-cubic oscillator (Duffing) with excitation amplitude $r_0 = 1$ and frequency ratio $\omega_E/\omega_0 = 1$

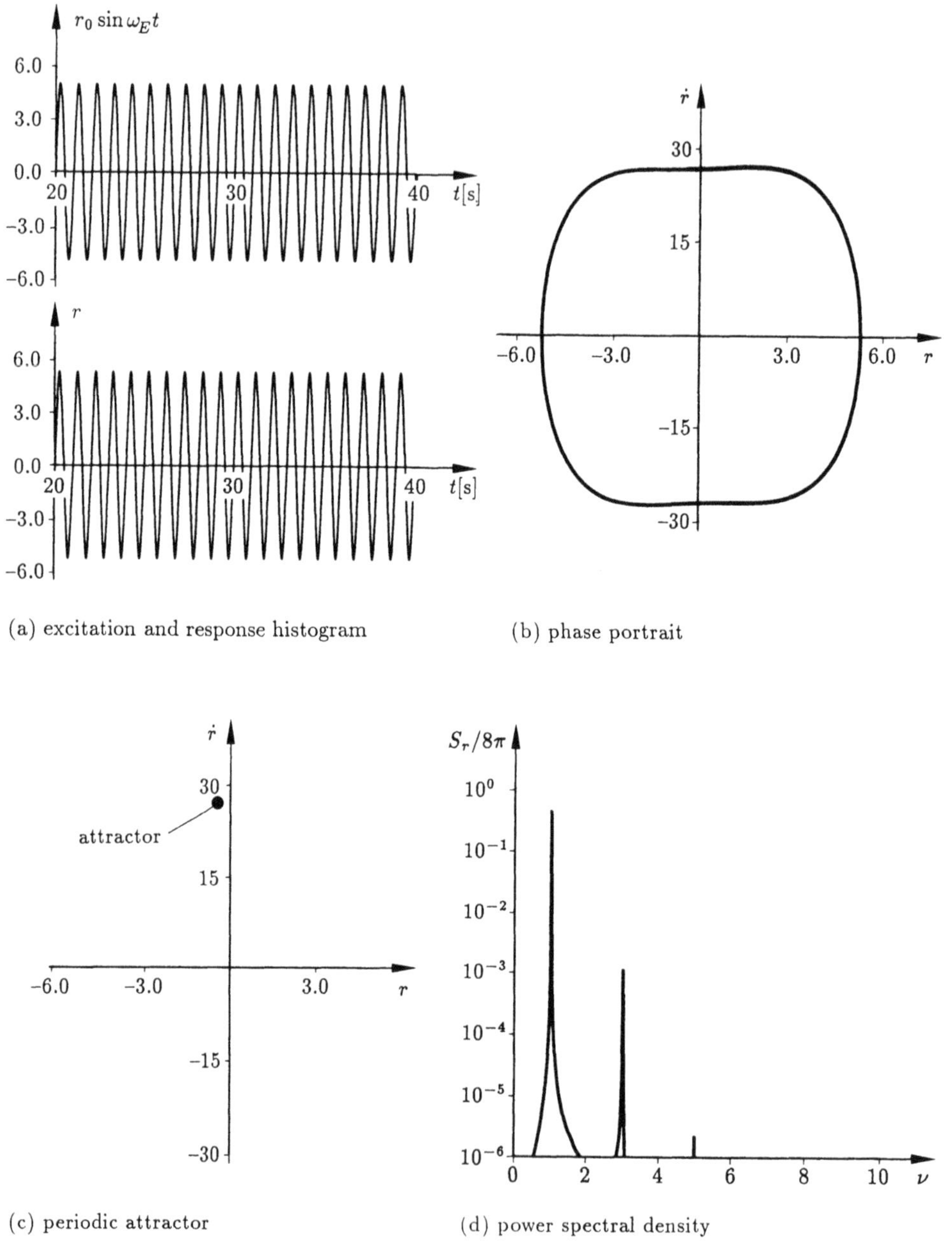

Fig. 11.1.19 Response of the linear-cubic oscillator (Duffing) with excitation amplitude $r_0 = 5$ and $\omega_E/\omega_0 = 1$

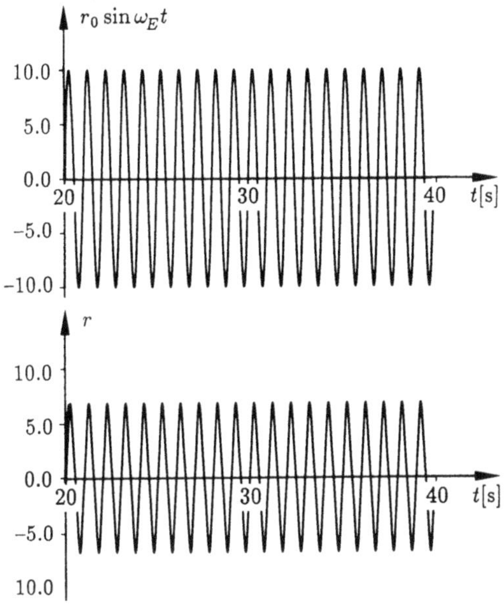

(a) excitation and response histogram

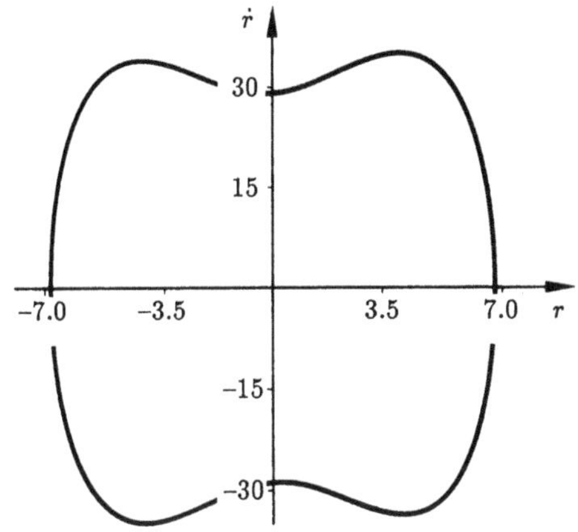

(b) phase portrait

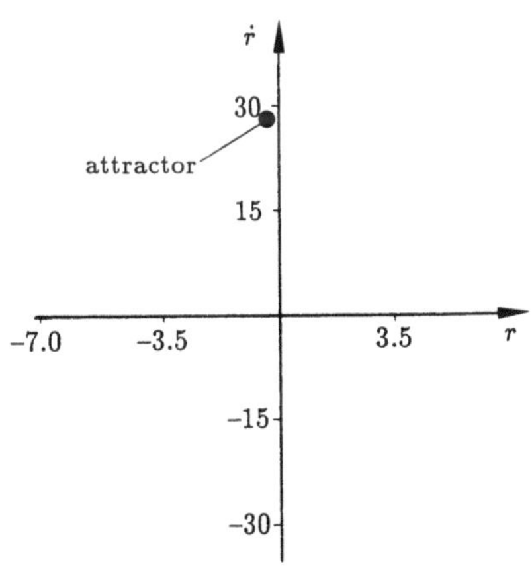

(c) periodic attractor

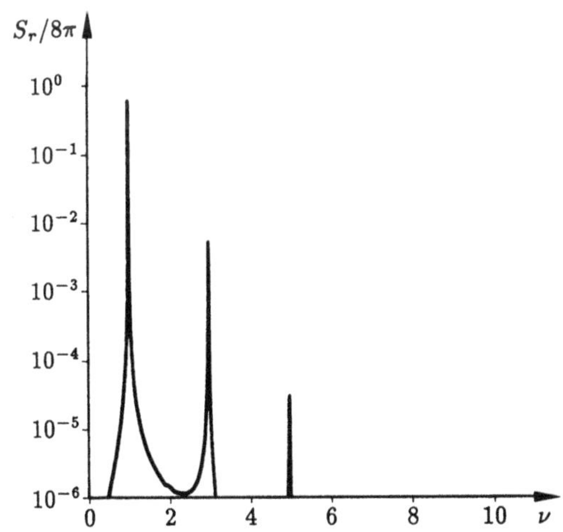

(d) power spectral density

Fig. 11.1.20 Response of the linear-cubic oscillator (Duffing) with excitation amplitude $r_0 = 10$ and $\omega_E/\omega_0 = 1$

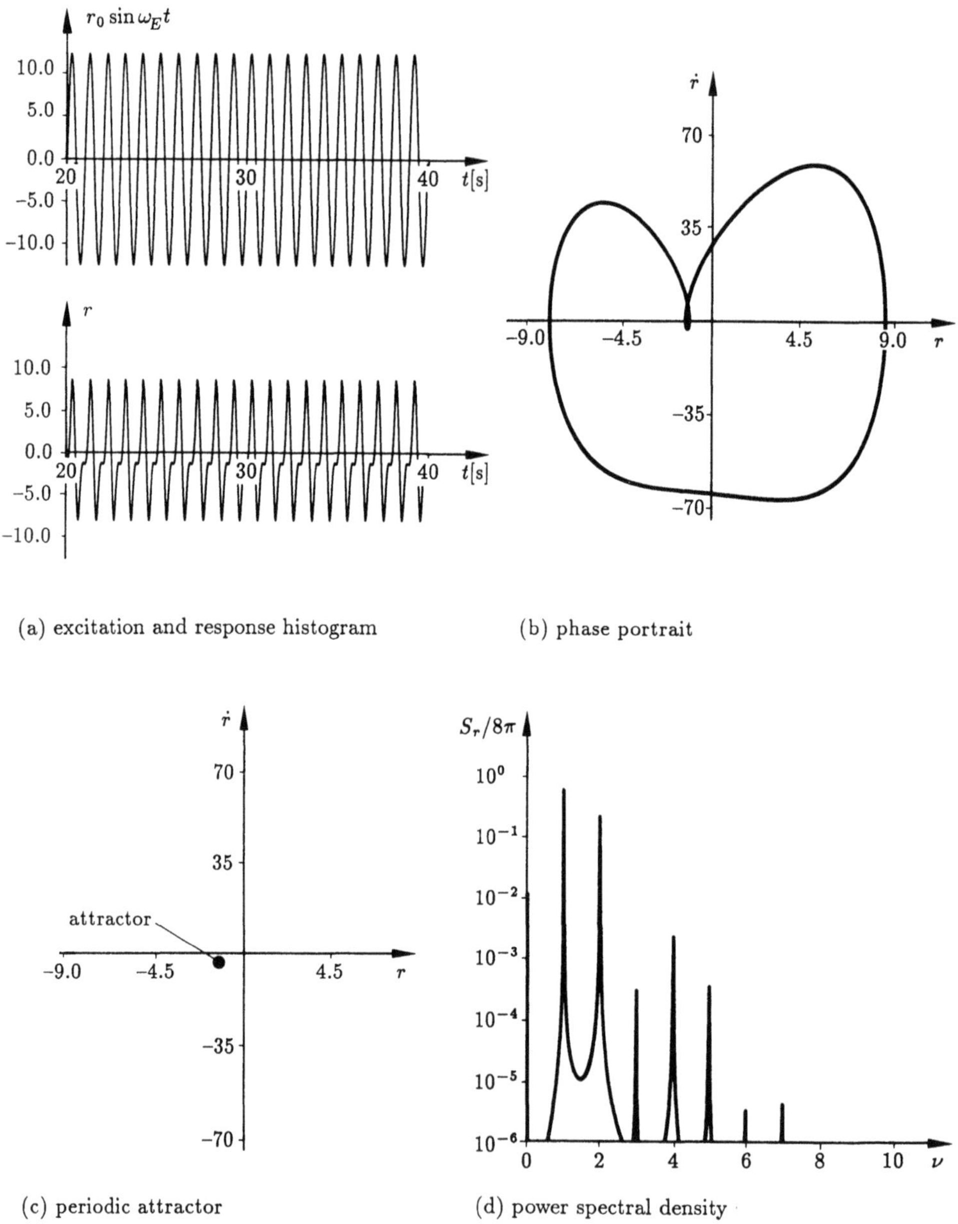

Fig. 11.1.21 Response of the linear-cubic oscillator (Duffing) with excitation amplitude $r_0 = 12.5$ and $\omega_E/\omega_0 = 1$

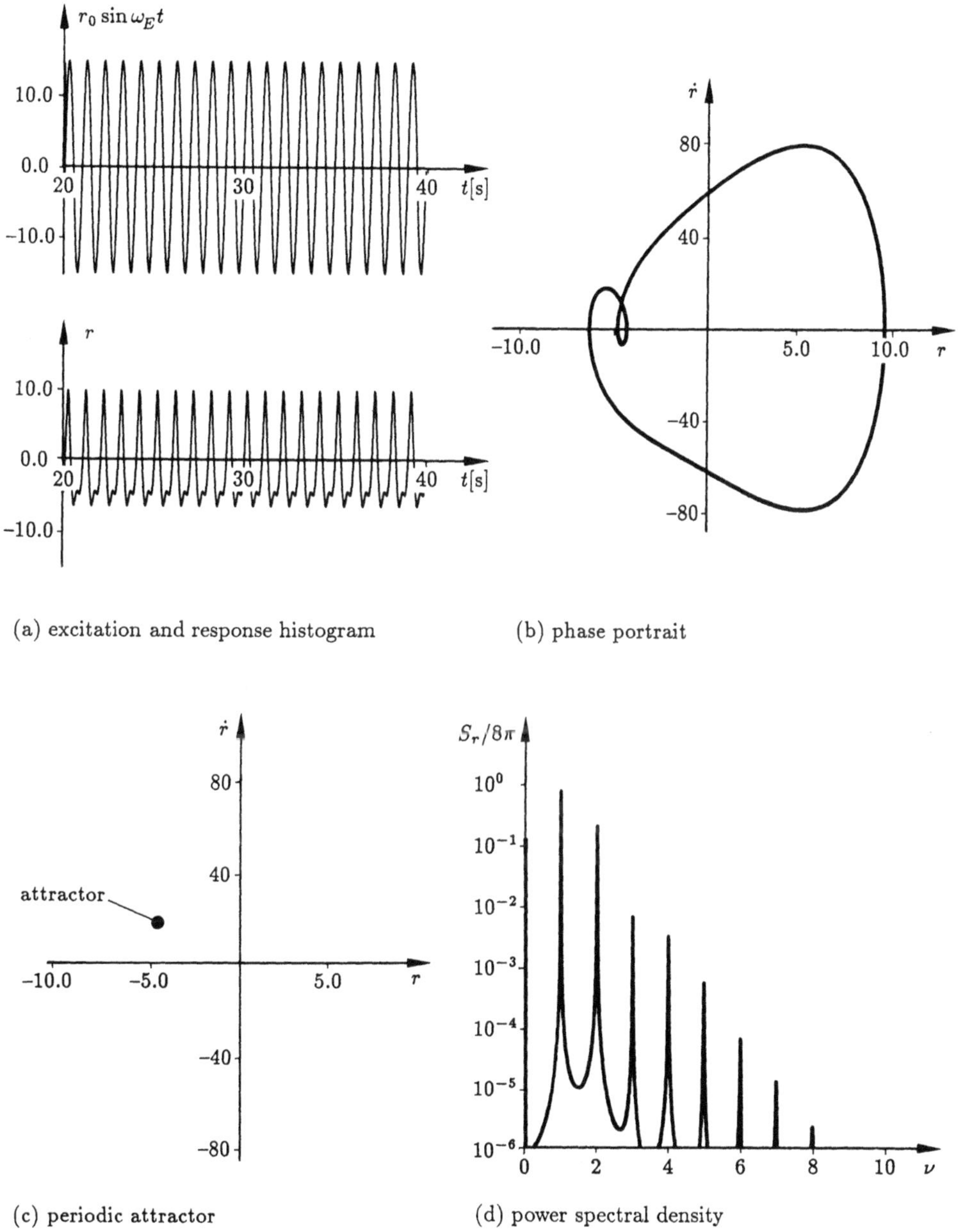

Fig. 11.1.22 Response of the linear-cubic oscillator (Duffing) with excitation amplitude $r_0 = 15$ and $\omega_E/\omega_0 = 1$

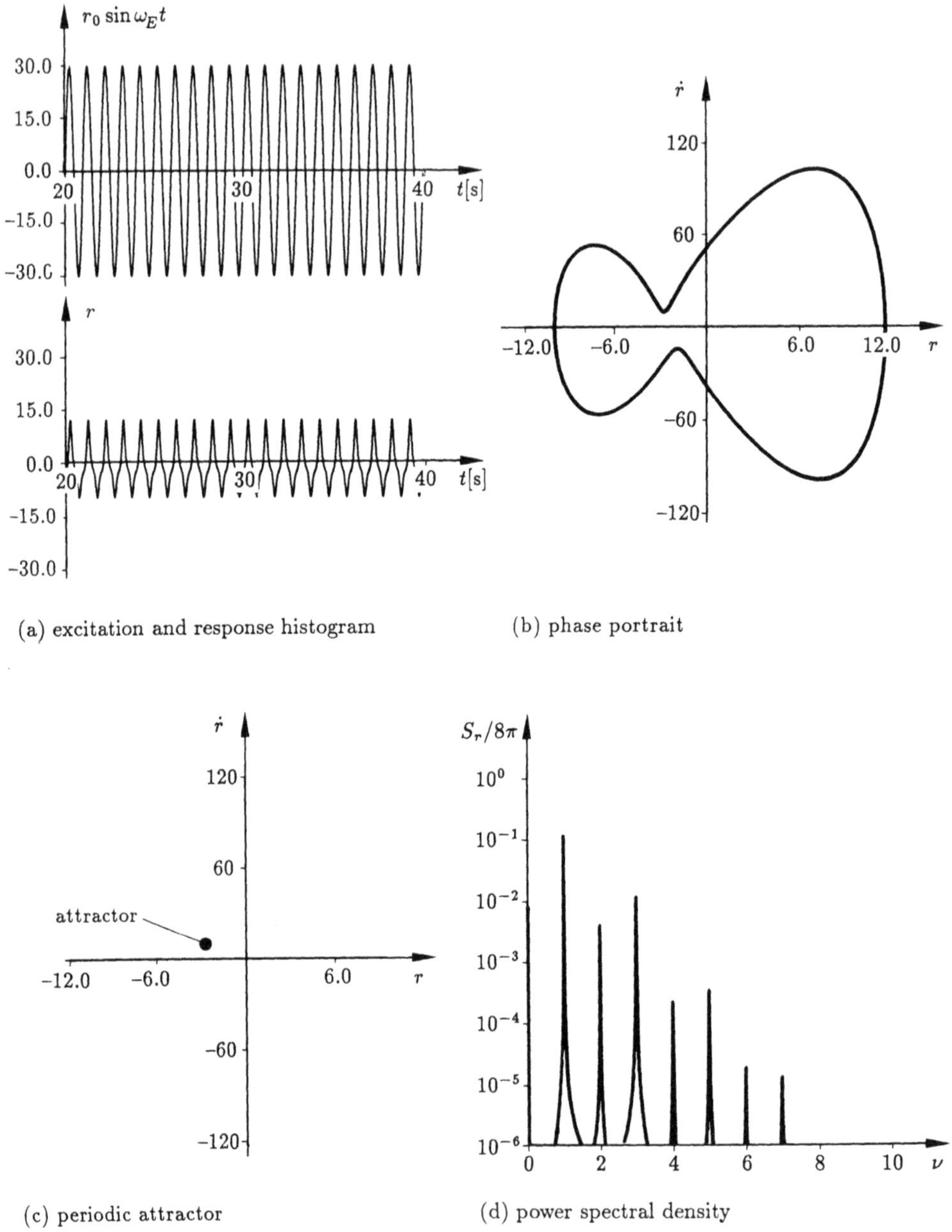

(a) excitation and response histogram

(b) phase portrait

(c) periodic attractor

(d) power spectral density

Fig. 11.1.23 Response of the linear-cubic oscillator (Duffing) with excitation amplitude $r_0 = 30$ and $\omega_E/\omega_0 = 1$

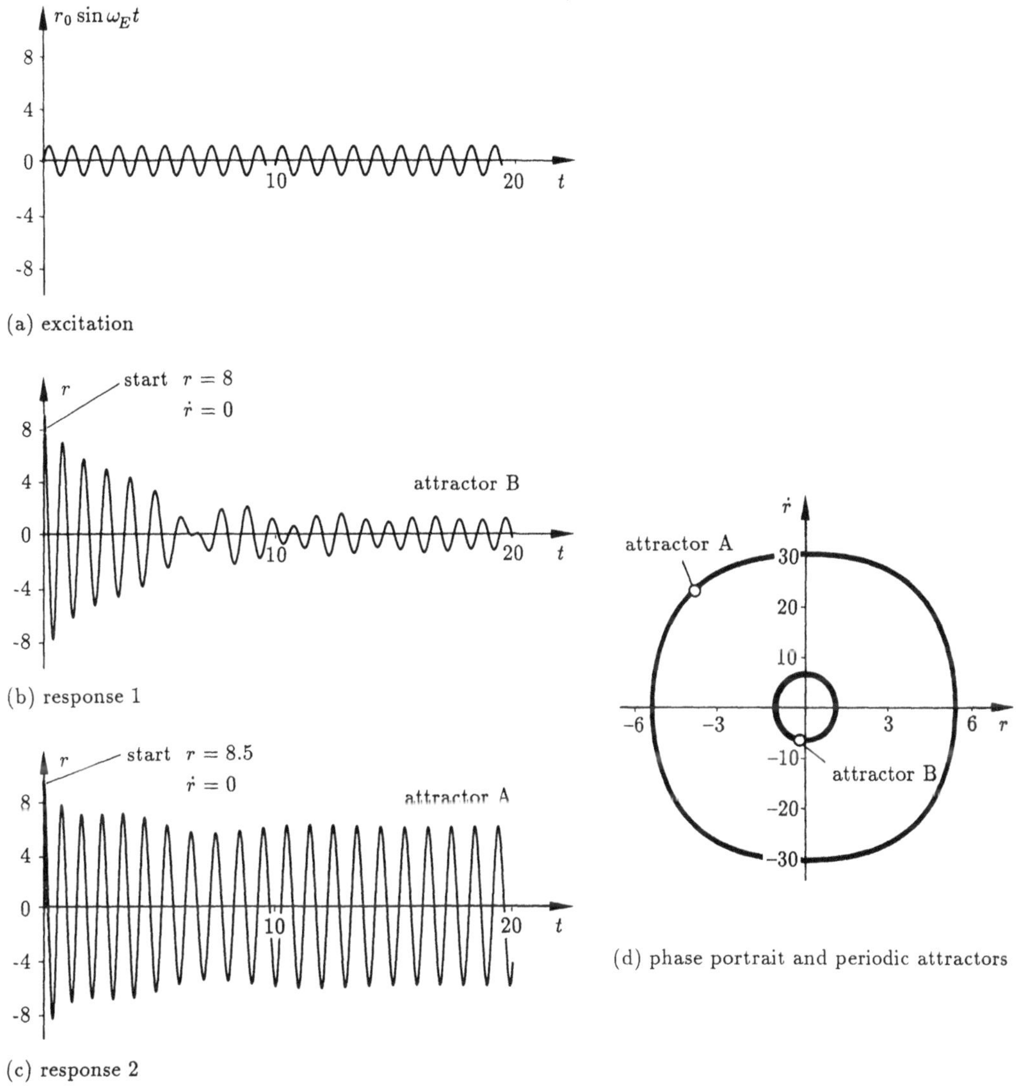

Fig. 11.1.24 Linear-cubic oscillator (Duffing) with varying initial conditions ($\zeta = 0.05$, $t_E = 1\,\mathrm{s}, \beta = \omega_E/\omega_0 = 1.4, \delta = 0.05$)

to very different amplitudes, phase portraits and attractors. In the phase space, each of these attractors possesses its own sphere of influence (basin of attraction) which we can imagine as being a territory surrounding the spiral shaped attractor in the phase space. Stability characterises the basin of attraction appertaining to an individual attractor. This means that solutions emanating from different initial conditions within the basin of attraction always lead to this attractor. Naturally, a surface or separatrix must also exist which separates the individual basins of attraction. In our case, this parting surface passes quite close by the attractors since we start from closely neighbouring initial points. This also demonstrates that we have to proceed with great care in the non-linear regime. When initial conditions are only imprecisely known, parameter studies become imperative in order to estimate the reliability of the solution. The variety of stable, alternative attractors – often more than two may occur in non-linear problems – is typical. In contrast, a single, unique attractor arises in the linear case. The aforementioned third amplitude, discernible in the amplitude-frequency-diagram, is, as has already been mentioned, unstable and thus cannot materialise in our numerical experiments.

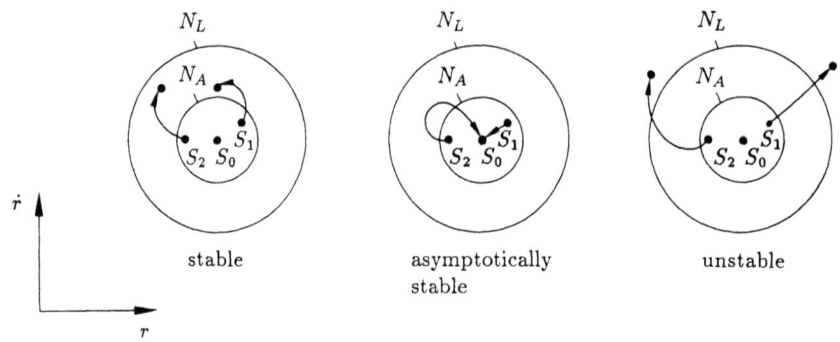

Fig. 11.1.25 Lyapunov's stability of stationary motions in the phase portrait

The present example offers an opportunity of briefly discussing the stability of non-linear oscillations. Here, we adopt the Lyapunov concept to obtain a definition for the stability of a steady-state motion. This is extremely simple: a stationary motion is stable if each additional solution generated in the immediate neighbourhood adheres for all the time closely to the first-named steady solution. These circumstances are demonstrated more precisely in fig. 11.1.25. Let S_0 be the phase portrait of a steady solution for which we wish to examine the stability. The surrounding encompassing the neighbouring solutions is bounded by N_A and the tolerance zone, which may still be referred to as the neighbouring zone, by N_L. If alternative solutions S_1 and S_2 remain within N_L, Lyapunov stability is ensured. If in the course of time t, alternative solutions converge to S_0, asymptotic stability is said to prevail. On the other hand, an escape from the N_L region indicates instability. These rather casual remarks should merely serve as an orientation in order to associate geometrical ideas with the concept of Lyapunov stability.

In concluding this account, we return once again to our computer laboratory and examine a modified Duffing equation discussed by Ueda in his recent publications (e.g. [11.12], [11.15]). To this purpose, we omit the linear stiffness term in our system and retain only the cubic term. From a physical point of view, this simply means that the initial tensional force N_0 is removed from our model of fig. 11.1.1. The restoring force then becomes in accordance with equ. (11.1.5)

$$R_e = EA\left(\frac{r}{a}\right)^3 \tag{11.1.108}$$

Hence, the present Duffing system is now defined by

$$\ddot{r} + \frac{c}{m}\dot{r} + \frac{EA}{ma^3}r^3 = \frac{R_0}{m}\sin\omega_E t \tag{11.1.109}$$

Following in the footsteps of Ueda, we select

$$f_k = \frac{c}{m} = 0.05 \tag{11.1.110}$$

$$\delta = \frac{EA}{ma^3} = 1 \tag{11.1.111}$$

$$\omega_E = 1 \tag{11.1.112}$$

and adapt the time-step to the period of the loading in such a manner that an equal number of load cycles are registered in the transient phase and in the ultimate stationary response. In the computer experiments, we vary the excitation term f_B in accordance with

$$f_B = \frac{R_0}{m} = \frac{R_0 a^3}{EA} \tag{11.1.113}$$

In fig. 11.1.26, we demonstrate a steady-state motion which, in principle, is already known to us from previous experiments. The periodic attractor appears once more as a point and the power spectral density distribution indicates a participation of a sequence of super-harmonic oscillations, with the 2nd order dominating. The increase of the excitation amplitude to $f_B = 5.0$ (fig. 11.1.27) heralds the emergence of special conditions. A steady-state periodic motion is seen to establish itself only with difficulties. Clearly, not only sub-harmonics but also ultra-subharmonics take part in the motion (see the power spectral density). This becomes apparent by the black spots of the periodic attractor, which now shows two definite centres. We now raise the excitation to $f_B = 7.5$ (fig. 11.1.28) and detect neither a periodic motion nor any regular motion. In this way, we reproduce Ueda's famous example [11.12, 11.15], which demonstrates a totally new phenomenon for an apparently well-documented differential equation. Both the phase portrait and the periodic attractor point to a totally irregular, erratic response which maintains this unexpected behaviour even when the lead time is extended (fig. 11.1.29). The power spectral density indicates a strong background noise of the motion. Thus, we face the at first surprising fact that a deterministic system may yield a stochastic response for a deterministic excitation. If the period of observation is extended, and the number of points in the periodic attractor diagram thereby increased, then a characteristic picture emerges typical for the phenomenon of chaos: namely the so-called strange attractor; to clarify the concept, see Ueda [11.15] and, in particular, Volume VII of this series [11.59]. If the excitation parameter is increased to $f_B = 10$, (fig. 11.1.30), a regular steady-state motion is once again restored. The phase portrait points to a limit cycle, now containing two loops, and the noise component in the power spectrum has disappeared. Raising f_B to 12.5, we evidently observe a re-emergence of a chaotic state (fig. 11.1.31).

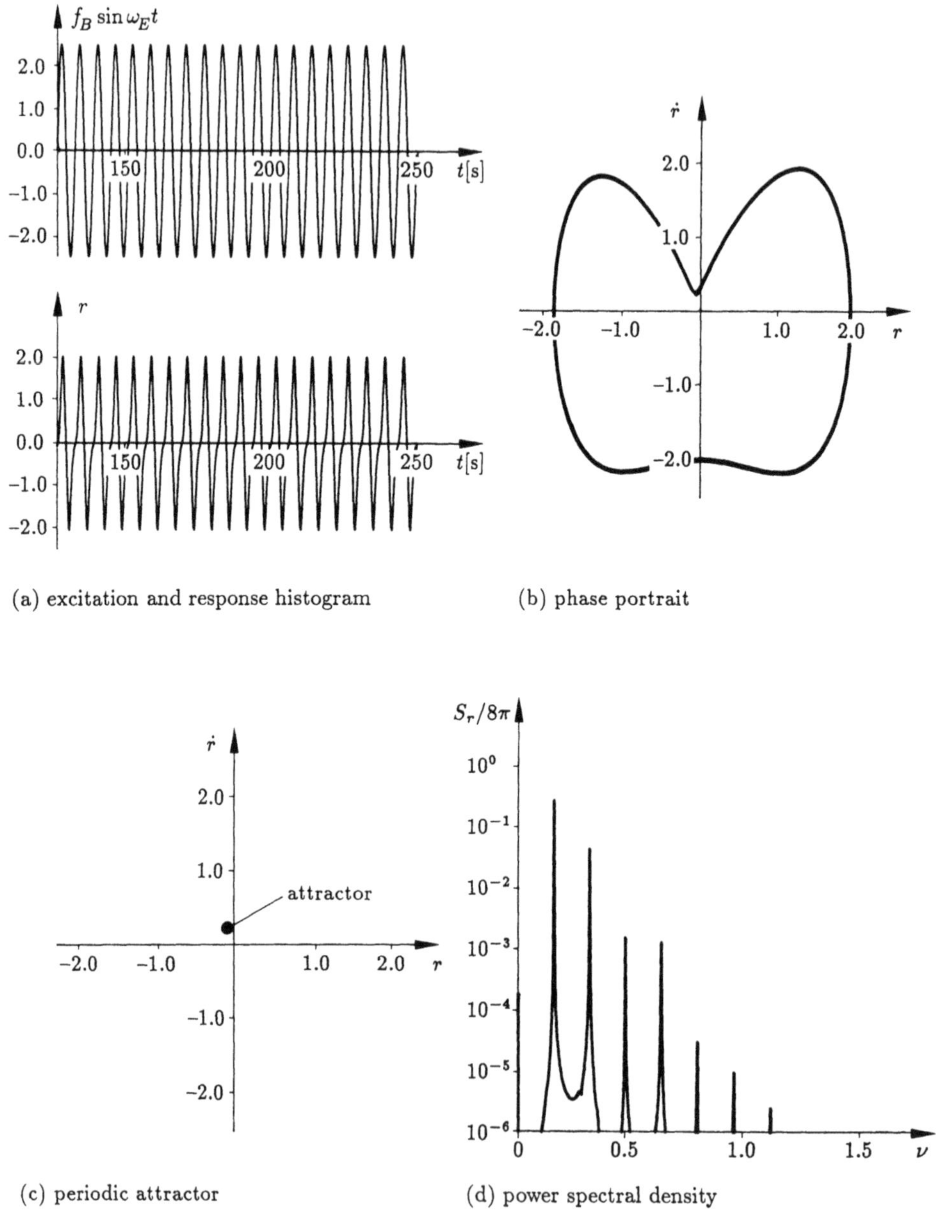

Fig. 11.1.26 The cubic Duffing oscillator (without linear stiffness term) with excitation amplitude $f_B = 2.5$

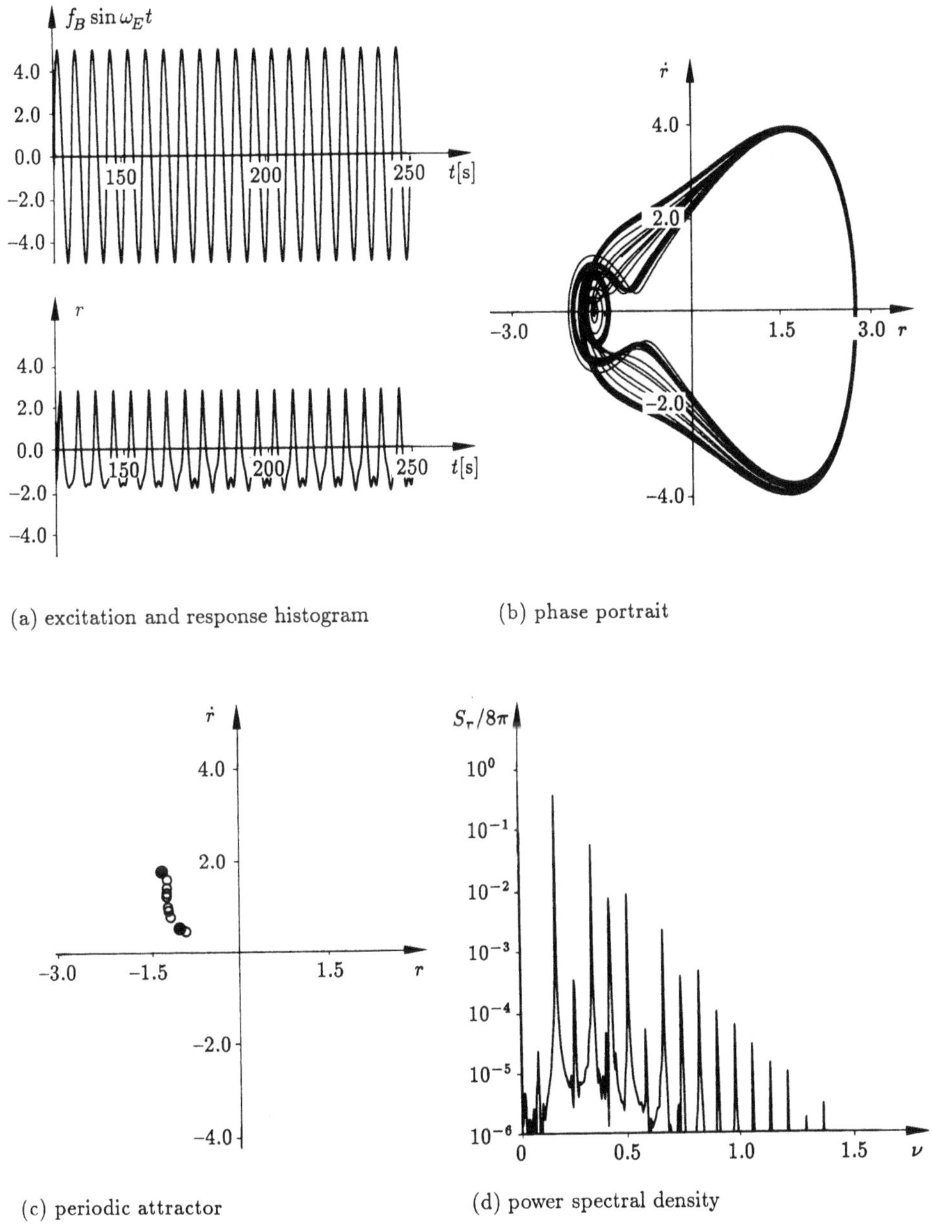

Fig. 11.1.27 The cubic Duffing oscillator with excitation amplitude $f_B = 5.0$

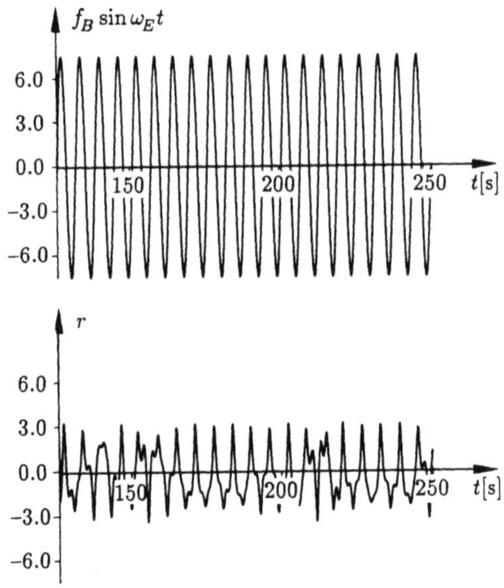

(a) excitation and response histogram

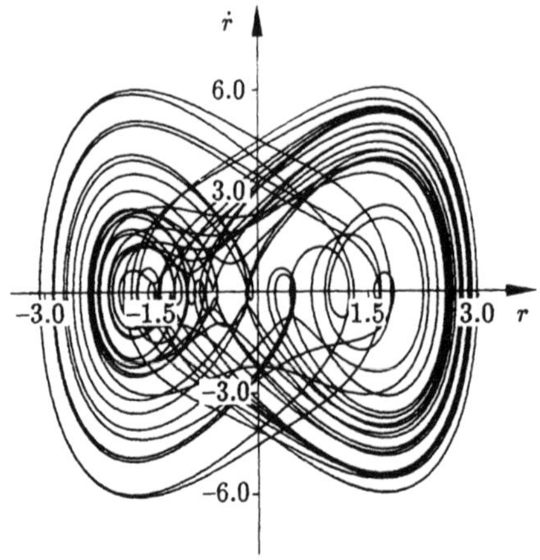

(b) phase portrait

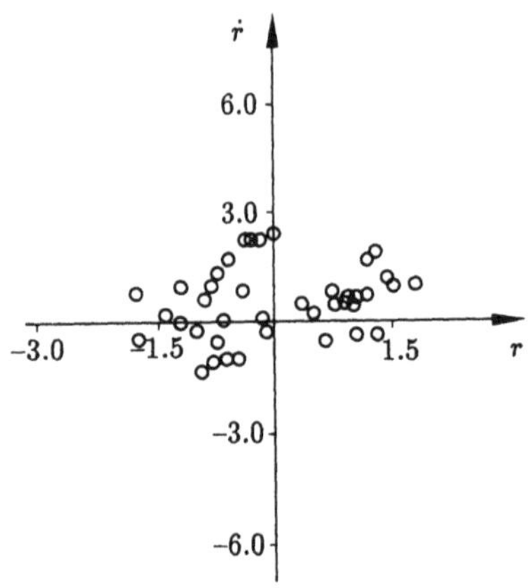

(c) periodic attractor

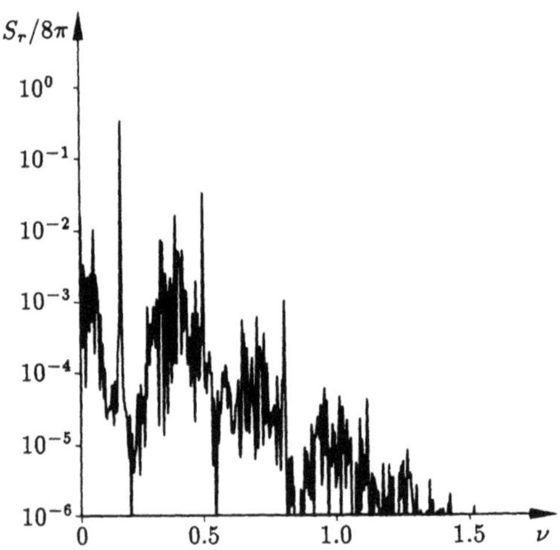

(d) power spectral density

Fig. 11.1.28 The cubic Duffing oscillator with excitation amplitude $f_B = 7.5$

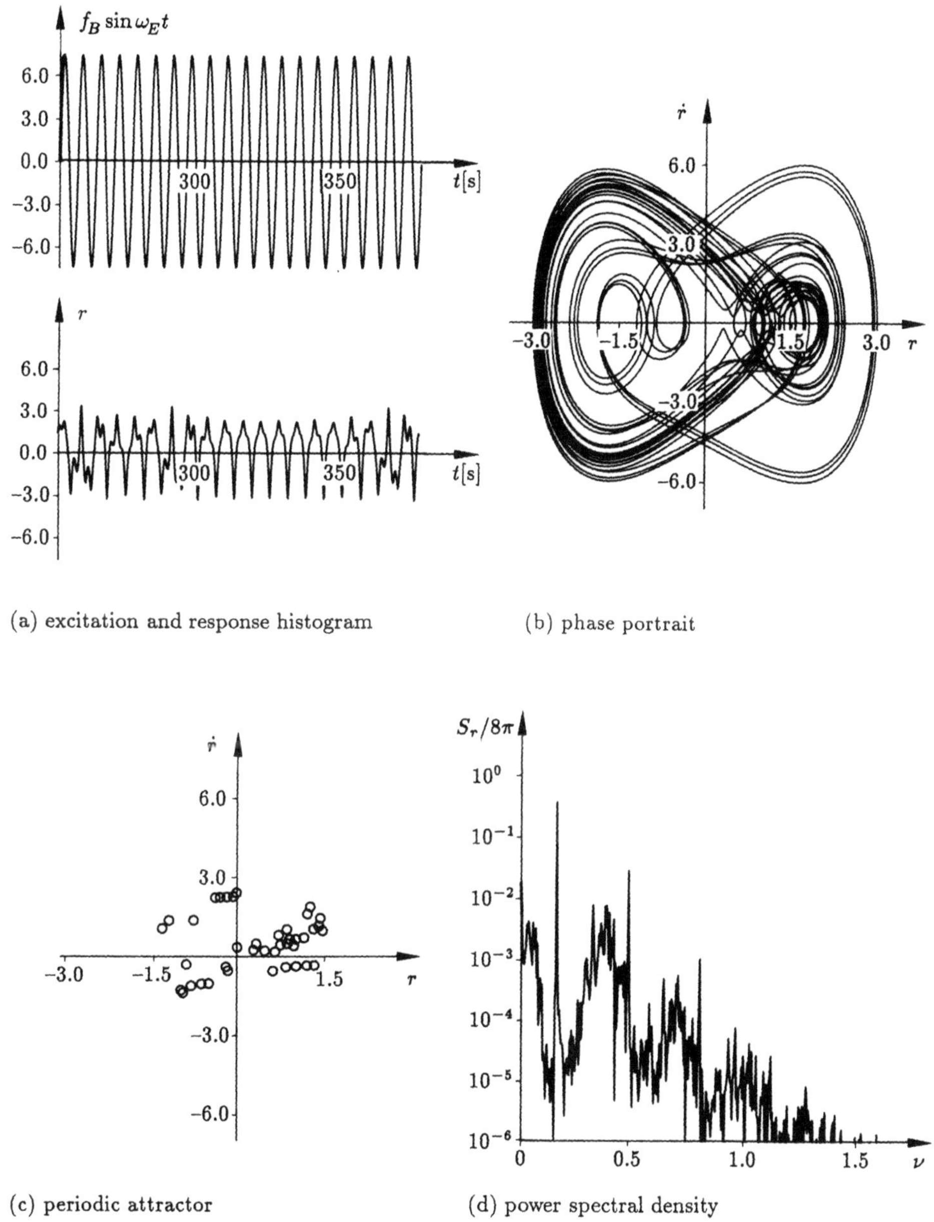

Fig. 11.1.29 The cubic Duffing oscillator with excitation amplitude $f_B = 7.5$ and with extended lead time

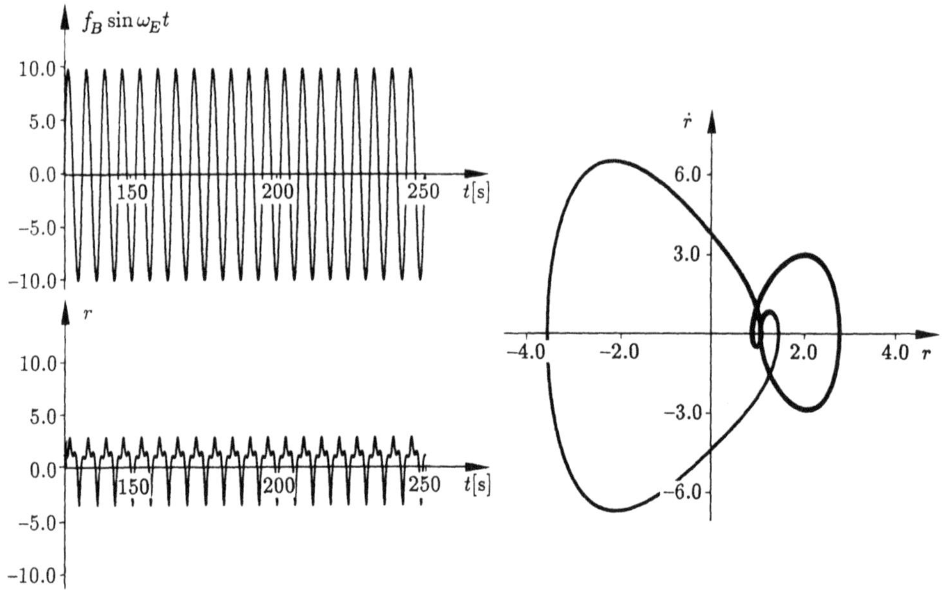

(a) excitation and response histogram

(b) phase portrait

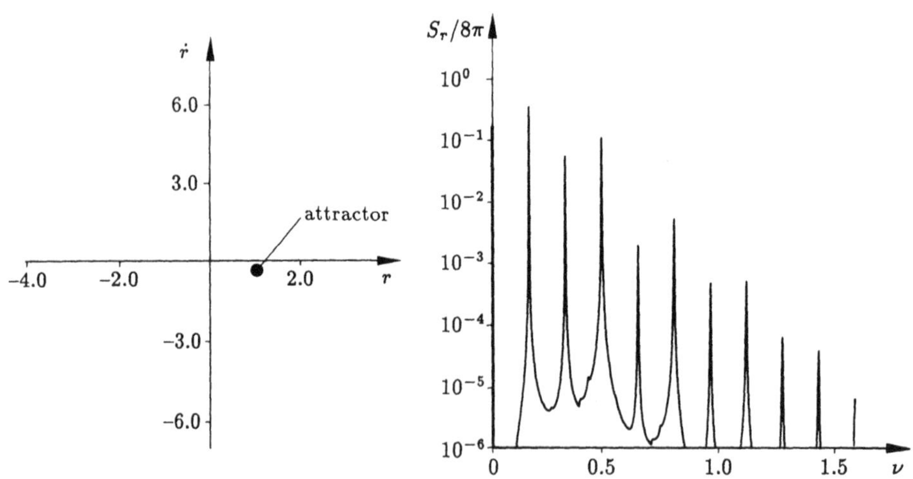

(c) periodic attractor

(d) power spectral density

Fig. 11.1.30 The cubic Duffing oscillator with excitation amplitude $f_B = 10$

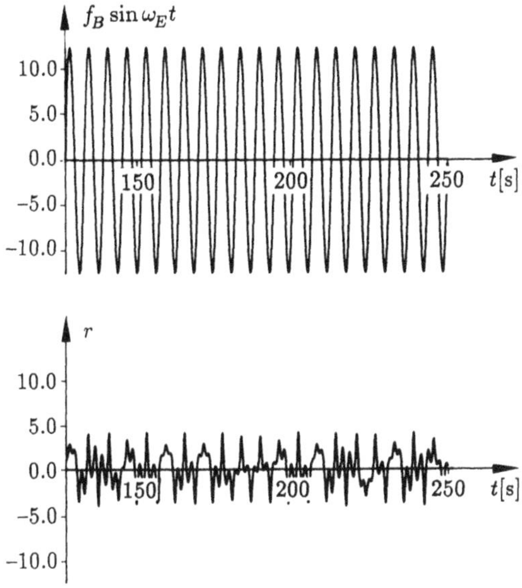

(a) excitation and response histogram

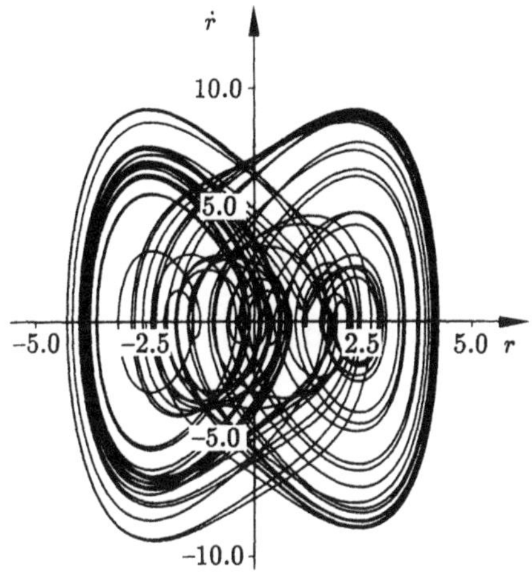

(b) phase portrait

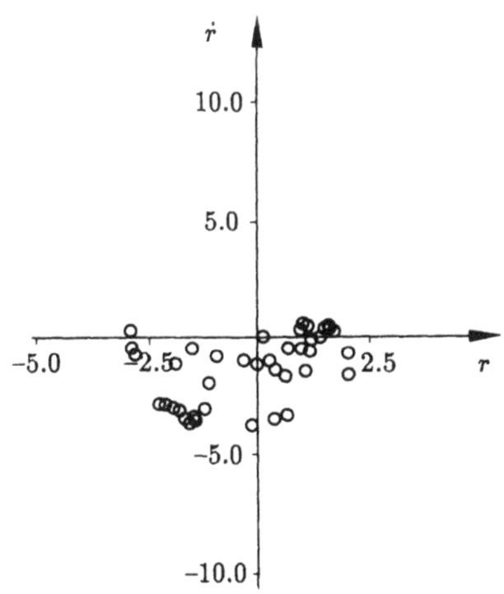

(c) periodic attractor

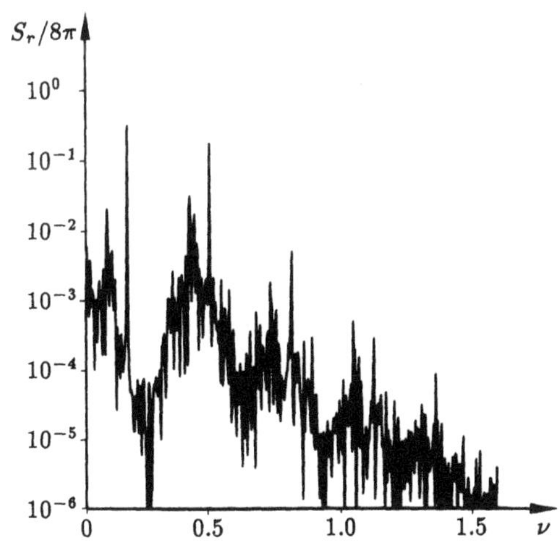

(d) power spectral density

Fig. 11.1.31 The cubic Duffing oscillator with excitation amplitude $f_B = 12.5$

With this in principle modest account, we conclude our introduction, having demonstrated some complex conditions prevailing in the field of non-linear dynamics. We refer, however, to Volume VII of this series, dedicated exclusively to chaos. Hopefully, our elementary account will induce in the reader a critical appraisal of the numerical integration procedures elaborated in the following sections, which, by virtue of their robust nature, always supply a result. Unfortunately, our experiments demonstrate that the practical engineer must apply extreme care in the acceptance of a 'solution', since in the case of an unstable response or a chaotic motion, slight pertubations in the initial conditions can result in dramatic changes in the solution.

11.2 Equations of motion for an arbitrary structure

Let us consider an idealised structure, discretised by means of the finite element method. We refer the reader to Chapter 10 in which we discussed the power *ansatz* for the space and the time variables, d'Alembert's principle and the principle of virtual displacements and ultimately obtained in this way the force equilibrium expression in dynamics, equ. (10.1.2). We prepared the ground for non-linear applications by introducing the incremental equilibrium expression and the associated secantial respectively tangential measures of the coefficient matrices. The authors now propose to expand upon these topics in this chapter. Non-linearities in structural dynamics are in the main the result of material and/or geometrical non-linearities. A varying mass (for example, due to burning in a rocket engine), stiffness or damping is in contrast rarer. Non-linearities of the material and time-dependence of structural characteristics pose for small deflections no difficulties in the interpretation of the force equilibrium given in equ.(10.1.2). In the presence of large deflections, however, (see, e.g., Vol. IV) we must adhere to a definite reference system and select suitable strain and stress measures. The reader should recollect here the total Lagrangian description where reference is made to the undeformed structure, and the Eulerian (convective) description where reference to the last recorded deformation state is used. Furthermore, the formulation of discrete force equilibrium is achieved by integration with respect to the reference state.

The specification of the relations between forces and displacements (respectively their temporal derivatives) permitted us to express in the linear case the force equilibrium through a system of ordinary linear differential equations. Here also, the intrinsically non-linear relations between forces and displacements respectively displacement derivatives will help us in the development of the theory. We thus initiate our account with the general equilibrium relation

$$\boldsymbol{R}_I(\boldsymbol{r},\dot{\boldsymbol{r}},\ddot{\boldsymbol{r}},t) + \boldsymbol{R}_D(\boldsymbol{r},\dot{\boldsymbol{r}},\ddot{\boldsymbol{r}},t) + \boldsymbol{R}_e(\boldsymbol{r},\dot{\boldsymbol{r}},\ddot{\boldsymbol{r}},t) = \boldsymbol{R}(\boldsymbol{r},\dot{\boldsymbol{r}},\ddot{\boldsymbol{r}},t) \tag{11.2.1}$$

The assumed functional dependence of the force components on displacement, velocity, acceleration and more explicitly on time must not be considered as inflated or exaggerated. This may be demonstrated immediately on the first group of forces, namely the inertial forces. When operating in a generally moving reference system, we confirm from equ. (2.3.15) that all functional dependences quoted above can occur, but fortunately, in the present case, in a linear form. In the following, we shall restrict ourselves to an inertial reference system for which

$$\boldsymbol{R}_I = \boldsymbol{M}\ddot{\boldsymbol{r}}(t) \tag{11.2.2}$$

holds. It is assumed that the mass matrix $\boldsymbol{M}$ does not depend on time and thus remains constant. This assumption can also be maintained in the presence of large strains and large deflections. Thus, for the simplex elements FLA2, TRIM3, TET4, we can adopt the elemental mass matrices established in sections 3.2 and 3.3 without change since only the element mass $(\rho\Omega t)$ but not the geometry enters into the corresponding formulae (e.g. equ. (3.3.4) for TRIM3), the overall element mass being considered invariant. In general, we shall limit the magnitude of the displacements so that changes of mass (elemental or global) cannot arise and consequently, the mass matrix has to be set up only once. We shall take up this topic once more in section 11.3. In section 11.8, the inertial forces will be discussed on a more general basis; in particular, rotating structures will be considered. This introduces additional dependences on displacements and velocities.

The specification of realistic damping mechanisms causes severe difficulties even in the context of classical linear theory and cannot easily be formulated within a consistent physical edifice. Thus, in many non-linear computations, we operate either with zero damping ($\boldsymbol{R}_D = \boldsymbol{o}$) or with the simplified viscous model adopted in the linear theory

$$\boldsymbol{R}_D = \boldsymbol{C}\dot{\boldsymbol{r}} \tag{11.2.3a}$$

Occasionally, a non-linear damping term has been incorporated in single-degree-of-freedom systems, e.g. in a pendulum with a damping proportional to the square of the velocity

$$R_D = (\operatorname{sign}\dot{r})\, c\, \dot{r}^2 \tag{11.2.3b}$$

However, a generalisation of this concept to multi-degree-of-freedom systems appears illusory. The only realistic approach to a formulation of a non-linear damping contribution can be achieved through a non-linear material law, i.e. a visco-elastic substance. In this case, it is advisable to combine the damping force $\boldsymbol{R}_D$ and the elastic force $\boldsymbol{R}_e$ into a single inner force $\boldsymbol{R}_i$

$$\boldsymbol{R}_i(\boldsymbol{r}, \dot{\boldsymbol{r}}) = \boldsymbol{R}_D + \boldsymbol{R}_e \tag{11.2.4}$$

Formally, we may then derive the tangential damping matrix

$$\boldsymbol{C}_t = \frac{\partial \boldsymbol{R}_i}{\partial \dot{\boldsymbol{r}}} \tag{11.2.5}$$

and the tangential stiffness matrix

$$\boldsymbol{K}_t = \frac{\partial \boldsymbol{R}_i}{\partial \boldsymbol{r}} \tag{11.2.6}$$

as linearised operators. In general, each of these matrices is a non-linear function of $\boldsymbol{r}$ and $\dot{\boldsymbol{r}}$. For the so-called symmetric systems, in addition to the mass matrix $\boldsymbol{M}$, the tangential damping $\boldsymbol{C}_t$ and the tangential stiffness $\boldsymbol{K}_t$ are presumed to be symmetric. We should not forget that the hypothesis of equ. (11.2.4), asserting exclusive dependence of the the inner forces $\boldsymbol{R}_i$ on displacements $\boldsymbol{r}$ and velocities $\dot{\boldsymbol{r}}$, is not the most comprehensive one. In the presence of plasticity and visco-plasticity, for example, $\boldsymbol{R}_i$ does not only depend upon the current values of $\boldsymbol{r}$ and $\dot{\boldsymbol{r}}$ but also on their histories. Altogether, we have to admit that we possess only restricted experience on this topic; the same also applies for unsymmetric systems to an even greater degree. In the following, we shall nearly always restrict ourselves to the simple assumption of equ. (11.2.2).

The central issue in most non-linear structural calculations concerns the elastic restoring forces $\boldsymbol{R}_e$. Here, an analysis is usually confined to a non-linear dependence on the displacements; this covers material and geometrical non-linearities. Analogous to equ. (11.2.6), the tangential stiffness can then be established using the relationship

$$\boldsymbol{K}_t = \frac{\partial \boldsymbol{R}_e}{\partial \boldsymbol{r}} \tag{11.2.7}$$

This tangential stiffness is of importance for a linearised interpretation of stability. However, in response calculations, the elastic forces are usually deduced directly from the stresses, e.g. following the updated Lagrangian description (see also Vol. IV)

$$^{1}\boldsymbol{R}_e = \int_{^{1}V} \boldsymbol{\alpha}_L^{\mathrm{t}}\left({}_{1}^{1}\boldsymbol{\sigma}\right) \mathrm{d}V \tag{11.2.8}$$

in which we use a simple formula well-known to us from linear theory. Equation (11.2.8) can immediately be deduced from the unit displacement method and the elemental strains $\boldsymbol{\alpha}_L \boldsymbol{\rho}_1$ associated with the relevant unit displacement pattern $\boldsymbol{\rho}_1$. Concerning the stresses ${}_{1}^{1}\boldsymbol{\sigma}$, the pre-set superior index 1 identifies the observed configuration as a reference state and the inferior index 1 the very same state of deformation to which the stresses are referred (for more details, see Vol. IV).

The external forces $\boldsymbol{R}$ need not necessarily be only dependent on time. For example, follower loads may arise which are dependent upon the displacement, and in the field of hydro- and aeroelasticity, the velocity and the acceleration of lifting surfaces are, in addition, an important factor. These will be explored to a limited extent in section 11.7, a more detailed discussion will be given in Chapters 1 to 9 of Volume VI. In the forthcoming sections, we shall restrict ourselves to the purely time-dependent external forces $\boldsymbol{R}(t)$. Thus, the equation of motion to be discussed in the following sections 11.3 to 11.6 can be expressed collectively as

$$\boldsymbol{M}\ddot{\boldsymbol{r}} + \boldsymbol{C}\dot{\boldsymbol{r}} + \boldsymbol{R}_e(\boldsymbol{r}) = \boldsymbol{R}(t) \tag{11.2.9}$$

with the given initial conditions $\boldsymbol{r}(0)$ and $\dot{\boldsymbol{r}}(0)$.

11.3 Explicit algorithms for non-linear response calculations

We now reconsider the explicit algorithms applied previously to linear systems, for example the Newmark method ($\beta = 0, \gamma = 1/2$), which is equivalent to the method of central differences . For explicit procedures, we derive displacement, velocity and acceleration at the end of the time-step from the known corresponding data at the beginning of this time-step, without having to solve a system of equations. Such a procedure appears particularly attractive for non-linear systems since the system matrices then possess tangential characteristics and alter continuously. In addition, such explicit algorithms can easily be programmed, their one disadvantage being the limitation on the step-size due to stability considerations. Bearing in mind our non-linear task, it is worthwhile casting a glance at these procedures.

Following equ. (10.2.10), the deflection at the end of the k-th time-step, $\boldsymbol{r}_{\mathrm{k}+1}$, can easily be derived – using the Newmark procedure with $\beta = 0$ – from the information available at the beginning of the time-step (i.e. $\boldsymbol{r}_{\mathrm{k}}$, $\dot{\boldsymbol{r}}_{\mathrm{k}}$, $\ddot{\boldsymbol{r}}_{\mathrm{k}}$). Thus,

$$\boldsymbol{r}_{\mathrm{k}+1} = \boldsymbol{r}_{\mathrm{k}} + \Delta t \dot{\boldsymbol{r}}_{\mathrm{k}} + \tfrac{1}{2}\Delta t^2 \ddot{\boldsymbol{r}}_{\mathrm{k}} \tag{11.3.1}$$

In any case, this step is fully explicit. Somewhat more problematic is the calculation of the velocity at the end of the calculation step according to equ. (10.2.11)

$$\dot{\boldsymbol{r}}_{\mathrm{k}+1} = \dot{\boldsymbol{r}}_{\mathrm{k}} + \Delta t(1-\gamma)\ddot{\boldsymbol{r}}_{\mathrm{k}} + \Delta t \gamma \ddot{\boldsymbol{r}}_{\mathrm{k}+1} \tag{11.3.2}$$

Here, the unknown accelerations at the end of the step do still appear in the formula. To determine them, we may use the expression for the dynamic equilibrium. From equ. (11.2.9), we have

$$\boldsymbol{M}\ddot{\boldsymbol{r}}_{\mathrm{k}+1} = \boldsymbol{R}_{\mathrm{k}+1} - \boldsymbol{C}\dot{\boldsymbol{r}}_{\mathrm{k}+1} - \boldsymbol{R}_{e|\mathrm{k}+1} \tag{11.3.3}$$

The force $\boldsymbol{R}_{\mathrm{k}+1}$ on the right-hand side can be considered known. Having computed $\boldsymbol{r}_{\mathrm{k}+1}$, the elastic forces $\boldsymbol{R}_{e|\mathrm{k}+1} = \boldsymbol{R}_e(\boldsymbol{r}_{\mathrm{k}+1})$ at the end of the step can be evaluated, for example, in geometrically non-linear problems through the sequence of operations

displacements $\boldsymbol{r}$
↓
Green's strains ${}^2_1\gamma$
↓
Piola-Kirchhoff stresses ${}^2_1\boldsymbol{\sigma}$
↓
elastic forces $\boldsymbol{R}_e$

(Here, the pre-set superior index 2 identifies the observed state as the current configuration and the inferior index 1 indicates the reference geometry to which the strains and stresses are referred.)

For the present, we assume the damping to be zero and consequently, the right-hand side of equ. (11.3.3) to be known. The mass matrix $\boldsymbol{M}$ may be considered constant in the non-linear calculation. However, the use of consistent mass matrices requires the solution of a sparsely populated system equivalent in its computational effort to a static calculation. At the same time, if the mass matrix $\boldsymbol{M}$ is constant, it can be triangularised *a priori*, and as a result, only a single forward and backward substitution remains within a time-step. This computing expenditure corresponds to that of the Trujillo algorithm (but in the latter case also involves the assembly of the damping matrix $\boldsymbol{C}$ and the stiffness matrix $\boldsymbol{K}$). Trujillo denotes his procedure as being explicit, but it would be more exact to term it pseudo-explicit. A *de facto* fully explicit technique is realised as in the linear case by a diagonal mass matrix, which can most easily be established by means of the lumping technique. In this way, however, we abandon the consistent application of the principle of virtual displacements and accept the vagaries of approximations. On the other hand, numerical experiments by Fujii [11.16], Krieg and Key [11.17] indicate that the errors of the central difference method – and thus also of the equivalent Newmark method for $\beta = 0$ – are opposed to those arising through mass lumping, and thus offer the possibility of a compensation.

Naturally, damping is present in any real motion. A first possibility for a solution technique is generated through substitution of equ. (11.3.3) into equ. (11.3.2) and a subsequent solution of the equation in terms of $\dot{\boldsymbol{r}}_{k+1}$, which is analogous to the linear case

$$[\boldsymbol{M} + \Delta t\gamma\boldsymbol{C}]\dot{\boldsymbol{r}}_{k+1} = \Delta t\gamma[\boldsymbol{R}_{k+1} - \boldsymbol{R}_{e|k+1}] + \boldsymbol{M}\dot{\boldsymbol{r}}_k + \Delta t(1-\gamma)\boldsymbol{M}\ddot{\boldsymbol{r}}_k \qquad (11.3.4)$$

Since our present discussion presumes $\boldsymbol{M}$ and $\boldsymbol{C}$ to be constant, the system matrix $[\boldsymbol{M} + \Delta t\gamma\boldsymbol{C}]$ can be triangularised *a priori* and our procedure then becomes at least pseudo-explicit. We obtain a completely explicit process if the matrices $\boldsymbol{M}$ and $\boldsymbol{C}$ are diagonal. Whilst a diagonal mass matrix is physically justifiable, this is difficult to accept for a diagonal damping. Following the method of central differences, we may now use the expedient of adopting a difference scheme for the damping term in equ. (11.3.3). Thus, we write

$$\dot{\boldsymbol{r}}_{k+1} \approx \frac{1}{\Delta t}[\boldsymbol{r}_{k+1} - \boldsymbol{r}_k] \qquad (11.3.5)$$

whereby full explicitness is achieved once more, even in the case of arbitrary viscous damping. For linear systems, we obtained the stability limit

$$\Delta t \leqslant \frac{2}{\omega_{max}} \qquad (11.3.6)$$

where ω_{max} is the highest circular eigenfrequency of the system. The errors inherent in the numerical integration procedure are of the order Δt^2. In principle, these observations are applicable to a consistently linearised non-linear system. Here, however, conditions change continuously since *inter alia*, the stiffness is undergoing a steady modification. Consequently, the adoption of a constant time-step in non-linear problems is often unsatisfactory from an economical point of view since the time-step then has to be very small. Hence, we generally choose a variable time-step, the practical realisation of which is more easily attained in the case of the Newmark procedure ($\beta = 0$) than in the equivalent method of central differences. In many practical engineering tasks, considerable differences arise in the stability limit of various parts of a finite element idealisation. If we were to operate with one global time-step for the whole network, we would have to adopt the least favourable condition. A further stratagem can also be used: when applying the so-called explicit-explicit methods, we may use different time-steps for various parts of the network; alternatively, in an explicit-implicit method, we may proceed partly explicitly and partly implicitly (see the following section 11.4). The crucial issue in such applications lies in the analysis of the domain boundaries.

If we view the explicit procedures from the point of view of non-linear dynamics, they still retain the following advantages known to us from the linear theory:

— less computational expenditure per step
— less backing storage
— simple programming.

In the non-linear case, we can in addition

— incorporate complex material laws in a straightforward manner
— consider discontinuities (such as contact problems or phase changes) relatively easily.

On the other hand, we note the disadvantages:

— a low stability bound

and, in connection with this,

— a high number of time-steps.

In general, the explicit procedures are more suitable for problems in which medium and high frequencies determine the response of the structure (wave propagation problems) or for time-dependent excitations, $\boldsymbol{R}(t)$, which can only be reproduced satisfactorily by a small time-step.

Until now, our present account was restricted to the Newmark method ($\beta = 0$) and the method of central differences. To expand the reader's horizon, we propose to conclude our discussion by drawing attention to some other explicit procedures. One of the most prominent techniques goes back to Runge-Kutta and is of the fourth order. It has played a significant role in numerical mathematics for over half a century [11.19]. The Fu method [11.20] possesses the same degree of accuracy, but unfortunately has an even stricter stability bound than the Newmark method ($\beta = 0$). Finally, there is the unconditionally stable (with respect to linear problems!) Trujillo method [11.21]. As has already been mentioned, this method is, however, only pseudo-explicit. Moreover, the limited accuracy of this method in effect only allows for time-steps of a size similar to those of the Newmark method ($\beta = 0$).

11.4 Implicit conditionally stable time-integration

11.4.1 Survey

As was already mentioned in Chapter 10 in the discussion of linear problems, implicit methods essentially possess two advantages as against explicit ones:

— higher accuracy
— improved stability.

However, the above assertion does not hold for all implicit algorithms. The Newmark procedure with $\beta = 1/6, \gamma = 1/2$, for example, possesses an increased stability limit (ϑ_{lim}=3.46 in comparison to ϑ_{lim}=2.00 for the Newmark method with $\beta = 0$), but the accuracy – when referred to the step-size required for a 3% limitation to the period elongation – proves for $\vartheta_{3\%} = 0.87$ hardly better than for the explicit Newmark procedure. The above assertion holds, on the other hand, for the Newmark procedure with $\beta = 0.10$ and still more for the cubic Hermitian integration technique.

The disadvantage of the implicit procedures arises from the necessity of having to solve a large system of equations within each time-step. For linear systems, this proves to be acceptable since the system coefficients are constant and thus, the computation effort involved in the time-step is very limited. In the case of non-linear problems, however, large non-linear systems of equations have to be solved, which is generally only iteratively possible. In fact, we are faced with a combination of the difficulties involved in a static non-linear analysis, i.e.

— selection of a suitable iterative non-linear solution procedure
— specification of a tolerance criterion for the dynamic equilibrium

and the difficulties involved in a step-wise dynamic calculation, i.e.

— selection of a suitable time-integration procedure
— specification of a time-step as influenced by physics, the adopted integration method and the non-linearity.

When contriving an economically efficient implicit procedure for non-linear applications, it is imperative to reduce the considerable expenditure involved in the integration over all time-steps. Therefore, the general aim will be to

— specify the admissible time-step as large as possible (see sections 11.5. and 11.6), thus confining the number of non-linear solution steps to a minimum
— design a most effective iterative solution within each time-step.

In principle, all the implicit procedures discussed in sections 10.2, 10.4 and 10.5 are also applicable to non-linear problems. In what follows, we select two typical procedures and discuss them in connection with specific non-linear applications.

11.4.2 The implicit Newmark procedures

In section 10.2, we attempted to familiarise the reader with a series of implicit Newmark variances involving, in particular, the procedure of average accelerations ($\beta = 1/4, \gamma = 1/2$), the linear acceleration method ($\beta = 1/6, \gamma = 1/2$), and the Fox-Goodwin variant ($\beta = 1/12, \gamma = 1/12$). In all these cases, the following expression holds for the displacements (see equ. (10.2.10))

$$\boldsymbol{r}_{\mathrm{k}+1} = \boldsymbol{r}_{\mathrm{k}} + \Delta t \dot{\boldsymbol{r}}_{\mathrm{k}} + \Delta t^2 (\tfrac{1}{2} - \beta) \ddot{\boldsymbol{r}}_{\mathrm{k}} + \Delta t^2 \beta \ddot{\boldsymbol{r}}_{\mathrm{k}+1} \tag{11.4.1}$$

and for the velocity (see equ. (10.2.11))

$$\dot{\boldsymbol{r}}_{\mathrm{k}+1} = \dot{\boldsymbol{r}}_{\mathrm{k}} + \Delta t (1 - \gamma) \ddot{\boldsymbol{r}}_{\mathrm{k}} + \Delta t \gamma \ddot{\boldsymbol{r}}_{\mathrm{k}+1} \tag{11.4.2}$$

A further essential relation is given by the dynamic equilibrium at the end of the time-step which we have already noted in equ. (11.3.3). In analogy to the static non-linear analysis, we can now specify dynamical out-of-balance forces in accordance with equ. (11.2.9)

$$\boldsymbol{R}_u = \boldsymbol{R}(t) - \boldsymbol{M}\ddot{\boldsymbol{r}} - \boldsymbol{C}\dot{\boldsymbol{r}} - \boldsymbol{R}_e(r) \tag{11.4.3}$$

These should vanish at the end of each incremental time-step

$$\boldsymbol{R}_u(\boldsymbol{r}_{\mathrm{k}+1}) = \boldsymbol{R}_{u|\mathrm{k}+1} = \boldsymbol{o} \tag{11.4.4}$$

In general, this can be realised only iteratively. We proceed from a known approximation $\boldsymbol{r}^{\mathrm{j}}_{\mathrm{k}+1}$ at the end of the time-step and postulate for the current iteration state j the linearised Taylor relation

$$\boldsymbol{R}_u(\boldsymbol{r}^{\mathrm{j}+1}_{\mathrm{k}+1}) = \boldsymbol{R}^{\mathrm{j}+1}_{u|\mathrm{k}+1} \approx \boldsymbol{R}^{\mathrm{j}}_{u|\mathrm{k}+1} + \boldsymbol{D}^{\mathrm{j}}_{\mathrm{k}+1}[\boldsymbol{r}^{\mathrm{j}+1}_{\mathrm{k}+1} - \boldsymbol{r}^{\mathrm{j}}_{\mathrm{k}+1}] \tag{11.4.5}$$

The out-of-balance forces may now be reduced as we proceed from j to j+1. The iteration matrix $\boldsymbol{D}$ of the dynamical process is deduced from

$$\begin{aligned} \boldsymbol{D}^{\mathrm{j}}_{\mathrm{k}+1} = \boldsymbol{D}(\boldsymbol{r}^{\mathrm{j}}_{\mathrm{k}+1}) &= \left.\frac{\partial \boldsymbol{R}_u}{\partial \boldsymbol{r}}\right|_{\boldsymbol{r}^{\mathrm{j}}_{\mathrm{k}+1}} \\ &= \left[\frac{\partial \boldsymbol{R}}{\partial \boldsymbol{r}} - \boldsymbol{M}\frac{\partial \ddot{\boldsymbol{r}}}{\partial \boldsymbol{r}} - \boldsymbol{C}\frac{\partial \dot{\boldsymbol{r}}}{\partial \boldsymbol{r}} - \frac{\partial \boldsymbol{R}_e}{\partial \boldsymbol{r}}\right]_{\boldsymbol{r}^{\mathrm{j}}_{\mathrm{k}+1}} \end{aligned} \tag{11.4.6}$$

whereby equ. (11.4.3) is used. The first term $\partial\boldsymbol{R}/\partial\boldsymbol{r}$ reproduces the effect of deflection-dependent external loads. At this point in our discussion, however, the forces are considered to be independent of the displacements and consequently, this term may be ignored. Nevertheless, we note that the presence of displacement-dependent loads induces an unsymmetric coefficient matrix $\boldsymbol{D}$ and thus can lead to a considerable increase in computational effort for the solution within the current time-step. As a result, some practitioners symmetrise such matrices and consider only this approximation which is better suited to the dynamic than the static case. We shall return to this point later on in our discussion.

We are already familiar with the last term in equ. (11.4.6) from statics: it represents the current tangential stiffness

$$\boldsymbol{K}^{\mathrm{j}}_{t|\mathrm{k}+1} = \left.\frac{\partial \boldsymbol{R}_e}{\partial \boldsymbol{r}}\right|_{\boldsymbol{r}^{\mathrm{j}}_{\mathrm{k}+1}} \tag{11.4.7}$$

Since the family of Newmark algorithms does not employ the explicit course of the displacements, velocities and accelerations within the time-step, we have to deduce the missing information in equ. (11.4.6), namely $\partial\ddot{\boldsymbol{r}}/\partial\boldsymbol{r}$ and $\partial\dot{\boldsymbol{r}}/\partial\boldsymbol{r}$, from the difference relations (11.4.1) and (11.4.2). To begin with, we obtain from equ. (11.4.1) the increment of the displacement due to an increase in the acceleration alone

$$\frac{\partial \boldsymbol{r}}{\partial \ddot{\boldsymbol{r}}} \approx \Delta t^2 \beta \tag{11.4.8}$$

and then from equ. (11.4.2) the increment of the velocity due to an increase in the acceleration only

$$\frac{\partial \dot{\boldsymbol{r}}}{\partial \ddot{\boldsymbol{r}}} \approx \Delta t \gamma \tag{11.4.9}$$

From equ. (11.4.8), we deduce by inversion

$$\frac{\partial \ddot{\boldsymbol{r}}}{\partial \boldsymbol{r}} = \frac{1}{\Delta t^2 \beta} \tag{11.4.10}$$

Combining equs. (11.4.9) and (11.4.10), we finally obtain

$$\frac{\partial \dot{\boldsymbol{r}}}{\partial \boldsymbol{r}} = \frac{\partial \dot{\boldsymbol{r}}}{\partial \ddot{\boldsymbol{r}}}\frac{\partial \ddot{\boldsymbol{r}}}{\partial \boldsymbol{r}} = \frac{\gamma}{\Delta t \beta} \tag{11.4.11}$$

Thus, we are now in a position to specify the iteration matrix $\boldsymbol{D}$ in detail. Using equ. (11.4.5), we find

$$\boldsymbol{D}(\boldsymbol{r}) = \frac{\partial \boldsymbol{R}}{\partial \boldsymbol{r}} - \frac{1}{\Delta t^2 \beta}\boldsymbol{M} - \frac{\gamma}{\Delta t \beta}\boldsymbol{C} - \boldsymbol{K}_t \tag{11.4.12}$$

Examining this expression, we observe immediately that the stiffness matrix $\boldsymbol{K}_t$ – known from statics and controlling the equilibrium iteration there – is, in dynamics, modified by mass and damping contributions. This also happened in the step solution for linear systems where the system matrix possesses some analogy to equ. (11.4.12). It is remarkable, however, that the factor $1/\Delta t^2$ causes the mass matrix to become dominant for small time-steps. If the mass matrix is positive definite and diagonal, then this term will exercise a stabilising effect upon the step solution. The rather stable behaviour of the out-of-balance loads has also been observed under more general conditions. Nevertheless, we should note

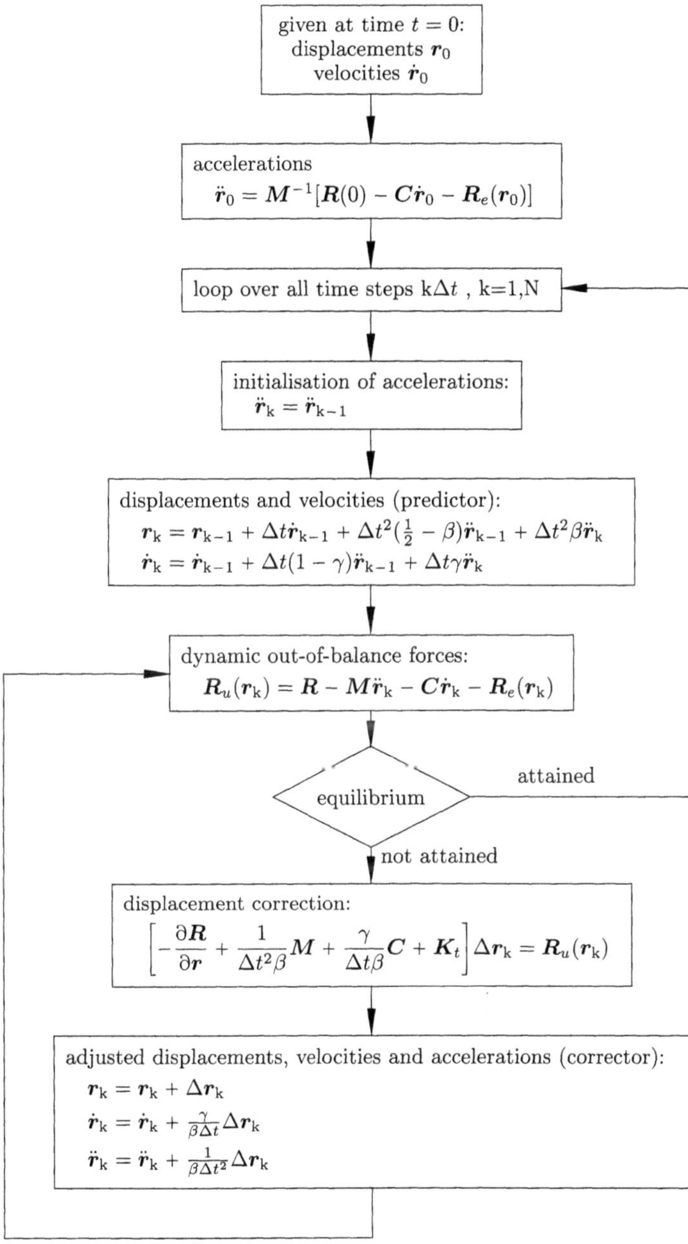

Fig. 11.4.1 Flow chart of the Newmark procedures applied to non-linear dynamic analyses

that a dynamical process involves strongly displacement-dependent effects and is consequently very sensitive with respect to break-off criteria in iterative operations. Hence, it is generally necessary to adopt a higher accuracy scheme than in statics.

Finally, we compose the non-linear Newmark algorithm using the above information (see fig. 11.4.1). Starting from the given initial conditions at the instant $t = 0$, $\boldsymbol{r}_0$ and $\dot{\boldsymbol{r}}_0$, the acceleration $\ddot{\boldsymbol{r}}_0$ can be deduced with relative ease from the force $\boldsymbol{R}_u$ at the instant $t = 0$ (see equ. (11.4.3)). Now we may proceed in our process along the time axis. To determine $\boldsymbol{r}_\mathrm{k}$ and $\dot{\boldsymbol{r}}_\mathrm{k}$ at the end of the time-step k from the Newmark formulae in equs. (11.4.1) and (11.4.2), we require the acceleration $\ddot{\boldsymbol{r}}_\mathrm{k}$ at the end of the time-step (we note in this connection that we proceed in fig. 11.4.1 from k$-$1 to k, but in equs. (24.4.1) and (24.4.2) from k to k+1, which is, however, equivalent). This operation is initialised by adopting either $\boldsymbol{o}$ or, even better, $\ddot{\boldsymbol{r}}_{\mathrm{k}-1}$ as an input. Using the Newmark predictor, we obtain first approximations to $\boldsymbol{r}_\mathrm{k}$, $\dot{\boldsymbol{r}}_\mathrm{k}$ and $\ddot{\boldsymbol{r}}_\mathrm{k}$ and are in a position to establish the out-of-balance forces from equ. (11.4.3). These are checked by an error tolerance, e.g. a check on forces analagous to statics

$$\frac{||\boldsymbol{R}_u||}{||\boldsymbol{R}||_{\max}} < \epsilon_R \tag{11.4.13}$$

in which the Euclidean norm of the out-of-balance forces is compared to the maximum norm of the given forces along the time axis. Alternatively, other expressions of residual bounds may be adopted using energy considerations, displacement residuals or different forms of force residuals, for example. If a satisfactory dynamic equilibrium is ultimately achieved, the operation within the time-step is completed. In general, however, this will not hold since $\ddot{\boldsymbol{r}}_\mathrm{k}$ was merely initialised. Consequently, following equs. (11.4.5) and (11.4.6), we evaluate a displacement correction using the equilibrium condition of equ. (11.4.4)

$$\boldsymbol{R}^{\mathrm{j}+1}_{u|\mathrm{k}+1} = \boldsymbol{o} \quad \rightarrow$$

$$\left[\boldsymbol{K}_t + \frac{1}{\Delta t^2\beta}\boldsymbol{M} + \frac{\gamma}{\Delta t\beta}\boldsymbol{C} - \frac{\partial \boldsymbol{R}}{\partial \boldsymbol{r}}\right]\left[\boldsymbol{r}^{\mathrm{j}+1}_{\mathrm{k}+1} - \boldsymbol{r}^{\mathrm{j}}_{\mathrm{k}+1}\right] = \boldsymbol{R}^{\mathrm{j}}_{u|\mathrm{k}+1} \tag{11.4.14}$$

In this way, we obtain a correction to the displacements and then, via equ. (11.4.11), also to the velocities

$$\dot{\boldsymbol{r}}^{\mathrm{j}+1}_{\mathrm{k}+1} - \dot{\boldsymbol{r}}^{\mathrm{j}}_{\mathrm{k}+1} = \frac{\gamma}{\Delta t\beta}\left[\boldsymbol{r}^{\mathrm{j}+1}_{\mathrm{k}+1} - \boldsymbol{r}^{\mathrm{j}}_{\mathrm{k}+1}\right] \tag{11.4.15a}$$

The correction to the accelerations is obtained by application of equ. (11.4.10)

$$\ddot{\boldsymbol{r}}^{\mathrm{j}+1}_{\mathrm{k}+1} - \ddot{\boldsymbol{r}}^{\mathrm{j}}_{\mathrm{k}+1} = \frac{\gamma}{\Delta t^2\beta}\left[\boldsymbol{r}^{\mathrm{j}+1}_{\mathrm{k}+1} - \boldsymbol{r}^{\mathrm{j}}_{\mathrm{k}+1}\right] \tag{11.4.15b}$$

Our discussion has now turned full circle and we may once again verify the equilibrium. We have described the algorithm in such a way as if Newton's method, already known to us from statics, forms part of the solution procedure within the step. This may suffice for the present, but alternatives will be considered in section 11.4.4.

11.4.3 The cubic Hermitian algorithm for non-linear use [11.22]

In section 10.2, we discussed the essentials of the cubic Hermitian algorithm in connection with linear applications. When proceeding to non-linear cases, no changes arise in equs. (10.2.47) and (10.2.48) for the determination of the displacement $\boldsymbol{r}_{\mathrm{k}+1}$ and the velocity $\dot{\boldsymbol{r}}_{\mathrm{k}+1}$. However, the equilibrium equations (10.2.51) and (10.2.52) are no longer applicable since they refer to linear conditions. They have to be replaced by equ. (11.2.9) and its derivatives with respect to time. As the authors wish to save the reader the trouble of looking up all these fundamental relations, we reproduce them here

$$\boldsymbol{r}_{\mathrm{k}+1} = \boldsymbol{r}_{\mathrm{k}} + \Delta t \dot{\boldsymbol{r}}_{\mathrm{k}} + \frac{\Delta t^2}{60}(21\ddot{\boldsymbol{r}}_{\mathrm{k}} + 3\Delta t \dddot{\boldsymbol{r}}_{\mathrm{k}} + 9\ddot{\boldsymbol{r}}_{\mathrm{k}+1} - 2\Delta t \dddot{\boldsymbol{r}}_{\mathrm{k}+1}) \tag{11.4.16}$$

$$\dot{\boldsymbol{r}}_{\mathrm{k}+1} = \dot{\boldsymbol{r}}_{\mathrm{k}} + \frac{\Delta t}{12}(6\ddot{\boldsymbol{r}}_{\mathrm{k}} + \Delta t \dddot{\boldsymbol{r}}_{\mathrm{k}} + 6\ddot{\boldsymbol{r}}_{\mathrm{k}+1} - \Delta t \dddot{\boldsymbol{r}}_{\mathrm{k}+1}) \tag{11.4.17}$$

$$\boldsymbol{M}\ddot{\boldsymbol{r}}_{\mathrm{k}+1} + \boldsymbol{C}\dot{\boldsymbol{r}}_{\mathrm{k}+1} + \boldsymbol{R}_{e|\mathrm{k}+1} = \boldsymbol{R}_{\mathrm{k}+1} \tag{11.4.18}$$

$$\boldsymbol{M}\dddot{\boldsymbol{r}}_{\mathrm{k}+1} + \boldsymbol{C}\ddot{\boldsymbol{r}}_{\mathrm{k}+1} + \dot{\boldsymbol{R}}_{e|\mathrm{k}+1} = \dot{\boldsymbol{R}}_{\mathrm{k}+1} \tag{11.4.19}$$

In principle, these four matrix equations represent a complete scheme to deduce the corresponding measures at the end of the step when $\boldsymbol{r}_{\mathrm{k}}, \dot{\boldsymbol{r}}_{\mathrm{k}}, \ddot{\boldsymbol{r}}_{\mathrm{k}}, \dddot{\boldsymbol{r}}_{\mathrm{k}}$ are given at the beginning of the time-step. In addition, we establish for the time-derivative of the elastic forces, $\dot{\boldsymbol{R}}_{e|\mathrm{k}+1}$, the relation

$$\begin{aligned} \dot{\boldsymbol{R}}_{e|\mathrm{k}+1} &= \left.\frac{\partial \boldsymbol{R}_e}{\partial t}\right|_{\boldsymbol{r}_{\mathrm{k}+1}} = \left.\frac{\partial \boldsymbol{R}_e}{\partial \boldsymbol{r}}\frac{\partial \boldsymbol{r}}{\partial t}\right|_{\boldsymbol{r}_{\mathrm{k}+1}} \\ &= \boldsymbol{K}_{t|\mathrm{k}+1}\dot{\boldsymbol{r}}_{\mathrm{k}+1} \end{aligned} \tag{11.4.20}$$

where $\boldsymbol{K}_{t|\mathrm{k}+1} = \boldsymbol{K}_t(\boldsymbol{r}_{\mathrm{k}+1})$ is the tangential stiffness of the non-linear system at the end of the time-step.

Application of equs. (11.4.16) to (11.4.20) yields an iterative process which is detailed in fig. 11.4.2 in the form of a flow chart and which we discuss in the following.

Once again, we assume that initial conditions for $\boldsymbol{r}_0$ and $\dot{\boldsymbol{r}}_0$ are given and determine with the help of the relations in equs. (11.4.18) and (11.4.19) the associated accelerations $\ddot{\boldsymbol{r}}_0$ and their derivatives $\dddot{\boldsymbol{r}}_0$. However, one has to take into consideration that, in general, the tangential stiffness may already be required here, since the structure may initially be pre-stressed or $\boldsymbol{r}_0$ itself be large. With this information, the initial set of pivotal values is completed. In order to compute the displacements and the velocities at the end of the time-step, we also require the accelerations and their derivatives at this instant – see equs. (11.4.16) and (11.4.17). These, however, are still unknown. Hence, we establish with the known information at the beginning of the time-step a Taylor expansion

$$\ddot{\boldsymbol{r}}_{\mathrm{k}+1} = \ddot{\boldsymbol{r}}_{\mathrm{k}} + \Delta t \dddot{\boldsymbol{r}}_{\mathrm{k}} \tag{11.4.21}$$

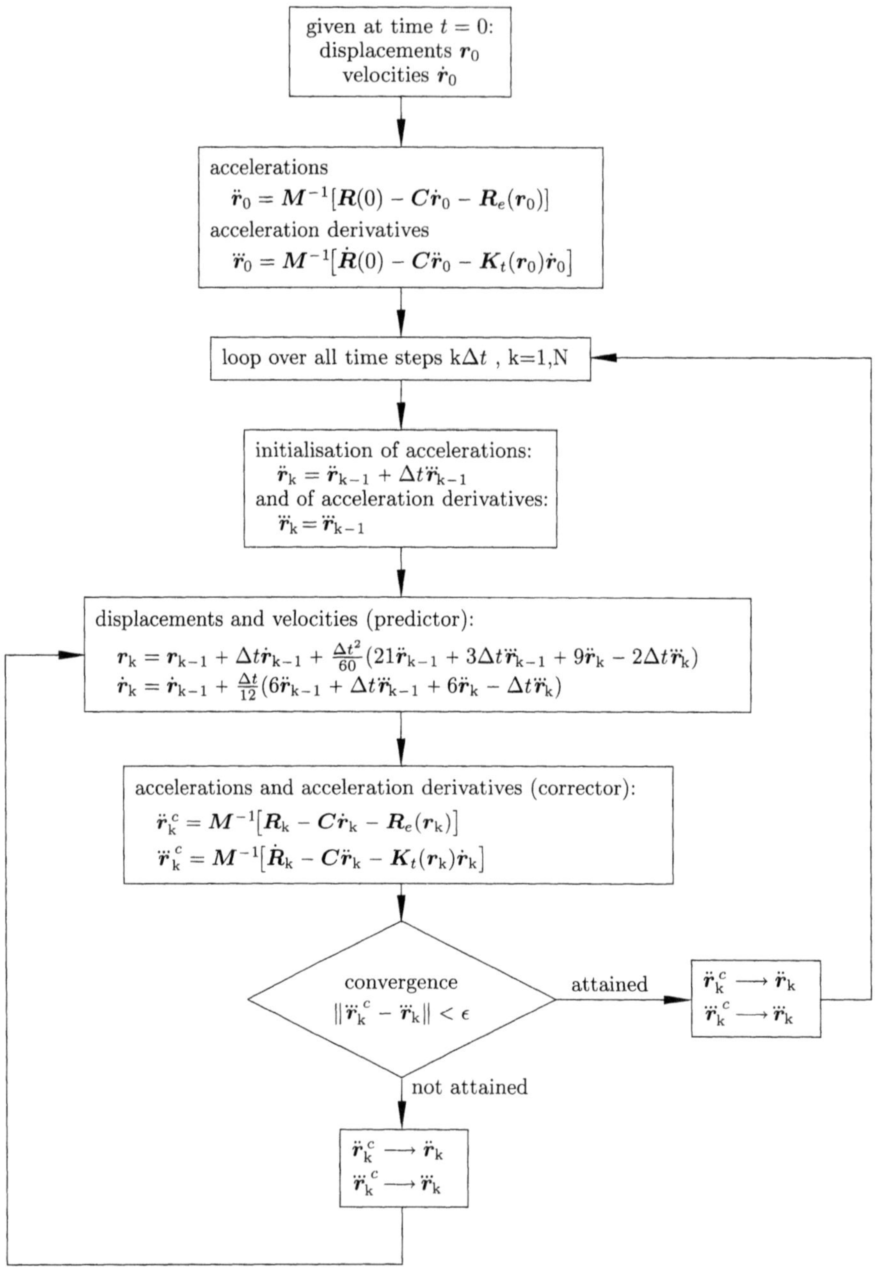

Fig. 11.4.2 Flow chart of the basic version of the conditionally stable cubic Hermitian algorithm when applied to non-linear dynamics

$$\dddot{\boldsymbol{r}}_{k+1} = \dddot{\boldsymbol{r}}_k \tag{11.4.22}$$

With the aid of this scheme, all displacements and velocities at the end of the step can be derived in conjunction with equs. (11.4.16) and (11.4.17). Since equs. (11.4.21) and (11.4.22) are only approximations, a correction has to be performed, for which we apply equs. (11.4.18) and (11.4.19). In this way, we obtain the corrected values (designated by a superscript c)

$$\ddot{\boldsymbol{r}}^c_{k+1} = \boldsymbol{M}^{-1}[\boldsymbol{R}_{k+1} - \boldsymbol{C}\dot{\boldsymbol{r}}_{k+1} - \boldsymbol{R}_{e|k+1}] \tag{11.4.23}$$

$$\dddot{\boldsymbol{r}}^c_{k+1} = \boldsymbol{M}^{-1}[\dot{\boldsymbol{R}}_{k+1} - \boldsymbol{C}\ddot{\boldsymbol{r}}_{k+1} - \boldsymbol{K}_{t|k+1}\dot{\boldsymbol{r}}_{k+1}] \tag{11.4.24}$$

Assuming the derivatives of the accelerations to form the most sensitive pivotal values, it is advisable to select the Euclidean norm of their variations as the convergence criterion

$$\|\dddot{\boldsymbol{r}}^c_{k+1} - \dddot{\boldsymbol{r}}_{k+1}\| < \varepsilon \tag{11.4.25}$$

although alternative measures are once again conceivable. If the convergence criterion in equ. (11.4.25) is fulfilled, then the iteration procedure of the time-step is concluded. If this is not the case, then we switch back to equs. (11.4.16) and (11.4.17) and determine the displacements and velocities on the basis of the new values.

Critical points – with respect to the computational effort – in the iterative operations within the step concern the evaluation of the elastic restoring forces $\boldsymbol{R}_{e|k+1}$ in equ. (11.4.23) and of the term $\boldsymbol{K}_{t|k+1}\dot{\boldsymbol{r}}_{k+1}$ in equ. (11.4.24). Furthermore, the existence of the damping matrix $\boldsymbol{C}$ in both equations plays an important role, even if we tacitly assume that the mass matrix $\boldsymbol{M}$ is diagonal. If the number of unknown displacements is high, at least $\boldsymbol{C}$ and $\boldsymbol{K}_t$ can no longer be stored in the central memory. The presence of a slow-speed backing store can cause a dramatic increase in the running time of the operation and may ultimately become unacceptable. It is thus advisable to reconsider the computational scheme.

The calculation of the elastic restoring forces can be reformulated without any difficulty in terms of vector operations. To this purpose, we use the relevant formula of Vol. IV of the series

$$\boldsymbol{R}_e = \int_V \boldsymbol{\alpha}^t_L \boldsymbol{\sigma}_c \mathrm{d}V \tag{11.4.26}$$

whereby $\boldsymbol{\sigma}_c$ are the Cauchy stresses in the structure (related to the deformation state $\boldsymbol{r}_{k+1}$) and the integration is applied over the deformed structure. Furthermore, $\boldsymbol{\alpha}_L$ is the well-known linear strain operator; see also the explanations to equ. (11.2.8). The integration over the total structure as it is seen from equ. (11.4.26) is – in a FE-discretisation – split up into an integration over the element volumina and a subsequent summation over the individual element contributions.

The Cartesian representation in equ. (11.4.26) has its natural complement (see Vol. III and IV)

$$\boldsymbol{R}_e = \sum_{\mathrm{g}=1}^{\mathrm{m}} \mathbf{a}_{\mathrm{g}}^{\mathrm{t}} \boldsymbol{a}_{N\mathrm{g}}^{\mathrm{t}} \int_{V_{\mathrm{g}}} (\boldsymbol{\alpha}_L^N) \boldsymbol{\sigma}_c \mathrm{d}V \tag{11.4.27a}$$

The summation is performed here over m finite elements. The matrix $\mathbf{a}_{\mathrm{g}}$ is a Boolean (incidence) matrix relating the Cartesian element displacements to global displacements. With the aid of $\boldsymbol{a}_{N\mathrm{g}}$, the incremental natural deformations can be derived from the infinitesimal increase of the Cartesian displacements at the state $\boldsymbol{r}_{\mathrm{k}+1}$. The matrix $\boldsymbol{\alpha}_L^N$ corresponds to the linear strain operator connecting Cartesian strains with natural deformations. As is shown in Vol. IV, the integral in equ. (11.4.27a) is equal to the natural forces $[\mathbf{P}_N + \mathbf{P}_{Ne}]_{\mathrm{g}}$ in the elements. We thus obtain

$$\boldsymbol{R}_e = \sum_{\mathrm{g}=1}^{\mathrm{m}} \mathbf{a}_{\mathrm{g}}^{\mathrm{t}} \boldsymbol{a}_{N\mathrm{g}}^{\mathrm{t}} [\mathbf{P}_N + \mathbf{P}_{Ne}]_{\mathrm{g}} \tag{11.4.27b}$$

where $\mathbf{P}_{Ne}$ exists only in the presence of loads distributed over the elements. For large displacements and small strains, $\mathbf{P}_N$ can be determined from

$$\mathbf{P}_N = \boldsymbol{k}_N \boldsymbol{\rho}_N \tag{11.4.28}$$

The explicit formation of the tangential stiffness matrix $\boldsymbol{K}_t(\boldsymbol{r}_{\mathrm{k}+1}) = \boldsymbol{K}_{t|\mathrm{k}+1}$ can be circumvented if the formula for its assembly is used

$$\begin{aligned} \boldsymbol{K}_{t|\mathrm{k}+1} \dot{\boldsymbol{r}}_{\mathrm{k}+1} &= \mathbf{a}^{\mathrm{t}} [\boldsymbol{k}_e + \boldsymbol{k}_G]_{\boldsymbol{r}_{\mathrm{k}+1}} \mathbf{a} \dot{\boldsymbol{r}}_{\mathrm{k}+1} \\ &= \mathbf{a}^{\mathrm{t}} [\boldsymbol{k}_e + \boldsymbol{k}_G]_{\boldsymbol{r}_{\mathrm{k}+1}} \dot{\boldsymbol{\rho}}_{\mathrm{k}+1} \end{aligned} \tag{11.4.29}$$

The elemental stiffnesses must be evaluated anyway. These are immediately multiplied by $\dot{\boldsymbol{\rho}}_{\mathrm{k}+1}$ and only the resulting pseudo-force need be stored. Thus, we have also ensured on the global level pure vector operations.

The incorporation of the damping matrix $\boldsymbol{C}$ is somewhat more problematic. If this matrix is built up element by element, we could in principle proceed in a manner analogous to equ. (11.4.29). Then, however, either the elemental damping matrices $\boldsymbol{c}_{\mathrm{g}}$ must be established anew at each step or a storage capacity is required equivalent to that for the matrix $\boldsymbol{C}$. As an alternative and as a suitable procedure for global $\boldsymbol{C}$ matrices not being assembled from elemental contributions, the method of 'lumping' the damping characteristics of the structure offers itself, although this is not altogether unproblematic, however.

In order to complete the historical account, we now consider a further possibility in the calculation of the elastic restoring forces. This method is developed in [11.22] and has already proven to be effective in practice. Consider the following relation

$$\boldsymbol{R}_{e|\mathrm{k}+1} = \boldsymbol{R}_{e|\mathrm{k}} + \boldsymbol{K}_s [\boldsymbol{r}_{\mathrm{k}+1}^t - \boldsymbol{r}_{\mathrm{k}}] \tag{11.4.30}$$

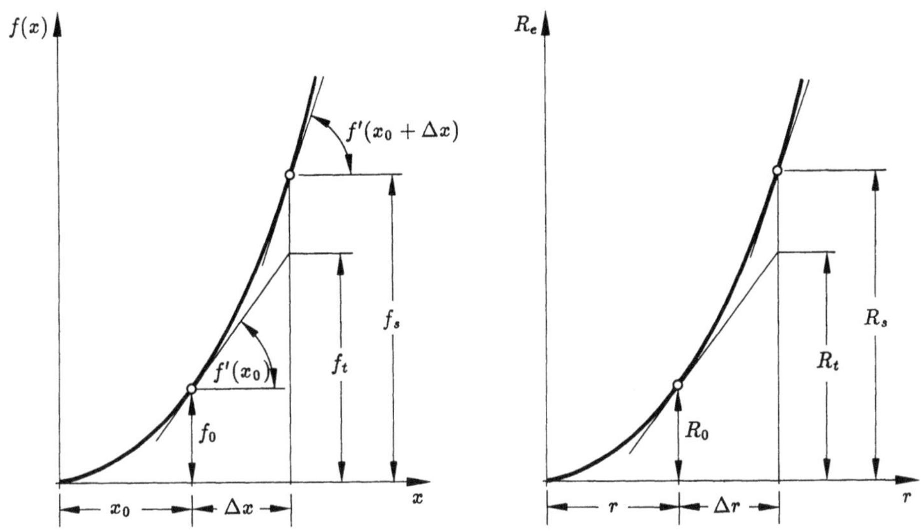

$$f_t = f(x_0) + f'(x_0)\Delta x \qquad \Longrightarrow \boldsymbol{R}^t_{e|k+1} = \boldsymbol{R}_{e|k} + \boldsymbol{K}_{t|k}\Delta\boldsymbol{r}$$

$$f_s = f(x_0) + \tfrac{1}{2}[f'(x_0) + f'(x_0 + \Delta x)]\Delta x \qquad \Longrightarrow \boldsymbol{R}_{e|k+1} = \boldsymbol{R}_{e|k} + \tfrac{1}{2}[\boldsymbol{K}_{t|k} + \boldsymbol{K}^t_{t|k+1}]\Delta\boldsymbol{r}$$

Fig. 11.4.3 Mathematical description and resulting secant stiffness having 2nd order accuracy

Here, $\boldsymbol{r}^t_{k+1}$ designates the iteratively established displacement vector $\boldsymbol{r}$ (at the instant k+1) which is derived by a purely tangential relation and $\boldsymbol{K}_s$ is the secant stiffness which is approximated to a second order accuracy (see also fig. 11.4.3) by

$$\boldsymbol{K}_s = \tfrac{1}{2}\left[\boldsymbol{K}_{t|k} + \boldsymbol{K}^t_{t|k+1}\right] \tag{11.4.31}$$

in which we set, in accordance with previous definitions, $\boldsymbol{K}^t_{t|k+1} = \boldsymbol{K}_t(\boldsymbol{r}^t_{k+1})$. In a preparatory step, $\boldsymbol{r}^t_{k+1}$ is calculated using the purely tangential relation

$$\boldsymbol{R}^t_{e|k+1} = \boldsymbol{R}_{e|k} + \boldsymbol{K}_{t|k}[\boldsymbol{r}^t_{k+1} - \boldsymbol{r}_k] \tag{11.4.32}$$

Hereafter, the evaluation proceeds with equ. (11.4.30) and the constant stiffness of equ. (11.4.31). For $\boldsymbol{K}_s$, no additional storage is required since the expression

$$\boldsymbol{K}_s = \tfrac{1}{2}\boldsymbol{K}_{t|k} + \tfrac{1}{2}\sum_{g=1}^{m} \mathbf{a}^t_g \boldsymbol{k}_t(\boldsymbol{\rho}^t_{k+1})\mathbf{a}_g \tag{11.4.33}$$

allows a direct accumulation. Theoretically, $\boldsymbol{K}_{t|k+1}$ would still be required in this process, see also equ. (11.4.24). In order to circumvent this, however, we use in the preparatory run $\boldsymbol{K}_{t|k}$ in place of $\boldsymbol{K}_{t|k+1}$ and in the subsequent runs $\boldsymbol{K}_s$; in this way, only one stiffness need be stored. Although this is somewhat problematic from the theoretical point of view, in practice it proves most effective.

11.4.4 Numerical examples[11.22]

Following the development of the conditionally stable cubic Hermitian algorithm in section 11.4.3, we next apply this technique in conjunction with the procedures of the preceding sections to a series of examples and compare the accuracy of the results with those obtained by alternative methods.

Pre-stressed string with one degree of freedom

We begin the series of applications with an example already discussed in detail in the introduction: a massless string with a central mass (fig. 11.4.4); the string is initially pre-stressed with $P_{N0} = 500$N. We do not use the approximation quoted in the introduction 11.4 and calculate the elastic restoring force "exactly" as

$$R_e = 2P_{N0}\frac{r}{\sqrt{l^2 + r^2}} + 2EA\frac{r}{l}\left(1 - \frac{l}{\sqrt{l^2 + r^2}}\right)$$

In a first calculation, the central mass is displaced 20cm and then allowed to oscillate freely without damping. After a period (here $T_0 = 0.264\,791\,269$ s as the *non-linear period*), the mass m should have returned to its initial position. In fig. 11.4.5, we plot the error along the time axis since this is directly related to the number of steps via

$$\Delta t = \frac{1}{\mathrm{N}}\frac{T_0}{4}$$

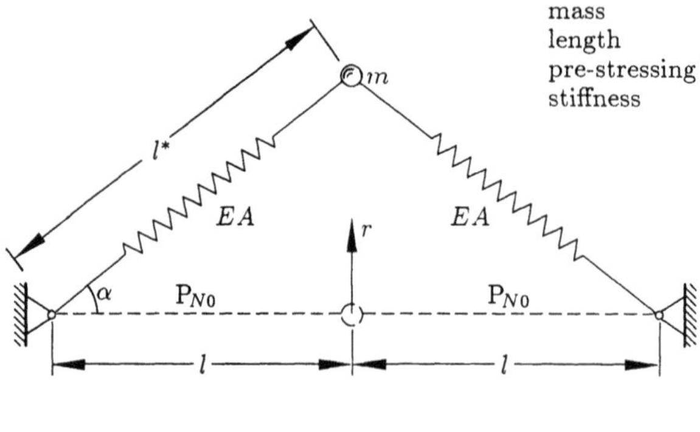

data:

mass	$m = 500$ kg
length	$l = 1$ m
pre-stressing load	$P_{N0} = 500$ N
stiffness	$EA = 10^7$ N

Fig. 11.4.4 Non-linear system with a single degree of freedom

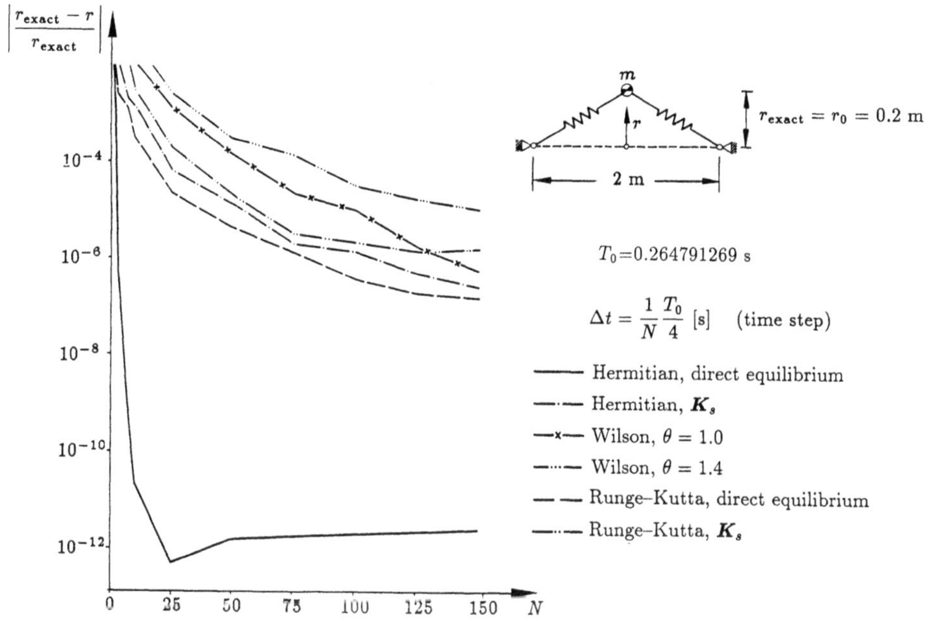

Fig. 11.4.5 Relative errors in the displacements after one period as a function of the step-size

The greater N is, the smaller the time-step and the smaller the integration error. Six different procedures are compared. The Hermitian algorithm in the basic version shown here (direct equilibrium) clearly leads the field. Under the same conditions and for the same time-step, Runga-Kutta is not so effective, but still in second place (this calculation should not be confused with the initial calculation which was executed with $\Delta t = 10^{-6}$ and yielded the aforementioned T_0). The $\boldsymbol{K}_s$-method naturally contains approximations which of necessity lead to a decrease in accuracy for both the Hermitian and the Runge-Kutta algorithms. Finally, two versions of Wilson's θ-method [11.23, 11.24, 11.25] are illustrated. The case $\theta = 1$ corresponds to the procedure of linear accelerations or to Newmark with $\beta = 1/6, \gamma = 1/2$. In the adopted time-step range of $0.25\,T_0$ to $0.00167\,T_0$, these procedures show the severest errors, although in such a comparison, the computational effort within the step ought to be taken into consideration; this is ignored here. In fig. 11.4.6, the error in the amplitude over a running time of 20 s is illustrated for a particular time-step $\Delta t = T_0/600$ resulting from N=150. Theoretically, about 80 oscillatory cycles should be executed in the specified time. In fact, due to integration and rounding errors, a (slight) rising error occurs during this running time; once again, the cubic Hermitian algorithm with direct equilibrium iteration yields the best results.

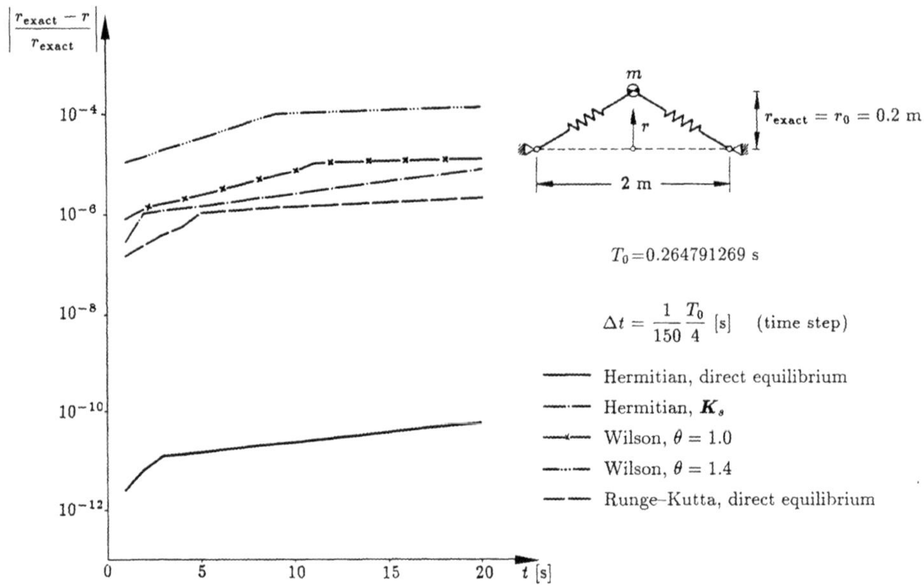

Fig. 11.4.6 Temporal evolution of the amplitude error at constant step-size

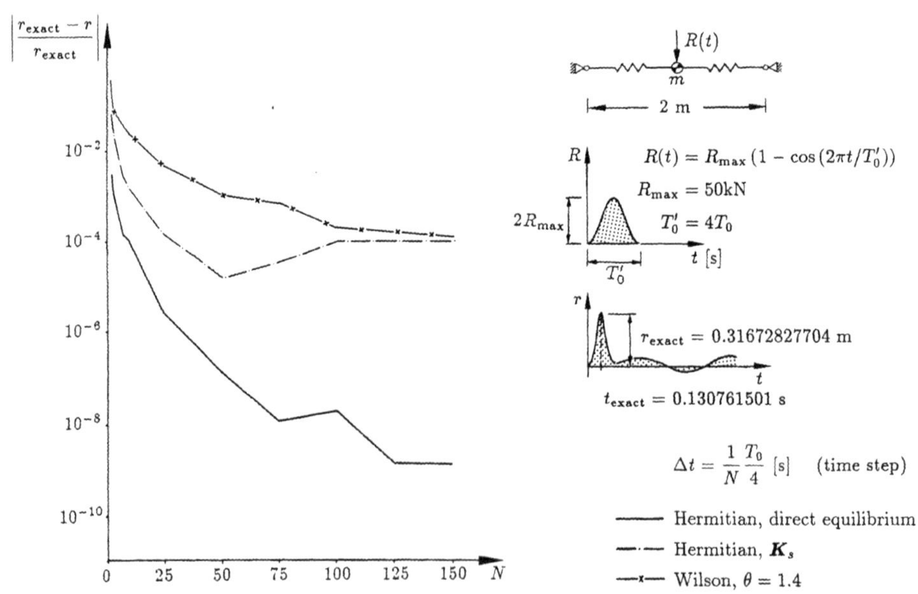

Fig. 11.4.7 Relative errors in the peak amplitudes as a function of the step-size

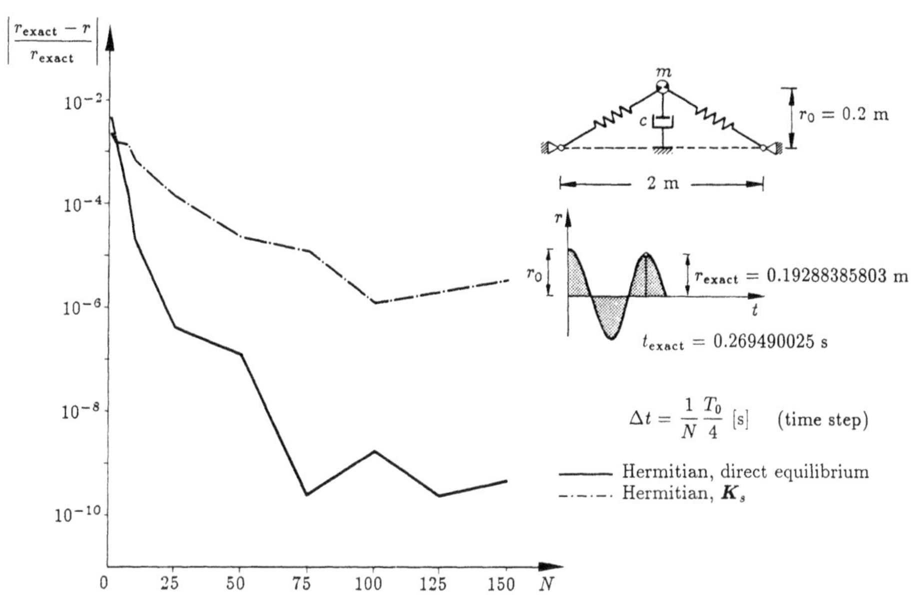

Fig. 11.4.8 Relative errors in the displacements after one damped non-linear period, plotted over the number of time-steps applied in a quarter period

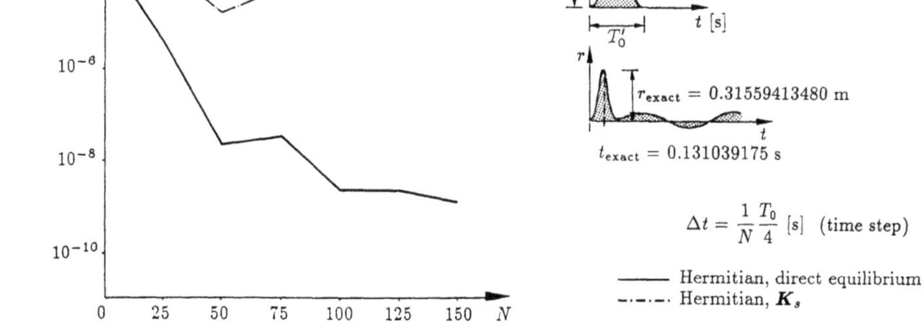

Fig. 11.4.9 Relative errors in the peak amplitudes within the transient phase, plotted over the number of time-steps applied in a quarter period

We now switch in this example to undamped, force-excited transient processes. In accordance with the legend to fig. 11.4.7, we apply a cosine excitation of limited duration to an oscillator. Since the loading action extends over four periods, this cannot be considered an impulse. Following an interval of $T_0' = 4\,T_0$, the assumed force progression is seen to lead to a free undamped natural vibration. However, we are interested in the maximum amplitude of the transient response which appears in the "exact" initial calculation at $t_{\mathrm{ex}} = 0.130\,761\,501$ s. In this case, two versions of the Hermitian algorithm and the Wilson θ-method are compared. Again, the Hermitian algorithm and in particular the direct iteration of the dynamic equilibrium yield the best results.

In two further tests (figs. 11.4.8 and 11.4.9), we introduce a damping of $c = 2\,\mathrm{Ns/m}$. In fig. 11.4.8, we calculate the deflection after one damped period and compare it with the "exact" value. Finally, in fig. 11.4.9, we reevaluate the maximum amplitude of the transient response; due to damping, this maximum response appears now with a slight time lag at $t_{\mathrm{ex}} = 0.131\,039\,175\,\mathrm{s}$. Both examples are further evidence of the good quality of the cubic Hermitian acceleration approximation. It goes without saying that even better results can be achieved with the conditionally stable quintic Hermitian algorithm.

Plane pre-stressed network

An example with one degree of freedom may serve as an amusing academic test, but it is not realistic. Therefore, we extend our investigations in a more practical direction and consider a wide-span cable network, initially taken to be plane, which is shown in fig. 11.4.10. The outer cables are pre-stressed with a load of about 180 000 N, the interior cables carry about 20 000 N. This plane system is subjected to a uniformly distributed transverse loading of 100 000 N *in toto*, which produces the sagging shape of the net shown in fig. 11.4.11. This static deflection, attained on a highly non-linear path and amounting to 1.62 m in the centre as compared with net dimensions of 12 m, represents the initial state for the subsequent oscillatory tests. The statically deflected structure is now released and allowed to oscillate freely. The $\boldsymbol{K}_s$-method and the Wilson θ-method are applied for two time-steps, $\Delta t = 0.0005\,\mathrm{s}$ and $\Delta t = 0.002\,\mathrm{s}$. In fig. 11.4.12, the motion of the central point A is plotted over the time t. During the course of the motion, it is possible that certain parts of the network experience a loss of their pre-stressing and become exposed in the computational procedure to compression, which is obviously physically impossible. Hence, a corresponding modification of the programme must be introduced providing for zero load in the case of compression strains. This realistic modification causes the breakdown of the calculation with the Wilson θ-method at the larger step of Δt=0.002 s (marked by "Div"). The other curves show a good to satisfactory consistency. For comparison, the calculation shown in fig. 13.4.13 was executed without the aforementioned modification, the rope elements being allowed to carry compression. After 0.5 s, we note clear deviations in the response. In this case, the Wilson θ-method continued to function up to a time-step of Δt=0.006 s (fig. 11.4.14). However, we discern clearly that the amplitude errors and, to a greater extent, the period elongation become no longer acceptable for such step-sizes.

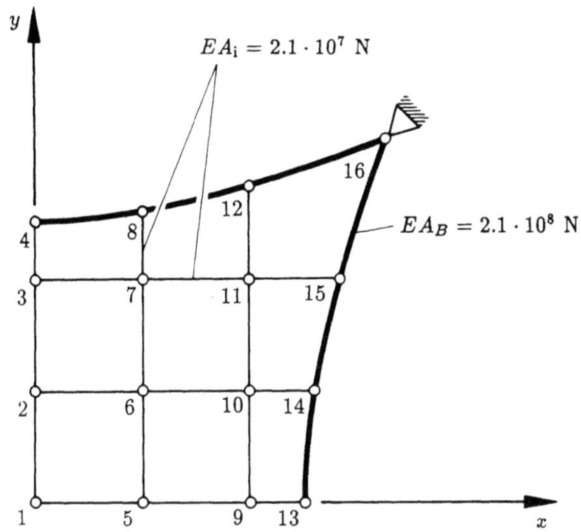

nodal points	x [m]	y [m]	mass [kg]
1	0.0000000000000	0.0000000000000	48.231548671863
2	0.0000000000000	2.0055650131893	48.197858694277
3	0.0000000000000	4.0111474799212	39.788641762874
4	0.0000000000000	4.6128398404689	245.288250325758
5	2.0055650131893	0.0000000000000	48.197858694277
6	2.0015054969710	2.0015054969710	48.163929366599
7	2.0052258136725	4.0029948413371	40.970658454237
8	1.9993209157720	4.8035476227263	248.274758583787
9	4.0111474799212	0.0000000000000	39.788641762874
10	4.0029948413371	2.0052258136725	40.970658454237
11	4.0104969345030	4.0104969345030	38.582265029508
12	3.9957683694213	5.2136537418985	259.384253827008
13	4.6128398404689	0.0000000000000	245.288250325758
14	4.8035476227263	1.9993209157720	248.274758583787
15	5.2136537418985	3.9957683694213	259.384253827008
16	6.0000000000000	6.0000000000000	

element	pre-stressing load [N]
1 – 2	$3.2107234141308 \cdot 10^4$
2 – 3	$3.2184288419281 \cdot 10^4$
3 – 4	$3.2435775092699 \cdot 10^4$
5 – 6	$1.8995475073131 \cdot 10^4$
6 – 7	$1.8921770313874 \cdot 10^4$
7 – 8	$1.8680148428175 \cdot 10^4$
9 – 10	$3.0928130275108 \cdot 10^4$
10 – 11	$3.1101045053774 \cdot 10^4$
11 – 12	$3.1606433700624 \cdot 10^4$
13 – 14	$1.7079500592538 \cdot 10^5$
14 – 15	$1.7343274915299 \cdot 10^5$
15 – 16	$1.8207756748960 \cdot 10^5$
4 – 8	$1.7079500592538 \cdot 10^5$
8 – 12	$1.7343274915299 \cdot 10^5$
12 – 16	$1.8207756748960 \cdot 10^5$
3 – 7	$3.0928130275204 \cdot 10^4$
7 – 11	$3.1101045053774 \cdot 10^4$
11 – 15	$3.1606433700624 \cdot 10^4$
2 – 6	$1.8995475073227 \cdot 10^4$
6 – 10	$1.8921770313874 \cdot 10^4$
10 – 14	$1.8680148428175 \cdot 10^4$
1 – 5	$3.2107234141308 \cdot 10^4$
5 – 9	$3.2184288419281 \cdot 10^4$
9 – 13	$3.2435775092699 \cdot 10^4$

Fig. 11.4.10 Plane pre-stressed network (one quarter); geometry and data

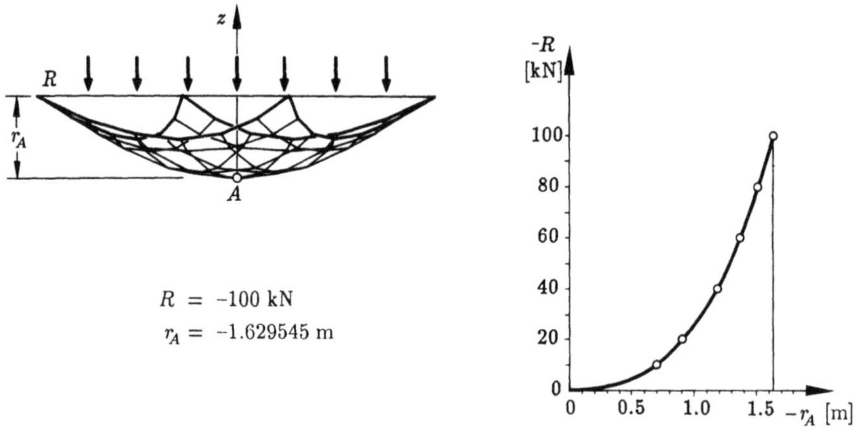

Fig. 11.4.11 Static deformation pattern of the pre-stressed network – initially plane – derived in a non-linear analysis

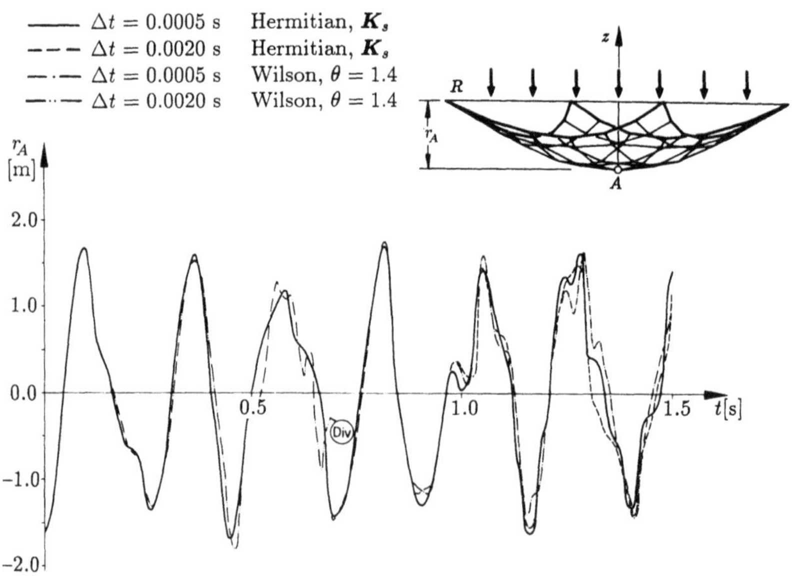

Fig. 11.4.12 Free undamped time-history response of centre node A; elements subject to compression are eliminated

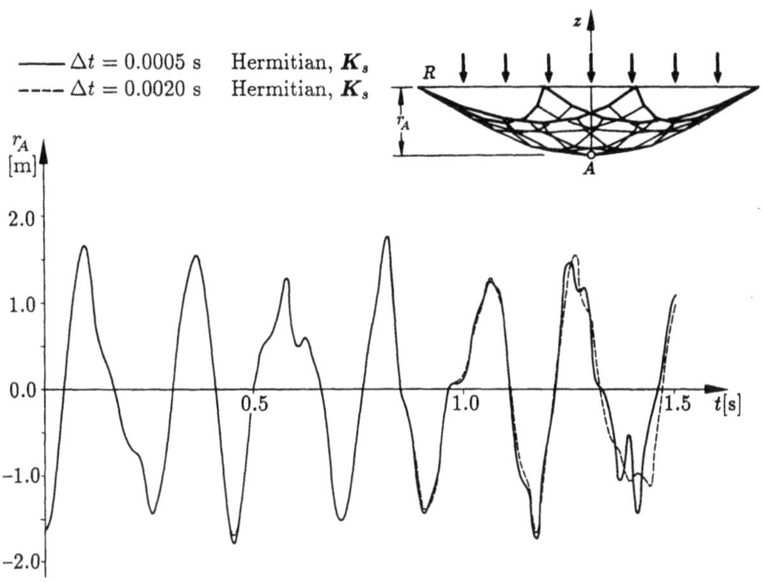

Fig. 11.4.13 Free undamped motion of centre node A; elements subject to compression not eliminated

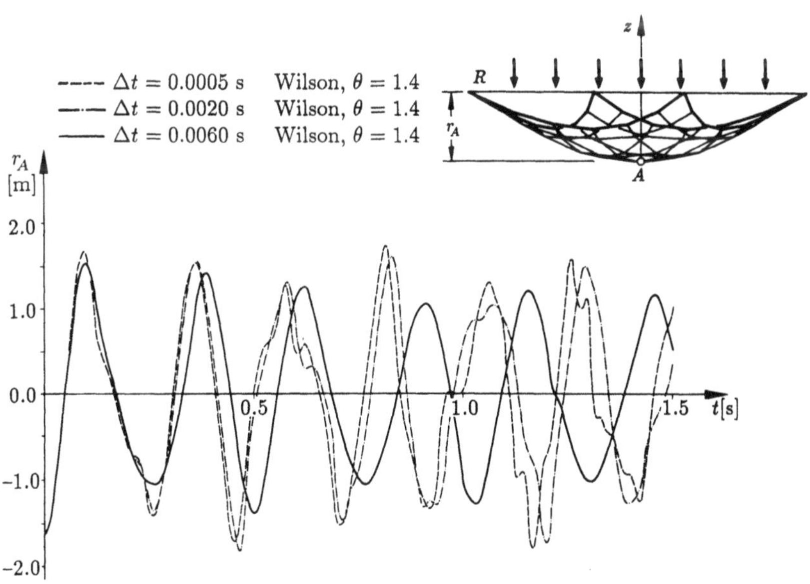

Fig. 11.4.14 Free undamped motion of centre node A evaluated with Wilson's procedure; elements subject to compression not eliminated

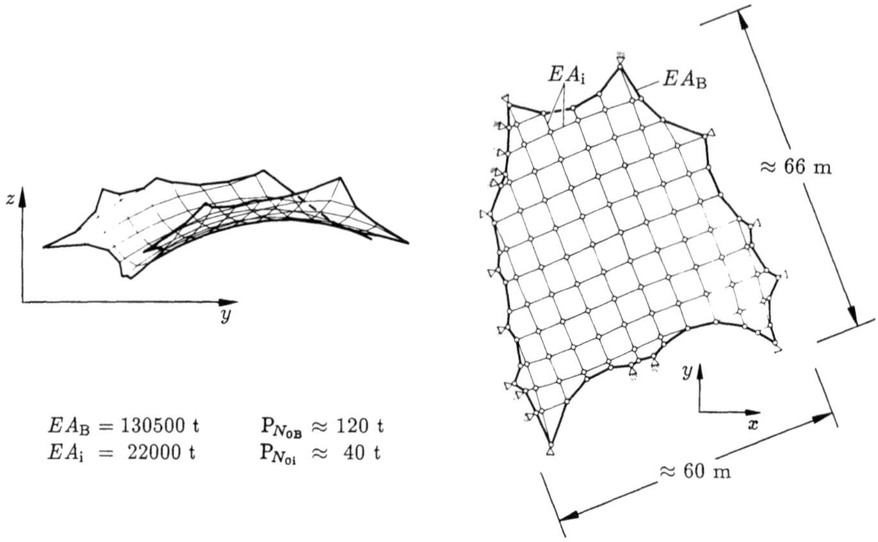

Fig. 11.4.15 Three-dimensional pre-stressed network (partial roofing of the Olympic Stadium in Munich)

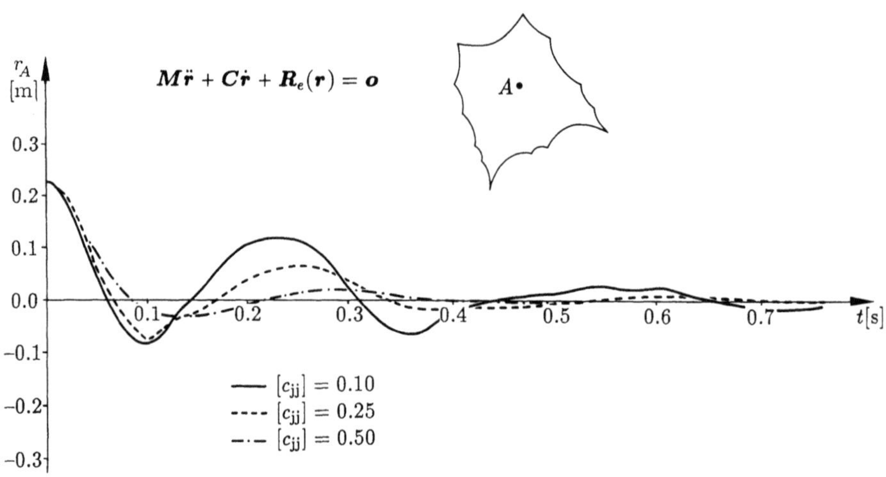

Fig. 11.4.16 Free damped response of nodal point A for different damping parameters

Three-dimensional pre-stressed network

This large scale, realistic problem formed part of the analysis of the roofing for the Olympic Stadium in Munich. The real net has a mesh spacing of 0.75 m which was converted to 6 m for the idealisation process (see fig. 11.4.15). The idealised structure possesses 288 unknown displacements, is geometrically extremely complex and strongly non-linear. Our investigation starts by deflecting the structure

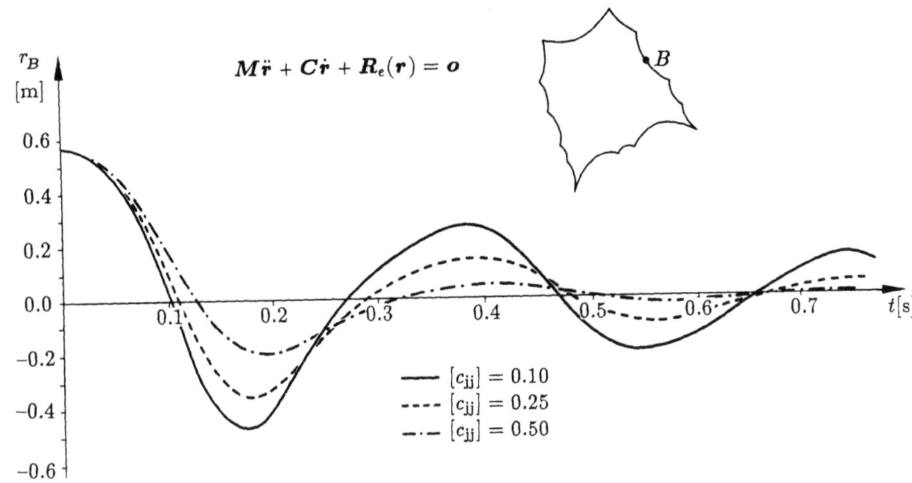

Fig. 11.4.17 Free damped response of nodal point B for different damping parameters

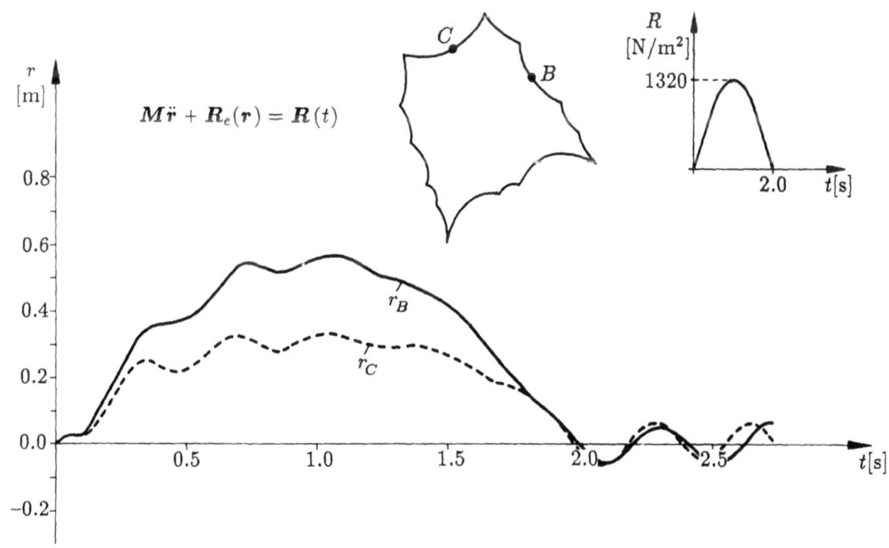

Fig. 11.4.18 Forced undamped transient response of nodal points B and C

statically due to a lateral wind thrust of 1320 N/m^2 and then releasing it into an oscillatory state. In fig. 11.4.16, we consider the motion of the point A, and in fig. 11.4.17 that of the point B. Damping effects are also taken into account and represented by a diagonal damping matrix; the influence of the damping parameters on the solution is shown. The chosen damping parameters are arbitrary and are not related to a specific physical argument. In fig. 11.4.18, damping is omitted,

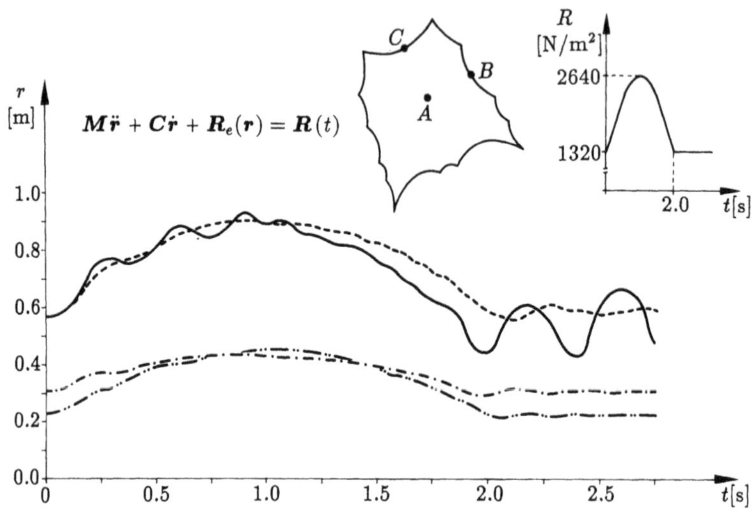

—— r_B , $[c_{jj}] = 0$

---- r_B , $[c_{jj}] = 0.25$

–·– r_C , $[c_{jj}] = 0.25$

–··– r_A , $[c_{jj}] = 0.25$

Fig. 11.4.19

Forced undamped transient response of nodal points A, B and C superimposed on a constant static deflection

and a lateral wind force in the form of a half sine wave with a maximum amplitude of 1320 N/m^2 is applied to the structure over 2 s. The maximum deflections in the transient phase are of particular interest. After 2 s, the force is removed and a free undamped vibration sets in. As the last test in this series, we combine in fig. 11.4.19 the dynamic application of the wind force with a static initial load. As a result, the structure is already statically deflected when the dynamic wind pressure is applied. Following the 2 s phase of wind loading, the oscillating system is again exposed solely to the initial static forces. The motion of point B without and with diagonal damping and the motion of points A and C, but only with damping, are illustrated. In the absence of damping, a free vibration about the statical equilibrium position will finally take place. In the presence of damping, the motion ultimately vanishes and the shape of the network coincides once more with that of the initial static case. Note that in an assumed linear response, the motion of point B (undamped) in the histogram of fig. 11.4.19 should be identical with that in fig. 11.4.18, apart from a vertical shift. However, due to the non-linear behaviour of the structure, the principle of superposition is no longer valid, so that apart from an overall similarity, clear deviations are observed.

Transmitter tower under wind loading

A 152.5 m high transmitter tower has a triangular cross-section space frame structure, stabilised by anchoring cables (backstays) (fig. 11.4.20). Despite its 737 elements and 549 degrees of freedom, the problem may be analysed fully in-core, due to the prevailing small bandwidth. Readers interested in the complete background to this problem may obtain all data in [11.26].

We first apply the dead weight of the structure as a constant dead load and then analyse the subsequent response of the system due to a gust of wind which is applied in the form of a half sine wave of 3 s duration (fig. 11.4.21). Figure 11.4.21 shows the average deflections of the cross sections A and B and fig. 11.4.22 those of the cross sections C and D. In addition, the displacements due to an assumed static application of the wind load are also indicated. In all cases, the conditionally stable cubic algorithm was applied.

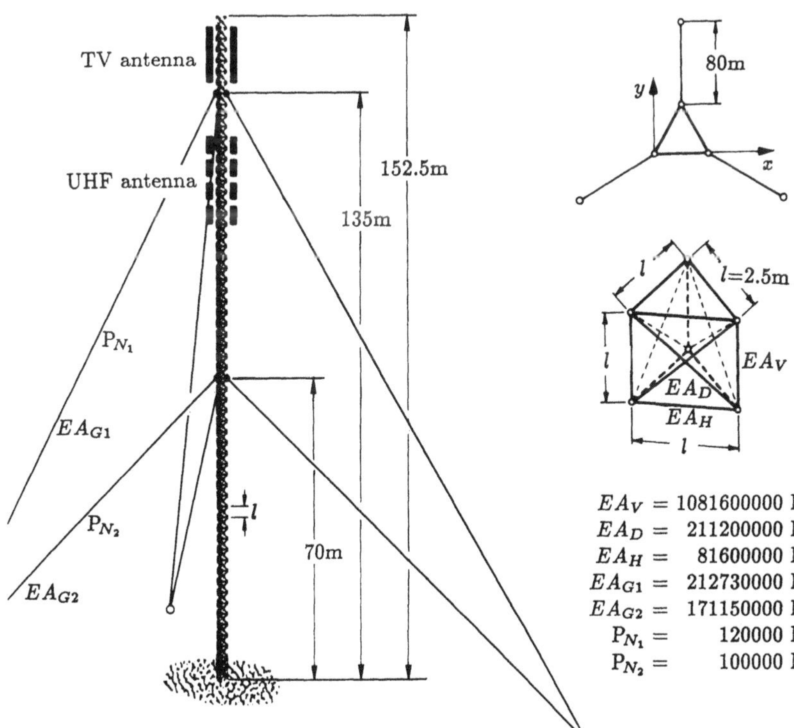

Fig. 11.4.20 Transmitter tower, geometry and idealisation resulting in 549 degrees of freedom

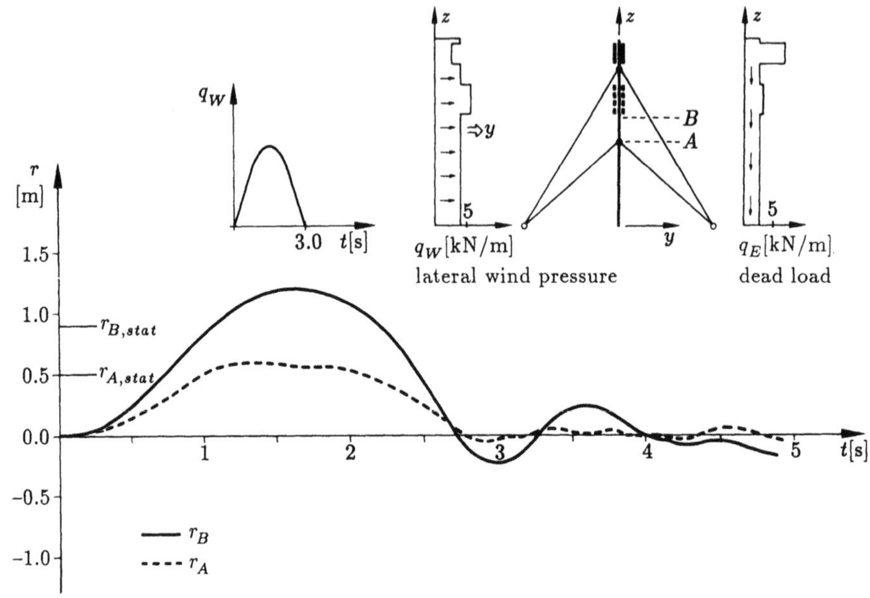

Fig. 11.4.21 Transmitter tower under wind loading; transient response of cross sections A and B

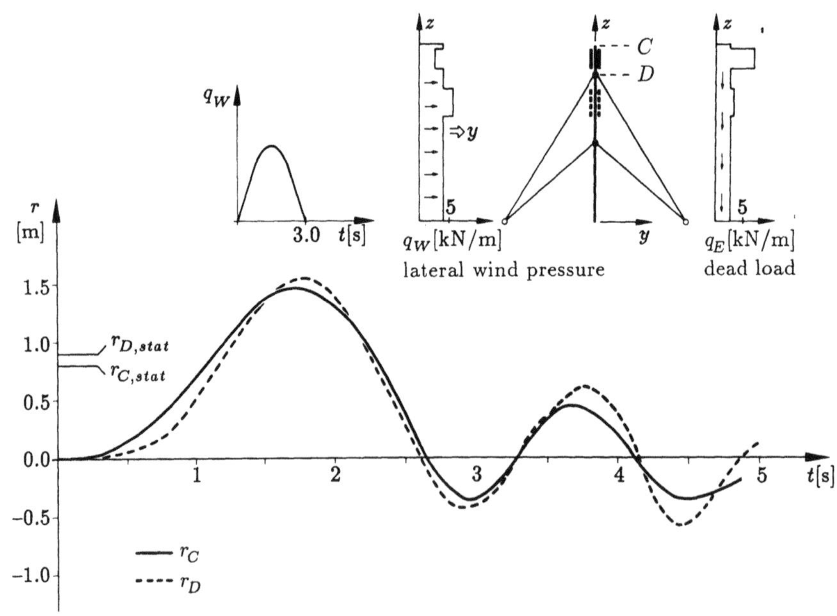

Fig. 11.4.22 Transmitter tower under wind loading; transient response of cross sections C and D

Transmitter tower under earthquake loading

As a (rough) simulation of an earthquake, the base of the tower including the cable supports is subjected to a horizontal acceleration in the form of a temporal sine wave, the amplitude of the acceleration being $g/2$ (=half the acceleration due to gravity). Some specific provisions must be made when considering the motion of the base. We first establish that we are working in an inertial system, so that the inertia forces are, as usual, expressed by $\boldsymbol{M}\ddot{\boldsymbol{r}}$, where $\boldsymbol{r}$ defines the motion with respect to the inertial system. In the present example, however, the base also moves. Thus, we may not suppress the base points as in the preceding calculation, but prescribe them within $\boldsymbol{r}$ as a subgroup (prescribed displacements $\boldsymbol{r}_u$). Thus, the total displacement vector $\boldsymbol{r}$ consists of the unknown displacements $\boldsymbol{r}_\sigma$ and the prescribed displacements $\boldsymbol{r}_u$, where the latter and their derivatives with respect to time, $\dot{\boldsymbol{r}}_u, \ddot{\boldsymbol{r}}_u, \dddot{\boldsymbol{r}}_u$, are assumed to be known

$$\boldsymbol{r} = \{\boldsymbol{r}_\sigma \quad \boldsymbol{r}_u\} \tag{11.4.34}$$

The predictor calculation proceeds now only in terms of $\boldsymbol{r}_\sigma$, while $\boldsymbol{r}_u$ is incorporated explicitly into the corrector operations. In place of equ. (11.4.23), we operate with a system of dynamic equilibrium relations split into hyperrows for $\boldsymbol{r}_\sigma$ and $\boldsymbol{r}_u$

$$\begin{bmatrix} \boldsymbol{M}_{\sigma\sigma} & \boldsymbol{M}_{\sigma u} \\ \boldsymbol{M}_{u\sigma} & \boldsymbol{M}_{uu} \end{bmatrix} \begin{bmatrix} \ddot{\boldsymbol{r}}_\sigma^c \\ \ddot{\boldsymbol{r}}_u \end{bmatrix}_{\mathrm{k}+1} = \begin{bmatrix} \boldsymbol{R}_\sigma \\ \boldsymbol{R}_u \end{bmatrix}_{\mathrm{k}} - \begin{bmatrix} \boldsymbol{C}_{\sigma\sigma} & \boldsymbol{C}_{\sigma u} \\ \boldsymbol{C}_{u\sigma} & \boldsymbol{C}_{uu} \end{bmatrix} \begin{bmatrix} \dot{\boldsymbol{r}}_\sigma \\ \dot{\boldsymbol{r}}_u \end{bmatrix}_{\mathrm{k}+1} - \begin{bmatrix} \boldsymbol{R}_{e\sigma|\mathrm{k}+1} \\ \boldsymbol{R}_{eu|\mathrm{k}+1} \end{bmatrix} \tag{11.4.35}$$

The first hyperrow yields a modified corrector for $\ddot{\boldsymbol{r}}_\sigma$

$$\ddot{\boldsymbol{r}}_{\sigma\mathrm{k}+1}^c = \boldsymbol{M}_{\sigma\sigma}^{-1}[\boldsymbol{R}_\sigma - \boldsymbol{M}_{\sigma u}\ddot{\boldsymbol{r}}_{u\mathrm{k}+1} - \boldsymbol{C}_{\sigma u}\dot{\boldsymbol{r}}_{u\mathrm{k}+1} - \boldsymbol{C}_{\sigma\sigma}\dot{\boldsymbol{r}}_{\sigma\mathrm{k}+1} - \boldsymbol{R}_{e\sigma|\mathrm{k}+1}] \tag{11.4.36}$$

Here, for typographical brevity, we have again introduced $\boldsymbol{R}_{e\sigma|\mathrm{k}+1} = \boldsymbol{R}_{e\sigma}(\boldsymbol{r}_{\mathrm{k}+1})$ etc.

The differentiation of this equation with respect to time leads by analogy to the transition from equ. (11.4.23) to equ. (11.4.24) to the supplementary corrector expression

$$\dddot{\boldsymbol{r}}_{\sigma\mathrm{k}+1}^c = \boldsymbol{M}_{\sigma\sigma}^{-1}[\dot{\boldsymbol{R}}_\sigma - \boldsymbol{M}_{\sigma u}\dddot{\boldsymbol{r}}_{u\mathrm{k}+1} - \boldsymbol{C}_{\sigma u}\ddot{\boldsymbol{r}}_{u\mathrm{k}+1} - \boldsymbol{K}_{t\sigma u|\mathrm{k}+1}\dot{\boldsymbol{r}}_{u\mathrm{k}+1} - \boldsymbol{K}_{t\sigma\sigma|\mathrm{k}+1}\dot{\boldsymbol{r}}_{\sigma\mathrm{k}+1}] \tag{11.4.37}$$

where

$$\boldsymbol{K}_{t\sigma u} = \frac{\partial \boldsymbol{R}_{e\sigma}}{\partial \boldsymbol{r}_u} \tag{11.4.38}$$

$$\boldsymbol{K}_{t\sigma\sigma} = \frac{\partial \boldsymbol{R}_{e\sigma}}{\partial \boldsymbol{r}_\sigma} \tag{11.4.39}$$

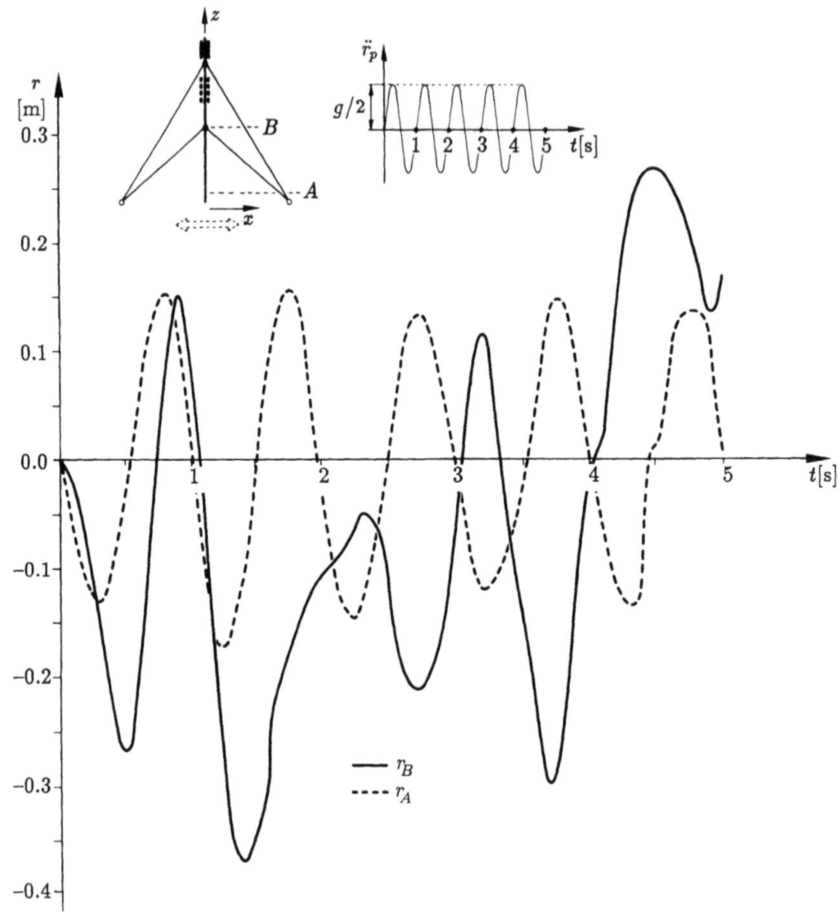

Fig. 11.4.23 Transmitter tower subject to an earthquake; time-history response of cross sections A and B

are submatrices of the total tangential stiffness matrix corresponding to the displacements $\boldsymbol{r}_\sigma$ or $\boldsymbol{r}_u$

$$\boldsymbol{K}_t = \begin{bmatrix} \boldsymbol{K}_{\sigma\sigma} & \boldsymbol{K}_{\sigma u} \\ \boldsymbol{K}_{u\sigma} & \boldsymbol{K}_{uu} \end{bmatrix}_t \tag{11.4.40}$$

When applying the $\boldsymbol{K}_s$-method, the procedure would in principle be the same; since it represents, however, an approximation of historical interest, it is not dealt with in detail here [11.22].

Figures 11.4.23 and 11.4.24 show the motions of the cross sections A, B, C and D of the tower. Section A oscillates as expected in phase with the motion of the ground. The calculation has to consider the possibility that the anchoring cables can be exposed computationally to compression. Consequently, these cables, idealised

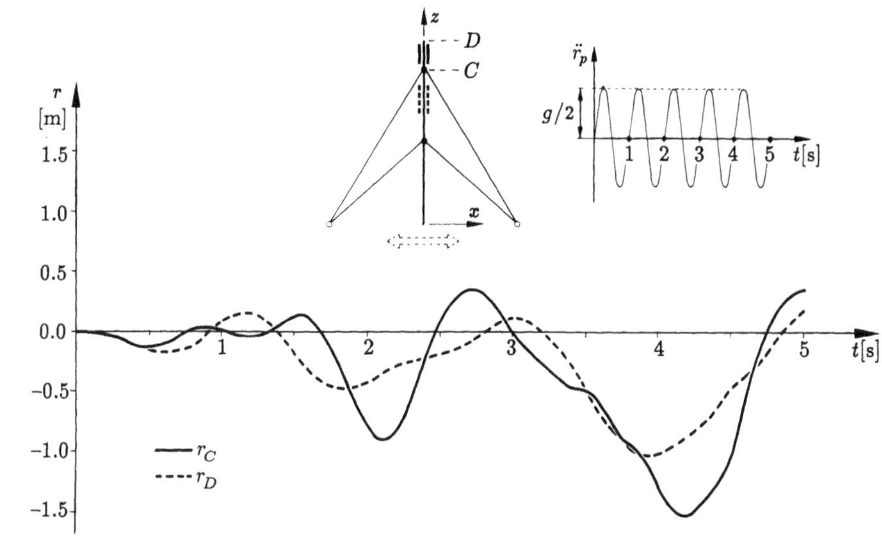

Fig. 11.4.24 Transmitter tower subject to an earthquake; time-history response of cross sections C and D

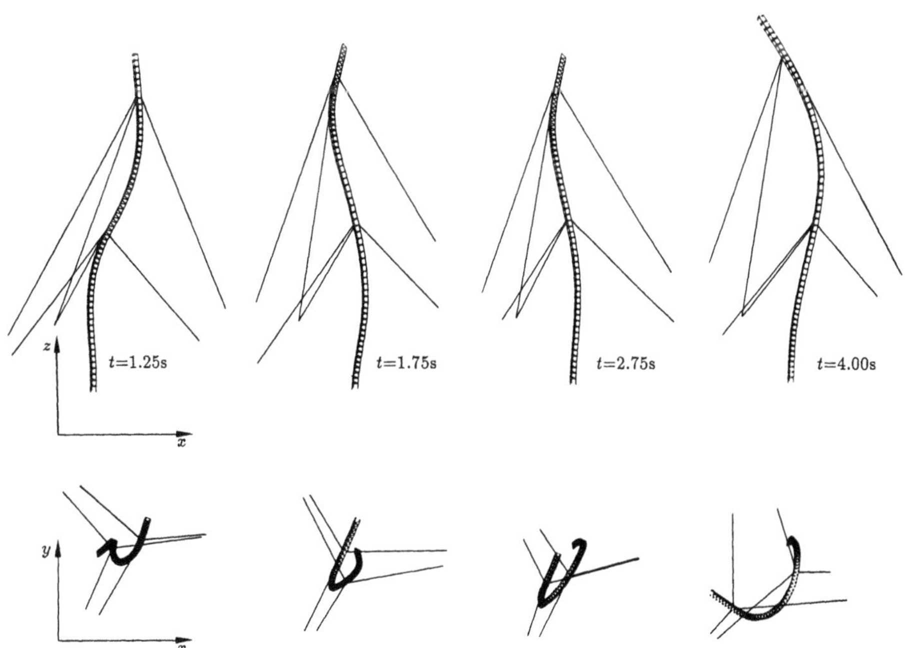

Fig. 11.4.25 Transmitter tower subject to an earthquake; deformation at four instants t

here as massless elements, have at times to be assumed inoperative if compression strains occur, but have to be reincorporated immediately into the calculation when tension re-emerges. Figure 11.4.25 clearly demonstrates the variety of deformations which take place in this problem.

Cosine impulse in an elastoplastic free-free bar [11.27]

Having considered a series of geometrically non-linear problems, we turn our attention to non-linear material behaviour and analyse the dynamic response under a cosine impulse of a bar of elastoplastic material which conforms to the Ramberg-Osgood material law and possesses kinematic hardening (fig. 11.4.26). Here, $\boldsymbol{K}_e$ must be replaced by $\boldsymbol{K}_{ep}$, the elastoplastic structural stiffness matrix discussed in Vols. III and IV of this series. In the case of the small deformations considered here, the contribution of $\boldsymbol{K}_G$ may be ignored. Figure 1.4.26 demonstrates the state of the bar after a period of 2μs for two differingly fine discretisations in time and space. We note a very good agreement and can thus limit the analysis to the coarser approximation. It should also be noted that the initial yield limit, which imposes an upper limit on the force, is already exceeded due to kinematic hardening. In fig. 11.4.27, we have also superimposed on the solution after 2μs the opposite (or antidromic) motion after 4μs run time. Moreover, by way of comparison, we show the elastic solution at 2μs and observe considerably higher forces.

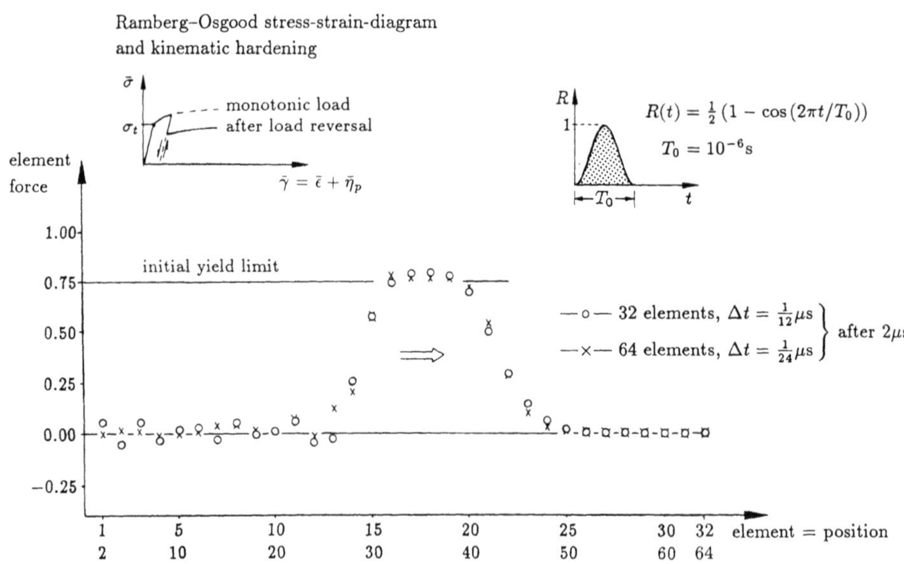

Fig. 11.4.26 Passage of a cosine impulse through an elastoplastic free-free bar; comparison of results after 2 μs for different spatial and temporal discretisations

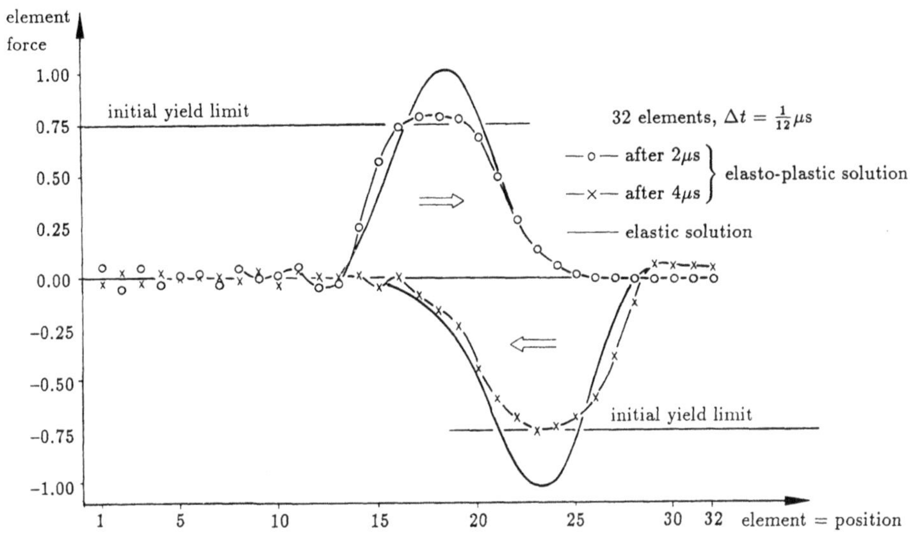

Fig. 11.4.27 Passage of a cosine impulse through an elastoplastic free-free bar; results at two different instants compared with the elastic solutions

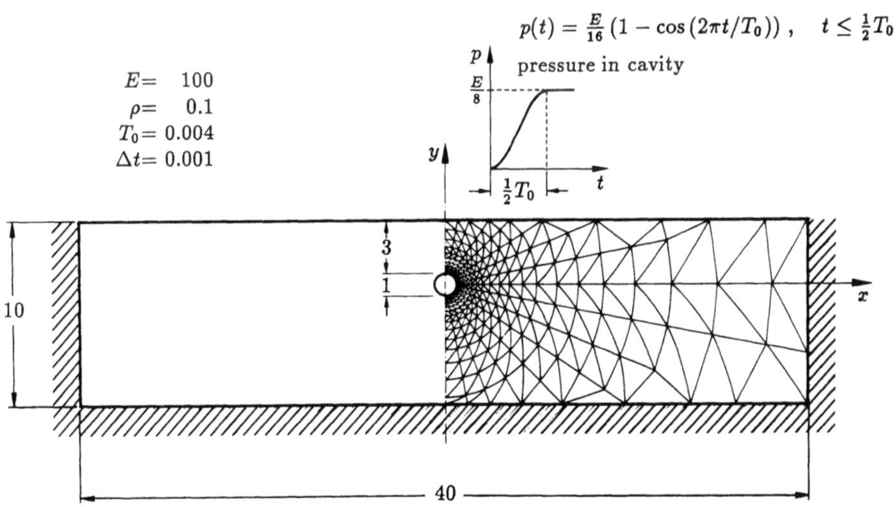

Fig. 11.4.28 Underground explosion with large strains; geometry, idealisation and data

Subterranean explosion with large strains

Figure 11.4.28 shows the idealisation of an underground area enclosing a cylindrical cavity. Adopting a simplified material law, we assume the validity of a quadratic Green strain energy. The pressure in the hollow space is taken to rise over an interval of 0.002 s in a cosine form to its ultimate peak. Figure 11.4.29 shows the motion of the point A at the top of the cavity. The upper cover is subjected to an oscillatory motion. In figure 11.4.30, the vertical displacements arising along the axis of symmetry are plotted for two instants. Note that after $t = 0.029\,\text{s}$, no reaction can be discerned on the surface, whereas after $t = 0.4\,\text{s}$, a bulging has taken place.

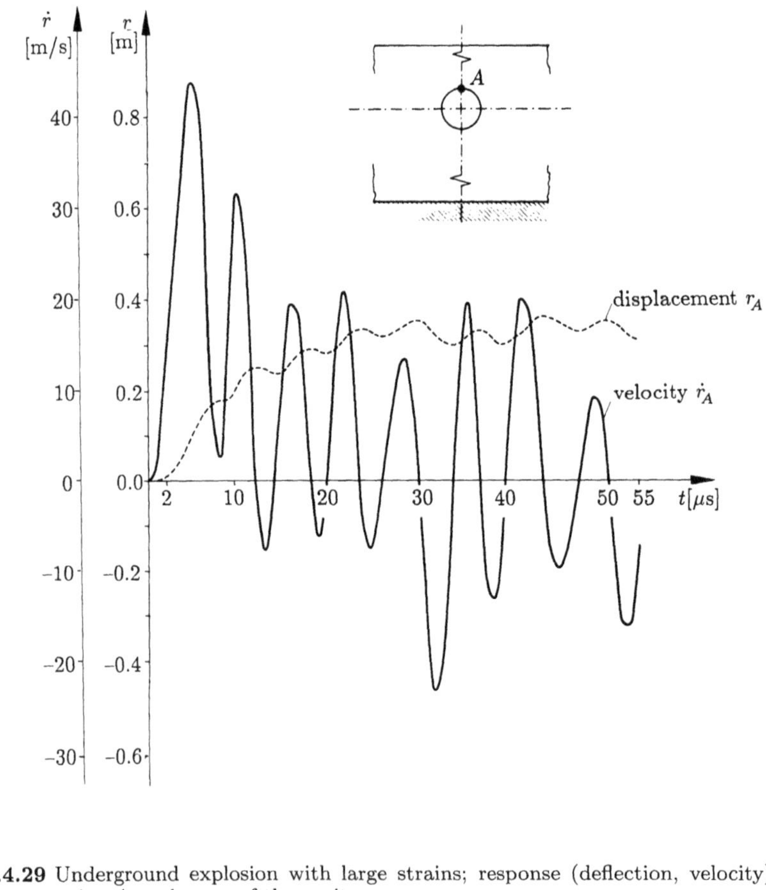

Fig. 11.4.29 Underground explosion with large strains; response (deflection, velocity) of the point A at the top of the cavity

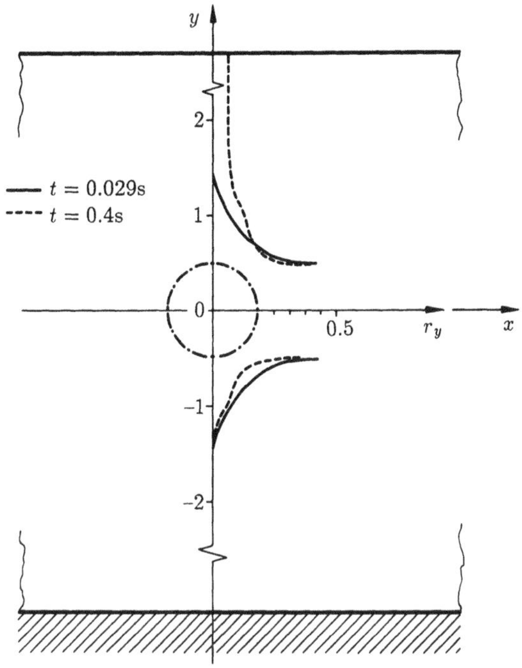

Fig. 11.4.30
Underground explosion with large strains; vertical deflections along the axis of symmetry plotted at two instants t

Calculations in celestial mechanics [11.27]

We conclude this series of examples on the application of conditionally stable algorithms with an excursion into celestial mechanics. Surprisingly enough, the field of celestial mechanics offers interesting possibilities for the application of the direct integration methods. In spite of the extraordinarily complex response of multi-body systems, the fundamental theory is essentially quite simple and has its origin in the revolutionary concept of Newton on the motion of celestial bodies, probably dating back to the year 1679 ([11.28],[11.29]). This is governed by his universal theory of gravity for mass points and postulates that every celestial object, e.g. P, possesses a positive scalar quantity, the gravitation mass, which does not depend on the position, motion or time. This gravitation mass is defined by the product of the gravitation constant γ and the inertia mass m_P of the object and has the dimension length^3/time^2. Any other celestial body, e.g. Q, is subject to an acceleration $\gamma m_P/r^2$ pointing to P, where r is the distance between P and Q. This acceleration takes place additionally and independently of whatever acceleration Q might experience in the absence of P. According to Newton's Law "Force equals mass times acceleration", it follows that two celestial bodies 1 and 2 attract one another with the force

$$\mathrm{P}_{N12} = -\gamma \frac{m_1 m_2}{l_{12}^2} \tag{11.4.41}$$

Our old friend, the natural force, is thus transferred into space. Consequently, a tangential natural stiffness arises between two celestial bodies

$$k_{Nt} = \frac{\mathrm{dP}_N}{\mathrm{d}l_{12}} = 2\gamma \frac{m_1 m_2}{l_{12}^3} \tag{11.4.42}$$

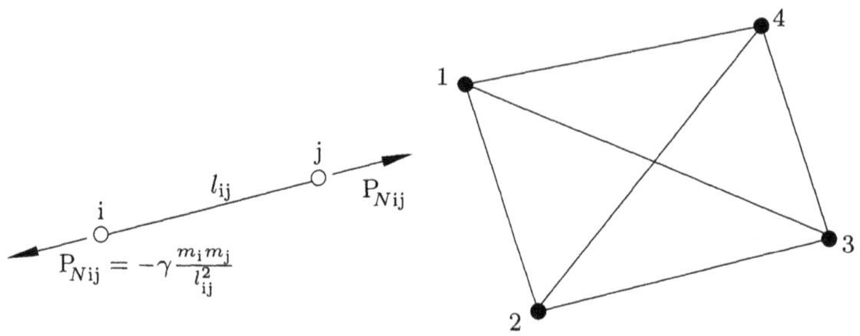

Fig. 11.4.31 Idealisation of a four-body problem with 6 fictitious non-linear bar elements

The physical characteristics of equs. (11.4.41) and (11.4.42) can be assigned to a fictitious (non-linear) FLA2 element. In a multi-body problem, we have to imagine "FLA2" elements connecting each mass point with all other mass points (fig. 11.4.31 displays this for n=4 bodies). Before applying this concept, we require a working hypothesis, since our investigations cannot account for an endless number of celestial objects. Thus, in all subsequent examples, we assume that the system in question extends over such a limited space that all inertia forces acting from outside sources are practically identical for all mass points. In fact, our analysis will then define a motion relative to the centre of gravity of the system. This in turn can itself move arbitrarily in space. For our solar system, the centre of gravity is practically located in the sun.

The celestial mechanics calculations demonstrated in fig. 11.4.32 were all executed with the high precision conditionally stable quintic Hermitian algorithm which, in contrast to the cubic Hermitian *ansatz*, has not yet been discussed in detail. The quintic procedure operates with the pivotal values $\ddot{\boldsymbol{r}}_k$, $\dddot{\boldsymbol{r}}_k$ and in addition also with $\ddddot{\boldsymbol{r}}_k$, so that in the case of two nodes, every degree of freedom is associated with six pivotal values for $\ddot{\boldsymbol{r}}$ and we can thus use a polynomial up to order 5 for the evolution of the acceleration within the time-step. The corresponding formula for the predictor can be established simply by means of a sampling procedure and the reader will be able to confirm the following expression without difficulty [11.22]

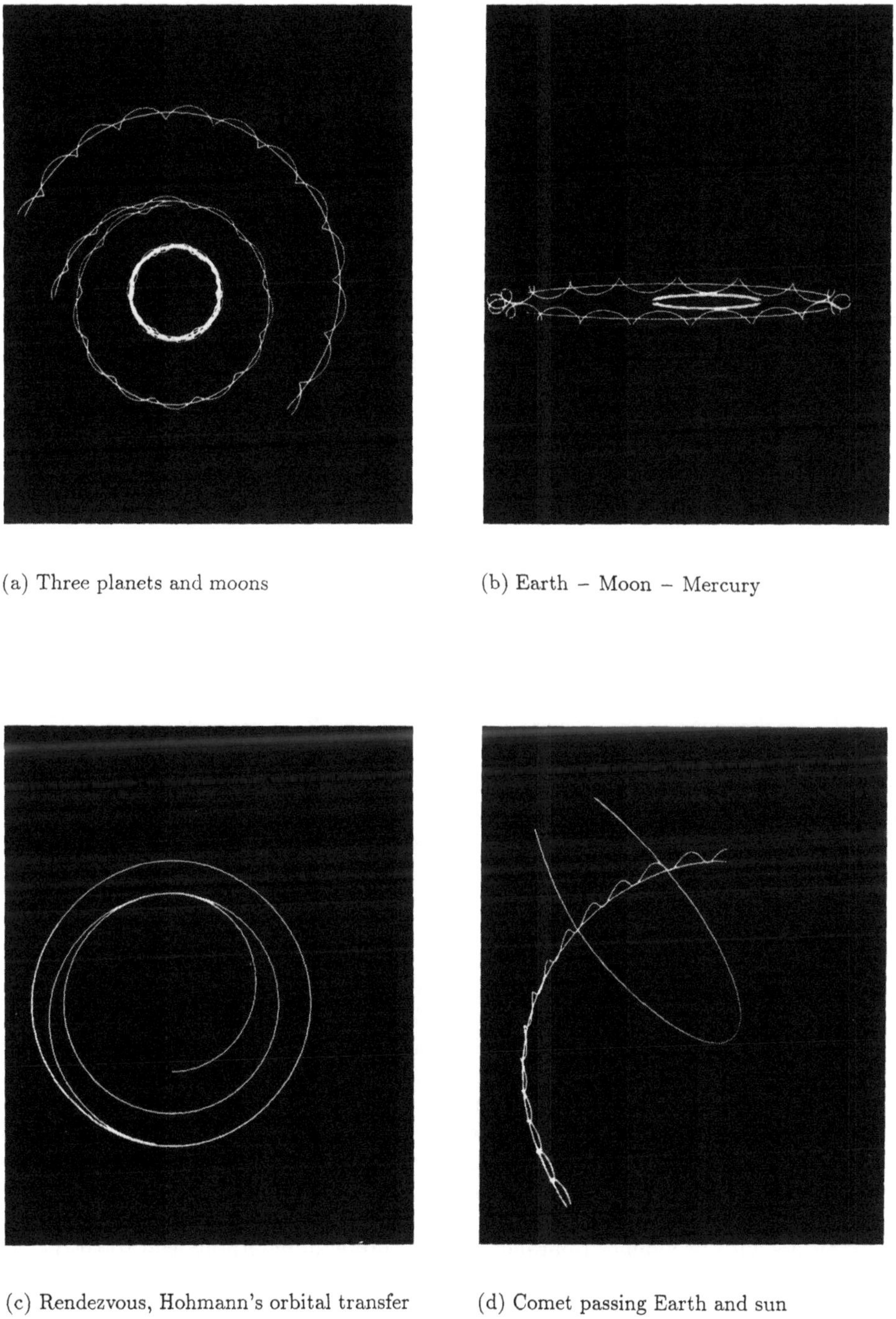

(a) Three planets and moons

(b) Earth – Moon – Mercury

(c) Rendezvous, Hohmann's orbital transfer

(d) Comet passing Earth and sun

Fig. 11.4.32 Calculations in celestial mechanics

$$\boldsymbol{r}_{\mathrm{k}+1} = \boldsymbol{r}_{\mathrm{k}} + \Delta t \dot{\boldsymbol{r}}_{\mathrm{k}} + \frac{\Delta t^2}{840}(300\ddot{\boldsymbol{r}}_{\mathrm{k}} + 52\Delta t \dddot{\boldsymbol{r}}_{\mathrm{k}} + 4\Delta t^2 \ddddot{\boldsymbol{r}}_{\mathrm{k}} + 120\ddot{\boldsymbol{r}}_{\mathrm{k}+1} - 32\Delta t \dddot{\boldsymbol{r}}_{\mathrm{k}+1} + 3\Delta t^2 \ddddot{\boldsymbol{r}}_{\mathrm{k}+1}) \tag{11.4.43}$$

$$\dot{\boldsymbol{r}}_{\mathrm{k}+1} = \dot{\boldsymbol{r}} + \frac{\Delta t}{120}(60\ddot{\boldsymbol{r}}_{\mathrm{k}} + 6\Delta t \dddot{\boldsymbol{r}}_{\mathrm{k}} + \Delta t^2 \ddddot{\boldsymbol{r}}_{\mathrm{k}} + 60\ddot{\boldsymbol{r}}_{\mathrm{k}+1} - 6\Delta t \dddot{\boldsymbol{r}}_{\mathrm{k}+1} + \Delta t^2 \ddddot{\boldsymbol{r}}_{\mathrm{k}+1}) \tag{11.4.44}$$

The associated corrector derives from the equilibrium at the end of the time-step and its first and second derivatives with respect to time. Thus,

$$\boldsymbol{M}\ddot{\boldsymbol{r}}_{\mathrm{k}+1} + \boldsymbol{C}\dot{\boldsymbol{r}}_{\mathrm{k}+1} + \boldsymbol{R}_{\mathrm{e}|\mathrm{k}+1} = \boldsymbol{R}_{\mathrm{k}+1} \tag{11.4.45}$$

$$\boldsymbol{M}\dddot{\boldsymbol{r}}_{\mathrm{k}+1} + \boldsymbol{C}\ddot{\boldsymbol{r}}_{\mathrm{k}+1} + \dot{\boldsymbol{R}}_{\mathrm{e}|\mathrm{k}+1} = \dot{\boldsymbol{R}}_{\mathrm{k}+1} \tag{11.4.46}$$

$$\boldsymbol{M}\ddddot{\boldsymbol{r}}_{\mathrm{k}+1} + \boldsymbol{C}\dddot{\boldsymbol{r}}_{\mathrm{k}+1} + \ddot{\boldsymbol{R}}_{\mathrm{e}|\mathrm{k}+1} = \ddot{\boldsymbol{R}}_{\mathrm{k}+1} \tag{11.4.47}$$

Here, $\boldsymbol{R}_e$ is deduced from the stresses, and its first derivative with respect to time, $\dot{\boldsymbol{R}}_e$, results, as before, from equ. (11.4.20). From this, $\ddot{\boldsymbol{R}}_e$ is immediately obtained

$$\ddot{\boldsymbol{R}}_{\mathrm{e}|\mathrm{k}+1} = \boldsymbol{K}_{t|\mathrm{k}+1}\ddot{\boldsymbol{r}}_{\mathrm{k}+1} + \dot{\boldsymbol{K}}_{t|\mathrm{k}+1}\dot{\boldsymbol{r}}_{\mathrm{k}+1} \tag{11.4.48}$$

While the determination of the tangential stiffness $\boldsymbol{K}_t$ represents a standard procedure in non-linear calculations, the derivation with respect to time of this matrix, $\dot{\boldsymbol{K}}_t$, naturally involves unpleasant additional effort. In the above calculations in celestial mechanics, no damping forces arise, of course, in equs. (11.4.45) to (11.4.47). All computations were performed with the conditionally stable Hermitian algorithm of the fifth order [11.22].

Summary

Our examples have demonstrated that non-linear dynamic calculations may be complicated, but are certainly practicable for problems with many degrees of freedom. The iterative solution process within a step has proved itself to be a most convenient tool. The price that has to be paid lies in a reduction of the time-step Δt by 60% in relation to the step required for the stability of the algorithm. This is advantageous to achieve a fast convergence within a step. The availability of super-computers like the Cray 2 and it successors and the continuous increase in the size of the central memory enable the practical solution of problems in non-linear dynamics with an ever-increasing size and complexity. The advent of computers with pronounced parallel architecture will inevitably lead to the generation of novel algorithmic and software concepts.

11.5 Time integration with large time-steps

11.5.1 Stability in non-linear dynamics

Let us now consider a non-linear dynamic system as exemplified by equ. (11.2.9) and attempt to apply formally the criteria of stability established in section 10.3 to a non-linear regime, using to this purpose the Newmark method. We observe that the predictor of equs. (11.4.1) and (11.4.2) remains completely unaffected in comparison with the linear case, whilst the corrector of equs. (11.4.3) or (11.4.4) can best be expressed in a linearised form as

$$\boldsymbol{M}\mathrm{d}\ddot{\boldsymbol{r}} + \boldsymbol{C}\mathrm{d}\dot{\boldsymbol{r}} + \boldsymbol{K}_t(\boldsymbol{r}_\mathrm{k})\mathrm{d}\boldsymbol{r} = \boldsymbol{R}(t_\mathrm{k}) - \boldsymbol{R}_{e|\mathrm{k}} \tag{11.5.1}$$

Assuming that the tangential stiffness does not change substantially within a step, equ. (11.5.1) is seen to apply approximately for the finite increments $\Delta\boldsymbol{r}, \Delta\dot{\boldsymbol{r}}, \Delta\ddot{\boldsymbol{r}}$ of a step. This is formally identical to the linear case; only the elastic stiffness matrix $\boldsymbol{K}$ must be replaced by the tangential stiffness $\boldsymbol{K}_t$ of the system. This has all been considered in a more general framework at the beginning of section 10.3. In order to assess the stability and accuracy in the case of linear systems, the eigenfrequencies of the structure were found to be relevant. For example, the maximum eigenfrequency determined the time-step in the case of conditionally stable algorithms. Equally, the linearised equilibrium equation (11.5.1) can be subjected to an eigenvalue analysis which, *inter alia*, yields the greatest eigenfrequency for small oscillations about $\boldsymbol{r}_\mathrm{k}$. With this information, it is possible to fix an upper limit to the time-step. Formally, this sounds quite simple, but is in fact complicated since the system frequencies vary with time. As a result, the tangential stiffness can only be assumed to remain constant over a few steps. In contrast to the linear case, a further analysis of the natural modes of vibration must then be carried out, or we have at least to obtain the upper and lower limits for the maximum eigenfrequency during the time-history response calculation. But what about the popular and unconditionally stable Newmark method ($\beta = 1/4, \gamma = 1/2$)? It would appear that it offers an ideal tool for non-linear problems since in this case, at least as far as the stability is concerned, there is no time-step limit. Unfortunately, however, this is jumping to conclusions, since equ. (11.5.1) can only represent an approximation for finite increments. The resulting conditions for stability prove, of course, necessary, but are sufficient in linear cases only. The observation that the unconditionally stable Newmark method ($\beta = 1/4, \gamma = 1/2$) is only conditionally stable in the non-linear case is a consequence of this argument ([11.27], [11.30]). Thus, the question arises whether a new, more powerful criterion can be found for non-linear analysis which could then lead to the development of algorithms for large time-steps. In [11.27], [11.30], [11.31], it is proposed to use the energy balance of dynamic systems as a suitable criterion. In a dynamic process, energy in the form of external work W is fed into the system

$$W = \int_0^{\boldsymbol{r}} \boldsymbol{R}^\mathrm{t}\mathrm{d}\boldsymbol{r} \tag{11.5.2}$$

and kinetic energy T

$$T = \frac{1}{2}\dot{\boldsymbol{r}}^{\mathrm{t}}\boldsymbol{M}\dot{\boldsymbol{r}} \tag{11.5.3}$$

and strain energy U

$$U = \iint\limits_{V0}^{\boldsymbol{\varepsilon}} \boldsymbol{\sigma}^{\mathrm{t}}\delta\boldsymbol{\varepsilon}\mathrm{d}V \tag{11.5.4}$$

are stored. We now stipulate that at an instant $t_1 \geqslant t_0$, the total energy $T+U-W$ does not exceed that at t_0

$$(T + U - W)_{t=t_1} \leqslant (T + U - W)_{t=t_0} \tag{11.5.5}$$

Were the total energy greater, then an inciting source of energy would have remained unnoticed in the process. An alternative formulation of equ. (11.5.5) is also

$$\Delta T \leqslant \Delta W - \Delta U \tag{11.5.6}$$

i.e. the increase in kinetic energy between t_0 and t_1 must be smaller than or equal to the work performed less the increase in strain energy.

We now firstly examine whether the Newmark algorithm ($\beta = 1/4, \gamma = 1/2$), which is unconditionally stable in the linear case and which according to literature may lose this characteristic in the non-linear case, satisfies this additional requirement or not. According to equs. (11.4.1) and (11.4.2), we obtain

$$\boldsymbol{r}_{\mathrm{k}+1} = \boldsymbol{r}_{\mathrm{k}} + \Delta t\dot{\boldsymbol{r}}_{\mathrm{k}} + \tfrac{1}{4}\Delta t^2[\ddot{\boldsymbol{r}}_{\mathrm{k}} + \ddot{\boldsymbol{r}}_{\mathrm{k}+1}] \tag{11.5.7}$$

$$\dot{\boldsymbol{r}}_{\mathrm{k}+1} = \dot{\boldsymbol{r}}_{\mathrm{k}} \qquad\qquad + \tfrac{1}{2}\Delta t[\ddot{\boldsymbol{r}}_{\mathrm{k}} + \ddot{\boldsymbol{r}}_{\mathrm{k}+1}] \tag{11.5.8}$$

and, with the help of equ. (11.5.8), equ. (11.5.7) can also be written as follows

$$\boldsymbol{r}_{\mathrm{k}+1} = \boldsymbol{r}_{\mathrm{k}} + \tfrac{1}{2}\Delta t[\dot{\boldsymbol{r}}_{\mathrm{k}} + \dot{\boldsymbol{r}}_{\mathrm{k}+1}] \tag{11.5.9}$$

With respect to this formulation, one speaks of a centre step algorithm. We now pre-multiply equ. (11.5.8) firstly by the mass matrix and then by $[\boldsymbol{r}_{\mathrm{k}+1} - \boldsymbol{r}_{\mathrm{k}}]^{\mathrm{t}}$

$$[\boldsymbol{r}_{\mathrm{k}+1} - \boldsymbol{r}_{\mathrm{k}}]^{\mathrm{t}}\boldsymbol{M}[\dot{\boldsymbol{r}}_{\mathrm{k}+1} - \dot{\boldsymbol{r}}_{\mathrm{k}}] = \tfrac{1}{2}\Delta t[\boldsymbol{r}_{\mathrm{k}+1} - \boldsymbol{r}_{\mathrm{k}}]^{\mathrm{t}}\boldsymbol{M}[\ddot{\boldsymbol{r}}_{\mathrm{k}} + \ddot{\boldsymbol{r}}_{\mathrm{k}+1}] \tag{11.5.10a}$$

We next substitute equ. (11.5.9) into the left-hand side and can thus eliminate the time increment Δt. Taking the symmetry of the mass matrix into consideration, we find

$$\tfrac{1}{2}[\dot{\boldsymbol{r}}_{\mathrm{k}+1}^{\mathrm{t}}\boldsymbol{M}\dot{\boldsymbol{r}}_{\mathrm{k}+1} - \dot{\boldsymbol{r}}_{\mathrm{k}}^{\mathrm{t}}\boldsymbol{M}\dot{\boldsymbol{r}}_{\mathrm{k}}] = \tfrac{1}{2}[\boldsymbol{r}_{\mathrm{k}+1} - \boldsymbol{r}_{\mathrm{k}}]^{\mathrm{t}}\boldsymbol{M}[\ddot{\boldsymbol{r}}_{\mathrm{k}} + \ddot{\boldsymbol{r}}_{\mathrm{k}+1}] \tag{11.5.10b}$$

The left-hand side of the equation records exactly the increase in kinetic energy in the step

$$\Delta T = \tfrac{1}{2}\left[\dot{\boldsymbol{r}}^{\mathrm{t}}_{\mathrm{k+1}}\boldsymbol{M}\dot{\boldsymbol{r}}_{\mathrm{k+1}} - \dot{\boldsymbol{r}}^{\mathrm{t}}_{\mathrm{k}}\boldsymbol{M}\dot{\boldsymbol{r}}_{\mathrm{k}}\right] \tag{11.5.11}$$

If we restrict ourselves for the moment to a free undamped non-linear oscillator, then

$$\boldsymbol{M}\ddot{\boldsymbol{r}} = -\boldsymbol{R}_e \tag{11.5.12}$$

also applies to the inertia forces on the right-hand side of equ. (11.5.10b), and thus

$$\Delta T = -\tfrac{1}{2}\Delta\boldsymbol{r}^{\mathrm{t}}\left[\boldsymbol{R}_{e|\mathrm{k+1}} + \boldsymbol{R}_{e|\mathrm{k}}\right] \tag{11.5.13}$$

Figure 11.5.1 illustrates the significance of the right-hand side for a single-degree-of-freedom system. In the linear case, the right-hand side is obviously identical with ΔU and the condition of equ. (11.5.6) is thus fulfilled. In the case of a hardening stiffness, the load deflection diagram indicates that the trapezoid area described by the right-hand side of equ. (11.4.13) is, however, greater than the area below the load displacement curve which defines the increase in strain energy

$$\Delta U < \tfrac{1}{2}\Delta\boldsymbol{r}^{\mathrm{t}}\left[\boldsymbol{R}_{e|\mathrm{k+1}} + \boldsymbol{R}_{e|\mathrm{k}}\right] \tag{11.5.14}$$

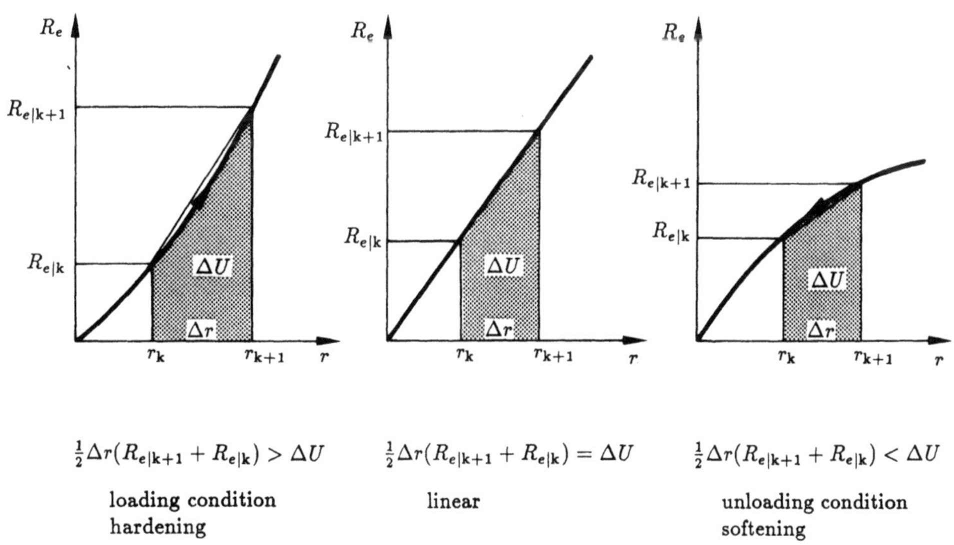

Fig. 11.5.1 Interpretation of the right-hand side in equ. (11.5.14)

The reverse condition holds in the case of unloading and a softening stiffness. In these cases, the requirement of equ. (11.5.6) is thus not fulfilled. In the course of consecutive loading and unloading operations, a certain compensation of the energy differences may occur, but even then, a limitation of the kinetic energy is not secured. Such a pathological energy increase has indeed been observed in particular cases. For non-linear cases, we consequently have to develop algorithms which retain the energy bounds of equ. (11.5.6) as also the fundamental equations in continuum mechanics prescribe. As a result, we establish at the same time procedures which allow large time-steps.

11.5.2 Modified unconditionally stable Newmark algorithm

The discrepancy registered in the energy balance of sub-section 11.5.1 is a result of the difference between the actual area below the load displacement curve and the trapezoid area, associated with the average value of the forces. We may conjecture that this discrepancy can be removed by a suitably defined mean value of the forces. For an undamped non-linear oscillator under excitation, we write

$$\boldsymbol{M}\ddot{\boldsymbol{r}} = \boldsymbol{R} - \boldsymbol{R}_e \tag{11.5.15}$$

Let $\overline{\boldsymbol{R}}$ be, in an energetical sense, such a "suitable" average value of the external forces in the time-step Δt, and $\overline{\boldsymbol{R}}_e$ that of the elastic forces. We introduce these improved average values into equ. (11.5.8) and obtain

$$\dot{\boldsymbol{r}}_{\mathrm{k}+1} = \dot{\boldsymbol{r}}_{\mathrm{k}} + \Delta t \boldsymbol{M}^{-1}[\overline{\boldsymbol{R}} - \overline{\boldsymbol{R}}_e] \tag{11.5.16}$$

As to equ. (11.5.9), it is kept unchanged and forms, together with equ. (11.5.16), the basis of the procedure. The relationship corresponding to equ. (11.5.13) for kinetic energy here is

$$\Delta T = \Delta \boldsymbol{r}^{\mathrm{t}}[\overline{\boldsymbol{R}} - \overline{\boldsymbol{R}}_e] \tag{11.5.17}$$

and reverts to the former for $\boldsymbol{R} = \boldsymbol{o}$ and $\overline{\boldsymbol{R}}_e = \frac{1}{2}[\boldsymbol{R}_{e|\mathrm{k}} + \boldsymbol{R}_{e|\mathrm{k}+1}]$. In order to set up the algorithm, we must specify precisely the adopted average value of the forces $\overline{\boldsymbol{R}}$ and $\overline{\boldsymbol{R}}_e$. In fact, we stipulate for the elastic forces

$$\begin{aligned} \Delta \boldsymbol{r}^{\mathrm{t}} \overline{\boldsymbol{R}}_e &= \int\limits_{\boldsymbol{r}_{\mathrm{k}}}^{\boldsymbol{r}_{\mathrm{k}+1}} \boldsymbol{R}_e^{\mathrm{t}} \mathrm{d}\boldsymbol{r} \\ &= \int\limits_{\boldsymbol{r}_{\mathrm{k}}}^{\boldsymbol{r}_{\mathrm{k}+1}} \left[\boldsymbol{R}_{e|\mathrm{k}} + \int\limits_{\boldsymbol{r}_{\mathrm{k}}}^{\boldsymbol{r}_{\mathrm{k}+1}} \boldsymbol{K}_t(\boldsymbol{r})\mathrm{d}\boldsymbol{r}\right]^{\mathrm{t}} \mathrm{d}\boldsymbol{r} \\ &= \Delta \boldsymbol{r}^{\mathrm{t}} \boldsymbol{R}_{e|\mathrm{k}} + \int\limits_{\boldsymbol{r}_{\mathrm{k}}}^{\boldsymbol{r}_{\mathrm{k}+1}} \int\limits_{\boldsymbol{r}_{\mathrm{k}}}^{\boldsymbol{r}_{\mathrm{k}+1}} \mathrm{d}\boldsymbol{r}^{\mathrm{t}} \boldsymbol{K}_t(\boldsymbol{r})\, \mathrm{d}\boldsymbol{r} \end{aligned} \tag{11.5.18}$$

where

$$K_t = \frac{\partial R_e}{\partial r} \tag{11.5.19}$$

is the known tangential stiffness of the system. The double integral in equ. (11.5.18) can now be evaluated, e.g. by means of varyingly precise Hermitian approximations. Thus, the information $R_{e|k}, K_{t|k}$ and $K_{t|k+1}$ allows a formulation which is quadratic in the displacements. In order to retain a physical clarity of the argument, we restrict our exposition to a single degree of freedom and obtain

$$R_e = R_{e|k} + K_{t|k}(r - r_k) + \frac{1}{2}\frac{K_{t|k+1} - K_{t|k}}{r_{k+1} - r_k}(r - r_k) \tag{11.5.20}$$

Correspondingly, equ. (11.5.18), first line, yields

$$\begin{aligned} \Delta r \overline{R}_e &= \Delta r R_{e|k} + \tfrac{1}{2} K_{t|k} \Delta r^2 + \tfrac{1}{6}(K_{t|k+1} - K_{t|k}) \Delta r^2 \\ &= \Delta r \left(R_{e|k} + \tfrac{1}{3} K_{t|k} \Delta r + \tfrac{1}{6} K_{t|k+1} \Delta r \right) \end{aligned} \tag{11.5.21}$$

or, upon a reversion to n degrees of freedom,

$$\overline{R}_e = R_{e|k} + \left[\tfrac{1}{3} K_{t|k} + \tfrac{1}{6} K_{t|k+1}\right] \Delta r \tag{11.5.22}$$

Higher approximations of the actual course of R_e within the step are conceivable if derivatives of the tangential stiffness are taken into consideration. Such derivatives proved necessary, for example, in the case of the quintic Hermitian algorithm, and they naturally increase the complexity considerably; we refer to the original publication [11.22]. The external work W does not constitute a critical quantity in the energy balance. Errors in the energy supply lead to neighbouring solution paths but certainly not to an unrestrained increase in kinetic energy. For this reason, the mean value

$$\overline{R} = \tfrac{1}{2}[R_k + R_{k+1}] \tag{11.5.23}$$

can certainly be used and, as an approximation of a higher order, the simple formula

$$\overline{R} = \tfrac{1}{2}[R_k + R_{k+1}] + \tfrac{1}{12}\Delta t[\dot{R}_k - \dot{R}_{k+1}] \tag{11.5.24}$$

can be adopted. Here, the average taken with respect to the deformation increment is replaced by the average value evaluated over the time increment Δt. Equations (11.5.22) and (11.5.24) complete the argument and we now deduce from equ. (11.5.16)

$$\dot{r}_{k+1} = \dot{r}_k - \tfrac{1}{2}\Delta t M^{-1}\left[R_{e|k} + \left[\tfrac{1}{3}K_{t|k} + \tfrac{1}{6}K_{t|k+1}\right][r_{k+1} - r_k]\right] + \tfrac{1}{2}\Delta t M^{-1}[R_k + R_{k+1}] \tag{11.5.25}$$

In conjunction with equ. (11.5.9), equ. (11.5.25) forms the foundation of an algorithm which, in the absence of external forces and for quadratically non-linear problems, is energy conserving and thus in a more precise sense unconditionally stable.

11.5.3 Modified unconditionally stable cubic Hermitian algorithm [11.27]

In section 10.4, we already discussed a modified cubic algorithm which proved unconditionally stable in the linear case. The necessary modifications are recorded in the predictor equs. (10.4.40) and (10.4.41). We start from these equations in our adaptation to non-linear problems. From equ. (10.4.40), we obtain

$$\tfrac{1}{12}\Delta t^2[\ddot{\boldsymbol{r}}_{\mathrm{k}} - \ddot{\boldsymbol{r}}_{\mathrm{k+1}}] = \dot{\boldsymbol{r}}_{\mathrm{k+1}} - \dot{\boldsymbol{r}}_{\mathrm{k}} - \tfrac{1}{2}\Delta t[\ddot{\boldsymbol{r}}_{\mathrm{k}} + \ddot{\boldsymbol{r}}_{\mathrm{k+1}}] \tag{11.5.26}$$

Substituting this into equ. (10.4.41), we find

$$\begin{aligned}\boldsymbol{r}_{\mathrm{k+1}} &= \boldsymbol{r}_{\mathrm{k}} + \Delta t\dot{\boldsymbol{r}}_{\mathrm{k}} + \tfrac{1}{12}\Delta t^2[4\ddot{\boldsymbol{r}}_{\mathrm{k}} + 2\ddot{\boldsymbol{r}}_{\mathrm{k+1}}] + \tfrac{1}{2}\Delta t[\dot{\boldsymbol{r}}_{\mathrm{k+1}} - \dot{\boldsymbol{r}}_{\mathrm{k}}] - \tfrac{1}{4}\Delta t^2[\ddot{\boldsymbol{r}}_{\mathrm{k}} + \ddot{\boldsymbol{r}}_{\mathrm{k+1}}]\\ &= \boldsymbol{r}_{\mathrm{k}} + \tfrac{1}{2}\Delta t[\dot{\boldsymbol{r}}_{\mathrm{k}} + \dot{\boldsymbol{r}}_{\mathrm{k+1}}] + \tfrac{1}{12}\Delta t^2[\ddot{\boldsymbol{r}}_{\mathrm{k}} - \ddot{\boldsymbol{r}}_{\mathrm{k+1}}]\end{aligned} \tag{11.5.27}$$

From this transformation, we observe that in the case of the unconditionally stable cubic Hermitian approximation, the displacements are evaluated in complete analogy to the velocities in equ. (10.4.41). The accelerations in equ. (11.5.27) can now be eliminated by means of equ. (11.5.15), i.e. the dynamic equilibrium. Thus,

$$\boldsymbol{r}_{\mathrm{k+1}} = \boldsymbol{r} + \tfrac{1}{2}\Delta t[\dot{\boldsymbol{r}}_{\mathrm{k}} + \dot{\boldsymbol{r}}_{\mathrm{k+1}}] + \tfrac{1}{12}\Delta t^2\boldsymbol{M}^{-1}[\boldsymbol{R}_{e|\mathrm{k+1}} - \boldsymbol{R}_{e|\mathrm{k}}] + \tfrac{1}{12}\Delta t^2\boldsymbol{M}^{-1}[\boldsymbol{R}_{\mathrm{k}} - \boldsymbol{R}_{\mathrm{k+1}}] \tag{11.5.28}$$

In order to specify the increase of the elastic forces, we use the exact secant formula

$$\boldsymbol{R}_{e|\mathrm{k+1}} = \boldsymbol{R}_{e|\mathrm{k}} + \boldsymbol{K}_s[\boldsymbol{r}_{\mathrm{k+1}} - \boldsymbol{r}_{\mathrm{k}}] \tag{11.5.29}$$

In the case of the second predictor equation (10.4.40), we also apply the dynamic equilibrium (11.5.19) and the secant formula (11.5.29). We thus obtain the two expressions

$$\boldsymbol{r}_{\mathrm{k+1}} = \boldsymbol{r}_{\mathrm{k}} + \tfrac{1}{2}\Delta t[\dot{\boldsymbol{r}}_{\mathrm{k}} + \dot{\boldsymbol{r}}_{\mathrm{k+1}}] + \tfrac{1}{12}\Delta t^2\boldsymbol{M}^{-1}\boldsymbol{K}_s[\boldsymbol{r}_{\mathrm{k+1}} - \boldsymbol{r}_{\mathrm{k}}] + \tfrac{1}{12}\Delta t^2\boldsymbol{M}^{-1}[\boldsymbol{R}_{\mathrm{k}} - \boldsymbol{R}_{\mathrm{k+1}}] \tag{11.5.30}$$

$$\begin{aligned}\dot{\boldsymbol{r}}_{\mathrm{k+1}} = \dot{\boldsymbol{r}}_{\mathrm{k}} &+ \tfrac{1}{2}\Delta t\boldsymbol{M}^{-1}[\boldsymbol{R}_{e|\mathrm{k}} + \boldsymbol{R}_{e|\mathrm{k+1}}] + \tfrac{1}{12}\Delta t^2\boldsymbol{M}^{-1}\boldsymbol{K}_s[\dot{\boldsymbol{r}}_{\mathrm{k+1}} - \dot{\boldsymbol{r}}_{\mathrm{k}}]\\ &+ \tfrac{1}{2}\Delta t\boldsymbol{M}^{-1}[\boldsymbol{R}_{\mathrm{k}} + \boldsymbol{R}_{\mathrm{k+1}}] + \tfrac{1}{12}\Delta t^2\boldsymbol{M}^{-1}[\dot{\boldsymbol{R}}_{\mathrm{k}} - \dot{\boldsymbol{R}}_{\mathrm{k+1}}]\end{aligned} \tag{11.5.31}$$

which are exactly equivalent to the relations in the original algorithm. For the free oscillator ($\boldsymbol{R} = \boldsymbol{o}$, $W = 0$), we now require that the upper bound rule of equ. (11.5.6) be fulfilled for the strain energy. We first check the algorithm and pre-multiply to this end equ. (11.5.31) by $[\boldsymbol{r}_{\mathrm{k+1}} - \boldsymbol{r}_{\mathrm{k}}]^{\mathrm{t}}\boldsymbol{M}$

$$[\boldsymbol{r}_{\mathrm{k+1}} - \boldsymbol{r}_{\mathrm{k}}]^{\mathrm{t}}\boldsymbol{M}[\dot{\boldsymbol{r}}_{\mathrm{k+1}} - \dot{\boldsymbol{r}}_{\mathrm{k}}] = \tfrac{1}{2}\Delta t[\boldsymbol{r}_{\mathrm{k+1}} - \boldsymbol{r}_{\mathrm{k}}]^{\mathrm{t}}[\boldsymbol{R}_{e|\mathrm{k}} + \boldsymbol{R}_{e|\mathrm{k+1}}] + \tfrac{1}{12}\Delta t^2[\boldsymbol{r}_{\mathrm{k+1}} - \boldsymbol{r}_{\mathrm{k}}]^{\mathrm{t}}\boldsymbol{K}_s[\dot{\boldsymbol{r}}_{\mathrm{k+1}} - \tag{11.5.32}$$

We next substitute equ. (11.5.30) for $[\boldsymbol{r}_{k+1} - \boldsymbol{r}_k]$ on the left-hand side and obtain for $\boldsymbol{R} = \boldsymbol{o}$ and the prescribed symmetry of $\boldsymbol{K}_s$ and $\boldsymbol{M}^{-1}$

$$\tfrac{1}{2}\Delta t[\dot{\boldsymbol{r}}_k + \dot{\boldsymbol{r}}_{k+1}]^t \boldsymbol{M}[\dot{\boldsymbol{r}}_{k+1} - \dot{\boldsymbol{r}}_k] + \tfrac{1}{12}\Delta t^2[\boldsymbol{r}_{k+1} - \boldsymbol{r}_k]^t \boldsymbol{K}_s[\dot{\boldsymbol{r}}_{k+1} - \dot{\boldsymbol{r}}_k]$$
$$= \tfrac{1}{2}\Delta t[\boldsymbol{r}_{k+1} - \boldsymbol{r}_k]^t[\boldsymbol{R}_{e|k} + \boldsymbol{R}_{e|k+1}] + \tfrac{1}{12}\Delta t^2[\boldsymbol{r}_{k+1} - \boldsymbol{r}_k]^t \boldsymbol{K}_s[\dot{\boldsymbol{r}}_{k+1} - \dot{\boldsymbol{r}}_k] \tag{11.5.33}$$

This immediately yields

$$\Delta T = \tfrac{1}{2}\Delta \boldsymbol{r}^t[\boldsymbol{R}_{e|k} + \boldsymbol{R}_{e|k+1}] \tag{11.5.34}$$

and thus leads us to the conclusion that the unconditionally stable cubic Hermitian algorithm for linear problems violates the rule of energy bounds in non-linear cases. At the same time, however, we can gather from this brief chain of demonstrations that the energy rule applies when

$$\tfrac{1}{2}[\boldsymbol{R}_{e|k} + \boldsymbol{R}_{e|k+1}]$$

is replaced by an average force $\overline{\boldsymbol{R}}_e$ which ensures that

$$\Delta U = \Delta \boldsymbol{r}^t \overline{\boldsymbol{R}}_e \tag{11.5.35}$$

is adhered to as closely as possible. To this purpose we proceed in exactly the same way as we did in the preceding section in connection with Newmark. Thus, as a modification of equ. (11.5.31) and retaining an unmodified equ. (11.5.30), the basic algorithm becomes

$$\dot{\boldsymbol{r}}_{k+1} = \dot{\boldsymbol{r}}_k + \Delta t \boldsymbol{M}^{-1}\overline{\boldsymbol{R}}_e + \tfrac{1}{12}\Delta t^2 \boldsymbol{M}^{-1}\boldsymbol{K}_s[\boldsymbol{r}_{k+1} - \boldsymbol{r}_k] + \boldsymbol{M}^{-1}\overline{\boldsymbol{R}} \tag{11.5.36}$$

where, interestingly enough, $\overline{\boldsymbol{R}}$ was already proposed in equ. (11.5.24) as an approximation of higher order of the loading within the time-step. If a quadratic formulation (based on $\boldsymbol{R}_{e|k}, \boldsymbol{K}_{t|k}, \boldsymbol{K}_{t|k+1}$) is chosen for $\boldsymbol{R}_e$, the corresponding average value may be deduced from equ. (11.5.22). The corresponding secant stiffness is then given by

$$\boldsymbol{K}_s = \tfrac{1}{2}[\boldsymbol{K}_{t|k} + \boldsymbol{K}_{t|k+1}] \tag{11.5.37}$$

It is, of course, possible to proceed to an approximation of higher order for $\boldsymbol{R}_e$, for example on the basis of $\boldsymbol{R}_{e|k}$ and the first and second derivatives at the interval ends, using a quartic *ansatz*. For the sake of conciseness, we demonstrate this once again, using as an example one with a single degree of freedom. Let $R_{e|k}$ as well as $R'_{e|k}, R'_{e|k+1}, R''_{e|k}, R''_{e|k+1}$ be given, where

$$R'_e = \frac{\mathrm{d}R_e}{\mathrm{d}r} = K_t \tag{11.5.38}$$

and

$$R_e'' = \frac{\mathrm{d}^2 R_e}{\mathrm{d}r^2} = \frac{\mathrm{d}K_t}{\mathrm{d}r} = K_t' \tag{11.5.39}$$

must be observed. The quartic *ansatz*

$$R_e = c_0 + c_1(r - r_\mathrm{k}) + \tfrac{1}{2}c_2(r - r_\mathrm{k})^2 + \tfrac{1}{3}c_3(r - r_\mathrm{k})^3 + \tfrac{1}{4}c_4(r - r_\mathrm{k})^4 \tag{11.5.40}$$

immediately yields by spot checks at sampling values of r

$$c_0 = R_{e|\mathrm{k}} \tag{11.5.41}$$

$$c_1 = K_{t|\mathrm{k}} \tag{11.5.42}$$

$$c_2 = K'_{t|\mathrm{k}} \tag{11.5.43}$$

and, with somewhat more effort,

$$c_3 = \ 3(K_{t|\mathrm{k}+1} - K_{t|\mathrm{k}})/\Delta r^2 - (2K'_{t|\mathrm{k}} + K'_{t|\mathrm{k}+1})/\Delta r \tag{11.5.44}$$

$$c_4 = -2(K_{t|\mathrm{k}+1} - K_{t|\mathrm{k}})/\Delta r^3 + (K'_{t|\mathrm{k}} + K'_{t|\mathrm{k}+1})/\Delta r^2 \tag{11.5.45}$$

where

$$\Delta r = r_{\mathrm{k}+1} - r_\mathrm{k} \tag{11.5.46}$$

We can now establish the secant stiffness as

$$\begin{aligned} K_s &= \frac{R_{e|\mathrm{k}+1} - R_{e|\mathrm{k}}}{\Delta r} = c_1 + \tfrac{1}{2}c_2\Delta r + \tfrac{1}{3}c_3\Delta r^2 + \tfrac{1}{4}c_4\Delta r^3 \\ &= \tfrac{1}{2}(K_{t|\mathrm{k}} + K_{t|\mathrm{k}+1}) + \tfrac{1}{12}(K'_{t|\mathrm{k}} - K'_{t|\mathrm{k}+1})\Delta r \end{aligned} \tag{11.5.47}$$

We are already familiar with the form of this equation from equ. (11.5.27). Moreover, equ. (11.5.18) yields the average force $\overline{R}_e$, which is correct from the energetical point of view

$$\begin{aligned} \overline{R}_e &= \frac{1}{\Delta r}\int_{r_\mathrm{k}}^{r_{\mathrm{k}+1}} R_e \mathrm{d}r \\ &= R_{e|\mathrm{k}} + (\tfrac{7}{20}K_{t|\mathrm{k}} + \tfrac{3}{20}K_{t|\mathrm{k}+1})\Delta r + (\tfrac{1}{20}K'_{t|\mathrm{k}} - \tfrac{1}{30}K'_{t|\mathrm{k}+1})\Delta r^2 \end{aligned} \tag{11.5.48}$$

We are acquainted with the coefficients from the cubic Hermitian formulation of equ. (11.4.16). Equations (11.5.47) and (11.5.48) can now be extended to the multi-dimensional case. We thus obtain

$$\boldsymbol{K}_s = \tfrac{1}{2}\left[\boldsymbol{K}_{t|\mathrm{k}} + \boldsymbol{K}_{t|\mathrm{k}+1}\right] + \tfrac{1}{12}\sum_{\mathrm{j}=1}^{\mathrm{n}}\left[\boldsymbol{K}_{t,\mathrm{j}|\mathrm{k}} - \boldsymbol{K}_{t,\mathrm{j}|\mathrm{k}+1}\right]\Delta r_\mathrm{j} \tag{11.5.49}$$

and

$$\overline{R}_e = R_{e|k} + \left[\tfrac{7}{20}K_{t|k} + \tfrac{3}{20}K_{t|k+1}\right]\Delta r + \left[\sum_{j=1}^{n}\left[\tfrac{1}{20}K_{t,j|k} - \tfrac{1}{30}K_{t,j|k+1}\right]\Delta r_j\right]\Delta r \tag{11.5.50}$$

Here, $K_{t,j|k}$, for example, represents – as has already been mentioned repeatedly – a typographically compact expression for the derivative of K_t with respect to r_j at the state r_k

$$K_{t,j|k} = \left.\frac{\partial K_t}{\partial r_j}\right|_{r\,=\,r_k} \tag{11.5.51}$$

The algorithm with K_s according to equ. (11.5.49) and $\overline{R}_e$ according to equ. (11.5.50) retains the exact energy balance of a free oscillator, as long as the load displacement relationship is one of the 4th order in r or the elastic strain energy is one of the 5th order in r. Algorithms of higher order are conceivable, but are problematic in that derivatives of higher order of the tangential stiffness are required; see [11.47], for example, which contains an extensive discussion of the unconditionally stable quintic algorithm.

11.5.4 Some comments on the implementation of the modified unconditionally stable cubic Hermitian algorithm; examples of application and some alternative procedures

The algorithm described in equs. (11.5.30) and (11.5.36), as well as in equs. (11.5.49) and (11.5.50), for example, can only compete with the conditionally stable procedure of section 11.4.3 if it is really applicable for large time-steps. Since K_s and $\overline{R}_e$ depend on r_{k+1}, the solution in the time-step is iterative in the sense that K_s and $\overline{R}_e$ repeatedly have to be set up anew. For every evaluation of K_s and $\overline{R}_e$, the vectors r and $\dot{r}$ must be calculated by means of a direct solution technique. The method is thus analogous to the unconditionally stable procedure in the linear case, on the understanding that that the matrices D_1 (see equ. (10.4.43)) and D_0 (see equ. (10.4.44)) can no longer be assumed to be independent of r_k and r_{k+1}.

The solution for a large time-step conforms to a great extent to the technique in non-linear statics when seeking a solution in a direct iteration without incrementation. In the case of problems that have to be solved by incrementation, the use of large time-steps is not very promising. The following rule of thumb is based on a number of examples: without adopting any special precautionary measures, convergence is achieved if the time-step is not greater than a quarter of the period of the component of the motion having a maximum amplitude. The time-step thus established is, of course, by no means generous compared with the possibilities available in linear applications. The active participation of high frequencies in a non-linear motion would exclude the application of an "unconditionally" stable procedure. Taking the average value of two consecutive iterates has proved a simple additional technique of achieving stabilisation. These tricks, however, have their limits, since the programming of the process is complex enough and further operations cast doubt on the profitability of the procedure.

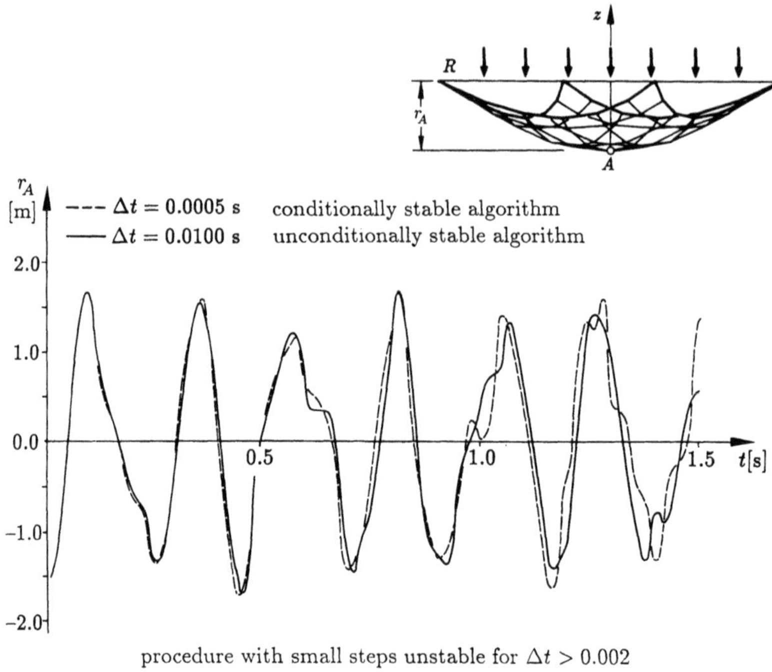

Fig. 11.5.2 Cubic Hermitian algorithm for small and large (modified version) time-steps

We have already applied the conditionally stable cubic Hermitian algorithm to the plane network shown in figs. 11.4.10 and 11.4.11. We revert to this example in fig. 11.5.2 and compare the "exact" curve, computed with a time-step of only $\Delta t = 0.0005$ s by means of the conditionally stable procedure, with the result obtained with the modified algorithm using a time-step of $\Delta t = 0.01$ s. For larger time-steps, the process no longer converges. This time-step corresponds to five times the limiting time-step of the conditionally stable procedure.

In fig. 11.5.3, we again apply the conditionally stable and the "unconditionally" stable cubic Hermitian algorithm on a simple oscillator with cubic stiffness. The system possesses no damping, the mass is deflected statically and then allowed to oscillate freely. After a time T_0 (= non-linear period obtained from a precise calculation), the oscillator returns to its original position. Figure 11.5.3 indicates the error in the maximum deflection and in the oscillatory period for an increasingly fine grading of the time T_0, i.e. for a decreasing time-step. The modified cubic Hermitian algorithm yields better results in all cases and still functions even below the limiting number N of steps allowed for the conditionally stable variant.

Finally, we refer to some alternative energy conserving procedures. The algorithm of Haug, de Rouvray and Nguyen [11.32] works with a cubic interpolation of the accelerations and conserves energy to the 5th degree in the displacements. Programming is inevitably somewhat more complex than in the case of the modified

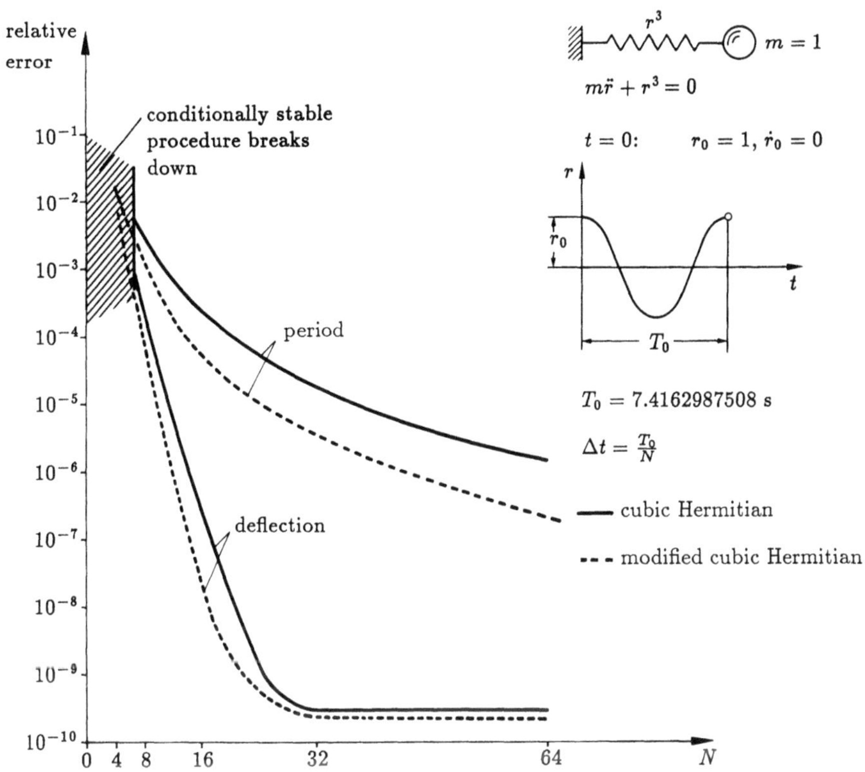

Fig. 11.5.3 Simple oscillator with cubic stiffness: free motion evaluated with the conditionally stable cubic Hermitian algorithm and with the modified version designed for large time-steps

cubic Hermitian algorithm, especially when the Newton-Raphson or the BFGS technique is used for the incremental solution within a step.

The method employed by Hughes, Caughy and Lin [11.33] secures conservation of energy by means of Lagrange's multipliers. The solution in the step can be achieved iteratively. Hughes points out in [11.34] that this technique can be considered a universal method of stabilising transient procedures.

11.6 The incorporation of displacement-dependent forces

11.6.1 Introduction and survey

Standard loading actions in statics and dynamics of structures and in other areas of applied mechanics encompass forces which are not dependent on the motion of the system, i.e. on displacements and velocities. In this case, the load vector $\boldsymbol{R}(t)$ is known at each point of time t and prescribed. On the other hand, we also meet forces in nature which depend on motion. Since in this case, displacements and velocities are not known *a priori* and first have to be calculated, we do not know the loads *a priori*, either. Moreover, when applying a load increment – departing from a known state of equilibrium – the change of load arising from the displacements and velocities must be taken into account. This, however, leads to force correction matrices which must be incorporated in the system of incremental dynamic equilibrium. Examples of such displacement-dependent loads (for velocity-dependent loads, see section 11.7 and the chapters on aeroelasticity in the companion Vol. VI) are centrifugal forces and pressure loads.

Before we deal with this field of mechanics in more detail, let us take a look at the classical theory and recall some concepts for classifying loading actions (see e.g. in Ziegler [11.35]). One classification of the forces (fig. 11.6.1) is based on the calculus of variations in mechanics which we have virtually ignored in this work. Thus, if the forces acting on a mechanical system can be deduced from a scalar function, the so-called potential function Π, then we speak of monogenetic forces. If this is not the case, then the forces are denoted as polygenetic. A second division of the forces is based on the functional character of W of the external loading. If the work W is independent of the path for every admissible displacement and is thus only dependent on the initial and final state of the system, then we speak of conservative forces, otherwise of non-conservative forces. This classification is one more suited to our engineer's philosophy.

As we can gather from fig. 11.6.1, the classifications of the external forces overlap, so that monogenetic forces may certainly also be non-conservative. Thus, all loads that are explicitly dependent on time are instationary or unsteady and non-conservative. This includes practically all the forces of excitation dealt with up to now in this treatise. This classification entails no consequences for the process of the numerical calculation as long as the forces are not dependent on the displacement or on the velocity. On the other hand, the velocity-dependent forces, which are normally non-conservative, include the special case of conservative gyroscopic forces for which $\mathrm{d}W = 0$ applies. Thus, in this special case, $\mathrm{d}W$ is again independent of the path. The Coriolis forces, which will be discussed briefly in the next section, the Lorentz forces and the gyroscopic moments appertain to this group of forces. Although they represent only marginal cases within the conservative classification, load correction matrices are necessary for their correct numerical interpretation. Velocity-dependent loads, however, can certainly be non-conservative, as in damping effects in liquids and gases.

Confronted with this confusing variety of definitions, we consider it most convenient to classify the external loads according to whether particular extensions to theory, like force correction matrices, are necessary or not. In this section, we restrict ourselves to purely displacement-dependent loads. Under certain conditions, these can be noncirculatory and thus conservative. In this case, we obtain symmetric force correction matrices. A characteristic example is given by centrifugal field forces. However, in the present account, we wish to direct our attention mainly to non-conservative circulatory loads which induce non-symmetric force correction matrices. Examples are follower loads, follower moments and pressure forces.

By way of a first elementary introduction to displacement-dependent forces, we consider some simple examples from the field of statics. Figure 11.6.2 shows a simple bar (single element) which is placed radially to the axis of rotation and is referred to a rotating convective coordinate system. Let the angular velocity be Ω. Mass is distributed uniformly over the element (ρ=mass density, A=area of

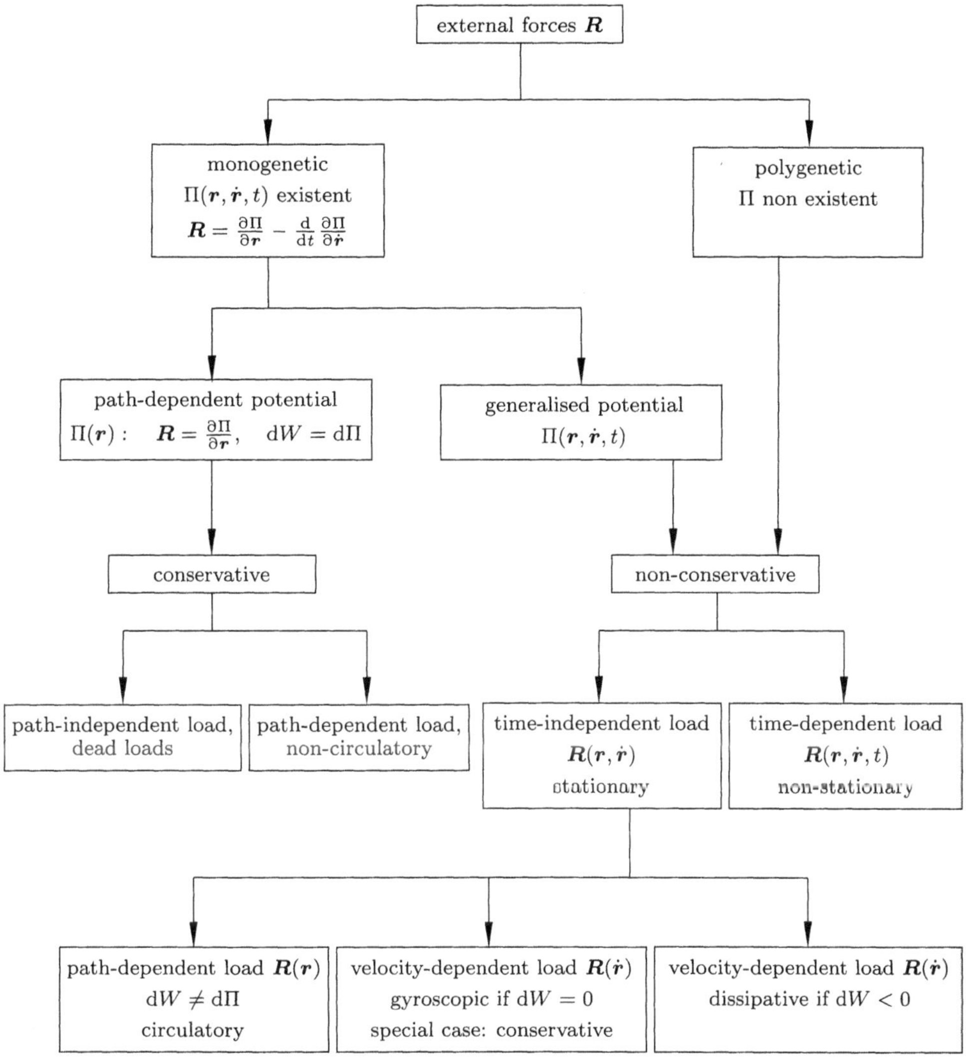

Π = scalar potential function
$\mathrm{d}W$ = work of $\boldsymbol{R}$ over $\mathrm{d}\boldsymbol{r}$

Fig. 11.6.1 Standard classification of external forces in analytical mechanics

cross section) which is consequently subject to distributed centrifugal forces. An infinitesimal part of the FLA2 element of the length $\mathrm{d}x$ experiences an infinitesimal inertia force in the direction x

$$\mathrm{d}U = \rho A \mathrm{d}x \Omega^2 x \tag{11.6.1}$$

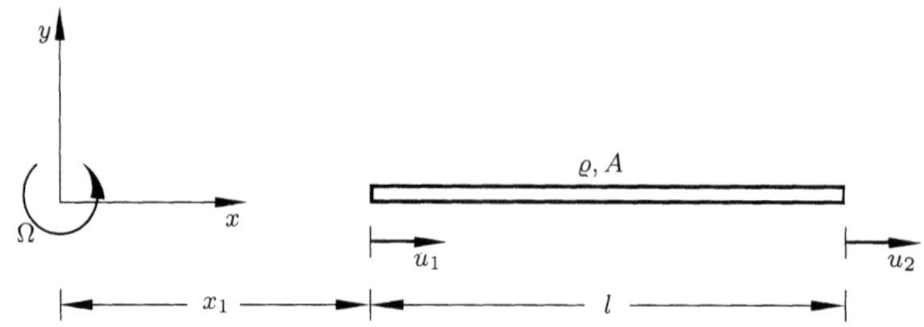

Fig. 11.6.2 Rotating flange element FLA2, placed radially to the axis of rotation

which yields

$$p_{Vx} = \frac{\mathrm{d}U}{A\mathrm{d}x} = \rho\Omega^2 x \tag{11.6.2}$$

for the local volume force at x where x represents the actual position of the increment $\mathrm{d}x$. This is determined by the initial position ${}^0x$ – at zero angular velocity, for example – and the displacement u

$$x = {}^0x + u \tag{11.6.3}$$

If we now deduce kinematically equivalent nodal forces from the distributed loads by the standard technique, we find, using a modal function $\boldsymbol{\omega}$,

$$\begin{aligned} \mathbf{P}_e &= \int_V \boldsymbol{\omega}^{\mathrm{t}} \boldsymbol{p}_V \mathrm{d}V \\ &= \int_x \boldsymbol{\omega}_x^{\mathrm{t}} \rho\Omega^2({}^0x + u) A\mathrm{d}x \end{aligned} \tag{11.6.4}$$

Interpolating in the usual manner the displacements in a FLA2 element, we obtain

$$\begin{aligned} u &= \boldsymbol{\omega}_x \boldsymbol{\rho} \\ &= l^{-1}\,[l - (x - x_1) \qquad (x - x_1)]\{u_1 \qquad u_2\} \end{aligned} \tag{11.6.5}$$

Thus, equ. (11.6.4) reduces to

$$\mathbf{P}_e = \int_x \rho A\omega^2({}^0x)\boldsymbol{\omega}_x^{\mathrm{t}} \mathrm{d}x + \left[\int_x \rho A\omega^2 \boldsymbol{\omega}_x^{\mathrm{t}} \boldsymbol{\omega}_x \mathrm{d}x\right]\boldsymbol{\rho} \tag{11.6.6}$$

The first integral supplies the traditional vector of the kinematically equivalent nodal forces with reference to the state of rest. Since it represents a known force, it may be transferred to the the right-hand side of the dynamic equilibrium. In contrast, the second integral is displacement-dependent and hence unknown. The constituent matrix

$$\boldsymbol{z} = \int_x \rho A \Omega^2 \boldsymbol{\omega}_x^{\mathrm{t}} \boldsymbol{\omega}_x \mathrm{d}x \tag{11.6.7}$$

describes a force correction matrix and is handled formally like a stiffness. It is denoted a centrifugal stiffness and makes due allowance for the variation of the load arising from the motion of the bar. This force correction matrix has of necessity to be added to the standard elastic stiffness during the course of the calculation and thus modifies the standard coefficients of the stiffness. Were it ignored, then the correct tangential direction of this problem would be falsified to a greater or lesser extent, leading not only to an eventual increased iterative effort but also to completely false results, for example in the case of instability problems. The work of the centrifugal forces clearly does not depend on the path, but only on the final position. Thus, these forces are conservative. As a result, we observe the symmetry of the matrix $\boldsymbol{z}$, which represents a special case of equ. (2.3.19). Here, we obtain

$$\boldsymbol{z} = \frac{m\Omega^2}{6} \begin{bmatrix} 2 & 1 \\ 1 & 2 \end{bmatrix} \tag{11.6.8}$$

where

$$m = \rho A l \tag{11.6.9}$$

is the total mass of the element. The general formula for an arbitrary finite element subject to a steady rotation is

$$\boldsymbol{z}_{\mathrm{g}} = m\Omega^2 \frac{1}{V_{\mathrm{g}}} \int_{V_{\mathrm{g}}} \boldsymbol{\omega}^{\mathrm{t}} [\boldsymbol{I}_3 - \boldsymbol{e}_\Omega \boldsymbol{e}_\Omega^{\mathrm{t}}] \boldsymbol{\omega} \mathrm{d}V \tag{11.6.10}$$

where $\boldsymbol{e}_\Omega$ represents a unit vector in the direction of the axis of rotation.

To continue our preparatory discussion, we now consider another elementary example also from the field of pure statics, namely a cantilever beam under the influence of a lateral force assumed to act continually in a normal direction to the deformed beam (follower load, fig. 11.6.3). This model case can be considered a highly simplified example of pressure loading. At a certain load level, P, the vector of the nodal loads becomes

$$\boldsymbol{R} = P\{-\sin\psi_1 \quad \cos\psi_1 \quad 0\} \tag{11.6.11}$$

i.e. the deformations enter in a non-linear form into this expression. If, in the process of an incremental calculation, the load level is increased from P to $P + \Delta P$ and, as a result, ψ_1 increases to $\psi_1 + \Delta\psi_1$, then at the end of the increment we have

$$\boldsymbol{R} + \Delta\boldsymbol{R} = (P + \Delta P)\{-\sin(\psi_1 + \Delta\psi_1) \quad \cos(\psi_1 + \Delta\psi_1) \quad 0\} \tag{11.6.12}$$

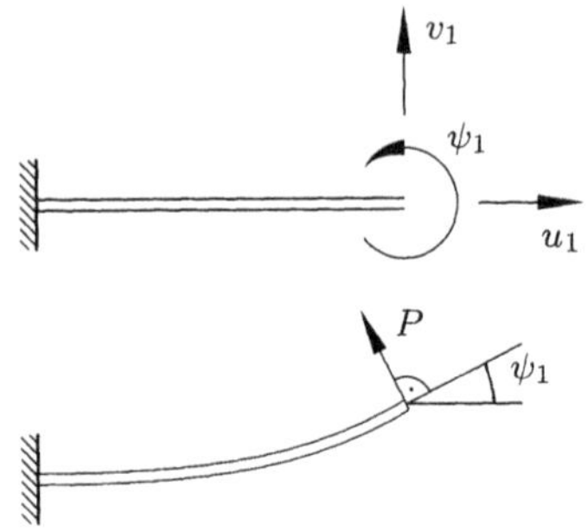

Fig. 11.6.3
Cantilever beam under "follower" transverse load

Taking the differences of equs. (11.6.12) and (11.6.11) and applying a linearisation, we obtain

$$\Delta \boldsymbol{R} = \Delta P\{-\sin\psi_1 \quad \cos\psi_1 \quad 0\} + P\begin{bmatrix} 0 & 0 & -\cos\psi_1 \\ 0 & 0 & -\sin\psi_1 \\ 0 & 0 & 0 \end{bmatrix}\begin{bmatrix} u_1 \\ v_1 \\ \psi_1 \end{bmatrix} \tag{11.6.13}$$

The incremental load vector contains two terms, the first of which describes a conventional load expression which may be transferred to the right-hand side of the equilibrium expression. The second term, however, represents a force correction matrix which records the change in the force direction in the incremental step. This force correction matrix, which again forms a pseudo-stiffness, is unfortunately now no longer symmetric. The loading is correspondingly non-conservative, i.e. the external work depends on the path. This is, of course, very unpleasant already in statics, since the standard and fastest solution algorithms assume the symmetry of the stiffness matrix. No few attempts – but to no avail – have been made to circumvent such force corrections or to enforce a symmetrisation. Inevitably, however, such attempts involve a falsification of the tangential direction which induces negative effects of varying significance into the computations.

11.6.2 General FEM formulations of displacement-dependent loads

We assume that there are three types of displacement-dependent loads: surface forces $\boldsymbol{p}_S$, volume forces $\boldsymbol{p}_V$, and discrete loads $\boldsymbol{U}_\mathrm{i} = \{U \ V \ W\}_\mathrm{i}$ acting at the positions $\boldsymbol{x}_\mathrm{i}$ of the structure. Each of these groups of forces is understood as the product of a loading factor $(\lambda_S, \lambda_V, \lambda_\mathrm{i})$ with corresponding vector functionals $\boldsymbol{f}_S, \boldsymbol{f}_V$ and $\boldsymbol{f}_\mathrm{i}$ depending on the position $\boldsymbol{x}$ and the displacements $\boldsymbol{u}$

$$\boldsymbol{p}_S = \lambda_S \boldsymbol{f}_S(\boldsymbol{x},\boldsymbol{u}(\boldsymbol{x})) \tag{11.6.14}$$

$$\boldsymbol{p}_V = \lambda_V \boldsymbol{f}_V(\boldsymbol{x},\boldsymbol{u}(\boldsymbol{x})) \tag{11.6.15}$$

$$\boldsymbol{U}_\mathrm{i} = \lambda_\mathrm{i} \boldsymbol{f}_\mathrm{i}(\boldsymbol{u}(\boldsymbol{x}_\mathrm{i})) \tag{11.6.16}$$

Using the standard modal displacement distributions for the displacements within the finite element

$$\boldsymbol{u}(\boldsymbol{x}) = \boldsymbol{\omega}(\boldsymbol{x})\boldsymbol{\rho} \tag{11.6.17}$$

which, as we know, is also applicable – at least in incremental form – in the presence of large displacements and strains, the equivalent nodal loads can be expressed as simply as in the linear case

$$\mathbf{P}_e = \lambda_S \int_S \boldsymbol{\omega}^{\mathrm{t}} \boldsymbol{f}_S \mathrm{d}S + \lambda_V \int_V \boldsymbol{\omega}^{\mathrm{t}} \boldsymbol{f}_V \mathrm{d}V + \sum_{\mathrm{i}=1}^{\mathrm{k}} \lambda_{\mathrm{i}} \boldsymbol{\omega}^{\mathrm{t}}(\boldsymbol{x}_{\mathrm{i}}) \boldsymbol{f}_{\mathrm{i}} \tag{11.6.18}$$

In the non-linear case, possibly associated with large deformations, it is only necessary to remember that the integration must be carried out over the current (instantaneous) geometry. It should also be observed that $\mathbf{P}_e$ is generally a highly non-linear function of the element displacements $\boldsymbol{\rho}$. The current displacements $\boldsymbol{\rho}$ are, of course, presumed known in equ. (11.6.18). Thus, equ. (11.6.18) forms the basis for the iterated equilibrium procedure in non-linear calculations. In order to set up the incremental operation, we require the incremental formulation of equ. (11.6.18). Variations in the equivalent element nodal forces $\mathrm{d}\mathbf{P}_e$ arise from incrementations of the load factors $(\mathrm{d}\lambda_S, \mathrm{d}\lambda_V, \mathrm{d}\lambda_{\mathrm{i}})$ and the nodal displacements $\mathrm{d}\boldsymbol{\rho}$; thus, we obtain here

$$\begin{aligned} \mathrm{d}\mathbf{P}_e =& \mathrm{d}\lambda_S \int_S \boldsymbol{\omega}^{\mathrm{t}} \boldsymbol{f}_S \mathrm{d}S + \mathrm{d}\lambda_V \int_V \boldsymbol{\omega}^{\mathrm{t}} \boldsymbol{f}_V \mathrm{d}V + \sum_{\mathrm{i}=1}^{\mathrm{k}} \mathrm{d}\lambda_{\mathrm{i}} \boldsymbol{\omega}^{\mathrm{t}}(\boldsymbol{x}_{\mathrm{i}}) \boldsymbol{f}_{\mathrm{i}} \\ &+ \left[\lambda_S \int_S \boldsymbol{\omega}^{\mathrm{t}} \frac{\partial \boldsymbol{f}_S}{\partial \boldsymbol{\rho}} \mathrm{d}S + \lambda_V \int_V \boldsymbol{\omega}^{\mathrm{t}} \frac{\partial \boldsymbol{f}_V}{\partial \boldsymbol{\rho}} \mathrm{d}V + \sum_{\mathrm{i}=1}^{\mathrm{k}} \lambda_{\mathrm{i}} \boldsymbol{\omega}^{\mathrm{t}}(\boldsymbol{x}_{\mathrm{i}}) \frac{\partial \boldsymbol{f}_{\mathrm{i}}}{\partial \boldsymbol{\rho}} \right] \mathrm{d}\boldsymbol{\rho} \end{aligned} \tag{11.6.19}$$

This formula corresponds closely in its formulation to that of our introductory example in equ. (11.6.13). The first part represents a known load increment which is transferred to the right-hand side of the equilibrium expression like any normal load. The second part contains the quadratic (but in general nonsymmetrical!) force correction matrix

$$\boldsymbol{k}_R = \lambda_S \int_S \boldsymbol{\omega}^{\mathrm{t}} \frac{\partial \boldsymbol{f}_S}{\partial \boldsymbol{\rho}} \mathrm{d}S + \lambda_V \int_V \boldsymbol{\omega}^{\mathrm{t}} \frac{\partial \boldsymbol{f}_V}{\partial \boldsymbol{\rho}} \mathrm{d}V + \sum_{\mathrm{i}=1}^{\mathrm{k}} \lambda_{\mathrm{i}} \boldsymbol{\omega}^{\mathrm{t}}(\boldsymbol{x}_{\mathrm{i}}) \frac{\partial \boldsymbol{f}_{\mathrm{i}}}{\partial \boldsymbol{\rho}} \tag{11.6.20}$$

The element correction stiffness $\boldsymbol{k}_R$ will generally be a non-linear function of the element displacements $\boldsymbol{\rho}$ and will depend, of course, on the preceding load history, as defined by the load factors $(\lambda_S, \lambda_V, \lambda_{\mathrm{i}})$. If we substitute equs. (11.6.19) and (11.6.20) into the incremental relation for dynamic equilibrium, we find

$$\boldsymbol{M}\mathrm{d}\ddot{\boldsymbol{r}} + \boldsymbol{C}\mathrm{d}\dot{\boldsymbol{r}} + [\boldsymbol{K}_e + \boldsymbol{K}_G]\mathrm{d}\boldsymbol{r} = \mathrm{d}\boldsymbol{R}_{e\lambda} + \mathrm{d}\boldsymbol{R}_{eu} \tag{11.6.21}$$

Using the standard Boolean assembly matrix $\mathbf{a}$, we find

$$\begin{aligned} \mathrm{d}\boldsymbol{R}_{e\lambda} &= \mathbf{a}^{\mathrm{t}}\mathrm{d}\mathbf{P}_{e\lambda} \\ &= \mathbf{a}^{\mathrm{t}}\left[\mathrm{d}\lambda_S\int_S \boldsymbol{\omega}^{\mathrm{t}}\boldsymbol{f}_S\mathrm{d}S + \mathrm{d}\lambda_V\int_V \boldsymbol{\omega}^{\mathrm{t}}\boldsymbol{f}_V\mathrm{d}V + \sum_{\mathrm{i}=1}^{\mathrm{k}}\mathrm{d}\lambda_{\mathrm{i}}\boldsymbol{\omega}^{\mathrm{t}}(\boldsymbol{x}_{\mathrm{i}})\boldsymbol{f}_{\mathrm{i}}\right] \end{aligned} \tag{11.6.22}$$

This corresponds to the standard load contribution in the incremental step. Also,

$$\begin{aligned} \mathrm{d}\boldsymbol{R}_{eu} &= \mathbf{a}^{\mathrm{t}}\mathrm{d}\mathbf{P}_{eu} \\ &= \mathbf{a}^{\mathrm{t}}\boldsymbol{k}_R\mathrm{d}\boldsymbol{\rho} = \mathbf{a}^{\mathrm{t}}\boldsymbol{k}_R\mathbf{a}\,\mathrm{d}\boldsymbol{r} \\ &= \boldsymbol{K}_R\,\mathrm{d}\boldsymbol{r} \end{aligned} \tag{11.6.23}$$

represents a linearised displacement-dependent contribution. Here, $\boldsymbol{K}_R$ is the force correction matrix on the global level. With its help, a modified tangential stiffness can be formed in the presence of displacement-dependent loads if the term $\mathrm{d}\boldsymbol{R}_{eu}$ in equ. (11.6.21) is transferred onto the left-hand side. This yields the compound stiffness

$$\boldsymbol{K}_t = \boldsymbol{K}_e + \boldsymbol{K}_G - \boldsymbol{K}_R \tag{11.6.24}$$

Thus, we obtain the same form of the incremental equilibrium as in the case of dead loads

$$\boldsymbol{M}\mathrm{d}\ddot{\boldsymbol{r}} + \boldsymbol{C}\mathrm{d}\dot{\boldsymbol{r}} + \boldsymbol{K}_t\mathrm{d}\boldsymbol{r} = \mathrm{d}\boldsymbol{R}_{e\lambda} \tag{11.6.25}$$

We do not want to conclude this section without a brief reference to the natural formulation which has often supported our interpretative desire, whether as a compact, elegant formulation or as a valuable physical insight.

In our preceding research publications, as also in Vol. III of this series, we observed that in the non-linear regime of large displacements, the Cartesian displacements $\boldsymbol{\rho}$ can be expressed by a linear transformation in the natural or generalised displacements $\widetilde{\boldsymbol{\rho}}'$ and vice versa

$$\boldsymbol{\rho} = \boldsymbol{A}_e\widetilde{\boldsymbol{\rho}}' \tag{11.6.26}$$

$$\widetilde{\boldsymbol{\rho}}' = \boldsymbol{a}_e\boldsymbol{\rho} \tag{11.6.27}$$

Correspondingly,

$$\widetilde{\boldsymbol{\rho}}' = \{\widetilde{\boldsymbol{\rho}}_0 \quad \widetilde{\boldsymbol{\rho}}_N\} \tag{11.6.28}$$

is construed so that $\widetilde{\boldsymbol{\rho}}_0$ contains to a first order approximation ($\mathrm{d}\widetilde{\boldsymbol{\rho}}_0$) the rigid body motion contributions ($\mathrm{d}\boldsymbol{\rho}_0$) and $\widetilde{\boldsymbol{\rho}}_N$ the elastic ones ($\mathrm{d}\widetilde{\boldsymbol{\rho}}_N = \mathrm{d}\boldsymbol{\rho}_N$). If we consider the generalised forces $\widetilde{\mathbf{P}}_e$ which can always be deduced via the principle of virtual displacements from $\mathbf{P}_e$, and vice versa,

$$\widetilde{\mathbf{P}}_e = \boldsymbol{A}_e^{\mathrm{t}} \mathbf{P}_e \tag{11.6.29}$$

$$\mathbf{P}_e = \boldsymbol{a}_e^{\mathrm{t}} \widetilde{\mathbf{P}}_e \tag{11.6.30}$$

then it is, of course, also feasible to differentiate with respect to the generalised displacements $\widetilde{\boldsymbol{\rho}}'$ when determining increments. Observing the modal forms

$$\widetilde{\boldsymbol{\omega}} = [\widetilde{\boldsymbol{\omega}}_0 \quad \widetilde{\boldsymbol{\omega}}_N] = \boldsymbol{\omega} \boldsymbol{A}_e \tag{11.6.31}$$

appertaining to $\widetilde{\boldsymbol{\rho}}'$, we obtain

$$\mathrm{d}\widetilde{\mathbf{P}}_e = \mathrm{d}\widetilde{\mathbf{P}}_{e\lambda} + \mathrm{d}\widetilde{\mathbf{P}}_{eu} \tag{11.6.32}$$

as an alternative to equ. (11.6.19) and note the standard load increment

$$\mathrm{d}\widetilde{\mathbf{P}}_{e\lambda} = \mathrm{d}\lambda_S \int_S \begin{bmatrix} \widetilde{\boldsymbol{\omega}}_0^{\mathrm{t}} \\ \widetilde{\boldsymbol{\omega}}_N^{\mathrm{t}} \end{bmatrix} \boldsymbol{f}_S \mathrm{d}S + \mathrm{d}\lambda_V \int_V \begin{bmatrix} \widetilde{\boldsymbol{\omega}}_0^{\mathrm{t}} \\ \widetilde{\boldsymbol{\omega}}_N^{\mathrm{t}} \end{bmatrix} \boldsymbol{f}_V \mathrm{d}V + \sum_{\mathrm{i}=1}^{\mathrm{k}} \mathrm{d}\lambda_{\mathrm{i}} \begin{bmatrix} \widetilde{\boldsymbol{\omega}}_0^{\mathrm{t}}(\boldsymbol{x}_{\mathrm{i}}) \\ \widetilde{\boldsymbol{\omega}}_N^{\mathrm{t}}(\boldsymbol{x}_{\mathrm{i}}) \end{bmatrix} \boldsymbol{f}_{\mathrm{i}} \tag{11.6.33}$$

within the step, as well as the additional linearised contribution

$$\begin{aligned} \mathrm{d}\widetilde{\mathbf{P}}_{eu} &= \widetilde{\boldsymbol{k}}_R \mathrm{d}\widetilde{\boldsymbol{\rho}} \\ &= \begin{bmatrix} \widetilde{\boldsymbol{k}}_{00} & \widetilde{\boldsymbol{k}}_{0N} \\ \widetilde{\boldsymbol{k}}_{N0} & \widetilde{\boldsymbol{k}}_{NN} \end{bmatrix}_R \begin{bmatrix} \mathrm{d}\widetilde{\boldsymbol{\rho}}_0 \\ \mathrm{d}\widetilde{\boldsymbol{\rho}}_N \end{bmatrix} \end{aligned} \tag{11.6.34}$$

where the submatrices are defined by

$$\begin{bmatrix} \widetilde{\boldsymbol{k}}_{00} \\ \widetilde{\boldsymbol{k}}_{N0} \end{bmatrix}_R = \lambda_S \int_S \begin{bmatrix} \widetilde{\boldsymbol{\omega}}_0^{\mathrm{t}} \\ \widetilde{\boldsymbol{\omega}}_N^{\mathrm{t}} \end{bmatrix} \frac{\partial \boldsymbol{f}_S}{\partial \widetilde{\boldsymbol{\rho}}_0} \mathrm{d}S + \lambda_V \int_V \begin{bmatrix} \widetilde{\boldsymbol{\omega}}_0^{\mathrm{t}} \\ \widetilde{\boldsymbol{\omega}}_N^{\mathrm{t}} \end{bmatrix} \frac{\partial \boldsymbol{f}_V}{\partial \widetilde{\boldsymbol{\rho}}_0} \mathrm{d}V + \sum_{\mathrm{i}=1}^{\mathrm{k}} \lambda_{\mathrm{i}} \begin{bmatrix} \widetilde{\boldsymbol{\omega}}_0^{\mathrm{t}}(\boldsymbol{x}_{\mathrm{i}}) \\ \widetilde{\boldsymbol{\omega}}_N^{\mathrm{t}}(\boldsymbol{x}_{\mathrm{i}}) \end{bmatrix} \frac{\partial \boldsymbol{f}_{\mathrm{i}}}{\partial \widetilde{\boldsymbol{\rho}}_0} \tag{11.6.35}$$

and

$$\begin{bmatrix} \widetilde{\boldsymbol{k}}_{0N} \\ \widetilde{\boldsymbol{k}}_{NN} \end{bmatrix}_R = \lambda_S \int_S \begin{bmatrix} \widetilde{\boldsymbol{\omega}}_0^{\mathrm{t}} \\ \widetilde{\boldsymbol{\omega}}_N^{\mathrm{t}} \end{bmatrix} \frac{\partial \boldsymbol{f}_S}{\partial \widetilde{\boldsymbol{\rho}}_N} \mathrm{d}S + \lambda_V \int_V \begin{bmatrix} \widetilde{\boldsymbol{\omega}}_0^{\mathrm{t}} \\ \widetilde{\boldsymbol{\omega}}_N^{\mathrm{t}} \end{bmatrix} \frac{\partial \boldsymbol{f}_V}{\partial \widetilde{\boldsymbol{\rho}}_N} \mathrm{d}V + \sum_{\mathrm{i}=1}^{\mathrm{k}} \lambda_{\mathrm{i}} \begin{bmatrix} \widetilde{\boldsymbol{\omega}}_0^{\mathrm{t}}(\boldsymbol{x}_{\mathrm{i}}) \\ \widetilde{\boldsymbol{\omega}}_N^{\mathrm{t}}(\boldsymbol{x}_{\mathrm{i}}) \end{bmatrix} \frac{\partial \boldsymbol{f}_{\mathrm{i}}}{\partial \widetilde{\boldsymbol{\rho}}_N} \tag{11.6.36}$$

This presentation will be used in the next section in order to establish the force correction matrices for a beam; in this connection, the important role of large rotations discussed in Vol. III will be elaborated again.

Before dealing with this range of problems, however, we consider a specific group of forces, namely loads acting directly on the nodes of a FE network. In principle, these are already contained in the concentrated loads $\boldsymbol{U}_{\mathrm{i}}$ acting on discrete points $\boldsymbol{x}_{\mathrm{i}}$ of the element. In the present case, the position $\boldsymbol{x}_{\mathrm{i}}$ simply coincides with a nodal point position. It is nevertheless inexpedient to analyse this special case via the roundabout way of a finite element. In analogy to equs. (11.6.14) and (11.6.16), we express the vector of the external forces, $\boldsymbol{R}$, as a product of load factors $\boldsymbol{\lambda}$ and a displacement-dependent basic distribution $\overline{\boldsymbol{R}}(\mathbf{r})$

$$\boldsymbol{R} = \boldsymbol{\lambda}\overline{\boldsymbol{R}}(\boldsymbol{r}) \tag{11.6.37}$$

Here, $\boldsymbol{\lambda}$ is a diagonal matrix so that it is theoretically possible to steer every single load component in the loading process individually. In analogy with equ. (11.4.16), we set up the incremental load expression and obtain

$$\mathrm{d}\boldsymbol{R} = \mathrm{d}\boldsymbol{\lambda}\overline{\boldsymbol{R}}(\boldsymbol{r}) + \boldsymbol{\lambda}\frac{\partial \overline{\boldsymbol{R}}}{\partial \boldsymbol{r}}\mathrm{d}\boldsymbol{r} \tag{11.6.38}$$

The force correction matrix on the structure or global level is hence given directly by

$$\boldsymbol{K}_R = \boldsymbol{\lambda}\frac{\partial \overline{\boldsymbol{R}}}{\partial \boldsymbol{r}} \tag{11.6.39}$$

Special conditions prevail if the line load $\boldsymbol{U}_{\mathrm{j}}$ and the moment vector $\boldsymbol{\Theta}_{\mathrm{j}}$ act on a node j and are only dependent on the local rotation vector $\boldsymbol{\vartheta}_{\mathrm{j}}$. The load measure $\overline{\boldsymbol{R}}$ which is then construed using this nodal load vector is

$$\overline{\boldsymbol{R}} = \{\overline{\boldsymbol{U}}_1 \quad \overline{\boldsymbol{\Theta}}_1 \quad \dots \quad \overline{\boldsymbol{U}}_{\mathrm{j}} \quad \overline{\boldsymbol{\Theta}}_{\mathrm{j}} \quad \dots \quad \overline{\boldsymbol{U}}_{\mathrm{np}} \quad \overline{\boldsymbol{\Theta}}_{\mathrm{np}}\} \tag{11.6.40}$$

where np indicates the number of nodes; this, in conjunction with the corresponding displacement vector

$$\boldsymbol{r} = \{\boldsymbol{u}_1 \quad \boldsymbol{\vartheta}_1 \quad \dots \quad \boldsymbol{u}_{\mathrm{j}} \quad \boldsymbol{\vartheta}_{\mathrm{j}} \quad \dots \quad \boldsymbol{u}_{\mathrm{np}} \quad \boldsymbol{\vartheta}_{\mathrm{np}}\} \tag{11.6.41}$$

yields a hyperdiagonal force correction matrix

$$\boldsymbol{K}_R = \boldsymbol{\lambda}\lceil \boldsymbol{K}_R^1 \quad \dots \quad \boldsymbol{K}_R^{\mathrm{j}} \quad \dots \quad \boldsymbol{K}_R^{\mathrm{np}} \rfloor \tag{11.6.42}$$

Here, the force corrector assigned to the point j has the form

$$\boldsymbol{K}_R^{\mathrm{j}} = \begin{bmatrix} \boldsymbol{O} & \dfrac{\partial \overline{\boldsymbol{U}}_{\mathrm{j}}}{\partial \boldsymbol{\vartheta}_{\mathrm{j}}} \\ \\ \boldsymbol{O} & \dfrac{\partial \overline{\boldsymbol{\Theta}}_{\mathrm{j}}}{\partial \boldsymbol{\vartheta}_{\mathrm{j}}} \end{bmatrix} \tag{11.6.43}$$

Our introductory example illustrated in fig. 11.6.3 could be dealt with in this way. In such cases, the force corrector can be set up directly on the global level and inserted into the tangential stiffness of the structure.

11.6.3 Load correction matrix for a beam element under displacement-dependent distributed loads [11.36]

We demonstrate the application of the theory developed in the preceding section on a straight cylindrical beam in space (fig. 11.6.4). This beam element is subjected to follower loads p_x, p_y, p_z distributed linearly along its length (fig. 11.6.5). The distribution is determined by the respective nodal point values, i.e. for example, p_x^1 and p_x^2. Here, however, it is more attractive to work with the mean values (m) and the differences (d)

$$\begin{aligned} p_{xm} &= \tfrac{1}{2}(p_x^2 + p_x^1) \qquad & p_{xd} &= p_x^2 - p_x^1 \\ p_{ym} &= \tfrac{1}{2}(p_y^2 + p_y^1) \qquad & p_{yd} &= p_y^2 - p_y^1 \\ p_{zm} &= \tfrac{1}{2}(p_z^2 + p_z^1) \qquad & p_{zd} &= p_z^2 - p_z^1 \end{aligned} \tag{11.6.44a}$$

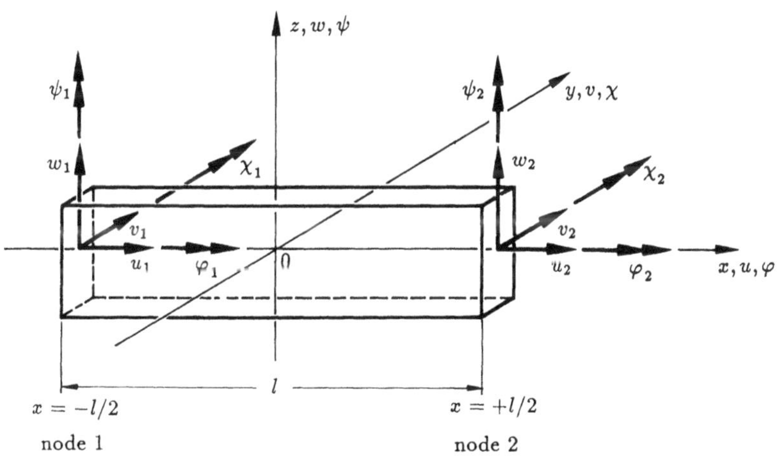

Fig. 11.6.4 Beam element in space – local system and nodal freedoms

where p_x is a follower load acting tangentially to the beam axis and p_y and p_z act normally (transversally) to it. If we combine the mean values of the load intensity in the vector

$$\boldsymbol{p}_m = \{p_{xm} \quad p_{ym} \quad p_{zm}\} \tag{11.6.44b}$$

and the difference values in

$$\boldsymbol{p}_d = \{p_{xd} \quad p_{yd} \quad p_{zd}\} \tag{11.6.45}$$

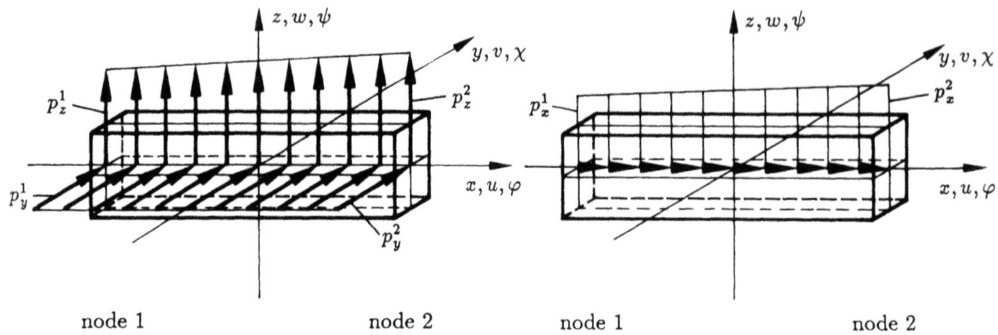

Fig. 11.6.5 Beam element under distributed follower loads: p_x retains its axial direction, p_y and p_z remain normal to the beam axis

then we obtain, using the dimensionless coordinate

$$\xi = \frac{x}{l} \tag{11.6.46}$$

the following expression for the distribution of load intensity in the three directions

$$\boldsymbol{p}_0 = \boldsymbol{p}_m + \xi \boldsymbol{p}_d = \{p_x \quad p_y \quad p_z\} \tag{11.6.47}$$

In the non-deflected state, this expression reproduces the forces in the directions of the axes (per unit length). If, however, the beam is deformed, then at the point ξ, a rotation, defined by the local rotation vector $\boldsymbol{\vartheta}$, will take place and the local force vector is deduced by a rotation transformation applied to the local load intensity. Thus,

$$\boldsymbol{p}(\boldsymbol{\vartheta}) = \boldsymbol{T}(\boldsymbol{\vartheta})\boldsymbol{p}_0 \tag{11.6.48}$$

The non-linear transformation matrix $\boldsymbol{T}$ can be set up using the classical Rodrigues formulae. The following elegant matrix formulation [11.36] offers an efficient tool. We set up

$$\boldsymbol{\vartheta} = \{\varphi \quad \chi \quad \psi\} \tag{11.6.49}$$

the local torsion vector with its Cartesian components φ, χ and ψ and the modulus

$$\vartheta = (\varphi^2 + \chi^2 + \psi^2)^{1/2} \tag{11.6.50}$$

We also require $\boldsymbol{\Theta}$, a skew-symmetric matrix (this designation should not be confused with that of the moment vector in section 11.6.2) also applied in Vol. III of this series

$$\boldsymbol{\Theta} = \begin{bmatrix} 0 & -\psi & \chi \\ \psi & 0 & -\varphi \\ -\chi & \varphi & 0 \end{bmatrix} \tag{11.6.51}$$

With this notation, $\boldsymbol{T}$ becomes

$$\boldsymbol{T} = \boldsymbol{I}_3 + \frac{\sin\vartheta}{\vartheta}\boldsymbol{\Theta} + \left(\frac{\sin\vartheta/2}{\vartheta/2}\right)^2\boldsymbol{\Theta}^2 \tag{11.6.52}$$

(see incidentally the detailed account in Vol. III and [11.59]). Applying the natural approach, the force correction matrix of the element may be expressed as a slight modification of equs. (11.6.35) and (11.6.36) (note the presence of line loads) in the form

$$\widetilde{\boldsymbol{k}}_R = l\int\limits_{-1/2}^{+1/2}\begin{bmatrix}\widetilde{\boldsymbol{\omega}}_0^{\mathrm{t}}\\ \widetilde{\boldsymbol{\omega}}_N^{\mathrm{t}}\end{bmatrix}\frac{\partial\boldsymbol{T}}{\partial\boldsymbol{\rho}}\boldsymbol{p}_0\mathrm{d}\xi \tag{11.6.53}$$

where, in addition, the relationship between the components of the local rotation vector and the natural displacements is of importance. Adopting the following expressions from Vol. I, we have (noting that $\rho_{0\mathrm{i}}$ defines a rigid body rotation)

$$\begin{aligned}\varphi(\xi) &= \rho_{04} + \xi\rho_{N6}\\ \chi(\xi) &= -w_{,x} = \rho_{05} + \xi\rho_{N4} - (3\xi^2 - \tfrac{1}{4})\rho_{N5}\\ \psi(\xi) &= v_{,x} = \rho_{06} - \xi\rho_{N2} + (3\xi^2 - \tfrac{1}{4})\rho_{N3}\end{aligned} \tag{11.6.54}$$

Restricting ourselves to relatively small rotations, equ. (11.6.52) may be approximated by

$$\boldsymbol{T} = \boldsymbol{I}_3 + \boldsymbol{\Theta} + \tfrac{1}{2}\boldsymbol{\Theta}^2 \tag{11.6.55}$$

We now assign to the components of the torsion vector $\boldsymbol{\vartheta}$ of the property of semi-tangential rotations , as expounded in [11.59]. With equs. (11.6.54) and (11.6.55), we find

$$\frac{\partial\boldsymbol{T}}{\partial\boldsymbol{\rho}_0}\boldsymbol{p}_0 = \begin{bmatrix}0 & 0 & 0 & 0 & p_z & -p_y\\ 0 & 0 & 0 & -p_z & 0 & p_x\\ 0 & 0 & 0 & p_y & -p_x & 0\end{bmatrix} \tag{11.6.56a}$$

as well as

$$\frac{\partial\boldsymbol{T}}{\partial\boldsymbol{\rho}_N}\boldsymbol{p}_0 = \begin{bmatrix}0 & \xi p_y & (\frac{1}{4} - 3\xi^2)p_y & \xi p_z & (\frac{1}{4} - 3\xi^2)p_z & 0\\ 0 & -\xi p_x & -(\frac{1}{4} - 3\xi^2)p_x & 0 & 0 & -\xi p_z\\ 0 & 0 & 0 & -\xi p_x & -(\frac{1}{4} - 3\xi^2)p_x & \xi p_y\end{bmatrix} \tag{11.6.56b}$$

In this connection, we should also observe equ. (11.6.47), i.e. the fact that p_x, p_y and p_z are linear functions of ξ. With some effort, the rather large (12×12)-matrix $\widetilde{\boldsymbol{k}}_R$ is obtained via an exact analytical integration. Following equ. (11.6.34) its four sub-matrices are given by

$$\widetilde{\boldsymbol{k}}_{R00} = \begin{bmatrix} \cdots & \cdot & lp_{zm} & -lp_{ym} \\ \cdots & -lp_{zm} & \cdot & lp_{xm} \\ \cdots & lp_{ym} & -lp_{xm} & \cdot \\ \cdots & \cdot & \cdot & \cdot \\ \cdots & -\frac{1}{12}l^2p_{yd} & \frac{1}{12}l^2p_{xd} & \cdot \\ \cdots & -\frac{1}{12}l^2p_{zd} & \cdot & \frac{1}{12}l^2p_{xd} \end{bmatrix} \tag{11.6.57}$$

$$\widetilde{\boldsymbol{k}}_{R0N} = \begin{bmatrix} \cdot & \frac{1}{12}lp_{yd} & \cdot & \frac{1}{12}lp_{zd} & \cdot & \cdot \\ \cdot & -\frac{1}{12}lp_{xd} & \cdot & \cdot & \cdot & -\frac{1}{12}lp_{zd} \\ \cdot & \cdot & \cdot & -\frac{1}{12}lp_{xd} & \cdot & \frac{1}{12}lp_{yd} \\ \cdot & \cdot & \cdot & \cdot & \cdot & \cdot \\ \cdot & \cdot & \cdot & \frac{1}{12}l^2p_{xm} & -\frac{1}{60}l^2p_{xd} & -\frac{1}{12}l^2p_{ym} \\ \cdot & -\frac{1}{12}l^2p_{xm} & \frac{1}{60}l^2p_{xd} & \cdot & \cdot & -\frac{1}{12}l^2p_{zm} \end{bmatrix} \tag{11.6.58}$$

$$\widetilde{\boldsymbol{k}}_{RN0} = \begin{bmatrix} \cdots & \cdot & \frac{1}{12}lp_{zd} & -\frac{1}{12}lp_{yd} \\ \cdots & -\frac{1}{12}l^2p_{zm} & \cdot & \frac{1}{12}l^2p_{xm} \\ \cdots & \frac{1}{120}l^2p_{zd} & \cdot & -\frac{1}{120}l^2p_{xd} \\ \cdots & \frac{1}{12}l^2p_{ym} & -\frac{1}{12}l^2p_{xm} & \cdot \\ \cdots & -\frac{1}{120}l^2p_{yd} & \frac{1}{120}l^2p_{xd} & \cdot \\ \cdots & \cdot & \cdot & \cdot \end{bmatrix} \tag{11.6.59}$$

$$\widetilde{\boldsymbol{k}}_{RNN} = \begin{bmatrix} \cdot & \frac{1}{12}lp_{ym} & -\frac{1}{60}lp_{yd} & \frac{1}{12}lp_{zm} & -\frac{1}{60}lp_{zd} & \cdot \\ \cdot & -\frac{1}{240}l^2p_{xd} & -\frac{1}{120}l^2p_{xm} & \cdot & \cdot & -\frac{1}{240}l^2p_{zd} \\ \cdot & \frac{1}{120}l^2p_{xm} & -\frac{1}{1680}l^2p_{xd} & \cdot & \cdot & \frac{1}{120}l^2p_{zm} \\ \cdot & \cdot & \cdot & -\frac{1}{240}l^2p_{xd} & -\frac{1}{120}l^2p_{xm} & \frac{1}{240}l^2p_{yd} \\ \cdot & \cdot & \cdot & \frac{1}{120}lp_{xm} & -\frac{1}{1680}l^2p_{xd} & -\frac{1}{120}l^2p_{ym} \\ \cdot & \cdot & \cdot & \cdot & \cdot & \cdot \end{bmatrix} \tag{11.6.60}$$

11.6.4 Incorporation of displacement-dependent forces into an incremental non-linear algorithm using the Newmark method with average accelerations (implicit), as in [11.37]

We demonstrate in what follows the inclusion of displacement-dependent loads into an algorithm which has also been used in [11.37]. With this in mind, we consequently choose the Newmark method with $\beta = 1/4, \gamma = 1/2$. According to equs. (10.2.10) and (10.2.11), the basic expressions of the method are

$$\boldsymbol{r}_{\mathrm{k}+1} = \boldsymbol{r}_{\mathrm{k}} + \Delta t \dot{\boldsymbol{r}}_{\mathrm{k}} + \frac{\Delta t^2}{4}(\ddot{\boldsymbol{r}}_{\mathrm{k}} + \ddot{\boldsymbol{r}}_{\mathrm{k}+1}) \tag{11.6.61}$$

$$\dot{\boldsymbol{r}}_{\mathrm{k}+1} = \dot{\boldsymbol{r}}_{\mathrm{k}} + \frac{\Delta t}{2}(\ddot{\boldsymbol{r}}_{\mathrm{k}} + \ddot{\boldsymbol{r}}_{\mathrm{k}+1}) \tag{11.6.62}$$

These are supplemented here by the non-linear dynamic equilibrium relation at the end of the step. Ignoring damping, we have

$$\boldsymbol{M}\ddot{\boldsymbol{r}}_{\mathrm{k}+1} + \boldsymbol{R}_{e|\mathrm{k}+1} = \boldsymbol{R}(\boldsymbol{r}_{\mathrm{k}+1}) \tag{11.6.63}$$

Note that the external loads now also depend on the displacements. In analogy with the algorithm for linear systems developed in section 10.2, the accelerations at the end of the step are deduced from equ. (11.6.61)

$$\ddot{\boldsymbol{r}}_{\mathrm{k}+1} = \frac{4}{\Delta t^2}[\boldsymbol{r}_{\mathrm{k}+1} - \boldsymbol{r}_{\mathrm{k}} - \Delta t \dot{\boldsymbol{r}}_{\mathrm{k}}] - \ddot{\boldsymbol{r}}_{\mathrm{k}} \tag{11.6.64}$$

Substituting equ. (11.6.64) into equ. (11.6.63), we find

$$\frac{4}{\Delta t^2}\boldsymbol{M}\boldsymbol{r}_{\mathrm{k}+1} + \boldsymbol{R}_{e|\mathrm{k}+1} = \boldsymbol{R}(\boldsymbol{r}_{\mathrm{k}+1}) + \frac{4}{\Delta t^2}(\boldsymbol{r}_{\mathrm{k}} + \Delta t \dot{\boldsymbol{r}}_{\mathrm{k}}) + \boldsymbol{M}\ddot{\boldsymbol{r}}_{\mathrm{k}} \tag{11.6.65}$$

This represents a non-linear set of equations for the calculation of the displacements $\boldsymbol{r}_{\mathrm{k}+1}$ at the end of the step deduced from the existing information $\boldsymbol{r}_{\mathrm{k}}, \dot{\boldsymbol{r}}_{\mathrm{k}}$ and $\ddot{\boldsymbol{r}}_{\mathrm{k}}$ at the beginning of the step. Having calculated $\boldsymbol{r}_{\mathrm{k}+1}$, we find $\ddot{\boldsymbol{r}}_{\mathrm{k}+1}$ via equ. (11.6.64) and then $\dot{\boldsymbol{r}}_{\mathrm{k}+1}$ from equ. (11.6.62). This completes formally the operations within the current time-step.

In order to solve the non-linear system of equ. (11.6.65) iteratively according to Newton-Raphson or a modified Newton-Raphson procedure, we require a linearisation of the system as we already expounded in Vol. III. In analogy to statics, we interpret equ. (11.6.65) in the form of a dynamic non-equilibrium force $\boldsymbol{R}_u$, which has to be reduced to zero. We have at the state $\boldsymbol{r}_{\mathrm{k}+1}$

$$\boldsymbol{R}_{u|\mathrm{k}+1} = \frac{4}{\Delta t^2}\boldsymbol{M}\boldsymbol{r}_{\mathrm{k}+1} + \boldsymbol{R}_{e|\mathrm{k}+1} - \boldsymbol{R}(\boldsymbol{r}_{\mathrm{k}+1}) - \frac{4}{\Delta t^2}(\boldsymbol{r}_{\mathrm{k}} + \Delta t \dot{\boldsymbol{r}}_{\mathrm{k}}) - \boldsymbol{M}\ddot{\boldsymbol{r}}_{\mathrm{k}}$$

$$= \boldsymbol{o} \tag{11.6.66}$$

Inevitably, we discover that for the j-th iterated value $\boldsymbol{r}^{\mathrm{j}}_{\mathrm{k}+1}$, $\boldsymbol{R}_u$ does not vanish

$$\boldsymbol{R}^{\mathrm{j}}_{u|\mathrm{k}+1} \neq \boldsymbol{o} \tag{11.6.67}$$

Consequently, we require an increment $\Delta \boldsymbol{r}^{\mathrm{j}}_{\mathrm{k+1}}$ for which $\boldsymbol{R}_u(\boldsymbol{r}^{\mathrm{j}}_{\mathrm{k+1}} + \Delta \boldsymbol{r}^{\mathrm{j}}_{\mathrm{k+1}})$ is reduced to zero. Applying a consistent linearisation, we have

$$\boldsymbol{R}_u(\boldsymbol{r}^{\mathrm{j}}_{\mathrm{k+1}} + \Delta \boldsymbol{r}^{\mathrm{j}}_{\mathrm{k+1}}) \approx \boldsymbol{R}^{\mathrm{j}}_{u|\mathrm{k+1}} + \frac{\partial \boldsymbol{R}_u}{\partial \boldsymbol{r}_{\mathrm{k+1}}} \Delta \boldsymbol{r}^{\mathrm{j}}_{\mathrm{k+1}} = \boldsymbol{o} \tag{11.6.68}$$

Considering now the displacement-dependent external forces, see e.g. equs. (11.6.38) and (11.6.39), equ. (11.6.66) yields the system matrix

$$\boldsymbol{K}_{\text{eff}} = \frac{\partial \boldsymbol{R}_u}{\partial \boldsymbol{r}_{\mathrm{k+1}}} = \frac{4}{\Delta t^2} \boldsymbol{M} + \boldsymbol{K}_e + \boldsymbol{K}_G - \boldsymbol{K}_R \tag{11.6.69}$$

which has to be evaluated at $\boldsymbol{r}^{\mathrm{j}}_{\mathrm{k+1}}$. The inclusion of the force correction matrix $\boldsymbol{K}_R$ generally induces a non-symmetric form of this matrix. Thus, a special equation solver has to be applied. In the case of small time-steps – especially for a diagonal mass matrix – a stabilising effect will be achieved by $\boldsymbol{M}$. The transition to the limiting case of statics is obtained effortlessly by the omission of inertia effects, i.e. the mass matrix. The relevant increment $\Delta \boldsymbol{r}^{\mathrm{j}}_{\mathrm{k+1}}$ is now deduced simply from

$$\boldsymbol{K}_{\text{eff}} \Delta \boldsymbol{r}^{\mathrm{j}}_{\mathrm{k+1}} = -\boldsymbol{R}^{\mathrm{j}}_{u|\mathrm{k+1}} \tag{11.4.70}$$

where the right-hand side is given by equ. (11.6.66). Noting the relation

$$\boldsymbol{r}^{\mathrm{j+1}}_{\mathrm{k+1}} = \boldsymbol{r}^{\mathrm{j}}_{\mathrm{k+1}} + \Delta \boldsymbol{r}^{\mathrm{j}}_{\mathrm{k+1}} \tag{11.6.71}$$

the algorithmic argument is evident.

11.6.5 A comment on the stability response of a system under displacement-dependent loads

In Vol. III, we already discussed stability problems, but restricted ourselves to purely static ones, dealing with linear or linearised non-linear buckling problems and applying the so-called equilibrium method. We asked ourselves whether only one definite state of equilibrium exists for a particular load factor λ, as in linear statics, or perhaps several (neutral point). The latter case resulted in the requirement

$$\boldsymbol{K}_t \Delta \boldsymbol{r} = \boldsymbol{o} \tag{11.6.72}$$

which can only be fulfilled if $\boldsymbol{K}_t(\lambda)$ is singular. This is indeed the case for a critical load factor λ_{crit}. The criterion in question can also be applied to displacement-dependent loads as long as the force correction matrix $\boldsymbol{K}_R$ incorporated into $\boldsymbol{K}_t$ is symmetric, i.e. the loads involved are conservative. However, we have already ascertained that non-conservative displacement-dependent loads generally give rise to non-symmetrical force corrections. Bolotin [11.38] and Ziegler [11.39, 11.35] have demonstrated that the so-called static procedures, like the equilibrium or the equivalent energy method, then no longer provide reliable information on a loss of

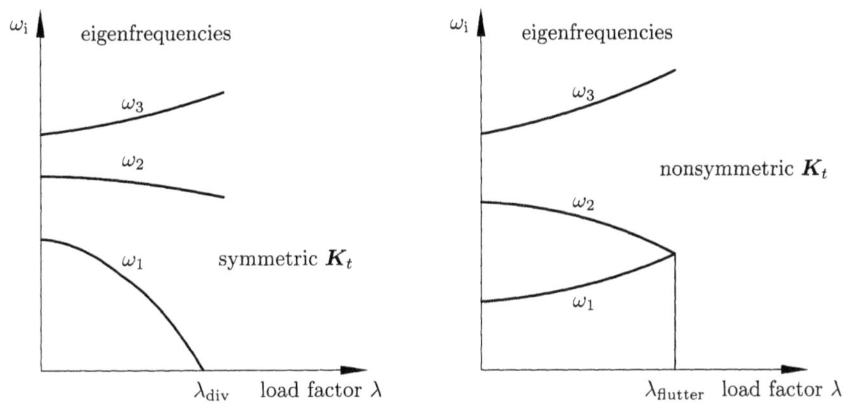

Fig. 11.6.6 Loss of stability for displacement-dependent loads

stability. In fact, such systems can become unstable due to bifurcation, but also due to flutter-like phenomena. In the latter case, static procedures necessarily fail to yield a reliable estimate of stability. Hence, in the presence of non-symmetric force correction matrices, we have to examine the equilibrium at a particular point of loading by dynamic criteria – even for purely static load applications. Let us assume that we have reached a state of equilibrium defined by $\boldsymbol{r}_k$ in the course of a loading process. Our structure is then allowed to execute small oscillations at a constant load intensity about this position of equilibrium. This can be expressed as

$$[\boldsymbol{K}_e + \boldsymbol{K}_G - \boldsymbol{K}_R - \omega_i^2 \boldsymbol{M}]\boldsymbol{x}_i = \boldsymbol{o} \tag{11.6.73}$$

This is formally the same equation as the one we already know for free undamped oscillation, formulated here, however, with the current tangential stiffness $\boldsymbol{K}_t = \boldsymbol{K}_e + \boldsymbol{K}_G - \boldsymbol{K}_R$ which may, moreover, be non-symmetric. How does an impending loss of stability manifest itself? Let us first consider the classical case of a symmetric tangential stiffness. Here, all eigenfrequencies are positive real. We know, however, that the tangential stiffness matrix becomes singular when approaching the point of instability. This means that at least one eigenfrequency tends to zero (fig. 11.6.6, left) and divergence occurs. In the case of a non-symmetric tangential stiffness matrix, the eigenfrequencies are real or complex. A loss of stability can take place when two consecutive eigenfrequencies coalesce. Beyond the critical load factor, complex conjugate eigenfrequencies arise and divergent oscillations occur (flutter).

11.6.6 Computational demonstrations

We divide the following examples into four groups

- simple non-linear calculations with displacement-dependent loads
- onset of instabilities with divergence
- onset of instabilities with flutter
- dynamic calculations with non-stationary displacement-dependent loads.

We begin with the first and simplest example.

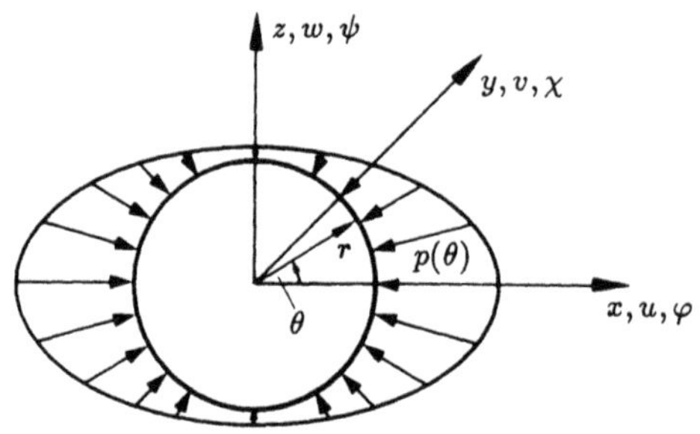

ring-idealisation: 18 elements, 53 unkown (1/4–ring)

geometrical data:

radius	$r = 100$ cm
moment of inertia	$I = 1.666667$ cm^4
cross-sectinal area	$A = 20$ cm^2

material data:

Young's modulus	$E = 2.1 \cdot 10^7$ N/cm^2
Poisson's ratio	$\nu = 0.3$

load:
non-conservative, nonuniform normal pressure $p(\theta) = p_0(1 + e\cos 2\theta)$

Fig. 11.6.7
Circular ring under nonuniform normal pressure: idealisation and data

Large deflections of circular ring under nonuniform normal pressure [11.36]

A quarter of the circular ring is idealised with 18 beam elements. The data of the structure and the loading are shown in fig. 11.6.7. The analysis presumes a very flexible ring which is treated as an elastica. In this case, the axial extension and the shear deformation of the ring are negligible and the bending moment is proportional to the change of curvature. The nonuniform pressure applied over the circumference is expressed as

$$p = p_0(1 + e\cos 2\theta)$$

where e defines the deviation from the constant load distribution; it is a follower load and thus non-conservative. In the course of the loading process, no instability occurs unless the load is distributed uniformly ($e = 0$). Consequently, we can adopt the assumption of double symmetry and need only idealise one quarter of the ring. Figure 11.6.8 demonstrates the highly non-linear character of this problem. The high quality of the results for a series of parameters e is demonstrated by a comparison with an analytical solution due to Seide-Jamjoom [11.40]. In the case of a small eccentricity ($e = 0.1$) and a value of $\bar{p}_0$ approaching 3, we observe a distinctly non-linear response. This load value represents the critical load when the pressure distribution becomes uniform. Figure 11.6.9 illustrates the considerable deformations which take place in this system. Significant changes in the direction of the local pressure loading are, of course, associated with this deformability.

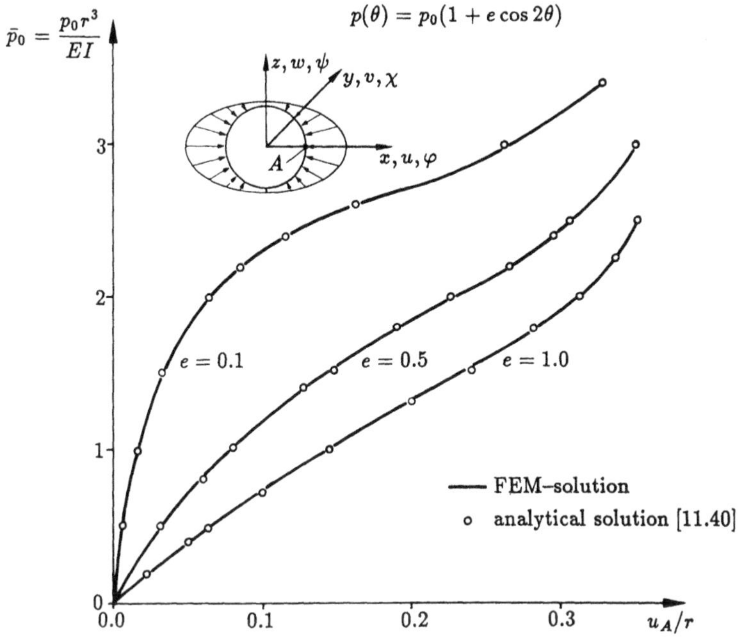

Fig. 11.6.8 Circular ring under normal pressure: load-displacement diagram of point A

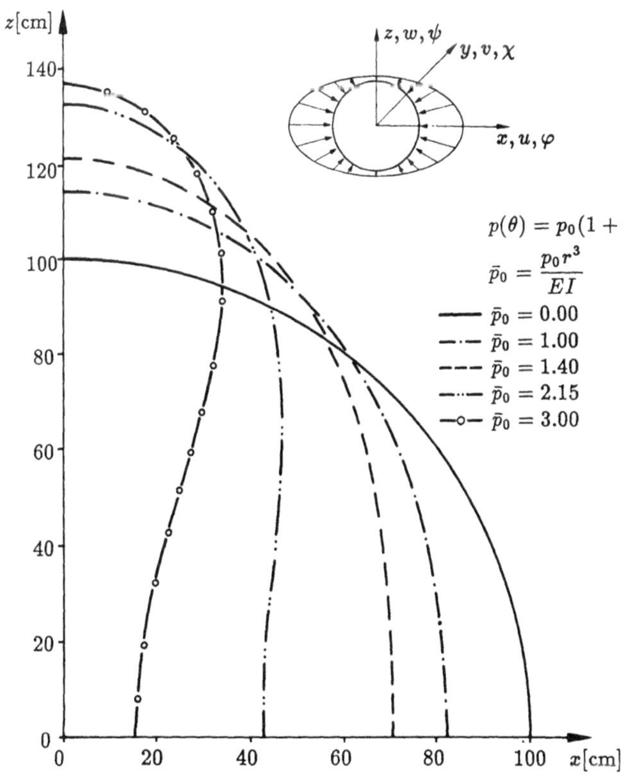

Fig. 11.6.9

Circular ring under normal pressure: deformed geometry of ring centre line for e=1.0

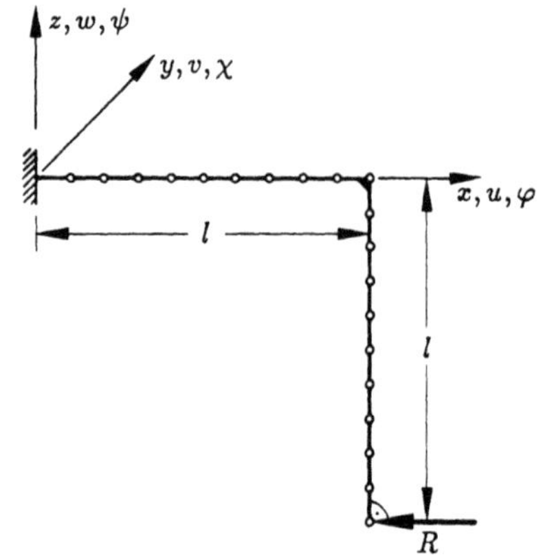

beam idealisation: 20 elements, 60 unknowns

geometrical data:

length	$l = 24$ cm
moment of inertia	$I = 0.135$ cm^4
cross-sectional area	$A = 0.18$ cm^2

material data:

Young's modulus	$E = 7.124 \cdot 10^6$ N/cm^2
Poisson's ratio	$\nu = 0.3$

load:

non-conservative, normal (follower) tip load R

Fig. 11.6.10 Right-angle frame under non-conservative transverse end load: geometry, idealisation and data

Large deflections of right-angle frame under non-conservative end force [11.36, 11.41]

Figure 11.6.10 shows a right-angle frame subjected to a follower load at the free end. The frame is idealised with 2×10 beam elements. Figure 11.6.10 illustrates the dimensions, elastic properties and loading. Figure 11.6.11 demonstrates the complicated non-linear load-deflection response of this structure by means of the three displacement components at node 1. The force correction matrix in this problem is, as we have already ascertained, non-symmetric; consequently, the tangential stiffness is, too. In order to calculate the incremental displacements, we need an equation solver for non-symmetric coefficient matrices, which is provided by the LARSTRAN 80 system developed at the ISD/ICA and used in the present case. This is often not available in traditional FEM systems; sometimes, the attempt is made to avoid the effort involved in solving nonsymmetrical equations. A standard method of symmetrising the tangential stiffness is

$$\boldsymbol{K}_{t\,\mathrm{appr}} = \tfrac{1}{2}[\boldsymbol{K}_t + \boldsymbol{K}_t^{\mathrm{t}}]$$

Figure 11.6.11 also shows how the enforced symmetrisation performs. Up to a load of $R = 2.8\,$kN, the aforementioned approximation appears to yield the same results as the procedure with the correct nonsymmetrical tangential stiffness. Figure 11.6.13 demonstrates, however, that this success can only be gained at the cost of an effort which increases proportionately to the load intensity: the number of equilibrium iterations increases continuously and is finally no longer acceptable. Moreover, the method of symmetrisation breaks down completely for loads over 2.8 kN and yields erroneous results. This is shown by the load-displacement diagram in fig. 11.6.11 and also by the deflected structure at various stages of loading in fig. 11.6.12. For a loading of about 5 kN, the symmetrised tangential stiffness simulates a bifurcation which does not in fact exist.

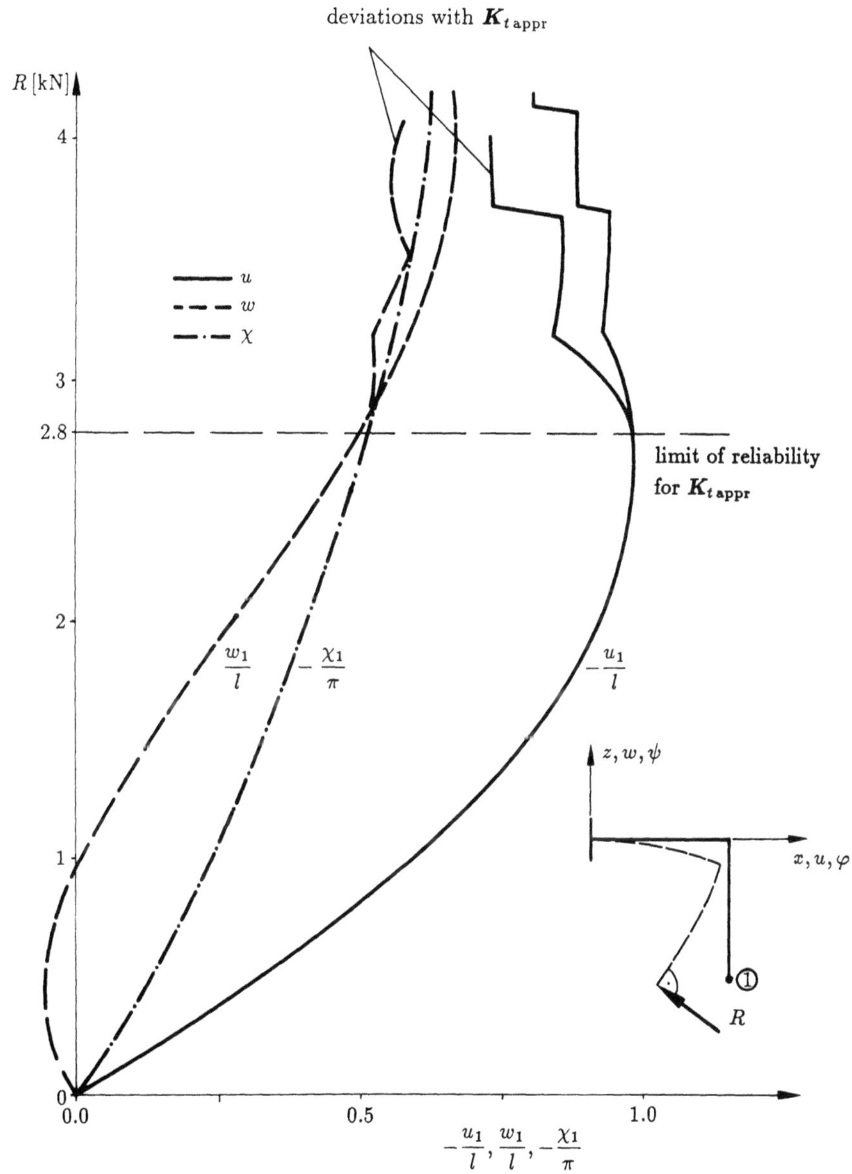

Fig. 11.6.11 Right-angle frame under non-conservative transverse end load: load-displacement diagram of point 1

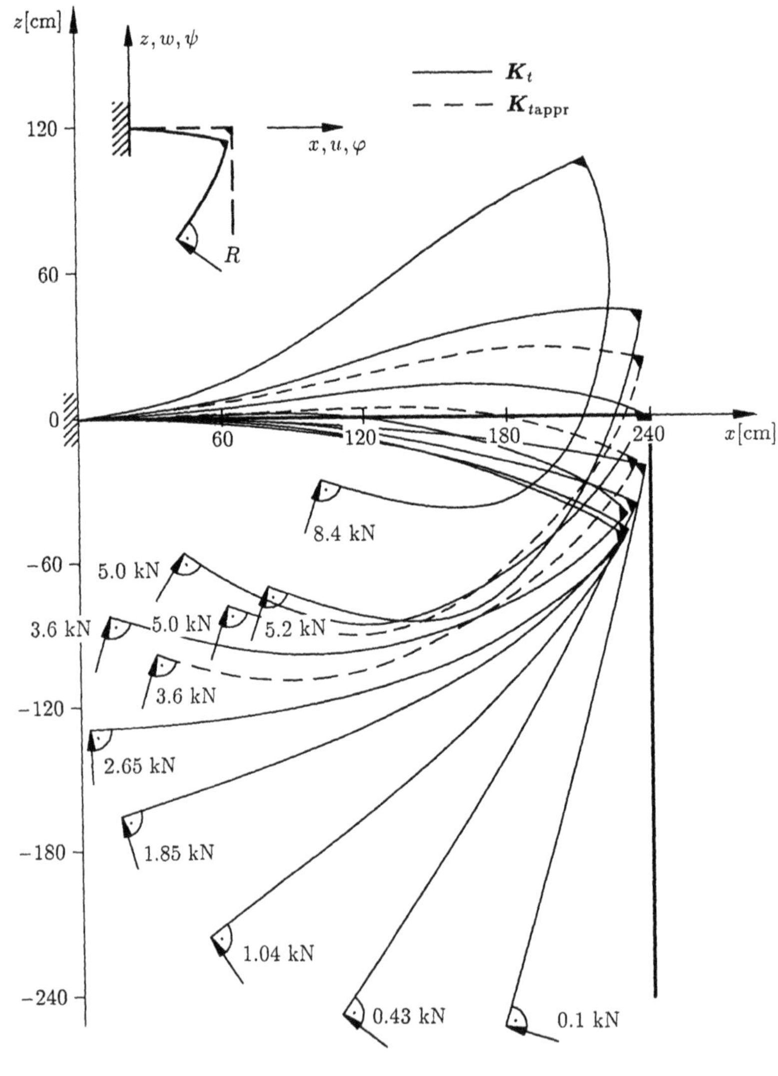

Fig. 11.6.12 Right-angle frame under non-conservative transverse end load: deformed geometry of centre line

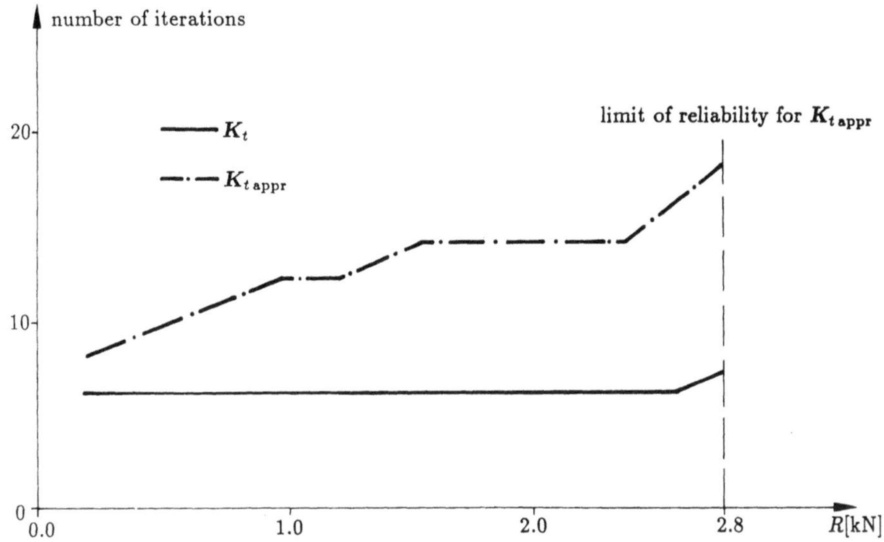

Fig. 11.6.13 Number of equilibrium iterations increasing with the load intensity

Large deflections and bifurcation instability of hinged right-angle frame under non-conservative transverse force [11.36, 11.41]

Figure 11.6.14 presents a hinged right-angle frame, idealised with 20 beam elements and loaded eccentrically with a follower force. From the start, the example demonstrates a non-linear response (fig. 11.6.16) and experiences a loss of stability by divergence for both conservative and follower loads. A non-conservative load apparently induces a stabilising effect on the system, as compared with a conservative one. In the case of a direction-preserving (conservative) load, divergence already occurs at $\lambda = 18.55$, but not until $\lambda = 36$ in the case of the follower load. This increase in the critical load can also be seen in fig. 11.6.17 which displays the displaced state of the structure at various stages of both conservative and non-conservative loading. The horizontal component of the follower force clearly has a pre-stressing effect on the horizontal beam, thus tending to pull the frame back to its original state. As a result, the deflections are considerably smaller in the case of a follower load than in the case of a conservative load. We now return to fig. 11.6.16 and note that the correct divergence point cannot be attained in the case of a symmetrised tangential stiffness, since convergence loss takes place beforehand. The same can be observed in fig. 11.6.15. In addition, with increasing load, we note again a marked increase in the number of equilibrium iterations when applying the symmetrised approximation.

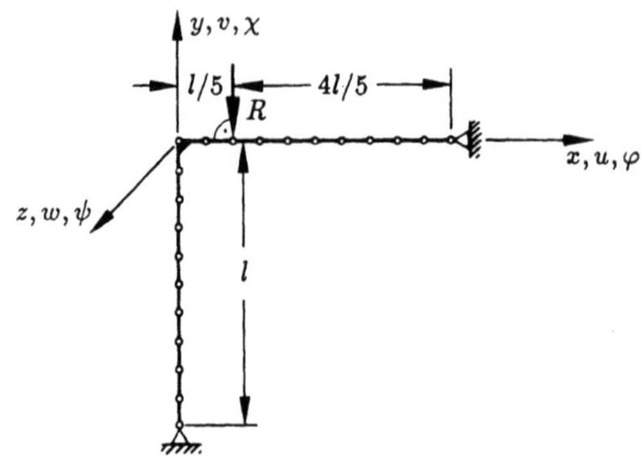

beam idealisation: 20 elements, 59 unknowns

geometrical data:
- length $l = 120$ cm
- moment of inertia $I = 2$ cm^4
- cross-sectional area $A = 6$ cm^2

material data:
- Young's modulus $E = 7.2 \cdot 10^6$ N/cm^2
- Poisson's ratio $\nu = 0.3$

load:
non-conservative normal (follower) load R, eccentrically applied

Fig. 11.6.14
Hinged right-angle frame under non-conservative transverse load: geometry, idealisation and data

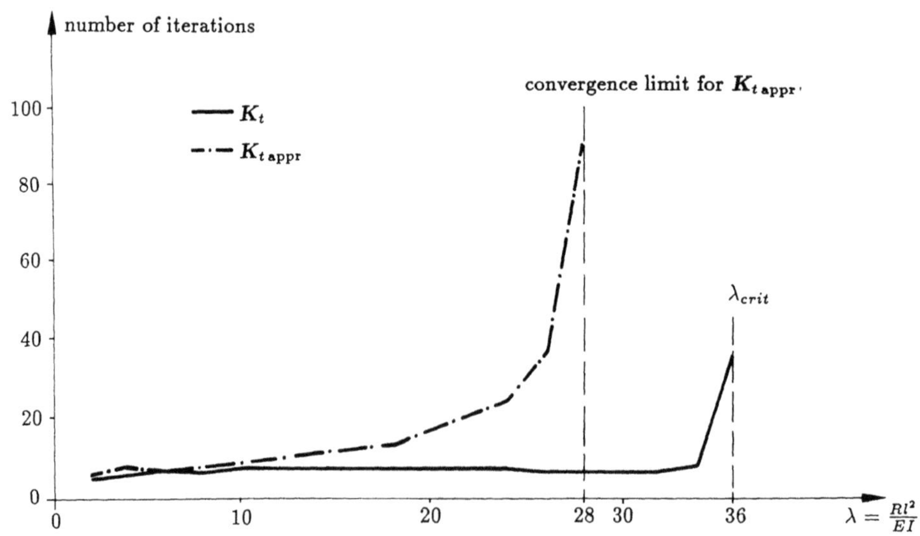

Fig. 11.6.15 Hinged right-angle frame under non-conservative transverse load; number of equilibrium iterations in dependence of the applied loading

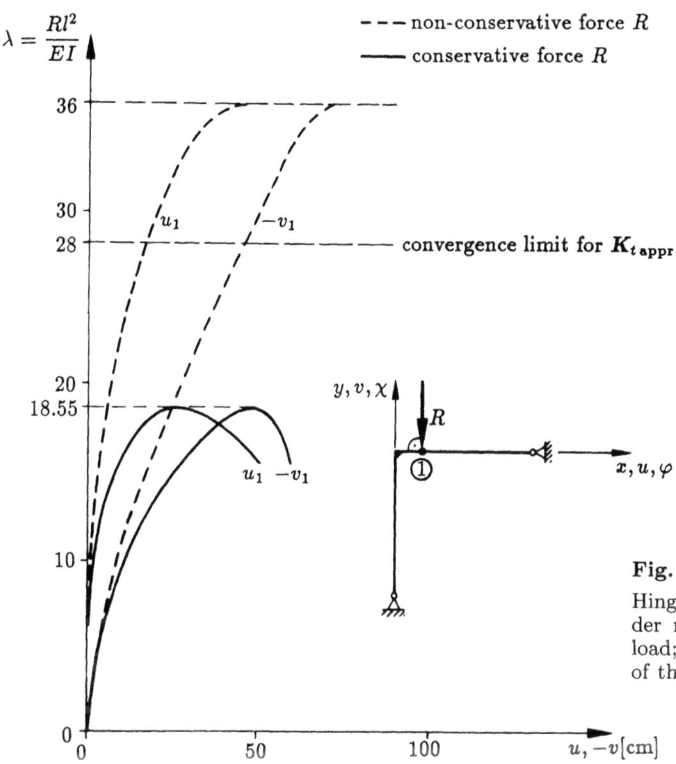

Fig. 11.6.16

Hinged right-angle frame under non-conservative transverse load; load-displacement diagram of the loading point

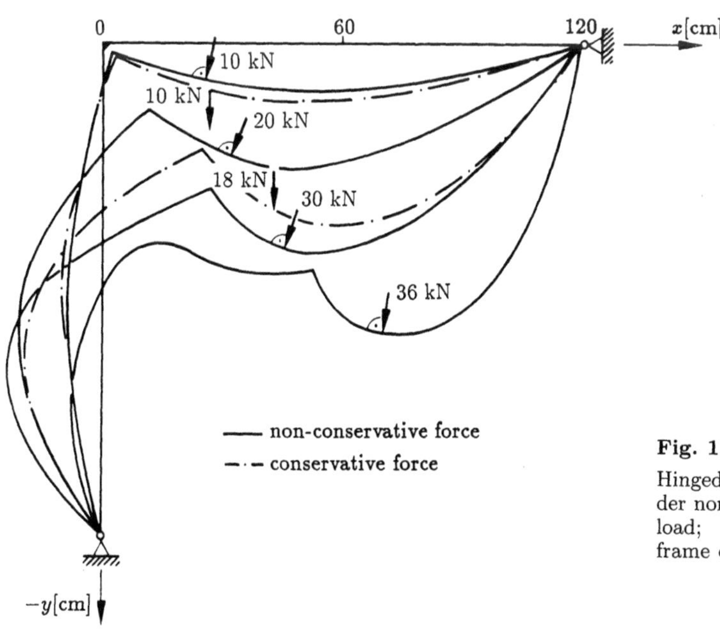

Fig. 11.6.17

Hinged right-angle frame under non-conservative transverse load; deformed geometry of frame centre line

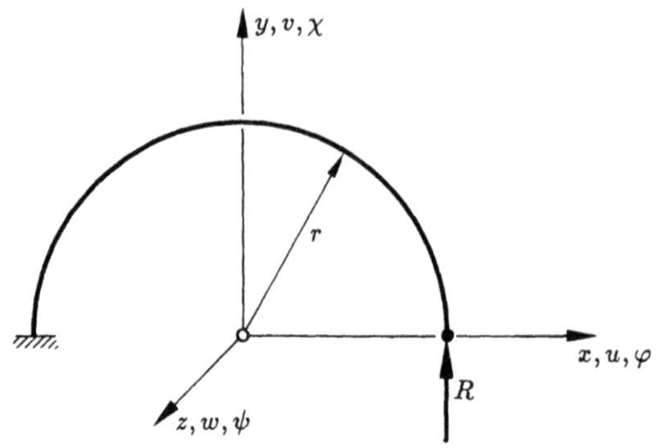

beam idealisation: 36 elements, 108 unknowns

geometrical data:
- radius $r = 50$ cm
- moment of inertia $I = 0.5$ cm^4
- cross-sectional area $A = 1$ cm^2

material data:
- Young's modulus $E = 7.2 \cdot 10^6$ N/cm^2
- Poisson's ratio $\nu = 0.3$

load:
- non-conservative tangential (follower) tip load R

Fig. 11.6.18

Semi-circular arch under tangential follower load: geometry, idealisation and data

Large displacements and flutter of a semi-circular ring under non-conservative tangential end load [11.36, 11.41]

We now move on into the realm of dynamics. Figure 11.6.18 shows a semi-circular cantilever ring under a tangential follower load at the free end. The arch-beam is idealised with 36 beam elements and fig. 11.6.19 demonstrates right away a highly non-linear static response. We observe to begin with that the method of a symmetrised tangential stiffness fails at a relatively early stage. In this example, the loss of stability does not occur due to a singularity of tangential stiffness (divergence), but due to flutter at $R = 2.4\,\text{kN}$. At this load, oscillations are initiated with increasing amplitude. Figure 11.6.20 shows the considerable modification of the geometry in the course of loading. When flutter arises, the half-ring is already bent right back.

We now proceed to the last group of examples, but keep to the semi-circular ring. So far, we have calculated a quasistatic pre-flutter path. However, static calculations are no longer valid when two consecutive eigenfrequencies merge, so that we must apply dynamic considerations in order to gain further information on post-flutter behaviour. This is analogous to linear buckling, where a linear analysis can yield the buckling load, whereas the post-buckling behaviour can only be analysed by means of a non-linear calculation. All the subsequent dynamic calculations were carried out with the Newmark method of averaged accelerations (see section 11.6.4) with correct nonsymmetrical tangential stiffness.

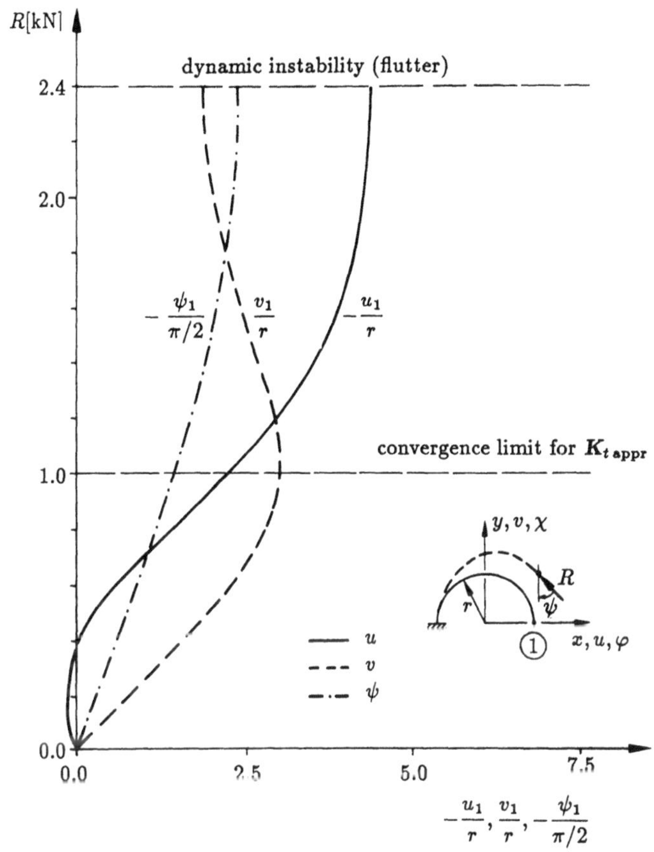

Fig. 11.6.19 Semi-circular arch under tangential follower load: load-displacement diagram of the loading point

Firstly, it is interesting to examine the dynamic behaviour before the onset of the critical flutter limit (fig. 11.6.21(a)). To this end, the load is increased from $R = 2.0\,\text{kN}$ to $R = 2.001\,\text{kN}$ and then kept constant. The load increase is executed in $\Delta t_0 = 1/(\sqrt{10}\ 10^5)$ s. The time increment of the numerical integration here and subsequently is $\Delta t = 1/(\sqrt{10}\ 10^3)$ s. The motion which sets in below the critical limit is obviously irregular, but not incited. We now repeat the operation, but this time on the flutter boundary $R = 2.431\,\text{kN}$ (fig. 11.6.2(b)). At first, an oscillation with increasing amplitude sets in, as could be predicted. Subsequently, however, the amplitude decreases and the whole process is repeated periodically. Thus, the system seems to stabilise itself dynamically in time. This is known as a limit cycle phenomenon and only takes place in the case of non-linear oscillators. The same behaviour occurs at a higher load level (fig. 11.6.21(c)). An additional harmonic excitation (fig. 11.6.21(d)) yields qualitatively the same picture. Thus, the critical flutter point by no means necessarily results in the destruction of the structure (note the analogy to the buckling of a shell, which also stabilises itself within an elastic regime). A really catastrophic build-up of amplitudes beyond the critical flutter point can only be achieved by means of a steadily increasing load R (fig. 11.6.21(e)).

Finally, in fig. 11.6.22, we investigate the effect of mass distribution on the results. While the preceding calculations presuppose lumped translational masses simulating a uniform mass distribution, a comparative calculation with a concentrated mass at the free end was also made (fig. 11.6.22(b)). The concentrated additional mass is ten times larger than the lumped masses. A constant load incrementation was also applied and initiated below the critical point. Both calculations display qualitatively the same response. Below the flutter limit, oscillations of small but constant amplitudes occur. In excess of the critical load, the vibration builds up. However, we observe a definite influence of the mass distribution on the flutter point.

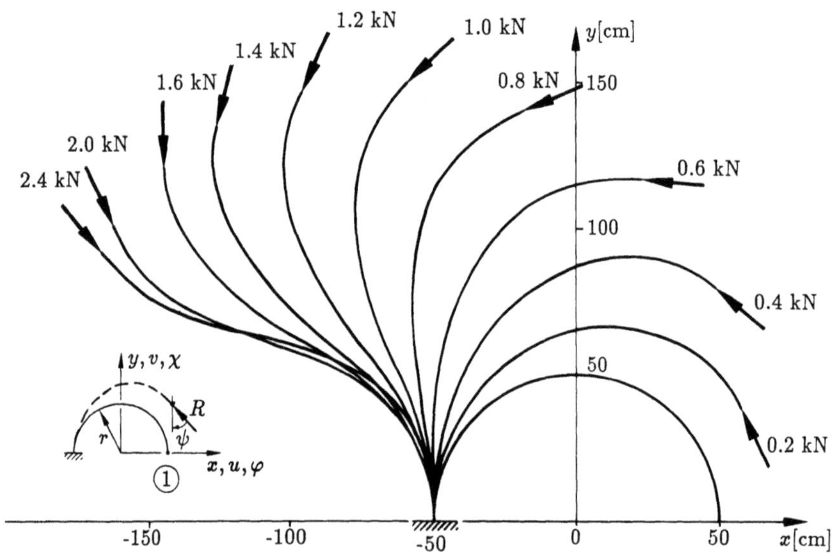

Fig. 11.6.20 Semi-circular arch under tangential follower load: deformed geometry at different load intensities

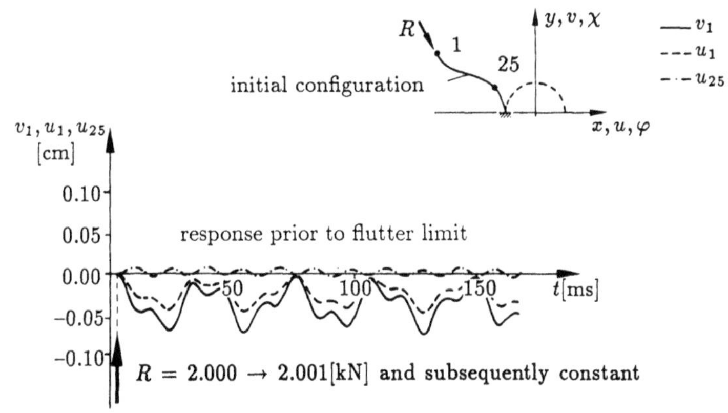

(a) dynamic behaviour generated by a small load increment below the flutter limit

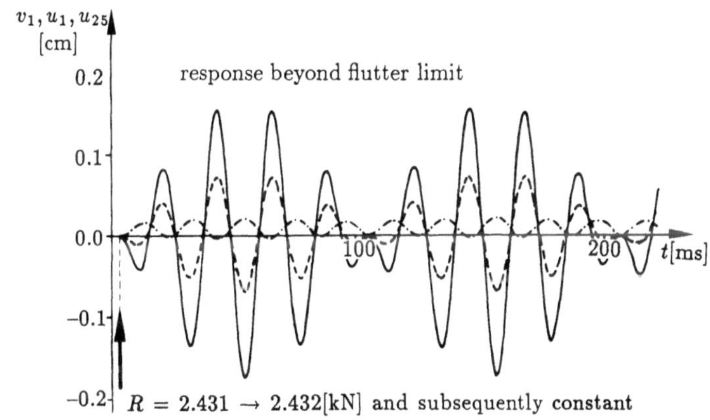

(b) application of small load increment when flutter limit is attained

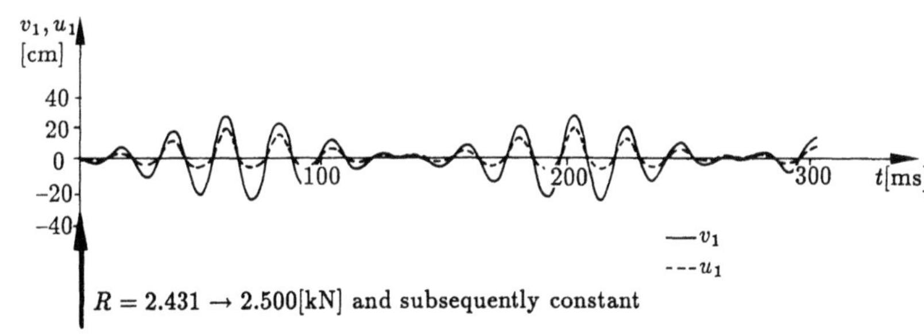

(c) application of larger load increment when flutter limit is attained

Fig. 11.6.21 Semi-circular arch under tangential follower load: some response histograms calculated with Newmark's method of average accelerations, *Part 1*

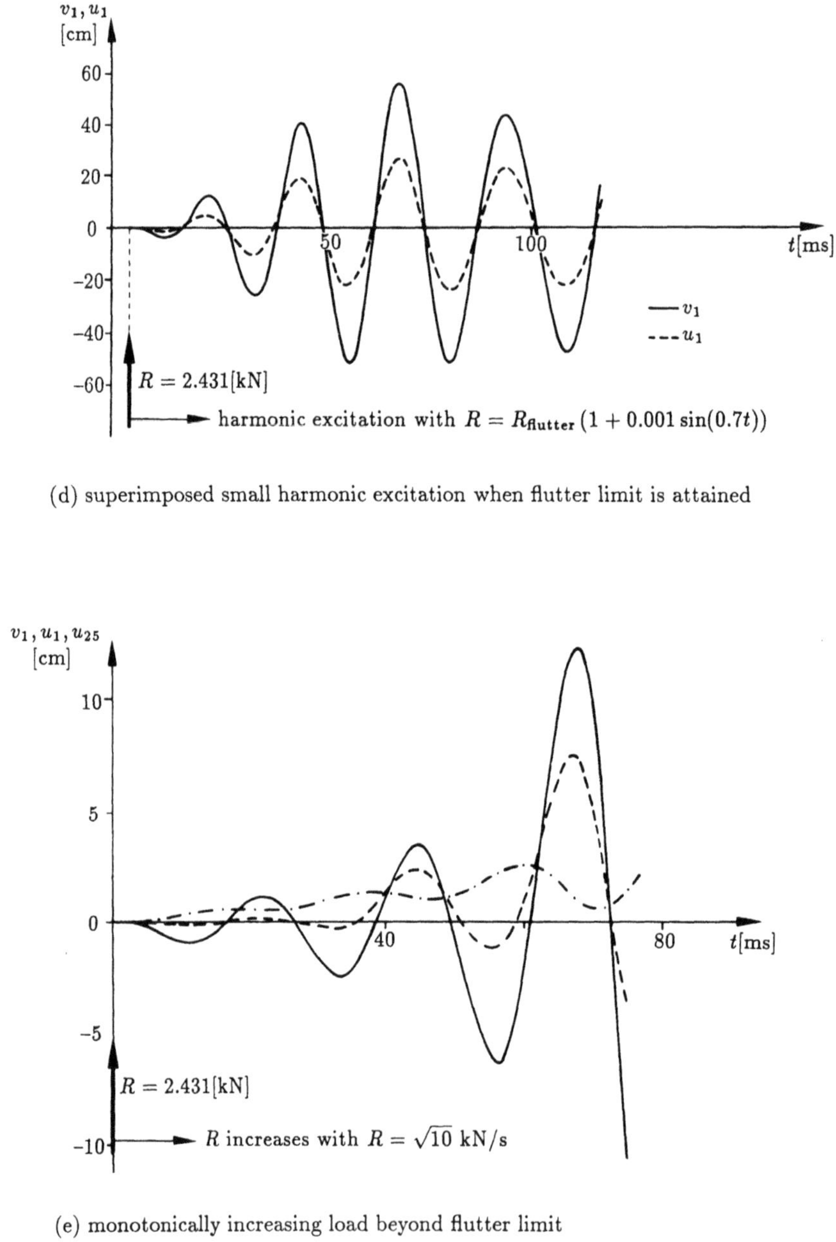

Fig. 11.6.21 Semi-circular arch under tangential follower load: some response histograms calculated with Newmark's method of average accelerations, *Part 2*

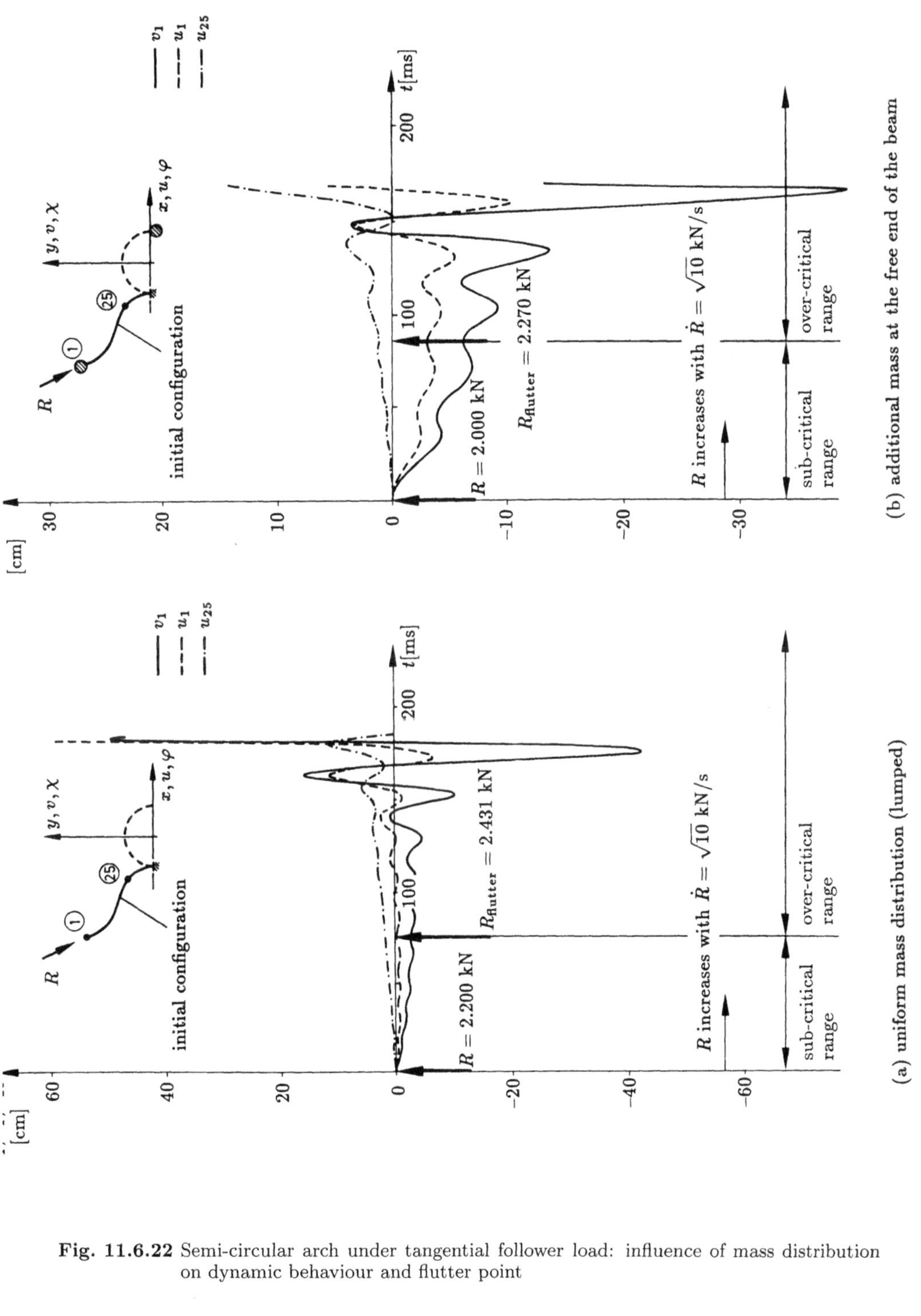

Fig. 11.6.22 Semi-circular arch under tangential follower load: influence of mass distribution on dynamic behaviour and flutter point

Lateral buckling of a cantilever subjected to non-conservative transverse force [11.36, 11.37, 11.44]

Up to the occurrence of instability, the cantilever, illustrated in fig. 11.6.23, does not give rise to large deflections. The differences in the critical loads are notable. In the case of a direction-preserving load, the loss of stability occurs in the form of divergence. With a follower load, flutter sets in at a considerably higher intensity. In the case of a linear stability analysis with follower load, we assume that the load only rotates about the x-axis, whereas in a non-linear calculation, a rotation of the loaded cross section, and thus of the load, can occur about all three axes. This also results in significant differences in the critical load.

As has already been mentioned, flutter of this system results from a coalescence of two eigenfrequencies, in this case a bending and a torsional mode (fig. 11.6.24). Figure 11.6.24 clearly indicates the behaviour of the imaginary and the real part of the complex circular frequencies as functions of time, and thus also of the load. When the flutter point is reached, a dynamic calculation must be applied. This was carried out with a steadily increasing load and a time-step of 10^{-3}s; in the post-flutter range, strongly diverging oscillations, typical of flutter, occur.

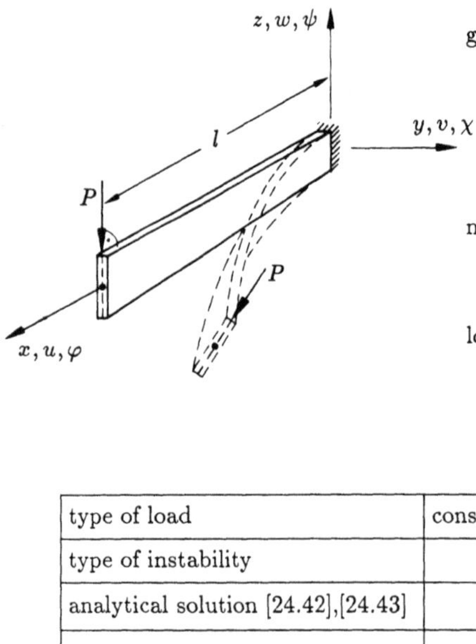

beam idealisation: 10 elements, 60 unknowns

geometrical data:
- length $l = 100$ cm
- moments of inertia $I_{yy} = 1$ cm^4, $I_{zz} = 0.125$ cm^4
- St.Venant torsion constant $J = 0.01$ cm^4
- cross-sectional area $A = 1$ cm^2

material data:
- Young's modulus $E = 10^4$ N/cm^2
- shear modulus $G = 0.5 \cdot 10^4$ N/cm^2

load:
- follower transverse force P at the free end

type of load	conservative tip load	non-conservative tip load
type of instability	divergence	flutter
analytical solution [24.42],[24.43]	0.100325	0.17475
FEM-solution	0.100563	linear 0.17494 non-linear 0.17082

Fig. 11.6.23 Lateral buckling of cantilever under non-conservative transverse end load: idealisation, data and some critical load values

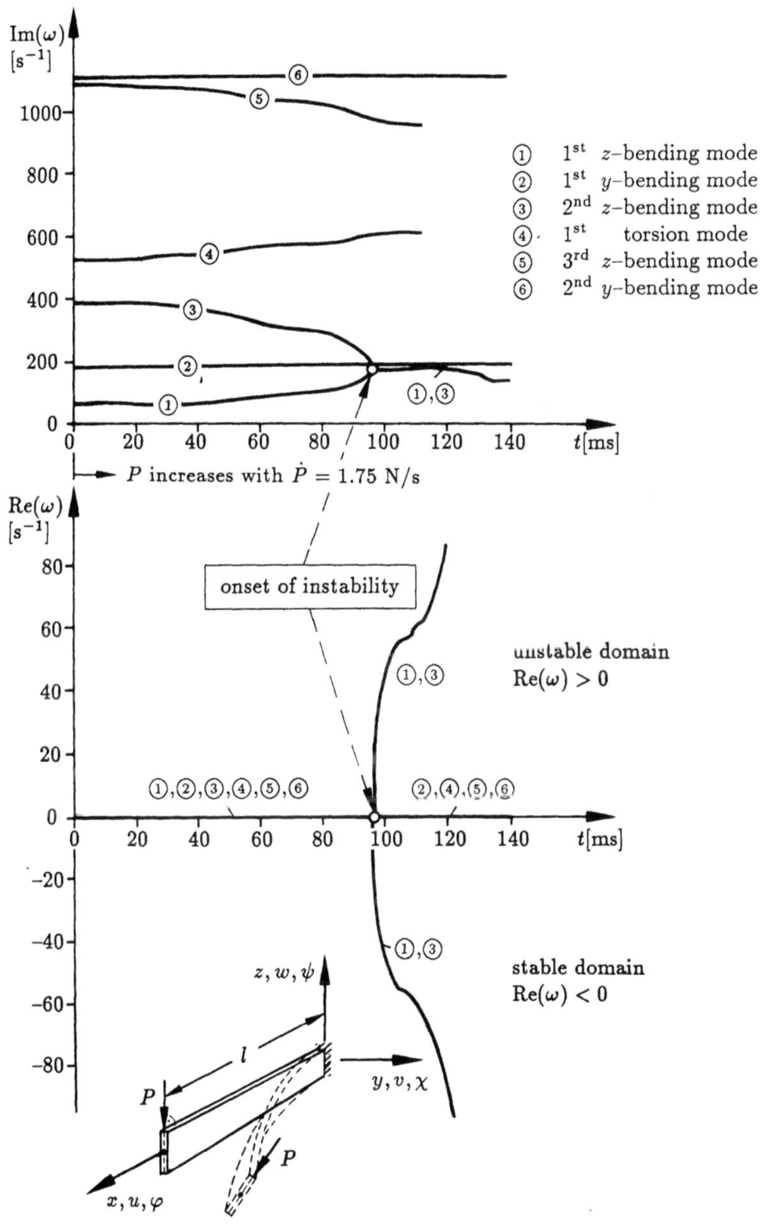

Fig. 11.6.24 Lateral buckling of cantilever under non-conservative transverse end load: evolution of eigenfrequncies

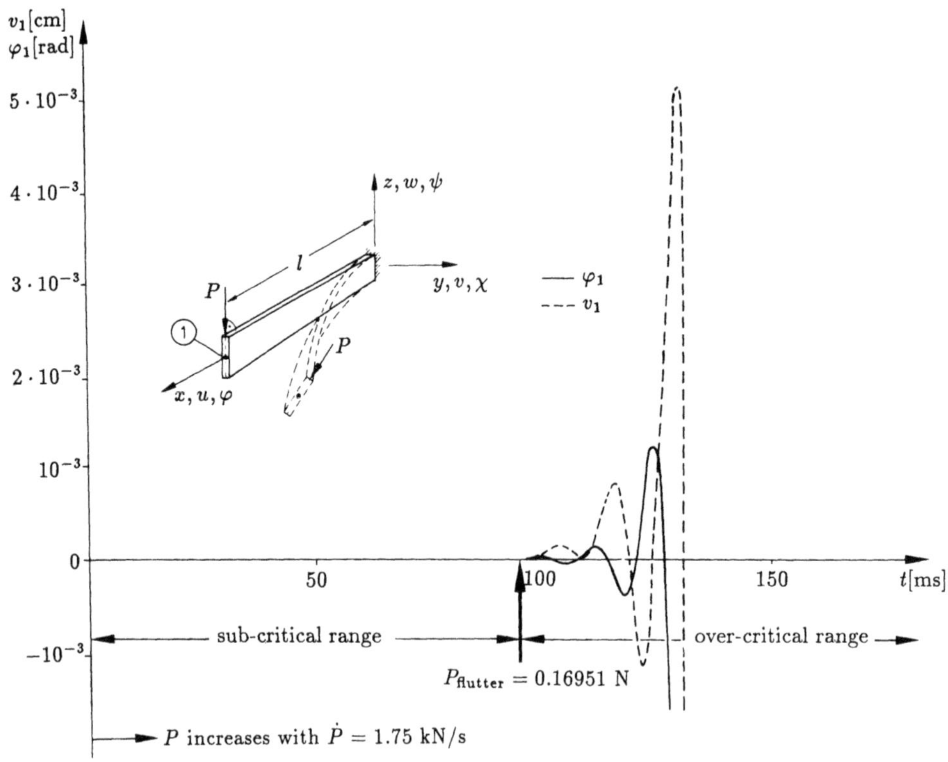

Fig. 11.6.25 Lateral buckling of cantilever under non-conservative transverse end load: response due to continuously increasing loading

In this final example, we extend the analysis to instationary follower loads (fig. 11.6.25). In such a case, the load intensity is made time-dependent; thus, in analogy to equ. (11.6.48), we have

$$\boldsymbol{P}(\boldsymbol{\vartheta}, t) = \boldsymbol{T}(\boldsymbol{\vartheta})\boldsymbol{P}_0(t)$$

where the load intensity vector is written

$$\boldsymbol{P}_0(t) = \{0 \quad 0 \quad -P_0 \sin \omega t\}$$

The excitation frequency $\omega = 100\,\text{Hz}$ lies between the two lowest eigenfrequencies. The load amplitude P_0 is taken to be equal to the critical flutter load in the stationary case, i.e. $P_0 = 0.17475\,\text{N}$. The time-step is 10^{-3}s as before. In fig. 11.6.26, we apply an impulse in the form of a torsion moment, $\Delta M_x = 10^{-5}$Ncm, which initiates the flutter. A highly diverging oscillation with an explosive increase of the amplitudes occurs which corresponds to the behaviour of a linear flutter model.

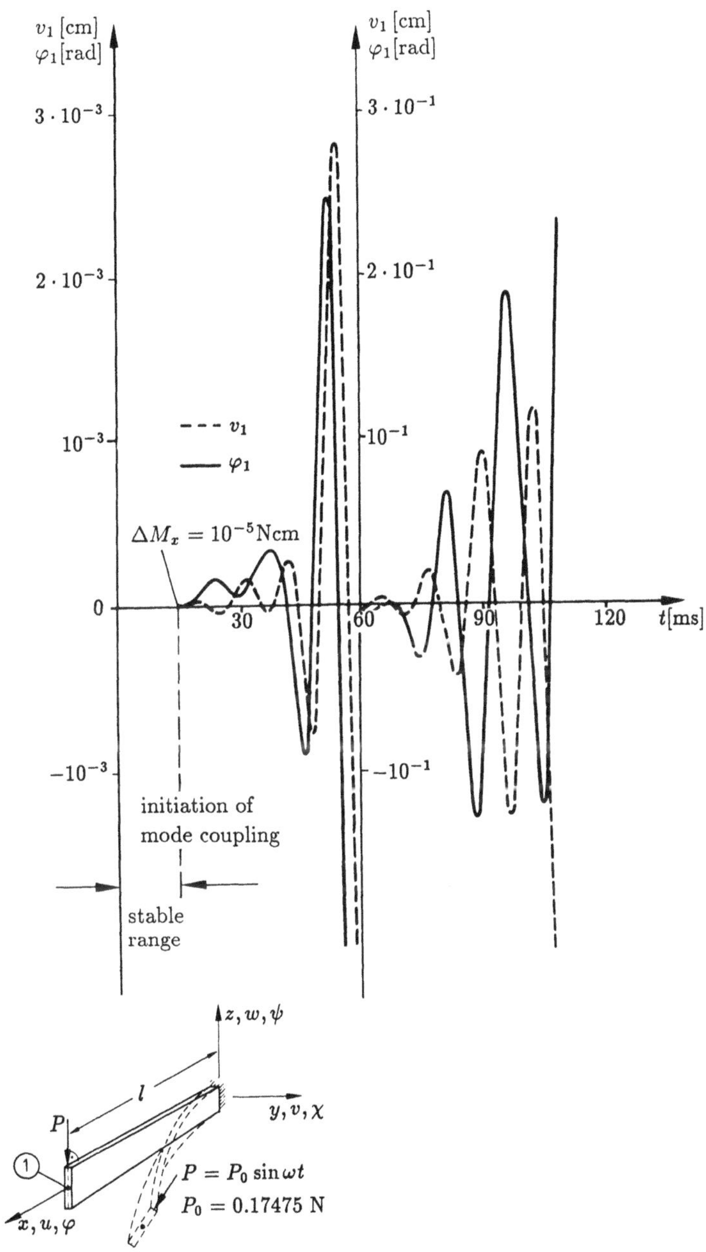

Fig. 11.6.26 Lateral buckling of cantilever under nonstationary non-conservative end load: post-flutter behaviour induced by an additional impulse loading

11.7 Velocity-dependent loads like gyroscopic effects

11.7.1 Introduction, general formulation and gyroscopic pseudo-damping for a beam

We begin our observations with the example of a system rotating uniformly about a fixed axis. Such a problem is a special case in our preceding discussion on generally moving systems given in Chapter 2. There, we observed that in a system under arbitrary motion, additional contributions to the forces of inertia may arise. Apart from a contribution to the external forces, $\boldsymbol{R}_{IO}(t)$, a displacement-dependent additional force emerges for uniformly rotating systems ($\dot{\boldsymbol{\omega}}_O = \boldsymbol{o}$)

$$\boldsymbol{Z}\boldsymbol{r}$$

where $\boldsymbol{Z}$ represents the so-called centrifugal field stiffness . This additional force was already discussed in the previous section in connection with displacement-dependent loads. A velocity-dependent additional force

$$\boldsymbol{G}\dot{\boldsymbol{r}}$$

also arises, and it is to this that we now turn our attention. The incremental expression of the dynamic equilibrium is given by

$$\boldsymbol{M}\ddot{\boldsymbol{r}} + [\boldsymbol{C} + \boldsymbol{G}]\dot{\boldsymbol{r}} + [\boldsymbol{K}_e + \boldsymbol{K}_G + \boldsymbol{Z}]\boldsymbol{r} = \boldsymbol{R}(t) - \boldsymbol{R}_{IO}(t) \tag{11.7.1}$$

(see equ. (2.3.31)). If in the course of a non-linear static analysis, linear calculations of natural modes of vibration are carried out at a state of equilibrium, as will be the case in the following examples, the right-hand side of the system vanishes. We already know from our discussion in Chapter 2 that the effective tangential stiffness is symmetrical. This also applies to the original damping matrix $\boldsymbol{C}$, but not to the Coriolis matrix $\boldsymbol{G}$. According to equ. (2.3.21)

$$\boldsymbol{G} = \int_V 2\rho\boldsymbol{\omega}^{\mathrm{t}}(\boldsymbol{\omega}_O \times \boldsymbol{\omega})\mathrm{d}V \tag{11.7.2}$$

is a skew-symmetric matrix. Naturally, this has consequences for the calculation of the eigenvalues. In the presence of a damping matrix $\boldsymbol{C}$, an effective general damping matrix $\boldsymbol{C} + \boldsymbol{G}$ arises, and a general eigenvalue procedure is necessary for non-symmetric matrices. When $\boldsymbol{C} = \boldsymbol{O}$, special algorithms are available which make use of the skew-symmetric form of $\boldsymbol{G}$. Special measures are necessary in any case in order to include the effect of $\boldsymbol{G}$. For the discretised structure, we replace the global integration over the whole system by an integration over the individual element volumes and a subsequent summation over the element contributions

$$\begin{aligned}\boldsymbol{G} &= \sum_{\mathrm{g}=1}^{\mathrm{m}} \mathbf{a}_{\mathrm{g}}^{\mathrm{t}}\left[\int_{V_{\mathrm{g}}} 2\rho\boldsymbol{\omega}^{\mathrm{t}}(\boldsymbol{\omega}_O \times \boldsymbol{\omega})\mathrm{d}V\right]\mathbf{a}_{\mathrm{g}} \\ &= \sum_{\mathrm{g}=1}^{\mathrm{m}} \mathbf{a}_{\mathrm{g}}^{\mathrm{t}}\boldsymbol{g}_{\mathrm{g}}\mathbf{a}_{\mathrm{g}}\end{aligned} \tag{11.7.3}$$

where

$$\boldsymbol{g}_{\mathrm{g}} = \int\limits_{V_{\mathrm{g}}} 2\rho\boldsymbol{\omega}^{\mathrm{t}}(\boldsymbol{\omega}_{O} \times \boldsymbol{\omega})\mathrm{d}V \tag{11.7.4}$$

is the gyroscopic element damping which, according to equ. (11.7.3), represents a building block for the construction of the global matrix $\boldsymbol{G}$. The assembly procedure is identical with that of $\boldsymbol{K}, \boldsymbol{C}$ and $\boldsymbol{M}$. Since the vector product $(\times)$ in equ. (11.7.4) may prove slightly disturbing to the reader, we recall our work on rotations in Vol. III and write

$$\boldsymbol{\omega}_{O} \times \boldsymbol{\omega} = \boldsymbol{\Theta}_{\Omega}\boldsymbol{\omega} \tag{11.7.5}$$

We recall that here, $\boldsymbol{\omega}$ is the deflection interpolation matrix in the element whilst $\boldsymbol{\omega}_{O}$ is the rotation vector of the system whose components are φ, χ, ψ

$$\boldsymbol{\omega}_{O} = \{\varphi \quad \chi \quad \psi\} \tag{11.7.6}$$

Thus, the matrix $\boldsymbol{\Theta}_{\Omega}$ becomes

$$\boldsymbol{\Theta}_{\Omega} = \begin{bmatrix} 0 & -\psi & \chi \\ \psi & 0 & -\varphi \\ -\chi & \varphi & 0 \end{bmatrix}_{\Omega} \tag{11.7.7}$$

and is, obviously, skew-symmetric. With equ. (11.7.4), we obtain

$$\boldsymbol{g}_{\mathrm{g}} = \int\limits_{V_{\mathrm{g}}} 2\rho\boldsymbol{\omega}^{\mathrm{t}}\boldsymbol{\Theta}_{\Omega}\boldsymbol{\omega}\mathrm{d}V \tag{11.7.8}$$

which confirms once more the skew-symmetrical form of $\boldsymbol{g}_{\mathrm{g}}$, and thus also of $\boldsymbol{G}$. We now follow the comments of Mai [11.45] who has dealt extensively with the analysis of gyroscopic shell and beam structures. In his work, he formulates, among other things, the Coriolis pseudo-damping. As the reader should already be familiar with interpolation modal forms for a straight beam, see e.g. Vol. I of this series, it suffices here to reproduce only the complete result in fig. 11.7.1, taken from [11.45]. The corresponding matrix $\boldsymbol{g}$ is evaluated in a local system, the x-axis of which points in the longitudinal axial direction of the beam. It is noticeable that all the linear interpolation forms (horizontal displacement u, torsion φ) are uncoupled. In the work quoted above, we also find the Coriolis matrix for simple plate elements like TRIB3 or TRUMP. However, since the modal forms $\boldsymbol{\omega}$ associated with these elements are detailed in Vol. I and equ. (11.7.8) can in an emergency also be integrated numerically, we may refrain here from developing the detailed analytical formulations.

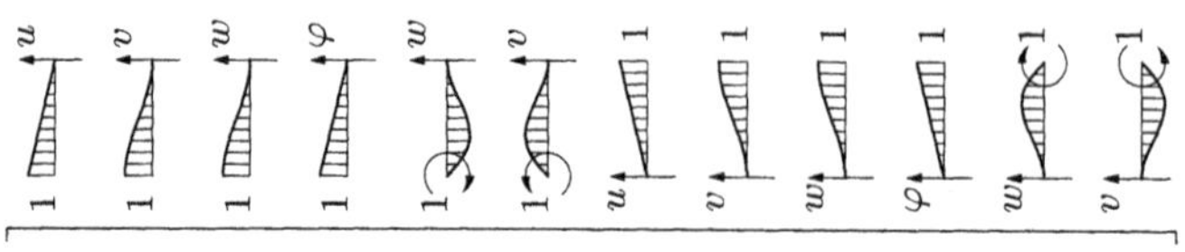

$$
\boldsymbol{g} = \frac{\rho A l}{420} \cdot \begin{bmatrix}
\cdot & -294\psi & 294\chi & \cdot & -42l\chi & -42l\psi & \cdot & -126\psi & 126\chi & \cdot & 28l\chi & 28l\psi \\
 & \cdot & -312\varphi & \cdot & 44l\varphi & \cdot & 126\psi & \cdot & -108\varphi & \cdot & -26l\varphi & \cdot \\
 & & \cdot & \cdot & \cdot & 44l\varphi & -126\chi & 108\varphi & \cdot & \cdot & \cdot & -26l\varphi \\
 & & & \cdot & \cdot & \cdot & \cdot & \cdot & \cdot & \cdot & \cdot & \cdot \\
 & & & & \cdot & -8l^2\varphi & 28l\chi & -26l\varphi & \cdot & \cdot & \cdot & 6l^2\varphi \\
 & & & & & \cdot & 28l\psi & \cdot & -26l\varphi & \cdot & -6l^2\varphi & \cdot \\
 & & & & & & \cdot & -294\psi & 294\chi & \cdot & 42l\chi & 42l\psi \\
 & & & & & & & \cdot & -312\varphi & \cdot & -44l\varphi & \cdot \\
 & & & & & & & & \cdot & \cdot & \cdot & -44l\varphi \\
 & & \text{skew symmetrical} & & & & & & & \cdot & \cdot & \cdot \\
 & & & & & & & & & & \cdot & -8l^2\varphi \\
 & & & & & & & & & & & \cdot
\end{bmatrix}
$$

Fig. 11.7.1 Gyroscopic pseudo-damping for the straight cylindrical beam [11.45]

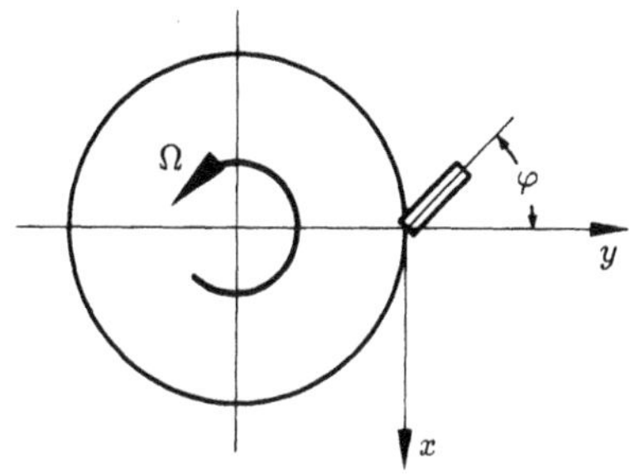

geometrical data:	
cross-sectional area	$A = 1.6{\cdot}10^{-3}$ in^2
moments of inertia	$I_{xx} = 4{\cdot}10^{-5}$ in^4
	$I_{zz} = 6{\cdot}10^{-5}$ in^4
St.Venant's constant	$J = 10^{-4}$ in^4
material data:	
Young's modulus	$E = 1.5 \cdot 10^7$ psi
Poisson's ratio	$\nu = 0.3$
density	$\rho = 2.5900{\cdot}10^{-4}$ lb s^2/in^4

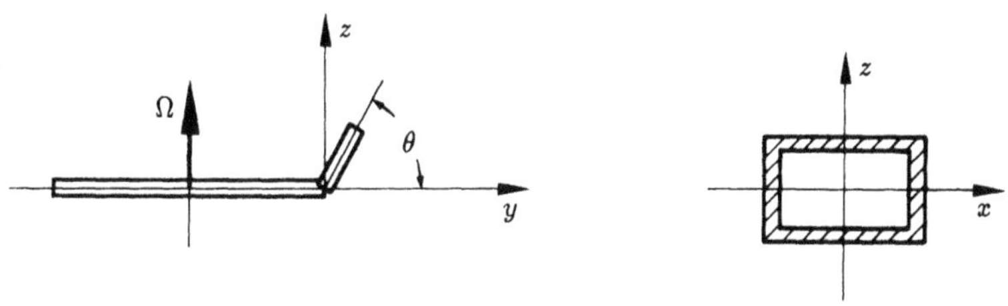

Fig. 11.7.2 Rotating beam: geometry and data

11.7.2 Coriolis effects in some calculation examples

Figure 11.7.2 shows a rigid circular ring with a radius $R = 80$ in onto which a very flexible beam of the length $l = 40$ in is mounted. The beam can be swivelled in the direction of the circumference (angle φ, in the following example 0) as well as in the axial direction (angle θ, in the following example 90°). The original example is taken from [11.46]. In this publication, however, the non-linear deformation behaviour in the revving-up range is not taken into account, although it is clearly significant. An eigenvalue analysis at zero rotational speed (fig. 11.7.3, left), which any reader could certainly carry out by hand, serves as a basis for comparison. Subsequently, the rotational velocity is increased in five increments to 41.8 rpm and – on the unrealistic assumption that only Coriolis forces are acting upon the structure – a further linearised eigenvalue analysis is carried out. In this way, we get a summary view of the qualitative effect of the Coriolis forces on the eigenvalue spectrum. As a typical effect, we note a spreading of the frequency spectrum: the first frequency falls distinctly and the second rises. We also note a coupling effect due to the skew-symmetric matrix $\boldsymbol{G}$: the uniaxial bending actions are all transformed into biaxial bending deformations. Each point of the beam follows an elliptical path whereby the semiaxes of the ellipse at any point are described by the deflections at this point shown in fig. 11.7.3 (b) and (c).

If the centrifugal forces are now introduced, as is only realistic, then naturally, considerable static pre-deformations are produced (fig. 11.7.3(c)), and the reference state of the oscillation is significantly modified. We also note that the influence of the centrifugal forces on the frequency spectrum is dominant and masks to a great extent the effects of the Coriolis forces. This phenomenon can be observed in a large series of real rotating structures, such as compressor blades or turbine

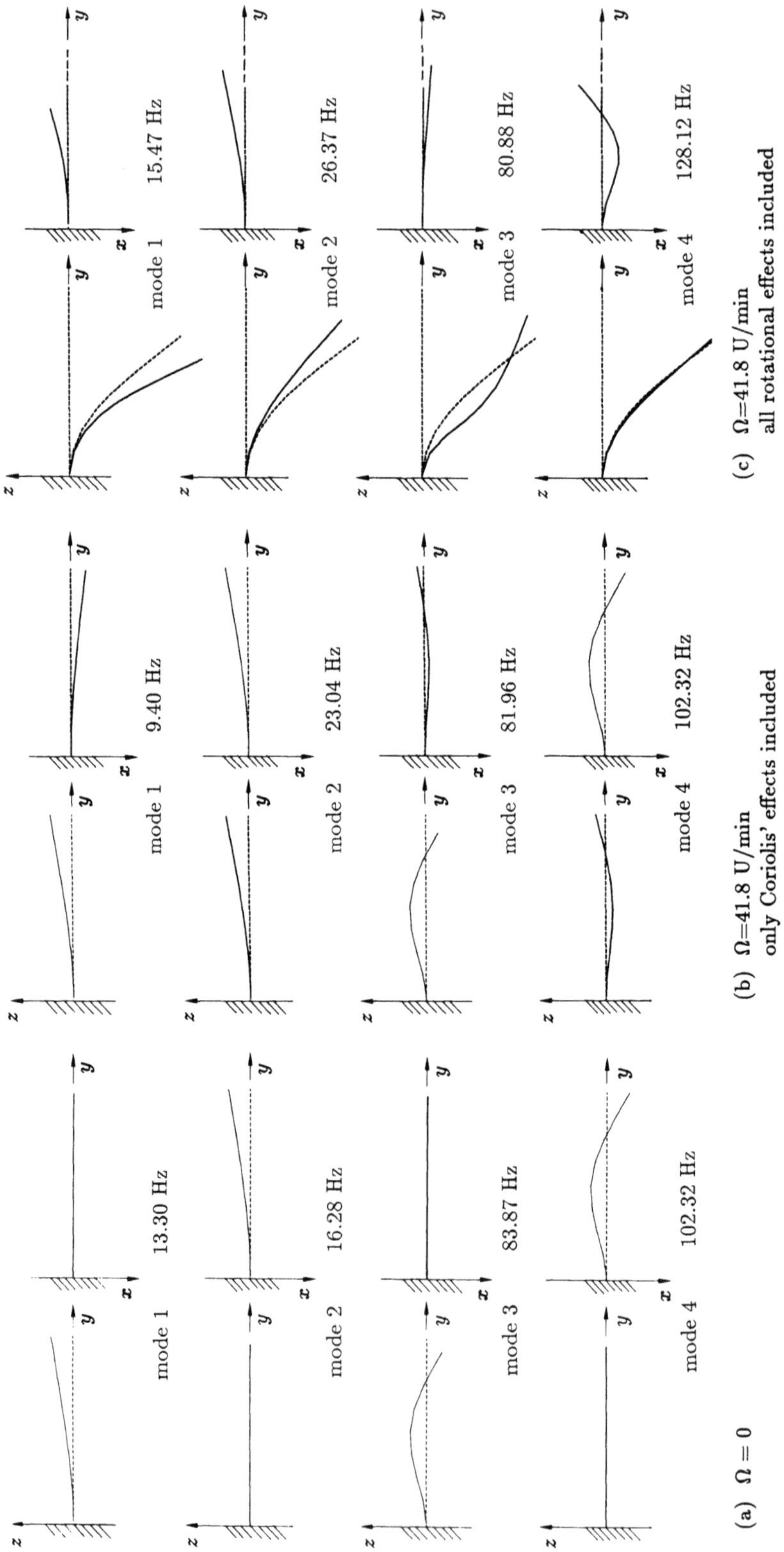

Fig. 11.7.3 Rotating beam with $\theta = 90^\circ$: eigenfrequencies and eigenmodes resulting from a linearised analysis at a non-linear state of equilibrium

blades. We continue to notice the influence of the Coriolis forces in one effect, however: the coupling of the mode shapes about different planes is still present.

Our second and final example, illustrated in fig. 11.7.4, is a long, slender, rotating circular cylinder which may represent a rotor hub or orbiter model [11.45]. The cylinder rotates with 398 rad/s about the longitudinal axis z. In this example, the centrifugal field forces have scarcely any influence on the behaviour of the natural modes of vibration. Figure 11.7.5 demonstrates the first three eigenfrequencies for both the rotating cylinder and the cylinder at rest. In order to improve the quality of the illustration, the radius is shown magnified by 6; thus, the cylinder appears considerably more stumpy than in reality. On the first bending mode, the Coriolis forces have a drastic effect, since the eigenfrequency falls by about 40%. For the second eigenmode when the cylinder is not rotating, we simply obtain a flexural oscillation about a different axis at the same frequency (the structure is doubly symmetrical). In the case of a rotating structure, this bending mode is replaced by a torsional vibration which displays a slight "respiration" oscillation. This torsional mode takes third place when the structure is at rest and has obviously hardly changed its frequency. The third eigenfrequency of the rotating structure involves now bending about the other axis (no. 2 in the case of the structure at rest), now combined with torsion. The increase in frequency amounts to circa 74%. Further comparisons of frequency can be found in the original work [11.45]. We can thus conclude that the Coriolis forces possess a considerable influence on this type of structure and should certainly be taken into consideration in the case of similar systems like complete turbine rotors with blades and spin-stabilised space structures.

geometrical data:

length	$l = 81$ in
radius	$r = 3.5$ in
thickness	$t = 0.43$ in

material data:

Young's modulus	$E = 30.700 \cdot 10^6$ psi
Poisson's ratio	$\nu = 0.3$
density	$\rho = 7.768 \cdot 10^{-4}$ lb s^2/in^4

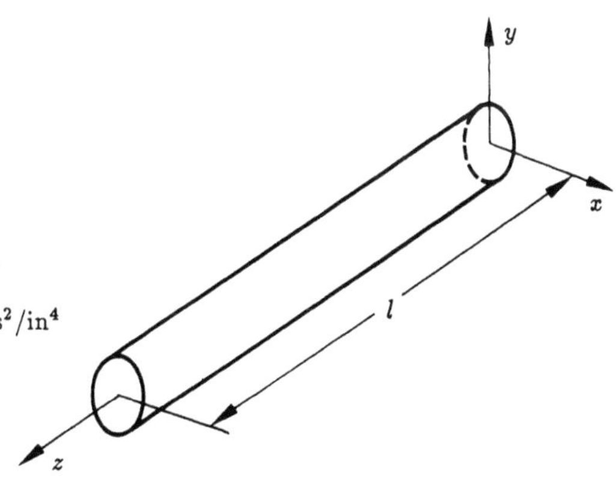

Fig. 11.7.4 Rotating cylinder: geometry and data

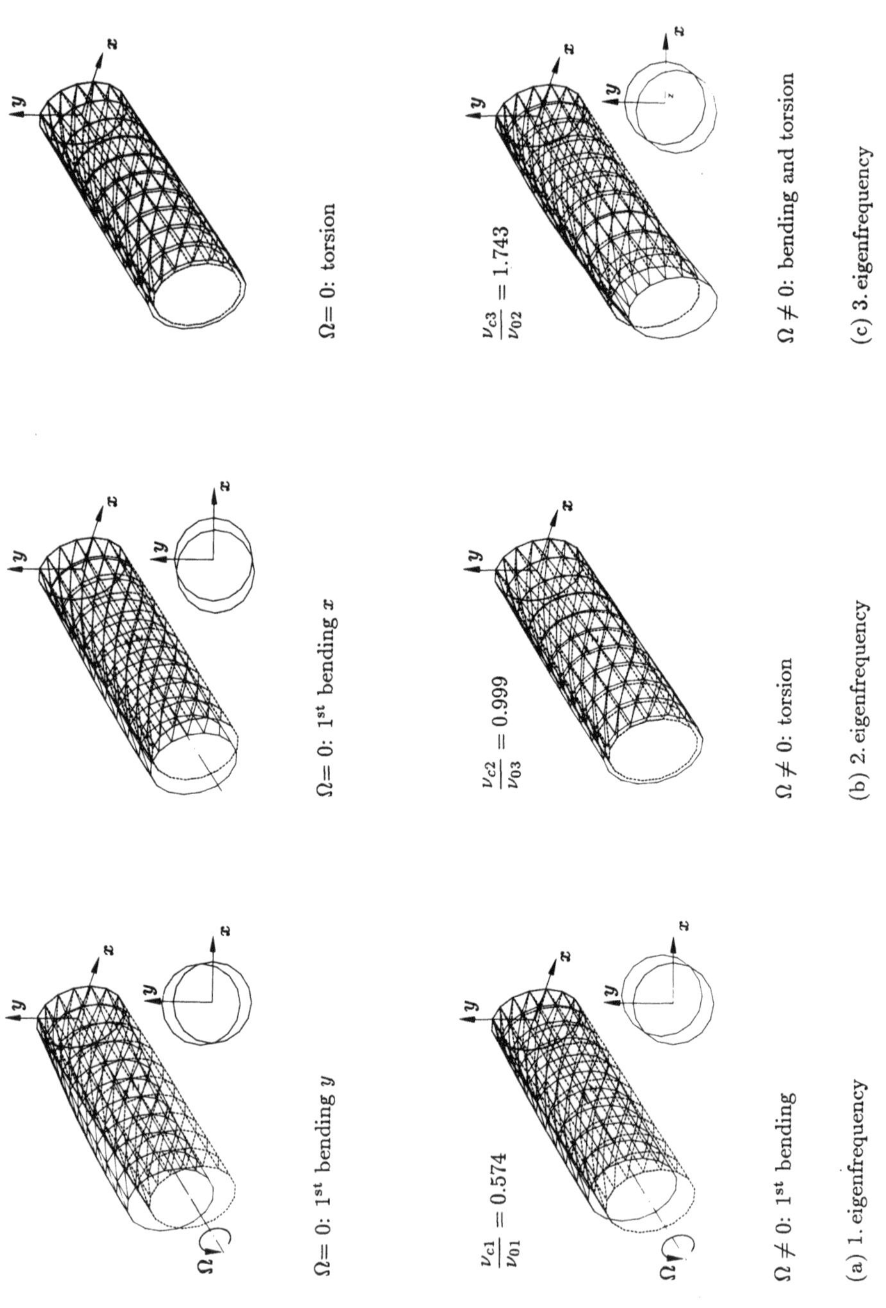

Fig. 11.7.5 Rotating cylinder: effect of Coriolis forces on eigenfrequencies [11.45]

11.8 Computational analysis of the crash process

Authors: John Argyris, Hans Balmer, Ioannis St. Doltsinis

So far in this chapter, we have, with one exception, dealt only with the non-linear dynamic properties of elastic structures. In the exceptional case quoted in section 11.4.4, we discussed the dynamic propagation of an impulse through an elastoplastic bar; in this case, the elastic matrix $\boldsymbol{K}_e$ had to be replaced by the elastoplastic stiffness matrix $\boldsymbol{K}_{ep}$.

In the following two sections, we consider two methodologies developed for non-linear material properties and large displacements and investigate as an example of the first procedure the crash response of a car shell. The second procedure, to be expounded in Section 11.9, is applied to the detonation loading generated by an explosive pad placed on a thin cylindrical shell. The impact velocities are thereby an order of magnitude higher than in a car crash.

In a crash the forces involved and the severe deformations of the structure are of great relevance to the safety of the car passengers. To satisfy the last requirement, we demand that the structure should absorb the energy produced at impact, causing deformations within limits acceptable to the passengers. In the past, industry mainly relied on expensive tests in this area of structural design. Experiments dealing with the crash response either of the whole structure or of susceptible parts are simulated and evaluated. Figure 11.8.1 shows the typical layout of such a test; it represents the front part of a car without the engine, but with bumper as well as mudguards. In this case, the deformation response of the rest of the construction is not of great importance, whereas its mass is. This is simulated by welding metal plates onto the bodywork, increasing the total impact mass – without engine to $M = 588\,\mathrm{kg}$. In the test, this structure on rails is catapulted at the required velocity against a solid concrete wall; the wall can be considered rigid. In the following example, the velocity at the instant of the crash is $v = 49.3\,\mathrm{km/h}$. The process of deformation caused by the crash results in significant permanent contributions to the overall strain which should absorb the kinetic energy as desired. In contrast, the elastic contributions to the overall strain are negligible, although their significance in the evaluation of stresses should not be underestimated.

It is obvious that a reliable computational analysis of the crash process – at least by the first procedure – is bound to be complex and costly. On the other hand, a successful computer simulation is less expensive than the test and supplies in addition an abundance of results which repeated tests could only approximately achieve. It will be plain to the experienced reader that the crash analysis of a sheet metal construction involves complicated simulations and calculations. Moreover, it would be unrealistic to expect that every detail of a pronounced deformation could be rendered by a calculation, considering the uncertainties of the precise original geometry and the material properties. For reasons of economy, therefore, it is advisable to aim at a technically feasible idealisation of the construction and to use only the material properties essential to the simulation in the calculation. In this first procedure, we consider it sufficient to assume for the metal an inelastic viscous law of an incompressible material as a constituent foundation for the determination of the deformations (not, however, for the determination of the actual stresses). This aspect is described in detail in section 11.8.1. In this process, it is now important to register that it is not the nodal displacement $\boldsymbol{r}$, but the

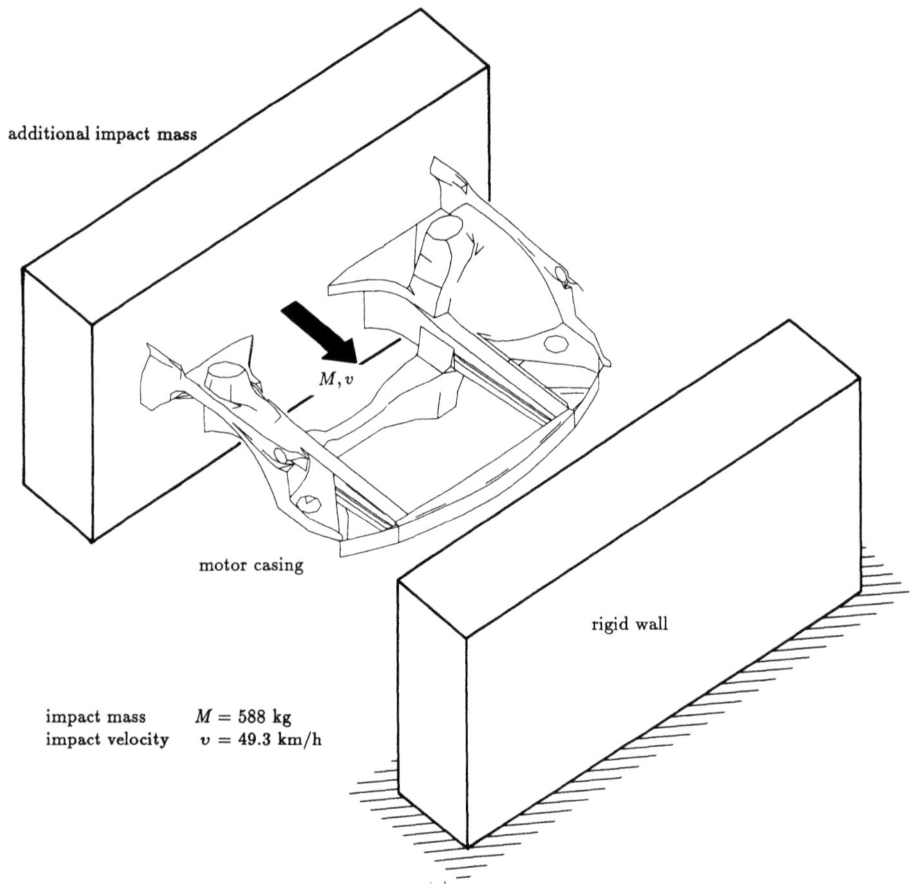

Fig. 11.8.1 Experimental layout for crash simulation with the front part of an automobile construction

velocities $\dot{\boldsymbol{r}}$ which enter as unknowns. In our example, the influence of the large displacements is taken into account using a modified form of the standard concept of geometric stiffness. At every node of the discretised structure, the resultants of the internal stresses are established directly for the current geometry. These resultants together with the mass forces and external forces must constitute a state of dynamic equilibrium; this establishes the equation of motion as a function of $\dot{\boldsymbol{r}}$. The principle of the procedure can be gathered from the following two subsections; further details can be found in [11.48]. Subsequently, we shall consider in section 11.9 the second, more economical, high-impact procedure.

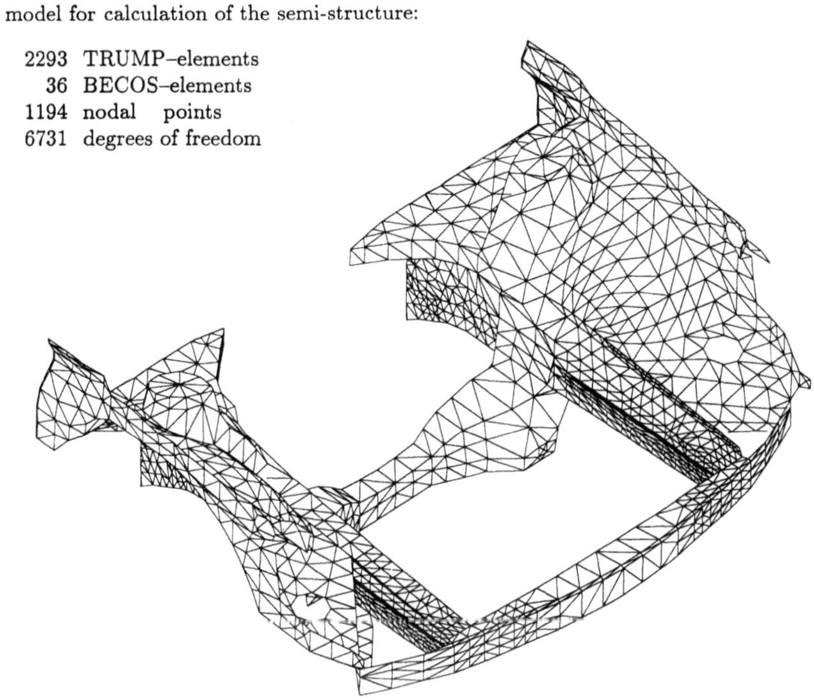

Fig. 11.8.2 Discretisation of the automobile front part

11.8.1 On the constitutive law of metal sheets

The nature of the adopted constitutive law of the structural material is reflected on the determination of the current stresses resulting from the deformation which in turn define the stress resultants at the nodal points of the discretised model.

The characteristics of the sheet metal used in the construction are shown for a uniaxial tension load in figs. 11.8.3 and 11.8.4. Figure 11.8.3 refers to standard steel, fig. 11.8.4 to the high strength steel of the main longitudinal beams. The stress-strain pattern displayed in the diagrams indicates a certain hardening of the material under increasing strain, the latter appearing almost entirely as inelastic strain. Interestingly enough, different stress-strain relations were observed during tension tests carried out at different velocities; it can be concluded that the material behaviour depends significantly on the strain rate. In order to convey to the reader a physical idea of this behaviour, figs. 11.8.3 and 11.8.4 contain the two limiting curves which mark the beginning and end of this dependence. For simplicity's sake, the material law is based here on a viscous incompressible material model. The mathematical formulation for the adopted isotropic dependence of the material refers to deviatoric stresses

$$\boldsymbol{\sigma}_D = 2\mu\dot{\boldsymbol{\gamma}} \tag{11.8.1}$$

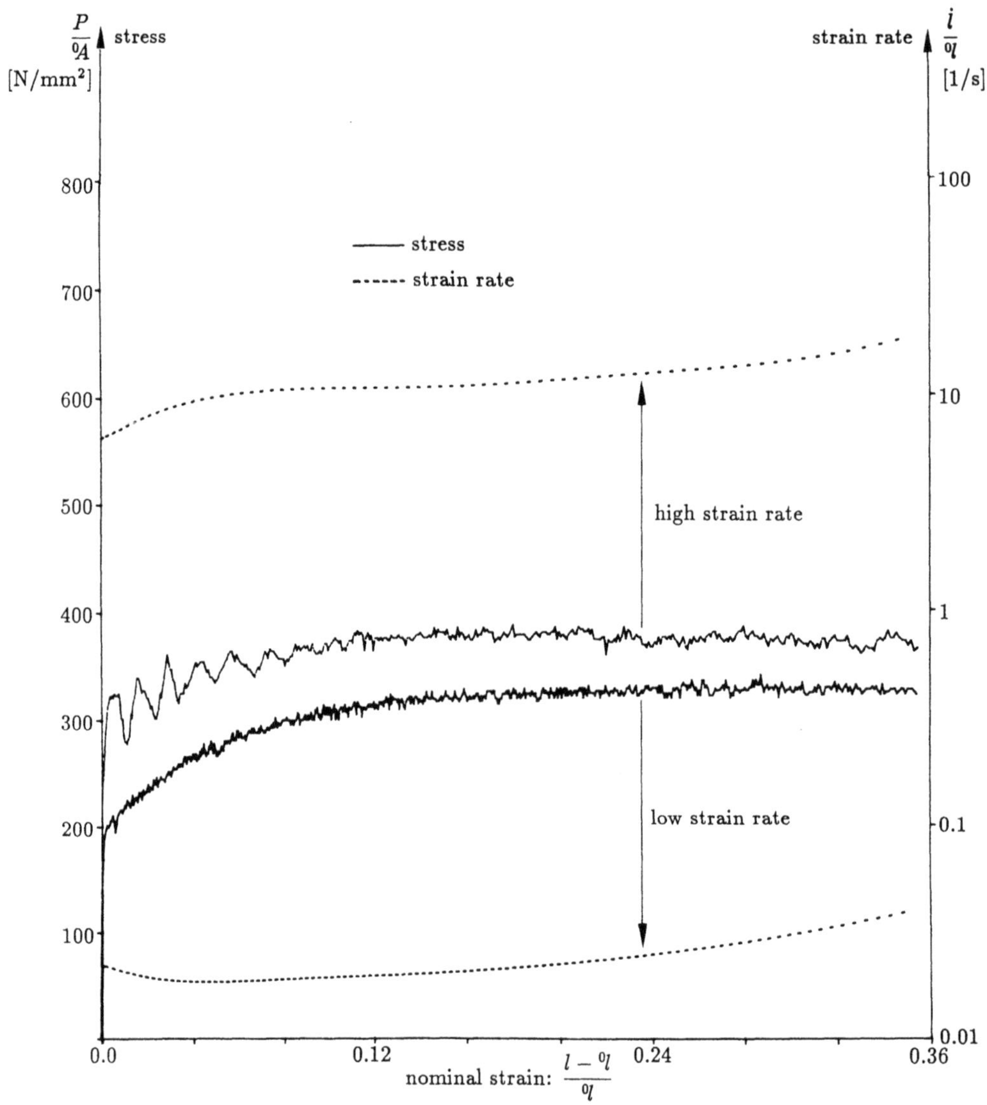

Fig. 11.8.3 Stress-strain relationship for a standard automobile steel sheet

where μ is the viscosity coefficient. The assumed incompressibility requires that the volumetric component of the strain velocity disappear and thus, the strain rate here is purely deviatoric. Hence, we have

$$\dot{\gamma} \equiv \dot{\gamma}_D \tag{11.8.2}$$

Due to the assumed incompressibility, it is in general not possible to express the hydrostatic stress by the strain rate. In the case in hand, however, we are dealing

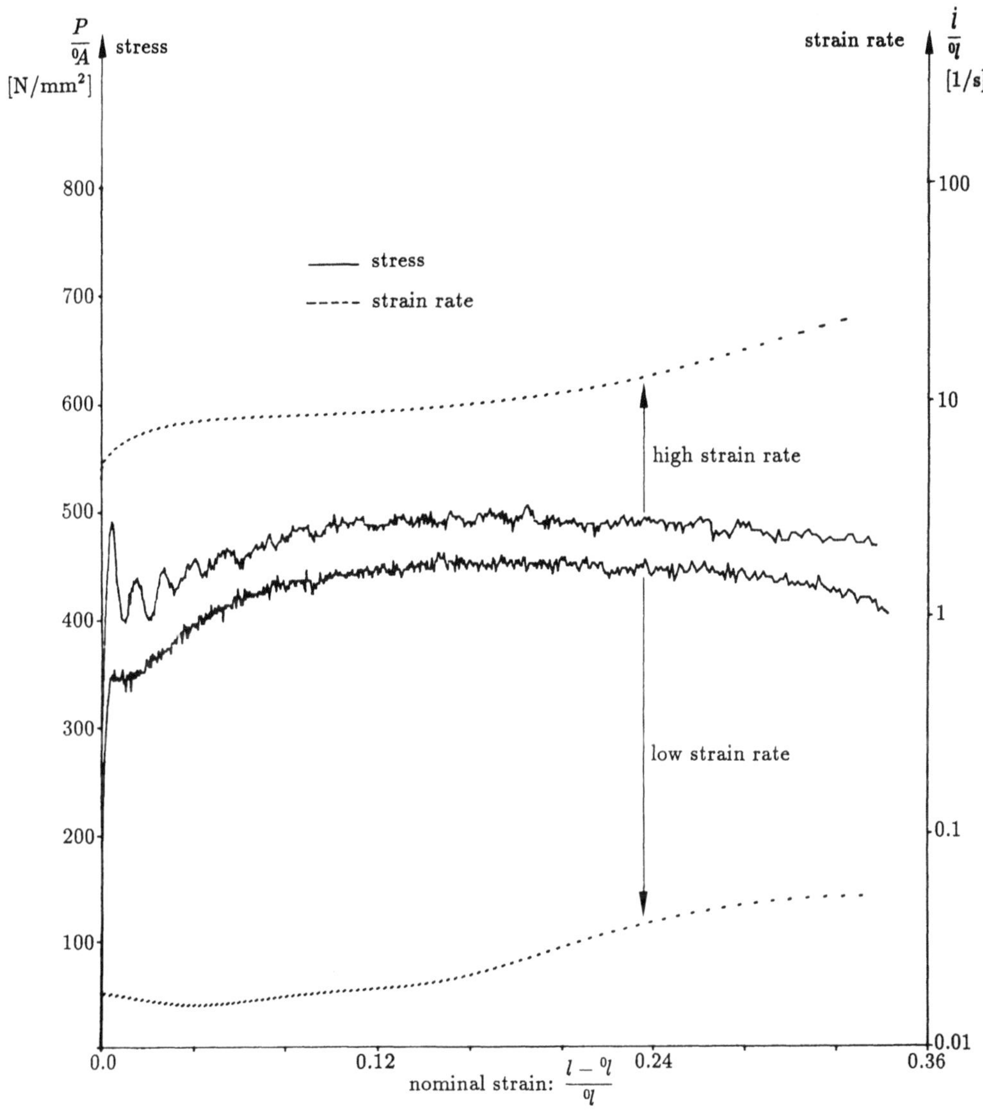

Fig. 11.8.4 Stress-strain relations for high strength steel (main longitudinal beams)

exclusively with plane states of stress, and the appropriate (reduced) stress and strain rates are related by

$$\boldsymbol{\sigma} = 2\mu\left[\boldsymbol{I}_3 + \boldsymbol{e}_{2,1}\boldsymbol{e}_{2,1}^{\mathrm{t}}\right]\dot{\boldsymbol{\gamma}} \tag{11.8.3}$$

where

$$\boldsymbol{e}_{2,1} = \{1 \quad 1 \quad 0\}$$

A comparison of equ. (11.8.3) with the analogous elastic expression shows that in the case of the viscous material used, the viscosity coefficient μ takes the place of the elastic shear modulus G and the inelastic strain velocity $\dot{\gamma}$ takes the place of the elastic strain $\boldsymbol{\epsilon}$. Hence, the viscosity coefficient μ is here the determining material parameter and has to be established from the available uniaxial experimental data. To this end, we deduce from equ. (11.8.1) an expression relating measures according to von Mises, stress $\overline{\sigma}$ and strain rate $\dot{\overline{\gamma}}$. In this way, we obtain

$$\overline{\sigma} = 3\mu\dot{\overline{\gamma}} \tag{11.8.4}$$

In the same way, the uniaxial material characteristic can be described by a functional connection between equivalent stress and strain rate. We write

$$\overline{\sigma} = f(\dot{\overline{\gamma}}, \overline{\gamma}) \tag{11.8.5}$$

where the equivalent strain itself takes the material hardening into account. A comparison between equs. (11.8.4) and (11.8.5) yields the viscosity coefficient

$$\mu = \frac{\overline{\sigma}}{3\dot{\overline{\gamma}}} = \frac{f(\dot{\overline{\gamma}}, \overline{\gamma})}{3\dot{\overline{\gamma}}} \tag{11.8.6}$$

The material law is thus established.

A speedy formulation of the theory is now made possible by the implied analogy with the response of elastic materials. The Cartesian stress resultants $\mathbf{P}_D$ at the nodal points of a finite element can be determined from the corresponding deformation velocities $\dot{\boldsymbol{\rho}}$ of the element nodes by applying the relevant expression

$$\mathbf{P}_D = \boldsymbol{c}\dot{\boldsymbol{\rho}} \tag{11.8.7}$$

where $\mathbf{P}_D$ represents the so-called element damping forces and the viscosity matrix $\boldsymbol{c}$ of the element is established in the same way as the stiffness matrix.

11.8.2 Formulation of the equations of motion of the structure

Following the preceding sub-section dealing with the characteristic behaviour of the chosen inelastic material, we can now turn to the actual analysis of the crash process. We refer once more to the discretisation of the construction in fig. 11.8.2. The problem can be considered symmetrical about the vertical medial plane, so that it is only necessary to analyse one half of the structure. The adopted model consists of 1194 nodal points comprising 2293 TRUMP plate elements as well as 36 BECOS beam elements. These element types, originally elastic, are modified to account for a non-linear response, including viscous and hardening effects. Not only the aforementioned symmetry but also the rigid connection of the structure to the end plate can be taken into consideration in the analysis. Even with these

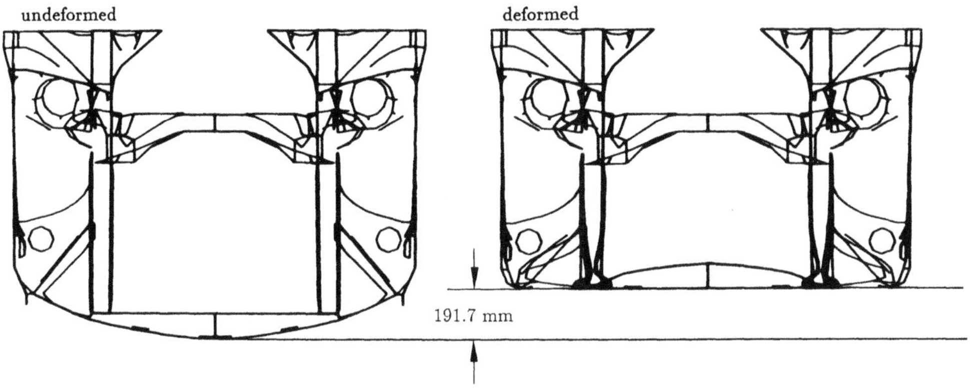

Fig. 11.8.5 Crash simulation: deformation pattern with a permanent shortening of 191.7mm; view from above

simplifications, 6731 unknowns remain and condition a non-linear problem of considerable complexity. The analysis was performed on a CRAY-1M, at the time the supercomputer installed at the University of Stuttgart.

The Boolean assembly process of the contributions of the individual elements of the discretised structure yields, at each instantaneous deformed geometry, the stress resultants at the nodal points which represent quasi-damping forces in the form

$$\boldsymbol{R}_D = \boldsymbol{C}(\boldsymbol{r})\dot{\boldsymbol{r}} \tag{11.8.8}$$

where $\boldsymbol{C}(\boldsymbol{r})$ is the viscosity matrix of the discretised structure. The equation of motion which expresses the dynamic equilibrium between external and internal forces thus becomes

$$\boldsymbol{M}\ddot{\boldsymbol{r}} + \boldsymbol{C}(\boldsymbol{r})\dot{\boldsymbol{r}} = \boldsymbol{R}(\boldsymbol{r}) \tag{11.8.9}$$

and has to be established at each step for the current geometry of the deformed structure. In this case, the influence of the usual geometric stiffness is contained in the instantaneous non-linear $\boldsymbol{C}(\boldsymbol{r})$. The contribution of the current geometry is contained in the determination of the viscosity matrix $\boldsymbol{c}(\rho)$ (equ. (11.8.7)) or $\boldsymbol{C}(\boldsymbol{r})$ (equ. (11.8.9)) as well as the specification of the load vector $\boldsymbol{R}(\boldsymbol{r})$.

The temporal integration of the equation of motion (11.8.9) is based on one of the algorithms already discussed. At the start of the computer simulation of the crash process, the whole discretised structure experiences the prescribed impact velocity except for the parts which collide with the rigid barrier and thus have zero velocity. The approximate temporal integration of the equations of motion is based on an adaption of Newmark's unconditionally stable algorithm with $\beta = 1/4, \gamma = 1/2$. The resulting incremental problem is non-linear and demands the repeated solution of a matrix equation. In the course of the analysis, the adopted time increment varied between 0.1 ms and 0.2 ms. Time increments up to 1ms could occasionally

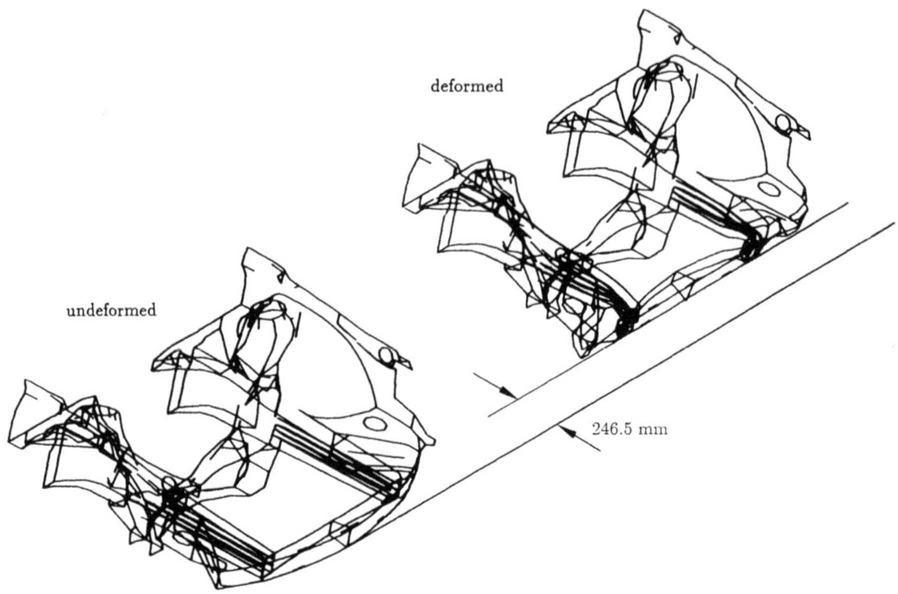

Fig. 11.8.6 Crash simulation: deformation pattern with a permanent shortening of 246.5mm; side view

also be used. Results of the analysis are shown in figs. 11.8.5 to 11.8.7. The impact initially leads to a flattening of the bumper. Subsequently, the main longitudinal beams are subjected to impact by contact with the wall. At first, they are deformed locally along the longitudinal axis so that folds and bumps are formed as shown in fig. 11.8.5. The view from above allows a comparison between the undeformed and the deformed state following a permanent shortening of the structure of 191.7 mm. Further motion results in a buckling of the main beams which begin to bend downward. The state following front end permanent displacement of 246.5 mm can be seen in a perspective view in fig. 11.8.6. The crash process was calculated up to a final front end displacement of 301.22 mm. Figure 11.8.7 shows this final state of deformation in a side view and contrasts this with the undeformed original structure. It is perhaps of interest to mention that this extensive crash analysis was the first of this magnitude. The results of the analysis were confirmed to a large extent by comparison with an experimental crash test. As mentioned in the beginning, however, a meaningful determination of the actual stresses requires that the elastic material properties be allowed for in a final reassessment of the results. It is not possible to deal with these details in this work.

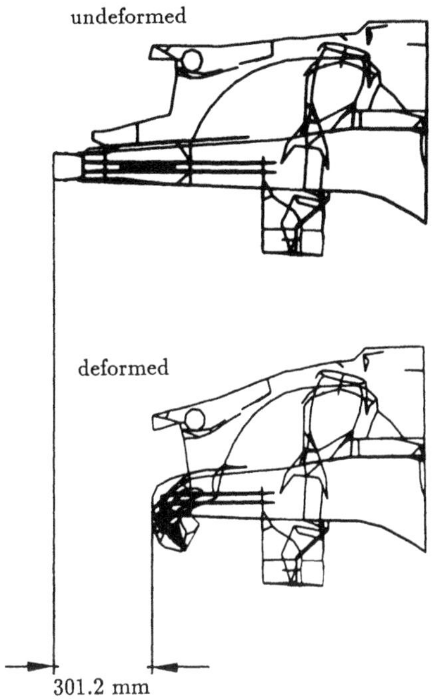

Fig. 11.8.7
Crash simulation: deformation pattern with a permanent shortening of 301.2mm: side view

11.9 An alternative analysis of shell-like structures under impact loading

Authors: John Argyris, Hans Balmer, Ioannis St. Doltsinis

11.9.1 Introductory remarks

As mentioned in the preceding section, the procedure on crash analysis developed therein proves, for all its simplicity, inevitably costly even on present-day supercomputers like the Cray 2. As a result, the authors researched into the possibility of a still simpler idealisation requiring only computer times of an order of magnitude lower than in the first procedure. To this end, significant spatial and temporal simplifications had to be adopted. The present section summarises the new theory which does not exclude elastic effects. Details are also given in [11.49], [11.50] and [11.51].

Comparative calculations of plates with the new mechanical model (which might also be denoted a natural difference model) and with the FE-technique have shown that the critical time-step for numerical stability in the case of the FE-model is approximately 20× smaller than in the case of the difference model. In the idealisation, the same mesh size and the same fully explicit difference algorithm with respect to time were used for a linear plate problem. Whereas the difference model does not explicitly contain rotational degrees of freedom, in the FE-method, these appear to increase the eigenfrequencies of the structure and influence the critical time-step substantially. In the non-linear case on the other hand, the eigenfrequencies may be controlled rather by the membrane effect and will be higher due to the increased stiffness of the structure.

For a comparable accuracy of the results, a coarser mesh can of course be selected in the FE-model than in the difference model. This advantage is counterbalanced by the fact that considerably longer computing time per time-step may be necessary for the FE-model. Moreover, since large plastic deformations usually occur at impact, the element size must be limited anyway. In the difference model and a non-linear calculation, which is our main concern here, the repeated setting-up of the various element matrices becomes unnecessary. In addition, the difference model displays good numerical behaviour.

The following, extremely simple model, which nevertheless retains all the essential static and dynamic characteristics of the shell, is suitable for an efficient elasto-plastic analysis with large displacements, but small strains. It follows naturally from an already existing beam model, the main characteristic of which is that the bending deformation is described by (finite) folding angles between two neighbouring elements – in contrast to the FE-method where the curvatures are defined by infinitesimal angles which are integrated over the element. To ease comprehension, the dynamic calculation scheme on which the analysis of both the beam and the shell model is based is first described using the beam.

11.9.2 The original beam model [11.52]

The beam, which is assumed to move in a plane, is represented by a chain of links which are rigid in bending but can act as extensible flanges (fig. 11.9.1). Thus, the extensional deformation of the beam is assigned to the flanges while the bending deformation is concentrated at the joints and is defined by the change of the angle between adjoining flanges. These angles can be related directly to second-order difference quotients in the displacements. Rotational degrees of freedom as such do not appear explicitly.

In the case of dynamic applications, the mass distribution is modelled by mass points at the joints, while the flanges are massless. The motion of the beam is described in terms of the trajectories of the individual mass points, starting from a given configuration at time t_0. The time integration of the differential equations of motion is based on a simple difference scheme which is also used for finite element idealisations.

The calculation scheme proceeds as follows:

(1) Given, for example, the initial velocities of the mass points, calculate the displacement increments at time $t_1 = t_0 + \tau$, where τ is the constant time increment.

(2) Knowing the displacement increments at time t_1, calculate the increments of the flange elongations and angular changes which jointly define the state of strain.

(3) With an appropriate constitutive relation, determine the stress increments which are added to (accumulated in) the already existent stresses.

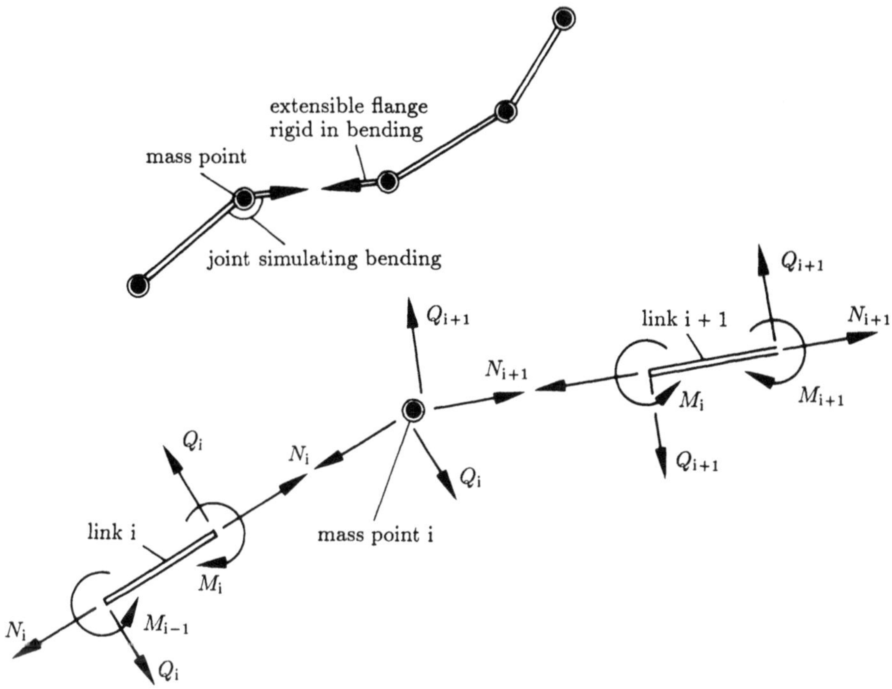

Fig. 11.9.1 Beam model

(4) Calculate the stress resultants, i.e. the axial (normal) forces and bending moments.

(5) Deduce the shear forces in the massless links from equilibrium relations.

(6) Using the known axial forces, shear forces and, if present, external forces, determine the accelerations of the mass points. If these are expressed in difference form, they will in turn supply the displacement increments at time t_2.

The current time cycle is thus completed and the process is repeated, beginning at step (2). It should be pointed out that the displacements themselves do not enter into the calculation directly. They are obtained as by-products of the summation of the increments and define the geometry of the structure which serves as the initial state for the next time-step.

11.9.3 Natural extension (generalisation) of the beam model to a shell model

The analysis of a shell follows exactly the same pattern as the beam analysis. The shell is idealised by a triangular lattice placed on its middle surface (fig. 11.9.2). The lattice is provided with nodal joints and composed of flanges rigid in bending. The basic shell pattern depicted in fig. 11.9.2 must always be adhered to, even in a topologically distorted form. We note that each nodal point may be considered the meet of *three* strings of flanges which are beam models as defined in section 11.9.2,

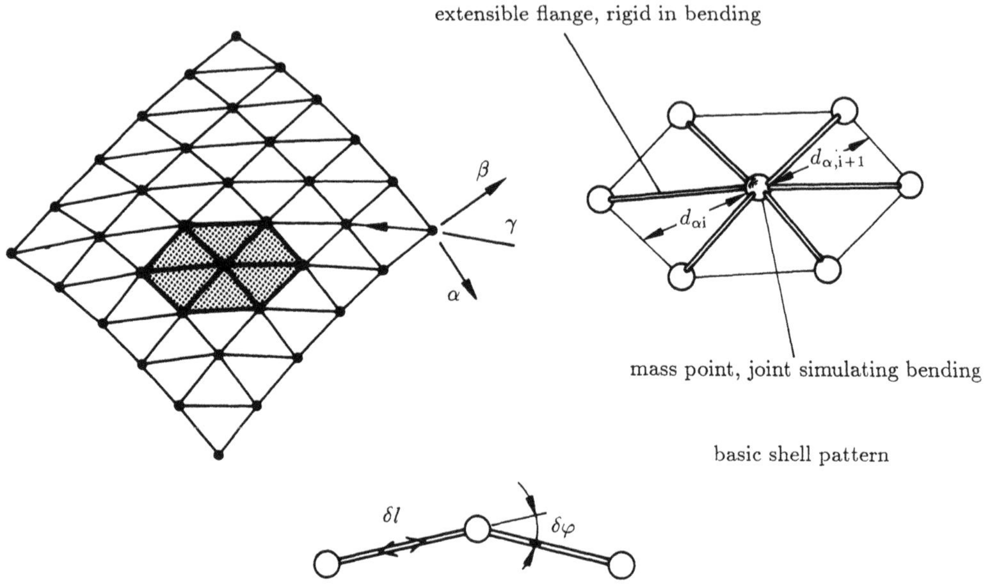

Fig. 11.9.2 Lay-out of shell model

pointing in different directions on the middle surface. The individual strings form three (supernumerary) families of surface curves (α, β, γ-curves).

The deformation of the shell is defined at the nodal point. While the membrane deformation is given by the elongations of the three concurrent flanges, the bending deformation is expressed for each of the strings by the change of angle between two adjoining flanges in the corresponding normal plane of the shell. Thus, in addition to the three middle surface strains, we obtain three curvature measures; together, these uniquely define the deformation state of the shell in a completely homogeneous way (only extension measures with no explicit shear expressions).

11.9.4 Kinematics of the model

A homogeneous stress or strain representation, as in the natural approach, does not entail any preferences in any of the three directions. Thus, it suffices to consider only a typical direction like α. The corresponding equations for the other directions follow by cyclic permutation.

Membrane deformation: the elongation of a flange is obtained in the usual way by projection of the difference of the nodal displacements on the axis of the flange. The elongation increments δl of the three concurrent flanges at a given nodal point are collected in the column vector

$$\delta \boldsymbol{l} = \{\delta l_\alpha \quad \delta l_\beta \quad \delta l_\gamma\} \tag{11.9.1}$$

The *total* natural membrane strain increments appertaining to this point may be expressed as

$$\delta\boldsymbol{\gamma}_t = \boldsymbol{l}_0^{-1}\delta\boldsymbol{l} \tag{11.9.2}$$

where the diagonal matrix

$$\boldsymbol{l}_0 = \lceil l_{0\alpha} \quad l_{0\beta} \quad l_{0\gamma} \rfloor \tag{11.9.3}$$

defines the lengths of the flanges in the undeformed state of the shell. This conforms with the assumption of small strains compatible with the large deflections admitted here.

At this stage, it should be noted that the strain increments as specified in equ. (11.9.2) actually appertain to the mid-point of the flanges and not to the common nodal point. However, as a result of the elasto-plastic material considered here, the state of stress in the shell cannot be divided into bending and membrane parts and therefore should be allocated to a common point for curvature and middle surface strain; most convenient in this case is the nodal point. Assigning the strain to one of the end points of the flange inevitably involves an approximation.

Bending deformation: as has already been mentioned, the angular change between two adjoining flanges serves to determine the curvature over a finite arc length. Thus, the *total* curvature increment in the direction of the string of flanges α is defined by

$$\delta\chi_{\alpha t} = \left[\frac{\delta(\Delta\varphi)}{\Delta s_0}\right]_\alpha \tag{11.9.4}$$

where Δs_0 defines a reference arc length (see below). The increment $\delta(\Delta\varphi)_\alpha$ is measured in a plane perpendicular to the shell middle surface, intersecting it along an α-line, and is given by the algebraic sum of the angular increments $\delta\varphi_{\mathrm{i}}$ and $\delta\varphi_{\mathrm{i}+1}$ of two adjacent flanges (fig. 11.9.3, top). These angular increments are taken as positive if they are directed towards the inner (negative) side of the shell. In this way, the curvature increment has a positive sign if a fibre layer is stretched on the positive side of the shell.

If we consider a single flange (fig. 11.9.3, bottom), the angular increments $\delta\varphi_1$, $\delta\varphi_2$ associated with its ends are construed by the contributions of the normal components δw_1, δw_2 of the nodal displacement increments. The normal to the shell surface is readily obtained by normalising the vector sum of the unit normals on the triangles adjacent to the flange. This is illustrated in fig. 11.9.4. Collectively, the angular increments may be written

$$\delta\boldsymbol{\varphi}_\alpha = \{\delta\varphi_1 \quad \delta\varphi_2\}_\alpha = \boldsymbol{a}_\alpha \delta\boldsymbol{w}_\alpha \tag{11.9.5}$$

where

$$\boldsymbol{a}_\alpha = \frac{1}{l_\alpha}\begin{bmatrix} 1 & -1 \\ -1 & 1 \end{bmatrix} \tag{11.9.5a}$$

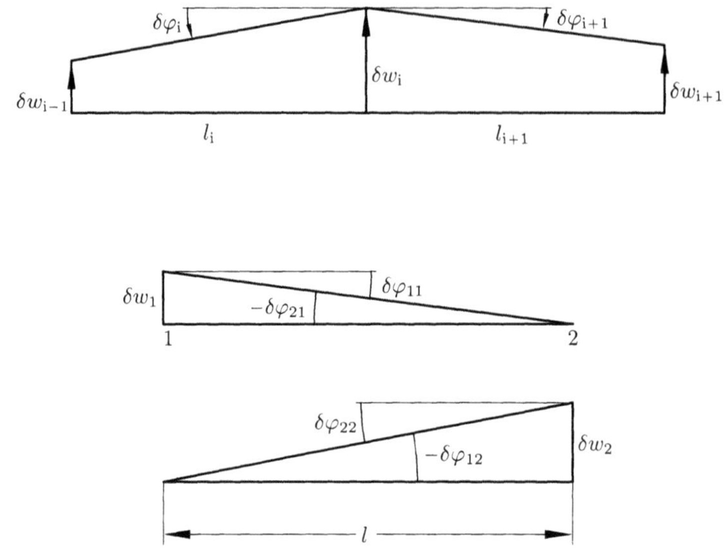

$$\delta\varphi_1 = \delta\varphi_{11} + \delta\varphi_{12} \qquad \delta\varphi_2 = \delta\varphi_{21} + \delta\varphi_{22}$$

Fig. 11.9.3 Bending kinematics for α, β, γ directions

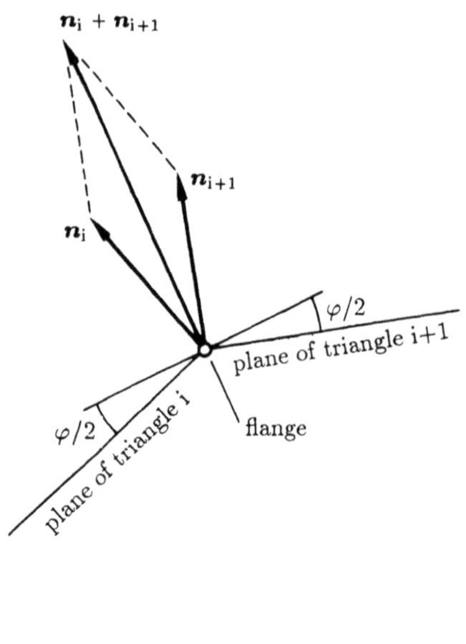

Fig. 11.9.4
On the construction of the shell normal

and l_α is the current length of the flange.

The reference arc length $\Delta s_{0\alpha}$ in equ. (11.9.4) is simply defined as the average length of the flanges i and i + 1 in the undeformed state. The (3×1) vector of the total natural curvature increments at a nodal point is thus given by

$$\delta\boldsymbol{\chi}_t = \{\delta\chi_{\alpha t} \quad \delta\chi_{\beta t} \quad \delta\chi_{\gamma t}\} \tag{11.9.6}$$

In conjunction with equ. (11.9.4), we note that its entries represent difference quotients in the directions α, β, γ of a triangular mesh. As in the case of the beam model, these can be related to second-order difference expressions in the displacements. This bending model can be considered a natural (homogeneous) version of the traditional two-dimensional Cartesian difference scheme.

We now consider the case where, as a result of the idealisation of the shell middle surface, the string of flanges forms a slightly kinked line and is no longer straight. In the evaluation of the curvature according to equ. (11.9.4), the angular contributions from the right and the left of the nodal point are no longer placed in the same normal plane, but in two planes inclined slightly towards one another. Despite the assumed kink in the string of flanges, we may, in a first approximation, proceed as before and simply add the contributions, thus ignoring the deviations of the angle and the reference length caused by the projection (i.e. we assume cosine ≈ 1). In the case of a larger (geodesic) curvature of the surface lines, this condition can only be justified by the adoption of a finer triangular mesh.

11.9.5 Statics of the model

The forces corresponding to the elongation increments of equ. (11.9.1) are the natural flange forces in the three directions α, β, γ (fig. (11.9.5))

$$\mathbf{P}_N = \{\mathrm{P}_\alpha \quad \mathrm{P}_\beta \quad \mathrm{P}_\gamma\} \tag{11.9.7}$$

The static (equilibrium) relations for the flange are dual to the kinematic (compatibility) relations (equ. (11.9.5))

$$\boldsymbol{Q}_\alpha = \boldsymbol{a}_\alpha^{\mathrm{t}} \boldsymbol{M}_\alpha \tag{11.9.8}$$

where

$$\boldsymbol{M}_\alpha = \{M_1 \quad M_2\}_\alpha \tag{11.9.9}$$

define the bending moments and

$$\boldsymbol{Q}_\alpha = \{Q_1 \quad Q_2\}_\alpha \tag{11.9.10}$$

the associated transverse shear forces at the ends of the flange. Equation (11.9.8) can also be established using the equilibrium diagram of fig. 11.9.5.

It should be mentioned that the three simple (one-dimensional) beam equilibrium relations of equ. (11.9.8), if applied independently to the three different directions, are consistent with the basic concepts of the natural method of *component* bending moments, component transverse forces and component stress flows. Torsional moments and shear flows do not appear explicitly, but are contained implicitly (cf. section *8.4.6.2, Vol. I(G).

Fig. 11.9.5
Equilibrium for α, β, γ directions

11.9.6 Constitutive relations

As already mentioned, the model has also to incorporate inelastic effects, e.g. plasticity. The easiest way of achieving this is by assigning to the shell a layered assembly across its thickness. Material hardening due to increasing strain is readily included by further subdivision of the layers into sublayers [11.53]. In this way, the mechanics of repeated loading and unloading in the inelastic domain which are particularly important in dynamic processes can be properly included.

We now consider a nodal point of the shell which is presumed to be thin. The *total* natural strain increment in the m-th layer at a distance z_{m} from the middle surface of the shell is

$$\delta\boldsymbol{\gamma}_{t\mathrm{m}} = \delta\boldsymbol{\gamma}_t + z_{\mathrm{m}}\delta\boldsymbol{\chi}_t \tag{11.9.11}$$

where $\delta\boldsymbol{\gamma}_t$ is the natural membrane strain increment of equ. (11.9.2) and $\delta\boldsymbol{\chi}_t$ is the natural curvature increment of equ. (11.9.6).

Before we can develop a suitable computational scheme for the constitutive relation in a single layer, we have to introduce some concepts used in the inelastic analysis of the *plane* stress state of a triangular membrane of thickness t. For simplicity's sake, we restrict ourselves to an elastic, perfectly plastic material. An extension to more complex constitutive relations is relatively simple due to the explicit representation adopted here.

In the following, we use, adopting good aeronautical tradition, the so-called natural normal flows, i.e. the *component* stresses multiplied by the thickness t

$$\boldsymbol{N}_c = \{N_{\alpha c} \quad N_{\beta c} \quad N_{\gamma c}\} = t\{\sigma_{\alpha c} \quad \sigma_{\beta c} \quad \sigma_{\gamma c}\} \tag{11.9.12}$$

In the case of isotropy, the corresponding material stiffness relating total strains to component stresses is given by equ. (5.3.45) of Vol. I(G)

$$\boldsymbol{\kappa}_{ct} = 2G\left[\boldsymbol{\mathcal{A}}^{-1} + \frac{\nu}{1-\nu}\boldsymbol{\mathcal{A}}^{-1}\boldsymbol{e}_3\boldsymbol{e}_3^{\mathrm{t}}\boldsymbol{\mathcal{A}}^{-1}\right] \tag{11.9.13}$$

where the matrix $\boldsymbol{\mathcal{A}}$ is specified in equ. (5.3.44) of Vol. I(G)

$$\boldsymbol{\mathcal{A}} = \begin{bmatrix} 1 & \cos^2\gamma & \cos^2\beta \\ \cos^2\gamma & 1 & \cos^2\alpha \\ \cos^2\beta & \cos^2\alpha & 1 \end{bmatrix} \tag{11.9.14}$$

and

$$\boldsymbol{e}_3 = \{1 \quad 1 \quad 1\} \tag{11.9.15}$$

In the case of an inelastic analysis, it is advantageous to split the stress into a hydrostatic and a deviatoric contribution. If we denote the membrane flows in a Cartesian reference system by $\boldsymbol{N}$, then, in accordance with equ. (5.3.33) of Vol. I(G),

$$\boldsymbol{N} = \{N_{xx} \quad N_{yy} \quad \sqrt{2}N_{xy}\} = \boldsymbol{\mathcal{B}}\boldsymbol{N}_c \tag{11.9.16}$$

applies; the transformation matrix $\boldsymbol{\mathcal{B}}$ is given in equ. (5.3.33) of Vol. I(G)

$$\boldsymbol{\mathcal{B}} = \begin{bmatrix} c_{x\alpha}^2 & c_{x\beta}^2 & c_{x\gamma}^2 \\ c_{y\alpha}^2 & c_{y\beta}^2 & c_{y\gamma}^2 \\ \sqrt{2}c_{x\alpha}c_{y\alpha} & \sqrt{2}c_{x\beta}c_{y\beta} & \sqrt{2}c_{x\gamma}c_{y\gamma} \end{bmatrix} \tag{11.9.17}$$

with

$$c_{x\alpha} = \cos(x, \alpha)$$

In accordance with equ. (12.2.112) of Vol. II(G), the hydrostatic membrane flow can be written as

$$N_H = \tfrac{1}{3}\boldsymbol{e}_{2,1}^{\mathrm{t}}\boldsymbol{N} \tag{11.9.18}$$

with

$$\boldsymbol{e}_{2,1} = \{1 \quad 1 \quad 0\} \tag{11.9.19}$$

Substitution of equ. (11.9.16) into equ. (11.9.18) yields the hydrostatic membrane flow in the natural system

$$N_H = \tfrac{1}{3}\boldsymbol{e}_{2,1}^{\mathrm{t}}\boldsymbol{\mathcal{B}}\boldsymbol{N}_c = \tfrac{1}{3}\boldsymbol{e}_3^{\mathrm{t}}\boldsymbol{N}_c \tag{11.9.20}$$

This simple relation, demonstrating that the hydrostatic flow is given by the average of the component flows, follows immediately from the special form of $\boldsymbol{\mathcal{B}}$. This formulation was first deduced in [11.54]. The hydrostatic state of stress is characterised by the fact that the normal flow in each direction is N_H. The corresponding vector is thus a total quantity

$$\boldsymbol{N}_{tH} = \boldsymbol{e}_3 N_H = \tfrac{1}{3}\boldsymbol{e}_3\boldsymbol{e}_3^{\mathrm{t}}\boldsymbol{N}_c \tag{11.9.21}$$

As a matter of fact, the component N_H normal to the plane is suppressed in this reduced form of $\boldsymbol{N}_{tH}$ since it is not relevant.

The deviatoric membrane flow is immediately deduced from the difference

$$\boldsymbol{N}_{tD} = \boldsymbol{N}_t - \boldsymbol{N}_{tH} \tag{11.9.22}$$

where $\boldsymbol{N}_t$ is the total membrane flow. In analogy to the spatial case expounded in Vol. II(G), (see equ. (12.2.136)), but with $\boldsymbol{\mathcal{A}}$ from equ. (11.9.14), we find

$$\boldsymbol{N}_t = \boldsymbol{\mathcal{A}}\boldsymbol{N}_c \tag{11.9.23}$$

(which can also be confirmed by an equilibrium consideration in the stress plane). Thus ultimately, we obtain

$$\boldsymbol{N}_{tD} = \left[\boldsymbol{\mathcal{A}} - \tfrac{1}{3}\boldsymbol{e}_3\boldsymbol{e}_3^{\mathrm{t}}\right]\boldsymbol{N}_c \tag{11.9.24}$$

The transverse component, which is of course also present here, is again ignored.

Finally, we need the equivalent membrane flow $\overline{N}$. In the case of the plane state of stress under consideration, it is most convenient to use the formula of equ. (12.2.149) in Vol. II(G) which generally applies to a spatial state of stress

$$\overline{N}^2 = \tfrac{3}{2}\boldsymbol{N}_{tD}^{\mathrm{t}}\boldsymbol{N}_c \tag{11.9.25}$$

This relation also applies for the reduced stress vector of equ. (11.9.12) since the transverse component of $\boldsymbol{N}_c$ is zero by definition and cannot contribute to the scalar product in equ. (11.9.25). With equ. (11.9.24), we obtain

$$\overline{N}^2 = \tfrac{3}{2}\boldsymbol{N}_c^{\mathrm{t}}\left[\boldsymbol{\mathcal{A}} - \tfrac{1}{3}\boldsymbol{e}_3\boldsymbol{e}_3^{\mathrm{t}}\right]\boldsymbol{N}_c \tag{11.9.26}$$

In accordance with equ. (12.2.119) of Vol. II(G), the yield function required for our simple constitutive relation may thus be expressed as

$$\Phi = \boldsymbol{N}_c^{\mathrm{t}}\left[\boldsymbol{\mathcal{A}} - \tfrac{1}{3}\boldsymbol{e}_3\boldsymbol{e}_3^{\mathrm{t}}\right]\boldsymbol{N}_c - \tfrac{2}{3}(\sigma_p t)^2 \tag{11.9.27}$$

where σ_p is the constant yield stress for perfectly plastic material.

In the following calculation scheme for a triangular membrane, we start from the usual rule that the stress increment involving plastic flow leads from one state of stress j$-$1 at the yield limit to a neighbouring state j at the yield limit (yield condition). We thus specify

$$\Phi(\boldsymbol{N}_c)_{\mathrm{j}} = \Phi(\boldsymbol{N}_c)_{\mathrm{j}-1} = 0 \tag{11.9.28}$$

The state of stress j can be deduced from that of j$-$1 via

$$\boldsymbol{N}_{c\mathrm{j}} = \boldsymbol{N}_{c,\mathrm{j}-1} + t\boldsymbol{\kappa}_{ct}\delta\boldsymbol{\varepsilon}_t \tag{11.9.29}$$

where $\boldsymbol{\kappa}_{ct}$ is the elastic material stiffness relating total strains and component stresses of equ. (11.9.13). The increment

$$\delta\boldsymbol{\varepsilon}_t = \delta\boldsymbol{\gamma}_t - \delta\boldsymbol{\eta}_t \tag{11.9.30}$$

is the elastic contribution in the (total) strain increment $\delta\gamma_t$. Following the standard flow rule, the plastic contribution $\delta\boldsymbol{\eta}_t$ is proportional to the deviatoric stress. This may be expressed as

$$\delta\boldsymbol{\eta}_t = \frac{1}{t}\delta\lambda \boldsymbol{N}_{tD} \tag{11.9.31}$$

Using equs. (11.9.30) and (11.9.31), equ. (11.9.29) can be written as

$$\boldsymbol{N}_{cj} = \boldsymbol{N}_c^* - \delta\lambda\boldsymbol{\kappa}_{ct}\boldsymbol{N}_{tD} \tag{11.9.32}$$

where

$$\boldsymbol{N}_c^* = \boldsymbol{N}_{c,j-1} + t\boldsymbol{\kappa}_{ct}\delta\gamma_t \tag{11.9.33}$$

may be designated a fictitious stress flow arising from $\boldsymbol{N}_{c,j-1}$ if the incremental deformation is purely elastic. Using equs. (11.9.13) and (11.9.24), the matrix product on the right-hand side of equ. (11.9.32) is reduced to

$$\boldsymbol{\kappa}_{ct}\boldsymbol{N}_{tD} = 2G\left[\boldsymbol{I}_3 - \frac{1-2\nu}{3(1-\nu)}\boldsymbol{\mathcal{A}}^{-1}\boldsymbol{e}_3\boldsymbol{e}_3^t\right]\boldsymbol{N}_c \tag{11.9.34}$$

In the algebraic simplification process leading to equ. (11.9.34), advantage is taken of the relation $\boldsymbol{e}_3^t\boldsymbol{\mathcal{A}}^{-1}\boldsymbol{e}_3 = 2$ (cf. the end of section 12.2.5 in Vol. II(G). By substitution of equ. (11.9.32) into equ. (11.9.27) and consideration of equ. (11.9.34), the yield condition (equ. (11.9.28)) in the stress state j can be expressed as a quadratic equation in the proportionality factor $\delta\overline{\lambda} = 2G\delta\lambda$. In this way, we obtain

$$a(\delta\overline{\lambda})^2 - 2b\delta\overline{\lambda} + c = 0 \tag{11.9.35}$$

with the coefficients

$$a = \boldsymbol{N}_c^t\left[\boldsymbol{\mathcal{A}} - \frac{1}{27}\left\{13 - \frac{2\nu}{(1-\nu)^2}\right\}\boldsymbol{e}_3\boldsymbol{e}_3^t\right]\boldsymbol{N}_c \tag{11.9.35a}$$

$$b = \boldsymbol{N}_c^{*t}\left[\boldsymbol{\mathcal{A}} - \frac{4-5\nu}{9(1-\nu)}\boldsymbol{e}_3\boldsymbol{e}_3^t\right]\boldsymbol{N}_c \tag{11.9.35b}$$

$$c = \boldsymbol{N}_c^{*t}\left[\boldsymbol{\mathcal{A}} - \tfrac{1}{3}\boldsymbol{e}_3\boldsymbol{e}_3^t\right]\boldsymbol{N}_c^* - \tfrac{2}{3}(\sigma_p t)^2 = \Phi(\boldsymbol{N}_c^*) \tag{11.9.35c}$$

The factor $\delta\overline{\lambda}$ is obtained as the smaller root of equ. (11.9.35) and may be written as

$$\delta\overline{\lambda} = \frac{c}{b + \sqrt{b^2 - ac}} \tag{11.9.36}$$

The calculation scheme for the m-th layer (thickness t_m) at a given nodal point is now completely analogous to that for the triangular membrane described above.

The three strings of flanges which meet at a nodal point have merely to be substituted in place of the three sides of the membrane triangle.

The scheme proceeds as follows:

(1) Starting from the state of stress j−1, estimate a stress flow using equs. (11.9.11) and (11.9.33)

$$\boldsymbol{N}_{\mathrm{cm}}^{*} = \boldsymbol{N}_{\mathrm{cm,j-1}} + t_{\mathrm{m}} \boldsymbol{\kappa}_{ct0} \delta \boldsymbol{\gamma}_{t\mathrm{m}} \tag{11.9.37a}$$

where $\boldsymbol{\kappa}_{ct0}$ is the elastic material stiffness and is assumed to be constant through the thickness of the layer. The index 0 denotes the initial configuration of the shell. The expression conforms to the assumption of small strains in the present large deflections analysis.

(2) Calculate the corresponding value of the yield function (equ. (11.9.35c))

$$\Phi(\boldsymbol{N}_{\mathrm{cm}}^{*}) = \boldsymbol{N}_{\mathrm{cm}}^{*\mathrm{t}} \left[\boldsymbol{\mathcal{A}}_0 - \tfrac{1}{3} \boldsymbol{e}_3 \boldsymbol{e}_3^{\mathrm{t}}\right] \boldsymbol{N}_{\mathrm{cm}}^{*} - \tfrac{2}{3} (\sigma_p t_{\mathrm{m}})^2 \tag{11.9.37b}$$

(3) If $\Phi(\boldsymbol{N}_{\mathrm{cm}}^{*}) \leqslant 0$, then

$$\boldsymbol{N}_{\mathrm{cmj}} = \boldsymbol{N}_{\mathrm{cm}}^{*} \tag{11.9.37c}$$

otherwise, application of equs. (11.9.32) and (11.9.34) provides the corrected stress flow

$$\boldsymbol{N}_{\mathrm{cmj}} = \boldsymbol{N}_{\mathrm{cm}}^{*} - \delta \overline{\lambda}_{\mathrm{m}} \left[\boldsymbol{I}_3 - \frac{1 - 2\nu}{3(1 - \nu)} \boldsymbol{\mathcal{A}}_0^{-1} \boldsymbol{e}_3 \boldsymbol{e}_3^{\mathrm{t}} \right] \boldsymbol{N}_{\mathrm{cm}} \tag{11.9.37d}$$

where $\delta\overline{\lambda}_{\mathrm{m}}$ is obtained from equ. (11.9.36).

In order to avoid formal matrix operations, it is advantageous to use the hydrostatic membrane flow of equ. (11.9.20) explicitly in the numerical evaluation of equs. (11.9.35) and (11.9.37d).

With respect to the membrane flow $\boldsymbol{N}_{\mathrm{cm}}$ in equs. (11.9.35a), (11.9.35b) and (11.9.37d), we observe that the statement $\boldsymbol{N}_{\mathrm{cm}} = \boldsymbol{N}_{\mathrm{cm,j-1}}$ is only correct if the stress state j −1 coincides with a point on the yield surface. If this is not the case, the expression

$$\boldsymbol{N}_{\mathrm{cm}} = \tfrac{1}{2} \left[\boldsymbol{N}_{c,\mathrm{j}-1} + \boldsymbol{N}_{c}^{*}\right]_{\mathrm{m}} \tag{11.9.38}$$

is more appropriate.

Once the normal flows in the layers $\boldsymbol{N}_{\mathrm{cmj}}$ have been evaluated, we obtain the natural forces $\mathbf{P}_N$ and bending moments $\boldsymbol{M}_N$ at the nodal point by summation over the layers and by subsequent multiplication by the average of the distances $d_{\alpha\mathrm{i}}, d_{\alpha,\mathrm{i}+1}$ between the nodal point i and the adjacent pair of strings (fig. 11.9.2). Finally, the α, β, γ-components of $\mathbf{P}_N$ and $\boldsymbol{M}_N$ are assigned to the adjacent flanges (section 11.9.5).

11.9.7 Integration of the differential equations of motion

We first determine the Cartesian nodal forces **P** of the flanges by projection of the natural forces $\mathbf{P}_N$ of equ. (11.9.7) and the transverse shear forces $\boldsymbol{Q}$ of equ. (11.9.10) onto the x, y, z-directions. The resulting nodal forces are obtained by the standard Boolean assembly procedure

$$\boldsymbol{S} = \mathbf{a}^{\mathrm{t}}\mathbf{P} \tag{11.9.39}$$

where **a** is the well-known Boolean incidence or assembly matrix.

The differential equations of motion at the node at the time t_{j} may be written as

$$\boldsymbol{R}_{\mathrm{j}} - \boldsymbol{S}_{\mathrm{j}} = \boldsymbol{M}\ddot{\boldsymbol{r}}_{\mathrm{j}} \tag{11.9.40}$$

where $\boldsymbol{R}_{\mathrm{j}}$ represents the vector of the external forces. These also include the additional forces arising due to a possible relative motion (centrifugal, Coriolis). The mass matrix $\boldsymbol{M}$ is diagonal, contains the point masses at the nodes and reproduces (approximately) the mass distribution in the shell. We now write the accelerations $\ddot{\boldsymbol{r}}$ in a difference form

$$\ddot{\boldsymbol{r}}_{\mathrm{j}} = \frac{\delta\boldsymbol{r}_{\mathrm{j}+1}/\tau - \delta\boldsymbol{r}_{\mathrm{j}}/\tau}{\tau} = \frac{\delta\boldsymbol{r}_{\mathrm{j}+1} - \delta\boldsymbol{r}_{\mathrm{j}}}{\tau^2} \tag{11.9.41}$$

where $\delta\boldsymbol{r}_{\mathrm{j}}$ are the displacement increments given by

$$\delta\boldsymbol{r}_{\mathrm{j}} = \boldsymbol{r}_{\mathrm{j}} - \boldsymbol{r}_{\mathrm{j}-1} \tag{11.9.42}$$

Substituting equ. (11.9.41) into equ. (11.9.40), we obtain

$$\delta\boldsymbol{r}_{\mathrm{j}+1} = \delta\boldsymbol{r}_{\mathrm{j}} + \tau^2\boldsymbol{M}^{-1}[\boldsymbol{R}_{\mathrm{j}} - \boldsymbol{S}_{\mathrm{j}}] \tag{11.9.43}$$

Finally, the new nodal displacement increments $\delta\boldsymbol{r}_{\mathrm{j}+1}$ are assigned to the adjacent α, β, γ-flanges (section 11.9.4). As already mentioned in section 11.9.2, the equations of motion are written directly in terms of displacement *increments* and not displacements as such. This proves to be numerically more advantageous. An estimation of the time increment τ for numerical stability is given in [11.49].

In order to estimate the dissipated work absorbed by the shell at impact, it is useful to calculate the instantaneous kinetic energy in the form

$$T = \frac{1}{2\tau^2}\left[\sqrt{\boldsymbol{M}}\delta\boldsymbol{r}\right]^{\mathrm{t}}\left[\sqrt{\boldsymbol{M}}\delta\boldsymbol{r}\right] \tag{11.9.44}$$

In the case of a collision with a rigid wall, for example, when the work of the external forces acting on the shell is zero, the dissipation at time t_{j} simply results from the difference between the initial and the current kinetic energy

$$D_{\mathrm{j}} = T_0 - T_{\mathrm{j}} \tag{11.9.45}$$

This holds as long as the elastic contribution to the work of deformation can be assumed to be negligible. This is usually true in the case of thin-walled shells which experience mainly plastic deformations.

11.9.8 Numerical test of the shell model

Structures under impact are often analysed by subjecting them to a sudden change of impulse in the form of an initial velocity distribution. As a simple quality test of our model, we calculate the eigenmode of vibration of a plate with the corresponding exact velocity distribution for zero deflection as an initial condition. The plate is square and simply supported on all four sides. Particular attention must be paid to the boundary (symmetry) conditions as a result of the oblique directions of the triangular difference mesh. This is illustrated schematically in fig. 11.9.6. Figure 11.9.7 reproduces the first symmetrical eigenmode of the plate and reproduces the evolution of the deflection in the centre as a function of time for a progressively diminishing mesh size. The results are normalised with respect to the exact amplitude and period of oscillation. The figure demonstrates convergence towards the exact solution.

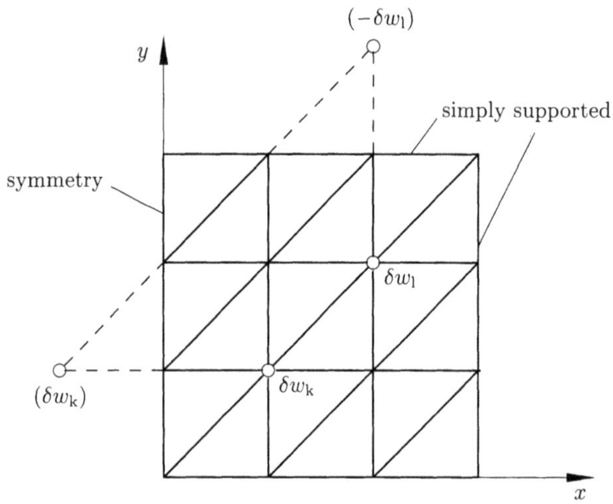

Fig. 11.9.6
A quarter of a square plate: definition of the boundary conditions in the natural difference mesh

11.9.9 Response of an explosively loaded panel

In this somewhat more ambitious example extracted from [11.55], calculation predictions of the present shell model are compared with experimental results obtained from a cylindrical panel exposed to the detonation of a thin high-explosive pad on the surface of the shell. Figure 11.9.8 shows the lay-out and data of the shell, fully built-in on all four sides. The symmetrically placed high-explosive pad covers a 60°-peripheral sector on the circumference of the cylindrical shell while the shell itself comprises a 120°-peripheral sector. The layer covers 10.21 inches of the 12.56 inch long shell. The impact caused by the detonation is simulated by an instantaneous uniform velocity distribution $v_0 = 5650$ in/s. To account properly for the stress distribution during plastic deformation, the shell is divided into four layers across its thickness. The response of the material, a 6061-T6 aluminium alloy, exhibits only weak strain-hardening and can be satisfactorily reproduced by an elastic, perfectly plastic stress-strain relation.

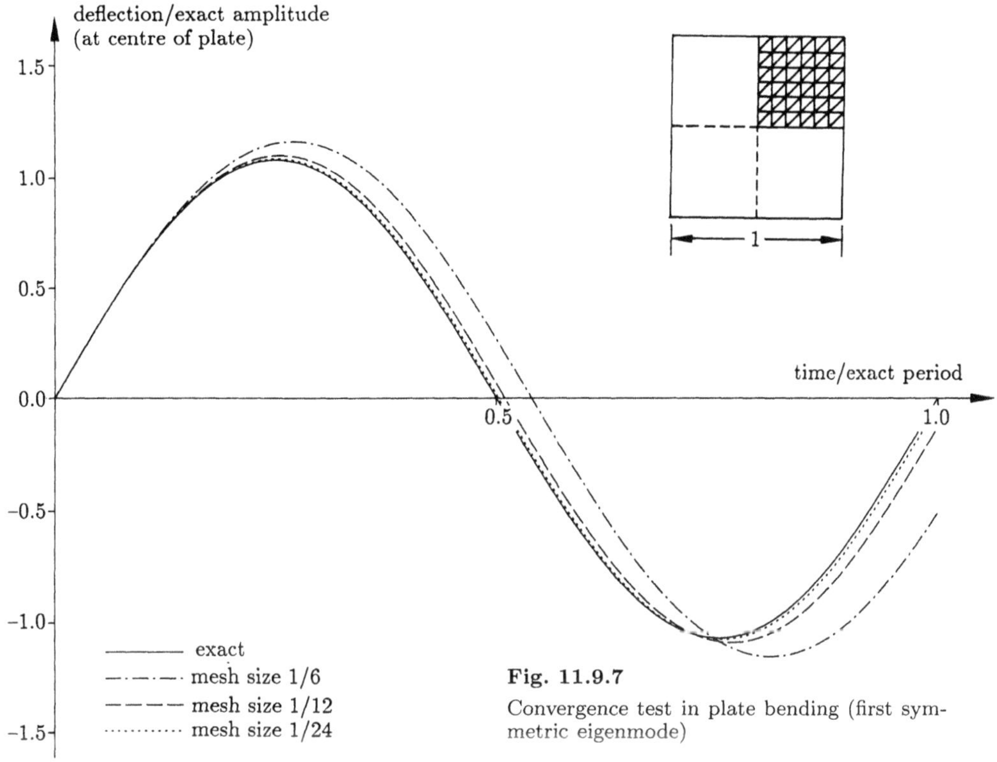

Fig. 11.9.7

Convergence test in plate bending (first symmetric eigenmode)

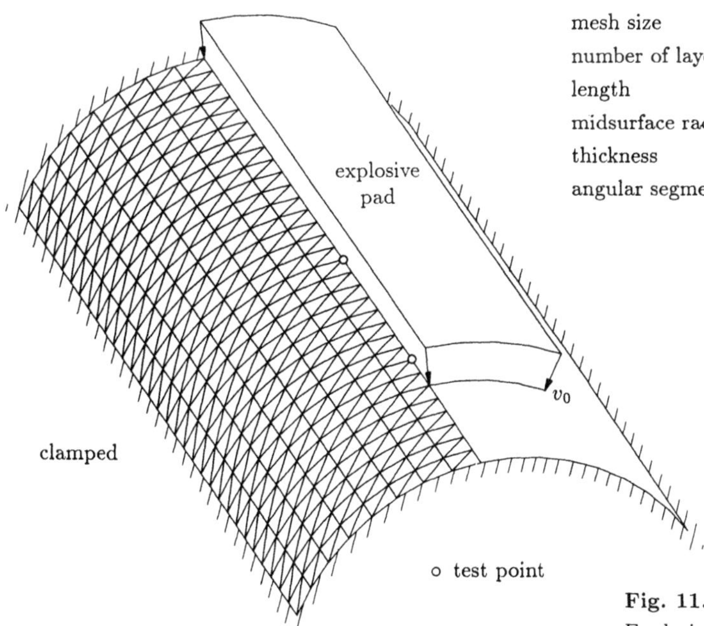

mesh size	8 × 32
number of layers	4
length	12.56 in
midsurface radius	2.9375 in
thickness	0.125 in
angular segment	120°

$v_0 = 5650$ in/s
$E = 10.5 \cdot 10^6$ psi
$\nu = 1/3$
$\sigma_y = 4.4 \cdot 10^4$ psi
$\rho = 2.5 \cdot 10^{-4}$ lb s^{2}in^{-4}

Fig. 11.9.8

Explosively loaded cylindrical panel. Problem description

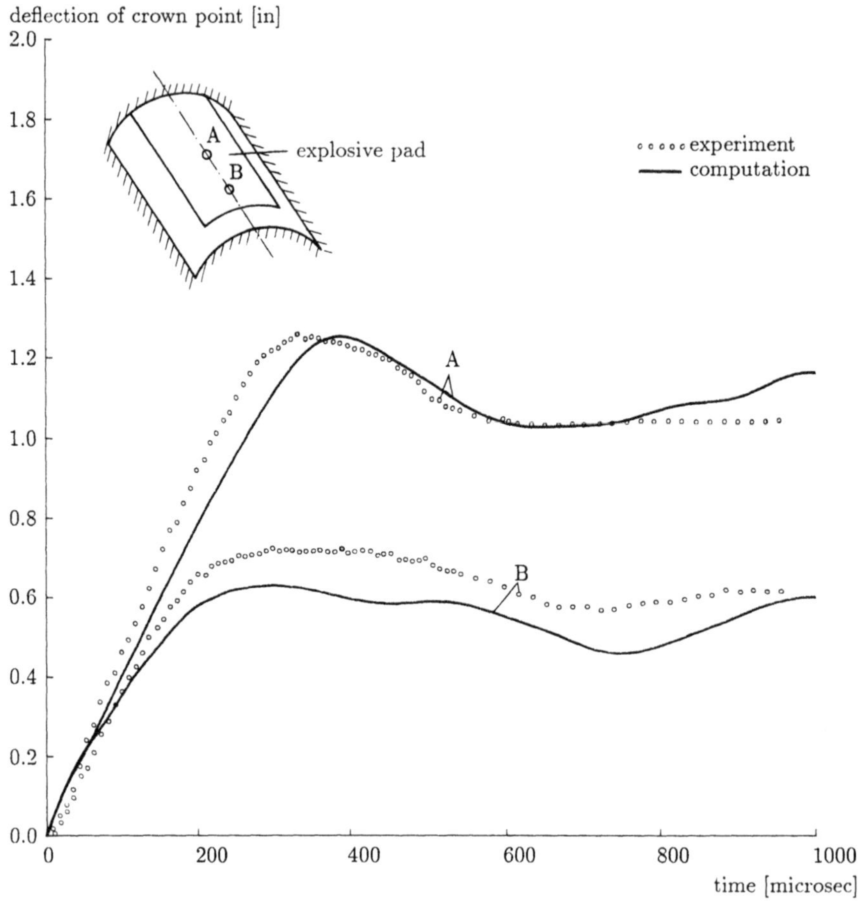

Fig. 11.9.9 Variation of crown point deflection with time. Comparison of calculated and experimental results

Figure 11.9.9 compares the calculated course of the deflection with experimental results at two test points, one at the centre of the panel, the other near the edge of the high-explosive layer. The agreement between theoretical and test results is remarkably good. Figure 11.9.10 illustrates the shell in its deformed state (actual geometry) after 100 μs and after 387 μs, the latter being the instant of maximum indentation. The pronounced plastic hinge at the upper edge which is in contact with the high-explosive layer is particularly conspicuous. On the other hand, on the opposite edge where no high-explosive is applied, the bend is much gentler. Furthermore, a comparison of the crosswise bending deformations along the crown shows that they change locally from convex to concave.

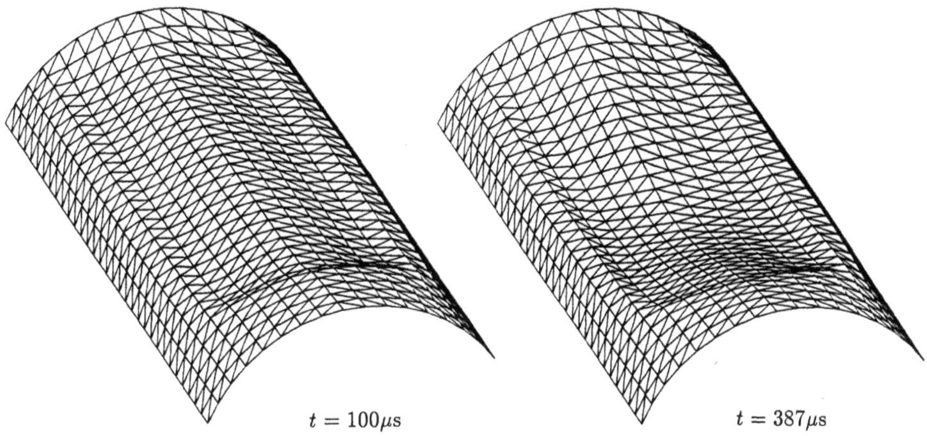

Fig. 11.9.10 Deformed shapes of panel at two instants t after detonation

Bibliography for Chapter 11

[11.1] *K. Klotter*; Nonlinear vibrations, in: Handbook of engineering mechanics, ed. W. Flügge (McGraw-Hill, New York, 1962).

[11.2] *J.J. Stoker*; Nonlinear vibrations in mechanical and electrical systems (Interscience, New York, 1950).

[11.3] *Ch. Hayashi*; Nonlinear oscillations in physical systems (McGraw-Hill, New York, 1964).

[11.4] *P. Hagedorn*; Non-linear oscillations (Clarendon Press, Oxford, 1981).

[11.5] *R.E. Mickens*; An introduction to nonlinear oscillations (Cambridge University Press, Cambridge, 1981).

[11.6] *J.M.T. Thompson, H.B. Stewart*; Nonlinear dynamics and chaos, geometrical methods for engineers and scientists (John Wiley and Sons, Chichester, 1986).

[11.7] *E. Kreuzer*; Numerische Untersuchung nichtlinearer dynamischer Systeme (Springer, 1987).

[11.8] *A. Kunick, W.H. Steeb*; Chaos in dynamischen Systemen (BI Wissenschaftsverlag, Mannheim, Wien,Zürich, 1986).

[11.9] *K. Straub, G. Vallianos*; ASKA Part II, linear dynamics, earthquake spectral response analysis, ASKA UM 221, ISD Stuttgart (1978).

[11.10] *J.W. Miles, W.T. Thompson*; Statistical concepts in vibration, in: Shock and vibration handbook, ed. by Harris and Crede (McGraw-Hill, New York, 1961).

[11.11] *L. Meirovitch*; Elements in vibration analysis (McGraw-Hill, New York, 1975).

[11.12] *Y. Ueda*; Explosion of strange attractors exhibited by Duffing's equation, in: Nonlinear dynamics, ed. by R.H.G. Helleman, New York Academy of Sciences, New York (1980) 422-434.

[11.13] *W.C. Hurty, M.F. Rubinstein*; Dynamics of structures (Prentice Hall, Englewood Cliffs, 1964).

[11.14] *R.W. Clough, J. Penzien* ; Dynamics of structures (McGraw-Hill, New York, 1975).

[11.15] *Y. Ueda*; Steady motions exhibited by Duffing's equation: a picture book of regular and chaotic motions, in: New approaches to nonlinear problems in dynamics, ed. by P.J. Holmes, SIAM, Philadelphia (1980) 311-322.

[11.16] *H. Fujii*; Finite element schemes: stability and convergence, in: Advances in computational methods in structural mechanics and design, ed. by J.T. Oden *et al.*, University of Alabama Press (1972) 201-218.

[11.17] *R.D. Krieg, S.W. Key*; Transient shell response by numerical time integration, Int. J. Num. Meth. Eng., Vol. 17 (1973) 273-286.

[11.18] *T. Belytschko, R. Mullen*; Explicit integration of structural problems, in: Finite elements in nonlinear mechanics, Vol. 2, Tapir (1977) 697-720.

[11.19] *P.J. van der Houwen*; Construction of integration formulas for initial value problems (North-Holland, Amsterdam, 1977).
[11.20] *C.C. Fu*; A method for the numerical integration of the equations of motion arising from a finite element anlysis, J. Appl. Mech. (1970) 599-605.
[11.21] *D.M. Trujillo*; The direct numerical integration of linear matrix differential equations using Padé approximations, Int. J. Num. Meth. Eng., 9 (1975) 259-270.
[11.22] *J. Argyris, P.C. Dunne, T. Angelopoulos*; Non-linear oscillations using the finite element technique, Comp. Meth. Appl. Mech. Eng., Vol. 2 (1973) 203-250.
[11.23] *I. Farhoomand* ; Non-linear dynamic stress analysis of two-dimensional solids, Ph.D. Thesis, University of California, Berkeley (1970).
[11.24] *R.W. Clough, E.L. Wilson*; Dynamic finite element analysis of arbitrary thin shells, Computers and Strutures 1 (1972) 33-35.
[11.25] *E.L. Wilson, I. Farhoomand, K.J. Bathe* ; Nonlinear dynamic analysis of complex structures, Earthquake Eng. and Struct.Dyn., Vol. 1 (1973) 241-252.
[11.26] *C. Petersen*; Abgespannte Maste und Schornsteine, Statik und Dynamik (Wilhelm Ernst und Sohn Verlag, Berlin, 1970).
[11.27] *J. Argyris, P.C. Dunne*; Some Contributions to nonlinear solid mechanics, Colloque INRIA, Versailles, 1973, Springer, Berlin (1974).
[11.28] *I. Newton*; Principia: Book III, General Scholium.
[11.29] *I. Newton*; The System of the World, § 26.
[11.30] *T.J.R. Hughes*; Stability, convergence and growth and decay of energy of the average acceleration method in nonlinear structural dynamics, Computers and Structures 6 (1976) 313-324.
[11.31] *T. Belytschko, D.F. Schoeberle*; On the unconditional stability of an implicit algorithm for nonlinear structural dynamics, J. Appl. Mech. 42 (1975) 865-869.
[11.32] *E. Haug, A.L. de Rouvray, Q.S. Nguyen*; An improved energy conserving implicit time integration algorithm for nonlinear dynamic structural analysis, 4th Int. Conf. SMIRT, San Francisco, Aug.(1977) Paper M5/3.
[11.33] *T.J.R. Hughes, T.K. Caughy, W.K. Liu*; Finite element methods for nonlinear elastodynamics which conserve energy, J. Appl. Mech. 45 (1978) 366-370.
[11.34] *T.J.R. Hughes*; Analysis of transient algorithms with particular reference to stability behaviour, in: Computational methods for transient analysis ed. by T. Belytschko and T.J.R. Hughes, North-Holland, Amsterdam (1983).
[11.35] *H. Ziegler*; Principles of structural stability (Blaisdell, Waltham/Massachusetts, 1968).
[11.36] *J. Argyris, Sp. Symeonidis*; Nonlinear finite element analysis of elastic systems under nonconservative loading – Natural formulation. Part I. Quasistatic problems, Comp. Meth. Appl. Mech. Eng., 26 (1981) 75-123.
[11.37] *J. Argyris, K. Straub, Sp. Symeonidis*; Nonlinear finite element analysis of elastic systems under nonconservative loading – Natural formulation. Part II. Dynamic problems, Comp. Meth. Appl. Mech. Eng. 28 (1981) 241-258.
[11.38] *V.V. Bolotin*; Nonconservative problems of the theory of elastic stability (Moskau, 1961), English translation (Pergamon Press, New York, 1963).
[11.39] *H. Ziegler*; Linear elastic stability, Z. angew. Math. Physik 4 (1953) 89-121;168-185.
[11.40] *P. Seide, T.M.M. Jamjoom*; Large deformations of circular rings under nonuniform normal pressure, ASME, J. Appl. Mech. 41 (1974) 192-196.
[11.41] *J. Argyris, Sp. Symeonidis*; A sequel to: Nonlinear finite element analysis of elastic systems under nonconservative loading – Natural formulation. Part I. Quasistatic problems. Comp. Meth. Appl. Mech. and Eng., 26 (1981) 377-383.
[11.42] *S.P. Timoshenko, J.M. Gere*; Theory of elastic stability, 2nd ed. (McGraw-Hill, New York, 1961).
[11.43] *M. Como*; Lateral buckling of a cantilever subjected to a transverse follower force, Internat. J. Solids and Structures 2 (1966) 515-523.
[11.44] *J. Argyris, K. Straub, Sp. Symeonidis*; Static and dynamic stability of nonlinear elastic systems under nonconservative forces. Natural approach, Comp. Meth. Appl. Mech. Eng. 32 (1982) 59-83.
[11.45] *M.M. Mai* ; Zur nichtlinearen statischen und dynamischen Berechnung gyroskopischer Schalen- und Balkenstrukturen mit der Finite Elemente Methode, Dissertation, Universität Stuttgart (1985).
[11.46] *R.M. Laurenson*; Modal analysis of rotating flexible structures, AIAA J., 14 (1976) 1444-1450.
[11.47] *J. Argyris, P.C. Dunne, T. Angelopoulos*; Dynamic response by large step integration, Earthquake Eng. Struct. Dyn., 2 (1973) 185-203.

[11.48] *J. Argyris, H.A. Balmer, I.St. Doltsinis, A. Kurz*; Computer simulation of crash phenomena, Int. J. Num. Meth. Eng., 22 (1986) 497-519.
[11.49] *J. Argyris, H. Balmer, I.St. Doltsinis*; Some thoughts on shell modelling for crash analysis, Comp. Meth. Appl. Mech. Eng. 71 (1988) 341-365.
[11.50] *J. Argyris, H. Balmer, I.St. Doltsinis*; A simple but subtle model for the analysis of shell-like structures, Comp. Meth. Appl. Mech. Eng. 85 (1991) 1-20.
[11.51] *J. Argyris, H. Balmer, I.St. Doltsinis*; On shell models for impact analysis, in: Analytical and computational models of shells, ed. by A.K. Noor *et al.*, CED-Vol. 3, ASME Winter Annual Meeting, San Francisco, Dec. (1989) 443-456.
[11.52] *E.A. Witmer, H.A. Balmer, J.W. Leech, T.H.H. Pian*; Large dynamic deformations of beams, rings, plates and shells, AIAA J., 1 (1963) 1848-1857.
[11.53] *H.A. Balmer, E.A. Witmer*; Theoretical-experimental correlation of large dynamic and permanent deformations of impulsively-loaded simple structures, Massachusetts Institute of Technology, Cambridge, MA, FDL-TDR-64-108 (1964).
[11.54] *J. Argyris, D.W. Scharpf, J.B. Spooner*; Die elastoplastische Berechnung von allgemeinen Tragwerken und Kontinua, Ing. Archiv 37 (1969) 326-352.
[11.55] *J.W. Leech*; Finite-difference calculation method for large elastic-plastic dynamically-induced deformations of general thin shells, Ph.D. Thesis, Dept. of Aeronautics and Astronautics, Massachusetts Institute of Technology, Cambridge, MA, (1966).
[11.56] *H.J. Flüh*; Einsatz einer modernen Prozeßrechneranlage zur experimentellen Schwingungs- und Dämpfungsanalyse an räumlichen Seilnetztragwerken, 2. Internat. Symposium "Weitgespannte Flächentragwerke", Stuttgart, 1979, Vorberichte S. 4.4 - 1.
[11.57] *J. Argyris, W. Aicher, H.J. Flüh*; Vergleich theoretischer und experimenteller Schwingungs- und Dämpfungsuntersuchungen an ebenen Seilnetztragwerken, Forsch. Ing. Wiss. Bd. 45 (1979) Nr. 2. S. 37-68.
[11.58] *H.J. Flüh*; Zur Dynamik von Flächentragwerken. Eine rechnergestützte experimentelle Analyse strukturdynamischer Probleme mit interaktiver Versuchsdurchführung, doctoral dissertation, Dr.Ing., University of Stuttgart (1981).
[11.59] *J. Argyris, G. Faust. M. Haase*; Das Abenteuer Chaos (Vieweg, Wiesbaden, 1991).
[11.60] *J. Argyris*; An excursion into large rotations, Comp. Meth. Appl. Mech. Eng. 32 (1982) 85-155.

Index

www.ingramcontent.com/pod-product-compliance
Ingram Content Group UK Ltd.
Pitfield, Milton Keynes, MK11 3LW, UK
UKHW052229270726
14060UKWH00004B/683

9 780444 890450